# Cave and Karst Systems of the World

**Series editor**

James W. LaMoreaux, Tuscaloosa, AL, USA

More information about this series at http://www.springer.com/series/11987

Sam Upchurch • Thomas M. Scott
Michael C. Alfieri • Beth Fratesi
Thomas L. Dobecki

# The Karst Systems of Florida

## Understanding Karst in a Geologically Young Terrain

Springer

Sam Upchurch
SDII Global Corporation
Land O' Lakes, FL, USA

Thomas M. Scott
SDII Global Corporation
Havana, FL, USA

Michael C. Alfieri
Water Resource Associates, LLC
Tampa, FL, USA

Beth Fratesi
Southwest Research Institute
San Antonio, TX, USA

Thomas L. Dobecki
Dobecki Geosciences, LLC
Mishawaka, IN, USA

ISSN 2364-4591 ISSN 2364-4605 (electronic)
Cave and Karst Systems of the World
ISBN 978-3-030-09903-9 ISBN 978-3-319-69635-5 (eBook)
https://doi.org/10.1007/978-3-319-69635-5



Printed on acid-free paper

This Springer imprint is published by the registered company Springer International Publishing AG part of Springer Nature.
The registered company address is: Gewerbestrasse 11, 6330 Cham, Switzerland

# Preface

We began considering development of this book in 2010 while we were preparing a guidebook for a karst geology field trip. Our discussions resulted in realization that there are few, if any, karst areas in the world where so much research and investigation existed, especially when one considers that Florida's near-surface carbonate rocks are eogenetic. That is, they have never been deeply buried or subjected to heat and pressure or had sufficient time to substantially alter their original textures and mineral contents. The youngest karstified carbonate rocks in Florida are only about 100,000 to 150,000 years old, and the oldest, near-surface karstic rocks are about 50 million years old. We concluded that it was time for a synthesis of the extensive literature and construction of a resource to help understand Florida's karst and the processes that have created our landscape.

Sinkholes are one of the major focuses of this book because of their importance to understanding the geomorphology of the state and to the welfare and economy of Florida's residents. The media commonly report on damaging sinkholes, and there is widespread fear or concern as a result. Florida has more limestone springs than any other state, and the springs are loved by all. At present, the springs are suffering from nutrient enrichment, loss of flow, and other problems. These, and many other issues, are addressed in this book.

The highest point in Florida is located at Britton Hill in Walton County, near the community of Lakewood about 0.4 km south of the Alabama line. This hill is 105 m above sea level. With this towering hill, Florida has the lowest high point elevation of any state in the USA. Elevations in Florida are generally less than 46 m, so it is a really flat state compared to the remainder of the country. Geologists who work in Florida, however, soon become acclimated to the subtleties of topography. It is amusing to take out-of-state colleagues on a tour of the state and watch their reactions to statements such as "the hill over there" or "the bottom of that depression." When one becomes attuned to the subtle topography and lack of obvious landmarks, it is amazing to see and understand the origins of the subtle contours of the Florida landscape. The geology of Florida is quite interesting, diverse, and important. The goal of this book is to help visualize Florida's karst topography in a subtle terrain through use of digital elevation maps, geographic information systems (GIS) analysis, and light detection and ranging (LIDAR) mapping. By means of these tools, we wish to highlight the diversity, complexity, and, especially, the interesting karst features that dot the state from the Florida Keys to the Alabama and Georgia state lines.

Most of the variations of the topography are a result of three processes: (1) sea level rising and falling over the last few tens of millions of years; (2) dissolution of carbonate-rich sediments (including shell beds), limestone, and dolostone to form karst; and (3) development of surface-water and groundwater drainage systems. The excursions of sea level have resulted in deposition of sandy and clayey cover sediments on top of the older carbonate sediments and rocks. Only Plio-Pleistocene carbonate rocks in southern Florida are predominately bare. Scattered and spatially limited areas of bare karst can be found elsewhere, but cover predominates. The siliciclastic sediment cover complicates interpretation of the karst and leads to sudden and unexpected sinkholes. Our emphasis on Florida karst includes in-depth considerations of the importance and interactions between geologically young, bare, and covered karst and the cover sediments. The interplay of these three processes forms the dominant subject of this book.

Because Florida karst is a concern to the lay public and the processes of karst development in geologically young sediment and rocks are of interest to the geological and engineering communities in Florida and worldwide, we began this book with several introductory chapters that set the scene for discussing karst and interactions with cover sediments. After an introduction to the importance and uniqueness of Florida karst and to the concept of eogenetic karst in Chap. 1, we begin in Chap. 2 by presenting relevant information on the geologic materials that are common in Florida sediments and rocks. Chapter 3 presents an introduction to the most recent geomorphologic characterization of the state. Chapter 3 also discusses the geologic history of Florida and introduces the origins, names, and ages of strata that underlie the state.

Florida depends on groundwater for much of its potable water supply, and the aquifers that supply water are largely karstic in nature. Chemical processes in these aquifers result in development of karst features, and the presence of karst features makes the aquifers highly productive in terms of water supply. It is our dependence on karstic aquifers that has resulted in the abundant literature on Florida karstic, carbonate aquifers. Chapter 4 presents the hydrogeology of Florida's aquifer systems, and Chap. 5 discusses the chemistry of surface water and groundwater in terms of potential for karst development and for interactions with the clay- and sand-rich cover materials.

Chapters 6, 7 and 8 describe the karst processes operative in Florida and provide information for recognition of karst landforms, such as sinkholes and caverns. Chapter 6 describes the controls on development of karst, such as time required, and controls on locations of preferential dissolution. It emphasizes the role of dual porosity in geologically young sediment and rock as a control on development of groundwater flow and dissolution in carbonate rocks. Chapter 7 is a summary of the karst processes and landforms resulting from interaction of the host strata with groundwater. This chapter deals with formation of caves and cave decorations and sinkholes. Detailed explanations are provided to show how the different forms of sinkholes develop and their risks. Overall sinkhole risk is also addressed in Chap. 7. Chapter 8 concludes the book by discussing (1) processes and interactions with surface water, including shallow, meteoric water and streams, and (2) hypogenetic, deep-seated karst related to dissolution at saltwater transition zones and in evaporite-bound strata.

There is some overlap in content from chapter to chapter in order for the chapters to stand alone to the extent possible. This allows the publisher to offer online purchases of specific chapters. Since we hope that lay readers will enjoy this book, this repetition may also allow for better understanding and reinforcement of the provided content.

Our intent is that this book serves two audiences. Florida includes the largest expanse of well-known, eogenetic karst in the world. Therefore, by integrating knowledge of the karst, we hope to share our understanding of the eogenetic karst of Florida with our colleagues worldwide. We also hope to further inform geologists, engineers, and lay persons interested in sinkholes, springs, caves, and other karst features and hazards in Florida. We have attempted to explain karst processes in such a way that interested lay persons can understand Florida karst science.

For those who wish to inspect karst features in the field, we have included geographic coordinates for many of the important and accessible karst features with which we are familiar. Many of the rock exposures and related karst features are either on private land or have long since been destroyed. Therefore, in some cases the karst features are best observed in historical aerial photographs, such as those provided on publicly distributed earth-imaging software. Please respect private property, and do not trespass without permission if you choose to visit these sites. Coordinates are not given for privately owned caves and properties in order to protect the caves and property owners.

We have, for many years, worked with cave divers, who routinely explore the many subaqueous caves in the state. Where possible, we have included discussions of their experiences, cave maps, and photographs of these beautiful, but dangerous, caves. Cave diving is extremely hazardous, and one should not attempt to explore underwater caverns without the proper certifications, equipment, and safety protocols. Novices and experts die each year in Florida caves, so be cautioned. The Cave Diving Section of the National Speleological Society (https://caves.org/) is an excellent source of information on cave diving and safety.

## Acknowledgments

We thank the many persons who instructed, supported, and collaborated with us over the years. There are way too many to acknowledge all. Our students and colleagues at SDII Global Corporation and the Florida Geological Survey were invaluable. In addition, we acknowledge colleagues who provided us with direct support for this book. Cave divers Mark and Annette Long, Bob Schulte, Jill Heinerth, Wes Skiles, and Pete Butt taught us about underwater cave diving and the beauty of Florida's deep caves. Mark Long and Jill Heinerth graciously provided photographs for use in this book, and Bob Schulte assisted in development of our concepts about the potentially hypogenetic, deep caves on Florida's west coast. Florida Geological Survey staff, including Jon Arthur, Harley Means, Clint Kromhout, Cindy Fischler, Christopher Williams, Rick Green, Tom Greenhalgh, and Alan Baker, assisted us in developing many of the concepts presented here and provided significant GIS support. Florida Department of Environmental Protection staff, Rick Copeland, Gary Maddox, and Debra Harrington, provided data on Florida springs. Our associates at Florida's water management districts, especially Eric DeHaven at the Southwest Florida Water Management District and David Hornsby at the St. Johns River Water Management District, also provided helpful data. John Mylroie and Jim LaMoreaux encouraged us to write the book and provided helpful suggestions in its early stages of development. Lee Florea and Arthur and Margaret Palmer provided much appreciated reviews of the entire text prior to publication. Finally, we thank SDII's Bob Windschauer and WRA's Pete Hubbell and Mark Farrell for providing time and resources for Upchurch and Alfieri, and we greatly appreciate the support of Springer Nature's editors and staff, especially for their patience as we attempted to perfect the manuscript.

Land O' Lakes, FL, USA — Sam Upchurch
Havana, FL, USA — Thomas M. Scott
Tampa, FL, USA — Michael C. Alfieri
San Antonio, TX, USA — Beth Fratesi
Mishawaka, IN, USA — Thomas L. Dobecki
2018

# Contents

# 1 Eogenetic Karst in Florida

**Abstract**

Florida's karst constitutes some of Earth's largest expanses of geologically young carbonate sedimentary deposits (shelly sediments, limestone and dolostone) with bare and covered karst. Because of the large population of Florida and the dependence of that population on carbonate aquifers, the karst of Florida is has been extensively investigated. This book synthesizes our knowledge about Florida karst, beginning with why Florida karst differs from older, telogenetic karst elsewhere in North America and the world.

Florida's highly productive aquifers are part of an exemplary karst landscape, an extensive, mantled, geologically young carbonate terrain with dozens of first magnitude springs. Florida's karst has provided water resources to an exploding population and fueled tourism, while creating or exacerbating problems such as sinkhole formation and saltwater intrusion.

The Florida Platform is low and flat and its geology has been dominated by sea-level fluctuations that have left behind a variety of carbonate strata and karst landforms ranging in age from Eocene to Recent. The eogenetic carbonates that comprise these aquifers were never deeply buried and, therefore, they have extensive primary porosity and well developed permeability, distinguishing them from older, telogenetic karst with little remaining primary porosity and very low matrix permeability. Florida's karst is polygenetic, resulting in a complex array of karst features. The overlying siliciclastic sediments create an environment ripe for damaging sinkholes and other hazards.

**Keywords**

Florida karst · Floridan Aquifer System · Biscayne Aquifer · Eogenetic karst · Carbonate aquifer

## 1.1 Introduction

Florida's carbonate aquifers and springs are world famous for the quality of their waters. Florida is also world famous for sinkholes and other artifacts of karst development. These conditions represent eogenetic karst. Florida includes the largest expanse of eogenetic karst in North America and one of the largest in the world. Because Florida is low-lying, it has been inundated and exposed many times by sea-level fluctuations. These sea-level excursions have alternately exposed carbonate rocks to karst-development processes and covered the karstic limestones and dolostones with sand and clay cover materials. The properties of Florida's eogenetic karst and importance of understanding the interactions of karstic sediments and rocks with cover materials are explained in this chapter.

## 1.2 Importance of Karst Processes

- By 1900, groundwater from Florida's limestone and dolostone aquifers was becoming the dominant source of potable water for its inhabitants. The water was touted worldwide for its purity and abundance. Tourist attractions had developed at many springs.
- In 1927, the U.S. Geological Survey's noted hydrogeologist O.E. Meinzer declared that Florida's Silver Springs was the largest spring in a carbonate aquifer in the United States, and that Florida had more first magnitude springs[1] than any other state in the U.S.
- In 1968, the Florida legislature authorized property insurance companies in Florida to offer *optional* insurance against damage caused by sinkhole development. This was the first attempt at legislating sinkhole insurance in

[1] First magnitude springs are springs that discharge at least 2.8 $m^3/s$ on average.


S. Upchurch et al., *The Karst Systems of Florida*, Cave and Karst Systems of the World, https://doi.org/10.1007/978-3-319-69635-5_1

the U.S. There was little public interest in this insurance coverage.

- In 1981, a massive sinkhole opened in the urban area of Winter Park, Florida. This sinkhole "swallowed" several Porsches, half of an Olympic-sized swimming pool, and other structures. The press around the world spread the word about Florida's sinkholes. Public interest was aroused.
- In 1981, the Florida legislature passed a law requiring property insurers in the state to provide *mandatory* insurance coverage for damage caused by sinkhole development. After 1981, all insured property in Florida was "protected" from sinkhole damage by insurance.
- In 1994, a sinkhole developed within a waste-gypsum disposal area, draining slightly acidic and radioactive water into the underlying Floridan aquifer. There was widespread public concern about pollution potential, but little evidence as to where the water went.
- In the late 1990s, concerns began to be expressed about springs, a major water and tourism resource, experiencing reductions in discharge and increases in algae and other indicators of eutrophication. In 1999, the Florida Department of Environmental Protection created the Florida Springs Task Force, a multi-agency group of experts on springs.
- In 2000, the Florida Springs Task Force concluded that Florida's springs are threatened by eutrophication caused by nitrate in their water. The Task Force recommended strategies for mitigation of flow and nitrate issues.
- In 2003, M.A. Bonn and F.W. Bell reported that Florida's springs constitute major sources of economic benefit to Florida with individual public springs generating as much as $23 million annually.
- In 2010, a deep freeze resulted in heavy pumping of the Floridan Aquifer. As a result, potentials in the aquifer plummeted and over 200 sinkholes developed near Plant City (Hillsborough County).
- By 2011, Florida's property insurance carriers were reporting large financial losses as a result of sinkhole claims and litigation.
- In 2012, heavy rainfall from Tropical Storm Debby caused hundreds of sinkholes in west-central and northern Florida.
- In 2013, a man was killed when a sinkhole beneath his house caused the collapse of his bedroom floor slab. He was in bed when the sinkhole developed. His body was never recovered.
- In 2016, another sinkhole developed in a waste gypsum disposal facility near the 1994 sinkhole. Public outcry was loud.
- By 2016, all of Florida's water management districts had issued "water use caution areas" in order to control permitting for consumptive groundwater use. They were developing water conservation measures, including development of alternative sources, water reuse, and water trading.
- In 2017 a large (80 m) diameter sinkhole developed in Land O'Lakes (Pasco County) destroying two homes and causing five others to be condemned. News of this sinkhole spread throughout the U.S. and Europe.

These, and many other events, have made Florida's 67 counties (Fig. 1.1) famous for sinkholes, springs and other issues related to karst. In order to understand the processes, hazards, and resources related to karst in Florida, one should have an understanding of karst and its origins in Florida. So what is karst and what are karst processes?

### 1.2.1 Definitions of Karst and Karst Processes

The term "karst" is derived from the Serbo-Croatian word "Kras" meaning barren, stony ground, which is a characteristic of some of the classical karst terrains in middle Europe (Cvijic 1893; Sweeting 1981). The term has come to mean land features that have been developed by dissolution of sediments and rocks that are soluble in water. In Florida, these materials are carbonate sediments and shell beds and carbonate rock, specifically limestone and dolostone. Karst most commonly forms in areas where limestone**,** a sedimentary rock composed of calcite (calcium carbonate, $CaCO_3$) or dolostone**,** a sedimentary rock composed of the mineral dolomite (calcium, magnesium carbonate, $CaMg(CO_3)_2$) are near the land surface or in contact with waters capable of causing mineral dissolution. These rock types are collectively known as carbonate rocks. In Florida, karst development is not restricted to just carbonate rocks. Shell beds and other unconsolidated or poorly consolidated, carbonate-rich sediments are also developing karst-like landforms.

Karst is found in many areas of the world, and it gives a distinctive appearance to landscapes wherever it exists. Karst processes include (1) the dissolution of sediments and rocks in natural ground or surface water and related movement of any sediments or rocks that overlie the voids created by dissolution and (2) precipitation of minerals from water in caves and on the land surface.

Karst landforms dominate many areas of the United States, especially Florida, Kentucky, Tennessee, Missouri, Texas, and many other states in the eastern and mid-continent U.S. (Fig. 1.2; Veni et al. 2001; Weary and Doctor 2014). Karst landscapes, or karst terrains, are areas characterized by a series of landforms that have been created by the chemical and physical actions of water on soluble rocks, such as limestone or dolostone. Chemical action generally relates to dissolution or chemical precipitation of soluble sediments and rocks of a karst terrain by surface water or groundwater. The

**Fig. 1.1** Locations of Florida's 67 counties

process of dissolution is the dominant cause of development of karst landforms, but precipitation of minerals and rocks in caves and springs is also part of the karst process. In addition, the movement of water that is transporting sediment can also contribute to destruction of sediments and rocks in karst terrains.

Physical landforms are natural, physical features of the landscape, such as hills and valleys, which make up the surface and shallow subsurface details of an area. Examples of landforms that form in karst terrains include sinkholes, caves, springs, sinking streams, towers and pinnacles, and arches.

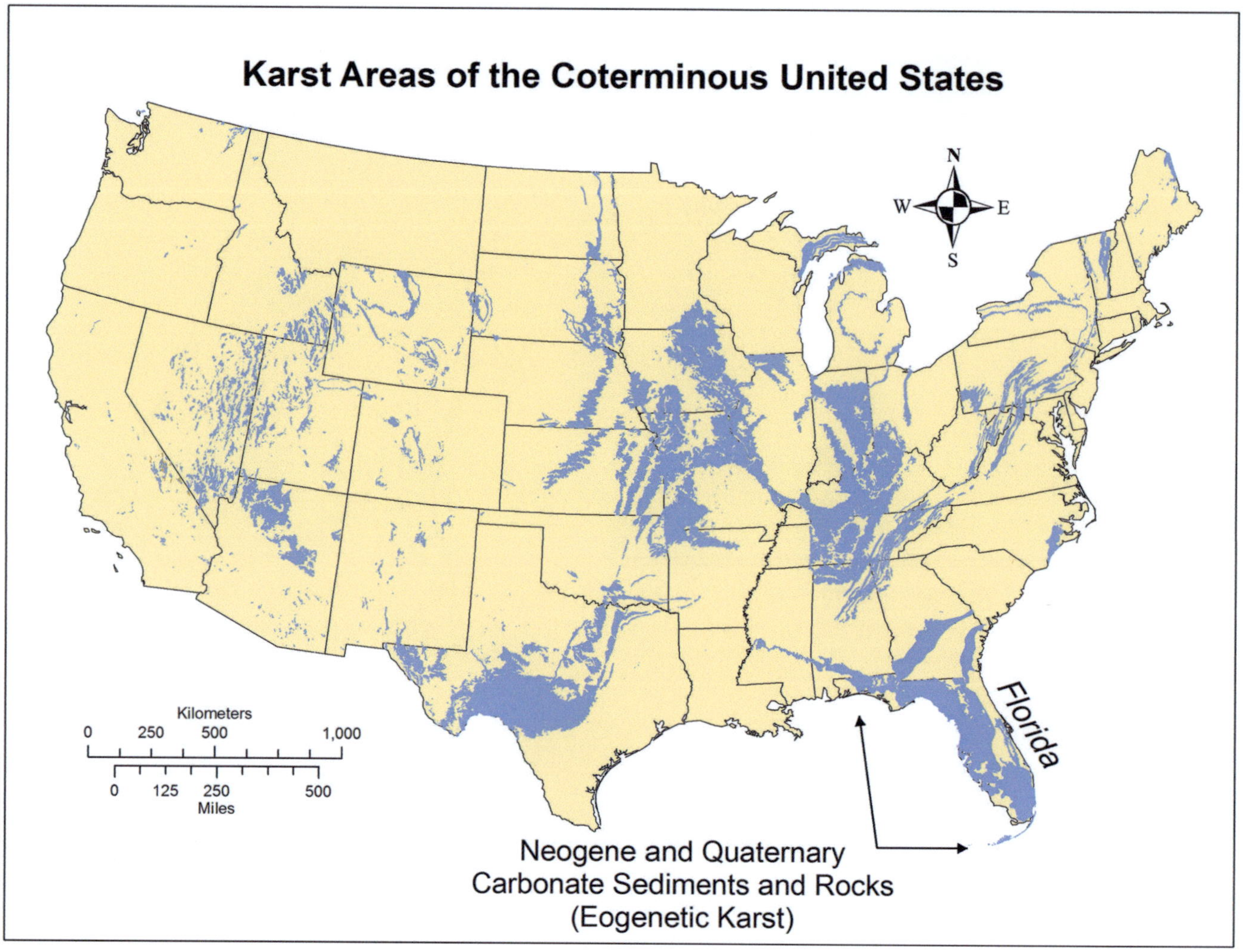


**Fig. 1.2** Karst distribution in the United States. Note the exposed or shallow karst in the Coastal Plain of Florida and adjacent states. (Modified from Weary and Doctor (2014))

This text is about the about how the karst features in Florida form and how to recognize them. However, the subject has significant, broader implications that should be of interest to karst scientists worldwide and to those interested in the social and political implications of karst as a geologic hazard or resource. Here's why.

- Florida and limited portions of Georgia and South Carolina (Fig. 1.2) contain the most extensive expanse of geologically young (Paleogene to Quaternary age), eogenetic karst (see Sect. 1.3) in North America, and it is one of the largest areas in the world where geologically young carbonate sediments and rocks have developed karst landforms. Eogenetic karst consists of landforms whose host rocks are geologically young and have never been deeply buried. As a result, the host rocks have not been subjected to pressure and heat that alter the fabric of the rocks and change their ability to transmit water.
- The sediments affected by karst processes in Florida include a complete range of carbonate strata from unconsolidated or poorly consolidated, carbonate-rich sediment, such as shell beds, to well-lithified rock. Mineral content ranges from aragonite and high-magnesium calcite, which are chemically unstable, to low-magnesium calcite and dolomite, which are chemically relatively stable.
- Karst landforms range from minor, newly formed and poorly understood small-scale features to large-scale landforms reminiscent of the well-known karst regions with ancient rocks that have been altered by deep burial, heat, and pressure, such as the carbonate rocks shown elsewhere in the continental United States (Fig. 1.2).
- The karstic carbonates of Florida include two of the most productive aquifers in the world, the Floridan Aquifer System and the Biscayne Aquifer. These aquifers are among the most studied in the world because of their extensive use for water supply.

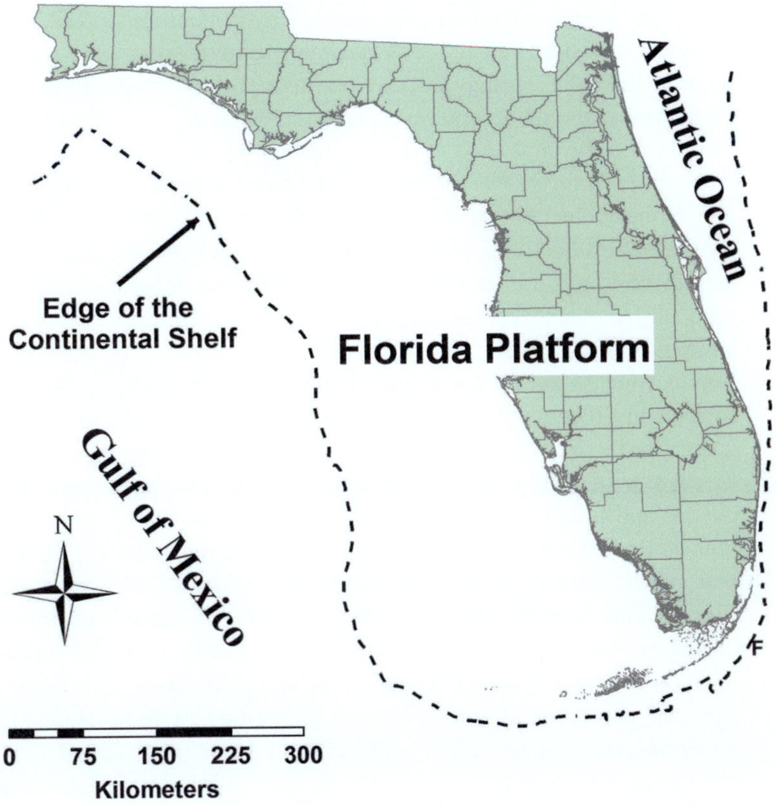


**Fig. 1.3** The extent of the Florida Platform, which includes both mainland Florida and the continental shelf

- Because of Florida's large, and growing, population, the effects of water extraction from the aquifers provides an excellent laboratory for studying rock and soil mechanics, hydraulics, hydrogeology, and hydrogeochemistry of karstic limestone and dolostone aquifers under stress.
- The highest point in Florida is only 105 m above sea level, and about half of the Florida Platform (Fig. 1.3), the area of modern land and continental shelf, is currently submerged with an approximate average water depth of about 75 m. Karst features are known to exist on the continental shelf and even in deeper water on the continental slope.
- Since the Florida Platform is topographically low and flat, Florida has been dramatically affected by the sea-level changes resulting from glaciers forming and melting over the past five million years or so.[2] High, interglacial sea stands resulted in deposition of sand, clay, and carbonate sediment over carbonate rock in many areas. Low sea stands during glacial episodes resulted in development of near-surface alteration of the sediment and rock, including development of karst features. Circulation of groundwater in Florida's aquifers during low sea stands also resulted in deeply buried karst features in the limestones and dolostones. As a result, Florida karst features and the processes that formed them are complex and multidimensional in both time and space.
- With its long coastline, saline-water intrusion into the freshwater aquifers is a regional problem in Florida, especially where the karstic aquifers are highly confined, and recharge from rainfall is limited. Much of the seminal research on saltwater intrusion was done in Florida.
- Florida's springs are world famous. It has been said that Florida contains more first-magnitude springs than any other state or nation. This statement is certainly true for springs in carbonate rocks (Meinzer 1927). These springs have been extensively studied for a variety of reasons, including concerns about water supply, maintenance of ecological and recreational values, and eutrophication and deterioration of environmental values.
- Unlike many areas of eogenetic karst elsewhere, most of Florida's karst is covered by mixtures of varying percentages of sand and clay that mask and exacerbate geological hazards normally present in karst terrains. These hazards include sinkholes, soil creep and slope movement, and the expansion and shrinkage of certain clays.
- Carbonate sediments are being formed at the present time in Florida. Older carbonate sediments have been converted into rock, which remains at shallow depths. Because of these circumstances, an extensive literature exists concerning carbonate sediment deposition and

[2] Glaciers during this timeframe extended as far south as southern Ohio in North America. The rise and fall of sea level due to the repeated melting and formation of the glaciers had a direct effect on depositional and erosional processes in Florida.

post-depositional alteration (diagenesis); hydrogeology; saltwater intrusion; soil and rock mechanics and development of sinkholes; springs and surface water/groundwater interactions; and other issues of concern in Florida.

- Florida was the first state to develop laws providing for property insurance coverage[3] as a result of sinkhole damage. In the context of this book, insurance claim investigations have resulted in thousands of subsurface test borings and/or soundings in search of evidence of sinkhole development. No other location has produced such a wealth of geological and geotechnical information.
- Finally, most texts on karst science focus on the landforms and processes that directly affect development of karst landforms. They do not deal in detail with the mechanics of failure and movement of sediments that overlie and interact with the karstic features of the carbonate. Because of the widespread sand and clay deposits that overlie carbonate sediments and rocks in much of Florida and because these cover sediments, in some instances, have the potential to collapse suddenly into voids in the underlying carbonate rocks and cause damage, the geotechnical and geological properties of the cover materials are addressed in this text.

Because of these factors, Florida has gained an international reputation for its water-filled caverns, springs, sinkholes, and other landforms associated with karst. Few have attempted to integrate and synthesize data from numerous different, but related, disciplines in order to present the story of Florida karst development and encourage understanding of eogenetic karst in a carbonate- and sand or clay- sediment-rich terrain that has been affected by interplay with changing sea levels. In their discussion of the evolution of the concept of carbonate aquifers with multiple forms of permeability, Vacher and Florea (2015) synthesized the development concepts of the evolution of permeability in Florida karst. Others have synthesized portions of the story of Florida karst, such as sinkhole-development processes. It is the goal of this text to synthesize this rich and diverse literature in order to understand eogenetic karst, the landforms and sediment or rock properties that have developed in Florida and provide an explanation as to how these features formed. The authors have added our insights and ideas where gaps in the literature exist.

[3]It is not the intent of this text to discuss sinkhole insurance and the sociopolitical issues that have resulted from the insurance coverage. These topics will be addressed in a later publication.

### 1.2.2 Why Study Florida Karst?

Florida includes the largest geographic area of geologically young carbonate sediments and rocks with well-developed karst features in North America, and it is one of the world's largest eogenetic karst areas. The extensive Yucatan Peninsula in Mexico has a similar karst landscape, but substantially less is known about the details and origins of its karst. As such, the thoroughly studied karst of Florida provides an excellent laboratory for understanding how karst landforms evolve within a few thousand to millions of years after deposition. With the exception of a few carbonate islands, such as the Bahamas and Bermuda, there are few areas where karst processes have been so intensely studied in areas where the rocks have not been deeply buried and primary porosity and permeability not destroyed.

A societal reason that the karst of Florida is of interest is that millions of dollars in damage occur each year as a result of sinkhole development. Understanding the risks and causes of sinkholes should be a priority for all Floridians.

This book synthesizes what we know and understand about (1) how Florida's geologically young karst has formed, and (2) how it has affected the Florida landscape. These landforms include sinkholes, caves, springs, and many other artifacts of dissolution and/or precipitation of soluble materials.

### 1.2.3 Geologic Hazards and Resources in Florida

In addition to a need to develop its karstic aquifers as water resources and protect the many scenic and environmental benefits of the carbonate terrain, karst development is a well-known geologic hazard in Florida. Inadequate foundation support, sinkholes, groundwater contaminant migration and other issues frequently confront Floridians. Construction continually has to deal with designing for the karst in the subsurface (Fig. 1.4); property is damaged by sinkholes and subsidence (Fig. 1.5); and groundwater contaminants travel in unpredicted pathways through caverns and fractures in the carbonate strata.

Karst features are also highly prized resources. Florida's springs (Fig. 1.6) are major tourist attractions and local recreational sites, sites for the development of bottled water plants, plant and animal refuges (Fig. 1.7), and sources of many streams. Sinkhole lakes are prime real estate with scenic water-front lots and swimming, boating, and wildlife observation and conservation opportunities.

Finally, most of Florida's potable water comes from its karstic aquifers. These resources are, without doubt, our most valuable karst resources of all.

**Fig. 1.4** Collapse failure of a landfill over a sinkhole at the Southeast Hillsborough Landfill near Wimauma, Hillsborough County in 2010. Site, which has been remediated, was located at 27.776°N, 82.188°W. People provide scale

### 1.2.4 Florida's Karstic Aquifers

The state's two major, karstic, carbonate aquifers are the Floridan Aquifer System (FAS) and the Biscayne Aquifer (Fig. 1.8; Table 1.1). Because of their high permeabilities, these aquifers are considered to be among the most productive in the world. Each is capable of providing millions of cubic meters of water to wells each day. In addition, there is a number of sub-regional limestone, dolostone, or shell aquifers that are also, to some degree, karstic (Fig. 1.8).

The Sand and Gravel Aquifer in the extreme western part of the Florida panhandle (Fig. 1.8) is the only Florida aquifer without a significant carbonate sediment component and that has not been affected by karst processes. It produces 401,000 $m^3/d$ of freshwater (Marella 2009; Table 1.1). The Sand and Gravel Aquifer only produced about 1.5% of the fresh water utilized by Florida in 2005, however (Marella 2009). As indicated in Table 1.1, about 60% of all freshwater consumed in Florida is derived from the karstic aquifers.

The FAS (Fig. 1.8) is the primary karstic, limestone and dolostone aquifer in Florida. Well over half of the state's population depends on the FAS for potable water. In southern Florida, the FAS is brackish to salty, so communities utilize shallower aquifers. In southeastern Florida, the Biscayne Aquifer, which is part of the Surficial Aquifer System (SAS), predominantly consists of limestone that is about 1 million to 125,000 years old. The Biscayne Aquifer, which is one of the most productive aquifers in the world, is the primary source of potable water for the Miami-Dade area and Florida Keys.

Because of the dependence of the population of Florida on groundwater derived from its karstic limestone and/or dolostone aquifers (Table 1.1), state and Federal agencies, local governments, industry, and academic institutions have sponsored numerous water-resource related studies of the karstic aquifers.

## 1.3 Eogenetic and Telogenetic Karst

In Sect. 1.1, the term eogenetic karst was freely utilized. In this section, the term is explained. It is because Florida's carbonate strata have not been subject to deep burial and the effects of heat and pressure that we can call the karst eogenetic, and it is because the karst is eogenetic that Florida has so many sinkholes, such productive aquifers, and so many large, limestone springs.

**Fig. 1.5** The 1981 Winter Park sinkhole (28.594°N, 81.362°W), located in Winter Park, Orange County, Florida. The largest sinkhole recorded in modern times in Florida captured the attention of the population and governments of Florida. (Photo courtesy of the Florida Geological Survey)

### 1.3.1 Primary and Secondary Porosity and Permeability

When carbonate sediments are originally deposited, they contain abundant void space within the sediment mass. Over time, this sediment may undergo diagenesis (physical and chemical changes that lead to conversion of sediment to rock) that result in development of additional porosity as solutions modify the sediment particles. When particles dissolve away, the resulting molds and other dissolution-related openings begin to touch each other, further enhancing porosity and permeability (Lucia 1995; Vacher and Mylroie 2002). The distinction between diagenesis and karst-related dissolution is often unclear, but the result is the same: a geologically young, carbonate-sediment deposit with enhanced porosity and permeability.

In Florida, the near-surface carbonate sediments are geologically young (less than about 40 million years old), have never been deeply buried (the shallow karstic carbonate strata in Florida have not been buried more than 100 m), and abundant pore space often remains between particles within the body of the carbonate mass. The early-stage pore space consists of the open spaces within and between fossils, sediment particles, and structures within the sediment. This original pore space is termed primary porosity. Unless the sediment is muddy, water can freely circulate between and within the sediment grains and porosity/permeability are enhanced by dissolution of sediment particles. Thus, Lucia's "touching vug" type of pore space, a form of secondary porosity (Vacher and Mylroie 2002) slowly emerged.

Because of the abundant pore space in Florida carbonate sediments and rocks, water movement is normally not restricted to fractures and other secondary openings. The result is a different pattern of pore and void space development in young carbonate strata as opposed to much older carbonate rocks. Geologically young rocks also develop fractures and poorly-developed bedding planes (Fig. 1.9b), so typically both primary and secondary porosity are present in rocks such as occur in Florida. This double set of open pore spaces and high permeability results in the high productivity of the aquifers, geologically rapid sediment and rock dissolution, and karst formation.

Much of the world literature on karst landforms and how they develop deals with areas where the carbonate strata were deeply buried in the geologic past where they underwent significant alteration as a result of the time, heat, and pressures that accompanied burial. As a result, this older, highly modified carbonate rock normally has little of the

**Fig. 1.6** Ichetucknee Head Spring, located in Ichetucknee State Park (29.984°N, 82.762°W), near Ft. White, Florida, is a popular swimming and tubing site. It is estimated that Ichetucknee Springs State Park alone brings $23 million to the economy of north Florida (Bonn and Bell 2003)

**Fig. 1.7** Biologist counting manatee seeking refuge during cold weather in the Blue Spring run, Blue Springs State Park (28.943°N, 81.341°W), near Orange City, Volusia County, Florida. There were 300 manatees in the spring and its run when the photo was taken

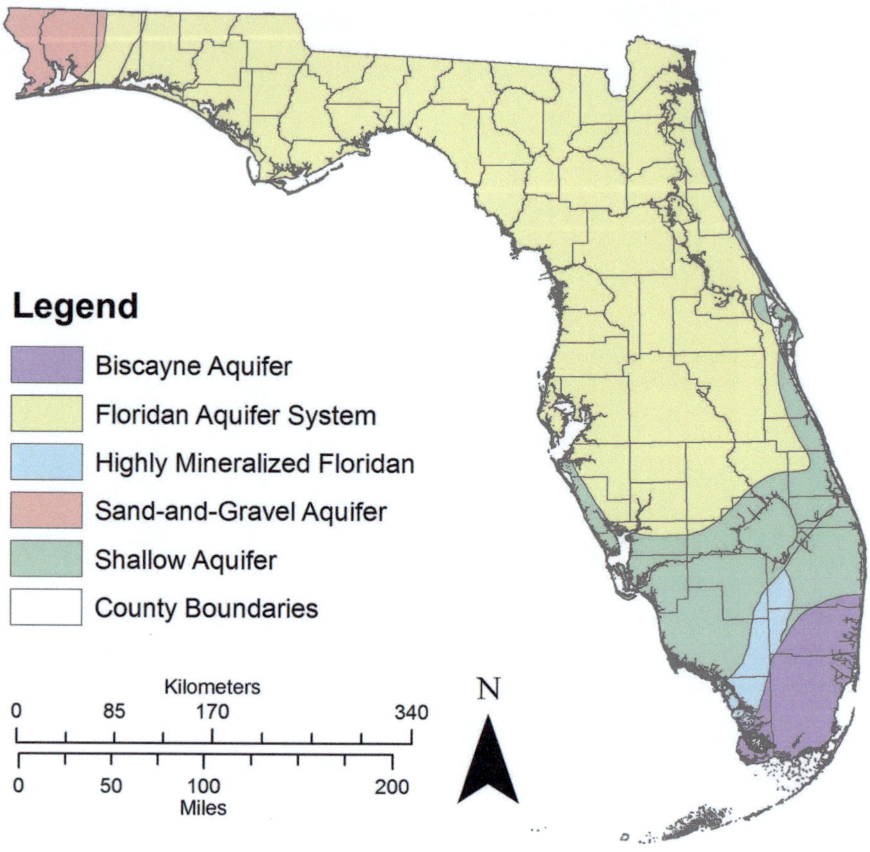


**Fig. 1.8** Florida's principal aquifers. The only aquifer without karst features is the Sand and Gravel aquifer of extreme west Florida

original pore space (porosity) remaining between the carbonate particles that originally accumulated and formed carbonate sediment and then carbonate. Permeability in these old carbonate deposits is typically developed along cracks and openings between sediment layers (Fig. 1.9a). There is little or no pore space within the matrix of the carbonate rock, so primary porosity in these carbonates is typically near zero.

After deep burial (hundreds to thousands of meters below the land surface) and for tens to hundreds of millions of years, the limestone (or other water-soluble rock) may be uplifted and exposed to near-surface conditions. With uplift, the old, low porosity carbonate rock is likely to be fractured and bedding planes (contacts between beds of sedimentary rocks that originally formed during deposition) may open. With the release of confining pressure and the onset of groundwater circulation, bedding planes become accentuated through dissolution. Groundwater and surface water seek these fractures and bedding planes because they are more permeable than the rock matrix. As the water flows through the fractures and bedding planes, they become enlarged due to the dissolution of the rock adjacent to these openings. As a result, much of the existing void space in these geologically old carbonate rocks is secondary porosity, or porosity that formed after deposition and burial of the carbonate.

**Table 1.1** Water use from surface water and all Florida aquifers in 2005

| | Daily water use ($\times 10^6$ m$^3$) | Percent of total |
|---|---|---|
| Saline water | 43.5 | 63 |
| Fresh water | 26.0 | 37 |
| **Total water use** | 69.5 | 100 |
| Fresh groundwater | 16.1 | 62 |
| Fresh surface water | 9.9 | 38 |
| **Total water use** | 26.0 | 100 |
| Saline groundwater | 0.042 | 0.01 |
| Saline surface water | 43.4 | 99.9 |
| **Total water use** | 43.5 | 100 |

Data from Marella (2009)

### 1.3.2 Comparison of Porosity in Geologically Ancient and Young Carbonate Sediments and Rocks

The most common forms of large openings along which water can flow in carbonate sediments and rocks are bedding planes, fractures, and, possibly, faults. In ancient, once deeply buried carbonate rocks, these structures dominate water movement. In young carbonate sediments and rocks that have never been deeply buried, intact and modified primary porosity is likely to remain and serve as a second set of

**Fig. 1.9** (**a**) Secondary porosity – solution-enlarged vertical fracture and horizontal bedding planes in an Ordovician[4] limestone near Nashville, Tennessee. Outcrop is about 6 m high. (**b**) The effects of primary and secondary porosity on the structure of the fractured Eocene Ocala Limestone in a quarry near Lowell, Marion County, Florida. Note the absence of well-developed bedding planes and rudimentary enlargement of the fracture. Quarry face is 15 m high and located near 29.322°N, 82.179°W

pathways for water movement (permeability) through the smaller, interstitial pore openings and touching vugs.

In ancient carbonates that have once been deeply buried and subjected to the effects of heat and pressure over millions of years, evidence can often be found in the body of the rock of the original, deposition-related composition, structure, and geologic history of the rock, but in the absence of small-scale, interstitial porosity and permeability related to these features, they do not normally substantially affect karst development. Permeability of the carbonate rock matrix is too low.

In many ways, the long and complex history of once deeply buried limestone or dolostone serves as a hydrogeologic simplifying process. The original properties of the carbonate rock matrix, which developed during deposition, early diagenesis (the physical and chemical changes in the sediment that result in transformation of the sediment into rock), and early karstification, are lost or obscured, resulting in carbonate strata with bulk properties that reflect burial rather than pre-burial diagenesis or karstification. Evidence of pre-burial diagenesis and/or karstification may be evident, but burial normally obliterates or significantly reduces primary and pre-burial secondary porosity and permeability, such as touching vug spaces.

The karst of Florida differs from the majority of the karst in the United States in that Florida's carbonate strata are significantly younger (all of the near-surface carbonate in Florida is less than 40 million years old, and near-surface limestone in southeastern Florida is 1 million to 125,000 years old) and has never been buried more than a hundred meters below the land surface. As a result, the original depositional features and primary porosity of the carbonate sediment or rock are often well-preserved and post-depositional alterations to the carbonate sediment or carbonate rock caused by early diagenesis are evident. In many areas of Florida, carbonate sediments have not even undergone sufficient diagenesis to result in well-lithified rock. In this setting, karstification and diagenesis may be synonymous, or at least they are complimentary and contemporaneous processes. Note that there are more deeply buried, karstic strata in Florida, but they are not involved in development of epikarst, the karst features that occur on or near the upper surface of the limestone or dolostone.

Compare the two exposures of limestone shown in Fig. 1.9. The limestone shown in Fig. 1.9a is over 400 million years old, has been deeply buried, subjected to burial diagenesis, and then uplifted to near the land surface. This section of Ordovician[4] limestone has been exposed to karst development for at least six million years (Middle Tennessee State University 2016). The limestone has a vertical fracture (joint) that has been enlarged by dissolution as groundwater passed through the rock. In addition, the bedding planes are well developed and many have also been enlarged by dissolution. The only significant pathways for groundwater circulation are through the secondary porosity of the fracture and

[4]The Ordovician Period extended from 485.4 to 443.8 million years ago (Mya).

along bedding planes. The matrix of the limestone is well-lithified (solidified into rock). As a result, the rock is hard and difficult to break or dig out of the rock face without a hammer or excavator.

Figure 1.9b is an exposure of the Ocala Limestone, which was deposited sometime between about 37 and 34 million years ago (Mya), has never been buried more than 100 m, and has not been subjected to geologically significant loading (application of weight from later sediments deposited on top of the limestone) or thermal stresses. This limestone exposure also has a fracture that has been enlarged by dissolution. The exposure is a quarry face, so weathering has not significantly accentuated the bedding planes. The limestone itself is often soft and easily excavated with a pick or shovel. In fact, in some areas of the exposure, one can dig the limestone out of the quarry face with bare hands. In this loose, poorly lithified limestone, pore spaces between the original particles of calcium carbonate (microscopic fossils, pellets, and other remains) are well preserved as primary porosity.

The differences in age, degree of burial diagenesis, and porosity and permeability in the two different limestones affect their appearance, how karst develops, and their engineering and hydrologic properties.

### 1.3.3 Eogenetic and Telogenetic Karst

In 2002, Vacher and Mylroie introduced the term eogenetic karst to describe karst features developed in geologically young, relatively unmodified carbonate sediments and rocks. Their use of the term eogenetic was derived from the classification of porosity development with time and depth of burial by Choquette and Pray (1970; Fig. 1.10).

According to the classification scheme developed by Choquette and Pray (1970), porosity in carbonate rocks develops in three "zones" – the eogenetic, mesogenetic, and telogenetic zones. The eogenetic zone reflects porosity development in a post-depositional regime before burial and deep-seated, heat- and pressure-related changes to the carbonate rock occur. The mesogenetic zone (Fig. 1.9) includes the environment of deep burial, where elevated temperature and pressure conditions exist. Burial and uplift of strata within the mesogenetic zone requires millions of years, so time is an important factor as compared to eogenetic processes. The telogenetic zone reflects porosity development after uplift and exposure to erosion at or near the land surface. Time, measured in tens to hundreds of millions of years, is required for a rock body to transition from eogenetic to mesogenetic and then to telogenetic zones.

Vacher and Mylroie (2002; Fig. 1.11) took this simple, rock-cycle-based porosity-development sequence and applied it to karst, suggesting that karst that develops shortly after deposition and early diagenetic alterations but before significant burial should be termed eogenetic karst. Telogenetic karst, in contrast, is karst developed on carbonate rocks that have been deeply buried, subjected to burial and tectonic pressures and heat over millions of years, and then uplifted to the land surface where they are exposed to subaerial and shallow subsurface erosion and dissolution processes. The mesogenetic zone is not subject to karstification because of depth of burial and high confining pressures. Note in Fig. 1.11 that the initial, primary porosity is typically lost during burial and that secondary porosity increases upon uplift and exposure to near-surface weathering, fracturing, and unloading.

As an example of the differences in the eogenetic karst of Florida and typical telogenetic karst, consider Table 1.2. This table compares some of the general properties of eogenetic and telogenetic karst based on Florida and the karst of mid-continent United States (Fig. 1.2). Note that there are substantial differences in the suite of host sediments and rocks

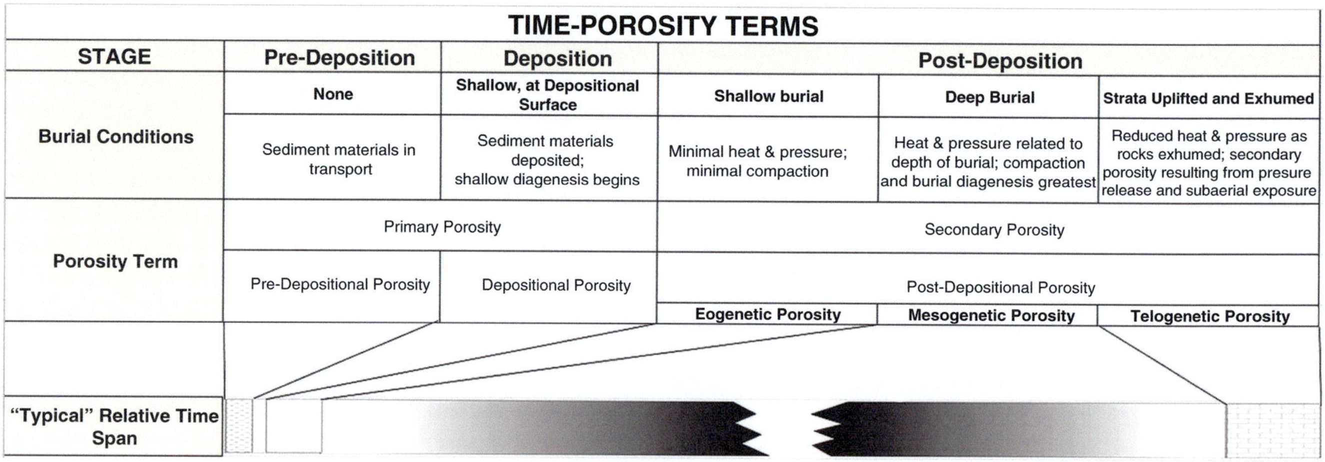


**Fig. 1.10** Time- and burial-related porosity terminology as suggested by Choquette and Pray (1970)

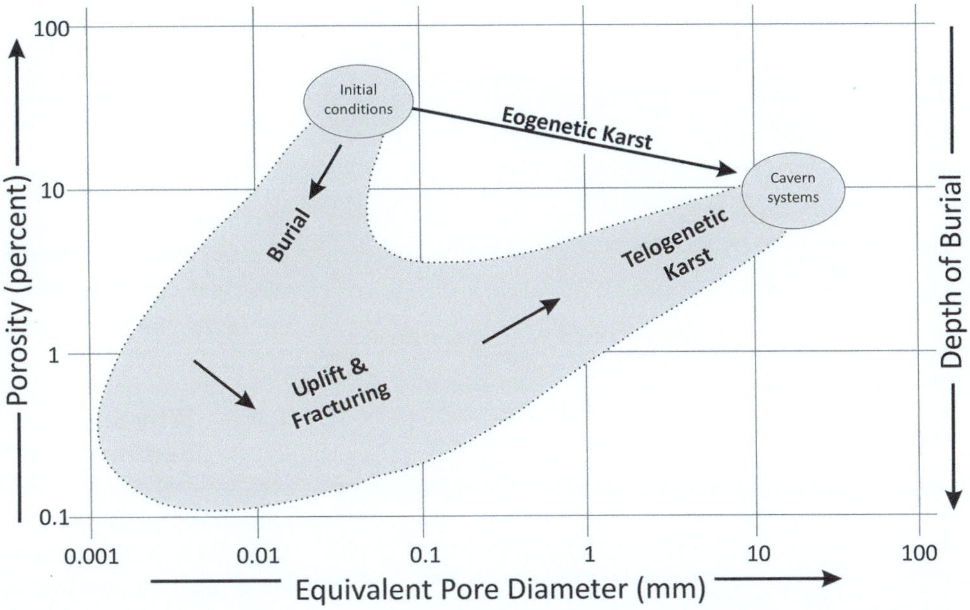


**Fig. 1.11** Evolution of porosity in eogenetic and telogenetic karst as a function of burial and time. (Modified from Vacher and Mylroie (2002))

and processes that affect karst development, including mineralogy, depth of burial, porosity distribution, relationship to sea level, and sediment/rock surface area available for chemical reactions with water.

We know from work in Florida and elsewhere around the world that near-surface, eogenetic karstification begins to develop upon exposure to subaerial and relatively shallow groundwater environments, so the eogenetic zone is not simply an environment of development of primary porosity as a result of deposition. Diagenesis and later karst processes related to dissolution in meteoric and groundwater systems can also result in secondary porosity, ranging from microscopic to very large pore spaces.

The depth of development of eogenetic and telogenetic karst is limited to those areas where groundwater can circulate. If karst features develop as a result of dissolution at or near the land surface, where groundwater is actively recharging and circulating through the shallow aquifer, the karst that develops is termed epigenetic, or karst that develops near the top of the soluble rock. Karst features that develop at depth as a result of acidic fluids migrating up from below or of mixing of saline and fresh ground water are termed hypogenetic.

Groundwater circulation is somewhat compartmentalized vertically. Within the upper part of the FAS, Biscayne Aquifer, and other shallow carbonate strata, karst is mostly epigenetic and groundwater circulation is limited to relatively shallow depths. Deeper circulation is limited by beds that prevent mixing with the shallower aquifers. This circulation may occur in strata to depths of at least 600 m in the FAS (Puri and Winston 1974; Smith and Griffin 1977). Any karst in these deeper strata either formed before the rocks were buried or is hypogenetic. For the most part, epigenetic karst in Florida is more-or-less restricted to the upper 100 m.

In Florida, the zone where eogenetic karst has formed has the following general physical and temporal constraints:

1. Both unconsolidated carbonate and carbonate-rich sediments and variably lithified carbonate rocks can be involved in karstification;
2. Host sediments and rocks are late Paleogene to Quaternary in age (sediments and rocks deposited between 40 Mya and about 10,000 years ago);
3. Karstification has been controlled by development of a groundwater flow system that responds to variations in sea level;
4. Eogenetic karst in Florida is likely multicyclic and polygenetic owing to changes in sea level of as much as 98 m (−90 to +8 m) over the last 120 thousand years and probably even more over the last 40 million years;
5. Landform development by epigenetic karst processes (development of sinkholes, caves and springs, etc.) is limited to approximately the upper 100 m, in sediments that have not been deeply buried or subjected to pressures equivalent to burial depths greater than 100 m;
6. Artifacts of deep karst (100 to over 600 m) consist of possible dissolution features developed as a result of groundwater mixing along the salt-water/fresh-water transition zone, prior to burial, or during extreme low sea stands; and
7. Owing to the youth and lack of burial to great depths, carbonate sediments and rocks within the upper 100 m in Florida retain texture, fabric, mineralogy, and structure related to their environments of deposition and early diagenesis.

Given this variable range of constraints, the eogenetic karst of Florida is complex. There are areas of Florida where

**Table 1.2** Comparison of some properties of eogenetic karst in Florida and in the telogenetic karst of the mid-continent United States

| Property | Eogenetic karst of Florida | Telogenetic karst of the mid-continent U.S. |
|---|---|---|
| Host materials | Carbonate-rich sediments, limestone, dolostone | Limestone, dolostone, calcitic and dolomitic marble |
| Age of host sediments and rocks | Tertiary and Quaternary | Primarily Paleozoic |
| Age of karst development | On-going, Eocene and younger | On-going, age variable |
| Influence by sea-level fluctuations | On-going karst development controlled by sea-level position and timing of deposition and subaerial exposure | No direct control; effects of sea level fluctuations absent for millions of years |
| Maximum depth of burial | Less than 100 m for subaerial and shallow karst | Thousands of meters |
| Karst development on the salt-water/ fresh water transition zone | On-going | Effects of sea-water/ fresh-water interaction absent; interaction with basin-derived brine possible |
| Sediment and rock mineral content | Aragonite, calcite, dolomite | Calcite, dolomite |
| Sediment and rock matrix primary porosity and permeability | Generally present | Generally absent |
| Sediment and rock secondary porosity and permeability | Poorly to well developed | Well developed |
| Relative surface area of sediment and rock in contact with groundwater | Very high | Low |

karstification and diagenesis are just beginning in geologically young sediments and there are areas where the eogenetic karst features strongly resemble features commonly associated with old, telogenetic karst (Table 1.2). For example, sinkholes develop in both eogenetic and telogenetic karst. As a result, the eogenetic karst of Florida stands as an excellent laboratory for documenting and synthesizing the range of eogenetic karst processes, a subject of concern to karst scientists, and for comparison to karst processes in older, telogenetic systems.

Florea (2012) summarized the explosion of investigations of eogenetic karst worldwide in a paper given at the 2012 Annual Meeting of the Geological Society of America. He concluded his abstract by saying

> In the decade following 2002, a wealth of research has highlighted the importance of eogenetic karst worldwide, and … stimulated revisions to the origin and morphology of paleokarst. …. [W]e may ultimately come to view eogenetic karst with equal importance to classical karst in older, diagenetically mature carbonates.

Synthesis of Florida karst science can contribute to this rise in importance of eogenetic karst, especially in areas where sand and clay have been deposited over the limestone or dolostone and sea level has risen and fallen multiple times.

## 1.4 Why Examine the Role of Cover Sediments in Florida?

One of the ways that this text differs from most traditional karst texts is emphasis on the materials and behavior of cover, the sediment that overlies the carbonate sediments or rocks. Throughout most of the state, the carbonate strata are covered by a wide range of sand and clay-rich sediments. It is this cover material that migrates into the void space within the carbonate sediment or rock and has the potential to cause unexpected damage through sinkhole development.

This text includes an extensive discussion of the role of the sediments that cover Florida limestone and dolostone because these cover sediments control karst development and, when sinkholes develop, the movement of cover materials that support structures usually causes the damage.

## 1.5 Summary

There is a prodigious amount of data available for piecing together the story of Florida karst. All totaled, we probably know more about Florida karst than any region of the United States. Unfortunately, most of these data have not been synthesized into a single story describing the formation of such a large and important expanse of eogenetic karst.

It is this wealth of data developed over many years by our friends and colleagues that we draw upon for this text. It is the authors' intent to characterize the eogenetic karst of Florida and assist those interested in the "sinkhole problem" in Florida to understand karst. Most important, the authors have attempted to synthesize and integrate these data into a "big picture" of eogenetic karst.

Notable conclusions from this chapter include:

- Shallow Florida karst can be considered eogenetic;
- Karst is developed in unconsolidated shell and other carbonate sediments and in well-lithified limestone and dolostone;

- Near-surface, eogenetic, karstic sediments and rocks have never been buried more than about 100 m, so they have not been subjected to significant burial pressures or temperatures;
- Near-surface karstic sediments and rocks, which range in age from Eocene to late Pleistocene, represent a variety of epigenetic karst landforms and aquifer characteristics;
- Deeper karst includes hypogenetic karst formed on saltwater/freshwater transition zones and in other environments;
- Because of the presence of primary and secondary porosity in the carbonate aquifers, they tend to be highly productive in terms of water supply; and
- Because of Pliocene to late Pleistocene sea-level fluctuations, many of the carbonate sediments and rocks have been covered by siliciclastic cover and subject to multiple cycles of karstification.

## References


Bonn MA and Bell FW (2003) Economic impact of selected Florida springs on surrounding local areas. Report prepared by Florida State University for the Florida Department of Environmental Protection. Tallahassee, Florida

Choquette PW and Pray LC (1970) Geologic nomenclature and classification of porosity in sedimentary carbonates. Bulletin, Amer. Assoc. Petrol. Geologists, 54:207–250

Cvijič J (1893) Das Karstphänomen. Geographische Abhandlungen, 5(3):215–319

Florea LJ (2012) Eogenetic karst aquifers after a decade of investigation. Geol. Soc. Amer. Abstracts with Programs, 44(7):205

Florida Springs Task Force (2000) Florida's springs: strategies for protection & restoration. Tallahassee, Florida Department of Environmental Protection. Available at http://www.dep.state.fl.us/springs/reports/files/SpringsTaskForceReport.pdf. Accessed 22 July 2014

Lucia FJ (1995) Rock fabric/petrophysical classification of carbonate pore space for reservoir characterization. Bulletin, Amer. Assoc. Petrol. Geologists, 79(9):1275–1300.

Marella RL (2009) Water withdrawals, use, and trends in Florida, 2005. USGS Sci. Inv. Report 2009-5125

Meinzer OE (1927) Large springs in the United States. USGS Water Supply Paper 557.

Middle Tennessee State University (2016) The history of the glades of the Central Tennessee Basin. Available at http://w1.mtsu.edu/gladecenter/gladehistory.php. Accessed December 15 2016

Puri HS and Winston GO (1974) Geologic framework of the high transmissivity zones in south Florida. Fla Bur Geology Sp. Pub. No. 20

Smith DL and Griffin GM (eds.) (1977) The geothermal nature of the Floridan Plateau. Fla Bur Geol. Sp. Pub. No. 21

Sweeting MM (1981) Karst geomorphology. Hutchinson Ross Pub. Co., Stroudsburg, Pennsylvania, Benchmark Papers in Geology, Vol. 59

Vacher HL and Florea LJ (2015) Quantitative hermeneutics: Counting forestructures on a path from W.M. Davis to the concept of multiple permeability karst aquifers. International Journal of Speleology, 44(3):207–230.

Vacher HL and Mylroie JE (2002) Eogenetic karst from the perspective of an equivalent porous medium. Carbonates and Evaporites, 17:182–196.

Veni G, DuChene H, Crawford NC, Groves, CG, Huppert GH, Kastning EH, Olson R, and Wheeler BJ (2001) Living with karst: A fragile foundation. Environmental Awareness Series, American Geological Institute

Weary DJ and Doctor DH (2014) Karst in the United States: A digital map compilation and database. USGS Open-File Report 2014-1156

# Geological Materials: An Overview

# 2

**Abstract**

In Florida's mantled, eogenetic karst, the sedimentology and mineralogy of siliciclastic and carbonate geological materials greatly impact karst development. Florida's cover materials are integral parts of the karst development story, and they can present hazards themselves. Cover materials are mostly allochthonous quartz, feldspars, clay minerals, while carbonate rocks consist of calcite with some aragonite and dolomite. Magnesium content of calcite affects diagenetic and karst processes very little; but aragonite dissolves preferentially, and dolostone deposits can be resistant to dissolution. Therefore, it is important to understand the mineralogy of geologically young, carbonate sediments and rocks.

Marine sediments in Florida are predominantly well-sorted, fine-to-medium sands, sourced from areas with limited grain sizes and deposited by waves, currents, and wind. Eolian sands are poorly consolidated with frosted grain surfaces. Grain shapes range from angular shells through variable sands to rounded pebbles and cobbles.

Under-consolidated clay beds in the Miocene Hawthorn Group (Chap. 3) act as confining layers despite high porosities, and the fine-grained sediments have the potential to flow if sinkholes develop. Smectites are common expansive clays in the Hawthorn Group. The Hawthorn clays were significantly altered during two periods of intense weathering in the Late Miocene-Early Pliocene and Late Pliocene-early Pleistocene. The first weathering event formed a prominent paleosol throughout central and northern Florida.

Florida's carbonate rocks were deposited in broad, shallow seas, creating flat, laterally extensive layers. The overlying siliciclastics, in contrast, represent complex environments resulting in sediment facies with great lateral and vertical variations. This chapter describes the origins and properties of these sediments.

**Keywords**

Carbonate sediments · Limestone · Dolostone · Paleosols · Depositional environments · Sand · Clay

## 2.1 Introduction

This chapter introduces the classification and chemical/mineralogical properties of geological materials involved with karst in Florida. This chapter is included for those readers who are unfamiliar with the basics of sediment classification and mineralogy or the terminology used to characterize such materials. In eogenetic karst systems, the mineralogy of geological materials strongly impacts karst development because of the presence of thermodynamically unstable minerals, mineral deposits that add to our knowledge of Florida karst, and evidence that minerals provide with respect to weathering, transport and deposition, and age.

Carbonate minerals and rocks are discussed first, because of their importance to formation of karst features. The properties of cover sediments that overlie carbonate rocks are discussed toward the end of this chapter. Note that geologic formation names are used in this chapter in order to optimize the utility of the chapter. Refer to Chap. 3 for a discussion regarding formation names and lithologies.

It is important to understand the materials that comprise Florida's eogenetic carbonate sediments and rocks and the materials that cover the carbonates. It is the intent of this chapter to introduce sediment and rock terminology, mineralogy, textures, and classification schemes utilized in this text. After discussing these issues, the concepts of facies, lithosomes, and depositional control on sediment textures and compositions are introduced. By understanding the nature of Florida carbonate sediments and rocks and siliciclastic sediments that commonly overlie the carbonates, the


S. Upchurch et al., *The Karst Systems of Florida*, Cave and Karst Systems of the World,
https://doi.org/10.1007/978-3-319-69635-5_2

concepts of sediment mixing as a result of deposition and of cover collapse can be evaluated.

### 2.1.1 Definitions of Soil, Sediment, and Rock

It is important to understand the terms soil, sediment, and rock in the context of Florida geology and engineering literature. Most engineers utilize the term "soils" to identify any geological material that consists of disaggregated, granular particles, regardless of grain size. This usage is fine when describing the mechanical (i.e., load bearing) properties of geologic materials, but it is too limiting for a discussion of the processes that form karst and alteration products resulting from chemical and physical weathering.

Physical weathering is the simple mechanical disaggregation of geologic materials to smaller grain sizes by breakage or separation of grains or crystals within a rock. Chemical weathering involves the chemical alteration of earth materials, usually as a result of interactions with water.

In this text, the term soil refers to the uppermost portion of the earth where chemical and physical alterations of the underlying sediments and rocks are, or have been, caused by chemical and physical weathering. It is within the soil column, consisting of disaggregated, weathered and altered materials, where most plant growth occurs. The term soil is utilized for disaggregated and altered materials that are currently at the land surface and undergoing pedogenesis, or soil formation. The term paleosol is utilized for ancient soils that have been preserved in the subsurface.

Therefore, in this text, the term soil is restricted to chemically altered and/or disaggregated materials that have developed as a result of weathering. Unlithified materials that were formed by depositional processes, as opposed to weathering, are termed sediments. Generally speaking, sediment refers to the materials deposited by earth-surface processes. Sediments consist of separate, discrete particles deposited by transport processes such as the wind, rivers and streams, glaciers, and the sea. Sediment can also include biogenic materials, such as peat and muck, coral reef frameworks, and seashells, and inorganic chemical precipitates, such as beds of rock salt or other minerals formed as water evaporates. These latter materials form *in situ* and are not necessarily transported.

Rock consists of aggregates of minerals that have been lithified, or solidified, regardless of origin. Rocks that have formed by cementation or compaction of sediments are termed sedimentary rocks.

### 2.1.2 Major Mineral Groups of Florida Carbonate Rocks and Sediments and of Cover Materials

The earth materials that are involved in formation of karst features in Florida are carbonate sediments and rocks. The most abundant carbonate minerals in Florida are composed of calcite or dolomite, both of which are slowly soluble in slightly acidic water. Dissolution of these minerals may lead to development of such karst features as sinkholes and caverns.

The earth materials that overlie the carbonate sediments or rocks in Florida are typically silica-bearing minerals that include silica ($SiO_4^{-4}$) as a primary chemical component. Some Florida minerals are silicate minerals that include silica coordinated with positively charged elements such as calcium or magnesium. Each silicon atom coordinates with four oxygen atoms to form a tetrahedron functional group known as the silica tetrahedron. If the silica tetrahedra are also coordinated with aluminum, which forms an octahedral arrangement with oxygen (aluminum octahedra), the minerals are termed aluminosilicates. Quartz, which forms the sand on most beaches and is a major constituent in the surficial sediments of Florida, is a silicate composed entirely of silica tetrahedra, so it has the composition $SiO_2$. Many of the minerals that comprise clay deposits in Florida are aluminosilicates. These silica-bearing minerals comprise the cover sediments that often overlie Florida's carbonate sediments and rocks.

### 2.1.3 Definition of Cover

In Florida, carbonate sediments and rocks are often covered by younger sediments composed of non-carbonate minerals. The blanket of non-carbonate sediments and soils that overlies the carbonate sediments and rocks is termed cover. This cover material is important for two reasons. First, cover sediments interact with the underlying carbonate rocks in a number of different ways, sometimes enhancing dissolution of the underlying carbonate material and sometimes preventing or lessening dissolution. Second, the movement of cover materials into voids in the underlying carbonate sediment or rock causes sinkholes and other karst features to develop within the cover.

Cover is an inclusive term used by karst scientists in reference to the overburden sediments that overlie karstic carbonate sediment or rock. In Florida, cover materials are highly variable, heterogeneous mixtures of unconsolidated to poorly indurated sand, clay, and shell. Organic-rich sediment (peat, muck) is also common as a cover material in southern Florida and locally elsewhere in the state.

There are three reasons for discussing cover materials in Florida.

1. Since most of the karst of Florida is covered with siliciclastic cover sediment, which masks and interacts with the processes of karst development, there is a need to understand the processes that affected deposition and subsequent alterations of the cover materials in order to recognize any artifacts of karst interactions, such as movement of cover into underlying void spaces.
2. Many of the cover materials (i.e., expansive clay, organic-rich sediments, and poorly compacted dune sands) in Florida constitute geological hazards in their own right. The properties of these sediments that make them hazards have been frequently and erroneously confused and intertwined with karst-related processes.
3. Finally, in order to identify the early stages of the development of a sinkhole, one must understand the properties of the materials that overlie the limestone or other soluble material. The process of identification of raveling or subsidence involves accurate characterization of variations in cover sediment strength, texture, and stratigraphy and separation of the artifacts of karst-related sediment movement from depositional and post-depositional processes.

Therefore, this chapter discusses the composition, origin, and post-depositional behaviors of the materials that comprise the cover over the carbonate strata in Florida.

Cover materials include sediment bodies containing many different minerals, including silicates, aluminosilicates, phosphates (minerals with phosphorus and oxygen as components of their crystal structures), and non-minerals, such as organic-rich sediments (i.e., peat, muck).

The term clastic refers to sediment particles that have been transported by wind, waves, currents, or other processes prior to deposition. Carbonate sediments can be clastic, or allochthonous, in which case the sediment particles formed at some distant location and were transported to the deposition site, or they can form in the location where they were preserved in the sediment. In the latter case, the sediments are non-clastic, or autochthonous. Both types of carbonate sediments exist in Florida.

For the most part, the sand- and clay-sized silicate- and aluminosilicate-bearing sediment particles in Florida were formed elsewhere and transported to Florida by waves, river or sea currents, or the wind. Therefore, they are allochthonous and are considered clastic because they have been transported to their sites of deposition. Most of these clastic sediments formed in the ancestral Appalachian Mountains, north of the state (Hine 2013), although southern Florida has siliciclastic sediments that were transported by the wind to Florida from Africa (Prospero et al. 1987, 2001, 2010; Prospero 1999; Shinn 2001; Muhs et al. 2007). These siliciclastic sediments comprise much of the cover over Florida's carbonate sediments and rocks. For an excellent discussion of the mechanisms and timing of the influx of these siliciclastic sediments, see Chap. 8 in Hine (2013).

## 2.2 Florida Carbonate Sediments and Rocks

### 2.2.1 Carbonate Minerals in Florida Sediments and Rocks

Minerals that are composed of carbonate ($CO_3^{2-}$) combined with one or more metals, such as calcium ($Ca^{2+}$) and magnesium ($Mg^{2+}$), are collectively called "carbonate" minerals, and carbonate sediments and rocks are primarily composed of carbonate minerals. This section describes the carbonate minerals that comprise limestone, a sedimentary rock that primarily consists of calcite, dolostone, a sedimentary rock composed of dolomite, and geologically young, shelly sediments that include the minerals calcite and aragonite, a calcium carbonate mineral that is chemically similar to calcite but has a different crystal structure.

In this chapter, special emphasis is placed on the composition of carbonate sediment as it is first deposited and the preservation potential of those minerals as they are subjected to post-depositional changes, such as weathering and the development of karst.

### 2.2.2 Calcite, Dolomite, and Aragonite

There are three carbonate minerals of importance to karst development in Florida. These are calcite, dolomite, and aragonite. Calcite and aragonite are polymorphs, minerals with the same chemical composition – calcium carbonate ($CaCO_3$) – but with different crystal structures. The differences in crystal structure result in differences in solubility of the minerals in water.

By far, the most widespread and abundant carbonate mineral in the limestones of mainland Florida is calcite. Aragonite is still present in dominant to trace amounts in the Pleistocene limestone and shell deposits of southeastern Florida (Bock et al. 1969).

Many late Tertiary and Quaternary shell and shelly sand beds and most modern carbonate sediments in Florida include mixtures of aragonite and calcite, with the proportions of the two minerals depending on the age and geologic setting of the strata. Figure 2.1 illustrates the mineral compositions of modern marine organisms that secrete carbonate-mineral hard parts in Bermuda. These biogenic hard parts are mineralogically similar to those

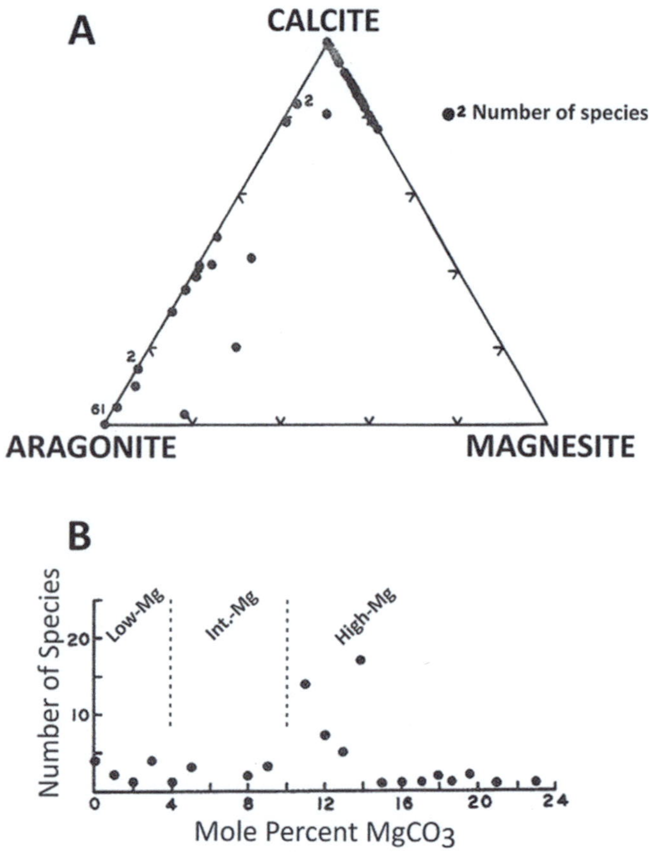


**Fig. 2.1** Compositions of calcareous hard parts in marine organisms from Bermuda. The ranges of low-, intermediate-, and high-Mg calcite (Morse et al. 1985) are shown for reference. Magnesite is the magnesium carbonate mineral $MgCO_3$, which is used as a surrogate to demonstrate the amount of Mg in the shell. (Modified from Upchurch (1970b))

currently being deposited in the Florida Keys and elsewhere offshore in Florida.

When lithified, calcite and aragonite form the rock limestone. The Pleistocene limestones of southern Florida often contain mixtures of calcite and aragonite (Ball 1967; Hoffmeister et al. 1967; Robinson 1967; Bock et al. 1969) with calcite as the dominant mineral and aragonite as a secondary constituent. Older limestones are composed entirely of calcite, often with some included dolomite.

Dolomite ($CaMg(CO_3)_2$) occurs in Recent sediment in the Florida Keys (Taft 1961; Deffeyes and Martin 1962; Shinn 1964a, b, 1968; Carballo et al. 1987) and as a minor to dominant constituent of Tertiary sediments and rocks throughout Florida. Rocks composed predominantly of dolomite are termed dolostone.

Because of their compositions, crystal structures, and grain size as sediments, these three minerals have variable solubilities in water. These differences in solubility affect the rates of formation and preservation of diagenetic and/or karst features, especially in the youngest carbonate sediments of Florida's eogenetic karst.

### 2.2.3 Effect of Magnesium on Calcite Solubility

Many marine organisms, notably certain mollusks and algae, precipitate calcite hard parts as they grow. Magnesium can substitute for calcium in the calcite crystal lattice, so some biogenic calcite contains small amounts of magnesium when precipitated by marine organisms (Fig. 2.1; Upchurch 1970b; Milliman 1974; Carter 1980a, b, c). Calcite with less than 4 mole percent magnesium, as $MgCO_3$, is termed low-Mg calcite, calcite with between 4 and 10 mole percent $MgCO_3$ is considered intermediate-Mg calcite, and over 10 mole percent is termed high-magnesium calcite (high-Mg calcite; Morse et al. 1985). As seen from the data from Bermuda organisms (Fig. 2.1), the mole percentages of $MgCO_3$ in subtropical to tropical, modern marine organisms with calcite hard parts can be as high as 24–25 mole percent.

The amount of magnesium substituted for calcium in the calcite crystal lattice affects the solubility of the calcite. Figure 2.2 illustrates the relative solubilities of low-, intermediate-, and high-Mg calcite in fresh water at atmospheric temperatures and pressures. Note that the minimum solubility of calcite is in the vicinity of 3–4 mole percent $MgCO_3$ and that the solubility of calcite then increases with increasing Mg content. Over time, high-Mg calcite is converted to low-Mg calcite, often without significant loss of shell volume or structure (Land 1966, 1967). As a result, losses of Mg and the resulting changes in mineral solubility are unlikely to cause significant development of karst landforms.

In addition to precipitation by organisms, inorganic forms of calcite are common in Florida's carbonate sediments. Inorganic calcite commonly occurs as cements in limestone and other sedimentary rocks and as cave decorations, such as stalactites and stalagmites. The calcite that forms in cave decorations, such as stalactites, is low-Mg. The metabolic activity of living organisms appears necessary for the hard parts of marine organisms to incorporate significant amounts of magnesium.

### 2.2.4 Aragonite Solubility

The other common form of calcium carbonate precipitated by marine organisms is aragonite (Fig. 2.1). Unlike biogenic calcite, aragonite does not incorporate significant amounts of magnesium, although it can contain small amounts of strontium through substitution for calcium. Aragonite is chemically unstable at earth-surface temperatures and pressures (Jamieson 1953; MacDonald 1956). Given the right conditions and sufficient time, biogenic, aragonitic sedimentary materials can either be destroyed by dissolution or altered with little or no loss of particle volume or structure

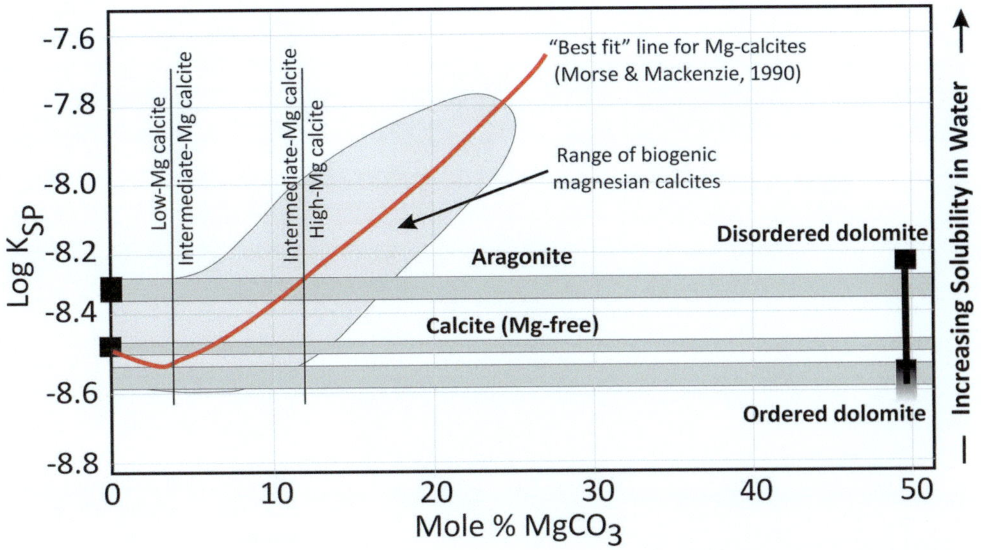


**Fig. 2.2** Solubilities of carbonate minerals as a function of magnesium content. Log $K_{SP}$ is the solubility product constant, a measure of the solubility of a mineral. (Modified from Railsback (2013))

to the more chemically stable calcite (Land 1966, 1967). As shown in Fig. 2.2, aragonite has the same approximate solubility in freshwater as high-Mg calcite with about 14 mole percent $MgCO_3$.

Carbonate sediments, such as those in late Pleistocene limestones in southeastern Florida and the Florida Keys, that originally contained high proportions of aragonite, are likely to undergo substantial volume reductions through dissolution of the aragonite when exposed at the land surface to rainfall. Consequently, these originally aragonitic carbonate sediments may contain diagenetically derived or even karstic landforms related to subaerial weathering and dissolution and/or precipitation of carbonate minerals. In Florida, these karst landforms are usually relatively small and unlikely to cause harm to structures built over them.

### 2.2.5 Dolomite Solubility

The third important carbonate mineral in Florida is dolomite, a mineral that makes up dolostone and occurs as an accessory mineral in other sediments and rocks. Dolomite is widespread throughout the state where it makes up regionally extensive dolostone beds, especially within the Miocene Hawthorn Group and in the Eocene Avon Park Formation (Chap. 3). Florida dolomite formed in many different environments and there is a long history of investigations as to the origin and distribution of dolomite origins in Florida. In Florida, dolomitization has been identified as a modern, authigenic process (Taft 1961; Deffeyes and Martin 1962; Shinn 1964a, b, 1968; Carballo et al. 1987) and as post-depositional (diagenetic) sediments that have replaced pre-existing carbonate sediments (Randazzo 1978; Thayer and Miller 1984; Randazzo and Bloom 1985). Dolostone and silt-sized dolomite sediments (dolosilt) from the Miocene Hawthorn Group have been attributed to both primary depositional (authigenic) and diagenetic processes (Scott 1988).

The solubility of dolomite in fresh water varies with the perfection of the mineral crystal structure (Fig. 2.2). Well-ordered (well-crystallized) dolomite is less soluble in fresh water than either calcite or aragonite. It is for this reason that karst is commonly less well developed in dolostone deposits than in limestone of the same age and diagenetic and tectonic history. Many geologically young, authigenic dolomites are poorly ordered (poorly crystallized) as a result of rapid or disrupted mineral nucleation processes. These poorly ordered dolomites are somewhat more soluble than well-ordered crystals (Fig. 2.2) and may be more susceptible to dissolution and formation of minor diagenetic or karstic features upon subaerial exposure. Preservation of well-ordered dolomite in karstic rock is much more likely than of poorly ordered crystals.

Because of the lower solubility of dolomite in meteoric and fresh groundwater (Fig. 2.2), development of karst features is limited in Florida dolostones. The dolostones are relatively brittle and are, therefore, often highly fractured (Navoy 1986; Safko and Hickey 1991; Duerr 1994; Winston 1995). Examination of borehole logs and down-hole videos (Navoy 1986; Safko and Hickey 1991; Duerr 1994) suggests that the deeper dolostones have some possible dissolution features, indicating that some dissolution has occurred either during lower sea stands, along the transition zone between the freshwater lens and underlying saline water, or by dissolution related to dilute sulfuric acid in the vicinity of evaporite (calcium sulfate) deposits at depth. By far, however, the greatest amount of porosity development in deeply buried dolostone appears to be a result of fracturing. Shallow dolostone deposits exhibit little karst development. However, sinkholes are known to develop on rare occasions in dolostones, and there is

**Fig. 2.3** Quartz sand weakly cemented by kaolinite and ferric hydroxides. From the reworked Cypresshead Formation in a sand pit (27.919°N, 81.584°W) near Lake Wales, Polk County, Florida

**Fig. 2.4** Thin bed of heavy mineral concentrates (arrow) in dune sand from the reworked Cypresshead Formation. Same location as the sample in Fig. 2.3

small-scale pitting and scalloping on the surfaces of dolostone deposits exposed along rivers in northern Florida.

## 2.3 Florida Siliciclastic Sediments

Siliciclastic sediments are typically composed of quartz, feldspars, clay minerals, other silicate or aluminosilicate minerals, and rock fragments. Throughout most of Florida, quartz is the dominant, coarse-grained mineral in siliciclastic sediments. Feldspars, notably microcline and orthoclase, are common in the Pliocene Cypresshead Formation. Weathering of these potassium feldspars has resulted in commercial kaolinite deposits and in kaolinite cements in the quartzose sandstones (Fig. 2.3; Fountain 2009).

Florida sand deposits often contain a suite of minerals that are termed "heavy minerals." The term heavy mineral comes from the fact that these mineral clasts have higher specific gravities (mass/unit volume) than quartz of equivalent grain size. As such, the heavy minerals tend to settle out in water or wind more rapidly than similar sized quartz grains and form concentrates (Fig. 2.4). Because of intense weathering toward the end of the Pliocene Epoch (Fountain 2009), many of the heavy minerals have been oxidized in the sand deposits that are currently exposed in Florida. Weathering of the iron-containing minerals released iron to groundwater and resulted in the widespread presence of ferric hydroxide ($Fe(OH)_3$), a common cement in Florida sand deposits.

Important heavy mineral deposits remain offshore and on Trail Ridge (Fig. 2.5). Trail Ridge, a prominent Pleistocene beach ridge complex that extends from southern Georgia to Putnam County (Fig. 2.5), is well known for its economically important heavy mineral deposits. The heavy minerals that occur in Trail Ridge sediments include: enstatite, epidote, garnet, ilmenite, rutile, kyanite, staurolite, and zircon (Miller 1945; Pirkle et al. 1977). The heavy mineral deposits of Trail Ridge are currently mined as a source of titanium and refractory silicate minerals, such as zircon and staurolite. Leucoxene, a common weathering product of the titanium minerals, is also abundant in Florida's heavy mineral deposits. Arthur et al. (1986), Nocita et al. (1990), and others have examined the shallow sand deposits offshore in Florida and found substantial deposits of similar heavy minerals.

Based on the presence of ferric-hydroxide staining in the surficial sand deposits of Florida, iron-bearing heavy minerals were apparently once widespread in the sandy sediments of Florida. Weathering of these iron-containing minerals has resulted in the brown and reddish colors common in Florida sand deposits (Fig. 2.6). The reddish and brown colors are a result of deposition of ferric hydroxide.

The thin bands of ferric hydroxide shown in Figs. 2.3 and 2.6 are termed cutans, pedogenic (soil) horizons formed by groundwater passing through the sediments. Ferric hydroxide does not normally form clasts, but it does form cements, typically at or near the water-table surface.

In addition to silicate and carbonate clasts, Florida cover sediments may contain trace to abundant amounts of other mineral and rock fragments, including phosphate minerals and rock fragments (phosphorite) and oxides of iron and titanium.

As discussed in Chap. 3, the Miocene Hawthorn Group is mined commercially in several areas of the state for phosphate, which is used in the production of agricultural fertilizer and other products. The Hawthorn Group phosphate deposits include both direct chemical precipitates of

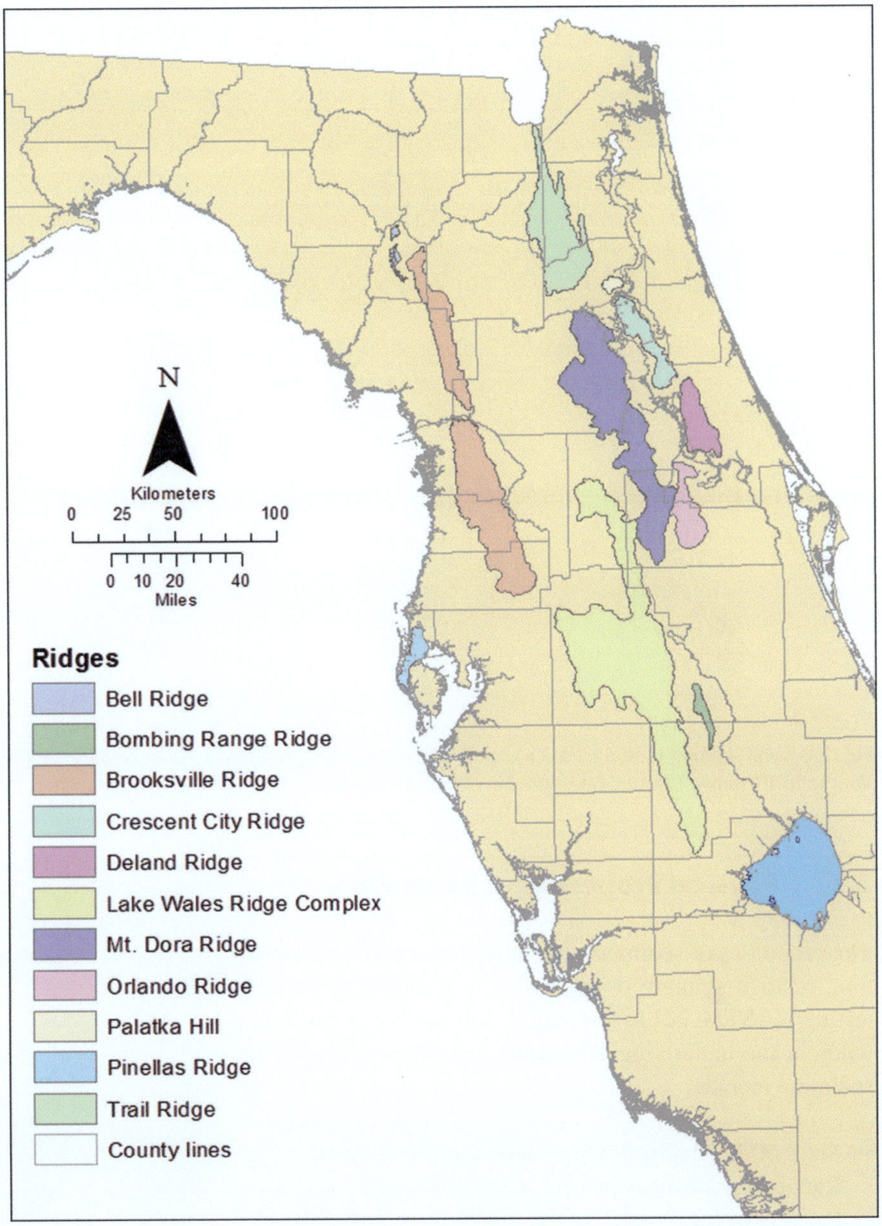


**Fig. 2.5** Prominent ridges in Florida. Based on geomorphic provinces of Williams et al. (in prep.)

phosphate minerals and replacements of pre-existing carbonate materials. Fragments of limestone and dolostone were commonly reworked by waves and currents at the time of deposition and formed clasts in the sand-rich, clay-rich sediments of the Hawthorn Group. The primary phosphate mineral in the marine phosphatic clasts is carbonate fluorapatite ($Ca_5(PO_4,CO_3)_3F$; Upchurch 1992).

## 2.4 Florida Evaporites

Evaporites are geologic strata composed of minerals formed by evaporation of seawater or some other saline solution. In Florida, the important evaporite minerals are gypsum ($CaSO_4 \cdot nH_2O$) and anhydrite ($CaSO_4$).

Gypsum crystals are occasionally found in clay-rich sediments of the Hawthorn Group (Upchurch 1992). These rare crystals are, in some cases a result of evaporation of sea water, but oxidation of the sulfur in pyrite ($FeS_2$) may also be responsible for formation of some shallow gypsum. In any event, these shallow gypsum deposits are unimportant relative to karst formation or cover behavior unless sulfuric acid ($H_2SO_4$) forms from these sulfate or sulfide sources.

Deeply buried evaporite deposits consisting of beds of gypsum and anhydrite may have been responsible for creating sulfuric acid and contributing to the hypogene dissolution of carbonate rocks in Florida (Herbert and Upchurch 2016; Upchurch 2017). This process is discussed in Chap. 8.

**Fig. 2.6** Iron staining in thin cutans and multiple soil horizons in Plio-Pleistocene dune sand in the reworked Cypresshead Formation near Lake Wales, Polk County, Florida. Location: 27.919°N, 81.584°W

## 2.5 Physical Properties of Sediments

The properties of sediments that affect the texture, composition, relative strength or resistance to standard penetration testing[1] (ASTM 2011c), mobility and ability to ravel into voids in the underlying limestone, and permeability of the sediment include:

- Sizes of the majority of the clasts in the sediment;
- Range of grain sizes present in the sediment;
- Packing of the grains into the sediment volume; and
- Angularity of sediment grains.

**Table 2.1** The Udden-Wentworth grain-size classification system

| Grain size (mm) | US standard sieve mesh (Openings/inch of screen) | Wentworth size classification |
|---|---|---|
| <256 | No standard sieve size | Boulder |
| 256–64 | | Cobble |
| 64–4 | 5 | Pebble |
| 4–2 | 10 | Granule |
| 2–1 | 18 | Very coarse sand |
| 1–0.5 | 35 | Coarse sand |
| 0.5–0.25 | 60 | Medium sand |
| 0.25–0.125 | 120 | Fine sand |
| 0.125–0.0625 | 230[a] | Very fine sand |
| 0.0625–0.002 | No standard sieve size | Silt |
| <0.002 | | Clay |

[a]The USCS uses the 200 mesh sieve (0.074 mm) as a lower size limit for classification of sand sizes (ASTM 2011a)

### 2.5.1 Sediment Grain Size

#### 2.5.1.1 Sediment Grain Size Classification

Classification of sediments into basic textural types, such as sand, gravel, silt, or clay is based on determining the proportions of different grain sizes (Table 2.1) in a sample. The sediment is named on the basis of the predominant size fraction(s) present. The most common classification scheme used in Florida follows ASTM methods (ASTM 2009, 2011a). The Unified Soil Classification System (USCS; ASTM 2011a) notation is commonly used by engineers and some geologists to describe the textures of sediments. Most geologists utilize a similar classification, but with different and more extensive descriptive modifiers.

It is important to understand that the grain-size classification applies to all siliciclastic, carbonate, and other sediments. It is simply a size classification and should not connote any mineral content or origin of the sediment. For example, the term sand is commonly used to connote quartz

[1]Standard penetration testing is the most common subsurface testing method for evidence of sinkhole development in Florida.

grains with diameters in the sand size category. In reality, the term sand can apply to carbonate, phosphate, heavy minerals, and other particles.

For visually classified soils in geotechnical projects, the Unified Soil Classification System (ASTM Method D2488 (ASTM 2009)) is a commonly utilized classification scheme. This method relies on visual inspection and sample manipulation for classification. ASTM Method D2487 (ASTM 2011a) is utilized for classification of soils and sediments according to the USCS and that have been subjected to Atterberg Limits testing and grain-size analysis.

For sandy sediments, the simplest method of determining grain-size distribution in a sample is by sieving the sample. For a full grain-size curve of sand-sized materials, the dried sample is passed through a nest of sieves that have standard sized openings. By weighing the fraction retained on each screen, the distribution, by weight, of particle sizes can be determined. Assuming that all of the particles in the sediment have the same mineral composition, the size distribution by weight is essentially the same as the volume proportions.

When fine-grained sediments are present in the sample, the USCS classification scheme utilizes washing a wet sediment sample through the 200-mesh sieve (0.074 mm; Table 2.1). The sediment is classified based on the proportions of sample retained on, and passing through, the 200-mesh screen. The proportion of the sediment fraction finer than the 200-mesh screen (−200 fraction) is utilized in the USCS soil classification system (ASTM 2007, 2011a) as a surrogate for determination of the proportions of the fine sand, silt, and clay fractions of the sample. If actual proportions of the fine sand, silt, and clay fractions are needed, the hydrometer method is utilized (ASTM 2011b).

#### 2.5.1.2 Relationship of Grain Size and Depositional Environment

As a general rule of thumb, the coarser the sediment particles, the higher the wave and current energies were at the time of deposition. However, in Florida, the range of sizes of siliciclastic sediments may be limited by source area and depositional environment. While granule- to cobble-sized, non-biogenic clasts are found on beaches and in certain strata in Florida, most of the marine sediments in Florida are predominantly fine- to medium-sand sized. Silt-sized sediment is generally limited in extent in Florida, and much of the silt-sized material in Miocene strata is dolomitic (Scott 1988). Finally, siliciclastic clay is common within the Miocene, Pliocene, and younger sediments.

#### 2.5.1.3 Grain-Size Distributions

The range and proportions of grain sizes in a sample represent the sample's grain-size distribution. Figure 2.7 illustrates some examples of grain-size distributions from Florida sands. The dominant grain size(s) determine the classification of the sample according to the USCS. For example, Fig. 2.7a would represent medium to coarse sand, 2.7b depicts fine sand, and 2.7c shows a bimodal shelly sand (the coarse mode consists of shell and the fine mode is quartz sand). However, note the differences in the total distributions of each of these samples. The size distribution of sediment is critical to understanding the mechanical behavior of the sediment and its depositional environment. The grain-size distribution is an important element in characterizing the sediments of the karst cover and for understanding their behavior during a subsidence event.

Granular, non-cohesive sediments (silty and sandy sediments) have different behaviors than cohesive, clayey sediments. These two groups of sediment types (cohesionless and cohesive sediments) are discussed separately below.

### 2.5.2 Sorting (Gradation) of Sediments

Sorting is a measure of the range of grain sizes in the sediment. If the range of grain sizes is limited, such as in the sample shown in Fig. 2.7b, the sediment is said to be "well sorted" (Leeder 2011). If, however, there is a wide range of different grain sizes in the sediment, it is "poorly sorted" (Fig. 2.7a). As suggested above, geologists are concerned with the wave and current energies present at the time of deposition. These energies sort, or select for transport, different grain sizes. Thus, geologists are concerned with sorting as a process.

A different terminology is utilized in soil mechanics for the same data. This is because there is concern for the ability of sediment to support loads and remain stable. In soil mechanics, poorly sorted sediments with a wide range of grain sizes are best for supporting loads. Therefore, the proportions of the different "grades", or sizes, of sediment are important, and

- The grain-size distribution curve is termed a gradation curve;
- "Poorly sorted" sediment is said to be "well-graded", because it includes a wide range of grain sizes (Fig. 2.7a);
- "Well-sorted" sediment is termed "poorly graded" sediment, because the range of sizes is limited (Fig. 2.7b); and
- "Bimodal" sediment is termed gap-graded, because there is a gap in grain sizes between the two size modes of sediment (Fig. 2.7c).

The difference in terminology reflects traditional concerns of the two disciplines. Geologists are typically concerned with depositional processes, especially wave and current energy at the time of deposition, and diagenetic changes that have occurred in the sediment during and after deposition,

**Fig. 2.7** Grain-size distributions of sand samples from Florida. (**a**) Poorly sorted (well-graded) medium sand from the Cypresshead Formation in Putnam County. (**b**) Well sorted (poorly graded) fine, dune sand from Bay County. (**c**) Bimodal (gap-graded) beach sand from Pinellas County

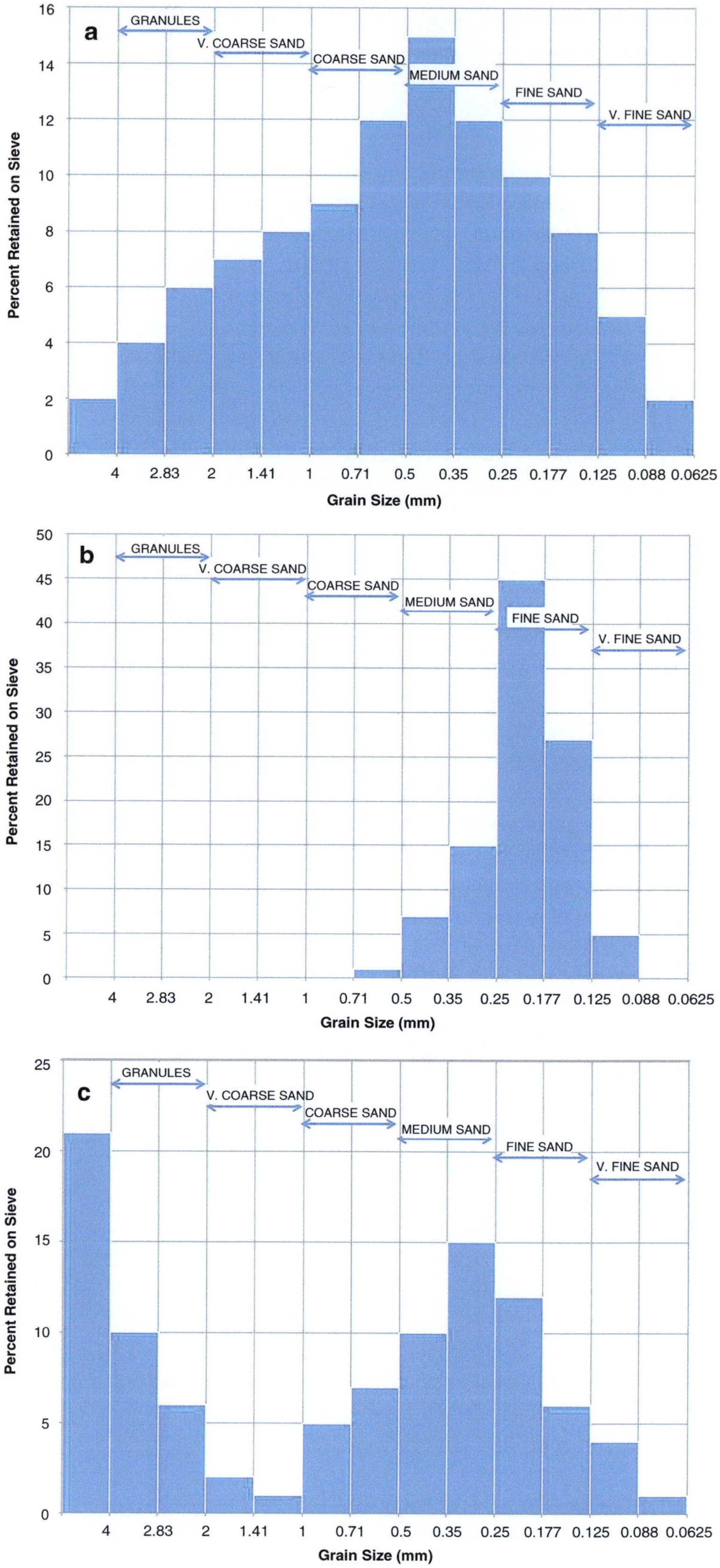

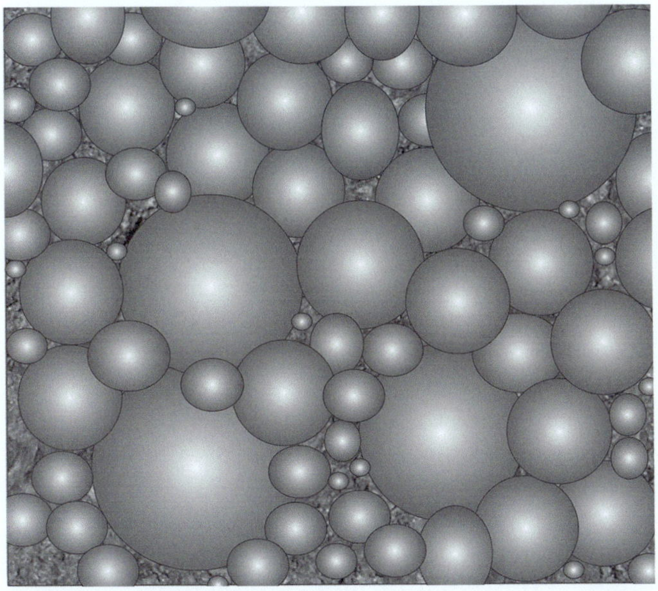

**Fig. 2.8** Hypothetical, close packed, poorly sorted (well-graded) sediment composed of spheres

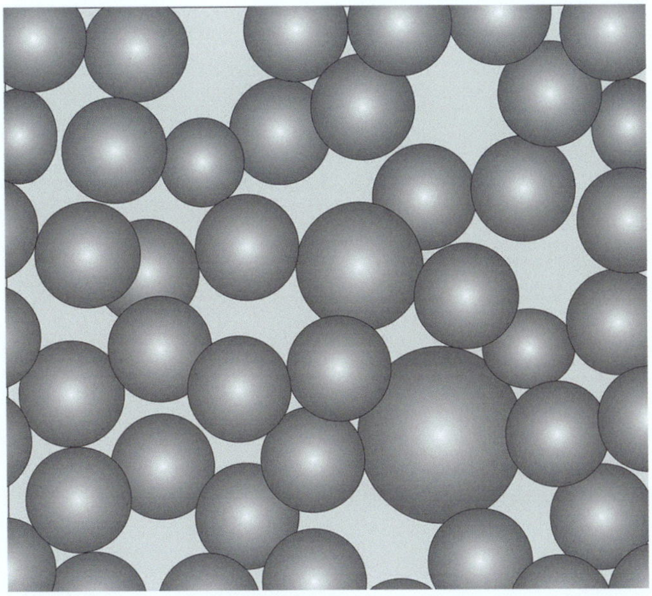

**Fig. 2.9** Hypothetical, open packed, well-sorted (poorly graded) sediment composed of spheres. Note the large amount of open space between grains

while engineers are more concerned with soil mechanics, especially the suitability of the sediment for construction. Gradation terminology is important for understanding relative sediment strengths and karst formation, so it is commonly used by both geologists and engineers in Florida when discussing the behavior of cover sediments in a karst-related scenario.

Because well-graded sand (Figs. 2.7a and 2.8) has a variety of different grain sizes, the larger grains are likely supported by smaller grains that fit in what would otherwise be pore space between the larger grains. This results in stronger sediment that is more likely to be suitable for construction. It is also more likely to have higher penetration resistances during standard penetration testing and resistance to raveling into voids in the underlying limestone as a result of the smaller grains interlocking and supporting the larger ones. Figure 2.7a is a grain-size distribution of sediment from a Pliocene beach preserved in the Cypresshead Formation near Davenport, Polk County, Florida. Granules are present, and in some areas even small cobbles and pebbles are known to occur. This sediment is poorly sorted or well-graded.

In contrast, Fig. 2.7b is fine sand with a limited range of grain sizes. This sample is dune sand reworked from the Cypresshead Formation near Lake Wales, Florida. It is well sorted, or poorly graded, owing to the limited range of grain sizes. Such well-sorted sand is generally less strong and somewhat less capable of supporting a structure than poorly sorted sediment with a diversity of different grain sizes. The lower relative strength is, in part, a result of the more uniform distribution of grain sizes (Fig. 2.9). The spaces between the grains are generally open and the grains are freer to move relative to each other. Well sorted (poorly graded) sand, especially wind-blown, dune sand is more likely to compact when a load is applied, exhibit creep down slopes, and fail if overlying a void – a sinkhole scenario.

Figure 2.7c is an example of a bimodally sorted grain-size distribution. In this case, the sediment distribution is of a sample from Caladesi Island, a modern barrier island in Pinellas County. The finer of the two modes (the distribution less than 1 mm in diameter) is fine to medium quartz sand with a minor shell fraction caused by breaking of the shells in the waves. The coarse mode (>1 mm) is shell derived from marine mollusks that lived offshore and whose shells were transported by waves and currents onto the beach. Such mixed sediments have variable strength behaviors depending on the proportions and sizes of the clasts. The sample shown in Fig. 2.7c has a high penetration resistance and is relatively strong, because of the coarse shell fraction and the fact that the sand-sized material supports the shell fragments.

Much of Florida's sand mantle is well-sorted (poorly graded) sand because of the limited range of different grain sizes that was available at the time of deposition. The quartz and other clasts that constitute the sand fraction are the result of the erosion in the ancestral Appalachian Mountains millions of years ago. The clasts were transported to Florida by waves, currents, and the wind. As they were transported, they were sorted, abraded, and rounded. By the time they arrived at their current locations, the siliciclastic grains had (and have) limited sizes.

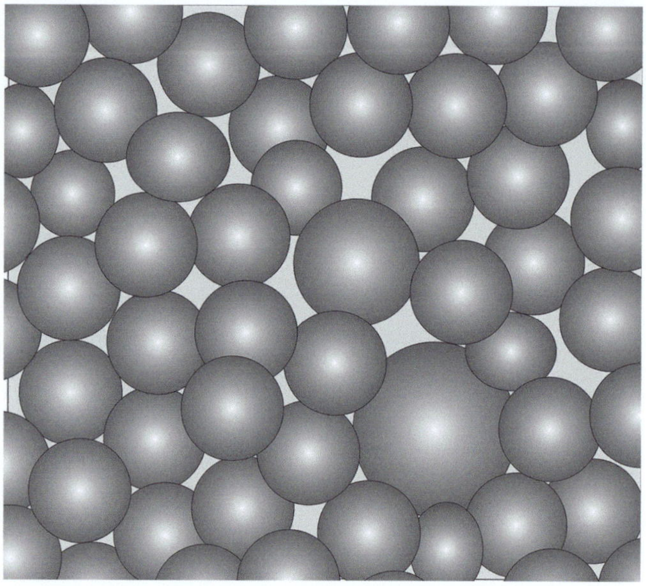

**Fig. 2.10** Hypothetical, well-sorted, close-packed arrangement of sediment grains represented by spheres

### 2.5.3 Sediment Packing

The concept of sediment packing is often overlooked when characterizing sediment, especially in sediment samples retrieved during penetration testing. Packing is a term used to describe the closeness of sediment grains to each other. Compare, for example, the sediment grains in Figs. 2.9 and 2.10. In Fig. 2.9, the grains are in contact and the sediment is grain-supported but there are large openings between the grains. Clearly the individual grains could be rearranged to fit together in a closer configuration. This is an example of open-packed sediment. Open-packed sediments typically form in lower energy, quiet depositional environments where the grains settle out of the wind or water and come to rest without the agitation necessary to cause them to come to rest in a close-packed arrangement.

The grains in the sediment sample shown in Fig. 2.10 are close packed. That is, the grains are arranged in such a way that the number of contact points between grains is maximized and the pore space is minimized. Close-packed sediments are typically deposited in higher energy environments (i.e., fast moving streams or wave-swept swash zones of a beach) where agitation causes the grains to seek the most efficient packing arrangement.

Open-packed sediments can be rearranged to a closer packed arrangement. In granular sediments this rearrangement is termed compaction. In order for sediment to naturally compact, time and great burial depths are required. Engineers utilize vibratory energy and artificial loading to enhance packing prior to construction on shallow, poorly compacted sediment. This anthropogenic compaction causes sediment grains to come closer together at shallow depths, where construction requires a stronger substrate. However, much more time and pressure is needed to cause sediments to compact at depth in natural systems. In natural systems, burial on the order of thousands of meters and times on the order of millions of years are needed to cause significant compaction.

Close-packed granular sediments cannot be compacted without grain reorientation, breakage, shattering, and/or, in some cases, dissolution of the grains at points of contact. This process is discussed below but suffice it to say that the heat and pressure associated with burial under thousands of meters of sediment are required for such events to occur (Hedberg 1926; Athy 1930; Weller 1959; Chilingarian and Wolf 1975, 1976).

### 2.5.4 Consolidation and Compaction

#### 2.5.4.1 Compaction of Granular Sediments

The transition from open packing to close packing requires that pore water must migrate from pore to pore in order to allow for the reduction in pore[2] space as compaction proceeds. In coarse, well-sorted granular sediments with abundant pore space, water can move freely to adjust for loss of pore volume during compaction. Filling of the pore space with fine-grained sediment clasts reduces the available pore space for water movement and reduces the potential for compaction.

#### 2.5.4.2 Consolidation of Fine-Grained Sediments

The transition from open-packed clay to close-packed clay, which naturally has very low permeabilities, is called consolidation. When deposited, clay-rich sediments often have very large amounts of pore space (over 50% by volume in many cases). Consolidation of clay deposits is dependent on burial pressure (load), time, and the availability of a mechanism and pathway for water to be forced out of the pore spaces within the clay (Hedberg 1936; Meade 1964; Rieke and Chilingarian 1974; Gordon and Flemmings 1998; Puttiwongrak et al. 2013). Clay deposits have low permeabilities because of electrochemical interactions of the clay particles and water molecules (Sect. 2.6.4). Therefore, pore water in poorly consolidated clay has a difficult time migrating out of the volume being consolidated; the incompressible water tends to resist consolidation. Burial pressures and long periods of time are typically required for water to migrate out of clay beds and for consolidation to occur.

[2] In this text, the term pore is utilized to describe the intergranular openings between particles in sediments. The term void is used to describe larger, dissolution-created openings.

Normally consolidated clay has the appropriate pore space and water content for the depth of burial and elapsed time since it was deposited. The shallow clays encountered in Florida have never been buried more than perhaps 100 m, so the amount of consolidation of clay beds is generally limited. In many areas of Florida, especially within the Hawthorn Group sediments, shallow clay deposits (depth < 100 m) can be under-consolidated. That is, the clay has not undergone consolidation consistent with its depth of burial because water could not migrate from the interstitial pores and allow for the packing of the clay to adjust to its load. These under-consolidated clays have been identified as a contributing factor in sinkhole development in west-central Florida (Upchurch et al. 2013).

#### 2.5.4.3 Matrix Porosity and Permeability in the Floridan Aquifer System

The Floridan Aquifer System (FAS; see Chap. 3) includes the Paleogene carbonate rocks of peninsular Florida. It is within this aquifer that most of Florida's well-developed karst exists. Because of its ability to supply high quality groundwater, the FAS has been intensely investigated.

Budd and Vacher (2004, p. 531) stated that "The Upper Floridan Aquifer of peninsular Florida retains most of its depositional porosity." This statement suggests that compaction or consolidation have been limited in the hundreds of meters of Eocene to Oligocene carbonate rocks of Florida. There have been, however, some alterations in the matrix permeability of the limestones and dolostones that comprise the aquifer. Budd (2001) noted a decrease in matrix permeability with depth that accounted for 22–31% of the variance in permeability data. Based on petrographic data, he concluded that this loss of permeability was a result of dolomitization and cementation, not compaction or consolidation. In grain-supported limestone, he noted evidence of mechanical compaction and onset of grain-to-grain pressure solution.

In other words, the matrix porosity and permeability of Florida's eogenetic limestone and dolostone reflect evidence of depositional conditions and only minor changes as a result of burial. Superimposed on the matrix conditions are the effects of fracturing, dissolution, and development of touching vug pore and void networks. As a result, the carbonate rock column has become one of the most productive aquifers in the world, and the carbonate rocks remain textbook examples of eogenetic, doubly porous and permeable strata.

#### 2.5.4.4 Effect of Sediment Texture on Penetration Resistance and Sediment Strength

As a general rule of thumb, granular and clay-rich, open-packed sediments are relatively weak under standard penetration testing. In contrast, close-packed sediments are stronger. Experience with standard penetration testing in Florida indicates that shallow Florida sediments have no or minor compaction or consolidation, which means that sediment strengths reflect the grain-size distributions and packing derived from the depositional environment more than consolidation or compaction.

Fig. 2.11 Photomicrograph of water-transported, fine sand from the Oligocene Suwannee or Eocene Ocala limestone in Hendry County. Note lack of surface frosting and angularity of the grains

### 2.5.5 Grain Shape

Grain shape also controls the strength and ability of granular sediments to undergo compaction and to migrate, either through raveling or soil creep. Shapes range from angular to well-rounded (see Leeder 2011) for classification of grain shapes). Figures 2.11 and 2.12 illustrate photomicrographs of selected Florida sandy sediments. Most of the grain shapes in both samples are subrounded to subangular. These are the most common grain shapes in the shallow quartz sands of Florida. Fragmented shell is commonly angular to subangular, while pebbles and cobbles are typically well rounded in Florida.

Figure 2.13 is a diagram showing the effect of grain angularity on the ability of grains to support a load. The sketch is of well-sorted (poorly graded) sand with angular shell fragments. The presence of angular grains tends to increase sediment resistance with loading and during standard penetration testing. This is because subangular clasts tend to lock together. If highly angular grains, such as the sliver-like shell fragments in the sketch shown in Fig. 2.13 are present, they tend to wedge against and between the more rounded grains, retarding compaction and increasing overall relative sediment strength.

**Fig. 2.12** Photomicrograph of aeolian dune sand from the reworked Cypresshead Formation. Note that the sand grains are somewhat more rounded than the grains in Fig. 2.11 and that they have frosted surfaces

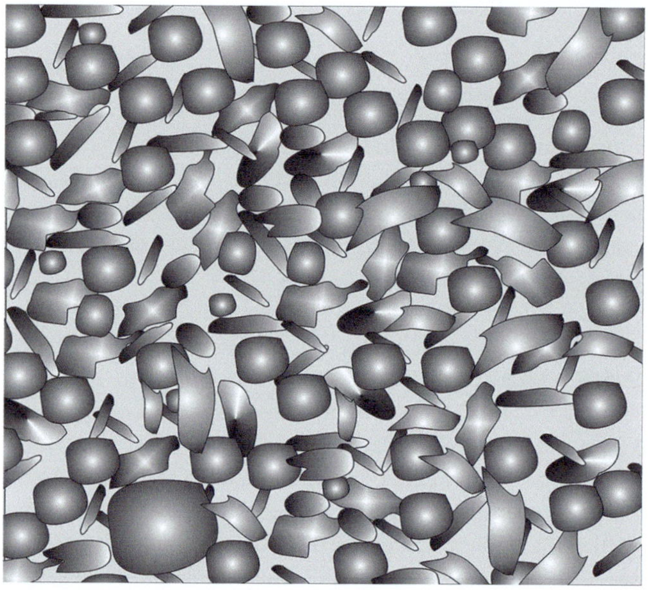

**Fig. 2.13** Poorly sorted, granular sediment with a variety of grain shapes. Note that the angular grains tend to wedge between better rounded grains and lock the grains together

**Fig. 2.14** Poorly sorted sand with frosted grain surfaces and kaolinite coatings, a weak cement. Reworked Cypresshead Formation from the Lake Wales Ridge, Polk County. Location: 28.182°N, 81.577°W

### 2.5.6 Grain Frosting

There is an abundance of loose, eolian (wind-blown) dune sand in Florida. If loose-packed, this sand can undergo some consolidation under the load of a poorly founded structure and, thus, cause damage. If the eolian sand is located on sloping ground, the sediment can creep down-slope and cause damage. Thus, poorly consolidated, eolian sand deposits may constitute a geologic hazard. Therefore, it is important to be able to identify eolian sand grains. Owing to the low density of air as compared to water, transport of sand grains by the wind is limited to certain grain sizes, usually less than 0.05 mm. Consequently, dune sands are normally very well-sorted (poorly graded) fine sand (Fig. 2.7b), the grains are subangular to rounded, and grain surfaces are often frosted (Fig. 2.14). The frosting is a result of grains striking each other without an intervening cushion of water. Grains transported in water do not abrade as quickly because the water cushions grain-to-grain impacts. Therefore, aqueous-transported quartz sand typically retains outer surfaces that are relatively bright and unabraded in appearance (Fig. 2.15).

### 2.5.7 Cements and Coatings

The presence of cements and grain coatings may affect the ability of cohesionless sediments to ravel, to transmit groundwater, and relative resistance during standard penetration testing. Cements are binding agents that essentially "glue" individual grains together. The degree of cementation may range from weak to very strong, and the cement may be localized and non-uniform within a sand section (Fig. 2.6) or it may constitute an aerially extensive horizon (Fig. 2.16).

The nature of the material cementing granular, host sediment can be instructive in terms of the genesis of any soil horizons superimposed on the host sediment and subsequent karst activity. Disruptions in soil horizons and sagging of cemented zones, for example, may suggest movement of the sediment related to karst processes in the underlying carbonate rock. Table 2.2 lists some common sedimentary cementing agents found in Florida strata.

Figures 2.6 and 2.16 illustrate ferric-hydroxide-cemented sand horizons. Colors vary, but are generally red or brown tones because the cementing agent is ferric hydroxide. These cements are sufficiently strong to allow for the mine faces to remain vertical and resist slumping. Figure 2.3 illustrates a sand sample that has been cemented by the clay mineral kaolinite. Kaolinite is a weathering product of potassium feldspars and is most common in the Pliocene Cypresshead Formation. Figure 2.17 is a well-cemented example of an organic hardpan formed at the water table in a wet soil. Cements of sediments with quartz, sedimentary opal, and aluminum-phosphate minerals are locally important in Florida sediments.

Coatings are mineral or organic films that encase individual grains in sandy sediments. While they do not cement sand bodies, they may cause local binding of grains together and cause the particles to resist raveling or slumping. Figure 2.14 is an example of poorly sorted sand from the reworked Cypresshead Formation. The grains have coatings of kaolinite, which cause the sand grains to weakly adhere to each other. Figure 2.15 depicts a sample of well-sorted, quartz sand with thin ferric hydroxide coatings.

**Fig. 2.15** Well-sorted quartz sand with ferric-hydroxide coating. Note that many of the grain surfaces are not frosted. Reworked Cypresshead Formation from the Lake Wales Ridge, Polk County. Location: 27.919°N, 81.584°W

## 2.6 Clay and Cohesive Sediments

Clay-rich sediments behave differently than the granular and cohesionless sediments discussed in Sect. 2.5. Because of these differences, it is important to characterize Florida's clay-rich sediments and the role they play in karst development and as geologic hazards in their own right.

Use of the word clay can result in confusion because there are two different definitions and uses of the term. One use of the term clay refers to very small sized sediment particles and the other use refers to a family of minerals (clay minerals) with well-defined crystal arrangements,

**Fig. 2.16** Layers of sand cemented by ferric hydroxides. Exposure is of the Cypresshead Formation in a mine (29.829°N, 81.968°W) near Keystone Heights, Putnam County

chemical compositions, and intergrain interactions. We begin with a discussion of the size and mineralogical aspects of clay in Florida.

### 2.6.1 Clay Size Definitions

As a grain-size designation, the term clay can be confusing because it has several very different definitions (Table 2.3). Today, 2 microns (μm) is utilized by most geologists and engineers to characterize the maximum dimension of clay-sized particles. However, many sedimentologists still utilize Krumbein's (1934) phi scale, which is based on conversion of sediment sizes to a $\log_2$ scale. Use of the phi scale normalizes certain grain-size distributions (Upchurch 1970a). The maximum grain size of clay according to the phi scale is 4 μm (Table 2.3).

**Table 2.2** Some common minerals, mineraloids, and organic materials that form cements in Florida sediments and rocks

| |
|---|
| Calcite |
| Ferric hydroxide |
| Kaolinite |
| Opal-A, opal-CT |
| Quartz (chalcedony) |
| Peat, muck |
| Aluminum-phosphate minerals (wavellite, crandallite, millisite, etc.) |

The widely utilized classification schemes of Udden (1898) and Wentworth (1922) have been combined by applying Wentworth's clay size definition to Udden's scale, which resulted in the Udden-Wentworth Scale (Table 2.1). See Krumbein and Pettijohn (1938) for a discussion of the evolution of these grain-size classification schemes.

The Unified Soil Classification System (ASTM 2011a) uses a complex definition based on the size and mechanical properties of the sediment (footnote, Table 2.3). In essence, the USCS classification is a hybrid classification that combines a grain-size component and the cohesive properties of the sediment. The latter is a function of the mineralogy and chemical conditions of the sediment. Even though the maximum size of "clay" according to the USCS is 75 μm, which incorporates fine sand and silt sizes, qualifying the clay designation by use of the Atterberg limits restricts the clay designation to cohesive sediment. The USCS terminology is advantageous in an engineering context as well as for discussion of geologic hazards, such as clay shrinkage and swelling.

In this text, the Udden-Wentworth Scale (Table 2.1) is used to define the maximum grain size of clay particles at 2 μm. However, the USCS terminology is used to describe clay-rich strata and discuss the mechanical behavior of clay deposits in the context of stratigraphy, shrink/swell behavior, and sinkhole development.

**Fig. 2.17** An organic (spodic) horizon in a clay pit (29.325°N, 82.178°W) near Lowell, Marion County. Note that the organics have cemented the sand so well that the exposure remains vertical with no slumping

**Table 2.3** Definitions of clay based on grain size

| Classification systems | Maximum clay size designation (microns; µm) | References |
|---|---|---|
| Udden Size Classification | 4 | Udden (1898) |
| Atterberg Scale | 2 | Atterberg (1905) |
| Wentworth Size Classification | 2 | Wentworth (1922) |
| Krumbein's Phi Scale | 4 | Krumbein (1934) |
| International Size Scale (ISO 14688-1) | 2 | ISO (2002) |
| Unified Soil Classification System (USCS; ASTM D 2487) | 75[a] | ASTM (2011a) |

[a]Size term based on sediment passing the No. 200 (75 µm) U.S. Standard Sieve and exhibiting plasticity. According to the USCS, clay has a plasticity index (PI) ≥ 4 and plots on a graph of PI versus liquid limit (LL) on or above the "A" line of Casagrande (1948)

**Table 2.4** Clay minerals and other materials that may be clay-sized in Florida cover materials

| Clay minerals with sheet structures | Other common, clay-sized (≤2 µm) minerals |
|---|---|
| Kaolinite<br>Illite<br>Smectite Group<br>Montmorillonite, Beidellite, Saponite, Nontronite<br>Chlorite Group<br>Mixed-Layer Clay Minerals | Carbonate Minerals<br>Calcite, Aragonite, Dolomite<br>Silicate and Silica Minerals<br>Feldspars (Albite, Potassium feldspars), Micas (Muscovite, Hydromuscovite), Opal-CT<br>Phosphate and Related Minerals<br>Apatite Group (Carbonate fluorapatite, Carbonate hydroxyapatite), Aluminum- Phosphate Minerals (Wavellite, Crandallite, Millisite) |
| **Fibrous clay minerals** | |
| Palygorskite<br>Sepiolite | Sulfides<br>Pyrite, Hydrotroilite<br>Oxides<br>Ferric oxyhydroxide, Goethite |

### 2.6.2 Clay Minerals

Clay minerals consist of a family of different aluminosilicate and silicate minerals with different crystallographic arrangements and/or chemical compositions. The physical properties of clay-sized particles vary depending on the crystallography, chemical composition, and size of the particles. Therefore, it is important to understand the chemical and crystallographic varieties of clay minerals that are common in Florida.

Clay minerals are normally clay sized (≤2 µm), but they may be coarser. Clay minerals have specific crystal structures that make them unique. Most of the clay minerals are hydrous aluminum phyllosilicates[3] with iron, magnesium, calcium, potassium and other metals incorporated in the crystal lattice. The word phyllosilicate suggests that the clay minerals typically form flat sheets, similar at the microscopic scale to sheets of mica.

The sheet structures of most clay minerals (Table 2.4) are caused by coordination of silica tetrahedra ($SiO_4^{-4}$) and aluminum octahedra ($AlO_6^{-6}$). Silica tetrahedra form the basic "building blocks" for all silicate minerals. The silica tetrahedron consists of a single silicon atom surrounded by four oxygen atoms (Fig. 2.18a). Each tetrahedron has triangular sides with oxygen atoms at the apices. The silica tetrahedron has four unsatisfied charges, one at each oxygen atom located at the points of the tetrahedron. These charges coordinate and share electrons with silicon atoms belonging to adjacent tetrahedra forming a sheet.

In the same fashion, the aluminum octahedron consists of an aluminum atom coordinated with six oxygen atoms to form an octahedron (Fig. 2.18b), a shape with eight triangular faces. The oxygen atoms at the points of each triangle also coordinate with adjacent aluminum atoms to form sheets of aluminum oxide. The clays with sheet structures consist of stacks of alternating sheets of silica tetrahedra and aluminum octahedra in different arrangements and with different metals substituted or coordinated within the crystal lattice (Grim 1953; Velde 1995).

Examination of Fig. 2.18a, b reveals an important result of building minerals with combinations of the two functional groups. Construction of these minerals results in outer mineral surfaces that are populated by oxygen atoms with a net negative charge. Thus, clay minerals have outer surfaces that can interact electromagnetically with surrounding minerals, water, and other substances.

#### 2.6.2.1 Mineralogy of Florida Clays: Overview

The most widespread group of Florida's clay minerals consists of sheets of silica and aluminum coordinated with oxygen bound together in different arrangements (Table 2.4). In addition to the clay minerals with sheet structures, the Miocene Hawthorn Group of sediments contains two important clay minerals that have fibrous structures (Table 2.4). Finally, a number of other minerals (not clay minerals) occur as clay-sized (≤2 µm) particles in Florida clay deposits. These, too, are listed in Table 2.4. The following sections discuss these minerals, beginning with the clays with sheet structures.

#### 2.6.2.2 Kaolinite

Kaolinite is the simplest of the aluminosilicate clay minerals. Each kaolinite crystal consists of a stack of alternating silica tetrahedra and aluminum octahedral sheets (Fig. 2.19), giving the mineral the formula $Al_2Si_2O_5(OH)_4$ (Back and Mandarino 2008).

[3]Phyllosilicates are silicate or aluminosilicate minerals with a "platy" or "sheet-like" crystal structure.

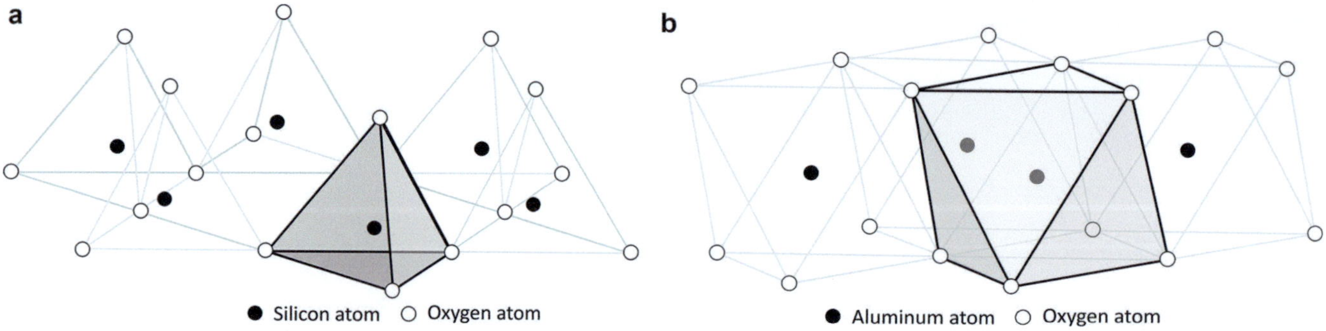


**Fig. 2.18** (**a**) A sheet of silica tetrahedra with a single, four-sided tetrahedron highlighted. (**b**) A sheet of aluminum octahedra with a single, eight-sided octahedron highlighted

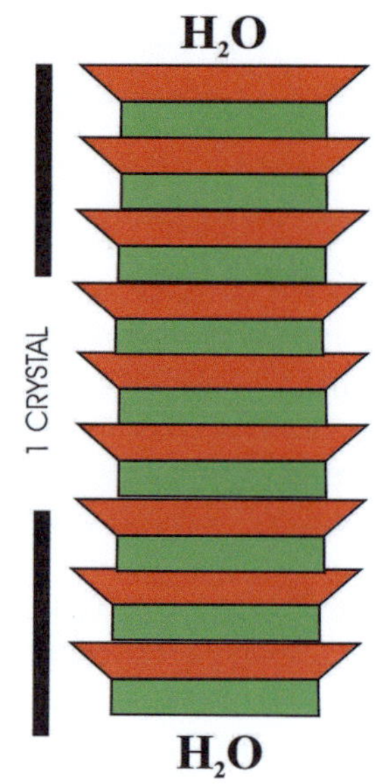


**Fig. 2.19** Generalized structure of kaolinite. Red represents the silica tetrahedra layers and green represents the aluminum octahedral layers. Individual crystals consist of repetitions of these two basic layers

**Table 2.5** Atterberg limits data from different clay minerals. Whole sample data only; data stratified by particle size are not included

| Mineral | Plastic limit (%) | Liquid limit (%) | Plasticity index (%) |
|---|---|---|---|
| Kaolinite | 29.8–39.3 | 35.0–71.6 | 5.1–32.3 |
| Illite | 23.9–35.7 | 29.1–111.2 | 5.2–58.3 |
| Smectite | 81.4–109.5[a]<br>10.7–51.0[b] | 117.5–700[a]<br>21.0–170.0[b] | 361–603[a]<br>3.0–124.0[b] |
| Palygorskite[c] | 116.6 | 177.8 | 61.2 |

Modified from White (1949) and from the senior author's data for Florida samples

[a]Includes volcanogenic clays from South Dakota

[b]Based on 143 native Florida clay samples processed by the senior author

[c]Based on one sample from Quincy, Florida

Note that there are no openings between the sheets within the crystal, so water and ionic constituents can only coordinate with the unsatisfied electrical charges at the outer edges of the stack of silica and aluminum sheets. This lack of locations where water, and ionic or electrically charged constituents dissolved in the water, can be coordinated within the crystal gives the clay low expansive properties and relatively low plasticity (Table 2.5).

Kaolinite is a fairly common clay mineral in certain areas of Florida, especially in northeastern Florida, where it is mined from deposits of the Cypresshead Formation (Fountain 2009; Figs. 2.20 and 2.21) and in soil weathering profiles in the post-Miocene paleosol (Van Kauwenbergh et al. 1990) in central Florida. Outside of the Cypresshead deposits and weathering profiles in the upper Hawthorn Group, kaolinite is usually a minor constituent.

Kaolinite is derived from weathering of feldspars (Fountain 2009), smectite (Amouric and Olives 1998), and other minerals (Van Kauwenbergh et al. 1990).

#### 2.6.2.3 Illite

Illite is a group of common clay minerals in sedimentary rocks, although seldom reported in Florida clay strata. As the general formula, $(K,H_3O)(Al,Mg,Fe)_2(Si,Al)_4O_{10}[(OH)_2.H_2O]$ (Back and Mandarino 2008), suggests, the composition of illite is complex, and the illite structure (Fig. 2.22) is often found interlayered with layers that have the structure and composition of smectite or chlorite (see below).

Illite is a weathering product of potassium-rich minerals, such as the potassium micas and potassium feldspars. Clay particles that consist of mixtures of illite and smectite layering are common in some partially weathered Florida clay deposits. Interlayering with smectite and other clay structures reduces the ability of the clay to shrink and swell. As a result, use of Atterberg limits data to quantify the plasticity of the clay and verify its swelling potential is recommended.

#### 2.6.2.4 Smectite Group

Smectite is a group of clay minerals that consists of a silica tetrahedra sheet coordinated with an aluminum octahedra

Fig. 2.20 Exposures of the Pliocene Cypresshead Formation that contains kaolinite, an abundant constituent in the Cypresshead sand deposits of northeastern Florida. Location: 29.829°N. 81.968°W, near Keystone Heights, Putnam County

Fig. 2.21 Kaolinite concentrate after the Cypresshead sand has been beneficiated to separate the kaolinite from the sand matrix

sheet and a second silica tetrahedra sheet (Fig. 2.23) forming a three-layer basic structure, much like a cream-filled cookie. This basic building block represents a single crystal that is separated and weakly attracted to adjacent crystals by the electrical charges of water molecules in the spaces between individual crystals.

On a volume basis, smectite minerals have extraordinary surface areas (Figs. 2.23 and 2.24a) onto which the electromagnetically polar water molecules can be attracted. This water is exchangeable, so the minerals of the smectite group have the ability to absorb[4] large amounts of moisture. When moisture is absorbed to the crystal surfaces, aggregates (beds) of smectite have the capacity to swell dramatically. Conversely, the clay sediment body will shrink as the water is removed from between the stacked crystals.

While all accumulations of clay minerals have some capacity to shrink and swell with changes in moisture content, smectite is by far the most active. This ability to shrink and swell with changes in moisture content can cause large changes in sediment volume and has the potential to damage surface structures affected by the volume change. Expansion and contraction of clay during wetting and drying cycles have been suggested as a mechanism of cover failure and sinkhole formation.

The term used for clays that have the capacity for large volume changes with moisture content is expansive. Expansive clays in arid environments are typically dehydrated (Chen 1988), so the problem with expansive clay in arid areas is typically damage caused by clays swelling as they hydrate during high soil-moisture events.

[4]A distinction is made between adsorption and absorption. If water simply coordinates with the external surfaces of a clay mineral, such as kaolinite, is it said to adsorb onto the mineral. If the water can enter into interlayers, such as in smectites, the water is said to absorb. If the process of hydration is unknown, the term sorption is utilized.

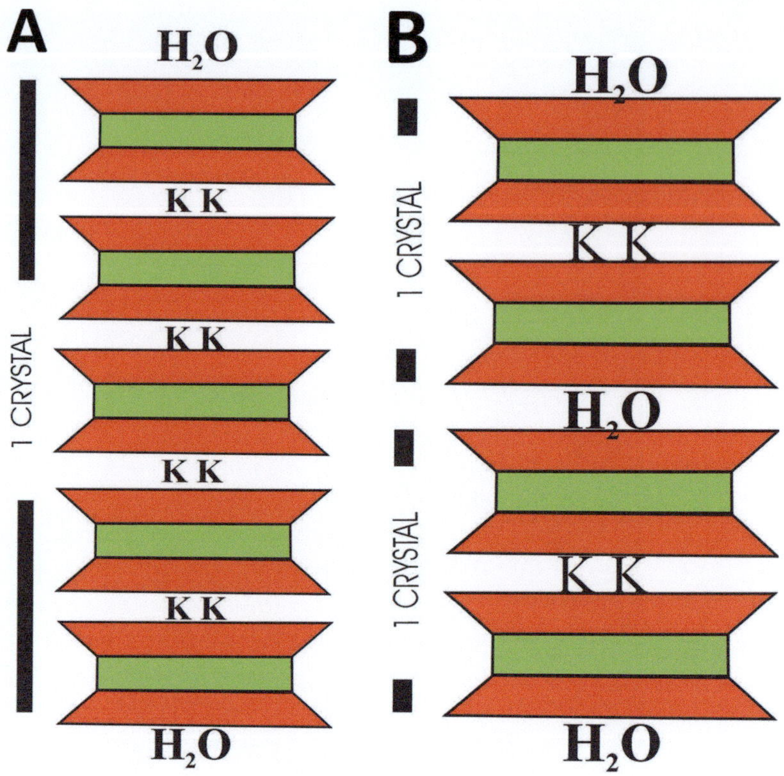


**Fig. 2.22** Examples of the structure of illite (**a**) and a mixed-layer illite (**b**). Red sheets represent silica tetrahedra layers and green represents aluminum octahedral sheets

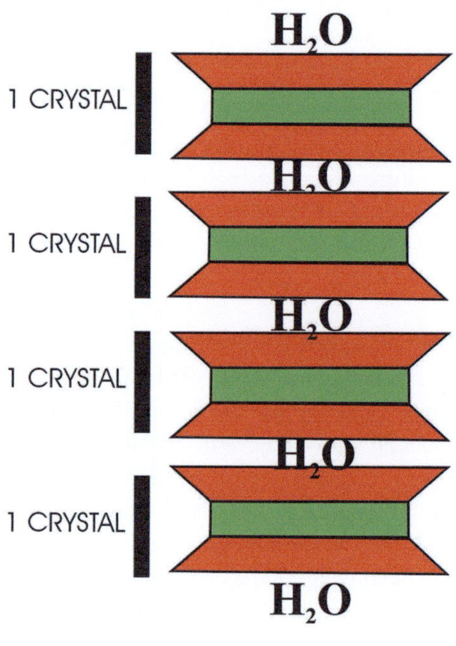


**Fig. 2.23** Example of the structure of smectite. Red sheets represent silica tetrahedra layers and green represents aluminum octahedral sheets

In Florida and elsewhere in the eastern United States, the climate is humid and water tables are high (Nelson and Miller 1997), so shrinkage of expansive clay as it dehydrates is a major cause of damage.

There is a dispute as to the depth of a bed of smectite at which shrinking and/or swelling can cause damage. Schmertmann and Crapps (1980) investigated the effects of expansive, smectitic clays in Florida and found that clay beds buried as deep as 6 m can cause damage when the clay dehydrates. Shrinking of clay allows overlying materials to subside in the direction of gravitational attraction. Therefore, damage caused by shrinkage can occur when the clay is as deep as 6 m. Conversely, if dehydrated clay is rehydrated and caused to swell, the upward pressure exerted by the clay is against the pull of gravity, so heaving caused by clay swelling may be limited to clay beds at shallower depths, typically about 1.5 m.

The "active zone" of expansive clay is the zone within which moisture changes are likely. These changes can be seasonal and driven by rainfall, they may be related to demands for water by plants, or they may result from human activities, such as modifications of drainage. In each case, repeated wetting and drying of expansive clay over limestone with voids in contact with the cover may induce cover failure by spalling and sinkhole development.

In Florida, at least four different smectite minerals have been identified. These are

- Beidellite – $(Na, Ca_{0.5})_{0.33}Al_2((Si, Al)_4O_{10})(OH)_2 \cdot nH_2O$, an aluminum-rich variety of montmorillonite;
- Montmorillonite – $(Na, Ca)_{0.33}(Al, Mg)_2Si_4O_{10}(OH)_2 \cdot nH_2O$, a sodium-calcium smectite (Fig. 2.25);
- Nontronite – $Na_{0.33}Fe_2^{+3}(Si, Al)_4O_{10}(OH)_2 \cdot nH_2O$, an iron-rich smectite; and

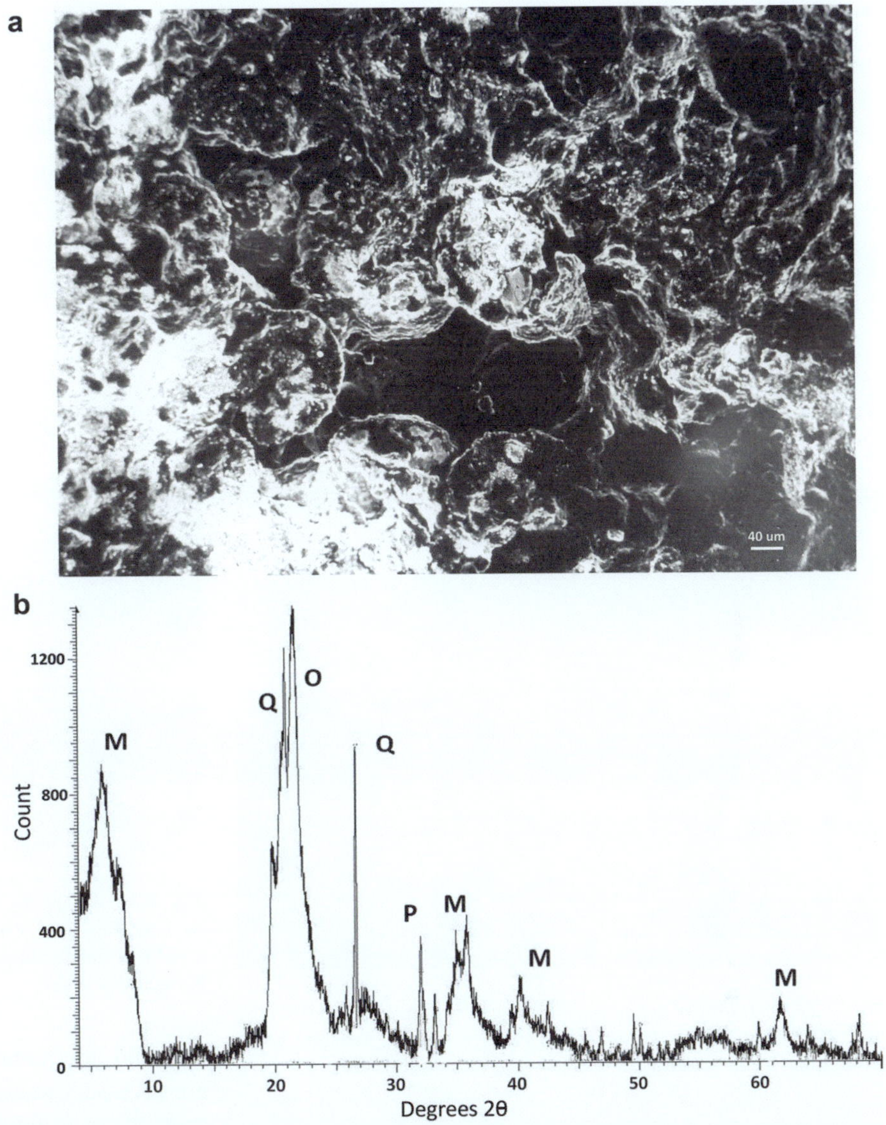


**Fig. 2.24** (**a**) SEM photomicrograph of smectite platelets in the Miocene Peace River Formation, Bartow, Polk County, Florida. (**b**) X-ray diffractogram of an expansive clay sample from Polk County, Florida. Sample was heat treated to force crystallization of the montmorillonite (M). Q = quartz, O = opal-CT, P = carbonate fluorapatite

- Saponite – $(Ca/2, Na)_{0.33}(Mg, Fe^{+2})_3 (Si, Al)_4O_{10}(OH)_2 \cdot 4H_2O$, a magnesium and iron smectite (mineral formulae from Back and Mandarino 2008).

The amount of true montmorillonite found in Florida is debatable, although it has been widely reported. Much of the material described as montmorillonite in Florida is probably beidellite (McClellan 2013), nontronite[5] (Strom and Upchurch 1985a), or saponite (RN Strom 1984 pers. comm.). Unless chemical and/or x-ray diffraction analyses (Fig. 2.24b) are undertaken to accurately characterize the clay mineral(s) in a sample, it is best to use the more encompassing, general classification of "smectite" to refer to the highly expansive clays in Florida. Note that all of the smectite minerals are expansive and X-ray diffraction analysis is not necessary to determine expansive properties if the Atterberg limits data indicate sufficient moisture sorption over a large range of plasticity.

Atterberg limits[6] (Atterberg 1905, 1911) data should be used as a minimum testing protocol to determine the

[5]The nontronite in Hawthorn Group clay of the Central Florida Phosphate District is a dioctahedral, iron-rich variety of smectite that is "about midway between montmorillonite and nontronite" (Strom and Upchurch 1985a, p. 118),

[6]Atterberg limits relate to the plastic behavior of clay as the moisture content of the sediment varies. There are two limits that are commonly identified for characterization of the properties of Florida clay. These are (1) the plastic limit, the moisture content below which cohesive sediment ceases to be plastic or moldable, and (2) the liquid limit, the moisture content above which plastic sediment becomes liquid. Note that calculation of these indices and moisture content are calculated as the ratios of the mass of moisture in a sample divided by the mass of the dry sediments.

Fig. 2.25 Exposure of green smectite at a construction site (28.0147°N, 82.2974°W) in Mango, Hillsborough County, Florida. Note the green coloration, a characteristic of unweathered Florida smectites

Fig. 2.26 Exposure of highly weathered clay at the same location depicted Fig. 2.24. The iron from the smectite is oxidized, resulting in a reddish tint to the minerals and ferric hydroxide cement. Note the hammer for scale

expansive properties of clay minerals and, in most cases, assign expansive clay minerals to the smectite group if they have low plastic limits and high plasticity indices. Table 2.5 presents some Atterberg limits data for selected clay minerals.

While it is occasionally necessary to perform x-ray diffraction analysis of a clay sample to quantify the amount of swelling of a clay sample, Atterberg limits data should suffice. If it is necessary to perform x-ray diffraction analyses, an experienced clay mineralogist should interpret the data. This is because Florida clays are often poorly crystallized (Fig. 2.24a), so simple x-ray diffraction may be insufficient to determine the abundance and crystal character of the clay. The smectite fraction often presents in x-ray diffractograms as a broad, low-magnitude "hump" of poorly crystallized material, thus rendering mineral identification by peak matching problematic. Diffraction analyses of heat-treated samples and samples treated with ethylene glycol to induce crystallization and expansion, respectively, of the clay and interpretation by an experienced clay mineralogist are necessary to establish the abundance and identity of clay minerals in a sample.

The major source of smectite in Florida is the Miocene Hawthorn Group, where thick, aerially extensive deposits of clay-rich sediments occur, especially in northern Florida. Relatively unweathered smectite has a greenish tint (Fig. 2.25) owing to the reduced iron ($Fe^{2+}$) in the crystal structure. Weathering typically causes oxidation of the ferrous iron ($Fe^{2+}$) to ferric iron ($Fe^{3+}$) and alters the colors of the clay deposit to reddish and brownish tones (Fig. 2.26). Oxidation of the iron may be accompanied by formation of ferric-hydroxide cements, but one should not assume that the ability of the clay to shrink and swell has been lost just because the iron has been oxidized. Many clay samples with brown coloration continue to demonstrate significant expansive behavior.

#### 2.6.2.5 Chlorite Group

The presence of chlorite group clay minerals is unusual in Florida. As a clay mineral, the composition is highly complex and variable (Bailey and Lister 1989). The general chemical formula for the chlorite group (Back and Mandarino 2008) is $A_{5\text{-}6}Z_4O_{10}(OH)_8$, where A = Al, $Fe^{+2}$, $Fe^{+3}$, Li, Mg, $Mn^{+2}$, N and Z = Al, B, $Fe^{+3}$, Si. This complex composition is a result of chlorite formation in metamorphic and weathering environments from a variety of aluminosilicate-rich rocks.

Florida chlorite was apparently not formed in Florida – it formed in North Africa! Chlorite has been found in the red clays of the Redland Agricultural District of Dade County by Kim (1984) and Estes (1987). They reported that the iron-stained soils of the Redland area include di-octahedral chlorite, a clay mineral that is often related to hydrothermal processes (Blatter et al. 1973; Lantenois et al. 2008). The clay in the Redland area was derived from extra-continental (African) dust blown into the area from the east (Prospero et al. 1987, 2001, 2010; Prospero 1999; Shinn 2001), so it is no wonder that the unusual clay is present in southern Florida.

#### 2.6.2.6 Mixed-Layer Clay Minerals

Mixed-layer clays are clay minerals that contain a mixture of different crystal structures such as illite and smectite. Particles that consist of mixtures of illite and smectite layering are common in some partially weathered Florida clay deposits. For example, the smectite group mineral beidellite may be a kaolinite/montmorillonite mixed layer (Weir and Greene-Kelly 1962; Yamada et al. 1991; Cuadros et al. 1994).

#### 2.6.2.7 Palygorskite and Sepiolite: Clay Minerals with Fibrous Habits

In addition to the sheet-structured clay minerals, Florida has important deposits of clay minerals that form acicular, or needle-like, structures (Fig. 2.27). These needle-like minerals, palygorskite and sepiolite, have crystal structures that consist of chains of silica tetrahedra rather than sheet-like structures (Grim 1953). While they may contain aluminum, these minerals are considered magnesium-silicate minerals rather than aluminosilicates.

The formula for palygorskite[7] is $(Mg,Al)_2Si_4O_{10}(OH)\cdot 4H_2O$ (Back and Mandarino 2008). Palygorskite's crystal structure is formed by double chains of silica tetrahedra. The formula for sepiolite is $Mg_4Si_6O_{15}(OH)_2{,}6H_2O$ (Back and Mandarino 2008). The sepiolite crystal structure consists of single chains of silica tetrahedra.

Palygorskite is locally abundant in Florida. Palygorskite is mined commercially to a limited extent in central Florida and extensively in northern Florida and in southern Georgia for use as absorptive media, such as "fuller's earth" and

[7]The disused term attapulgite is sometimes applied to palygorskite. The mineral name attapulgite is derived from deposits located near Attapulgus, Georgia, just over the state line from Florida.

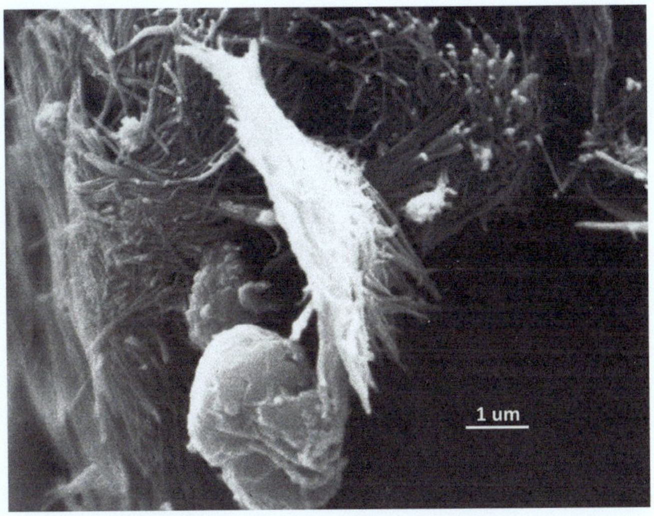


Fig. 2.27 Scanning electron micrograph of a bundle of palygorskite crystals from a Peace River Formation clay bed near Bartow, Polk County, Florida. Note the acicular (needle-like) crystals in the foreground and in the background matrix

"kitty litter." Large, commercial deposits of palygorskite with smectite can be found in Gadsden County, Florida (Figs. 2.28 and 2.29; Kerr 1937; Sowers and Kennedy 1964; Patterson 1974; Weaver and Beck 1977; Olive et al. 1989; McClellan 2013) and palygorskite-rich clay has been noted in Marion County, Florida (Espenshade 1958). Palygorskite/smectite-rich clay has been mined for kitty litter in at least one pit near Lowell, Marion County (Scott and Upchurch, personal observations). Unusual strata consisting of nearly pure palygorskite have been documented in west-central Florida (Upchurch et al. 1982; Strom and Upchurch 1985a, b), as well. All of these strata are within the Miocene Hawthorn Group.

Sepiolite is also present in the Hawthorn Group, but is typically found only in trace quantities. The northern Florida mineral deposits are typically mixtures of palygorskite and beidellite (McClellan 2013). Essentially pure deposits of palygorskite (Fig. 2.30) have been identified in Polk and Hillsborough counties by Upchurch et al. (1982) and Strom and Upchurch (1985a, b).

Both palygorskite and sepiolite have high absorptive capacities for water, neutral organic molecules, and metal cations (Kerr 1937; Patterson 1974; Galan 1996; Shariatmadari et al. 1999).

X-ray diffraction analysis is required to identify the presence of palygorskite and sepiolite. Atterberg limits data may provide for a qualitative indication of the presence of palygorskite and/or sepiolite, however. Based on limited data from White (1949), samples consisting of either clay mineral appear to have high liquid and plastic limits, and low plasticity indices (Table 2.5). As a result, the clays are not

**Fig. 2.28** Exposure of palygorskite-rich Hawthorn Group strata at the BASF Little River Mine (30.636°N, 84.461°W) near Havana, Gadsden County. The working face of the mine is about 25 m high. Photograph courtesy of the Florida Geological Survey

**Fig. 2.29** Example of the palygorskite-smectite (beidellite)-rich clay at the BASF Mine near Havana, Gadsden County. Note that the Hawthorn Group clay has the typical greenish gray to gray coloration, blocky fracture, and well-developed bedding-plane partings

highly expansive even though they have the ability to absorb a variety of chemicals, such as water, urea, oils, and greases.

Palygorskite deposits are typically normally consolidated in both north and central Florida. Unlike the clay deposits that are predominantly smectite mixed with other clay- and silt-sized constituents, palygorskite-rich strata typically have well-developed partings that parallel bedding-plane structures (Figs. 2.29 and 2.30). Both palygorskite and smectitic clays develop blocky fracture when desiccated. In contrast, the smectite-rich clays rarely illustrate the bedding-plane partings common to the palygorskite-rich strata (compare Figs. 2.25 with 2.29 or 2.30). Both smectite- and palygorskite-rich deposits commonly exhibit blocky fracture.

The origins of the north Florida deposits were investigated by Patterson (1974) and Weaver and Beck (1977). Weaver and Beck concluded that the north Florida deposits were

**Fig. 2.30** X-ray pure palygorskite from the Clear Springs Mine (27.866°N, 81.787°W), Polk County. This Peace River Formation (Hawthorn Group) deposit was a "bed clay" deposit located immediately below the phosphatic ore deposit at the mine. The hand holding the sample gives scale

formed in perimarine water bodies that had been enriched in magnesium. Upchurch et al. (1982) and Strom and Upchurch (1985a, b) investigated the origin of the central Florida deposits. They, too, concluded that the palygorskite/sepiolite strata originated in Mg-rich, perimarine (lagoonal) environments during the Miocene. In addition, the abundance of silicification (void-filling and replacement chalcedony and rock-crystal quartz, and opal-CT within and below the clay beds) indicated that the lagoonal water was alkaline.

### 2.6.3 Other Minerals Likely to Occur in Clay Sizes

In addition to the silicate and aluminosilicate clay minerals, there are other common minerals that form clay- and/or silt-sized clasts in Florida sediments. As indicated in Table 2.4, these include carbonate minerals, other silicates, sulfides, and a suite of phosphate-related minerals. For the purposes of this text, the following discussion is limited to those silt- and clay-sized materials encountered in strata where karst features may be encountered.

#### 2.6.3.1 Carbonate Minerals (Calcite, Aragonite, Dolomite)

Calcite, aragonite, and dolomite are common clay- and silt-sized minerals in Florida. These minerals occur in sediments that range in age from modern to the most ancient rocks in the subsurface. Because these minerals may be soluble in meteoric and groundwater, the strata that contain them are involved in karst processes if they occur near land surface or within groundwater-mixing zones.

Aragonite appears to be restricted to southern Florida, where it is a dominant constituent in Recent and some Plio-Pleistocene mud deposits. For example, a dominant constituent of the carbonate muds currently being deposited in Florida Bay, north and west of the Florida Keys, is aragonite derived from the disarticulation of calcareous algae, such as *Penicillus* and *Halimeda* (Gray 1974, 1978; Multer 1977; Burns and Swart 1992). Aragonitic mud of similar origin is locally preserved in the Plio-Pleistocene strata of southern Florida, although diagenetic processes have altered much of the originally aragonitic mud to micrite (microcrystalline calcite).

Silt- and clay-sized calcite is found in many of the clay-rich deposits of Florida. Much of this calcite was derived from calcite-secreting marine organisms (Bathurst 1966; Milliman 1974), micritization of aragonitic sediments (Bathurst 1966, 1975), and post-depositional cementation. Most, if not all, of the calcite found in fine-grained, cohesive sediments in Florida is not the result of weathering of limestone. Rather, it is a result of primary depositional and diagenetic processes.

Silt- and clay-sized dolomite clasts are common in the fine-grained sediments of Florida, as well. Dolomitization is an on-going, but limited, process in perimarine sediments of southern Florida. For example, Shinn (1964a, b, 1968) documented dolomite forming in the Florida Keys and Upchurch et al. (1983) documented euhedral dolomite crystals in organic-rich, mangrove sediments of the Ten Thousand Islands, Monroe County, Florida.

Two deposits that include silt- and clay-sized, dolomitic sediments are important from a karst perspective in Florida. These "dolosilt" (Scott 1988) deposits are found in the plastic clay beds of the Miocene Hawthorn Group and in the non-plastic Eocene Avon Park Formation (see Chap. 3 for a discussion of these strata). Dolosilt consists of silt-sized, euhedral crystals of dolomite that are perfectly preserved in the clay matrices of the Hawthorn clay beds (Figs. 2.31, 2.32, and 2.45). The rhombohedral microcrystals clearly indicate that they were formed by crystallization at their current locations during or shortly after deposition of the clay beds. For example, Fig. 2.33 illustrates silt-sized crystals of dolomite with included palygorskite, which indicates that the dolomite crystals formed after deposition of the clay.

The dolosilt is neither a transported, clastic deposit nor a weathered carbonate stratum. The dolosilt was deposited in quiet water, possibly in lagoons similar to the perimarine lagoons described by Weaver and Beck (1977), Upchurch et al. (1982), and Strom and Upchurch (1985a, b). The Hawthorn dolosilt deposits also contain abundant smectite and palygorskite (Fig. 2.33), which give the Miocene dolosilt deposits a greenish color and its cohesive properties.

The Avon Park Formation also contains dolosilt beds that consist of limpid, euhedral dolomite crystals. The same

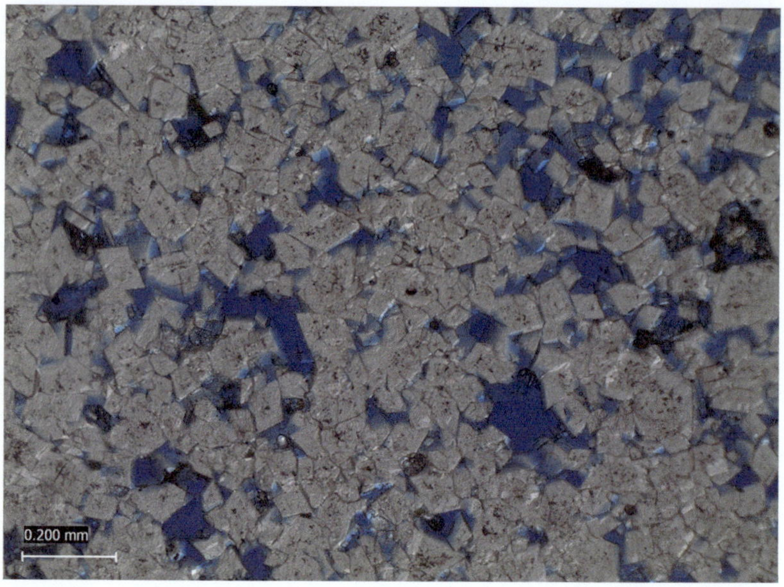


**Fig. 2.31** Dolosilt in thin section (plane polarized light; blue areas are pore spaces). Sample from the Peace River Formation, Polk County. Image courtesy of the Florida Geological Survey

**Fig. 2.32** Scanning electron micrograph of lithified dolostone composed of dolosilt from an exposure on the Alapaha River, Hamilton County. From McFadden (1982)

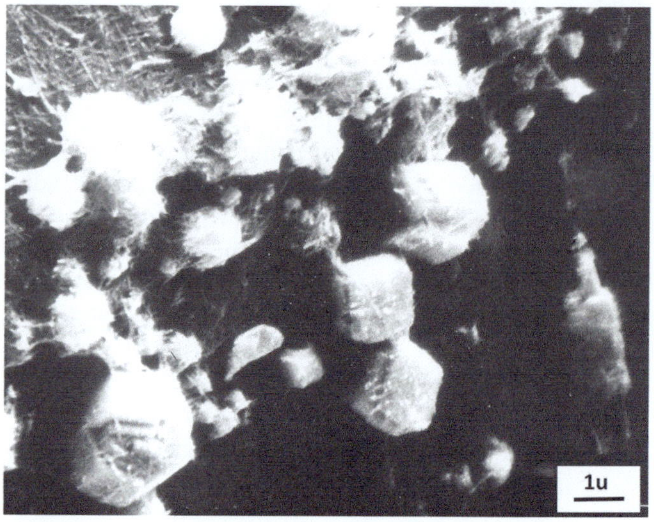


**Fig. 2.33** Scanning electron micrograph of clay-sized, rhombohedral, dolomite crystals penetrated by palygorskite needles. Sample is from a submerged, Miocene Hawthorn Group alkaline lake deposit (27.886°N, 82.437°W) near Ballast Point, Tampa Bay, Hillsborough County

arguments about lack of transport or weathering can be made as for the Hawthorn dolosilts; however, the Avon Park deposits are devoid of true clay minerals, so they are minimally cohesive, if at all. The Avon Park dolosilt is found in lenses enclosed within well-developed dolostone (Fig. 2.34). The deposits have either never been cemented or weak cementation has been ineffective in lithification of the sediment.

During standard penetration testing, weight-of-rod and/or –hammer conditions often occur in the dolosilt beds. The unconsolidated, silt- and clay-sized sediment usually washes out during mud rotary drilling. As a result, the dolosilt beds in the Miocene Hawthorn Group and Eocene Avon Park Formation are frequently misclassified and reported as "voids" or "weathered limestone." Weak dolosilt/clay deposits in the Hawthorn and dolosilt beds in the Avon Park have been misidentified as karst features in Florida.

#### 2.6.3.2 Silicate and Silica Minerals (Feldspars, Micas, Quartz, Opal-CT)

Sand- and silt-sized grains of the feldspars (albite, microcline and orthoclase) and micas (muscovite, hydromuscovite) commonly occur within Florida's clay-rich sediments (Van Kauwenbergh et al. 1990; Ardaman and Associates 1982a, b; Fountain 2009). While small quantities of these minerals may occur in any Florida sediment, they are most commonly encountered in the Miocene Hawthorn Group and the

Fig. 2.34 Dolostone with washed-out areas of dolosilt and post-depositional calcite cementation (see text) in the Avon Park Formation at the Dolime Mine (29.011°N, 82.653°W), Citrus County. Width of area photographed is 3 m

Pliocene Cypresshead Formation (Van Kauwenbergh et al. 1990; Ardaman and Associates 1982a, b; Fountain 2009). While the feldspars occur primarily as sand-sized clasts, they weather to clay-sized minerals.

The primary episodes of intense weathering associated with breakdown of silicate minerals occurred at the end of the Miocene and beginning of the Pliocene Epochs (the "post-Miocene unconformity") and in the Late Pliocene and early Pleistocene (Carr and Alverson 1959). During these episodes of weathering, the available feldspars were largely altered to kaolinite (Fountain 2009) and other clays. Muscovite, a potassium mica, was also altered to hydromuscovite by leaching of potassium and partial degradation of the crystal lattice.

Opal-CT, a sedimentary amorphous silica ($SiO_2 \cdot nH_2O$) phase, is common in the Hawthorn Group in central and northern Florida. Opal-CT gets its name from its x-ray diffraction pattern, which consists of a broad hump that represents poorly ordered, amorphous silica plus two distinctive peaks (cristobalite [101] and tridymite [200]) that reflect incipient ordering of the silica (Upchurch et al. 1982).

Opal-CT was deposited at scattered locations in northern Florida (McFadden 1982) and in west-central Florida (Figs. 2.35 and 2.36; Strom and Upchurch 1985a, b; Upchurch et al. 1982), as the clay-rich strata of the Hawthorn Group were being deposited. Opal-CT is a product of silica mobilization associated with perimarine alkaline lakes. The opal was deposited in the alkaline lake deposits of the Hawthorn Group and groundwater transported it into underlying limestone deposits. As a result, opal-CT occurs in the following forms in Florida sediments.

- Opal-CT and quartz chert occur as replacements of limestone and dolostone within (Fig. 2.36) and below Hawthorn Group sediments. The occurrences within the Hawthorn Group appear limited to the Peace River Formation in central Florida and the Statenville Formation in northern Florida. They are apparently contemporary with the palygorskite and sepiolite deposits in the alkaline, perimarine lake deposits.
- The replacements that occur stratigraphically below the Peace River Formation are typically found in the Tampa Member of the Arcadia Formation (Hawthorn Group), Suwannee Limestone, and Ocala Limestone. The opaline and quartzose chert is associated with the unconformable contact with the underlying limestone and were formed either (1) concurrently with alkaline lakes or (2) as a result of dissolution of opal and chemical alteration of clays in the Hawthorn Group and reprecipitation from groundwater in the underlying limestone. Figure 2.36 illustrates a replacement of a paleosol and breccia at an unconformity at the top of the Tampa Member of the Arcadia Formation (Hawthorn Group).
- Opal-CT (Fig. 2.38) is a common cement in clay beds associated with the perimarine, alkaline lake deposits of the Miocene Hawthorn Group (Fig. 2.35). Where

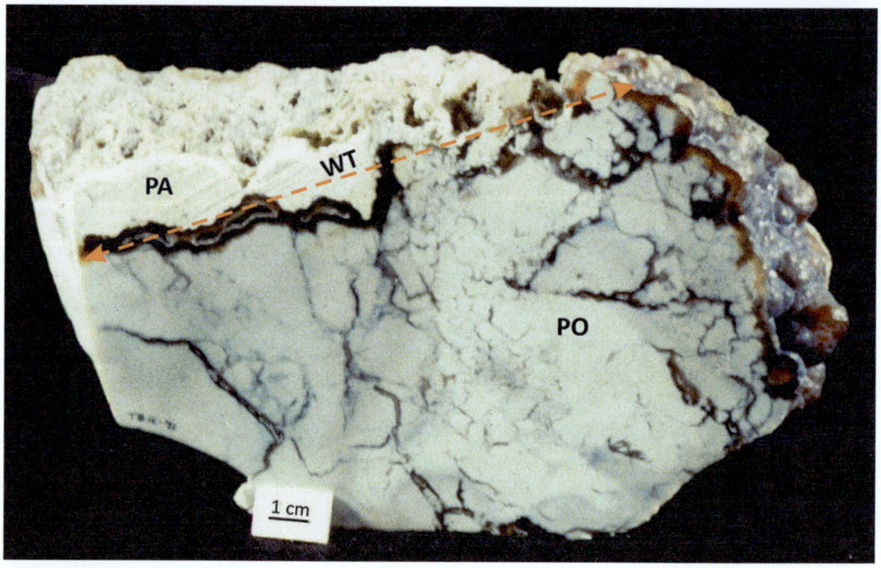


**Fig. 2.35** X-ray pure palygorskite cemented by opal-CT to form an opaline chert/claystone known as porcelainite (PO). WT denotes the location of the water table at the time of formation. Above the arrow denoting the paleo water table, there is no cementation and the sediment is pure palygorskite (PA). C is chalcedony filling desiccation cracks in the sediment. Specimen is from Hawthorn Group sediments dredged from Tampa Bay, Hillsborough County, Florida. Scale is 1 cm

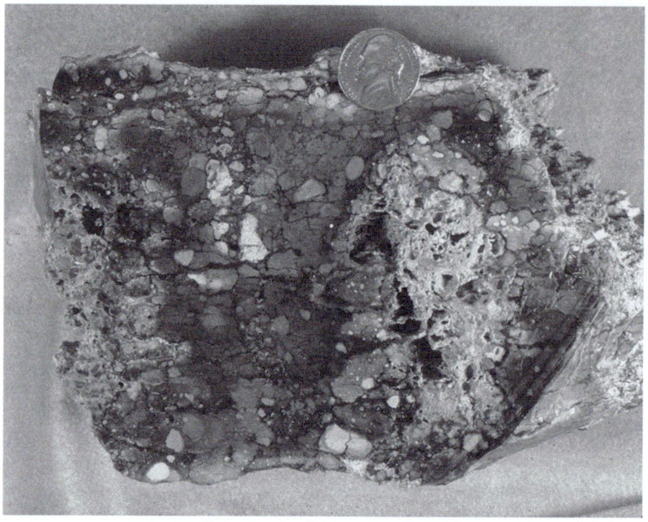

**Fig. 2.36** Chert formed by replacement of a limestone breccia and paleosol at the top of the Tampa Member, Arcadia Formation, by opal-CT and quartz. This deposit was formed adjacent to or under an alkaline, perimarine lake. Specimen is from northeastern Tampa, Hillsborough County, Florida

cementation is complete (Fig. 2.35) the resulting opaline claystone is termed porcelainite. Porcelainite is a hard, brittle, opaline variety of chert that can be difficult to penetrate during standard penetration testing.

- Opal-CT and quartz also occur as void fillings (Figs. 2.35 and 2.37). In most casts created in fossil molds and voids, the outer rim of the cast, which was deposited first, was opal-CT while the majority of the void fill is quartz (chalcedony and/or rock crystal). Outer surfaces of opaline void-filling crusts often exhibit clusters of microscopic balls of platy opal-CT termed lepispheres (Fig. 2.38).

Where present, opal-CT is generally relatively massive porcelainite. It is most likely to be termed cemented clay or siliceous clay in boring logs. However, opal is also common as small, disseminated flakes, lepispheres, and thin, discontinuous lenses in clay. These will probably not be noticed or considered consequential during penetration testing, but the presence of opal in the clay may affect the Atterberg limits and will likely be evident if the clay is subjected to x-ray diffraction.

#### 2.6.3.3 Phosphate Minerals (Apatite, Wavellite, Millisite, Crandallite)

The post-Miocene unconformity is represented in central and north Florida by paleosol development. Intense Late Miocene and Early Pliocene weathering of the phosphatic sediments of the Peace River Formation (Hawthorn Group) resulted in development of a number of important secondary, clay-sized phosphate minerals. The paleosol is best developed and has been most studied in the Central Florida Phosphate District. Within this mining district, the post-Hawthorn paleosol is known as the "leached zone."

The origin and mineral composition of the leached zone has been extensively studied because (1) it contains concentrations of uranium and uranium progeny, which have served as a mineral resource and may constitute a health hazard, and (2) after mining, separation of clay from the phosphate ore and subsequent disposal is a major problem for the phosphate industry. Important reports concerning the origin and mineralogy of the leached zone include Altschuler et al. (1951, 1956), Cathcart and Davidson (1952), Carr and Alverson (1959), Ketner and McGreevy (1959), Cathcart (1963a, b, 1964, 1966), Puri and Vernon (1964), Van Kauwenbergh et al. (1990), and Ardaman & Associates (1982a, b).

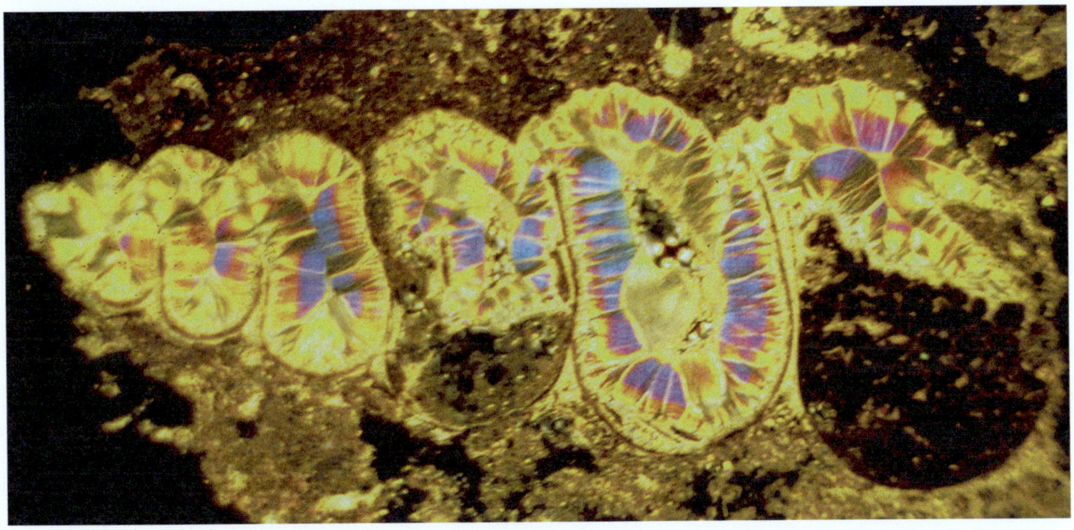

**Fig. 2.37** Cast of a Miocene gastropod from near Ballast Point, Tampa, Hillsborough County, Florida. The cast consists of void-filling, fibrous chalcedony. The matrix is clay composed of clay minerals and calcite that has been replaced by opal-CT. Gastropod is 1 cm in length; crossed polarized light

**Fig. 2.38** Opal-CT lepispheres in opaline chart from the Peace River Formation near Bartow, Polk County, Florida. Lepispheres are minute spheres consisting of platy, poorly formed crystals

While the leached zone paleosol in the Central Florida Phosphate Mining District has received most of the attention, studies elsewhere in Florida (Blanchard 1972; Blanchard and Denahan 1966; Upchurch 1992) demonstrate the widespread presence of secondary phosphate mineralization at or near the top of the Miocene Hawthorn Group within the post-Miocene paleosol and unconformity.

Phosphate minerals that commonly occur in the post-Hawthorn paleosols include carbonate fluorapatite (a.k.a. francolite; $Ca_3(PO_4,CO_3)_3F$), carbonate hydroxyapatite (a.k.a. dahlite; $Ca_5(PO_4,CO_3)_3(OH)$), wavellite ($Al_3(PO_4)_2(OH,F)_3 \cdot 5H_2O$), millisite ($(Na,K)CaAl_6(PO_4)_4(OH)_9 \cdot 3H_2O$), and crandallite ($CaAl_3(PO_4)_2(OH)_5 \cdot H_2O$). Carbonate fluorapatite is the primary phosphate mineral in the Hawthorn Group sediments. The others are products of leaching of the carbonate fluorapatite, silicates, and clay minerals included within the Miocene Hawthorn Group.

As a result of the leaching process, phosphate and other constituents were removed from the upper part of the leached zone and concentrated in the lower part. This chemical process was a result of intense weathering in the late Miocene, Pliocene, and perhaps early Pleistocene (Carr and Alverson 1959). Groundwater quality data from the leached zone (Oural et al. 1986), suggest that minor leaching is continuing today.

The leaching process resulted in a light-colored, clay-rich paleosol with weak, "frothy" textures (Puri and Vernon 1964). The open, pore-filled structure of the paleosol results in common occurrences of weight-of-rod or –hammer events during standard penetration testing. Because the pore spaces in the leached zone are not often interconnected, during standard penetration testing losses of drilling fluid circulation are uncommon.

In many areas on the flanks of the Brooksville Ridge (Fig. 2.5), the Late Miocene and Plio-Pleistocene weathering events resulted in an interesting, karst-related phosphate deposit, known as "hard-rock phosphate." As a result of weathering of the Miocene Hawthorn Group sediments, chemical precipitates and silt-and clay-sized detrital carbonate hydroxyapatite (hard-rock phosphate) were deposited in ancient sinkholes that had developed in the area (Sellards 1913a, b; Scott 1988; Upchurch 1992).

**Fig. 2.39** Pyrite framboids in the Oligocene Suwannee Limestone. SEM micrograph courtesy of R.E. Price

#### 2.6.3.4 Sulfides

Finally, sulfide mineralization is common in most strata in Florida. The dominant sulfide mineral is pyrite ($FeS_2$; Fig. 2.39). Pyrite is commonly disseminated through both limestones and clays (Price and Pichler 2002, 2006). Both individual crystals and spherical crystal clusters (framboids) are common. Pyrite has also been observed forming thin coatings on Pliocene and Pleistocene shells in the shell beds of southern Florida (Scott and Upchurch, personal observations). Hydrotroilite (a colloidal, hydrous ferrous sulfide) is common in trace quantities in Recent marine sediments where it stains shells and sediment dark gray to black.

The volume of these sulfide minerals in Florida sediment is small, but significant. Oxidation of the pyrite results in ferric hydroxide, the red, yellow, or brown stains that are common in both soils and sediments in Florida. The reaction

$$2FeS_{2pyrite} + 7O_{2gas} + 2H_2O_{water} \rightarrow 2Fe^{2+}{}_{aqueous} + 4SO_4{}^{2}{}_{aqueous} + 4H^{+}{}_{aqueous}$$

can be used to represent these reactions. Note that the ferrous iron ($Fe^{2+}$) is chemically unstable in oxygen-rich environments and it quickly oxidizes to ferric iron ($Fe^{3+}$), where it combines with oxygen and water to form the ferric hydroxide ($Fe^{+3}OH_3 \cdot nH_2O$) and other iron stains so common in Florida sediments. The other product of this reaction is weak sulfuric acid ($H_2SO_4$), which is represented in the sample reaction by the sulfate ($SO_4{}^{2-}$) and hydrogen ion ($H^+$) products of the reaction.

Sulfuric acid formed by sulfide mineral oxidation and other reactions is known to have formed significant karst systems elsewhere in the world (Ford and Williams 2007; Palmer 1991; Klimchouk 2007). It is probable that oxidation of pyrite in Florida sediments has contributed, albeit in a minor way, to shallow karst development in Florida. Figure 2.40 illustrates a situation where pyrite oxidation has been accelerated by dewatering and mining of a sandy, clayey shell bed in south-central Florida. Here, iron staining of the sediment and reddish runoff indicate development of iron oxidation products and acidic drainage. This same shell bed has developed widely separated, shallow cover-subsidence sinkholes, the formation of which may be related, in part, to the pyrite oxidation process.[8]

### 2.6.4 Clay Surface Properties: Cohesion, Permeability, and Sorption

Clay-rich deposits exhibit cohesion, low permeabilities, and shrink/swell actions because of the interaction between water and the clay mineral surfaces. This section further explains this interaction and its effects on the bulk properties of clays.

#### 2.6.4.1 Interaction of Water with Clay Minerals

Because of the oxygen that forms the outermost atoms of both silica tetrahedron and aluminum octahedron sheets in clays, the outer surfaces of clay minerals have net negative electromagnetic charges (Fig. 2.41).

The water molecule consists of an oxygen atom bound to two hydrogen atoms. The configuration of oxygen and hydrogen atoms in the water molecule has a "Y-shaped" arrangement with the oxygen atom at the base of the Y and the two hydrogen atoms at the two upper prongs of the Y (Fig. 2.41). This arrangement causes water to be a dipolar

[8] It is important to note that sulfuric acid derived from gypsum and/or anhydrite strata in the subsurface in Florida has likely also contributed to karst development, See Chap. 8 for a discussion of this process.

**Fig. 2.40** Iron staining and acidic runoff derived from oxidation of pyrite in Pleistocene shell beds at the Ruck's Mine (27.524°N, 80.755°W), Ft. Drum, Okeechobee County, Florida

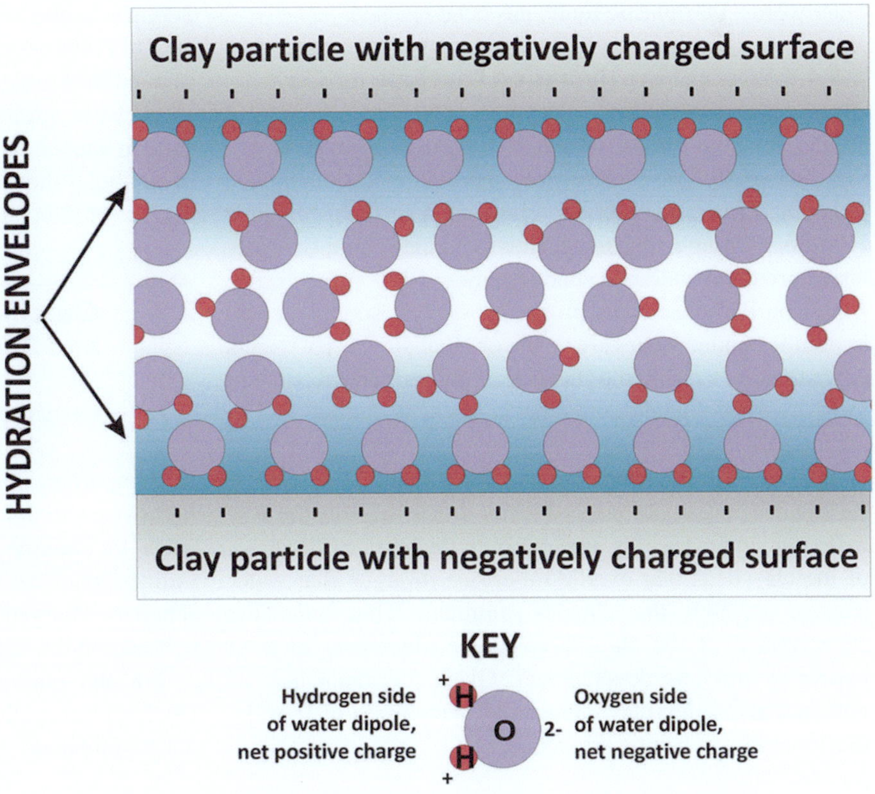


**Fig. 2.41** Arrangement of water molecules in clay interlayer and outer surface spaces

molecule, with a net negative electromagnetic charge on the oxygen side and net positive charge on the hydrogen side.

The positive poles of water molecules, therefore, are attracted to the negatively charged surfaces of the clay minerals, forming a hydration envelope (Fig. 2.41) on the outer surfaces of such minerals as kaolinite and illite and outer surfaces plus interlayer surfaces in smectites (Fig. 2.41). The ability of water to coordinate with the clay mineral and move in or out of the electromagnetic fields at the clay surfaces determines cohesion, permeability, and shrink/swell properties of clays.

Also, the higher the surface area of the clay, the more water the clay can sorb. It is for this reason that the smectites have such high sorption capacities and shrink/swell properties.

#### 2.6.4.2 Effect of the Hydration Envelope on Clay Cohesion

The electromagnetic interactions of the water and negatively charged clay surfaces cause cohesion. As long as the electromagnetic attractions can bind the clay particles together, cohesion exists. If the water concentration becomes so great that the water molecules cannot maintain electromagnetic bonding across the spaces between particles, the clay begins to behave as a liquid and the "liquid limit" of the clay has been reached. Conversely, if the moisture content in a clay mass becomes so low that water is no longer able to electromagnetically bridge the interparticle and interlayer spaces, the clay ceases to be plastic and behaves as a solid. In other words, the "plastic limit" has been reached.

#### 2.6.4.3 Effect of the Hydration Envelope on Clay Permeability

While clay beds typically have high porosities (void ratios), they generally have very low hydraulic conductivities (permeabilities). In Florida, clay strata in the Miocene Hawthorn Group and residual Hawthorn clay strata typically have horizontal hydraulic conductivities of $10^{-5}$ to $10^{-8}$ cm/s. Based on the anisotropic properties of clay strata, vertical hydraulic conductivities are on the order of $10^{-7}$ to $10^{-10}$ cm/s or smaller. As a result, the Hawthorn clay beds are considered to be hydrogeological confining layers with limited ability for the passage of water.

Hydrogeologists and karst professionals conducting subsidence investigations in Florida often rely on the low hydraulic conductivities, especially in the vertical direction, of the clay beds that overlie limestone to argue that water leakage through the clay is minimal. This, plus the cohesiveness of the clay, is considered by many to be a barrier to sinkhole development. Given the abundance of penetrations in thin clay deposits in many areas of Florida, this assumption may be erroneous.

The very low horizontal and vertical hydraulic conductivities of clay deposits are not representative or indicative of low porosity or even small pore throats. Porosities of clay beds often exceed 50%. Rather, the inability of clay to transmit water is because of the electromagnetic interactions of the water dipole and clay surface. As shown in Fig. 2.41, the electromagnetic attraction between the clay mineral surface and the water dipole causes the water molecules to coordinate with the clay rather than flow through the medium. As a result, clay-rich strata exhibit very low permeabilities.

#### 2.6.4.4 Effect of Low Clay Permeability on Consolidation and Penetration Testing

The difficulty that water has passing through a clay deposit is critical to consolidation of the clay and to the behavior of clay strata during standard penetration and cone penetrometer testing.

Suffice it to say, clay-rich sediments cannot consolidate under geostatic (gravitational) loads unless there is a mechanism that allows the included, uncompressible pore water to escape. Assuming that normally compacted, shallow (less than 30 m burial) sand has a porosity of about 25–30%, and if clay completely fills the intergranular space in the sand, then clayey sand with as little as 25–30% clay, can also resist dewatering under geostatic compression.

Depending on the clay content and distribution in the stratum, therefore, clayey sand, sandy clay, and clay strata may remain under-consolidated in shallow sediments for long periods of time. Upchurch et al. (2013) identified clay deposits in the Miocene Hawthorn Group in the Plant City area of Hillsborough County that had moisture contents near, or even in excess, of liquid limits. They concluded that these clays were under-consolidated and suggested that either sudden consolidation or flow of the clay was a possible cause of sinkholes induced by sudden declines in the potentiometric surface of the underlying Upper Floridan Aquifer (UFA).

## 2.7 Classification of Carbonate Sediments and Rocks

### 2.7.1 Dunham's Classification Scheme

It is important to understand the terms used to describe the limestone and dolostone of Florida. There are many different ways to classify carbonate sediment and rock. This text utilizes one of the most widely utilized classification schemes, Dunham's (1962) classification (Table 2.6), which is based on the texture and structure of the limestone.

For the purposes of this text, we take liberties with Dunham's classification in that we apply the classification to unconsolidated carbonate sediments, limestone, and

**Table 2.6** Dunham's (1962) limestone classification system based on sediment texture

| Dunham's classification according to depositional texture | | | | |
|---|---|---|---|---|
| Original components not bound together during deposition | | | | Original components were bound together during deposition… as shown by intergrown skeletal matter, lamination contrary to gravity, or sediment-floored cavities that are roofed over by organic or questionably organic matter and are too large to be interstices |
| Contains mud (particles of clay and fine silt size) | | | | |
| Mud supported | | | | |
| Less than 10 percent grains | More than 10 percent grains | Grain supported | Lacks mud and is grain supported | |
| **MUDSTONE** | **WACKESTONE** | **PACKSTONE** | **GRAINSTONE** | **BOUNDSTONE** |

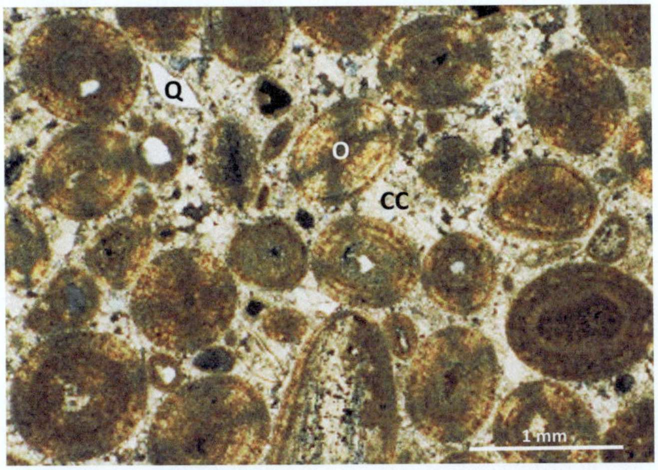


**Fig. 2.42** Photomicrograph of an oolitic grainstone. Oolite facies of the Miami Limestone, Miami (Dade County). O = ooliths, Q = quartz sand, CC = calcitic cement, crossed polarizers, scale is approximate. (Image modified from Lidz et al. (2005))

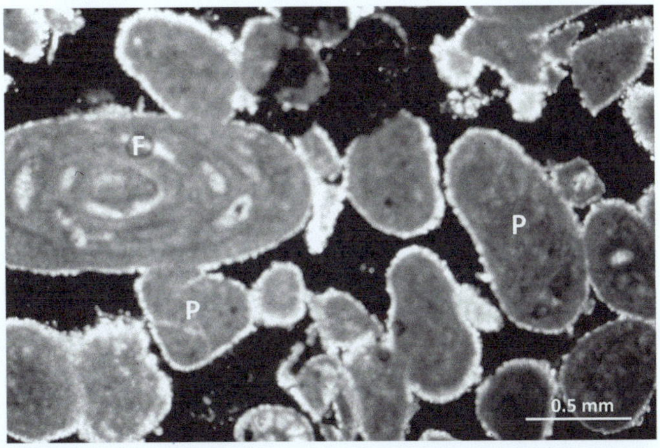


**Fig. 2.43** Photomicrograph of a poorly cemented fossiliferous grainstone from the Ocala Limestone in Marion County, Florida. Allochems labeled P are peloids and F are fossil foraminifera. Solid black areas are interstitial pore space. Crossed polarizers

dolostone. We also do not limit the "mud" fraction to carbonate silt- and clay-sized particles. We also apply Dunham's terminology to carbonate-rich sediments with included siliciclastic mud.

Figures 2.42 and 2.43 are examples of grainstones according to Dunham's classification. Note that the spaces between allochems (ooids, foraminifera, and peloids) are not filled with fine-grained sediment and the sediment is grain supported with individual grains in contact. The oolite shown in Fig. 2.42 has a crystalline calcite cement and the centers of some of the ooids have been dissolved away, leaving molds of the spherical sediment particles. This rock would be termed an oolitic grainstone.

In contrast, the rock shown in Fig. 2.44 is a packstone in Dunham's classification. Here the spaces between the fossils and peloids are filled with micrite. This rock would have lower primary porosity than the grainstones.

Figure 2.45 is an example of a mudstone according to Dunham's classification. It is from the Miocene Peace River Formation (Hawthorn Group) in Polk County. The phosphate peloid (P, Fig. 2.45) is sand sized. The remainder of the matrix consists of silt- and clay-sized micrite and euhedral crystals of dolomite. Permeability of such sediments is typically low, which protects them from dissolution to some extent.

### 2.7.2 Allochems, Mud, and Wave and Current Energy

The materials that make up the carbonate fraction of a sediment stratum are typically autochthonous; that is, they were formed at or near the site of deposition as a result of biological or other processes. Individual, locally derived carbonate sediment particles are termed allochems. Allochems include fossils (Figs. 2.43 and 2.44), ooids (Fig. 2.42), peloids (pellet-like particles, commonly of fecal origin; Fig. 2.43), and other discrete particles that formed at, or near, the site of deposition.

The matrix of the sediment may include clay- and silt-sized carbonate mud. When this material recrystallizes as calcite, it is termed micrite for microcrystalline calcite (Fig. 2.44).

Figure 2.43 is a photomicrograph of a grainstone from the Eocene Ocala Limestone from near Belleview, Marion County. The grainstone consists of grain-supported allochems, such as foraminifera and peloids. Note in Fig. 2.43 that the peloids have little or no internal structure. They most probably formed as fecal pellets. Also, note that the foraminifera retain some internal structure which appears "fuzzy." The loss of distinct boundaries in the foraminifera is a result of micritization, or conversion of the test (shell) of

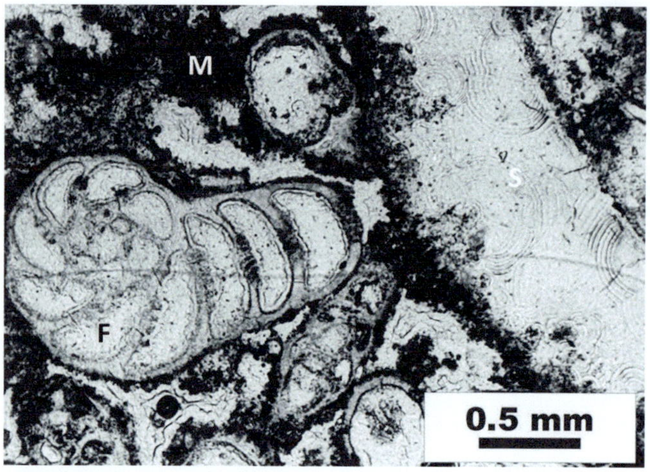


**Fig. 2.44** Photomicrograph of a fossiliferous packstone from the Miocene Tampa Member of the Arcadia Formation in Hillsborough County, Florida. F = foraminifera, S = shell fragment, M = micrite. Crossed polarizers

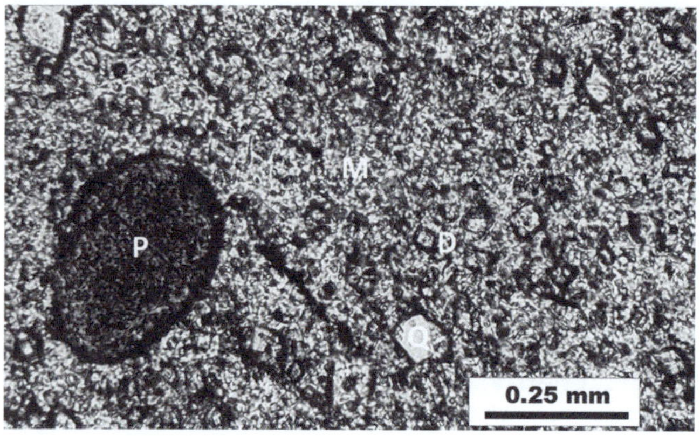


**Fig. 2.45** Photomicrograph of a mudstone from the Peace River Formation (Hawthorn Group) in Polk County. P = phosphate peloid, D = euhedral, silt-sized dolomite rhombohedra (dolosilt), M = micrite. Crossed polarizers

the foraminifera to microcrystalline calcite. These allochems have not been firmly cemented together, and there is only a thin calcite rim cement around each allochem.

A common diagenetic or post-depositional change in the texture and preservation of allochems in limestone is micritization. In the case shown in Fig. 2.43, the skeletal remains of the foraminifera have been "blurred" by conversion to micrite, or silt- and clay-sized calcite. Micritization is common in fossils that were originally composed of high-Mg calcite and aragonite. Note in the example shown in Fig. 2.43 that there has been no change in volume of the allochems as they have been micritized. In some cases, however, the conversion of coarser particles to micrite may represent a loss of sediment volume.

The intact outer margins of the allochems (Fig. 2.43) indicate that they have not been subject to significant dissolution. This limestone sample is an Eocene carbonate sand deposit that has apparently never been subjected to significant dissolution, cementation, or compaction. While micritized, the individual grains are not crushed or even closely packed together. The sample shown in Fig. 2.43 might have low penetration resistances during standard penetration testing and, owing to the large amount of primary, intergranular porosity, this rock would serve as a location of matrix groundwater flow.

Figure 2.44 illustrates a sample of a packstone from the Miocene Tampa Member of the Arcadia Formation (Hawthorn Group) limestone. Again, the allochems have not been weathered and no significant compaction has occurred. However, the presence of the micrite filling the interstitial spaces renders this rock relatively less permeable when compared to the Ocala Limestone in Fig. 2.43.

Most carbonate strata in Florida were deposited on broad, shallow sea bottoms as sea level rose and fell. As such, the carbonate strata may be laterally extensive and, while there are textural and biological differences, the beds generally do not change appreciably over short distances. An exception occurs in portions of the Miocene Hawthorn Group sediments where lenses of limestone and dolostone may be highly localized with transitions from carbonate rock to calcareous clay in distances of a few meters (Scott 1988).

The textures of the carbonate sediments and rocks derive from the wave and/or current energy present at the deposition site. Grainstones (Table 2.6) were deposited in high wave or current energy environments, and the included allochems may have been transported over short distances prior to final deposition. The high wave and/or current energy prevented deposition of the carbonate mud that fills interstitial openings in packstone. In a packstone (Table 2.6), the wave or current energy during deposition was less, but sufficient to sort the allochems. Mud accumulated in the more protected interstices between the allochems. Wackestone and mudstone textures represent wave or current energies that were so low that direct deposition of carbonate mud on the sea floor was possible. Any allochems present in these textures usually reflect organisms that lived and died at the site of deposition.

Therefore, in carbonate rocks (and sediments), application of Dunham's limestone classification (Table 2.6) allows one to understand something about how the materials were deposited and the relative amount of wave or current energy present at the approximate time of deposition. It also allows one to consider the ability of the rock to transport water by interstitial, or matrix, flow.

## 2.8 Classification of Siliciclastic Sediments

There are many classifications for sediments that rely on grain size and composition. Some of these have been discussed above. For the purposes of this text, the classification utilized is based on the Wentworth (1922) classification as modified by incorporation of Udden's (1914) clay size criterion (Table 2.1). Where mixtures of sizes occur, we utilize the less important constituent as a modifier (i.e., sandy clay, silty sand).

However, realize that most standard penetration test boring logs are prepared using the ASTM version of the USCS (ASTM 2009, 2011a). As a result, the term clay on the boring logs represents use of the USCS classification presented in Table 2.3.

## 2.9 Description of Sedimentary Strata

In addition to characterizing sediment and sedimentary rock samples on the basis of composition and texture, it is important to understand the terminology used to characterize strata composed of sediments. This section describes two very important concepts and associated terminology used in sedimentology and stratigraphy.

### 2.9.1 Facies and Walther's "Law"

#### 2.9.1.1 Facies

Sedimentologists have developed a very useful way of thinking about sedimentary environments and strata. Sedimentologists use the term facies to describe the different sedimentary environments and the sediment types and compositions they reflect. The word facies describes a body of sediment or rock with specified characteristics, such as composition, texture, fossil content, and sedimentary structures. Ideally, a facies is a distinctive sediment or rock type that formed under relatively uniform conditions of sedimentation and reflects a specific sedimentary process or environment.

The term facies was introduced by Gressly (1838). In the context of sedimentology and stratigraphy, a specific facies may consist of a sediment texture (i.e., sand facies, clay facies), composition (i.e., limestone facies, dolomitic facies, evaporite facies), fossil content (i.e., coral facies, *Chione*[9] facies) or depositional environment (i.e., reef facies, dune facies). In each case, naming the facies is intended to connote depositional, textural, or compositional conditions. The term has since been expanded into other areas of geology to characterize igneous and metamorphic rocks and water quality. In this text, Chap. 5 deals with hydrochemical facies to assist in characterizing groundwater quality and the chemical processes of diagenesis and karstification. See Krumbein and Sloss (1963, pp. 316–331) for an excellent discussion of the facies concept and how it developed.

[9]*Chione* is a common genus of bivalve mollusk in Tertiary and Quaternary sediments of Florida.

Fig. 2.46 Sedimentary facies in the Pliocene Caloosahatchee Formation of southwestern Florida

#### 2.9.1.2 Walther's "Law"

Geologist Johannes Walther (1860–1937) developed the facies concept in stratigraphy and sedimentation into a principle that is widely used to help understand how sedimentary facies relate to each other (Middleton 1973). Basically, Walther's "Law"[10] states that the vertical succession of facies, such as observed in a road cut, quarry, or sediment core reflects lateral changes in the depositional environment. In other words, the principle developed by Walther states that when a depositional environment "migrates" laterally, as would be caused by sea level rising or falling, sediments of one depositional environment become stacked, one on top of another. Sedimentary facies often grade or interfinger laterally into another facies by changes in textural or other characteristics (Fig. 2.46).

For example, the sediments of the Hawthorn Group show extremely complex transitions and interfingering of lenses of sand, clay, and carbonate sediment (Fig. 2.47). Because the Hawthorn Group sediments represent many different depositional environments in a complex, low gradient system that included deltaic, coastal, perimarine mudflats, and shallow marine areas, the sediment facies

[10]In modern terminology, Walther's statement of the relationships of facies in sediments and rocks is not considered to be a law, it is a principle.

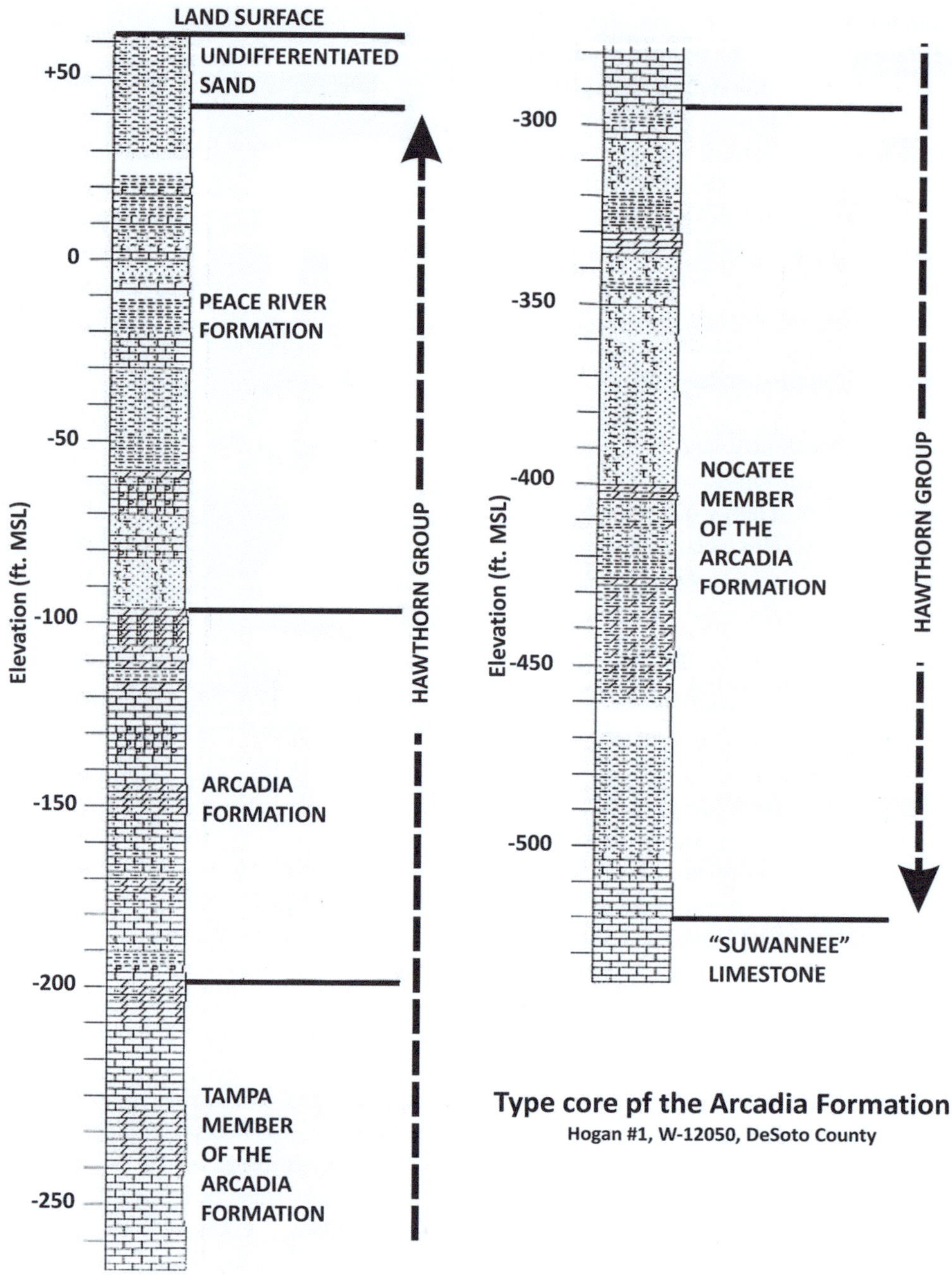


**Fig. 2.47** Example of the complex facies changes found within the Miocene Hawthorn Group. Each change in symbol on this boring log represents a facies change. This log is of the type core of the Arcadia Formation in DeSoto County, Florida. From Scott (1988, Fig. 34)

change rapidly in both the vertical and horizontal directions. Lenses of a particular sediment type commonly have limited areal extents, so it is not uncommon to encounter dolostone in one boring and sand or some other facies within a few meters at the same depth. These changes in sediment facies in the vertical sense (Fig. 2.47) reflect the changes in horizontal distribution of complex, small-scale depositional environments as the Hawthorn sea level changed.

Examine Fig. 2.48, an image from space of southern Florida, to get a perspective on the scale of facies changes in an area with many contemporaneous environments of deposition. Most of the area in the view is undergoing active sedimentation at the present time. The general types of sediments being deposited in each area are indicated on the figure. Sediments range from coarse shell, algae, and coral fragments along the reef tract to aragonite and calcite mud in Florida Bay, to calcareous mud and peat in the mangrove

**Fig. 2.48** South Florida from space showing some of the modern and contemporaneous depositional environments. 1 = coral reef (coralgal boundstone, shelly sand and gravel), 2 = Hawk Channel (back-reef lagoon; coralgal and shelly sand and gravel with scattered patch reefs), 3 = Card Sound (coralgal and shell sand mixed with quartz sand), 4 = Florida Bay (carbonate mud with varying amounts of shell), 5 = Everglades sawgrass prairie (carbonate mud and peat), 6 = Mangrove swamp (peat with varying amounts of shell and sand), 7 = Whitewater Bay (carbonate and organic mud with varying proportions of shell), 8 = Cape Sable beaches (quartz and carbonate sand mixed with shell), 9 = Gulf of Mexico (offshore quartz and carbonate sand predominates). Arrows delineate mud mounds in Florida Bay

fringe and Everglades, to quartz sand mixed with carbonates north of Card Sound, and quartz sand with shells on the beaches along the Gulf coast. All of these different sediment types are being deposited today, so they represent sets of contemporary sedimentary facies that grade from one to another and/or interfinger.

Within each major depositional environment, there are smaller variations in depositional environments. For example, narrow, sinuous mud mounds (arrows, Fig. 2.48) wind their way through Florida Bay (Area 4, Fig. 2.48). All of the sediments being deposited in Florida Bay are carbonates (aragonite and calcite). Between the mud mounds, the sediments consist of shelly, sandy mud (silt- and clay-sized sediment) and the mud mounds are almost wholly carbonate mud (silt- and clay-sized aragonite and calcite). The facies transitions are rapid and occur over short distances. Similar rapid facies changes occur throughout all of the major depositional environments shown in the figure.

Over geologic time and as sea levels fluctuate, these different sediment packages migrate laterally as the depositional environments shift. The result is a thickness of sediments that may remain uniform in texture and composition, or they may gradually change textures and compositions, or they may migrate and interfinger or become stacked, one sediment type over another. It is by these mechanisms that sediment facies become stacked, as shown in the type core of the Arcadia Formation shown in (Fig. 2.47).

### 2.9.2 Lithosomes

Another useful term used by sedimentologists to describe sediments is lithosome. A lithosome is a uniform mass of sediment or rock that has a different texture and/or composition from the surrounding sediment or rock. For example, the alternating limestone, dolostone, sand, and clay strata depicted in Fig. 2.47 represent different lithosomes. The best use of the term lithosome is to describe specific strata within a larger sedimentary package that represents a related set of sedimentary events. For example, one could describe deltaic

**Table 2.7** The hierarchy of lithostratigraphic and chronostratigraphic nomenclature adopted by the North American Commission on Stratigraphic Nomenclature (2005)

| Lithostratigraphic Units | Chronostratigraphic Units |
|---|---|
| Supergroup | Eonothem (Eons) |
| Group | Erathem (Supersystem or Era) |
| *Formation* | *System* |
| Member (or Lens or Tongue) | Series |
| Bed(s) or Flow(s) | Stage |
| | Chronozone |

facies (sediments deposited in a delta environment) as containing interfingering sand and clay lithosomes.

### 2.9.3 Stratigraphic Nomenclature

Finally, it is appropriate in this introductory chapter to present the formal nomenclature used by geologists to describe and classify geological strata for the purpose of mapping. The North American Commission on Stratigraphic Nomenclature (2005) provides guidelines for naming stratigraphic units. The basic, fundamental stratigraphic unit is the formation, which is a mappable body of sediment or rock (guidelines also exist for soils, igneous and metamorphic rocks, and other mappable geologic media). Sedimentary strata are classified according to the Commission's hierarchy for lithostratigraphic units (Table 2.7). Using this hierarchy, formations can be subdivided into smaller mappable units (members) and/or combined into larger units (groups).

Similarly, time can be divided into chronostratigraphic (temporal) units with the system as the fundamental time unit (Table 2.7). These units are utilized in Chap. 3. Reference is made to the publication by the North American Commission on Stratigraphic Nomenclature (2005) for the rules for naming lithostratigraphic and chronostratigraphic units.

## 2.10 Summary

In summary, the depositional environment, sources of sediment, packing, size and shape of clasts, groundwater deposits, and cements and grain coatings control the properties of Florida sediments and sedimentary rocks. These properties have profound effects on permeability, dissolution, and strength or penetration resistance. The natural variability of siliciclastic and carbonate sediment deposits is nearly infinite, and one simply cannot assume that all sediments behave in a simple, constant manner. For additional information on sediment grain sizes, sorting, packing and environments of deposition, see a good sedimentology text (i.e., Davis 1992; Leeder 2011). Engineering texts that offer excellent introductions to gradation and consolidation/compaction include Terzaghi and Peck (1948), Holtz and Kovacs (1981), and Wesley (2010).

Important conclusions of this chapter are:

- One must understand depositional environments in both carbonate and siliciclastic sediments and rocks in order to understand the processes of development of Florida karst and cover materials;
- In shallow, geologically young sediments, understanding depositional history is often most important in determining (1) the mineralogy, textures, compositions, and strengths of sediments and rocks and (2) their mode of response to karst processes;
- An understanding of the cover materials and how they were deposited and diagenetically altered is necessary in order to differentiate normal sedimentary processes from disruptions that result from sinkhole and other karst activities;
- Compaction and consolidation of shallow sediments have been minimal owing to the lack of sufficient lithostatic loading; and
- Understanding the classification of sediments and their geotechnical and hydrogeologic properties is important for understanding karst processes.

## References

Altschuler ZS Jaffe EB Dwornik E (1951) The stratigraphy of the upper part of the Bone Valley Formation and its relation to the leached zone. US Geol Surv, Trace Elements Memo Rept 237

Altschuler ZS Jaffe EB Cuttitta F (1956) The aluminum phosphate zone of the Bone Valley Formation, Florida, and its uranium deposits. US Geol Surv Prof Pap 300, pp. 495–504

Amouric M Olives J (1998) Transformation mechanisms and interstratification in conversion of smectite to kaolinite; an HRTEM study. Clays and Clay Min 46:521–527

Ardaman & Associates (1982a) Evaluation of phosphatic clay disposal and reclamation methods: Volume 1, Index properties of phosphatic clays. Bartow, Fla Inst Phosphate Res Pub No 02-002-003

Ardaman & Associates (1982b) Evaluation of phosphatic clay disposal and reclamation methods: Volume 2, Mineralogy of phosphatic clays. Bartow, Fla Inst Phosphate Res Pub No 02-002-004

Arthur JD Applegate J Melkote S Scott TM (1986) Heavy mineral reconnaissance off the coast of the Apalachicola River delta, northwest Florida. Fla Geol Surv Rept Inves No. 95

ASTM (2007) Standard test method for particle-size analysis of soils. Amer Soc Test and Mat Method D 422

ASTM (2009) Standard practice for description and identification of soils (visual-manual procedure). Amer Soc Test and Mat Method D 2488

ASTM (2011a) Standard practice for classification of soils for engineering purposes (Unified Soil Classification System). Amer Soc Test and Mat Method D 2487

ASTM (2011b) Standard test method for dispersive characteristics of clay soil by double hydrometer. Amer Soc Test and Mat Method D 4221

ASTM (2011c) Standard test method for standard penetration test (SPT) and split-barrel sampling of soils. Amer Soc Test and Mat Method D 1586

Athy LF (1930) Density, porosity, and compaction of sedimentary rocks. Bull Amer Assoc Petrol Geol 14(1):1–24

Atterberg A (1905) Die rationelle Klassifikation der Sande und Kiese. Chem Zeit 29:195–198

Atterberg A (1911) Plasticity in clays. Intern Repts Soil Sci 1:40–43 (Reproduced in English by the US Joint Public Res Serv for the US Army Cold Reg Res and Eng Lab, September, 1974)

Back M Mandarino JA (2008) Fleischer's glossary of mineral species 2008. Tucson, Mineral Rec Inc., 10th Ed

Bailey SW Lister JS (1989) Structures, compositions, and X-ray diffraction identification of dioctahedral chlorites. Clays and Clay Mins 37:192–202

Ball MM (1967) Carbonate sand bodies of Florida and the Bahamas. Jour Sed Pet 37:556–591

Bathurst RGC (1966) Boring algae, micrite envelopes and lithification of molluscan biosparites. Geol Jour 5:15–32

Bathurst RGC (1975) Carbonate sediments and their diagenesis. Amsterdam, Elsevier Sci Publ Co, 2nd Ed.

Blanchard FN (1972) Physical and chemical data for crandallite from Alachua County,Florida. Amer Mineral 57:472–484

Blanchard FN Denahan SA (1966) Wavellite-cemented sandstones from northern Florida. Quar Jour Fla Acad Sci 29:248–256

Blatter CL Roberson HE Thompson GR (1973) Regularly interstratified chlorite-dioctahedral smectite in dike intruded shales, Montana. Clays and Clay Mins 21:207–212

Bock WD Moore DR Neumann AC Supko PR (comp) (1969, reprinted 1994) Late Pleistocene geology in an urban area. Miami Geol Soc field trip guidebook, Available at http://sofia.usgs.gov/publications/reports/mgs_bock1969/mgs_bock1969.pdf. Accessed March 20, 2012

Budd DA (2001) Permeability loss with depth in the Cenozoic carbonate platforms of west-central Florida. Bull Amer Assoc Petrol Geol, 85:1253–2172

Budd DA Vacher HL (2004) Matrix permeability of the confined Floridan Aquifer, Florida, USA. Jour. Hydrogeology, 12(5):531–549.

Burns SL Swart PK (1992) Diagenetic processes in Holocene carbonate sediments: Florida Bay mud banks and islands. Sedimentology 39:285–304

Carballo JD Land LS Miser DE (1987) Holocene dolomitization of supratidal sediments by active tidal pumping, Sugarloaf Key, Florida. Jour Sed Pet 57:152–165

Carr WJ Alverson DC (1959) Stratigraphy of middle Tertiary rocks in part of west-central Florida. US Geol Surv Bull 1092

Carter JG (1980a) Environmental and biological controls of bivalve shell mineralogy and microstructure. In: Rhoads DC Lutz RA (eds.), Skeletal growth of aquatic organisms - Biological records of environmental change. New York, Plenum Press, pp. 69–113

Carter JG (1980b) Bivalve shell mineralogy and microstructure. Part A. Selected mineralogical data for the Bivalvia. In: Rhoads DC Lutz RA (eds.), Skeletal growth of aquatic organisms - Biological records of environmental change, New York, Plenum Press, pp. 627–643

Carter JG (1980c) Bivalve shell mineralogy and microstructure. Part B. Guide to bivalve shell microstructures. In: Rhoads DC Lutz RA (eds.), Skeletal growth of aquatic organisms - Biological records of environmental change, New York, Plenum Press, pp. 642–673

Casagrande A (1948) Classification and identification of soils. Trans Amer Soc Civil Eng 113:901–930

Cathcart JB (1963a) Economic geology of the Chicora Quadrangle, Florida. US Geol Surv Bull 1962-A

Cathcart JB (1963b) Economic geology of the Keysville Quadrangle, Florida. US Geol Surv Bull 1128

Cathcart JB (1964) Economic geology of the Lakeland Quadrangle, Florida. US Geol Surv Bull 1962-G

Cathcart JB (1966) Economic geology of the Fort Meade Quadrangle, Florida. US Geol Surv Bull 1207

Cathcart JB Davidson DF (1952) Distribution and origin of phosphate in the Land Pebble Phosphate District of Florida. US Geol Surv TEI-212

Chen FH (1988) Foundations on expansive soils. New York, Elsevier

Chilingarian GV Wolf KH (eds.) (1975) Compaction of coarse-grained sediments, I. Amsterdam, Elsevier Sci Publ Co, Develop in Sedimentology 18A

Chilingarian GV Wolf KH (eds.) (1976) Compaction of coarse-grained sediments, II. Amsterdam, Elsevier Sci Publ Co, Develop in Sedimentology 18B

Cuadros J Delgado A Cardenete A Reyes E Linares J (1994) Kaolinite/montmorillonite resembles beidellite. Clays and Clay Min 42(5):642–651

Davis RA Jr. (1992) Depositional systems: An introduction to sedimentology and stratigraphy. Englewood Cliffs NJ, Prentice Hall, 2nd Ed

Deffeyes KS Martin EL (1962) Absence of carbon-14 activity in dolomite from Florida Bay. Science, 136: 782

Duerr AD (1994) Types of secondary porosity of carbonate rocks in injection and test wells in southern peninsular Florida. US Geol Surv Water-Res Inves Rept 94–4013

Dunham RJ (1962) Classification of carbonate rocks according to depositional texture. In: Ham WE (ed.), Classification of carbonate rocks, Amer Assoc Petrol Geol Memoir. 1. pp. 108–121

Espenshade GH (1958) Geologic features of areas of abnormal radioactivity south of Ocala, Marion County, Florida. US Geol Surv Bull 1046-J, pp. 205–219

Estes CJ (1987) Chemical and structural analysis of an aluminum hydroxy-interlayered clay from terra rossa soil, south Florida. Master's Thesis, Univ South Florida, Tampa

Ford DC Williams PW (2007) Karst hydrogeology and geomorphology. New York, John Wiley & Sons

Fountain KB (2009) Origin and stratigraphic significance of kaolinitic sediments from the Cypresshead Formation: A sedimentological, mineralogical, and geochemical investigation. Ph.D. diss, Univ Fla, Gainesville

Galan E (1996) Properties and applications of palygorskite –sepiolite clays. Clay Min 31:442–453

Gordon DS Flemmings PB (1998) Generation of overpressure and compaction-driven fluid flow in a Plio-Pleistocene growth-faulted basin, Eugene Island 330, offshore Louisiana. Basin Res 16(10):177–196

Gray GW (1974) The origin and evolution of the sediments of Butch Key, an emergent mud mound in Florida Bay. Master's thesis, Wooster, OH, Coll of Wooster

Gray JR (1978) Genesis of a Holocene carbonate mud mound in Florida Bay. Master's thesis, Univ South Florida, Tampa

Gressly A (1838) Observation géologiques sur le Jura Soleurois. Neue Denksch der Allgem Schweiz Gesells für die Gesamn Naturwiss 2:1–112

Grim RE (1953) Clay mineralogy. New York, McGraw-Hill

Hine AC (2013) Geologic history of Florida: Major events that formed the Sunshine State. Gainesville, Univ Press Fla

Hedberg HD (1926) The effect of gravitational compaction on the structure of sedimentary rocks. Amer Assoc Petrol Geol Bull 10(11):1035–1072

Hedberg HD (1936) Gravitational compaction of clays and shales. Amer Jour Sci, Ser 5, 31:241–287

Herbert TA Upchurch SB (2016) The potential role of hypogene speleogenesis in the lower Floridan aquifer and Sunniland oil trend, south Florida, U.S.A. In: Chavez T Reehling P (eds), Proceedings of DeepKarst 2016: Origins, Resources, and Management of

Hypogene Karst, Carlsbad, New Mexico, April 11-14, 2016, Nat Cave and Karst Res Inst, Symp 6, pp. 119–129

Hoffmeister JE Stockman KW Multer HG (1967) Miami Limestone of Florida and its Recent Bahamian counterpart. Bull Geol Soc Amer 78:175–190

Holtz RD Kovacs WD (1981) An introduction to geotechnical engineering. Englewood Cliffs, NJ, Prentice Hall

ISO (2002) Geotechnical investigation and testing -- Identification and classification of soil -- Part 1: Identification and description. Internat Stand Org ISO 14688-1

Jamieson JC (1953) Phase equilibrium in the system calcite-aragonite. Jour Chem Phys 21(8):1385–1390

Kerr PF (1937) Attapulgus clay. Amer Mineral 22(5):534–550

Ketner KB McGreevy LJ (1959) Stratigraphy of the area between Hernando and Hardee counties, Florida. US Geol Surv Bull 1074-C

Kim JJ (1984) Genesis of dioctahedral chlorites in the Miami formation. Masters thesis, Univ South Florida, Tampa

Klimchouk A (2007) Hypogene speleogenesis: Hydrogeological and morphogenetic perspective. Carlsbad, NM, Nat Cave and Karst Res Inst

Krumbein WC (1934) Size frequency distributions of sediments. Jour Sed Pet 4:65–77

Krumbein WC Pettijohn FJ (1938) Manual of sedimentary petrography. New York, Appleton-Century-Crofts

Krumbein WC Sloss LL (1963) Stratigraphy and sedimentation. San Francisco, W.H. Freemen Co, 2nd Edition

Land LS (1966) Diagenesis of metastable skeletal carbonates. Ph.D. diss, Lehigh PA, Lehigh Univ

Land LS (1967) Diagenesis of skeletal carbonates. Jour Sed Pet 37:914–930

Lantenois S Champallier R Bény J-M Muller F (2008) Hydrothermal synthesis and characterization of dioctahedral smectites: A montmorillonites series. Appl Clay Sci 38(3):165–178

Leeder MR (2011) Sedimentology and sedimentary basins: From turbulence to tectonics. New York, Wiley-Blackwell, 2nd Ed

Lidz BH Reich CD Shinn EA (2005) Systematic mapping of bedrock and habitats along the Florida reef tract: Central Key Largo to Halfmoon Shoal (Gulf of Mexico). US Geol Surv Prof Pap 1714

MacDonald GJF (1956) Experimental determination of calcite-aragonite equilibrium relations at elevated temperatures and pressures. Amer Mineral 41:744–756

McClellan GH (2013) Summary of geology and mineralogy of palygorskite-sepiolite in northwest Florida and southwest Georgia. In: Southeastern Geol Soc, Field Trip Guidebook No 58, pp. 4–19

McFadden M (1982) Petrology of porcellanites in the Hawthorn Formation, Hamilton County, Florida. Master's thesis, Univ South Florida, Tampa

Meade RH (1964) Removal of water and rearrangement of particles during the compaction of clayey sediments – Review. US Geol Surv Prof Pap 497-B

Middleton GV (1973) Johannes Walther's Law of the correlation of facies. Geol Soc Amer Bull 84:979–988

Miller R III (1945) The heavy minerals of Florida beach and dune sands. Amer Mineral 30:65–75

Milliman JD (1974) Marine carbonates. New York, Springer-Verlag, Recent Sed Carb Part 1

Morse J Mackenzie FT (1990) Geochemistry of sedimentary carbonates, New York, Elsevier

Morse JW Zullig JJ Bernstein LD Millero FJ Milne P Mucci A Choppin GR (1985) Chemistry of calcium carbonate-rich shallow water sediments in the Bahamas. Amer Jour Sci 285:147–185

Muhs DR Budahn JR Prospero JM Carey SN (2007) Geochemical evidence for African dust inputs to soils of western Atlantic islands: Barbados, the Bahamas, and Florida. Jour Geoph Res, v. 112, F02009

Multer HG (1977) Field guide to some carbonate rock environments: Florida Keys and western Bahamas. Dubuque IA, Kendall/Hunt Pub Co

Navoy AS (1986) Hydrogeologic data from a 2,000-foot deep core hole at Polk City, Green Swamp area, central Florida. US Geol Surv Water-Res Invest Rept 84-4257

Nelson J Miller DJ (1997) Expansive soils: Problems and practice in foundation and pavement engineering. New York, Wiley-Interscience

Nocita BW Kohpina P Papetti LW Olivier MM Grosz AE Snyder S Campbell KM Green RC Scott TM (1990) Sand, gravel and heavy-mineral resources potential of surficial sediments offshore of Cape Canaveral, Florida. Fla Geol Surv Open File Rept No. 35

North American Commission on Stratigraphic Nomenclature (2005) North American Stratigraphic Code: Amer Assoc Petrol Geol Bull 89(11):1547–1591

Olive WW Chleborad AF Frame CW Schlocker J Schneider RR Schuster AL (1989) Swelling clays map of the coterminous United States. US Geol Surv Misc Inves Series Map I-1940

Oural CR Brooker HR Upchurch SB (1986) Source of gross-alpha radioactivity anomalies in recharge wells, central Florida phosphate district. Bartow FL, Fla Inst Phosph Res Pub No. 05-014-034

Palmer AN (1991) The origin and morphology of limestone caves. Geol Soc Amer Bull 103:1–21

Patterson SH (1974) Fuller's earth and other industrial mineral resources of the Meigs-Attapulgus-Quincy District, Georgia and Florida. US Geol Surv Prof Pap 828

Pirkle EC Pirkle WA Yoho WH (1977) The Highland heavy-mineral sand deposit on Trail Ridge in northern peninsular Florida. Fla Geol Surv Rept Invest No. 84

Price RE Pichler T (2002) Naturally occurring arsenic in the upper Floridan aquifer, southwest Florida: Implications for aquifer storage recovery. Amer Water Res Assoc, Proc AWRA Summer Specialty Conf, Ground Water/Surface Water Interactions, pp. 387–392

Price RE Pichler T (2006) Abundance and mineralogical association of arsenic in the Suwannee Limestone (Florida): Implication for arsenic release during water-rock interaction. Chem Geol 228:44–56

Prospero JM (1999) Long-range transport of mineral dust in the global atmosphere: Impact of African dust on the environment of the southeastern United States. Proc Nat Acad Sci, USA, 96:3396–3403

Prospero JM Olmez I Ames M (2001) Al and Fe in PM 2.5 and PM 10 suspended particles in south-central Florida: The impact of the long range transport of African mineral dust. Water, Air, and Soil Poll 125:291–317

Prospero JM Nees RT Uematsu M (1987) Deposition rate of particulate and dissolved aluminum derived from Saharan dust in precipitation at Miami, Florida. Jour Geophy Res 92(D12):14722–14731

Prospero JM Landing WM Schulz M (2010) African dust deposition to Florida: Temporal and spatial variability and comparisons to models. Jour Geophy Res 115:D13304–D13304

Puri HS Vernon RO (1964) Summary of the geology of Florida and a guidebook to the classic exposures. Fla Geol Surv Sp Pub 5 (revised)

Puttiwongrak A Honda H Matsuoka T Yamada Y (2013) Compaction curve with consideration of time and temperature effects for mudstones. Southeast Asian Geotech Soc, Geotech Engin Jour 44(1):34–39

Railsback LB (2013) Some fundamentals of mineralogy and geochemistry. Available at http://www.gly.uga.edu/railsback/FundamentalsIndex.html#Carbonates. Accessed February 6, 2013

Randazzo AF (1978) Dolomitization in the Floridan aquifer. Amer Jour Sci 278:1177–1184

Randazzo AF Bloom JI (1985) Mineralogical changes along the freshwater/saltwater interface of a modern aquifer. Sed Geol 43: 219–239

Rieke HH Chilingarian GV (1974) Compaction of argillaceous sediments. Amsterdam, Elsevier Sci Pub Co, Develop Sedimentology 16

Robinson RB (1967) Diagenesis and porosity development in Recent and Pleistocene oolites from southern Florida and the Bahamas. Jour Sed Pet 37:355–365

Safko PS Hickey JJ (1991) A preliminary approach to the use of borehole data, including television surveys, for characterizing secondary porosity of carbonate rocks in the Floridan aquifer system. US Geol Surv Water-Res Invest Rept 91–4168

Schmertmann JH Crapps DK (1980) Slope effect on house shrink-swell movements. Jour Geotech Engin Div, Proc Amer Soc Civil Engin 106(ST12): 1357–1343

Scott TM (1988) The lithostratigraphy of the Hawthorn Group (Miocene) of Florida. Fla Geol Surv Bull 59

Sellards EH (1913a) Map of Florida showing topography, hard rock and land pebble phosphate deposits, and areas of artesian flow. Fla Geol Surv

Sellards EH (1913b) Origin of the hard rock phosphates of Florida. Fla Geol Surv, 5th Ann Rept, pp. 23–80

Shariatmadari H Mermut AR Benke MB (1999) Sorption of selected cationic and neutral organic molecules on palygorskite and sepiolite. Clays and Clay Min 47(1):44–53

Shinn EA (1964a) Formation of Recent dolomite in Florida and the Bahamas (abs). Amer Assoc Petrol Geol Bull 48:547

Shinn EA (1964b) Recent dolomite, Sugarloaf Key. In: Ginsburg RN (ed), South Florida Carbonate Sediments, Geol Soc Amer Ann Meeting Field Trip Guidebook, Field Trip 1, pp. 62–67

Shinn EA (1968) Selective dolomitization of Recent sedimentary structures. Jour Sed Pet 38:612–616

Shinn EA (2001) African dust causes widespread environmental distress. US Geol Surv Open-File Rept 01-246

Sowers GF Kennedy CM (1967) High volume change clays of the Southeastern Coastal Plain. Third Pan-American Congress of Soil Mechanics and Foundation Engineering, Caracas, Venezuela, Volume II, pp. 98–120

Strom RN Upchurch SB (1985a) Palygorskite distribution and silicification in the phosphatic sediments of central Florida. In: Snyder S Riggs S Partin B Mallette P Walker R (eds.), Eighth Inter Field Workshop and Symp (Southeastern United States), Guidebook, Inter Geol Congress, Inter Geol Correl Prog, Project 156 - Phosphorites, pp. 118–126

Strom RN Upchurch SB (1985b) Palygorskite distribution and silicification in the phosphatic sediments of central Florida. In: Scott TM Cathcart JB (eds.), Florida land-pebble phosphate district. Geol Soc Amer, 1985 Ann Meet Field Trip Guidebook, pp. 68–75

Taft WH (1961) Authigenic dolomite in modern carbonate sediments along the southern coast of Florida. Science 134: 561–562

Terzaghi K Peck RB (1948) Soil mechanics in engineering practice. New York, John Wiley & Sons

Thayer PA Miller JA (1984) Petrology of lower and middle Eocene carbonate rocks, Floridan aquifer, central Florida. Trans Gulf Coast Assoc Geol Soc, 34:421–434

Udden JA (1898) Mechanical composition of wind deposits. Augustana Lib Pub No. 1

Udden JA (1914) Mechanical composition of clastic sediments. Geol Soc Amer Bull 25:655–744

Upchurch SB (1970a) Mixed-population sediment in nearshore environments. Inter Assoc Great Lakes Res, Proc 13th Conf Great Lakes Res, pp. 768–778

Upchurch SB (1970b) Sedimentation on the Bermuda Platform. Detroit, Lake Surv Dist, US Army Corps Eng Res Report 2-2

Upchurch SB (1992) Quality of waters in Florida's aquifers. In: Maddox GL Lloyd JM Scott TM Upchurch SB Copeland R (eds.), Florida Ground Water Quality Monitoring Program -- Volume 2, Background Hydrogeochemistry, Fla Geol Surv Sp Pub No. 34, Ch. IV, pp. 12–52, 64–84, 90–347

Upchurch SB (2017) Hypogene speleogenesis on the Floridan Platform. In: Klimchouk AB Palmer AN Waela JD Auler AS Audra P (eds.), Hypogene Karst Regions and Caves of the World, Springer Inter Publishing, Ch. 49, pp. 735–744

Upchurch SB Strom RN Nuckels MG (1982) Silicification of Miocene rocks from central Florida. In: Scott TM Upchurch SB (eds.), Miocene of the Southeastern United States, Fla Bur Geol Sp Pub No. 25, pp. 251–284

Upchurch SB Strom RN Williams MJ (1983) Preservation of dolomite in coastal peats of the Ten Thousand Islands area, Florida. In: Raymond R Jr. Andrejko MJ (eds.), Mineral matter in peat: Its occurrence, form, and distribution, New Mexico, Los Alamos Nat Lab LA-9907-OBES, UC-11, pp. 215–224

Upchurch SB Dobecki TL Scott TM Meiggs SH Fratesi SE Alfieri MC (2013) Development of sinkholes in a thickly covered karst terrane. Proceedings, Thirteenth Multidisciplinary Conference on Sinkholes and the Engineering and Environmental Impacts of Karst, Carlsbad NM, Nat Cave and Karst Res Inst, pp. 272–277

Van Kauwenbergh SJ Cathcart JB McClellan GH (1990) Mineralogy and alteration of the phosphate deposits of Florida. US Geol Surv Bull 1914

Velde B (1995) Origin and mineralogy of clays: Clays and the environment. Berlin, Springer-Verlag

Weaver CE Beck KC (1977) Miocene of the southeastern United States: A model for chemical sedimentation in a peri-marine environment. Amsterdam, Elsevier Sci Pub Co, Devel in Sediment vol. 22

Weir AH Greene-Kelly R (1962). Beidellite. Amer Mineral 47:137–146

Weller JM (1959) Compaction of sediments. Bull Amer Assoc Petrol Geol 43(2):273–310

Wentworth CK (1922) A scale of grade and class terms for clastic sediments. Jour Geol 30:377–392

Wesley LD (2010) Fundamentals of soil mechanics for sedimentary and residual Soils. New York, John Wiley & Sons

White WA (1949) Atterberg plastic limits of clay minerals. Amer Mineral 34(7–8):508–512

Williams CP Scott TM Upchurch SB Paul DT Hannon LM (in prep.) Florida geomorphology. Fla Geol Surv Bull 72

Winston GO (1995) The boulder zone dolomites of Florida Volume 1: Paleogene & upper Cretaceous zones of the southeastern peninsula and the Keys. Miami Geol Soc, Available at http://sofia.usgs.gov/publications/reports/mgs_winstonv11995/, accessed February 2, 2013

Yamada H Nakazawa H Yoshioka K Fujita T (1991) Smectites in the montmorillonite-beidellite series. Clay Min 26:359–369

# 3 Geologic Framework of Florida

**Abstract**

The Florida Platform is a series of Middle Jurassic to Quaternary evaporite, carbonate, and siliciclastic sediments deposited on the relatively stable, passive margin of the North American Plate. Faulting of pre-Middle Jurassic basement rocks created structures that have affected deposition through the Cenozoic. The Georgia Channel System, a structural trough, limited siliciclastic transport from the north to the peninsula until the Miocene.

On the Florida Platform, siliciclastic deposition predominated in the north, while carbonate and evaporite sediments formed to the south. The Cretaceous saw carbonate deposition advance northward across the Platform, with the Eocene Avon Park and Ocala limestones and Oligocene Suwannee Limestone deposited as carbonate production was at its most widespread. Deposition of the Miocene Hawthorn Group marked the progress of siliciclastic sediment southward across the Platform once again. The siliciclastic supply from the north waned slightly in the Quaternary, freeing the southern Platform for deposition of the Pleistocene Miami and Key Largo limestones.

Eocene and Oligocene limestones form the Floridan Aquifer System, a highly productive karstic aquifer. Miocene siliciclastics form the Intermediate Aquifer System and Confining Unit. The Surficial Aquifer System (including the Biscayne Aquifer of southeastern Florida) is comprised of Pliocene-Holocene siliciclastics and carbonates.

The geomorphology of the Florida Platform consists of east–west trending highlands in the north and west of the state, and north–south trending highlands down the middle of the peninsula, flanked by coastal lowlands. In every geomorphic district of Florida except one, karst features are ubiquitous but vary in form and degree of development.

It is necessary to understand the geology of Florida, including the ages of carbonate strata, in order to place Florida's karst in context. This chapter introduces the stratigraphy, aquifer systems, and geomorphology of Florida. The ages of major geologic events and deposits in the following paragraphs are based on the geologic time scale by Walker et al. (2012).

**Keywords**

Florida Platform · Florida stratigraphy · Carbonate sediments · Geomorphological districts · Geomorphological provinces · Florida aquifers

## 3.1 Introduction

This chapter provides a statewide presentation of Florida's geology with the intent of providing a context for understanding Florida's karst features. Florida's eogenetic karst varies in form and degree of development as a result of the age of the strata, their exposure to dissolution processes, and the relative development of primary and secondary porosity and permeability.

Florida's geologic history actually begins when the area was part of the supercontinent now known as Pangea. The area that became Florida was located near the center of a large tectonic plate, and it shared many geologic features with today's northwestern Africa. Throughout much of Earth's history, tectonic plates have been continually shifting, changing position, very slowly colliding with one another to form supercontinents and splitting apart again, forming new ocean basins. During the Paleozoic Era, the


S. Upchurch et al., *The Karst Systems of Florida*, Cave and Karst Systems of the World,
https://doi.org/10.1007/978-3-319-69635-5_3

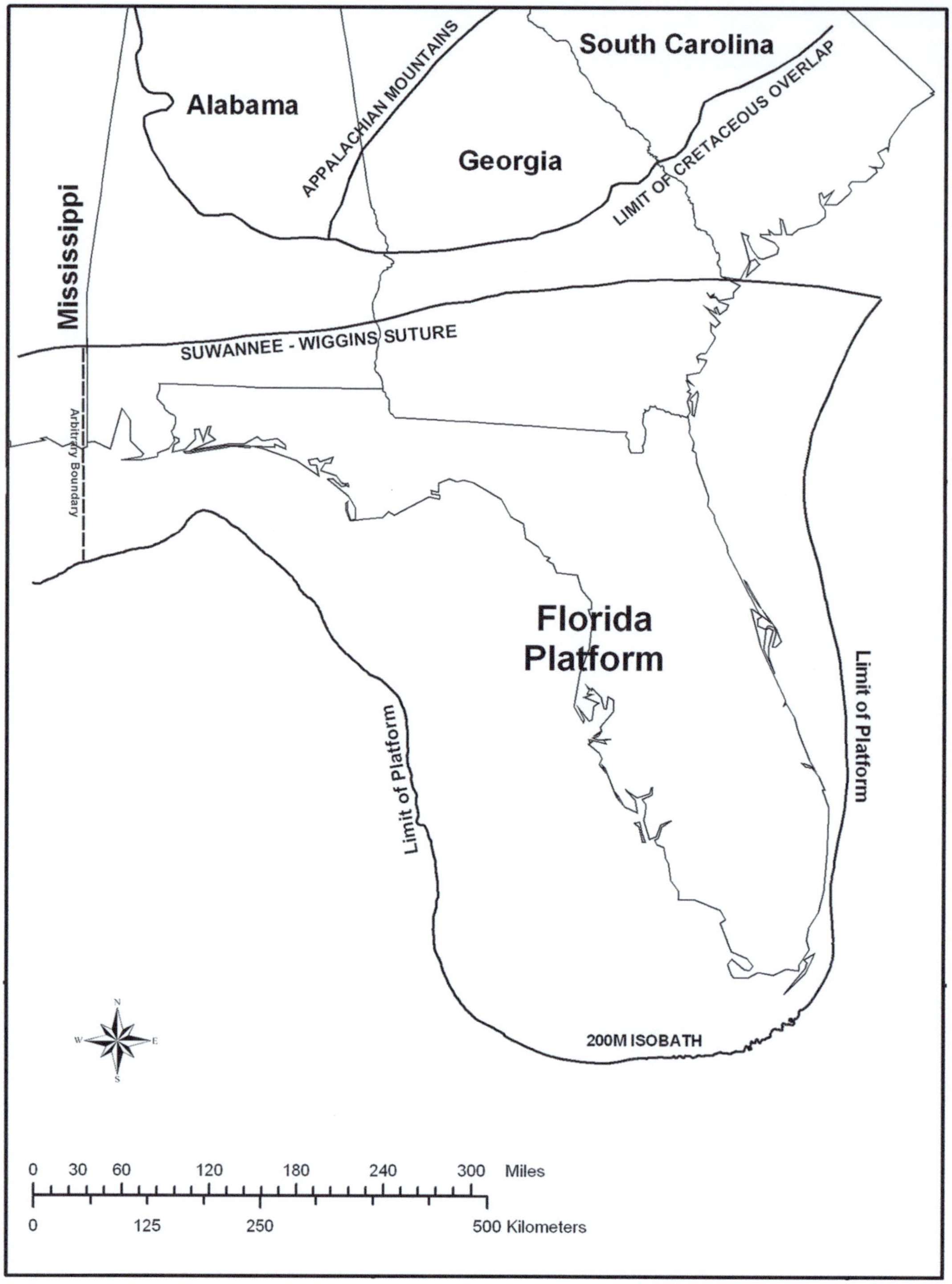


**Fig. 3.1** Limits of the Florida Platform

plates amalgamated into the supercontinent of Pangea. Pangea existed from approximately 335 million years ago (Mya) in the late Paleozoic until approximately 175 Mya (early to middle Mesozoic). The piece of Earth's crust that is now Florida was tucked in between portions of the plates that became North America, Africa, and South America. When Pangea began to break up in the early to middle Mesozoic Era, a piece of continental crust from Africa remained attached to the North American Plate.

This piece of continental crust, and the massive accumulations of limestone and siliciclastic sediments that built up on top of it in the following millions of years, make up what is known as the Florida Platform. The Florida Platform (Fig. 3.1) is delimited by the 200 m isobath at the continental shelf break and extends inland and northward to the approximate location of the Paleozoic, tectonic suture beneath southern Georgia and Alabama (Fig. 3.1). The Suwannee–Wiggins Suture (Thomas et al. 1989) is the proposed location where terranes with African affinities are welded to the North American Plate (Chowns and Williams 1983; McBride and Nelson 1988; Woods et al. 1991). These are the deepest known rocks in Florida. They are appropriately referred to as the basement rocks of Florida.

The basement rocks of the Florida Platform range in age from the late Precambrian to the Middle Jurassic (Barnett 1975) and are predominantly igneous and metamorphic rocks, which are overlain by sandstones and shales. Excellent reviews of the geology of the basement are provided by Smith (1982), Arthur (1988), Smith and Lord (1997), and Heatherington and Muller (1997). The basement is separated from overlying rocks by the "sub-Zuni surface" (Barnett 1975), a horizon representing a period of erosion sometime before the Middle Jurassic. Barnett's (1975) structure-contour map and interpretation of the basement surface has it occurring as shallow as approximately 915 m below mean sea level in central-northern peninsular Florida. The basement surface dips west and southwest toward the Gulf of Mexico basin, to the south into the South Florida Basin, and to the east into the Atlantic basin. The basement surface reaches depths of more than 5180 m below mean sea level in southern Florida (Barnett 1975).

Deposited on top of the erosional surface of the basement, the Florida Platform is constructed of Middle Jurassic to Quaternary evaporite, carbonate, and siliciclastic sediments deposited on a relatively stable, passive margin of the North American Plate. Age assignments for geologic formations are often made on the basis of the types of fossils within those formations and the periods in which they were known to live. In Florida, the age assignments for the Middle Jurassic to Quaternary formations are, at times, tentative due to limited, or lack of, paleontological evidence in some formations. The age determinations for some of the younger units, for example the Pliocene Tamiami Formation, are based on ubiquitous paleontological evidence. This inconsistency, in part, is responsible for varying and differing interpretations of when, where, and how much sediment was deposited across the Florida Platform over geologic time (compare Salvador 1991 and Randazzo 1997).

The Florida Platform is predominantly a carbonate platform that formed in clear tropical and subtropical shallow seas. Carbonate sedimentation generally occurs away from eroding mountain ranges and the muddy waters of areas undergoing siliciclastic sedimentation. The Bahamas Platform, which is often discussed with Florida as one large carbonate environment, is the major modern example of a carbonate platform. The tiny, calcareous skeletons of trillions of small animals and plants create carbonate sediments on both Platforms. These carbonate particles have accumulated over millions of years to form the Florida Platform.

Siliciclastic sediments are derived from the erosion of sedimentary, igneous or metamorphic rock. As mountain ranges and land masses dominated by these types of rocks erode away over millions of years, they provide a source of sediments that are deposited into surrounding low areas. During some periods of Florida's geologic history, siliciclastic sediments from the continent spilled onto the Florida Platform. The interplay between these siliciclastic sediments and the carbonate sediments of the Florida Platform is crucial to the development of Florida karst.

## 3.2 Geologic Structure of the Florida Platform

Geologic structures are rock deformations (folds, faults, sedimentary basins, uplifts) caused by tectonic movements of the Earth's crust and other processes that cause stress and strain in rock. It is important to understand the structure of the Florida Platform and the presence of features such as faults and fractures on the exposure of carbonate strata to erosion and development of karst. For example, some of Florida's geologic "structures" are actually features caused by erosion of pre-existing surfaces, not by tectonic movements of the strata.

The Florida Platform has been a relatively stable portion of the trailing edge of the North American Plate since the Middle Jurassic. Trailing edge plate margins develop when a tectonic plate splits. The two plates are initially uplifted along the line of their separation. As the two plates move apart, the rocks beneath cool and the margins of the plates begin to subside. As they subside, sedimentary basins develop and sediments accumulate. Both the Atlantic Ocean and Gulf of Mexico formed by spreading and the continental margins that face the spreading centers have subsided as a result of spreading. Spreading continues in the Atlantic Ocean, with the spreading zone located at the Mid-Atlantic Ridge. The Gulf of Mexico is no longer spreading but the margins of the continents facing the Gulf are slowly subsiding over geologic time.[1] The Florida Platform is sandwiched between the Gulf and Atlantic plate spreading zones and the margins of the Platform are faulted and have subsided as is normal for a trailing edge of a continental margin. The center of the Platform has subsided less that its margins, thus creating the appearance of a tectonic arch.

[1]Throughout this text the term "geologic time" is utilized to describe the durations of geologic processes that are so slow that they are not perceptible. Typically, the results of the processes are not easily detectible over the lifespan of an individual human being but are detectible in the geologic record.

Winston (1991) stated that the Mesozoic and Cenozoic Era structural movement on the Florida Platform was entirely downward, which is consistent with movements associated with rifting and separation along plate margins. Florida's arches, or structural highs, were not formed by uplift, but as the result of subsiding more slowly than the flanking basins. However, faulting of the basement rocks created many of the structural features recognized on the pre–Middle Jurassic surface (Barnett 1975; Smith and Lord 1997).

Faults disrupting the Upper Jurassic sediments have been identified in northwestern Florida; some fault displacements exceed 305 m (Lloyd 1989). Miller (1986) recognized a number of known or suspected Cenozoic faults that affect the Floridan Aquifer System (FAS). Duncan et al. (1994) identified faulting in the Lower to Middle Eocene Oldsmar Formation. A number of hydrogeologic and geomorphic investigations have proposed the existence of faults in later strata in Florida (Wyrick 1960; Leve 1966; Lichtler et al. 1968; Pirkle 1970; White 1970). The faults in the Cenozoic section have very limited displacement, generally less than 31 m and are difficult to identify due to limited displacement, poor well control, few "marker" beds, erosional disconformities, and karstification.

Little has been documented regarding folding of post–Middle Jurassic sediments on the Florida Platform. Missimer and Maliva (2004) believe that folding is more widespread on the Florida Platform than is presently recognized, due to the limited amount of detailed subsurface data. They recognized folding with associated fracturing and faulting in Eocene to Pliocene sediments on the southern portion of the Platform. They postulated that the interaction of the Caribbean and North American tectonic plates in the Late Miocene to Pliocene produced the folds, fractures, and faults (Missimer and Maliva 2004).

The oldest features recognized as having affected deposition of post–Middle Jurassic sediments on the Platform are expressed on the pre–Middle Jurassic surface (Arthur 1988). The Mesozoic structural features that influenced sediment deposition include a series of basins or embayments and arches (Fig. 3.2). Some of these features affected deposition into the middle Cenozoic (i.e., the South Florida Basin; Scott 1988). Other features influenced deposition into the late Cenozoic (i.e., the Apalachicola Embayment; Schmidt 1984). The Peninsular Arch (Fig. 3.2) affected deposition from the mid-Jurassic through the Cretaceous and was intermittently a positive structural feature during the Cenozoic (Miller 1986). The Cenozoic structural features that influenced sedimentation are depicted on Fig. 3.3.

One of the more interesting structural features of the Florida Platform is a southwest-to-northeast trending trough-like, tectonic low that has influenced deposition from the Middle Jurassic until at least the Middle Miocene. This feature has an extended list of names that have been applied to all or parts of it. Excellent reviews of the names applied to the feature were presented by Schmidt (1984) and Huddlestun (1993). However, Georgia Channel System is the name that has been applied to the entire sequence (Huddlestun 1993) (Figs. 3.2 and 3.3).

The Georgia Channel System had its origin in the formation of the South Georgia Rift in the Triassic–Jurassic (?) (Huddlestun 1993). From the late Cretaceous through the Paleocene, this area was the boundary between carbonate deposition to the south and siliciclastic deposition to the north. By the Eocene, the ancestral Appalachian Mountains had been highly eroded leaving relatively low hills and significantly reduced siliciclastic sediment transport via streams and rivers. In the Eocene and Oligocene as the result of a greatly reduced siliciclastic supply, carbonate deposition extended northward, across the Georgia Channel System. The channel system was then infilled by predominantly siliciclastic sediments in the Late Oligocene to the Early Miocene in response to uplift in the Appalachians (Scott 1988). Gallen et al. (2013) postulated that geomorphic changes in the southern Appalachians resulted from a "post-orogenic regional uplift in the Miocene." This rejuvenation would have resulted in a large amount of siliciclastic sediments reaching the marine environment in the southeastern United States and spilling southward, onto the Florida Platform. This influx of siliciclastic sediments was a crucial event in Florida's geologic and karstic history, the Miocene and younger siliciclastics being a key element in the nature of the cover materials overlying Florida's karstified carbonate sediments and rocks.

## 3.3 Depositional Environments of the Florida Platform

All of Florida's sediments above the basement rock were deposited in environments at or near sea level. The initial depositional environments affecting the Florida Platform were environments in which influx of water from the sea was often restricted, allowing for intense evaporation and the development of evaporites (halite, anhydrite, and gypsum) in limited areas. As the Gulf of Mexico, a region of tectonic plate separation, continued to expand and sea levels rose, siliciclastic and carbonate depositional environments began to cover more of the Florida Platform. Continued sea-level rise through the Cretaceous eventually caused the exposed land area in northern Florida to be submerged. The Florida Platform sediments were deposited in a complex interplay of siliciclastic, carbonate, and evaporite facies as a result of sea-level fluctuations (Randazzo 1997). Siliciclastic deposition predominated on the northern part of the Platform closest to the source of the sediments, while carbonate and evaporite sediments formed to the south (Randazzo 1997).

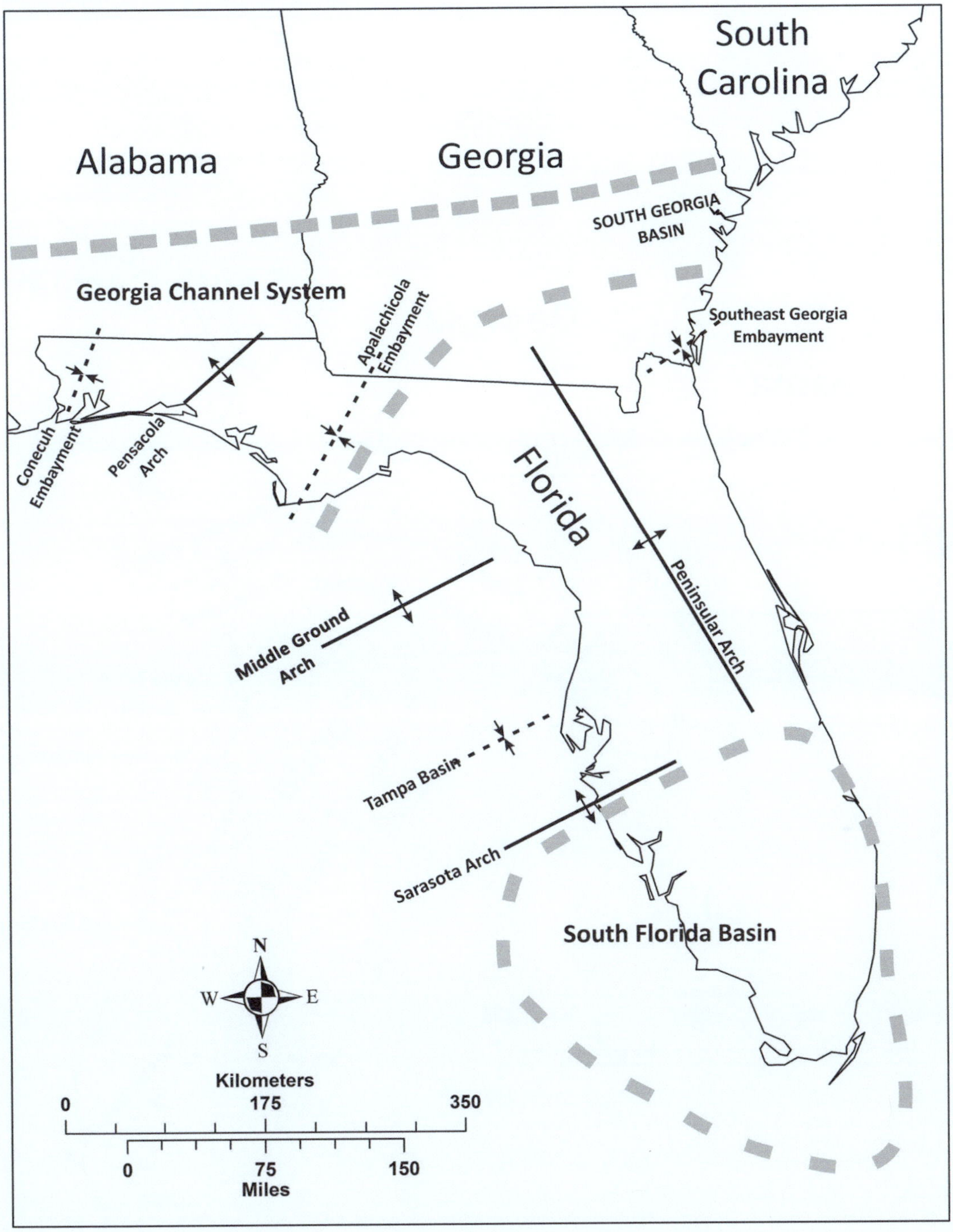


**Fig. 3.2** Structures affecting the Mesozoic and early Cenozoic deposits. (Modified after Lloyd 1997)

In the early Cenozoic (Paleogene), the siliciclastic sediment supply was limited due to the highlands of the ancient Appalachian Mountain trend having been reduced by erosion, and carbonate deposition expanded to cover the entire Florida Platform and beyond by the Oligocene. The carbonate platform, which began as a rimmed shelf in the Jurassic, evolved to a distally-steepened, carbonate ramp sequence by the early Cenozoic (Winston 1991; Randazzo 1997). As carbonate production reached its maximum extent on the Florida Platform and beyond, epeirogenic uplift occurred in the Appalachians providing a renewed supply of siliciclastic sediments (Scott 1988; Brewster-Wingard et al. 1997; Gallen et al. 2013). This influx of siliciclastic sediments in the Neogene replaced most carbonate deposition

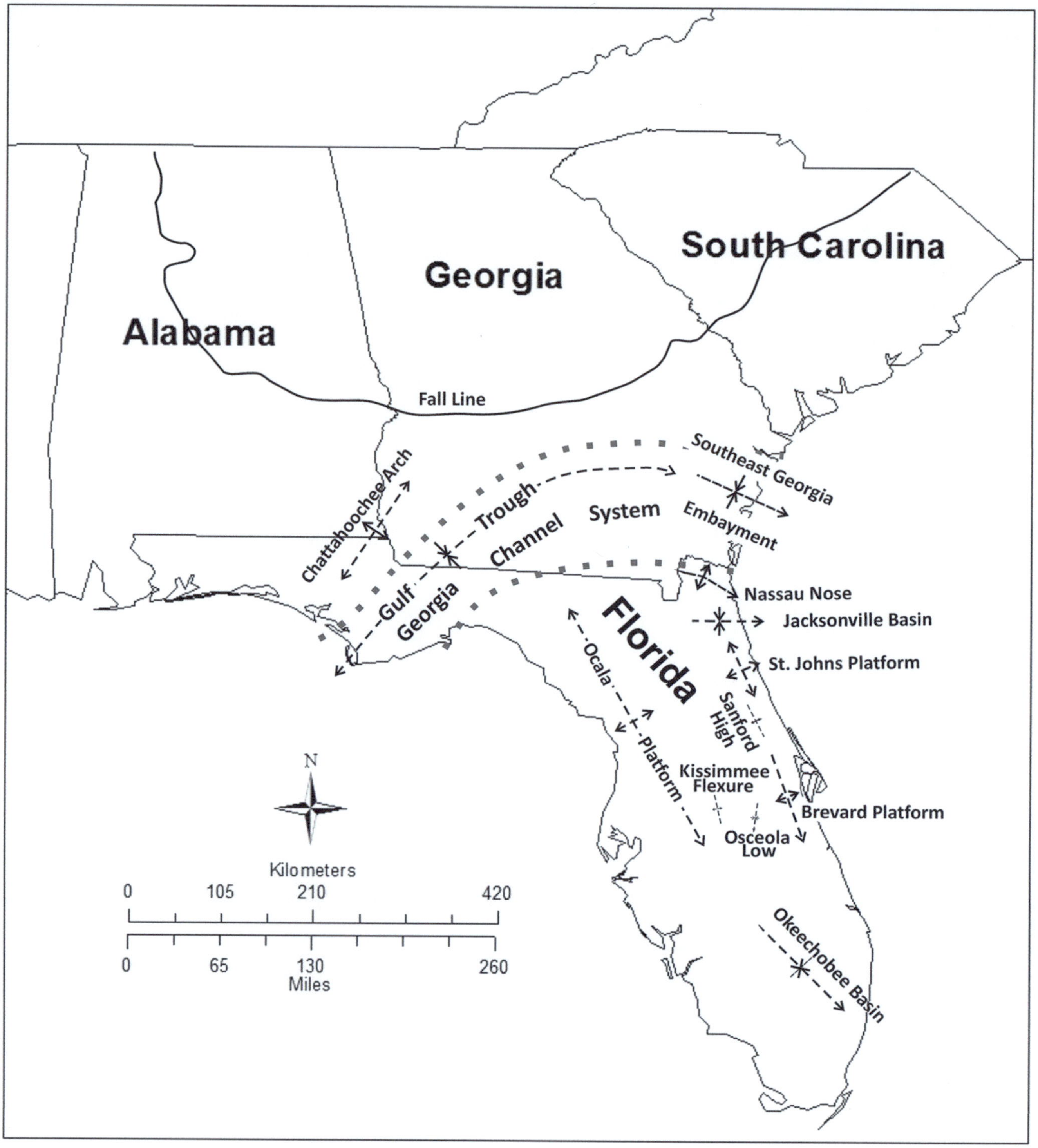


**Fig. 3.3** Structures affecting post-early Cenozoic deposits. (Modified after Scott 1988)

on the Florida Platform by the middle Pliocene. As sea level rose in the late Pleistocene, siliciclastic sedimentation decreased and carbonate deposition increased on the southern Florida Platform. The interplay of the carbonate and siliciclastic sediments with fluctuating sea level and changing climate created complex depositional environments (Scott 1988; Missimer 2002). The interaction of the carbonates and siliciclastics on the Florida Platform has been investigated and discussed by a number of authors (Warzeski et al. 1996; Cunningham et al. 1998, 2003; Guertin 1998; Guertin et al. 2000; Missimer et al. 2000; Missimer 2001, 2002). The siliciclastic sediments are important as cover materials and fill in sinkholes in relatively mature areas of karst. The geologically young, shallow carbonate sediments in southern Florida are in the early stages of karstification.

## 3.4 Stratigraphy of the Florida Platform

### 3.4.1 Pre-Mesozoic Era Stratigraphy

The pre-Mesozoic (Proterozoic Era and Paleozoic Era) framework of the Florida basement consists of igneous, sedimentary, and metamorphic rocks. These rocks have been penetrated by a limited number of oil exploration boreholes, so knowledge of the pre-Mesozoic stratigraphy of Florida is limited. A number of researchers have investigated the pre-Mesozoic rocks including Smith (1982), Chowns and Williams (1983), Dallmeyer et al. (1987), Arthur (1988), and others. Refer to Smith and Lord (1997) for a summary of the research on the basement rocks. The granitic igneous rocks, which occur in east-central Florida, have been dated at approximately 530 Ma (Early Paleozoic) (Dallmeyer et al. 1987; Smith and Lord 1997). Dallmeyer et al. (1987) recognized that these rocks, the Osceola Granite, were part of a complex that is also found in northwestern Africa. A metamorphic sequence, located on the southern flank of the Osceola Granite, indicates metamorphism was associated with the emplacement of the granite pluton.

Sedimentary rocks are found in two areas of the basement, a small area in the Panhandle near the junction of Alabama, Florida, and Georgia, and in the northern peninsula north of a line connecting Tampa Bay in the southwest and a point between St. Augustine and Jacksonville in the northeast (Jones 1997; Smith and Lord 1997). With the exception of sediments encountered in a few wells, the sandstone, siltstone, and shale are usually sparsely fossiliferous. Ages derived from the fossil assemblages range from Early Ordovician to Middle Devonian (approximately 485 to 393 Mya; Jones 1997). Opdyke et al. (1987) identified pre-Mesozoic shale in northern Florida that exhibited low-grade metamorphism. There is currently no evidence of karst features in these strata.

### 3.4.2 Mesozoic Era Stratigraphy

Stratigraphically, the Florida Platform is underlain by pre-mid-Mesozoic sedimentary, igneous, and metamorphic rocks overlain by post-mid-Mesozoic and Cenozoic sedimentary rocks and sediment. The post-mid-Mesozoic sediments consist predominantly of siliciclastics, except in central and southern Florida where carbonates and evaporites predominate.

#### 3.4.2.1 Pre-Middle Cretaceous Stratigraphy

The pre-mid-Mesozoic (Triassic and Jurassic) rocks (Table 3.1) occur nearest to the land surface in northern peninsular Florida. These rocks dip in the subsurface to the south, under the exposed portion of the Florida Platform, to the east under the Atlantic Ocean and west into the Gulf of Mexico (Puri and Vernon 1964). Consequently, the Mesozoic and Cenozoic sediments thicken in these areas to over 5180 m in southern Florida. Pre-Cretaceous rocks are not known to be involved in the formation of karst in Florida, so they will not be discussed in detail here. Refer to Scott (2011) for a discussion of the pre-Cretaceous stratigraphy.

#### 3.4.2.2 Cretaceous Period

By the beginning of the Cretaceous, a limited portion of the northern Florida Peninsula remained above sea level (McFarlan and Menes 1991). As sea level rose through the Early Cretaceous, more of the Platform was submerged (McFarlan and Menes 1991; Randazzo 1997). Deposition on the northwestern Florida Platform was dominated by marine and non-marine siliciclastics. On the northern portion of the Platform a mix of siliciclastics and carbonates was deposited. Carbonates and evaporites with minor amounts of siliciclastics covered the southern portion of the Platform (McFarlan and Menes 1991; Winston 1987, 1991). The thickness of the Lower Cretaceous sediments reaches more than 1830 m on the northwestern and 2740 m on the southern portions of the Florida Platform (Randazzo 1997).

During the Early Cretaceous, carbonates and evaporites of the Ocean Reef Group, Sunniland Formation (Table 3.1) and associated units were deposited. The Sunniland sediments became the reservoir rocks for Florida's first oil discovery in 1943 (Lloyd 1997). Herbert and Upchurch (2016) and Upchurch (2017) suggested that karst development may have enhanced porosity in the Sunniland Formation oil fields.

In the early portion of the Late Cretaceous, deposition on the northern portion of the Florida Platform continued to be dominated by siliciclastics, while carbonates were being deposited in southern Florida (Sohl et al. 1991; Winston 1991). By the mid-Late Cretaceous, carbonates, including chalk, with limited siliciclastics were deposited over the entire Florida Platform (Sohl et al. 1991). The Upper Cretaceous sediments are more than 915 m thick in northwestern and southern Florida (Randazzo 1997).

At the end of the Cretaceous, a large bolide (i.e., meteorite, asteroid, or comet) collided with the Earth in the Gulf of Mexico–Caribbean region (Hildebrand et al. 1991). This event is referred to as the Chicxulub impact, the infamous "dinosaur-killer" that caused a major mass extinction. The bolide impacted at an oblique angle, spreading ejecta to the north and west (Schultz 1996). It is thought that 100–300 m high tsunamis spread across the Gulf of Mexico (Bourgeois

**Table 3.1** Mesozoic stratigraphic columns

| PANHANDLE | | N. FLORIDA | | S. FLORIDA | | | SYSTEM | ERATHEM | AGE (Ma) |
|---|---|---|---|---|---|---|---|---|---|
| Selma Group | | Pine Key Formation | Lawson Formation | Lawson Formation | Pine Key Formation | Rebecca Shoals Dolomite | Cretaceous | Mesozoic | 66.0 |
| Eutaw Formation | | | | | | Orlando Sound Dolomite | | | |
| Tuscaloosa Group | | Atkinson Formation | | Atkinson Formation | | | | | |
| Unnamed Sediments | | Unnamed Sediments | | Marquesas Supergroup | Naples Bay Group | | | | |
| | | | | | Big Cypress Group | | | | |
| | | | | | Ocean Reef Group | Sunniland Fm. | | | |
| Unnamed Sediments | Sligo-Hosston Formation | | | | Glades Group | | | | |
| | | | | | Pumpkin Bay formation | | | | |
| | | | | | Bone Island formation | | | | |
| Cotton Valley Formation | | Cotton Valley Formation | | | Ft. Pierce Formation | Wood River Formation | Jurassic | | 145 |
| Haynesville Formation | | | | Unnamed Volcanic Complex | | | | | |
| Smackover Formation | | | | | | | | | |
| Norphlet SS | | | | | | | | | |
| Louann Salt / Werner Anhydrite | | Diabase | | | | | | | |
| Diabase | | | | | | | Triassic | | 201 |
| Eagle Mills Formation | | Newark Group | | | | | | | 252 |

Modified from Braunstein et al. (1988)

Note that the differences between the three columns illustrate the differences in depositional history of the regions of Florida. Ages are from Walker et al. (2012)

et al. 1988; Matsui et al. 1999; Kring 2000). Discussions with a number of geologists investigating the Chicxulub impact suggest that the Florida Platform should have been influenced by the event.[2] However, no evidence of the impact or tsunamis has been discovered on the Florida Platform to date. The lack of cores across the Cretaceous-Tertiary (KT) boundary, the limited number and widely spaced distribution of petroleum-exploration wells penetrating the KT boundary, and the general poor quality of the cuttings from the wells hinder the search for evidence.

### 3.4.3 Cenozoic Era Stratigraphy

In the Cenozoic, the Paleogene sediments are predominantly carbonates with some mixed carbonate-siliciclastic

[2]Discussions with T.M. Scott at the Chicxulub planning meeting–Group on Mesozoic–Cenozoic stratigraphy and the Cretaceous–Tertiary (KT) boundary, Puerto Vallarta, Mexico, 1993.

sediments. The Neogene and Quaternary sediments (Table 3.2) are predominantly siliciclastics (Braunstein et al. 1988).

#### 3.4.3.1 Paleocene, Eocene, and Oligocene Epochs (Paleogene Period) Stratigraphy

Carbonate sedimentation dominated during the Paleocene, Eocene, and Oligocene Epochs (Paleogene Period) on the Florida Platform. A significant change in sedimentation occurred in the early Neogene Period when siliciclastic sediments began to replace carbonates as the dominant sediment types. As siliciclastic sediments encroached on the Florida Platform, carbonate sedimentation was forced further and further to the south through the Miocene and Pliocene Epochs.

Carbonate and evaporite deposition dominated much of the Florida Platform during the Paleocene Epoch (Miller 1986). The carbonate and evaporite sediments graded to the northwest into shallow marine, fine-grained siliciclastic sediments across the Georgia Channel System. The main carbonate-producing area was interpreted as being rimmed by a reef system, which restricted flow of seawater into and out of the area, an environment necessary for restricted marine circulation and evaporite deposition (Winston 1991). Paleocene sediments cover the entire Florida Platform and have a maximum thickness of more than 670 m. Thick anhydrite beds in the Cedar Keys Formation (Table 3.2) form the regionally extensive lower confining unit of the FAS (Miller 1986, 1997).

The evaporite content of the Lower to Middle Eocene sediments declined in response to sea-level rise and resulted in the development of a more open, carbonate-ramp depositional system on the Platform. At this time, evaporites were deposited primarily as pore filling material (Miller 1986). The carbonate sediments grade into siliciclastic sediments in the Georgia Channel System (Miller 1986). Lower to Middle Eocene sediments cover the entire Platform, ranging to maximum thickness of more than 945 m. Middle Eocene carbonates (Avon Park Formation) are the oldest sediments exposed at the land surface in Florida (Scott et al. 2001). These sediments are exposed on the crest of the Ocala Platform (Fig. 3.3). Karst features are limited in this area due to the dolomitic nature of the Avon Park Formation sediments in western-central Florida. The Lower to Middle Eocene limestone and dolostone, in part, form the lower production zones of the FAS (the Lower Floridan Aquifer, LFA) while, in some areas, these sediments are part of the Middle Confining Unit (MCU) of the FAS (Miller 1986, 1997).

Carbonate deposition covered virtually the entire Florida Platform in the Late Eocene. Carbonates were deposited north of the Georgia Channel System, nearly to the Fall Line (limit of Cretaceous onlap), beyond the limits of the Florida Platform (Fig. 3.1). The carbonate ramp had a well-developed slope with no vestige of a rim to restrict seawater flow. As a result, evaporites have not been found or are very rare in Upper Eocene limestone or dolostone. The Upper Eocene carbonates grade into siliciclastics on the northwestern-most part of the Platform. These carbonates range in thickness to more than 213 m, but are absent in several areas of the Platform due to erosion (Miller 1986; Scott 1992, 2001). In a large area on the southern part of the Platform, the Upper Eocene sediments are absent, probably due to erosion by ocean currents similar to episodes identified in the Oligocene to Pliocene in this region (Guertin et al. 2000). On the areas of the Platform where the Oligocene carbonates are absent, the Upper Eocene carbonates form the upper portion of the FAS (Miller 1986, 1997). The most prominent limestone formation that was deposited during this time interval is the Ocala Limestone (Table 3.2). In addition to other karst features, the upper few meters of the Ocala Limestone have an extensively developed epikarst in much of Florida. Karst features (i.e., caves and cavernous porosity) are common throughout the Ocala Limestone.

Lower Oligocene carbonate deposition occurred as far landward as did the Upper Eocene deposition. The carbonates grade into siliciclastics on the northwestern-most part of the Platform. Very minor amounts of siliciclastics are incorporated into these carbonates; however, beds of fine quartz sand occur in the Early Oligocene of southern Florida (Missimer 2002). Whether or not the carbonate deposition covered the Platform is open to conjecture. The most prominent, karstic limestone formation that was deposited during this time interval is the Suwannee Limestone (Table 3.2). Similar to the Ocala Limestone, the upper few meters of the Suwannee Limestone have a well-developed epikarst, where it is near the land surface and in the subsurface. As with the Ocala Limestone, karst features are common throughout the Suwannee Limestone.

The Lower Oligocene sediments range in thickness to more than 200 m, but are absent over large portions of the Platform (Miller 1986; Scott 1992, 2001). These sediments are missing due to nondeposition or erosion, or both, in a large area on the eastern flank of the Ocala Platform in an area referred to as the paleo-Orange Island (Bryan 1991). Where the Lower Oligocene sediments are present, they form the upper portion of the FAS (Miller 1986, 1997).

In the subsurface, the upper surfaces of the Ocala and Suwannee limestones often exhibit high porosity and, for equivalent thicknesses of rock, produce substantially more water to wells than the remainder of the formations. These enhanced porosity zones appear to be examples of paleo-epikarst, or ancient, karstic weathering surfaces and conduits.

Chert (silicified limestone) occurs from the upper portion of the Middle Eocene through the Lower Oligocene carbonates. The chert itself formed long after the carbonates were

**Table 3.2** Cenozoic stratigraphic columns in Florida

| PANHANDLE | N. FLORIDA | S. FLORIDA | SERIES | SYSTEM | ERATHEM | AGE (Ma) |
|---|---|---|---|---|---|---|
| Undifferentiated Holocene - Pleistocene Sediments | Undifferentiated Holocene - Pleistocene Sediments | Undifferentiated Sediments | Holocene | Quaternary | Cenozoic | 0.01 |
| Undifferentiated Holocene - Pleistocene Sediments<br>Citronelle Formation | Undifferentiated Holocene - Pleistocene Sediments<br>Nashua Formation | Anastasia Formation; Miami Limestone; Key Largo Limestone<br>Fort Thompson Formation<br>Bermont Beds<br>Caloosahatchee Formation | Pleistocene | Quaternary | Cenozoic | 2.6 |
| Citronelle Formation; Intracoastal Formation; Miccosukee Formation<br>Coarse Clastics; Jackson Bluff Formation | Nashua Formation<br>Cypresshead Formation | Tamiami Formation<br>Long Key Formation; Stock Island Formation | Pliocene | Neogene | Cenozoic | 5.3 |
| Pensacola Clay<br>Choctawhatchee Formation<br>Intracoastal Formation; Shoal River Formation<br>Bruce Creek Limestone; Oak Grove Sands<br>Alum Bluff Group<br>Chipola Formation<br>Torreya Formation (Dogtown Member; Sopchoppy Member)<br>Chattahoochee Formation; St. Marks Formation | Hawthorn Group:<br>Charlton Member<br>Coosawhatchie & Statenville Formations<br>Marks Head Formation<br>Penney Farms Formation | Bone Valley Member<br>Long Key Formation; Stock Island Formation<br>Peace River Formation<br>Tampa Member<br>Arcadia Formation<br>Hawthorn Group | Miocene | Neogene | Cenozoic | 23.0 |
| Chickasawhay Limestone; Suwannee Limestone<br>Bucatunna Clay Member; Byram Formation; Bridgeboro Limestone; Marianna Limestone | Suwannee Limestone | Nocatee Member; Arcadia Formation; Hawthorn Group<br>Suwannee Limestone | Oligocene | Paleogene | Cenozoic | 33.9 |
| Ocala Limestone; Bumpnose Limestone Member<br>Lisbon Formation<br>Tallahatta Formation | Ocala Limestone<br>Avon Park Formation<br>Oldsmar Formation | Ocala Limestone<br>Avon Park Formation<br>Oldsmar Formation | Eocene | Paleogene | Cenozoic | 55.8 |
| Wilcox Group Undifferentiated | Cedar Keys Formation | Cedar Keys Formation; Rebecca Shoals Dolomite | Paleocene | Paleogene | Cenozoic | 65.5 |

Modified from Braunstein et al. (1988)
Ages are from Walker et al. (2012)

deposited, as the result of the weathering of the overlying clay-rich Miocene sediments that covered the Platform (Scott 1988). Weathering of the clays releases large amounts of silica into the groundwater and, in the appropriate geochemical environment, replaces limestone. Groundwater beneath the present-day fluviokarst escarpment near Lake City in northern Florida is supersaturated with respect to opal-CT and slightly saturated with respect to quartz due to weathering of the clays (Upchurch 2002, 2005), signaling that this is still a favorable environment for the formation of chert. Fossils including foraminifera and corals are often preserved in the chert. Often, the chert is concentrated at or just below the upper, karstic horizon of each formation.

Regression of the sea in the Late Oligocene restricted deposition to portions of southern and northwestern Florida (Missimer 2002). Although absent over much of the Platform, these sediments may exceed 135 m in thickness (Braunstein et al. 1988) in southern Florida. Because of this, the stratigraphic section in southern Florida may represent the most complete Upper Oligocene section in the southeastern United States (Brewster-Wingard et al. 1997). In very limited areas, the Upper Oligocene carbonates may form the top of the FAS (Miller 1986, 1997).

Cross sections showing the distribution of the Paleogene sediments are shown in Fig. 3.4. Figure 3.4a illustrates the Ocala Platform as a broad dome of Avon Park Formation surrounded by strata that dip away from the center of the Platform. The generalized geologic map of Florida (Fig. 3.5) shows the location of exposures of the Avon Park Formation in west-central part of the Florida Peninsula. The Paleogene lithostratigraphic units occurring in the surface and shallow subsurface of the panhandle, northern, and southern portions of Florida are shown in Fig. 3.5.

#### 3.4.3.2 Stratigraphy of the Miocene, Pliocene Epochs (Neogene Period) and Quaternary Period

Significant depositional changes occurred in the late Paleogene to early Neogene. Several factors were responsible for the changes including epeirogeny in the Appalachians that caused uplift of the highly eroded and reduced mountain range (Stuckey 1965; Schlee et al. 1988; Gallen et al. 2013). The rejuvenated mountain range again became a source of sediment due to increased erosion, and the siliciclastic sediments were transported downslope by streams and rivers. Marine currents transported the sediment southward onto the

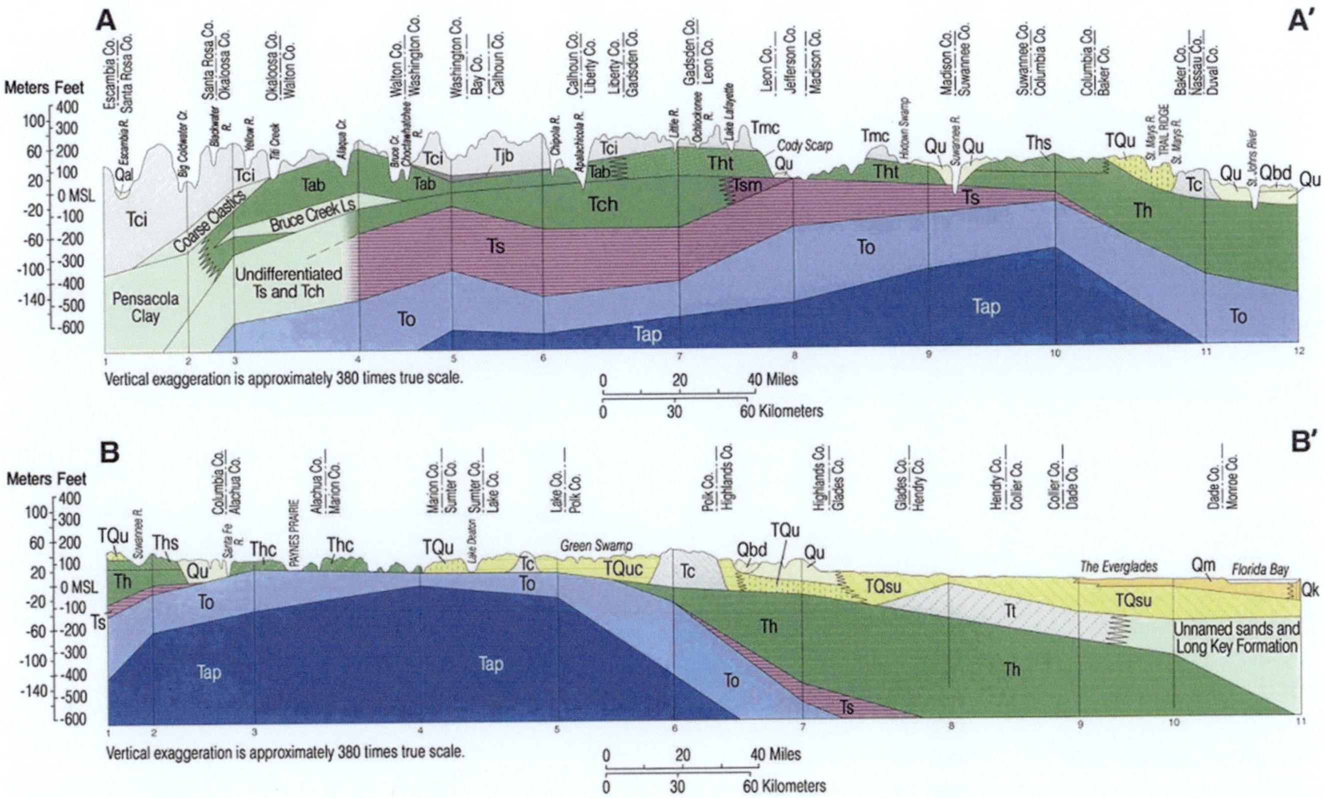


Fig. 3.4 (a) Cross sections showing the shallow subsurface and surface distributions of Paleogene, Neogene, and Quaternary lithostratigraphic units (Scott et al. 2001). There are no Mesozoic or older rocks at or near the land surface in Florida, so they are absent from the geologic map and accompanying cross sections. (b) Locations of cross sections shown in Fig. 3.4a. From Scott et al. (2001). (c) Legend for cross sections (Fig. 3.4a) and geologic map (Fig. 3.5). (From Scott et al. 2001)

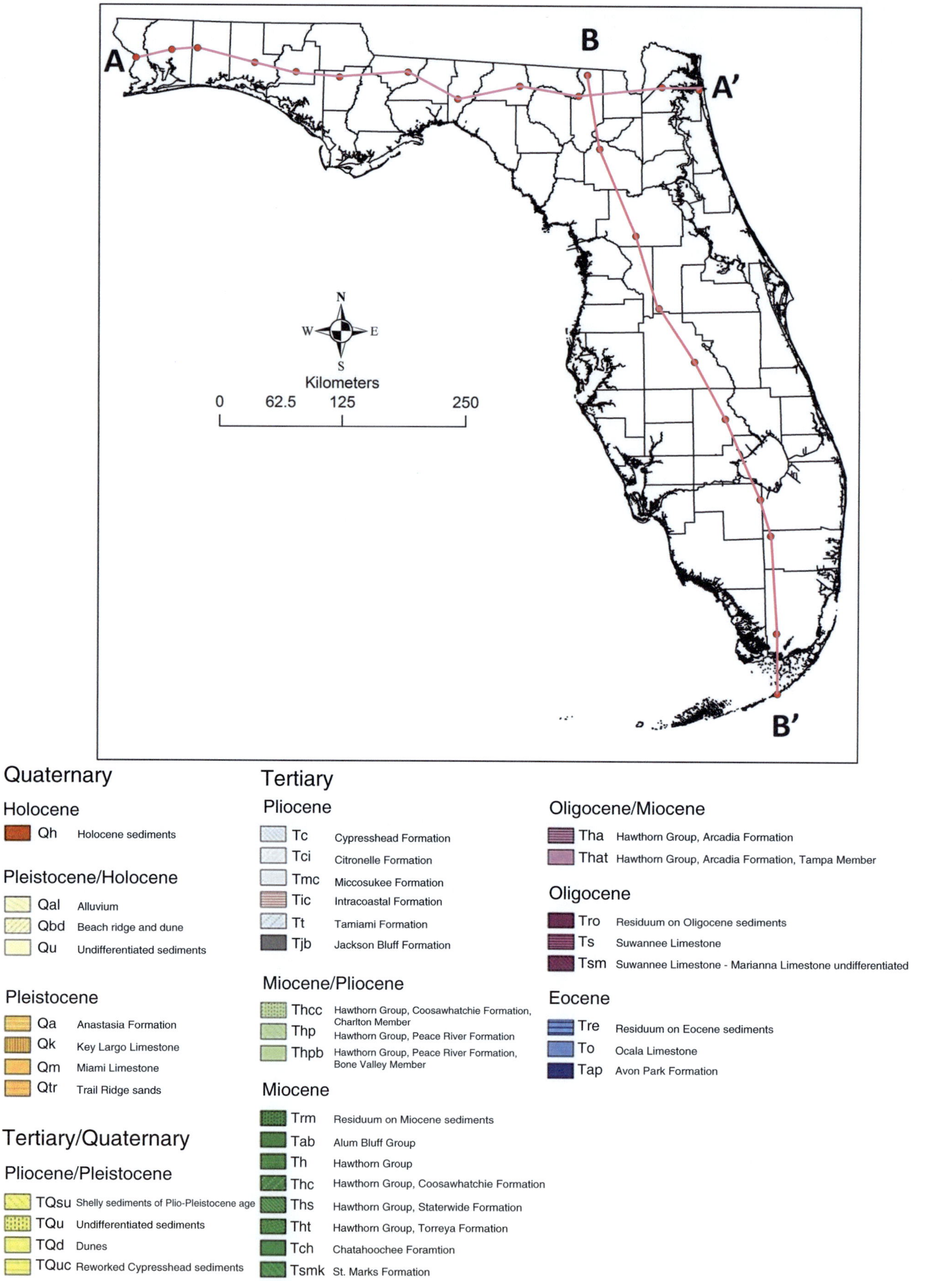


**Fig. 3.4** (continued)

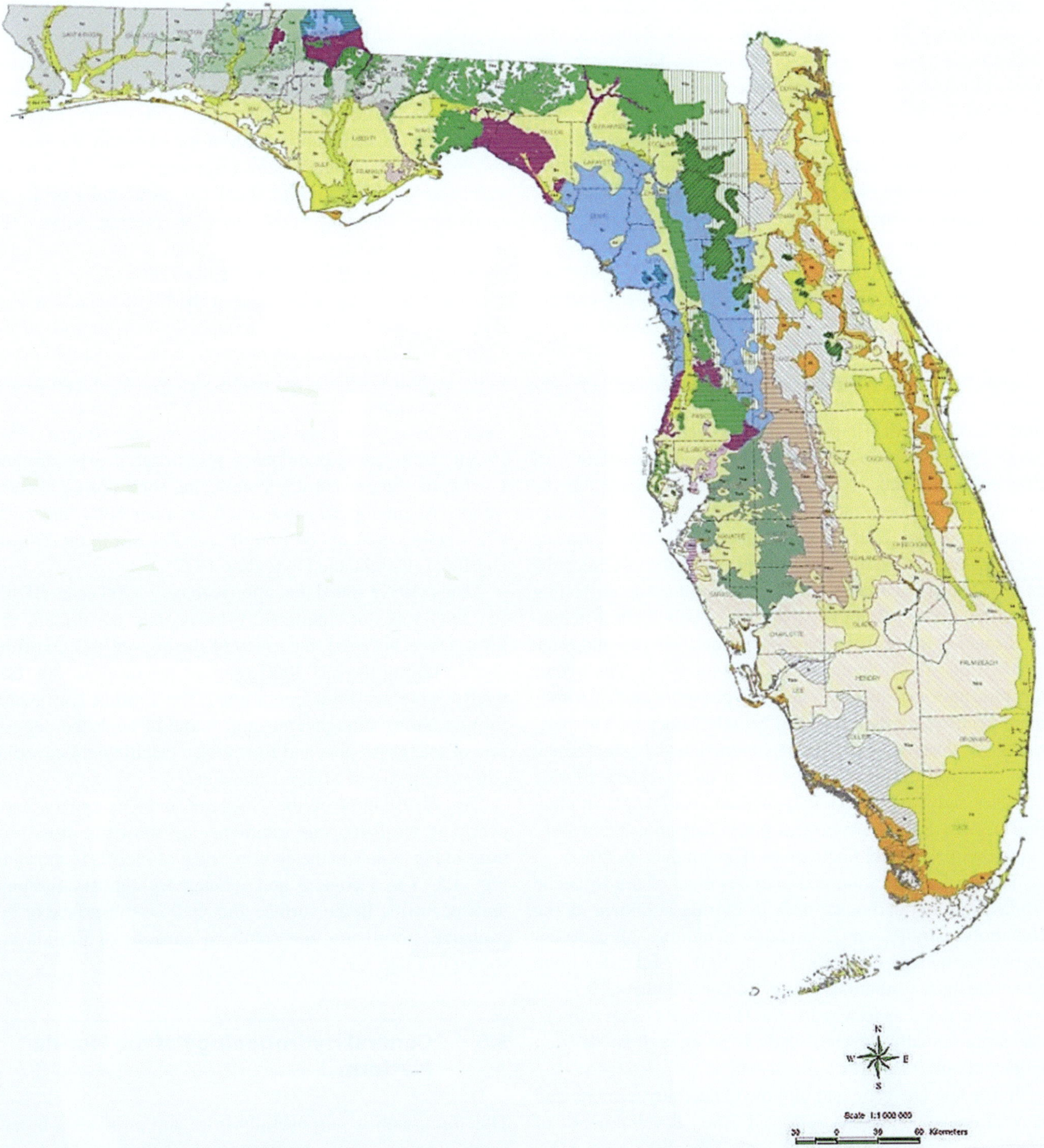

**Fig. 3.5** Geologic map of Florida (Scott et al. 2001). See Fig. 3.4c for a key to formations shown on the map

Florida Platform. Deposition of these siliciclastics was dependent on sea-level position. Sea level rose through the Middle Miocene then began fluctuating significantly until the end of the Pleistocene. Beginning near the end of the Pleistocene and continuing to the present, sea level rose to present levels transporting and depositing the siliciclastic sediments and sculpting the land surface in the process.

Initially, in the Early Miocene, the siliciclastics were deposited as strata interbedded and mixed with carbonates in northern Florida, while carbonates continued to dominate in southern Florida (Scott 1988). By the Middle Miocene, with continued sea-level rise, siliciclastic deposition generally replaced carbonate deposition (Scott 1988; Missimer 2002). Carbonate deposition continued only in the southernmost portions of the Platform; siliciclastic sediments continued to be transported further south and, ultimately, dominated the depositional system on most of the Florida Platform by the Early Pliocene. Carbonates continued to be produced, but on a limited scale. In the late Neogene, carbonates most often occurred solely as matrix materials between the siliciclastic grains. Siliciclastic sediments encroached onto the southernmost portion of the Platform during the Pliocene forming the foundation for the northern half of the Florida Keys (Cunningham et al. 1998). In the Quaternary, siliciclastics dominated over much of the Platform. However, in the late Quaternary, with a reduction in siliciclastic supply, carbonate deposition began to occur over the southernmost portion of the Florida Platform.

Sediments deposited during the Miocene covered the entire Platform; however, subsequent erosion and redeposition created the distributional pattern seen today (Fig. 3.5; Scott 1988; Scott et al. 2001). The initial distribution of Pliocene sediments is not known, but can reasonably be inferred to have been more extensive than the present occurrence (Scott et al. 2001). Unusual depositional environments are also recorded on the Florida Platform in the late Neogene as the result of sea-level fluctuations and marine upwelling of bottom waters. Major phosphorite and palygorskite deposits formed as the result of these conditions (Weaver and Beck 1977; Riggs 1979; Scott 1988; Compton 1997). The age of the phosphorites indicates that the phosphogenic environment occurred in the Early and Middle Miocene. The peri-marine environments where the palygorskite deposits formed also occurred during the Miocene in northwestern Florida (Weaver and Beck 1977). Palygorskite also formed in alkaline lakes in the western-central part of the peninsula in association with sea-level fluctuations (Upchurch et al. 1982).

The geologic stratum that includes most of the Miocene siliciclastic and carbonate beds in Florida is known as the Hawthorn Group, which contains a number of different named formations (Table 3.2; Scott 1988). Sinkholes often penetrate these formations as they collapse; however, it is not well understood as to whether the Hawthorn Group carbonate strata actually develop voids large enough to develop large, collapsed features.

In the late Neogene and into the Quaternary, climate and depositional conditions allowed the development of extremely fossiliferous molluscan-bearing lithologic units. Many of the formations defined within the late Neogene and early Quaternary contain some of the most diverse faunas in the world. How these units formed has been a source of discussion (Allmon 1992; Scott and Allmon 1992). Due to the abundance and diversity of the molluscan fossils, paleontologists have been drawn to study these sediments for more than a century (Scott 1997).

As sea level rose in the Pleistocene, sediments were deposited over that portion of the Platform that is below 18.3–30.5 m above sea level. The Pleistocene sea level rose no higher than this level (Colquhoun et al. 1968). The rising sea level in the late Pleistocene and increased carbonate production on the southern portion of the Platform allowed for the development of the Miami Limestone (Fig. 3.5), a broad carbonate bank and oolite shoal complex, and Key Largo Limestone, the paleo-reef tract of Florida Keys. The Neogene-Quaternary sediments range from a negligible thickness to more than 914 m (Miller 1986).

During the last glacial stage of the Pleistocene, sea level dropped approximately 122 m exposing vast portions of the Florida Platform which are presently beneath marine waters of the Gulf of Mexico and Atlantic Ocean. Submerged stream and river channels (Donahue et al. 1999; Cronin et al. 2007; Suthard et al. 2009) and archaeological sites (Ruppé 1980; Faught 2004) provide evidence for subaerial exposure and erosion on the present-day continental shelf during low sea stands. Archeological sites dated between 8000 and 6000 years before present (BP) are found offshore in the Florida Big Bend (Faught and Donoghue 1997).

Davis (1997) stated that approximately 3000 years BP the sea level was not significantly lower than the present sea level. Davis believed that much of the present-day coastline formed during the last 3000 years as the result of the relatively stable sea-level conditions. The Florida Everglades formed during this general time frame through the deposition of mangrove peat and freshwater, calcitic mud covering a broad expanse of Miami Limestone.

The distribution of the Neogene and Quaternary units overlying the Paleogene sediments are shown in cross sections in Fig. 3.4a and the geologic map of Florida is shown in Fig. 3.5. The Neogene and Quaternary lithostratigraphic units occurring in the surface and shallow subsurface of the panhandle, northern, and southern portions of Florida are shown in Table 3.2.

## 3.5 General Hydrogeology of the Florida Platform

This discussion is purposefully brief and provides a broad-brush overview of the hydrogeology of the Florida. An in-depth discussion on hydrogeology and its relationship with karst is found in Chap. 4.

The Cenozoic sediments of Florida form a series of aquifer systems, which provide greater than 90% of the drinking water (Berndt et al. 1998) and about 60% of all water used in the state (Table 1.1; Marella 2009). The prominent aquifer systems in Florida include: the Floridan; Intermediate; and

**Table 3.3** Hydrostratigraphic nomenclature of Florida

| | | PANHANDLE FLORIDA | | NORTH FLORIDA | | SOUTH FLORIDA | |
|---|---|---|---|---|---|---|---|
| **SYSTEM** | **SERIES** | FORMATION | HYDROSTRATIGRAPHIC UNIT | FORMATION | HYDROSTRATIGRAPHIC UNIT | FORMATION | HYDROSTRATIGRAPHIC UNIT |
| QUATERNARY | HOLOCENE<br>PLEISTOCENE | Undifferentiated | *sand and gravel aquifer*<br>*surficial aquifer system* | Undifferentiated<br>Anastasia Formation | *surficial aquifer system* | Undifferentiated<br>Miami Limestone<br>Key Largo Limestone<br>Anastasia Formation | *surficial aquifer system*<br>*Biscayne aquifer* |
| TERTIARY | PLIOCENE | Citronelle Formation<br>Miccosukee Formation<br>Jackson Bluff Formation<br>Intracoastal Formation<br>Alum Bluff Group<br>Coarse Clastics | | Undifferentiated<br>Miccosukee Formation<br>Cypresshead Formation | | Undifferentiated<br>Tamiami Formation<br>Long Key Formation<br>Hawthorn Group | |
| | MIOCENE | Coarse Clastics<br>Alum Bluff Group<br>Pensacola Clay<br>Intracoastal Formation<br>Hawthorn Group<br>Chipola Formation<br>Bruce Creek Limestone<br>St. Marks Formation<br>Chattahoochee Formation | *intermediate aquifer system or intermediate confining unit* | Hawthorn Group<br>St. Marks Formation | *intermediate aquifer system or intermediate confining unit* | Hawthorn Group | *intermediate aquifer system or intermediate confining unit* |
| | OLIGOCENE | Bucatunna Clay<br>Chickasawhay Limestone<br>Marianna Limestone<br>Suwannee Limestone | *Floridan aquifer system* | Suwannee Limestone | *Floridan aquifer system* | Suwannee Limestone | *Floridan aquifer system* |
| | EOCENE | Ocala Limestone<br>Avon Park Formation<br>Lisbon Formation<br>Tallahatta Formation<br>Claiborne Group Undiff. | | Ocala Limestone<br>Avon Park Formation<br>Oldsmar Formation | | Ocala Limestone<br>Avon Park Formation<br>Oldsmar Formation | |
| | PALEOCENE | Wilcox Group<br>Midway Group | *undifferentiated aquifer systems and confining units* | Cedar Keys Formation | *undifferentiated aquifer systems and confining units* | Cedar Keys Formation | *undifferentiated aquifer systems and confining units* |
| CRETACEOUS AND OLDER | | Undifferentiated | | Undifferentiated | | Undifferentiated | |

Modified from Southeastern Geological Society (1986)

Surficial aquifer systems (Fig. 1.8, Table 3.3; Southeastern Geological Society 1986). See Miller (1986) and Arthur et al. (2008) for more detailed discussions of Florida's aquifer systems.[3]

### 3.5.1 Floridan Aquifer System

The FAS is composed of Paleogene to early Neogene carbonates with highly variable permeability. This aquifer system, which is widespread in the Coastal Plain of the southeastern United States, is one of the most productive carbonate aquifers in the world (Miller 1986; Berndt et al. 1998). Budd and Vacher (2004) characterize the FAS as a multi-porosity aquifer: a fractured, porous aquifer where confined, and a karstic, fractured, porous aquifer where unconfined. The FAS covers the entire Florida Platform. The base of the FAS occurs in the lower Paleogene rocks where evaporites restrict the permeability (Miller 1986; Southeastern Geological Society 1986). The top of the FAS occurs where the carbonates are overlain by relatively impermeable sediments of the Intermediate Aquifer System/Confining Unit (IAS/ICU). In many areas of the state, the FAS can be subdivided into two aquifers (Table 3.3) – the Upper Floridan Aquifer (UFA) and Lower Floridan Aquifer (UFA). A middle confining unit (MCU) separates these aquifers.

### 3.5.2 Intermediate Aquifer System and Confining Unit

The IAS/ICU (Table 3.3; Southeastern Geological Society 1986) is composed of permeable and impermeable sediments deposited during the Neogene. The siliciclastics flooding onto the Florida Platform during the Miocene and Pliocene contained an abundance of clay. Deposition of the clayey sediments on the Paleogene carbonates created a relatively impermeable sequence of confining beds (Miller 1986, 1997). Permeable carbonate and siliciclastic sediments are, in some areas, interbedded with the impermeable units creating regionally limited, water-producing zones (Miller 1986,

[3]In this book the aquifer names are capitalized to facilitate use and abbreviation. Properly, there is no formal naming convention for hydrostratigraphic units, and only proper names (i.e., Floridan, Biscayne) should be capitalized.

1997). The base of the IAS/ICU occurs at the top of the regionally extensive, permeable carbonates of the FAS (Southeastern Geological Society 1986). The top of the IAS/ICU is considered to be at the top of the laterally extensive and vertically persistent lower permeability beds (Southeastern Geological Society 1986). The IAS/ICU is absent over much of the Ocala Platform.

### 3.5.3 Surficial Aquifer System

The Surficial Aquifer System (SAS) is composed of Pliocene through the Pleistocene-Holocene, permeable siliciclastic and carbonate sediments with some zones of more clayey, less-permeable sediments (Berndt et al. 1998). In two areas of the state, the SAS is particularly important since the FAS does not contain potable water. In these areas, the westernmost Panhandle and southeastern peninsula, the SAS is the primary source of drinking water. In the western panhandle, the SAS is a thick sequence (up to 152 m) of siliciclastic sediments. This aquifer is known as the Sand and Gravel Aquifer. In the southeastern peninsula, the SAS consists of highly permeable, interbedded carbonates and siliciclastics, which underlie some of the most populous metropolitan areas in Florida. This aquifer is known as the Biscayne Aquifer. The base of the SAS occurs at the top of the laterally extensive and vertically persistent lower-permeability beds of the IAS/ICU (Southeastern Geological Society 1986). By definition, the SAS is absent on the crest of the Ocala Platform where the FAS is unconfined and sandy sediments that overlie the unconfined FAS are saturated with FAS water (Southeastern Geological Society 1986).

## 3.6 Geomorphology of Florida

### 3.6.1 Introduction

The Florida Platform extends southward from the continental United States separating the Gulf of Mexico from the Atlantic Ocean. The exposed portion of the Platform, the Florida Peninsula, constitutes approximately one-half of the Florida Platform (Fig. 3.1). The remainder of the Platform is submerged. The margin of the outer continental shelf is delineated by the 200 m isobath, and the continental shelf is considered to be part of the Florida Platform. The axis of the Platform extends northwest to southeast approximately along the present-day west coast of the peninsula. The Florida Peninsula, from the Georgia State Line at the St. Mary's River to Key West, measures nearly 725 km. From the Alabama–Florida state line on the west eastward to the Atlantic coastline is approximately 595 km.

Florida lies entirely within the Coastal Plain Physiographic Province as defined by Fenneman (1938) and is one of two states in the United States that fall completely within the Coastal Plain. Much of the surface of Florida shows the influence of the marine processes that transported and deposited Cenozoic sediments. Fluvial processes, although more important in the panhandle and northern Florida, have helped sculpt the entire state, particularly during the low stands of sea level, redistributing the marine sediments.

Karst processes have had a dramatic effect on the Florida landscape due to the near-surface occurrence of soluble carbonate rocks. Middle Eocene to Pleistocene carbonate sediments are affected by karstification over large areas of the state. Siliciclastic sediments overlie the karstified carbonates in many areas of the state.

The general geomorphology of the Florida Platform consists of east–west trending highlands in the northern and western portions of the state and north–south trending highlands extending approximately two-thirds the length of the peninsula, flanked by coastal lowlands occurring between the highlands and the coastline that wraps around much of the state.

The highest point in the state, 105 m above sea level, occurs in the Western Highlands near the Alabama–Florida state line in Walton County (Fig. 1.1). There are several hilltops in the highlands in the center of the state that exceed 91 m msl in elevation. Florida has the distinction of having the lowest highest point of any state in the United States.

Many of the landforms of interest to residents and visitors are related to karst development. For example, more than 1000 springs are recognized in Florida with the major springs occurring within the karstic areas of the state (Scott et al. 2004). These springs, as well as sinkhole lakes, caves, and other karst features, are important, revenue-generating resources for Florida.

The geomorphic districts (Fig. 3.6) and provinces (Fig. 3.7) utilized in this text are based on the work of Scott (2005) and Williams et al. (in prep.) because they relate so well to karst processes in Florida.

### 3.6.2 Geomorphological Studies in Florida

White et al. (1964) and White (1958, 1970, Fig. 3.7) delineated the geomorphic subdivisions that most geologists working in the state utilize at the present time. See Schmidt (1997) for a review of the evolution of the geomorphology of Florida.

At the present time, a revised geomorphic map of Florida is in preparation by the Florida Geological Survey (Williams et al. in prep.). A preliminary version of this geomorphic map was released in a 2005 Southeastern Geological Society

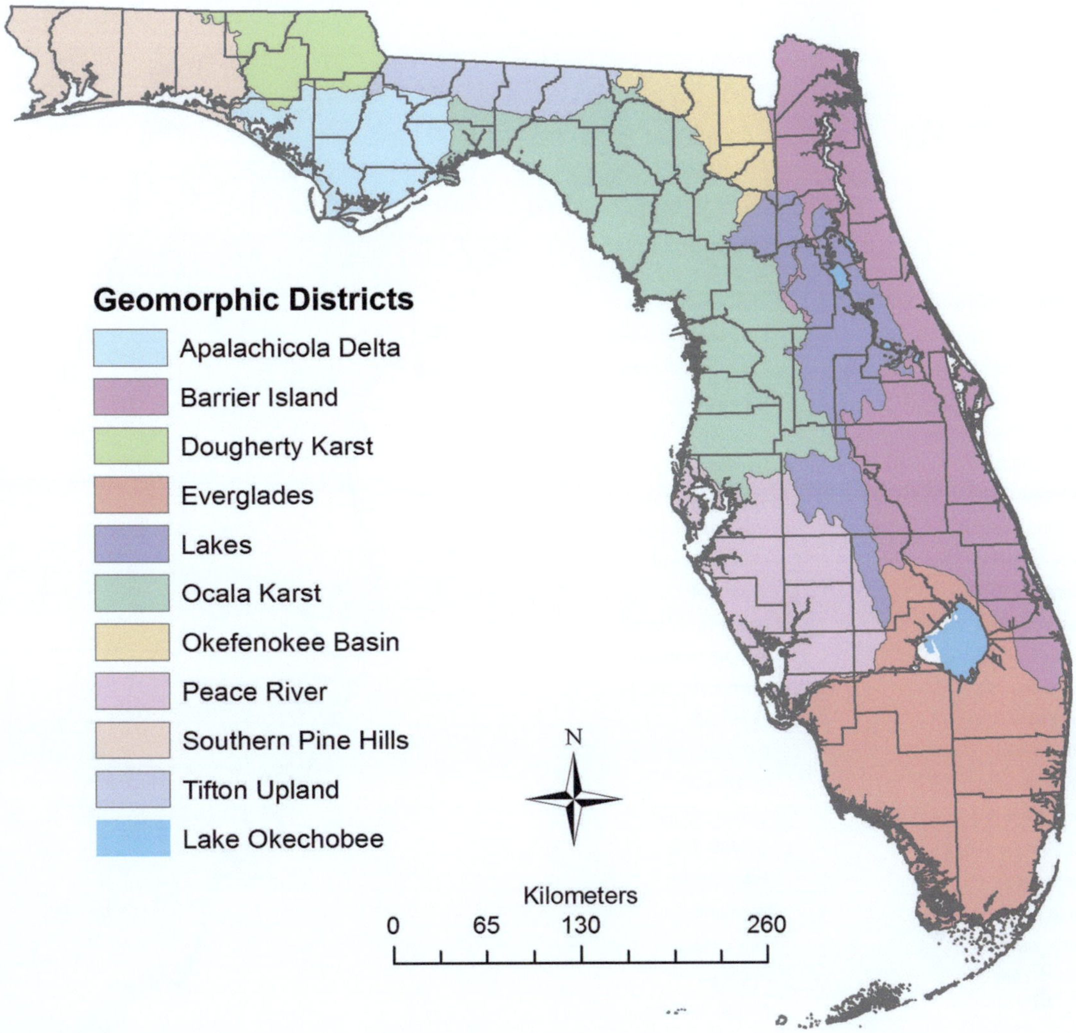


**Fig. 3.6** Revised geomorphic districts of Florida. (From Williams et al. in prep.)

field trip proceeding (Scott 2005). The geomorphic districts delineated by Scott (2005) and Williams et al. (in prep.) are more genetically related areas than the broad, generalized geomorphic zones of White (1970), partly as a result of (1) detailed geologic and topographic mapping since White developed his geomorphic provinces and (2) development of geographic information systems (GIS) and LIDAR (Light Detection and Ranging) technology. Williams et al. (in prep.) developed the new geomorphic map of the state, largely based on the GIS and LIDAR data. The Florida Geological Survey provided GIS files for the authors' use in this text because the revised geomorphological delineations best fit the patterns of karst in Florida.

The discussion below is an overview of the revised geomorphological presentation by Williams et al. (in prep.) and should be considered the most current interpretation of the geomorphology for Florida. The districts are discussed from west to east in the panhandle and peninsula and include the Southern Pine Hills, Dougherty Karst, Apalachicola Delta, Tifton Upland, Okeefenokee Basin, Ocala Karst, Lakes, Barrier Island, Peace River, and Everglades Districts (Fig. 3.6). Each district has been subdivided into geomorphic provinces based on geology and landforms. These provinces, or geomorphic terrains, have different karst-related features and hydrogeologic settings. For example, the vast majority of the springs and well-known karst landforms are located in the Ocala Karst District, the Lakes District, and the Dougherty Karst District (Fig. 3.6; Scott 2005).

### 3.6.3 Southern Pine Hills District

The Southern Pine Hills District is located in the westernmost portion of the Florida Panhandle (Fig. 3.6). The district occurs in Escambia , Santa Rosa, Okaloosa counties and part of Walton and Bay counties. This district grades into the

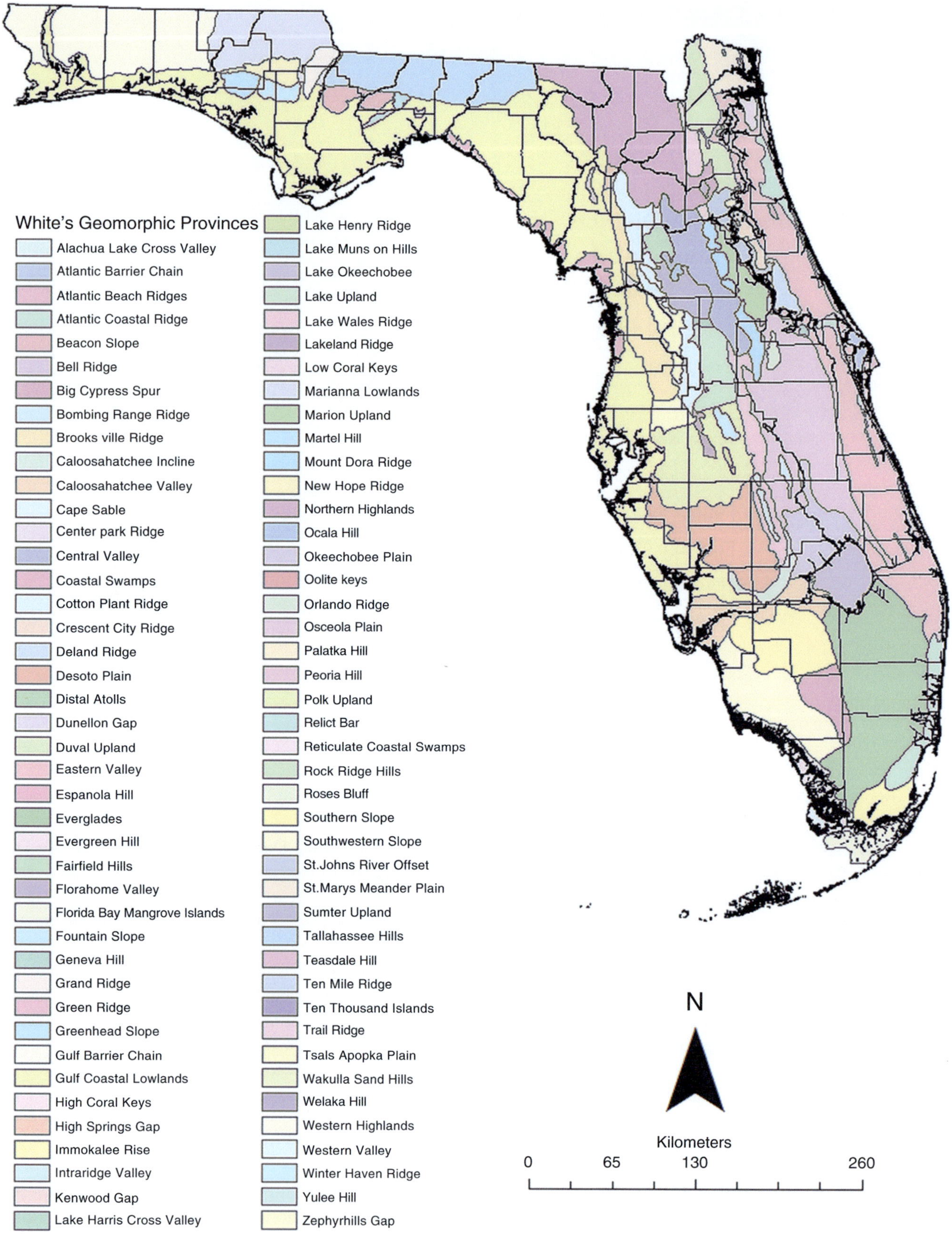


**Fig. 3.7** Geomorphological provinces in northern peninsular Florida as delineated by White (1970)

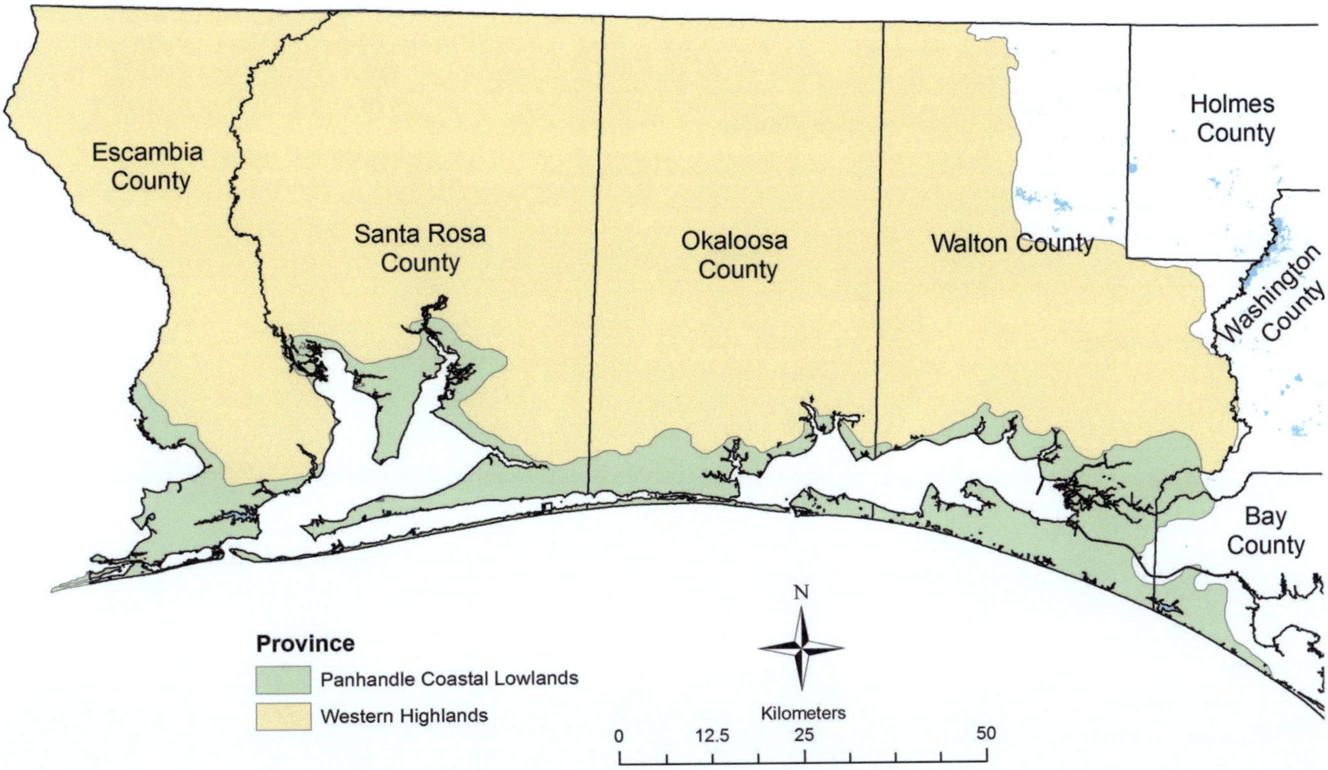


**Fig. 3.8** Physiographic provinces within the Southern Pine Hills Geomorphic District

Dougherty Karst Plain District to the east and the Apalachicola Delta District to the southeast. The two geomorphic provinces recognized within the Southern Pine Hills District are the Western Highlands and Panhandle Coastal Lowlands provinces (Fig. 3.8). White et al. (1964) identified the Western Highlands and the Gulf Coastal Lowlands within this district.

The river valleys of the Escambia and Yellow Rivers form major topographic breaks in the Southern Pine Hills District. The Escambia River Valley and the Yellow River Valley extend from the coastline to the Florida State line.

The land surface of the Southern Pine Hills District consists of rolling hills separated by streams and is a southward-sloping, stream- and river-dissected plain that developed on Miocene estuarine to marine sediments and Pliocene siliciclastics in Alabama and Florida (Table 3.2). Elevations within the district range from sea level to 105 m NGVD. The highest elevation in Florida, 105 m NGVD, occurs within this district. Coastal features occur below elevations of 15 m NGVD. Over much of the Southern Pine Hill District, the hill tops are gently rounded; however, in those areas where the elevation exceeds 76–82 m NGVD, remnants of a flat plain are obvious.

Sediments of the Southern Pine Hills District at or near the land surface are predominantly composed of siliciclastic sediments belonging to the Citronelle Formation, the Alum Bluff Group, and undifferentiated Quaternary sediments (Scott et al. 2001). The Panhandle Coastal Lowlands are underlain by undifferentiated sediments of the Pleistocene and Holocene (Scott et al. 2001). These sediments consist of fine to coarse-grained quartz sands with varying percentages of silt, clay and organic matter. In part, these sediments may contain locally and regionally reworked Citronelle sediments.

Karst features are rare to non-existent within most of the Southern Pine Hills District. As the district transitions into the Dougherty Karst Plain District, scattered karst features occur.

### 3.6.4 Dougherty Karst District

The Dougherty Karst District, located to the east of the Southern Pine Hills District and abutting the Apalachicola Delta District to the south and the Tifton Uplands District to the east, is a flat to gently rolling lowland with topography that is interrupted by the occurrence of karst features. The Dougherty Karst District occupies a portion of the central Panhandle covering Jackson and Holmes counties and parts of Bay, Calhoun, Walton, and Washington counties. Based on the characteristics of the karst landscape, the district can be subdivided into six distinct provinces: the Defuniak Springs Karst Hills, Bonifay Karst Hills, Marianna Karst Plain, Vernon Karst Hills, Compass Lake Karst Hills, and the

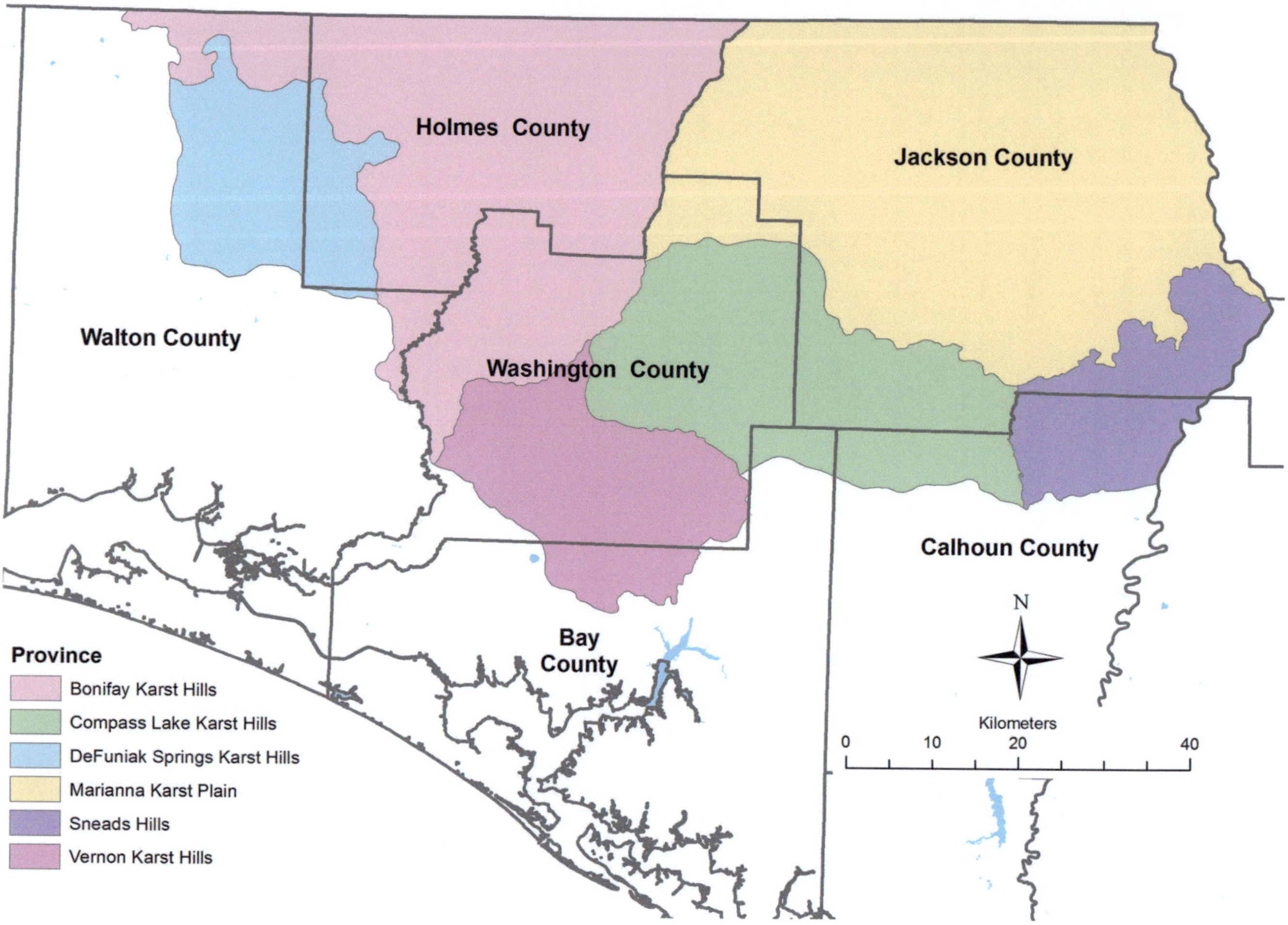


**Fig. 3.9** Physiographic provinces of the Dougherty Karst Geomorphic District

Sneads Hills (Fig. 3.9). White et al. (1964) identified the Marianna Lowlands, Greenhead Slope, Grand Ridge and a number of smaller outlier hills within this district. Elevations range from 6 m along the Choctawhatchee River and Holmes and Econfina Creeks to 100 m NGVD on the highest hilltops near Falling Waters State Park in the Compass Lake Karst Hills in eastern Washington County (Williams et al. in prep.). The river valleys of the Choctawhatchee, Chipola, and Apalachicola Rivers form major topographic breaks in the district.

Sediments at or near the land surface in the district range from Eocene carbonate rocks to Plio-Pleistocene siliciclastics (Table 3.2; Scott et al. 2001). The geology of the region controls the development of the karst landscape. Structurally, the Chattahoochee "anticline"[4], a high on the Tertiary limestone, brings the soluble Eocene to Miocene carbonates near the surface. Note that Froede et al. (2015) recently argued that the Chattahoochee Anticline is not a subsurface structural feature. Rather, they concluded that it is a result of differential erosion.

The maturity of the karst development varies across the Dougherty Karst District. Dissolution of the carbonates has created distinct landforms that characterize the district, including: caves, sinkholes, numerous springs, and few streams that sink into the ground by way of swallets and emerge at resurgences down gradient. In those areas near the Southern Pine Hills and Apalachicola Delta Districts, the karst is younger, less mature and has more relief. The development of karst features within the Dougherty Karst District is largely influenced by the thickness of the cover material overlying the carbonate rock.

Karst landscapes are better developed on the Marianna Karst Plain and less well developed in the karst hills areas. Springs and caves are common within the Dougherty Karst District. The best known are the caves in the Florida Caverns State Park, located just north of Marianna in Jackson County. There are numerous sinkholes that form basins for lakes and

[4]The term anticline has historically been applied to this positive tectonic feature. The Oligocene and other strata are near the land surface because they have not subsided or been eroded as much as the surrounding areas.

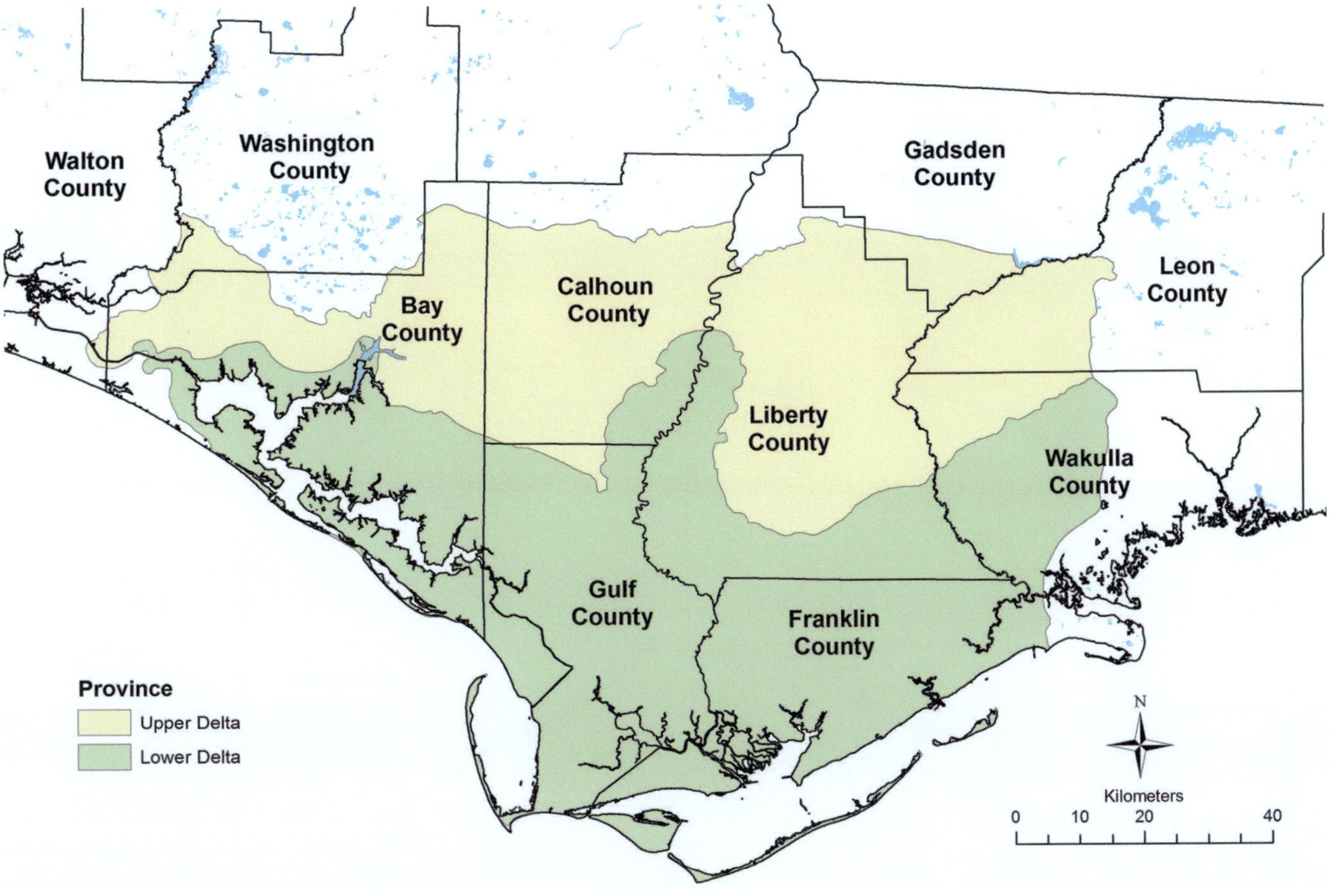


**Fig. 3.10** Physiographic provinces within the Apalachicola Delta Geomorphic District

swamps. The karst features form in the Eocene Ocala Limestone, the Oligocene Suwannee/Marianna Limestone and the Miocene Chattahoochee Formation (Table 3.2). The irregular surface of the carbonate rock is covered with residuum where younger units have been removed by erosion.

### 3.6.5 Apalachicola Delta District

Located south of the Dougherty Karst and Tifton Uplands Districts, the Apalachicola Delta[5] District extends from central Bay County east to the western third of Wakulla County, where it merges with the western-most edge of the Ocala Karst District. To the west, the Apalachicola Delta District merges with the Southern Pine Hills. The southern terminus of the district includes the barrier island complex developed along the Gulf coast. The district can be subdivided into two distinct provinces: the Upper Delta and Lower Delta provinces (Fig. 3.10). This area equates with the Gulf Coastal Lowlands, the Fountain Ridge, part of the New Hope Ridge and the Beacon Slope of White et al. (1964). The land surface of the Apalachicola Delta District varies from well drained in the higher elevations on the Upper Delta Province to swampy and poorly drained in the Lower Delta Province. The surface becomes more poorly drained and very swampy in the lower elevations toward the coast. Near the coastline, paleo-beach ridges are evident on the topographic maps and in LIDAR data. These features contain swampy swales and the crests of the ridges reach above 6 m NGVD.

Elevations in the Apalachicola Delta District range from more than 78 m NGVD along the northern edge to sea level at the coast. In the more northerly, higher portion of the district, the Upper Delta Province, streams exhibit mixed trellis and dendritic drainage patterns. Tanner (1982) interpreted the drainage systems to be controlled by high level marine terraces. The Apalachicola River occupies a rather broad valley across the delta district. Within the valley are many landforms related to the river's fluvial processes. Natural levees, point bar deposits, ox bow lakes and meander cutoffs are identifiable within the valley.

Sediments ranging from Miocene-Pliocene siliciclastics with limited carbonates to Pleistocene-Holocene

[5]The term "delta" is widely used in the Florida panhandle to describe wedges of fluvial sediment deposited by rivers and streams that originated to the north.

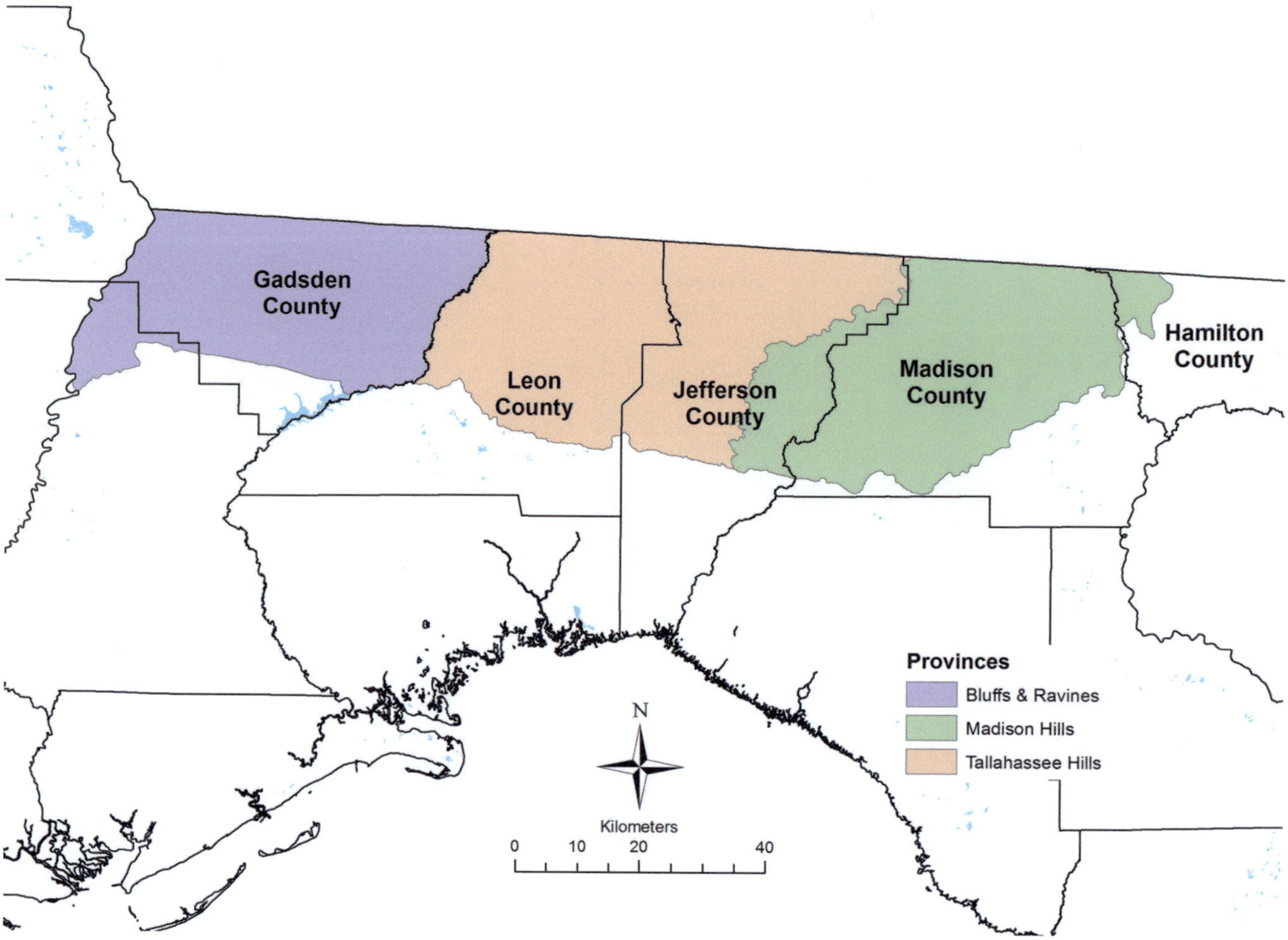


**Fig. 3.11** Physiographic provinces within the Tifton Upland Geomorphic District

siliciclastics occur at the surface within the Apalachicola Delta District (Table 3.2) (Scott et al. 2001). Citronelle Formation siliciclastics cover much of the Upper Delta province with Miocene and Pliocene units cropping out in very restricted areas. Within the Lower Delta Province, the sediments are mapped as undifferentiated Quaternary throughout with limited areas of Pliocene sediments exposed in creeks. Karst features are uncommon within much of the district. Karst features become more common in the transition to the Ocala Karst District to the east.

### 3.6.6 Tifton Upland District

The Tifton Upland District is located to the east of the Dougherty Karst District, west of the Okeefenokee Basin District, and abuts the Apalachicola Delta and Ocala Karst districts to the south. The Tifton Upland District can be subdivided into three distinct provinces based on the character of the topography, which is controlled by the underlying sediments: the Bluffs and Ravines Province; the Tallahassee Hills Province; and the Madison Hills Province (Fig. 3.11). The Tifton Upland District, in northern Florida, encompasses the area referred to as the Tallahassee Hills by White et al. (1964). The district extends from the Apalachicola River in Gadsden and Liberty Counties eastward to Hamilton County at the Withlacoochee River.

The Tifton Upland's topography is characterized by broad, undulating hills, many of which have well-developed dendritic drainage patterns. Dissection of the landscape increases near the southern boundary of the Tifton Uplands in the Tallahassee Hills. Elevations range from 12 m NGVD in rivers and stream valleys where the Tifton Upland District gives way to the Apalachicola Delta District and the Ocala Karst District to approximately 96 m NGVD in the Bluffs and Ravines Province. Elevations decrease toward the southern limit of the district. The Tifton Upland District merges with the northern portion of the Apalachicola Delta District at elevations around 61 m NGVD along the western portion of the Cody Scarp.

The Cody Scarp is an erosional escarpment that extends from northeastern Liberty County in the Florida Panhandle to near Gainesville in Alachua County. The escarpment was identified by Puri and Vernon (1964). In the eastern panhandle, there is evidence that the escarpment is predominantly marine in origin. However, as one moves to the east and then south to Alachua County, the escarpment becomes dominated by karst and fluvial erosion (Upchurch 2002, 2014).

The Cody Scarp marks the transition from the Tifton Upland District to the Ocala Karst District to the south. Elevations in the Cody Escarpment range from around 30 m NGVD down to approximately 15 m NGVD. The uplands merge with the flat, swampy Okeefenokee Basin District in central Hamilton County. In general, the land surface elevations decline very gradually from west to east across the Tifton Upland.

Sediments at or near the land surface are predominantly Miocene and Pliocene siliciclastics (Table 3.2; Scott et al. 2001). Oligocene and Miocene carbonates occur within 30 m of the surface in much of the district. Karst features are present within this district where the carbonates of the St. Marks Formation and Suwannee Limestone occur near the surface.

Within the Bluffs and Ravines Province, steepheads, which occur as the result of headward erosion of streams that form in response to seeps developed where clay beds underlie highly permeable sands, are abundant. Karst features are common in portions of the Tifton Uplands. In general, karst feature frequency increases as the Tifton Upland District transitions into the Ocala Karst District. Lake Jackson, within the Tallahassee Hills Province, is known world-wide for its history of intermittently and dramatically draining through sinkholes in the lake bed. Other lakes within the province have similar, but less well-known, histories.

### 3.6.7 Okeefenokee Basin District

The Okeefenokee Basin District occurs from the Georgia-Florida state line southward to Alachua County. The Okeefenokee Basin District lies between the Tifton Upland, Ocala Karst, and Barrier Island Districts and north of the Lakes District. The Okefenokee Swamp occupies much of the basin and forms the head waters of the Suwannee and St. Mary's Rivers. The district is divided into three provinces: the Ocean Pond Plain Province, Lake City Ridge Province, and Raiford Ridges Province (Fig. 3.12). The Lake City Ridge Province forms a separation between the northern, Ocean Pond Plain Province and southern, Raiford Ridges Province (Fig. 3.12). White (1970) referred to this district as a portion of the Northern Highlands.

Elevations generally vary within the district from approximately 30 m NGVD to more than 58 m NGVD. Where the Suwannee, St. Marys, Alapaha, and Santa Fe Rivers occur within the district, elevations along the rivers may be less than 15 m NGVD.

Miocene Hawthorn Group sediments to Plio-Pleistocene siliciclastics (Table 3.2) are at or near the land surface within the district (Scott et al. 2001). Karst features are relatively uncommon throughout most of the Okeefenokee Basin District. Karst features occur primarily in the transition zone (the Cody Scarp) with the Ocala Karst District to the west and in the transition zone with the Lakes District, where the Miocene Hawthorn Group sediment cover is thin and breached by sinkholes. The swampy wetland areas within the district are the result of low-permeability, clay-rich Miocene Hawthorn Group sediments, which have a low leakance and hamper the formation of karst.

### 3.6.8 Ocala Karst District

The Ocala Karst District (Fig. 3.13) extends from Leon and Wakulla Counties to Hillsborough and Pinellas Counties. The Ocala Karst District is developed on Paleogene and Neogene carbonate strata on the crest of the Ocala Platform, near the present day western coastline of the peninsula. Separated by the Cody Scarp, the Ocala Karst District lies to the south of the Tifton Upland District. It is bounded to the east by the Okeefenokee Basin and Lake districts. The southern margin of the district occurs where the impermeable Miocene Hawthorn Group sediments retarded the development of karst features in the Peace River District and where streams and rivers become more common.

The geomorphic provinces recognized within the Ocala Karst District include: the Alachua Karst Hills, Bell Ridge, Branford Karst Plain, Brooksville Ridge, Chiefland Karst Plain, Crystal River Karst Plain, Fairfield Karst Hills, Green Swamp, Gulf Hammock, Land O'Lakes Karst Plain, Ocala Karst Hills, Perry Karst, San Pedro Bay, Tsala Apopka Plain, Waccasassa Flats, Williston Karst Plain, and Woodville Karst Plain Provinces (Fig. 3.13). White (1970) recognized the following major geomorphic features within the Ocala Karst District: the Gulf Coastal Lowlands, Gulf Coastal Swamps, Bell Ridge, Brooksville Ridge, Western Valley, Fairfield Hills, cotton Plant Ridge, Ocala Hill, Sumter Upland, Tsala Apopka Plain, Lake Upland and part of the Polk Upland.

Elevations within the district range from sea level along the coast to over 91 m NGVD on the Brooksville Ridge. The topography over much of the district is gently rolling with only minor relief. However, on the Brooksville Ridge and the karst hills, the terrain looks more like that of the Lakes District and relief may exceed 61 m NGVD.

Sediments in the district range from Eocene carbonates to Pliocene-Pleistocene siliciclastics. Carbonate sediments ranging from the Middle Eocene Avon Park Formation to the Oligocene-Miocene Tampa Member of the Arcadia Formation, Hawthorn Group and the Miocene St. Marks Formation lie near the land surface (Table 3.2; Scott et al.

**Fig. 3.12** Physiographic provinces located within the Okeefenokee Basin Geomorphic District

2001). Maturity of karst development varies across the Ocala Karst District. Dissolution of the carbonates has created distinct landforms that characterize the district, including: caves, sinkholes, numerous springs, and streams that sink into the ground by way of swallets and emerge at resurgences down gradient. In those areas near the Tifton Upland, Okefenokee Basin and the Lakes districts, the karst is younger, less developed, and has more relief. Further away from these areas, in general toward the Gulf Coast, the karst is more mature and has less relief due to more extensive dissolution and erosion of the overlying siliciclastics. The development of karst features within the Ocala Karst District is largely influenced by the thickness of the cover material overlying the carbonate rock.

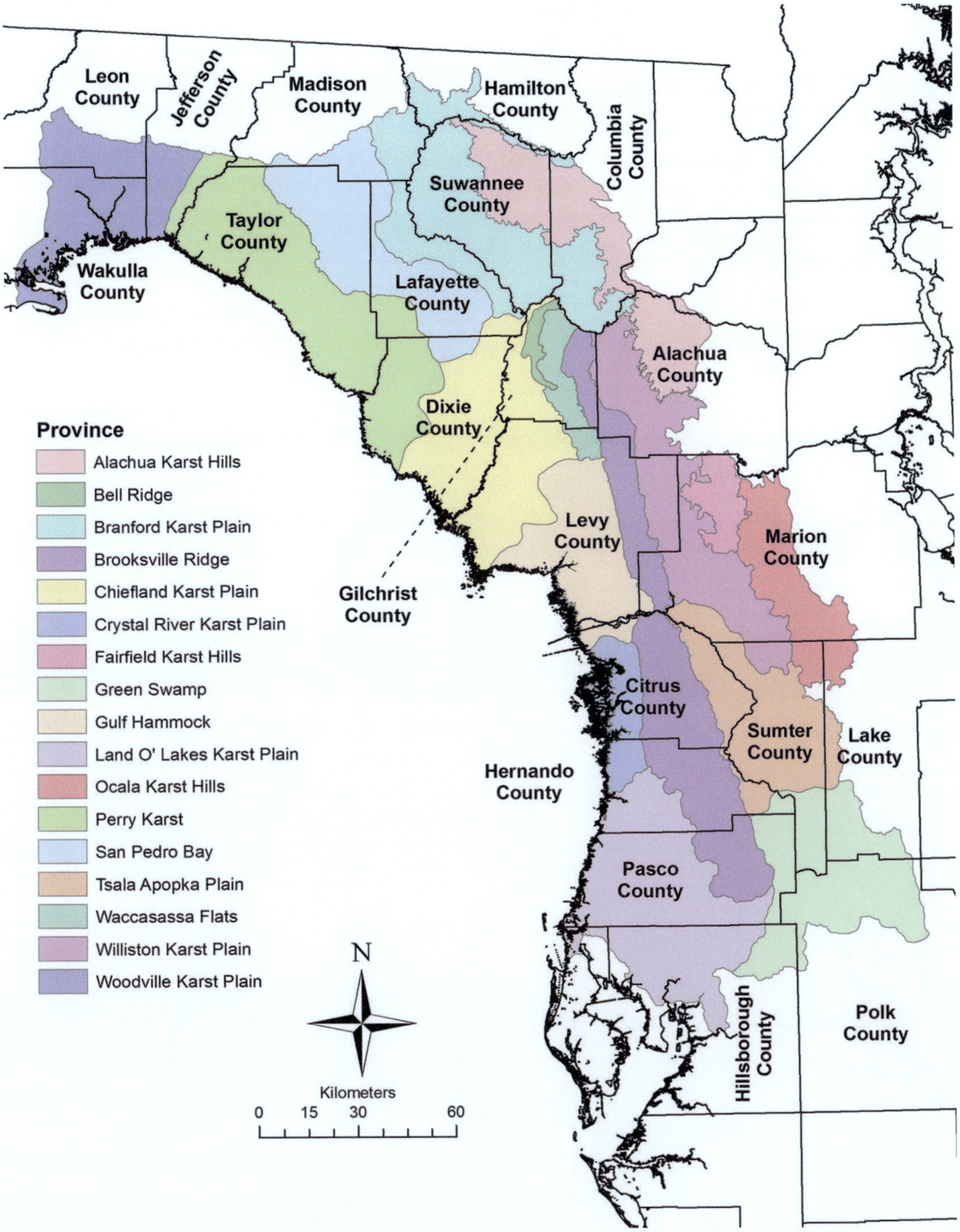


**Fig. 3.13** Physiographic provinces within the Ocala Karst Geomorphic District

### 3.6.9 Lakes District

The Lakes District extends from eastern Alachua County, southeastern Bradford County and southern Clay County to southernmost Highlands County. The Lakes District lies east and south of the Ocala Karst District, south of the Okeefenokee District, east of the Barrier Island District, and north of the Peace River District. The district is bounded on the east and west by erosional scarps. The district is divided into 14 distinct geomorphic provinces: Crescent City Ridge, Deland Ridge, Ft. McCoy Plain, Hawthorne Lakes, Lake Plain, Lake Wales Ridge Complex, Mt. Dora Ridge, Ocklawaha River Valley, Orlando Ridge, Palatka Hill, St. Johns River Offset Valley, Tavares Lakes, Trail Ridge, and Wekiva Plain Provinces (Fig. 3.14). Many of the geomorphic features recognized by White (1970) continue to be

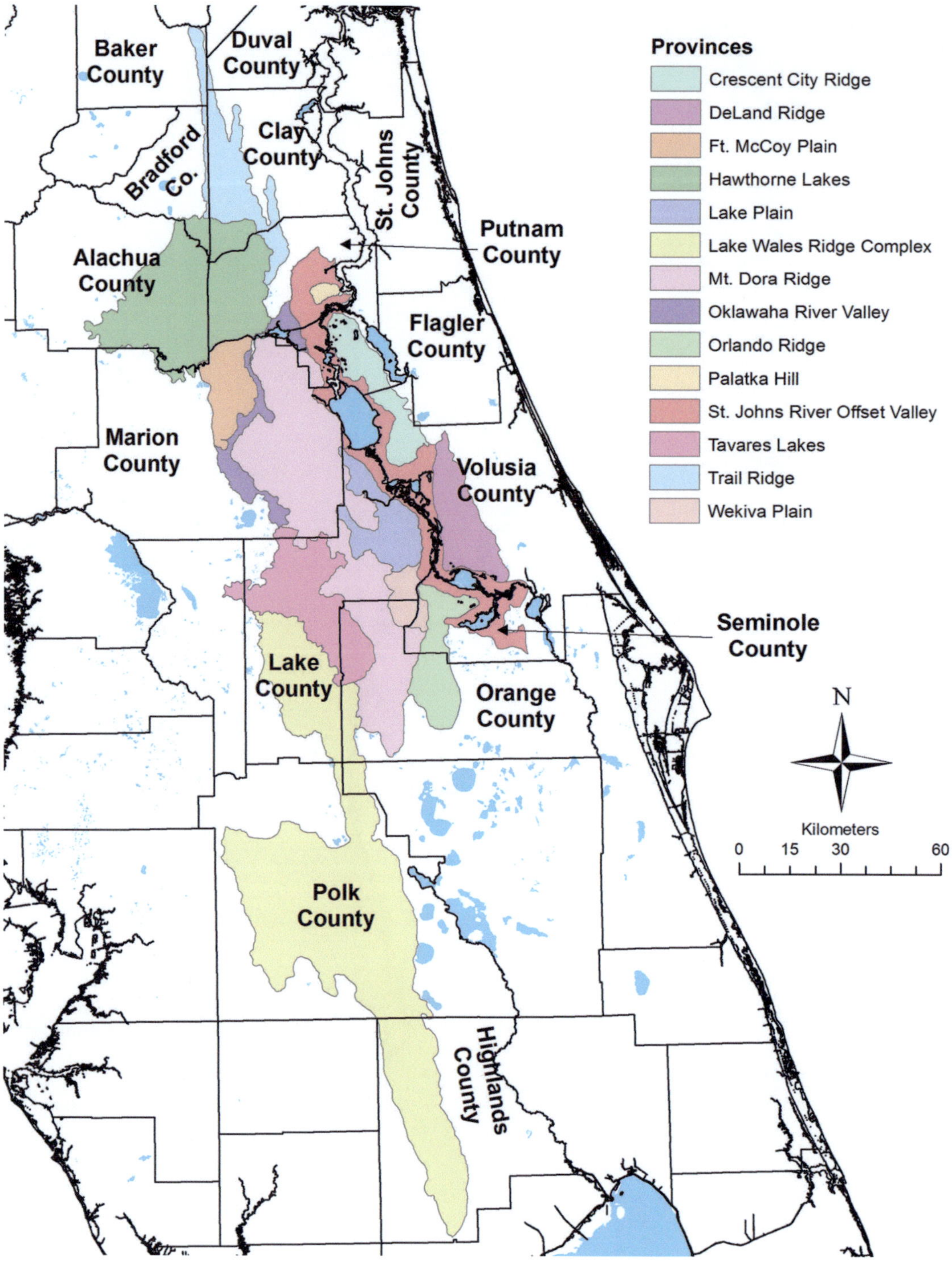


**Fig. 3.14** Physiographic provinces within the Lakes Geomorphic District

recognized in the Lakes District. White (1970) recognized the Lake Wales Ridge, Lakeland Ridge, Lake Henry Ridge and Winter Haven Ridge. These features are part of the Lake Wales Ridge Complex Province in the Lakes District.

Elevations in the Lakes District range from less than 1 m NGVD in the St. Johns River Offset to 95 m NGVD on Sugarloaf Mountain (28.6490°N, 81.7334°W), on the northwestern side of Lake Apopka. Sugarloaf Mountain is one of the highest points in the peninsula.

Sediments at or near the land surface include Miocene Hawthorn Group, the Pliocene Cypresshead Formation, and undifferentiated sediments (Table 3.2; Scott et al. 2001). A thick (up to at least 61 m) sequence of siliciclastic and carbonate sediments overlies the Eocene Ocala Limestone in much of the district. The Suwannee Limestone is absent throughout much of the district. However, the Oligocene Suwannee Limestone is present in western Polk and Hardee Counties and occurs between the Ocala Limestone and the overlying sediments.

Dissolution of the limestone and subsequent subsidence or collapse has created the characteristic sinkhole lakes and dry sinks that dominate the landscape. The distinctive circular to sub-circular shapes of the lakes can easily be seen to show individual and coalescing sinkholes in bathymetric maps. As a result of the numerous sinkholes, the Lakes District is primarily internally drained through these karst features, rather than by streams. The largest lakes in the Lakes District occur in the Tavares Lakes Province.

### 3.6.10 Barrier Island District

The Barrier Island District extends into Florida from Georgia and covers most of the eastern peninsula southward to Palm Beach County (Fig. 3.15). The district is bound on the west by the Lakes District and to the south by the Everglades District. The district is divided into nine provinces: Atlantic Coastal Complex, Bombing Range Ridge, Duval Upland, Geneva Hill, Lower St. Johns River Valley, Osceola Plain, Sea Islands, St. Marys Plain, and Upper St. Johns River Valley Provinces (Fig. 3.15). In this district, the major geomorphic features White (1970) recognized include: the Duval Upland, St. Marys Meander Plain, Eastern Valley, Atlantic Coastal Ridge, Atlantic Beach Ridges, Atlantic Barrier Chain, Atlantic Coastal Lagoons, Orlando Ridge and Osceola Plain,

The district is characterized by beach ridges, dunes and paleo-lagoons. Beach ridge plains (strand plains) occur in several areas of the Barrier Island District at elevations ranging from sea level to more than 23 m NGVD. These occur on the Duval Upland, Atlantic Coastal Complex, and the Osceola Plain Provinces. Drainage patterns in these areas often are strongly controlled by the relict beach ridges forming, in some cases, a distinct trellis drainage pattern. The beach ridge swales are often swampy and control the development of lakes on portions of the Osceola Plain and the Atlantic Coastal Complex. Elevations range from sea level to more than 41 m NGVD on the Duval Upland in the northern portion of the district and more than 45 m NGVD on the Bombing Range Ridge to the south.

The surficial and shallow subsurface sediments of the Barrier Island District were deposited during the Plio-Pleistocene and overlie sediments ranging from the Middle Eocene Avon Park Formation to the Miocene Hawthorn Group (Table 3.2; Scott et al. 2001). Plio-Pleistocene sea-level fluctuations deposited sediments forming step-like terraces, barrier islands, and beach ridges that dominate the topography of this area. The area has been slightly to moderately dissected by fluvial processes. Swamps are common in the low lying, poorly drained areas.

Karst is very limited in the district. Shallow depressions, referred to as sag ponds by Schmidt and Scott (1984), which are generally only a about a meter deep, are found throughout much of the Barrier Island District. These depressions are most numerous in the southern-most part of this district and decrease in abundance toward the north. The depressions are the result of dissolution of mollusk shells incorporated in the sediments at or near the surface. The percentage of shell in the sediment is greatest in southern Florida. Also, in the southern portion of the state, the sediment contains more fine-grained carbonate and less clay. As a result, the sediment is more permeable and relatively more soluble. The depressions formed as the cover subsided over tens of thousands to hundreds of thousands of years, as the uppermost shell material dissolved, letting the land surface subside slowly.

### 3.6.11 Peace River District

The Peace River District covers much of west-central, peninsular Florida extending from Hillsborough and Pinellas Counties southward to the Caloosahatchee River in Lee and Glades Counties (Fig. 3.16). The Peace River District lies south of the Ocala Karst District, west of the Lakes District and north of the Everglades District. The recognized provinces include: the Hardee Upland, Peninsula Coastal Lowlands, and Pinellas Ridge (Fig. 3.16). Within the Peace River District, White (1970) recognized the Gulf Coastal Lowlands, Gulf Barrier Chain, Gulf Coastal Lagoons, Polk Upland, Desoto Upland and the Caloosahatchee Incline.

The landscape of the Peace River District is relatively flat with limited relief. Elevations range from sea level along the coastline to more than 58 m NGVD.

The surficial and shallow subsurface sediments of the Peace River District were deposited during the Plio-Pleistocene and overlie sediments ranging from the Oligocene Suwannee Limestone to the Miocene Hawthorn Group (Table 3.2; Scott et al. 2001).

The district is characterized by the occurrence of numerous rivers and streams, which are more common than in any other geomorphic district in the State. The prominent rivers and their tributaries in the Peace River District include: the Alafia, Peace, Manatee, Braden, Myakka, and Caloosahatchee rivers. Clay-rich, siliciclastic sediments of the Miocene Hawthorn Group, which have a relatively low leakance, control the Peace River District drainage and landforms. This relatively impermeable substrate has contributed to the development of the numerous rivers and streams.

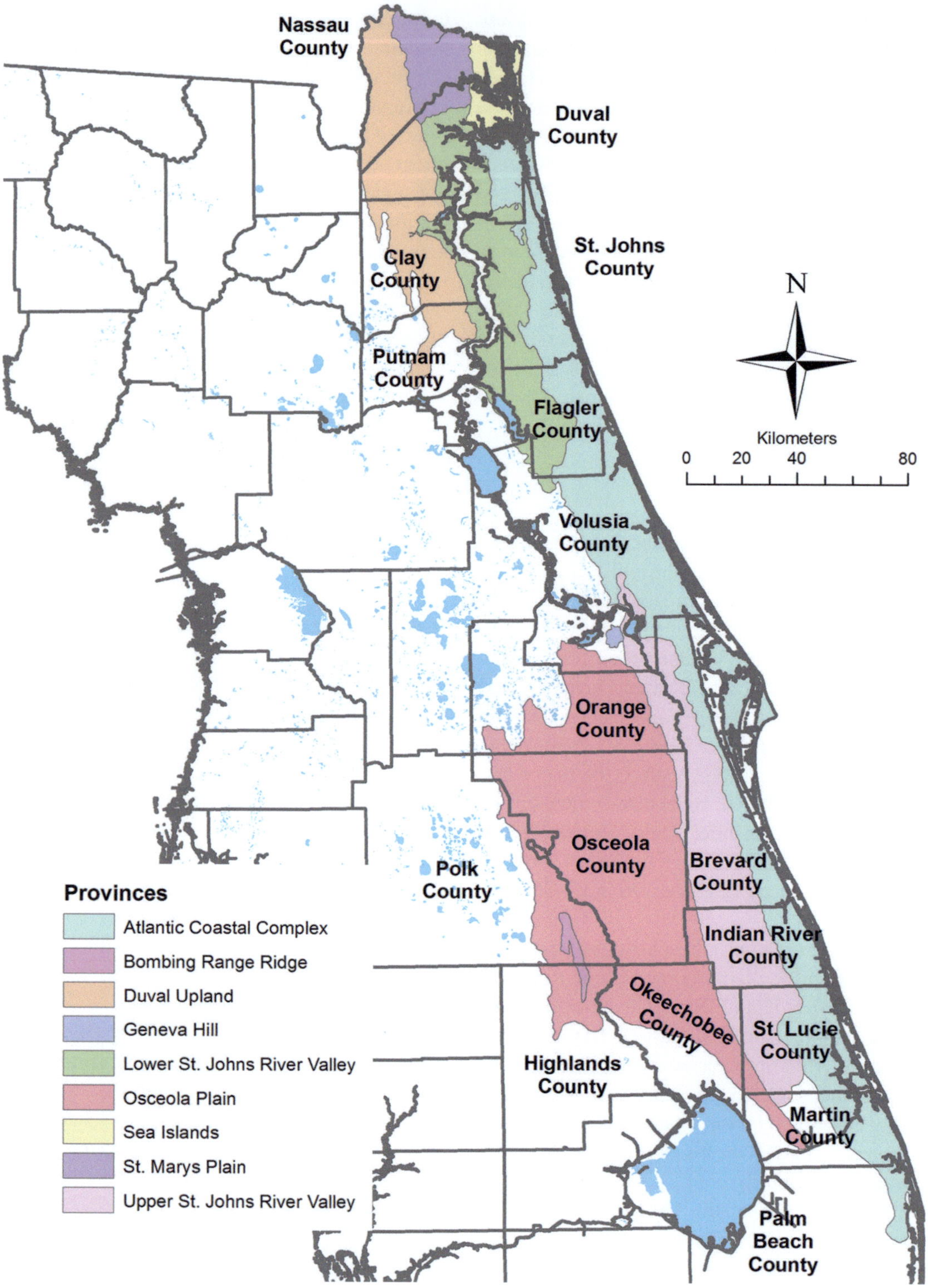


**Fig. 3.15** Physiographic provinces within the Barrier Island Geomorphic District

Surficial karst features are not common within the district. There are scattered sinkholes, swallets and few springs in the headwaters of the Peace River in Polk County. While uncommon, notable karst features include deep cenotes, such as Warm Mineral Springs and Little Salt Springs in Sarasota County. These deep, rock-collapse sinkholes penetrate the Hawthorn Group and younger sediments.

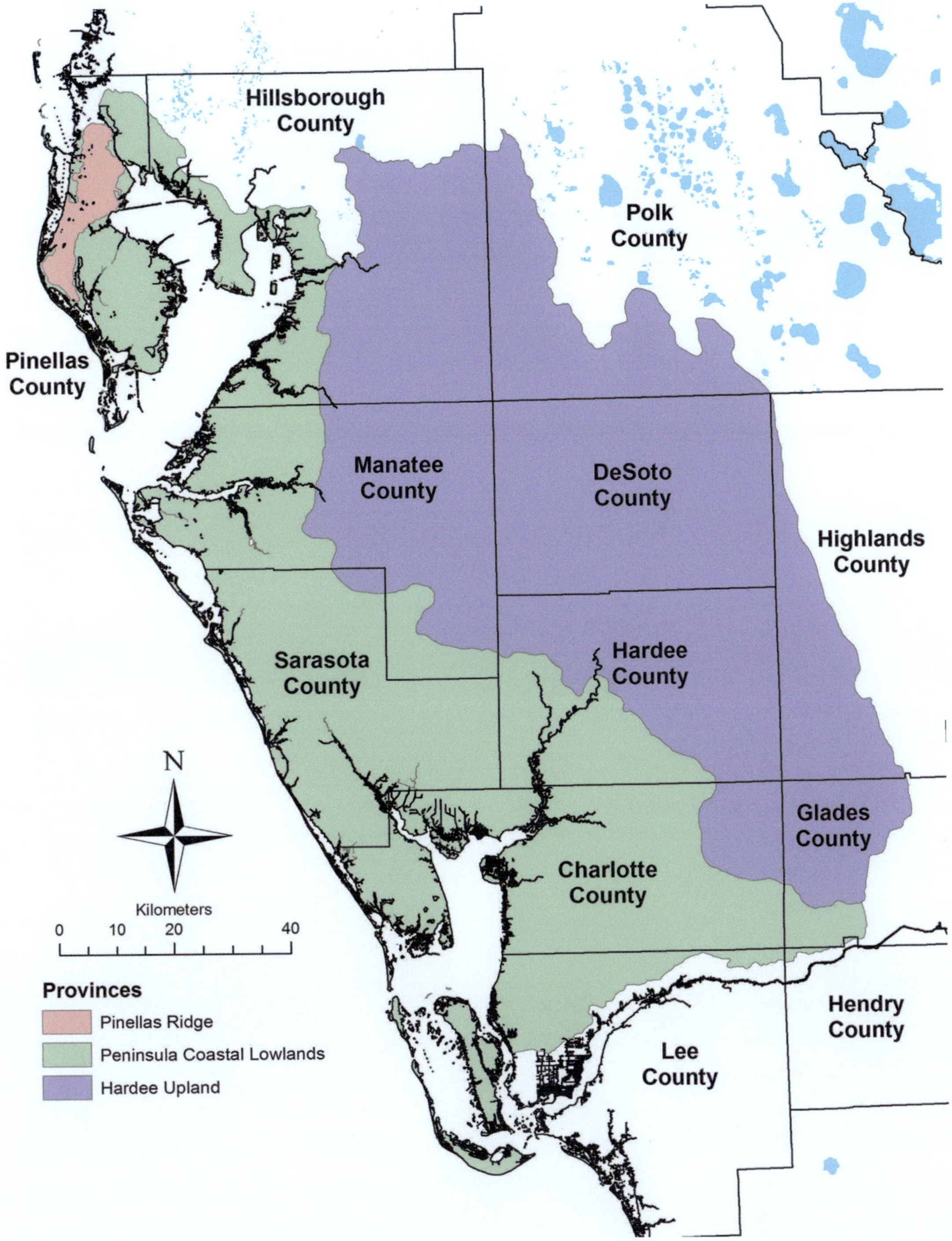


**Fig. 3.16** Physiographic provinces in the Peace River Geomorphic District

### 3.6.12 Everglades District

The Everglades District covers the southern one-third of the Florida peninsula and includes the Florida Keys (Fig. 3.17). The district extends from the Atlantic Coast to the Gulf Coast and incorporates Florida Bay. The district occurs from Lee County eastward to Palm Beach County and from Okeechobee and Highlands Counties southward to Monroe County. The Everglades District lies south of the Peace River District on the west and north, the Lakes District on the north and the Barrier Island District on the north and east.

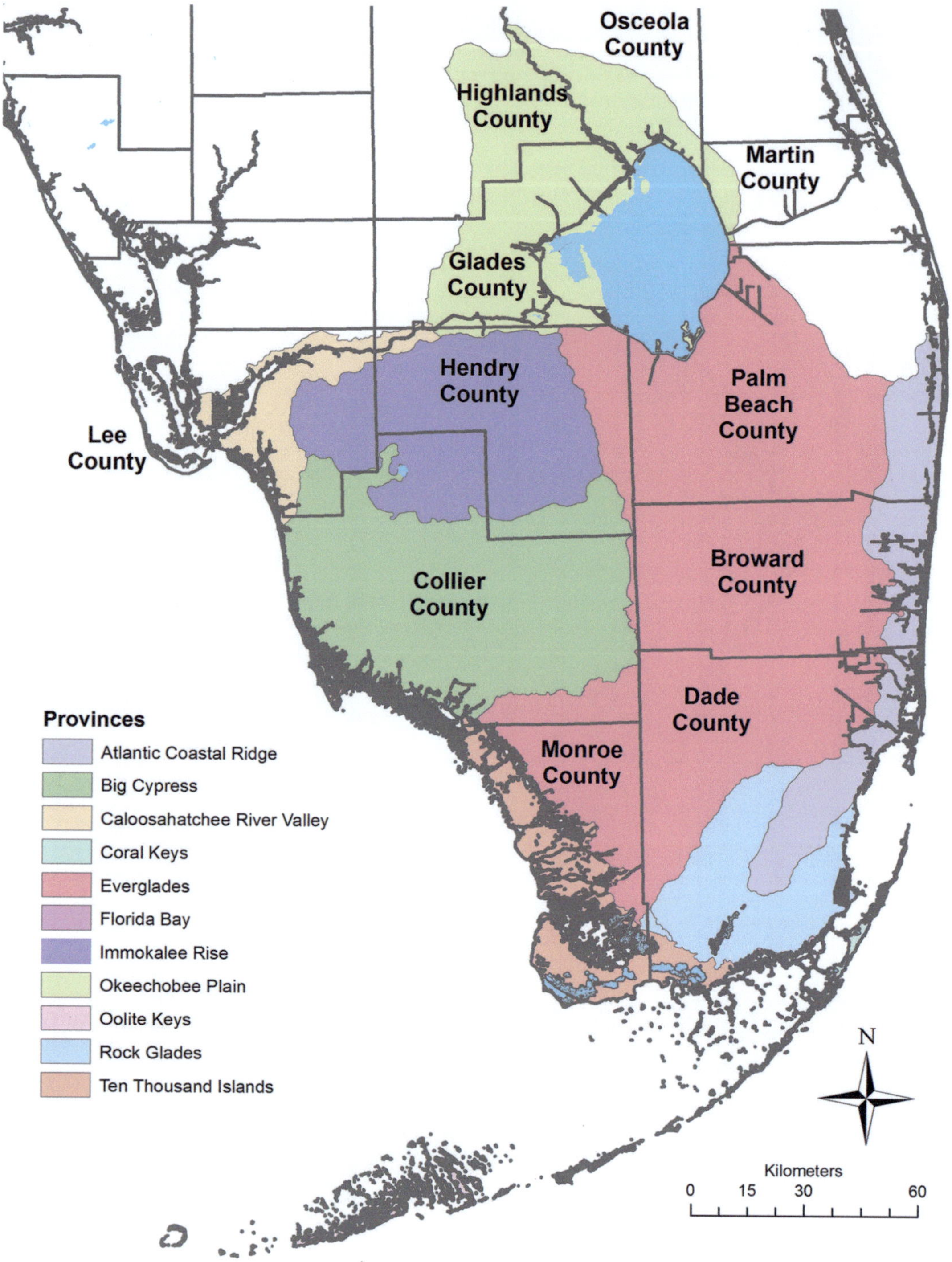


**Fig. 3.17** Physiographic provinces within the Everglades Geomorphic District

Geomorphic provinces within the district include: the Atlantic Coastal Ridge, Big Cypress, Caloosahatchee Valley, Cape Sable, Everglades, Florida Bay, Immokalee Rise, Lower Keys, Okeechobee Plain, Rock Glades, Ten Thousand Islands, and Upper Keys Provinces (Fig. 3.17). In this district, White (1970) recognized the Caloosahatchee Valley, Okeechobee Plain, Southwestern Slope, Immokalee Rise, Big Cypress Spur, Everglades, Gulf Coastal Lagoons, Reticulated Coastal Swamps, Ten Thousand Islands, Atlantic Coastal Ridge, Southern Slope, Florida Bay Mangrove Islands, Cape Sable, High Coral Keys, Low Coral Keys, Oolite Keys, and Distal Atolls.

Elevations within the Everglades District gradually decrease from north to south. The highest elevations generally occur along the northern boundary where elevations reach 15 m NGVD. Much of the district lies at elevations below 6 m NGVD.

Plio-Pleistocene sediments, ranging from quartz sand to limestone, underlie the entire district (Table 3.2; Scott et al. 2001). Karst landforms which have formed within the Everglades District are considered to be the youngest in the state and the karst features within much of the district are uncommon and small. The predominant karst features are microkarst and solution channels. Collapse sinkholes are uncommon, and cover subsidence and solution sinkholes are best developed in the western half of the district.

In contrast, the Atlantic Coastal Ridge and Rock Glades provinces, which are underlain by late Pleistocene limestone, have undergone considerable karst-related dissolution, producing solution pits, small caves, a few sinkholes, and a complex, solution-derived upper bedrock surface (Bock et al. 1969; Halley and Evans 1983; Cressler 1993; Florea and Yuellig 2007; Cunningham and Florea 2009).

## 3.7 Summary

The geomorphology of Florida is often subtle and relatively difficult to decipher in the field. However, topographic mapping and, more recently, LIDAR imaging have resulted in characterization of areas with distinctive combinations of landforms that can be divided into geomorphic districts (areas with related origins and landforms) and, within districts, provinces (areas with similar topographic attributes). The Florida Geologic Survey is currently revising Florida geomorphology using modern LIDAR technology and concepts of landscape development (Williams et al. in prep.).

The stratigraphy of Florida has affected development of Florida landforms, including development of karst. Every geomorphic district of Florida has some karst features. In the western panhandle (the Southern Pine Hills District), karst is limited to the extreme eastern part of the district, where it abuts the Dougherty Karst Plain. Elsewhere, karst features are ubiquitous but vary in degree of development.

The following chapters discuss Florida's hydrology, aquifers, and aquitards and the processes of water interactions with these materials to form the karst landscapes.

## References


Allmon WD (1992) Whence southern Florida's Plio-Pleistocene shell beds? In: Scott TM Allmon WD (eds) Plio-Pleistocene Stratigraphy and Paleontology of Southern Florida. Fla Geol Surv Sp Pub 36

Arthur JD (1988) Petrogenesis of the early Mesozoic tholeiite in the Florida basement and an overview of Florida basement geology. Fla Geol Surv Rept Inves 97

Arthur JD Fischler C Kromhout C Clayton J Kelley M Lee RA Li L O'Sullivan M, Green R Werner C (2008) Hydrogeologic framework of the Southwest Florida Water Management District. Fla Geol Surv Bull 68

Barnett RS (1975) Basement structure of Florida and its tectonic implications. Trans Gulf Coast Assoc Geol Soc 25:122–142

Berndt MP Oaksford ET Mahon GL (1998) Groundwater. In: Fernald EA Purdum ED (eds) Water resources atlas of Florida. Tallahassee, Fla State Univ

Bock WD Moore DR Neumann AC Supko PR (compilers) (1969, reprinted 1994). Late Pleistocene geology in an urban area. Miami Geol Soc, field trip guidebook, accessed at http://sofia.usgs.gov/publications/reports/mgs_bock1969/mgs_bock1969.pdf on 3/20/2012

Bourgeois J Hansen TA Wiberg PL Kauffman WG (1988) A tsunami deposit at the Cretaceous-Tertiary boundary in Texas. Science, 241:567–570

Braunstein J Huddlestun P Biel R (1988) Gulf coast region, correlation of stratigraphic units of North America (COSUNA) project. Tulsa, Amer Assoc Petrol Geol

Brewster-Wingard GL Scott TM Edwards LE Weedman SD Simmons KR (1997) Reinterpretation of the peninsular Florida Oligocene: An integrated stratigraphic approach. Sed Geol 108:207–228

Bryan JR (1991) Stratigraphic and paleontologic studies of Paleocene and Oligocene carbonate facies of the eastern Gulf Coastal Plain. Ph.D. diss, Knoxville, Univ Tenn

Budd DA Vacher HL (2004) Matrix permeability of the confined Floridan Aquifer, Florida, USA. Jour Hydrogeol 12(5):531–549

Chowns TM Williams CT (1983) Pre-Cretaceous rocks beneath the Georgia Coastal Plain: Regional implications. In: Gohn GS (ed) Studies related to the Charleston, South Carolina, earthquake of 1886: tectonics and seismicity. US Geol Surv Prof Paper 1313l

Colquhoun DJ Herrick SM Richards HG (1968) A fossil assemblage from the Wicomico Formation in Berkeley County, South Carolina. Bull Geol Soc Amer, 79:1211–1220

Compton JS (1997) Origin and paleogeographic significance of Florida's phosphorite deposits. In: Randazzo AF Jones DS (eds) The geology of Florida, Gainesville, Univ Press Fla, Ch 12, pp 195–216

Cronin T Edgar NT Brooks G Hastings D Larson R Hine A Locker S Suthard B Flower B Hollander D Wehmiller J Willard D Smith S (2007) Sea level rise in Tampa Bay. Eos, 88:117–118

Cunningham KJ Florea LJ (2009) The Biscayne aquifer of southeastern Florida. In: Palmer AN Palmer MV (eds.), Caves and karst of the USA. Huntsville, Ala, Nat Speleol Soc, pp. 196–199

Cunningham K McNeill D Guertin L Ciesielski P Scott T de Verteuil L (1998) New Tertiary stratigraphy for the Florida Keys and southern peninsula of Florida. Bull Geol Soc Amer, 110(2):231–258

Cunningham K Locker SD Hine AC Bukry D Barron JA Guertin LA (2003) Interplay of late Cenozoic siliciclastic supply and carbonate response on the southeast Florida Platform. Jour Sed Res 73(1):31–46

Dallmeyer RD Caen-Vachette M Villeneuve M (1987) Emplacement age of post-tectonic granites in southern Guinea (West Africa) and the peninsular Florida subsurface: Implications for the origins of southern Appalachian exotic terranes. Bull Geol Soc Amer, 99:87–93

Davis RA Jr (1997) Geology of the Florida coast. In: Randazzo AF Jones DS (eds) The geology of Florida, Univ Press Flor, Ch 10, pp. 155–168

Donahue BT Hine AC Tebbens SF Locker SD Twichell DC Hafen M (1999) Mouth of Tampa Bay side-scan sonar mosaic map. U.S. Geol Surv Open-File Rept 99-445

Duncan JG Evans WL III Taylor KL (1994) Geologic framework of the lower Floridan Aquifer System, Brevard County, Florida. Fla Geol Surv Bull 64

Faught MK (2004) Submerged Paleoindian and Archaic sites of the Big Bend, Florida. Jour Field Archeology, 29(3–4):273–290

Faught MK Donoghue JF (1997) Marine inundated archaeological sites and paleofluvial systems: Examples from a karst-controlled conti-

nental shelf setting in Apalachee Bay, northeastern Gulf of Mexico. Geoarchaeol 12(5):417–458
Fenneman NM (1938) Physiography of eastern United States. New York, McGraw-Hill
Florea LJ Yuellig A (2007) Everglades National Park – surveying the southernmost cave in the continental United States. Inside Earth, A Newsletter of the Cave & Karst Programs of the National Park Service, 10(1):3–5
Froede CR Jr. Reed JK Means GH (2015) The end of the line for the Chattahoochee Anticline, Gordon Anticline, and Decatur Arch, southeastern coastal plain (USA). Southeastern Geol 51(2):65–83
Gallen SF Wegmann KW Bohnensteih DR (2013a) Miocene rejuvenation of topographic relief in the southern Appalachians. GSA Today, 23(2):4–10
Cressler A (1993) The caves of Dade County, Florida. Georgia Underground, 30(3):9–16
Guertin LA (1998) A late Cenozoic mixed carbonate/siliciclastic system, south Florida: Lithostratigraphy, chronostratigraphy and sea-level record. Ph.D. diss, Univ Miami
Guertin LA Missimer TM McNeill DF (2000) Hiatal duration of correlative sequence boundaries from Oligocene-Pliocene mixed carbonate/siliciclastic sediments of the south Florida Platform. Sed Geol 134:1–26
Halley RB Evans CC (1983) The Miami Limestone: A guide to selected outcrops and their interpretation. Miami Geol Soc field trip guidebook, accessed on 3/20/2012 at http://sofia.usgs.gov/publications/reports/mgs_halley1983/index.html.
Heatherington AL Muller PA (1997) Geochemistry and origin of Florida crustal basement terranes. In: Randazzo AF Jones DS (eds) The geology of Florida, Gainsville Univ Press Fla Ch 3, pp. 27–37
Herbert TA Upchurch SB (2016) The potential role of hypogene speleogenesis in the lower Floridan aquifer and Sunniland Oil Trend, south Florida, U.S.A. In: Chavez T and Reehling P (eds) Proceedings of DeepKarst 2016: origins, resources, and management of hypogene karst, Carlsbad NM Nat Cave Karst Res Inst, Symp 6, pp. 119–129
Hildebrand AR Penfield GT Kring DA Pilkington M Camargo ZA Jacobsen SB Boynton WV (1991) Chicxulub crater: a possible Cretaceous/Tertiary boundary impact crater on the Yucatan Peninsula, Mexico. Geology, 19:867–871
Huddlestun PF (1993) A revision of the lithostratigraphic units of the Coastal Plain of Georgia: The Oligocene. Ga Geol Surv Bull 105
Gallen SF Wegmann KW Bohnensteih DR (2013b) Miocene rejuvenation of topographic relief in the southern Appalachians. GSA Today, 23(2):4–10.
Jones DS (1997) The marine invertebrate fossil record of Florida. In: Randazzo AF Jones DS (eds) The geology of Florida. Univ Press Fla, Gainesville, Ch 7, pp 89–117
Kring DA (2000) Impact events and their effect on the origin, evolution and distribution of life. GSA Today, 10(8):1–7
Leve GW (1966) Ground water in Duval and Nassau Counties, Florida. Fla Geol Surv Rept Inves 43
Lichtler WF Anderson W Joyner BF (1968) Water resources of Orange County, Florida. Fla Geol Surv Rept Inves 50
Lloyd JM (1989) 1986 and 1987 Florida petroleum production and exploration. Fla Geol Surv Inf Circ 106
Lloyd JM (1997) 1994 and 1995 Florida petroleum production and exploration. Fla Geol Surv Inf Circ 111
McBride JH Nelson KD (1988) Integration of COCORP deep reflection and magnetic anomaly analysis in the southeastern United States: Implications for the origin of the Brunswick and East Coast magnetic anomalies. Geol Soc Amer Bull, 100(3):436–445.
McFarlan E Menes LS (1991) Lower Cretaceous. In: Salvador A (ed), The Gulf of Mexico Basin. The geology of North America, Volume J, Geol Soc Amer, Ch. 9, pp. 181–204.
Marella RL (2009) Water withdrawals, use, and trends in Florida, 2005. US Geol Surv Sci Inves Rept 2009-5125
Matsui T Imamura F Tajika E Nakano Y Fujisawa Y (1999) K-T impact tsunami. Houston Texas, Lunar and Planet Inst, Lunar and Planet Sci XXX, Abst #1527
Miller JA (1986) Hydrogeologic framework of the Floridan Aquifer System in Florida, and in parts of Georgia, Alabama and South Carolina. US Geol Surv Prof Pap 1403-B
Miller JA (1997) Hydrogeology of Florida. In: Randazzo AF Jones DS (eds.), The geology of Florida, Gainesville, The Univ Press Fla Ch 6, pp. 69–88
Missimer TM (2001) Siliciclastic facies belt formation and the Late Oligocene to Middle Miocene partial drowning of the southern Florida Platform. Gulf Coast Assoc Geol Soc Trans 51:229–238
Missimer TM (2002) Late Oligocene to Pliocene evolution of the central portion of the South Florida Platform: mixing of siliciclastic and carbonate sediments. Fla Geol Surv Bull 65
Missimser TM Maliva RG (2004) Tectonically induced fracturing, folding and groundwater flow in south Florida. Gulf Coast Assoc Geol Soc Trans 54:443–459.
Missimer TM Maliva RG Walker CW Owosina E (2000) Anatomy of a nearshore mixed siliciclastic-carbonate deposit, the Plio-Pleistocene of southern Broward County, Florida. Gulf Coast Assoc Geol Soc Trans 50:111–128
Opdyke ND Jones DS MacFadden BJ Smith DL Muller PA Schuster RD (1987) Florida as an exotic terrane: Paleomagnetic and geochronologic investigation of the lower Paleozoic rocks from the subsurface of Florida. Geol 15:900–903.
Pirkle WA (1970) The offset course of the St. Johns River, Florida. Southeast Geol 13:39–59
Puri HS Vernon RO (1964) Summary of the geology of Florida and a guidebook to the classic exposures. Fla Geol Surv Sp Pub No 5 (revised)
Randazzo AF (1997) The sedimentary platform of Florida: Mesozoic to Cenozoic. In: Randazzo AF Jones DS (eds), The geology of Florida, Univ Press Fla, Gainesville, Ch 4 pp. 39–56
Riggs SR (1979) Phosphorite sedimentation in Florida – a model phosphogenic system. Econ Geol 74:285–314.
Ruppé R (1980) The archaeology of drowned terrestrial sites: A preliminary report. Fla Bur of Historic Sites and Properties Bull 6:35–45
Salvador A (1991) Triassic-Jurassic. In: Salvador A (ed), The Gulf of Mexico basin. Geol Soc Amer, The geology of North America Vol J Ch 8 pp. 131–180
Schlee JS Manspeizer W Riggs SR (1988) Paleo-environments: Offshore Atlantic U.S. margin. In: Sheridan RE Grow JA (eds), The Atlantic continental margin. Geol Soc Amer, The geology of North America, Vol. I–2, pp. 365–385
Schmidt W (1984) Neogene stratigraphy and geologic history of the Apalachicola Embayment. Fla Geol Surv Bull 58
Schmidt W (1997) Geomorphology and physiography of Florida. In: Randazzo AF Jones DS (eds), The geology of Florida, Univ Press Fla, Gainesville, Ch 1, pp. 1–12
Schmidt W Scott TM (1984) Florida karst: Its relationship to geologic structure and stratigraphy. In: Beck BF (ed.), Sinkholes: Their Geology, Engineering and Environmental Impact. Proc First Multidisc Conf on Sinkholes, Orlando, Florida, October 15–17, 1984, p. 11–16.
Schultz PH (1996) Cretaceous-Tertiary (Chicxulub) impact angle and its consequences. Geology, 24(11):963–967
Scott TM (1988) The lithostratigraphy of the Hawthorn Group (Miocene) of Florida. Fla Geol Surv Bull 59
Scott TM (1992) A geological overview of Florida. Fla Geol Surv Open File Rept 50
Scott TM (1997) Miocene to Holocene history of Florida. In: Randazzo AF Jones DS (eds), The geology of Florida, Univ Press Fla, Gainesville, Ch 5, pp. 57–67
Scott TM (2001) Text to accompany the geologic map of Florida. Fla Geol Surv Open File Rept 80

Scott TM (2005) Revisions to the geomorphology of Florida focusing on north-central Florida and the eastern Panhandle. In: Copeland R (comp), Geomorphic influence of scarps in the Suwannee Basin, Southeast Geol Soc Field Trip Guidebook 44, pp. 18–36.

Scott TM (2011) Geology of the Florida Platform – Pre-Mesozoic to Recent. In: Buster NA Holmes CW (eds), Gulf of Mexico: Origins, water, and biota, Vol 3 Geology, Texas A&M Univ Press, College Station, TX, Ch 2, pp. 17–31.

Scott TM Allmon WD (n.d.) (eds), Plio-Pleistocene stratigraphy and paleontology of southern Florida. Fla Geol Surv Sp Pub 36

Scott TM Means GH Meegan RP Means RC Upchurch SB Copeland RE Jones J Roberts T Willet A (2004) Springs of Florida. Fla Geol Surv Bull 66

Scott TM Campbell KM Rupert FR Arthur JD Green RC Means GH Missimer TM Lloyd JM Yon JW Duncan JG (2001) Geologic map of the State of Florida. Fla Geol Surv Map Series 146

Smith DL (1982) Review of the tectonic history of the Florida basement. Tectonophy 88:1–22

Smith DL Lord KM (1997) Tectonic evolution and geophysics of the Florida basement. In: Randazzo AF Jones DS (eds), The geology of Florida. Gainesville, Univ Press Fla, Ch 2, pp. 13–26

Sohl NF Martinez E Salmeron-Urena P Soto-Jaramillo F (1991). Upper Cretaceous. In: Salvador A (ed), The Gulf of Mexico basin, Geol Soc Amer, Geol North America, Vol J, Ch 10, pp. 205–244

Southeastern Geological Society (1986) Hydrogeological units of Florida. Fla Geol Surv Sp Pub 28

Stuckey JL (1965) North Carolina: Its geology and mineral resources. N Carol Dept Conserv and Nat Res, Raleigh

Suthard BC Hine AC Duncan DS Locker SD Morton RA (2009) A siliciclastic-filled sedimentary basin in a mid-carbonate platform setting, Tampa Bay, Florida. Tampa Bay Regional Planning Council, Tampa BASIS 5, conf presentation accessed 11/26/2014 at https://www.tbeptech.org/BASIS/BASIS5/DAY1-Session2-Geology_Sediments/3_Suthard_TB_Sinkhole_Geologic_Formation.pdf

Tanner WF (1982) High marine terraces of Mio-Pliocene age, Florida panhandle. In: Scott TM Upchurch SB (eds.), Miocene of the southeastern United States, Fla Bur Geol Sp Pub 25, p. 200–209

Thomas WA Chowns TM Daniels DL Neathery TL Glover III L Gleason RJ (1989) The subsurface Appalachians beneath the Atlantic and Gulf Coastal Plains. In: Hatcher RD Thomas WA Viele GW (eds), The Appalachian-Ouachita orogen in the United States, Geol Soc Amer, Geol North America, Vol F-2, pp. 445–458.

Upchurch SB (2002) Hydrogeochemistry of a karst escarpment. In: Martin JB Wicks CM Sasowsky ID (eds), Hydrogeology and biology of post-Paleozoic carbonate aquifers. Charles Town, WV, Karst Waters Inst Sp Pub 7, pp. 73–75.

Upchurch SB (2005) Geochemical and geomorphic evolution of a karst escarpment. In: Copeland R (comp), Geomorphic influence of scarps in the Suwannee River basin. Tallahassee, Southeast Geol Soc Field Trip Guidebook 44, pp. 38–42

Upchurch SB (2014) An introduction to the Cody Escarpment, north-central Florida. In: Lawn A (comp), Karst hydrogeology of the upper Suwannee River basin, Alapaha River area, Hamilton County, Florida. Tallahassee, Southeast Geol Soc Field Trip Guidebook 63, pp. 13–28

Upchurch SB (2017 in press) Hypogene speleogenesis on the Florida Platform, U.S.A. In: Klimchouk A. (ed), Hypogene karst of the world, Dordrecht, Springer Science

Upchurch SB Strom RN Nuckels MG (1982) Silicification of Miocene rocks from central Florida. In: Scott TM Upchurch SB (eds), Miocene of the Southeastern United States, Fla Bur Geol Sp Pub 25 pp. 251–284.

Walker JD Geissman JW Bowring SA Babcock LE (comp) (2012) Geologic time scale, Boulder, CO, Geol Soc Amer v 4.0

Warzeski ER Cunningham KJ Ginsburg RN Anderson JB Ding ZD (1996) Neogene mixed siliciclastic and carbonate foundation for the Quaternary carbonate shelf, Florida Keys. Jour Sed Res 66:788–800.

Weaver CE Beck KC (1977) Miocene of the southeastern United States: A model for chemical sedimentation in a peri-marine environment. Sed Geol 17:1–234

White WA (1958) Some geomorphic features of central peninsular Florida. Fla Geol Surv Bull 41

White WA (1970) The geomorphology of the Florida peninsula. Fla Geol Surv Bull 51

White WA Puri HS Vernon RO (1964) Physiographic setting. In: Puri HS Vernon RO, Summary of the geology of Florida and a guidebook to the classic exposures. Fla Geol Surv Sp Pub 5 (revised)

Williams CP Scott TM Upchurch SB Paul DT Hannon LM (n.d.) (in prep.) Florida geomorphology, Fla Geol Surv Bull 72

Winston GO (1987) Generalized stratigraphy and geologic history of the South Florida Basin. In: Maurrasse FJ-MR (ed), Symposium on south Florida geology. Miami Geol Soc Memoir 3

Winston GO (1991) Atlas of structural evolution and facies development on the Florida-Bahama Platform – Triassic through Paleocene. Miami Geol Soc

Woods RD Salvador A Miles AE (1991) Pre-Triassic. In: Salvador A (ed), The Gulf of Mexico basin, Geol North Amer, Vol J, Geol Soc Amer, Ch 7, pp. 109–129.

Wyrick GG (1960) The ground water resources of Volusia County, Florida. Fla Geol Surv Rept Invest 22

# Hydrogeology of Florida

# 4

**Abstract**

Recharge to Florida's three aquifer systems is controlled by prominent wet/dry seasons; multi-year wet/dry cycles; differences in weather patterns between northern and southern parts of the state; high evapotranspiration rates; and local degree of aquifer confinement.

The Surficial Aquifer System (SAS) comprises local, largely Plio-Pleistocene water-table aquifers, including the siliciclastic Sand and Gravel Aquifer and the extremely productive, carbonate Biscayne Aquifer, and undifferentiated surficial aquifer.

The Miocene Hawthorn Group hosts the Intermediate Aquifer System and Confining Unit (IAS/ICU). The IAS and ICU are defined in different areas based on the productivity of the very heterogeneous Miocene Hawthorn Group sediments. The ICU confines the underlying UFA via low-permeability clays and tortuous flow paths through interlayered sediments. Some karst development is evident in the IAS/ICU, especially near karst escarpments.

The Floridan Aquifer System (FAS) is subdivided in some areas into the Upper Floridan Aquifer (UFA), the Middle Confining Units (MCUs), and the Lower Floridan Aquifer (LFA). The UFA is highly productive, consisting of Middle Eocene to Early Miocene carbonates with well-developed karst as a result of dissolution by meteoric water and by freshwater/saltwater mixing dissolution at various elevations during past sea level fluctuations. Springs, sinkholes, and other karst features are common in unconfined areas. The LFA consists primarily of Paleocene-Eocene dolostones. It contains Boulder Zones (horizons of very high hydraulic conductivity) as well as possible deep karst probably developed along freshwater/saltwater mixing zones. The deep, high productivity zones may be the result of hypogenetic speleogenesis.

Groundwater flow in Florida's aquifers is largely lateral, gravity-driven flow recharging in unconfined areas and discharging in freshwater lenses and wedges at coastlines and rivers, with shallow aquifers exhibiting Tothian flow and geothermally driven circulation in the LFA.

**Keywords**

Floridan aquifer system · Freshwater lens · Florida climate · Karst aquifer · Biscayne aquifer · Hypogenetic speleogenesis · Epigenetic speleogenesis

## 4.1 Introduction

The hydrogeology and hydrogeochemistry of Florida karst are characterized by the interactions of carbonate sediment and rock with freshwater systems, deep saline groundwater, and sea water. This chapter is organized to follow the general pathways of water in the water cycle in Florida. The chapter begins with an overview of water resources in Florida, continues with a discussion of Florida hydrogeology, and ends with a detailed discussion of each of Florida's major aquifers. The geochemical processes that dominate karst aquifers are discussed in depth in Chap. 5.

Karst aquifers are unique in the degree to which the water within the aquifer can potentially restructure the aquifer itself. With Florida's eogenetic karst, the amount of restructuring, including development of sinkhole areas of focused recharge, karst conduits, and springs at discharge locations, varies. The variations in development of karst features result from the age of the carbonate rocks, amount of cover, and amount of confinement.


S. Upchurch et al., *The Karst Systems of Florida*, Cave and Karst Systems of the World,
https://doi.org/10.1007/978-3-319-69635-5_4

## 4.2 Overview of Florida Hydrogeology

### 4.2.1 Water Use in Florida

As noted in Chaps. 1 and 3, Florida is blessed with three major aquifer systems (Fig. 1.8): the Surficial Aquifer System (SAS, including the Sand and Gravel Aquifer in the Panhandle and the Biscayne Aquifer in southeastern Florida), the Intermediate Aquifer System and Confining Unit (IAS/ICU) and the Floridan Aquifer System (FAS).

As a result of Florida's karst, many areas of the state have little or no surface water upon which they can rely for public supply. Elsewhere, the surface waters are often characterized by natural water-quality issues, such as coliform bacteria, nitrate/nitrite, pesticides, herbicides from storm water runoff, that render treatment for public supply expensive. In both situations, Floridians have gravitated to use of groundwater from these aquifer systems.

In 2005, about 62% of the freshwater utilized in the state was obtained from Florida's aquifers (Marella 2009). Of the 62%, over half was for public supply. The majority of this groundwater primarily came from the Upper Floridan Aquifer (UFA) and the Biscayne Aquifer. These two aquifers (1) consist of carbonate rocks, (2) have high transmissivities, and (3) have generally excellent water quality. As such, they have been capable of supporting Florida's major metropolitan areas and the majority of Floridians.

Florida's other sources of groundwater are generally less productive and may have water quality issues. For the most part, these less productive aquifers are limited in areal extent. Where the UFA is saline, groundwater is usually withdrawn from the SAS, including the Sand and Gravel Aquifer, Biscayne Aquifer, and undifferentiated sand and/or carbonate aquifers within the SAS. With the possible exception of the Sand and Gravel Aquifer, these less extensive aquifers also contain carbonate-sediment components and contain karst features.

### 4.2.2 Florida's Water Budget

Florida is famous (or infamous) for its wetlands, a groundwater table that is often close to the land surface, and numerous lakes and streams, all of which suggest that Florida has a surplus of surface and ground water. While there is abundant rainfall, Florida's karstic limestone aquifers and sandy surficial sediments are characterized by rapid infiltration and dynamic surface and groundwater flow systems. This results in a dearth of surface water in many areas and shallow soil moisture in many areas of the state. It is also responsible for many of the processes that have formed karst features in Florida.

In order to understand Florida karst, therefore, it is important to first understand the water cycle in Florida. Inputs to the Florida water cycle include precipitation and surface and groundwater inflow from Alabama and Georgia. All totaled, inflow of water into Florida is approximately $6.7 \times 10^8$ $m^3$/day (based on data from Fernald and Patton 1984).

Water leaves the system through evapotranspiration, as runoff in streams and rivers discharging to the ocean, through consumptive use (pumping wells), or as groundwater discharge at the coastline. Water lost is also approximately $6.7 \times 10^8$ $m^3$/day, of which about $2.6 \times 10^8$ $m^3$/day is surface and groundwater.

### 4.2.3 Precipitation and Temperatures

#### 4.2.3.1 Atmospheric Temperatures

Statewide, Florida atmospheric temperatures average 16.6 °C (Florida Climate Center 2011b). The northern part of the state experiences more continental temperatures with brief intervals of temperatures below freezing and hot summers. Snow accumulation is rare, however. The southern part of the state has a sub-tropical to tropical climate, and freezing temperatures are rare. As a result, precipitation can be considered to consist entirely of rainfall.

#### 4.2.3.2 Rainfall

Statewide, Florida receives an average (1895–2011) of 1.37 m of precipitation a year (Florida Climate Center 2011a). Owing to Florida's warm climate, this precipitation is essentially all rainfall. This relatively high amount of rainfall is not evenly distributed over the state owing to a diversity of origins, including (1) continentally derived weather fronts, (2) local, convective storms, and (3) tropical storms and hurricanes.

Throughout the state, rainfall is seasonal with distinctive wet and dry seasons. Year to year, rainfall can vary tremendously. Extreme wet years are associated with tropical depressions and hurricanes, while drought years reflect an absence of tropical storms and stagnation of dry, high atmospheric-pressure cells over the state.

The highest annual precipitation in the state is in the Panhandle (average annual rainfall is as great as 1.7 m) and extreme southeastern Florida (up to 3.7 m). These highest precipitation areas result from the tropical storms and hurricanes that strike the state in those areas. Coastal areas tend to have slightly higher rainfall than the interior of the state as a result of convective thunderstorms that form near the coasts.

The month-to-month variation in rainfall is as important for understanding Florida hydrology as is annual rainfall. Fernald and Purdum (1998) summarized the monthly rainfall patterns throughout the state. Rainfall patterns in central and southern Florida are distinctly seasonal, with maximum rainfall in the "rainy season" from late May through early October. In northern Florida, there is a second rainy period in the late winter and early spring as a result of weather fronts originating north of the state.

Rainfall patterns throughout the state reflect the influence of the Atlantic Ocean and Gulf of Mexico bounding the eastern, western, and southern parts of the state, the proximity of the main North American landmass to the north, and the location of the southern part of the state in the subtropics. A maritime climate is characterized by convective thunderstorms in the summer. A continental rainfall pattern is relatively uniform throughout the year compared to a tropical rainfall pattern, which has a distinct wet and dry season ("monsoonal" pattern). Tropical and maritime patterns dominate in the south, resulting in a wet and dry season characterized by convective thunderstorms in the summer. This pattern gradually transitions northward to a continental pattern, which dominates north of the Suwannee River drainage basin. Rainfall patterns in the northern parts of the state are characterized by precipitation that is more evenly distributed between the warmer and cooler months. While average annual rainfall in the north is lower than the south, the average winter rainfall is somewhat higher than in the south. Winter precipitation events are a result of mid-latitude frontal weather systems with individual rainfall events that are usually widespread and associated with cold fronts. In addition to these climate influences, Florida's location on the southeastern coast of North America makes the entire state vulnerable to tropical storms and hurricanes spawned off of the coast of Africa, which account for a large amount of the rainfall volume in the summer months.

Heath and Conover (1981) recognized the existence of a "climatic river basin divide" in Florida that separates these different rainfall patterns and resulting stream-discharge patterns. Streams north and west of the climatic divide exhibit high flows in the late winter and early spring as a result of rainfall associated with continental weather fronts. Lower precipitation and low stream flows occur in late spring and fall north of the divide. Streams south of the climatic divide exhibit high flows in the late summer and fall, with spring low flows. This pattern reflects the summer convective storms that dominate the climate of peninsular Florida. Streams lying near the climatic divide tend to exhibit a mixture of both of these rainfall patterns (a "bimodal" pattern of floods in the spring and fall). More recently, Kelly (2004) and Kelly and Gore (2008) confirmed these hydrologic patterns in streams in Florida, which they termed the "northern river pattern" (spring flooding), the "southern river pattern" (fall flooding), and the "bimodal pattern" (both spring and fall flooding) near the divide.

These temporal rainfall and streamflow patterns may affect karst development if they persist through geologic time. For example, high rainfall during the summer months, when evapotranspiration is at a maximum, may result in a minimal amount of water available to interact with carbonate rock materials. In contrast, high rainfall during the cooler months is likely to be characterized by higher recharge, opportunity to interact with sediment or rock components, and potential for formation of karst landforms.

Year-to-year rainfall is rarely comparable to the average annual rainfall (Fernald and Purdum 1998; Winsberg 2003). For example, in the area covered by the National Oceanic and Atmospheric Administration's North Florida Climatic Division, annual (calendar year) rainfall has varied from a low of 90.2 cm in 1955 to a high of 198 cm in 1964. Such extremes are normal in Florida. As a result, Florida tends to suffer from droughts and torrential rainfall events on a regular basis.

For more information on the distribution of precipitation in Florida, see Hughes et al. (1971), Fernald and Purdum (1998), and Winsberg (2003).

### 4.2.4 Evapotranspiration

Evaporation is the loss of moisture to the atmosphere from a body of water such as a lake, the sea, or the moisture in the soil. Transpiration is the loss of moisture to the atmosphere through plants. Together, these processes are termed evapotranspiration or ET. Rates of evapotranspiration in Florida have been estimated with a variety of direct measurements and/or computational methods with varying degrees of accuracy.

Fernald and Purdum (1998) compared monthly rainfall and potential ET for 11 stations in Florida. ET is the estimated potential ET based on humidity and temperature data. By comparing rainfall and potential ET, it is clear that there are seasonal episodes of net rainfall surplus and/or deficit. For example, Pensacola, which is located in the extreme western Panhandle, tends to have surplus rainfall during both the winter and summer rainy periods. Only during the transition months is the rainfall less than ET, and it is during these periods that soil moisture is utilized to sustain crops. In contrast, Tampa and the other stations in the middle of the state only have surplus rainfall during the summer rainy season. At all other times, estimated ET exceeds rainfall. The balance between rainfall and ET in the Miami area has a similar pattern, but Key West, an island about 260 km south of Miami, never has a surplus of rainfall.

The rainfall/ET patterns depicted by Fernald and Purdum (1998) are important, because they show that, while Florida has abundant rainfall, only a small portion of that rainfall is

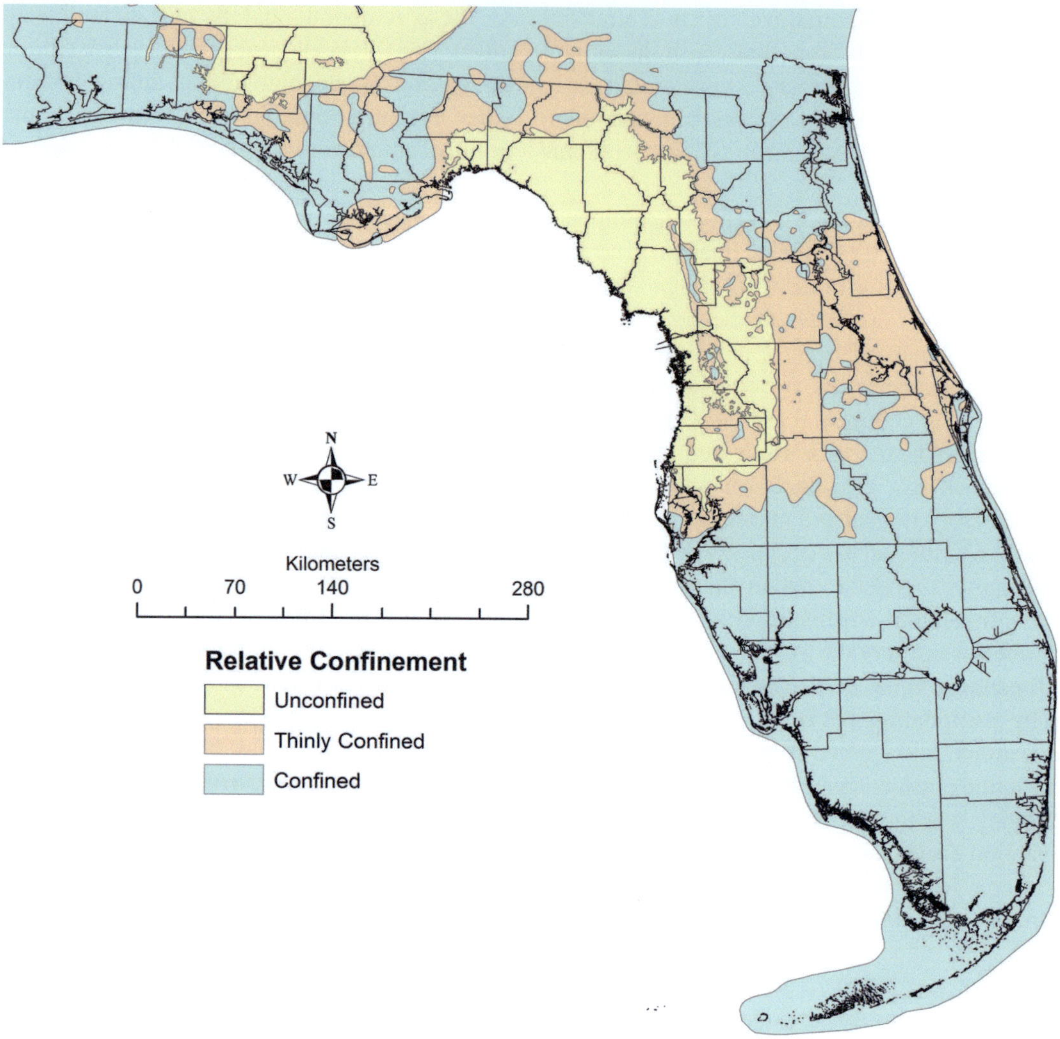


**Fig. 4.1** Relative confinement of the upper Floridan aquifer. (Modified from Williams and Kuniansky (2015, Fig. 21))

available for recharge and development of karst through dissolution of carbonate sediments and rocks.

For more information on the distribution of ET in Florida, see Visher and Hughes (1969), Fernald and Purdum (1998), and Winsberg (2003).

## 4.2.5 Runoff and Recharge

Once precipitation reaches the ground, it has three possible fates: evapotranspiration, runoff on the land surface, or recharge to Florida's aquifers. Runoff and recharge are closely related. Water can recharge an aquifer and then discharge to a stream, so it has a recharge and a runoff component. In the context of Florida karst, it is the recharge component that is most important. Exposed carbonate sediment and rock can undergo dissolution by surface water, but the complex, three-dimensional fabric of Florida aquifers is the most important to Florida water resources.

### 4.2.5.1 Runoff Versus Aquifer Confinement

The highest runoff is in areas where the UFA is well confined and significant recharge to the UFA is limited by the relatively impervious IAS/ICU. Figure 4.1 illustrates the relative confinement of the UFA in Florida. Streams are best developed in the areas where the UFA is indicated to be confined.

In the Everglades, Big Cypress, and other provinces of southern Florida, sheet flow through swamps and marshes (Fig. 4.2a) dominates, and recharge to the sediments of the surficial aquifer is related to the amount of water stored in the underlying aquifer. Runoff is characterized by high ET superimposed on a sluggish, sheet-flow system with a high surface area. The surficial aquifer is recharged by rainfall and this surface water. When excess water is available, the water table of the surficial aquifer rises, resulting in flowing water on the land surface (Fig. 4.2a). During the dry season and especially periods of drought, surface water may disappear as the water table drops below the land surface (Fig. 4.2b).

**Fig. 4.2** (**a**) The Everglades, a region of south Florida dominated by sheet flow of organic-rich surface water, during the rainy season. Everglades National Park, Dade County, Florida. (**b**) The Everglades during a drought. Note the mud-cracked calcitic sediment. Everglades National Park, Dade County, Florida

Low runoff areas in central and north Florida occur where the underlying UFA is unconfined or poorly confined (Fig. 4.1). These unconfined areas are important because they reflect karst terrains where drainage is internal.

In most circumstances, runoff in the karst terrains is restricted to rivers that originate in low recharge areas where the IAS/ICU is well developed. These rivers then cross and interact with the karst terrains. For example, the Suwannee River begins in the Okefenokee Swamp in southern Georgia. Here, the Miocene Hawthorn Group is thick and recharge to the underlying UFA is limited. As a result, any precipitation remaining after losses to ET is directed out of the Okefenokee as runoff through rivers, such as the Suwannee and St. Marys rivers.

The Suwannee, Santa Fe, and Withlacoochee rivers cross a vast karst area (the Branford and Chiefland karst plains; Figs. 3.13 and 4.3) in Florida. The Suwannee and Withlacoochee rivers originate in the Tifton Upland District, and the Santa Fe originates in the Raiford Ridges Province. In all three rivers, headwaters occur where the IAS/ICU is an effective aquitard that limits recharge, allowing for runoff to develop stream systems. In the karst plains (Fig. 4.3), sinkholes and other

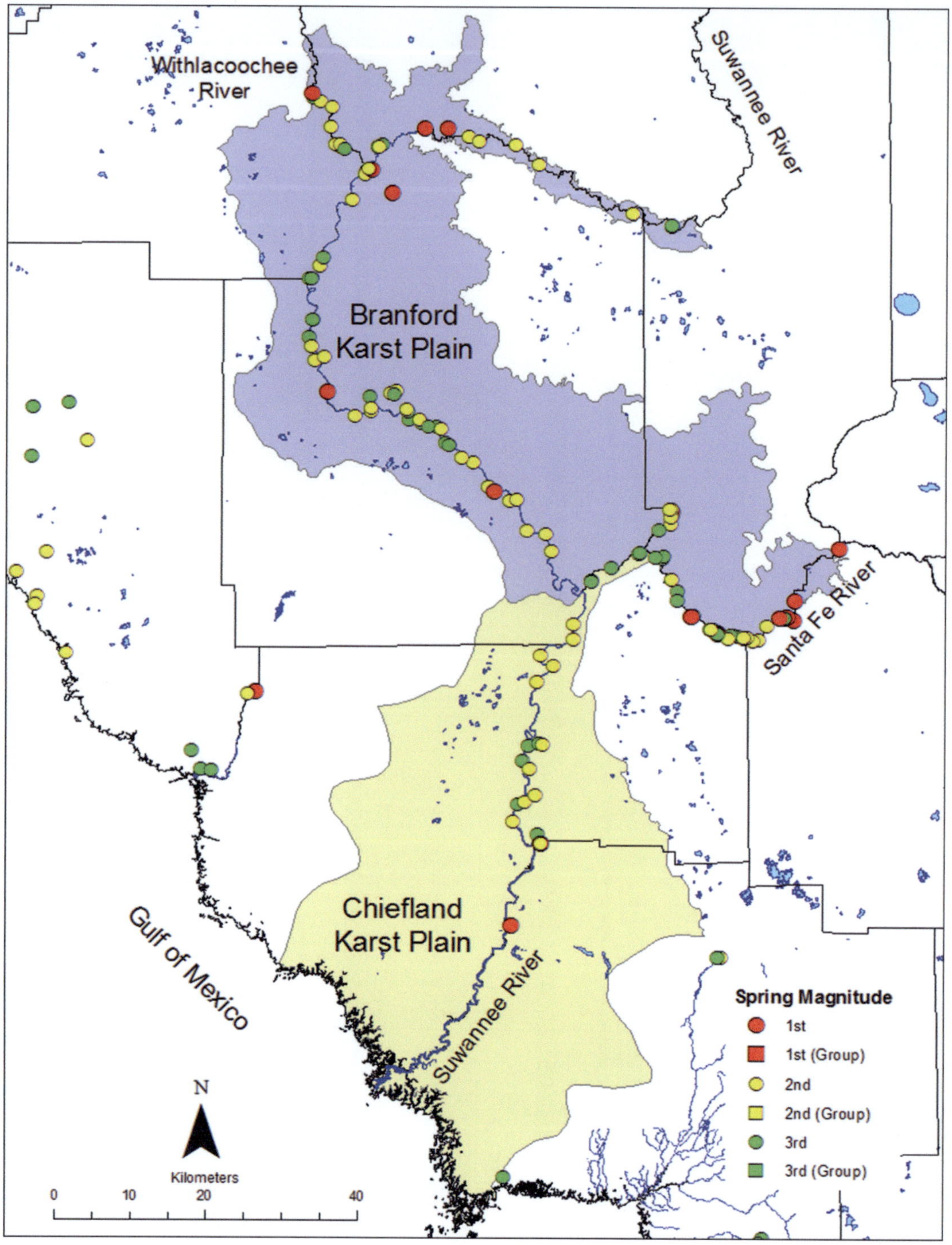


**Fig. 4.3** Major rivers that cross the Branford and Chiefland karst plains

karst features catch rainfall, which drains as groundwater internally into the unconfined or poorly confined UFA. That groundwater flows to the river corridors where it discharges by way of springs and seeps (Fig. 4.3). The Suwannee and Santa Fe rivers are fed by many springs that sustain flow in the rivers during droughts and the dry season.

Most of the rainfall in the low-runoff, karst regions is lost to evapotranspiration, leaving about 0.13–0.17 m annually for aquifer recharge and interaction with carbonate sediments and rocks (Fernald and Purdum 1998; Delfino and Heaney 2004). Owing to the presence of soils and sediments with high infiltration rates, moderate to high UFA permeabilities, and numerous internally drained basins (sinkholes and related karst features), recharge to the carbonate aquifers can be quite efficient.

The carbonate aquifers store vast amounts of water owing to their thicknesses, but they also drain efficiently. As a result, the potentiometric surface of the UFA is sensitive to the

**Fig. 4.4** An internally drained depression with a sinkhole lake at its center. Location: Loyce, Pasco County

amount of recharge and drainage through outflow or groundwater extraction. For example, water levels in sinkhole lakes that are well connected to the underlying carbonate-rock aquifer may be vulnerable to droughts and the effects of groundwater withdrawals. Figure 4.4 illustrates a sinkhole lake that is located near a major wellfield. The photograph was taken during a period of severe drought. Groundwater withdrawals at the wellfield have caused water levels in the lake and underlying UFA to decline. Note the dock as an indicator of water levels when potentials in the underlying UFA are high.

#### 4.2.5.2 Recharge and Discharge from the Floridan Aquifer System

Clearly, runoff and recharge to Florida's carbonate aquifers is a function of the degree of confinement. In the case of the FAS, the pattern of recharge and discharge from the UFA can be enlightening with respect to confinement and potential for karst development.

Figure 4.5 illustrates "pre-development" estimates of UFA recharge and discharge from Miller (1986, 1990) and Bush and Johnston (1988). The recharge estimates were derived from a groundwater flow modeling program (RASA or Regional Aquifer-System Analysis) in the mid-1980s. The calibrated, steady state model included groundwater withdrawals by human activity. By turning off the withdrawal component and allowing the model to adjust to solely natural recharge and discharge conditions, the "pre-development" estimate was developed.

There are several informative features of this "pre-development" map. First, note that highest UFA recharge areas are located on the Ocala Platform crest and the karst plains identified by Williams et al. (in prep.). Intermediate recharge amounts occur where the IAS/ICU is leaky, largely as a result of having been penetrated by ancient sinkholes. Areas of minimal to no recharge to the UFA are located where confinement by the IAS/ICU retards the downward vertical movement of water from the overlying SAS through the ICU to the UFA. Areas of highest UFA discharge are located along the coast, bordering the high recharge karst plains. These high discharge areas represent springs and seeps that are common along the coasts and some rivers. The area of UFA discharge offshore of the strip from Duval to Volusia counties (just below the north arrow) reflects a series of offshore springs that were of concern during model development.

For more information on the distribution of runoff and recharge in Florida, see Hughes (undated) and Fernald and Purdum (1998).

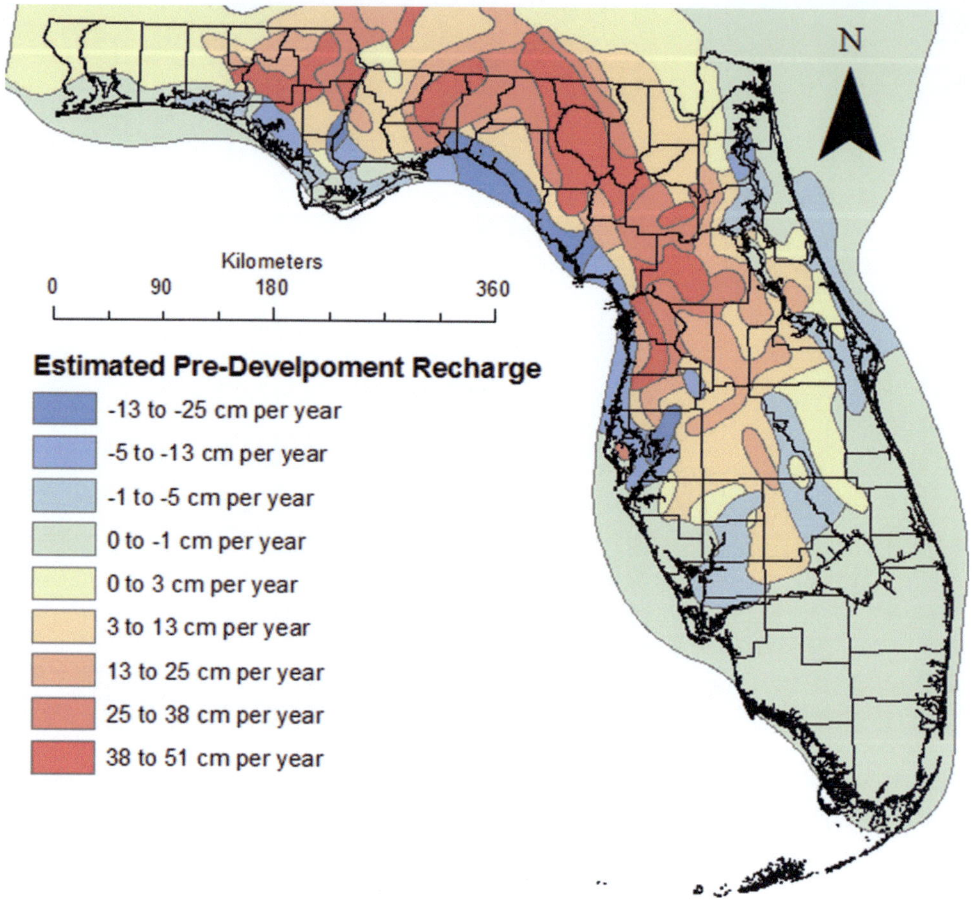


**Fig. 4.5** Estimated pre-development recharge and discharge from the upper Floridan aquifer. From geospatial data provided in Bellino (2011)

## 4.2.6 Groundwater

### 4.2.6.1 Vadose and Phreatic Conditions

Once rainfall soaks into near-surface soils or sediments, it becomes meteoric soil moisture. Much of the water that enters the surficial soils and sediments flows short distances (mostly downward), through the vadose zone, the part of the aquifer above the water table, where the pore spaces are mostly filled with air. The water table in a simple, unconfined aquifer represents the upper surface of the zone where pore and void space in the aquifer is saturated with water. Below the water table, in the phreatic zone, the aquifer pore spaces are filled with water that flows more or less horizontally under the influence of pressure or "potential" differentials. These pressure differentials are usually referred to as "hydraulic head", which can be conceptualized as the level to which water rises in a well.

### 4.2.6.2 Aquifer Confinement

In an unconfined aquifer, the heads measured in wells generally match up with the actual elevations of the water table in the aquifer, and the water table is a free surface in contact with the atmosphere. A confined aquifer is an aquifer that is overlain by a low vertical hydraulic conductivity stratum so that the hydraulic pressure field in the aquifer is unable to cause groundwater to rise to its potential elevation unless a conduit is present through the confining bed. When the confining stratum is penetrated by a well or other conduit, the water rises to an elevation that reflects its pressure field and potential. Thus, in confined aquifers, heads can extend above the top of the aquifer or above the ground surface. When this happens, the aquifer is said to be artesian. Strata that effectively confine movement of water from the aquifer are termed aquicludes. Strata that limit or retard movement of water from the aquifer are termed aquitards.

Figure 4.1 illustrates the general degree of confinement of the UFA. Where it is confined, artesian conditions can be expected and, where the UFA is unconfined, water levels in the aquifer represent the water table and potentiometric surface of the aquifer. Where it is semi-confined, either condition can exist.

### 4.2.6.3 Groundwater Flow Patterns

Groundwater generally flows from areas of higher hydraulic head to areas of lower hydraulic head. By mapping the spatial distribution of elevations (hydraulic heads) of water in an aquifer (the water table in unconfined aquifers and potentiometric surface in confined aquifers), one can create a map of likely groundwater flow paths.

Note that in isotropic and homogeneous aquifers[1] the groundwater flow paths are perpendicular to the isopotential contour lines of the potentiometric surface map. In fractured rock and rocks with cavernous porosity, the actual flow paths of groundwater likely follow the paths of least resistance – the fractures and karst conduits. Under conditions of fracture flow, the groundwater flows generally down gradient, but locally the path many be more tortuous, because it follows the fractures and conduits. Viewed from a large enough (regional) scale, these local deviations may not be detectable, Instead, these local flow deviations may blend into the general, regional, down-gradient trend. It is for this reason that scale is extremely critical in determining how a groundwater flow within karst system can be evaluated and modeled.

The concept that regional groundwater flow models that assume homogeneous aquifers can be used for fractured and/or cavernous strata is a subject of debate (Hickey 1984; Vacher and Mylroie 2002; Scanlon et al. 2003). The assumption that, when a large area is modeled, the effects of conduit flow are minimized has led to the concept of "equivalent porous media", which postulates that the conduit flow paths are integrated such that the regional flow appears to behave as it would in an homogeneous aquifer. In Florida, this concept has been widely adopted, but the reader is warned that modeling of groundwater flow systems in small areas may not replicate the actual, conduit flow system and require extensive geophysical and geological investigations verifying these systems prior to modeling. Additionally, these types of models may match heads relatively well regionally, but not travel times or pathways on a local scale.

The eogenetic karst of Florida's carbonate rocks and sediments has resulted in dual-porosity aquifers. That is, there is (1) interconnected, primary pore space between the individual particles of the carbonate-rock matrix and (2) between secondary, post-depositional fractures and karst conduits. Thus, there may be two different flow systems in a dual-porosity carbonate aquifer. The dominant flow system is likely through the fractures and karst conduits but there is also a secondary flow system through the intergranular space in the matrix of the aquifer. Owing to the differences in hydraulic conductivity of the matrix and conduits, water can flow quickly through the conduits and slowly through the matrix.

#### 4.2.6.4 Infiltration and Fate of Rainwater

From the perspective of a water budget, not all rainfall that enters the ground ends up in the phreatic zone. This moisture has several possible fates:

- It may be utilized by plants and ultimately released to the atmosphere as transpiration;
- It may evaporate directly from the land surface and from within porous and permeable soils or sediments;
- It may migrate into nearby streams as runoff or interflow through shallow sediments;
- It may become incorporated into the SAS and then discharge into nearby streams under small-scale, shallow flow systems with Tóthian flow (Tóth 1962, 1963); or
- It may pass more-or-less vertically through the surficial soils and sediments and recharge the underlying freshwater lens within the limestone aquifer.

The following section deals with the fate of water that recharges the soils and sediments that constitute Florida's aquifer systems.

#### 4.2.6.5 Florida's Aquifer Systems

As noted in Sect. 3.4, Florida has three regional aquifer systems (Southeastern Geological Society 1986; Fig. 4.6):

1. The Surficial Aquifer System (SAS), which includes the Sand and Gravel Aquifer of extreme western Florida, the Biscayne Aquifer of southeastern peninsular Florida, and undifferentiated sub-regional aquifers throughout the state;
2. The Intermediate Aquifer System (IAS) and Intermediate Confining Unit (ICU), collectively, the IAS/ICU, which includes sub-regional aquifers wherever permeable sediments of the Miocene Hawthorn Group of sediments occur. The IAS is an important and complex aquifer system in southwestern Florida with three identified production zones; and
3. The Floridan Aquifer System (FAS), which includes
   - The Upper Floridan Aquifer (UFA);
   - The Middle Confining Unit (MCU), a semi-confining unit at the base of the UFA; and
   - The Lower Floridan Aquifer (LFA), which lies below the MCU where the MCU is present.

#### 4.2.6.6 Involvement of Aquifer Systems in Karst Development

The following sections discuss the general recharge and flow systems of these aquifers in order to set the stage for understanding karst development. With the exception of the Sand and Gravel Aquifer in western Florida (Fig. 4.6), all of Florida's aquifers contain at least some carbonate-rich sediment and, as a result, have the potential for development of karst landforms and flow systems (Table 4.1).

The following sections describe the theoretical configurations of fresh groundwater in Florida. Much of the theory has been developed for islands with freshwater lens but can be directly extended to Florida aquifers given their proximity to the coasts.

[1] Aquifer in which the properties of permeability, porosity, and hydraulic conductivity are equal in all directions.

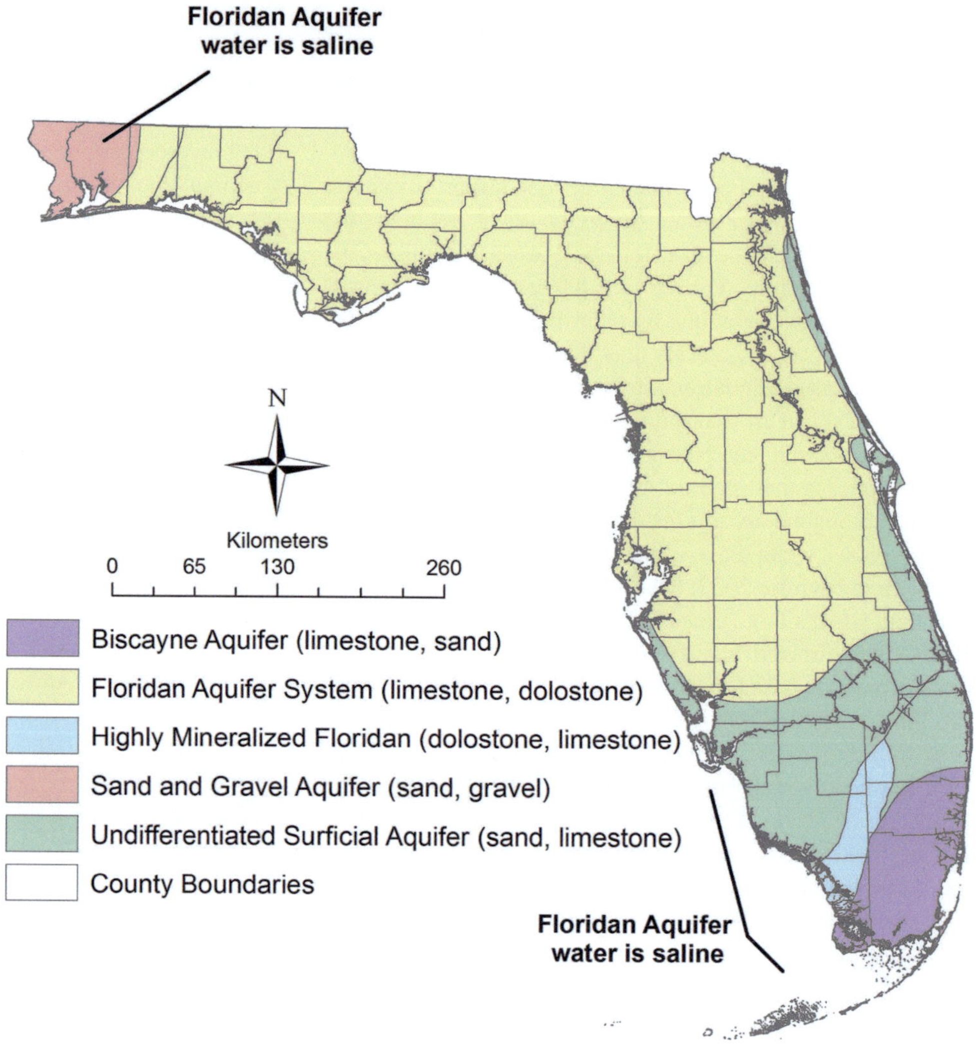


**Fig. 4.6** Lithologies of the principal aquifers of Florida

### 4.2.7 The Freshwater Lens

In the phreatic zone, below the water table, of coastal aquifers containing both freshwater and saltwater, the volume of freshwater generally occupies a lens-shaped body overlying the saltwater portion of the aquifer (Fig. 4.7). The freshwater is derived either from rainfall filtering downward through the vadose zone over the freshwater lens, or from aquifer water traveling from areas further inland. Within the freshwater lens, water generally flows seaward under the influence of gravity.

The freshwater body tends to take on a lens shape whose upper surface is the water table and whose lower surface is the freshwater/saltwater transition zone (A, Fig. 4.7) or a confining layer (B, Fig. 4.7).

It is convenient to consider the freshwater portion of Florida's aquifers as a freshwater lens. In reality, Florida aquifers are complex, span several hydrostratigraphic horizons with different hydraulic properties, and, because of Plio-Pleistocene sea-level excursions, contain water with varying chemical conditions. For the purpose of introducing groundwater flow systems and how they affect karst development, the following sections refer to simple models of freshwater lens development based on our understanding of simplified island and strip-shaped aquifers.

#### 4.2.7.1 Freshwater and Sea Water Circulation Patterns

When meteoric water enters an aquifer, it forms a mound (A, Fig. 4.7) and contributes to the pressure field that forms the potentiometric surface. Near the center of this mound, freshwater is likely to flow downward and then laterally to the coast. Recharge on the flanks of the mound also flows to the coast as shallower portions of the flow system. Shallow flow may be interrupted by discharge to streams, but the general pattern is one of movement of the water to the coast.

The strata that contain freshwater overlie strata that are saturated with saline water, which is denser than freshwater. Thus, a lens-shaped body of freshwater forms on top of the saline water. As will be explained below, the lens is thickest inland, under the crest of the freshwater mound, and thins to zero near the coast.

The saline water that underlies the freshwater lens also circulates, largely as a result of sympathetic movement with the freshwater. In Florida, it has also been suggested that the deep saline water of the LFA under southern Florida may be circulating as a result of geothermal heating (Kohout 1965, 1967, 1975).

#### 4.2.7.2 Geometry of the Freshwater Lens and the Mixing Zone

In freshwater lenses and wedges, the transition between fresh water and seawater may rarely comprise a relatively sharp boundary or, more commonly, a broad transition zone (or "mixing zone") containing brackish water of gradually varying salinity (Fratesi 2013). Flow in the freshwater portion of the wedge is generally towards the sea, with groundwater velocities increasing as the cross-sectional area of the lens or wedge decreases in the vicinity of the coastline. As Vacher (1988a, 1997) points out, in most depictions of freshwater-saltwater systems, including the ones in this chapter, the vertical scale is greatly exaggerated for ease of illustration and the transition zone is drawn as a sharp interface. This tends to overemphasize vertical flow in freshwater lenses and wedges. In reality, horizontal flow strongly dominates within freshwater wedges or lenses, except near the discharge zone (Fig. 4.7). Furthermore, use of a line to depict the position of a hypothetical interface diminishes the importance of the mixing zone, which may occupy a substantial volume of water near the coast. For example, Fig. 4.8 presents the distribution of chloride across a broad mixing zone in the Biscayne Aquifer near Miami (Dade County) in the 1950s. Note the thickness of the mixing zone and its changes in dip, which are a function of the hydraulic conductivities of the strata through which the water is passing.

**Table 4.1** Potential for karst development in Florida's aquifers

| Aquifer | Potential for Karst Development |
|---|---|
| **Surficial Aquifer System** | |
| Sand and Gravel Aquifer | No potential |
| Biscayne Aquifer | Carbonate aquifer; geologically young; potential for small karst features |
| Undifferentiated surficial aquifer | Mixtures of sand, shell, and local lenses of limestone; low potential (limited to small, shallow depressions caused by shell dissolution) |
| **Intermediate Aquifer System** | |
| Carbonate strata | Moderate in limestone; low to moderate in dolostone |
| Other strata | Very low to no potential |
| **Floridan Aquifer System** | |
| Upper Floridan Aquifer | High in limestone; low to moderate in dolostone |
| Middle Confining Unit | Dissolution of gypsum and anhydrite has been documented; otherwise, low |
| Lower Floridan Aquifer | Karst development limited to possible dissolution at salt-water/fresh-water interface and during former low sea levels; possible hypogenetic karst |

#### 4.2.7.3 The Ghyben-Herzberg Principle

Because of the difference in density (ρ) of fresh and sea water, the less dense freshwater ($\rho \approx 1.000$ g/cm$^3$ at earth surface temperatures and pressures) tends to override the denser ($\rho \approx 1.025$ g/cm$^3$) saline water as the freshwater flows to the coast. As the two water types interact, mixing creates a transition zone, primarily as a result of dispersion.

The relationship between the two water types and depth to the transition zone can be estimated based on the differences in density of the two end-member water types. This relationship was first characterized by Badon Ghyben (1889) and Herzberg (1901). The relationship described by Ghyben and Herzberg is known as the Ghyben-Herzberg principle. The Ghyben-Herzberg principle can be represented in terms of the relative densities of the water masses by:

$$z = \frac{\rho_f}{\left(\rho_s - \rho_f\right)} h, \quad (4.1)$$

where z = thickness of freshwater below sea level, $\rho_f$ = density (g/cm$^3$) of the freshwater, $\rho_s$ = density (g/cm$^3$) of the saline water, and $h$ = thickness of freshwater above sea level

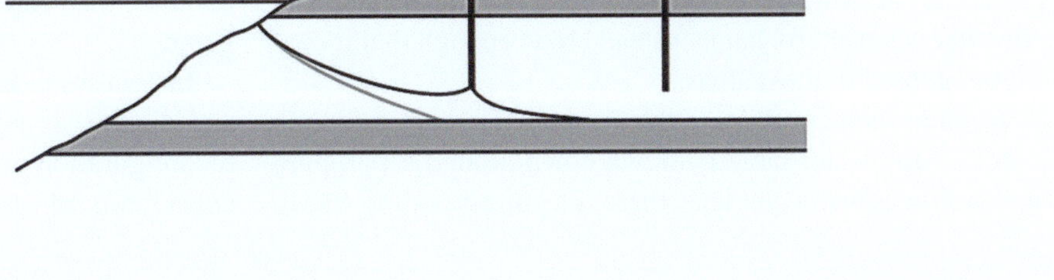


**Fig. 4.7** Cross section examples of a (**a**) an unconfined freshwater lens, as commonly found on a small island and (**b**) a freshwater lens in a confined aquifer, a configuration commonly seen along Florida's coasts. Both lenses are distorted by pumping from wells represented by vertical lines. (From Fratesi 2013)

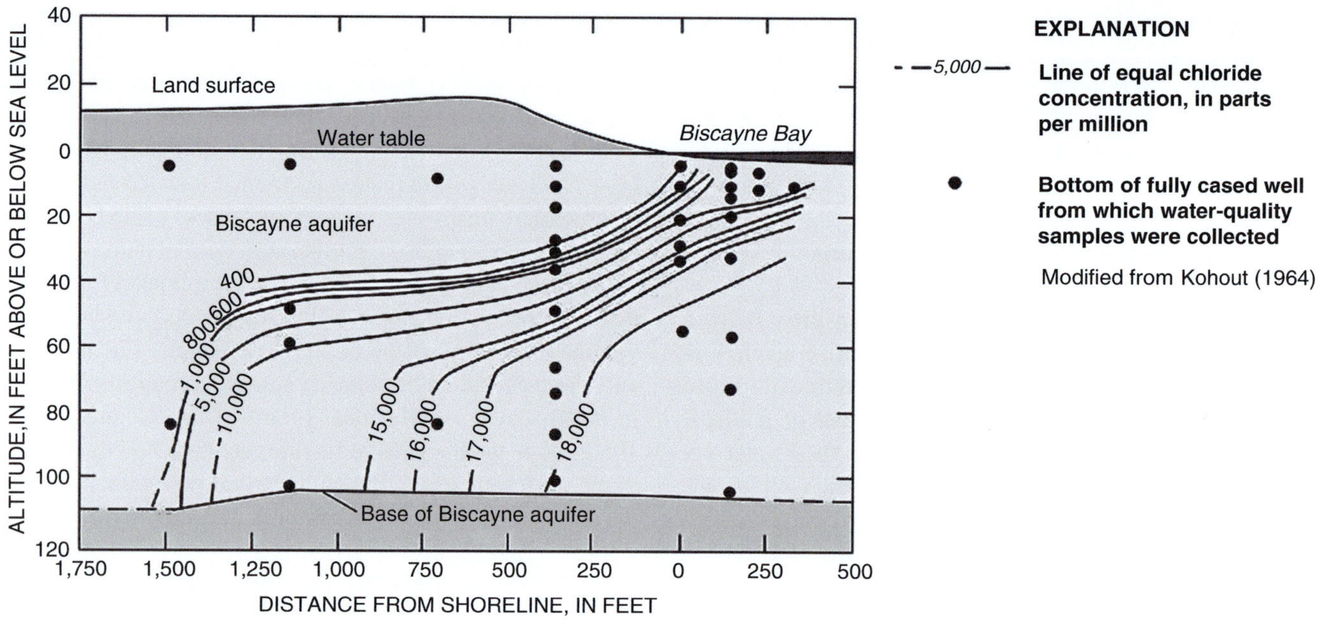


**Fig. 4.8** Cross section (perpendicular to coastline) through the Biscayne Aquifer in the Silver Bluff area of Miami showing the saline-water transition zone on November 2, 1954. Contours represent chloride isochlors in mg/l (Kohout 1964; Barlow 2003)

at 20 °C. Assuming that the densities are as presented above, Eq. 4.2 can be reduced to:

$$z = 40h \tag{4.2}$$

(Wentworth 1947; Freeze and Cherry 1979; Domenico and Schwartz 1997; Fetter 2001).

Equation 4.2 states that for every meter of freshwater (or freshwater head in a confined aquifer) above sea level in the freshwater lens, there should be 40 m of freshwater below sea level. This relationship creates a lens-shaped cross section of freshwater in an aquifer surrounded by sea water (such as on an island or peninsula) and a wedge-shaped cross section of freshwater in an aquifer with sea water on one side (such as on a continental coastal aquifer) (Fig. 4.7).

Equations 4.1 and 4.2 mathematically describe a sharp interface. Given that a sharp interface is improbable owing to dispersion and mixing, one must consider the position of the estimated saline-water interface as a surface that should be near the center of the freshwater/seawater transition zone.

This rule of thumb is useful for studying freshwater bodies using idealized geometries based on the simplifying assumptions that: (1) the boundary between freshwater and saltwater is sharp; (2) flow within the freshwater lens is purely horizontal; and (3) water in the seawater portion of the aquifer is static (i.e., Todd 1980; Fetter 2001; Vacher 1988a). In actuality, the boundary between freshwater and saltwater is not sharp, but is instead broad enough that it cannot be ignored in most areas.

An important corollary to the Ghyben-Herzberg principle is that, if the water table is drawn down around a pumping well within a freshwater lens, there is a corresponding rise in the freshwater/saltwater transition zone (Fig. 4.7), a process called upconing. Upconing causes vertical and lateral saltwater intrusion into freshwater portions of the aquifer and is an important source of freshwater contamination in Florida and other coastal areas of the world.

Many early investigations of freshwater lenses coupled analysis of the water-table configuration with some form of the Ghyben-Herzberg relationship to give general descriptions of the shape of the freshwater lens based on different sets of variables, such as recharge, permeability, island (and lens) width, and sea level (e.g., Fetter 1972; Glover 1964; Vacher 1978; Van der Veer 1977). Application of the Ghyben-Herzberg principle yields results that compare relatively favorably to those produced by more advanced analytical treatments except near the coastline, where flow has a more pronounced vertical component and the presence of an outflow face offsets the freshwater-saltwater interface somewhat (Vacher 1988b).

#### 4.2.7.4 Geometry of Freshwater Lens

In general, the maximum thickness of the freshwater lens is proportional to island or landmass size and recharge, and inversely proportional to hydraulic conductivity. However, in karstic rocks, hydraulic conductivity has the potential to vary by several orders of magnitude (Vacher and Mylroie 2002) and can, therefore, account for significant variability in lens geometry.

In limestone undergoing early karstification, dissolution and diagenesis tend to increase permeability over time by reconfiguration of porosity through selective dissolution, conversion of aragonite to calcite, and recrystallization

(Hanshaw and Back 1979; Vacher and Mylroie 2002). Freshwater lenses therefore tend to be thicker in younger, less permeable limestones than in older, more permeable limestones.

If recharge and hydraulic conductivity vary spatially within a landmass, the shape of the freshwater lens is modified accordingly (Fratesi 2013). Differences in permeability, for instance a high-permeability layer beneath a lower permeability layer, might cause the lens to be distorted or truncated (i.e., atoll islands (Buddemeier and Holladay 1977) and the Florida Keys (Vacher et al. 1992)). Lateral differences in permeability, evapotranspiration or recharge create an asymmetrical, irregular lens, or fragmented (i.e., lateral permeability changes in Bermuda (Vacher 1978), evaporative lakes in the Bahamas (Vacher and Wallis 1992), and possibly the Florida Keys). In addition to geographic variations in recharge, temporal variations are also observed in many areas where wet and dry seasonal patterns in rainfall are commonly seen, such as in the tropics and subtropics (see for example Buddemeier and Oberdorfer 1997; Halley et al. 1997; Humphrey 1997). Groundwater extraction also distorts the lens.

Often, such broad-scale variations can be treated analytically using simplified geometries, and such models have been crucial for interpreting field data and building conceptual models. However, additional complexities such as irregular island geometries, temporal variations in recharge, tidal signals, and density-driven flow usually require the use of numerical modeling to fully characterize the configuration of the freshwater body (i.e., Ayers and Vacher 1983; Essaid 1986; Hunt 1979; Ghassemi et al. 1993, 1996; Gingerich 1992; Langevin 2003; Contractor 1983; Oberdorfer et al. 1990; Underwood et al. 1992).

#### 4.2.7.5 Approximating the Freshwater-Saltwater Transition Zone

Although the theoretical treatments discussed to this point have dealt with the freshwater/saltwater boundary as a sharp interface, the mixing zone can be thick (Fig. 4.8), such that most of the meteoric water in the aquifer is mixed with seawater to form brackish water. In this case, the mixing zone may extend almost to the water table. Near the coastline, or in highly mixed lenses, seawater or brackish water may occupy the entire water column. Such appears to be the case in the LFA of southern Florida (Sect. 4.5.4).

At the transition zone, freshwater and saltwater flow parallel to one another toward the coast with the freshwater/saltwater mixing zone separating them. Mixing may be caused by:

1. Dispersion due to spatial variations in aquifer hydraulic conductivity;
2. Diffusion caused by chemical gradients within the mixing zone;
3. Migration of the water masses due to tidal oscillations;
4. Seasonal fluctuations in recharge or discharge; or
5. Perturbations caused by groundwater extraction (Cooper 1964; Wentworth 1948).

In locales with relatively homogeneous aquifers, the tidal signal propagates laterally inland from the coast, and attenuates with distance from the coast, causing a corresponding thinning of the mixing zone with distance inland (Vacher 1978; Whitaker 1992). The mixing zone is thicker in more transmissive karstified limestones, largely because of the efficient propagation of the tidal signal through these units (Cooper 1964; Whitaker and Smart 1997). In coastal locations with dual aquifer systems, there is no lateral trend in mixing zone thickness because tidal signal is transmitted relatively efficiently through the more transmissive limestones and vertically through the younger limestones to the lens (Buddemeier and Oberdorfer 1997; Hunt and Peterson 1980; Wheatcraft and Buddemeier 1981). In these situations, the tidal efficiency (ratio of tidal amplitude in wells to that in ocean) increases with depth and tidal lag (time difference between ocean and well tidal signals) decreases with depth (Underwood et al. 1992).

Many studies have characterized the freshwater portion of the aquifer as being bounded by an interface defined by drinking-water standards (<600 mg/L $Cl^-$ or < 250 mg/L $Cl^-$ [World Health Organization 1971, 2010, respectively]). There is no discrete boundary to the mixing zone; salinity theoretically varies with depth across this zone (Bear and Todd 1960; Cooper 1959; Kohout 1960). However, salinity profiles often deviate from this pattern, exhibiting deflections at hydrogeologic or lithologic boundaries (Buddemeier and Oberdorfer 1997; Harris 1967).

Below the mixing zone, the groundwater in the deeper, saltwater portion of the aquifer is generally not static. In southern Florida, Cooper (1959) and Kohout (1965) described landward flow of saltwater below the freshwater zone. This landward flow is counter to the seaward flow of fresh and brackish water (Fig. 4.9) near the freshwater/saltwater transition zone. Cooper (1959) attributed this flow to the fact that saltwater is entrained into the mixing zone where it moves out to sea. Saltwater, then, flows landward below the transition zone to replace the seaward flow near the mixing zone, creating a forced circulation system. Kohout (1965) cited geothermal heating and convection as a cause of this circulation in the LFA. Whitaker and Smart (1990) identified large-scale saltwater circulation in the Great Bahama Bank and attributed it to sea-level head differences and to density-driven flow of seawater concentrated by evaporation. Beddows et al. (2007) found widespread and persistent inflow of saltwater through cave systems in the Yucatan caused by gradual cooling of warmer water flowing inland from the coast.

Movement of saltwater through the deeper portions of the aquifer impacts the shallow flow systems. For instance, in

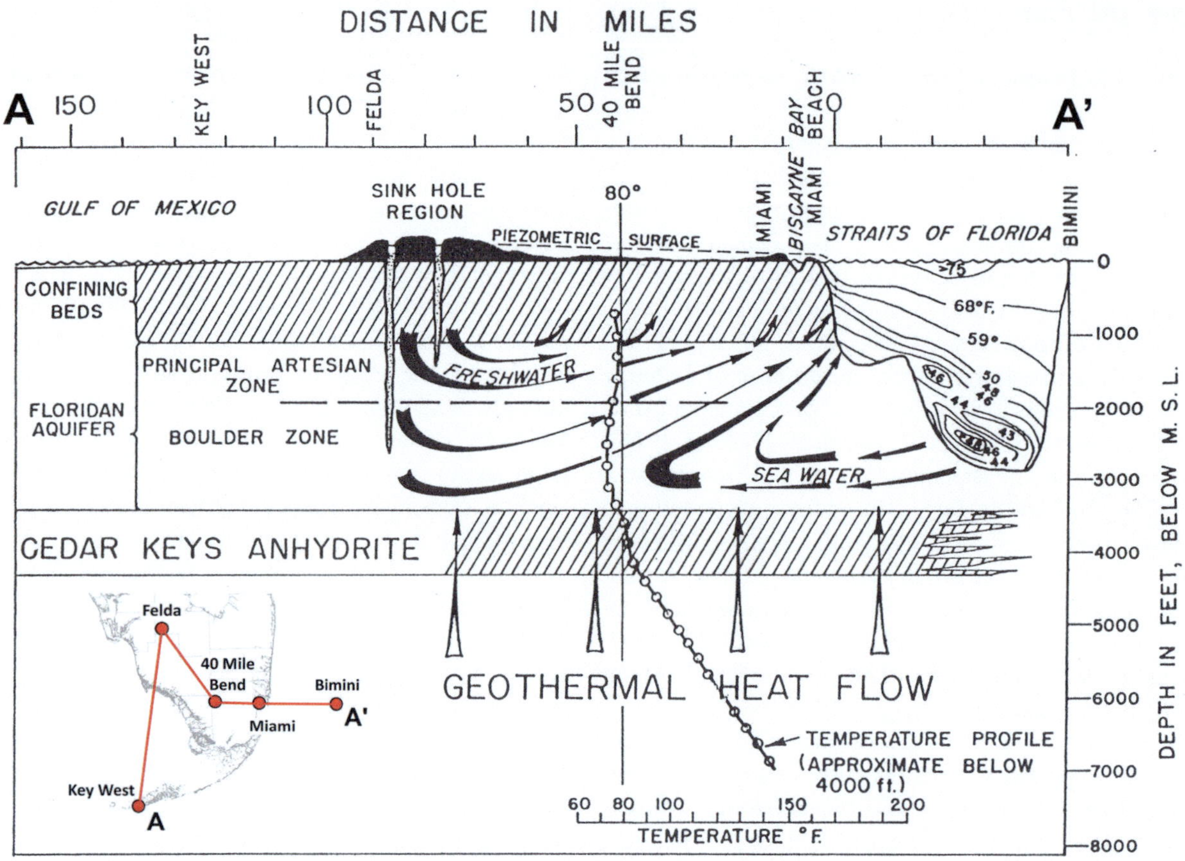


**Fig. 4.9** Hypothetical flow regime modified from Kohout (1967, 1975) and Kohout et al. (1977) for the Floridan aquifer in southern Florida. Note that the authors equate the "boulder zone" to the lower Floridan aquifer and that circulation is shown to cross the middle confining unit (see Sect. 4.5 for a discussion of these components of the Floridan aquifer). Note also the sinkholes drawn into the Boulder Zone

the Yucatan Peninsula, the freshwater lens is 40% thinner than predicted by Ghyben-Herzberg theory. Moore et al. (1992) attribute this to convectional flow of seawater. In southern Florida, deep flow paths bring brackish water and saltwater into contact with gypsum and anhydrite beds, adding sulfate compounds to the geochemical processes occurring in the shallow system (Copeland et al. 2011).

#### 4.2.7.6 Saltwater Intrusion

The shape of the freshwater body and the position of the freshwater/saltwater boundary are the result of a balance of the flux of freshwater towards the sea and aquifer-characteristic variables, such as hydraulic conductivity. Changing any of these variables can substantially change the configuration of the freshwater lens or wedge. The most common anthropogenic cause of saline-water intrusion is reduction in the freshwater flux or head by groundwater withdrawals. At the aquifer scale, decreased net recharge also causes the volume of the freshwater lens to decline, allowing the freshwater/saltwater boundary to "intrude" upward and/or inland. Seawater intrusion may also be caused by effectively fragmenting or decreasing the size of the land mass and freshwater lens by building seawater canals or other major land excavations (Bear et al. 1999; Langevin et al. 1998).

In areas where freshwater vertically overlies saltwater, drawdown of the water table or potentiometric surface by pumping wells or other means can cause the freshwater/seawater boundary to migrate vertically upward in general accordance with the Ghyben-Herzberg principle. This phenomenon is called upconing in reference to the fact that the up-drawn interface mirrors (with a 40:1 exaggeration) the drawdown cone around the well (Schmorak and Mercado 1969).

Reclamation of well fields contaminated with seawater is very difficult. It is possible to restore freshwater recharge to the aquifer; however, it takes only a small amount of seawater to make freshwater non-potable. Considering that virtually all of the residual seawater must be flushed from the aquifer to reclaim a saltwater-contaminated wellfield and the time it takes to do that, the only practical solution to seawater intrusion is prevention (Bear et al. 1999).

The largest saltwater intrusion reclamation effort in Florida was for the Biscayne Aquifer. In the 1940s and early

1950s, water from wells near the Atlantic coast began to show increasing salinity. The U.S. Geological Survey began studying this saline-water intrusion issue, and monitoring and study continue (Renken et al. 2005). When it was discovered that the intrusion was a result of lowering of the water table (the Biscayne Aquifer is unconfined) by constructing drainage canals through the Atlantic Coastal Ridge, drainage of water levels in the Everglades, and groundwater extraction in the populous areas of Miami, steps were taken to maintain water levels by construction of salinity dams on the canals. These dams raise the water levels in the canals and Everglades and help prevent saltwater intrusion within the Biscayne. The dams have been effective in dealing with the saltwater intrusion problem and partially restoring the freshwater lens in the Miami-Dade area.

The experiences in Florida with reversing saltwater intrusion are instructive with respect to groundwater quality in Florida's aquifers over geologic time. Sea level fluctuated more than 120 m as much as twenty times during the glacial events of the Pliocene and Pleistocene. Each high sea stand resulted in saltwater intrusion and each low sea stand led to flushing by freshwater. In addition, the freshwater/saltwater transition zone migrated up and down with each sea-level excursion. In Chap. 5, we will discuss the consequences of these excursions in terms of groundwater quality and karst development.

#### 4.2.7.7 Formation of Florida Karst Is Dynamic Over Time

This chapter discusses the present-day hydrogeology of Florida. One should not assume that present-day conditions reflect hydrogeologic conditions and karst development in the past. Sea levels, climate, and erosion and depositional patterns have changed dramatically in the past and will continue to change. This concept is a major "take-away" for this chapter. Changes in the geometry of the freshwater lens and freshwater/saltwater transition zone with time have created complex patterns of aquifer development, karstification, and water quality.

Depending on sea-level position, development of karst landforms near the water table and along the freshwater/saltwater transition zone varies in intensity and location (Gulley et al. 2013; Gulley and Florea 2016). Just in the last few thousand years, sea level has risen, and along with this change in sea level, groundwater levels and the thicknesses of Florida's aquifers have evolved. These changes continue as a result of anthropogenically induced and natural climate change. See Hine et al. (2016) for an excellent discussion of past and future sea levels and the effects of climate change in Florida.

The climate has changed dramatically in the last few million years. Some of the effects of climate changes on karst development are discussed in the following chapters. Suffice it to say here that Florida climate has fluctuated from dry and cool to warm and wet from the Miocene to the present. The time interval near the end of the Miocene and extending into the Pliocene was a period of intense weathering and soil formation (Carr and Alverson 1959). In the late Pleistocene and early Holocene, climate changed less dramatically but was nevertheless sufficient enough to affect karst development. For example, Watts and Stuiver (1980), Watts et al. (1992), and Watts and Hansen (1994) documented changes in climate during the late Pleistocene and Holocene based on pollen content in sediment from several Florida sinkhole lakes. The resulting changes in rainfall, meteoric-water drainage, and position of the freshwater lens affected development of karst, including speleogenesis and sinkhole development.

Variations in sea level and climate over the last few million years have resulted in significant changes to the geomorphology and sediment distribution in Florida. For example, the episode of intense weathering at the Mio-Pliocene boundary resulted in stripping of the Miocene Hawthorn Group sediments off the higher portions of the Ocala Platform (Scott 1981), including the karst plain provinces of the Ocala Karst District (Fig. 3.13). This episode of intense weathering resulted in extensive sinkhole development as demonstrated by the fossil content of sinkhole fill materials (McFadden 1997). Events such as this provide time frames for karst development and allow for estimates of the rates of carbonate-rock dissolution.

In summary, one should extrapolate today's climate and on-going karst-related processes into the past with care. Changes in sea level and climate continue to cause variations in patterns of karst development and exposure of carbonate-rich strata. Erosion and deposition that resulted from these events affected the probability of karst development and created limitations as to the time available for development of the eogenetic karst features present in Florida today.

## 4.3 Surficial Aquifer System

### 4.3.1 Aquifer Distribution

The SAS is the regional unconfined, water-table aquifer system in Florida. By definition (Southeastern Geological Society 1986; Table 3.3), the SAS can only exist where the intermediate confining unit (ICU) is present and where the ICU serves to separate the SAS and underlying UFA (Fig. 4.10). Therefore, in areas where the UFA is unconfined and exists under water-table conditions, the SAS does not exist, by definition. If the potentiometric surface of the unconfined UFA lies within the surficial sands that lie above the limestone, the water-saturated portion of the sand mantle is considered to be part of the UFA.

Throughout its extent, the SAS occurs within the near-surface, Plio-Pleistocene marine terrace sand deposits and equivalent shell and limestone in southern Florida.

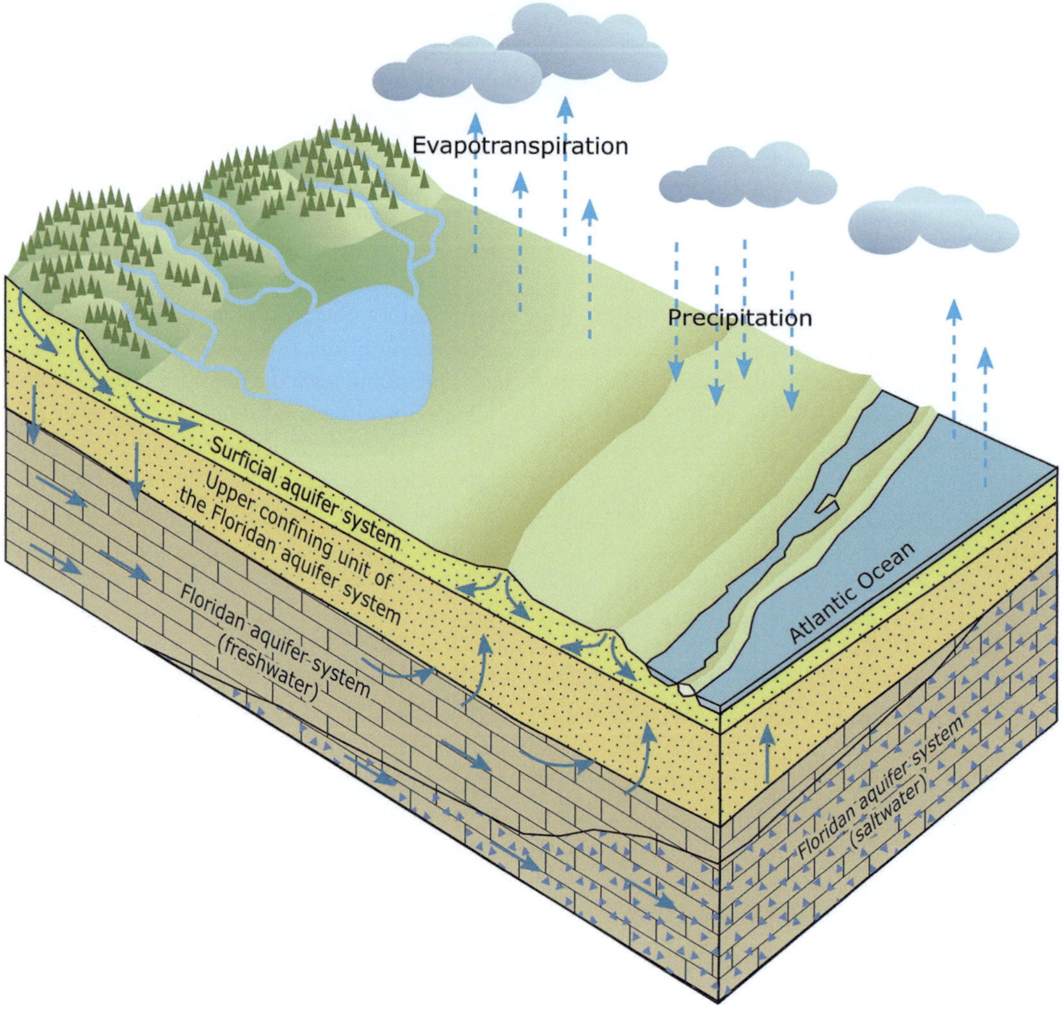


**Fig. 4.10** Flow systems in the Surficial, Intermediate, and Floridan aquifer systems

As shown in Fig. 4.6, the SAS only serves as a major water-supply source in the Sand and Gravel Aquifer of Escambia and Santa Rosa counties in the western panhandle of Florida and in Dade and portions of Broward counties in southeastern Florida, where the Biscayne Aquifer is the primary water-supply source. Elsewhere, the aquifer is commonly utilized for small-volume, domestic and public supply, and for irrigation.

Water quality in the Sand and Gravel and Biscayne aquifers is generally excellent; however, color and turbidity are common issues for treatment. Where the SAS is shallow and consists of sand, color, turbidity, iron, odor and other issues limit its utility without treatment.

Throughout Florida, color, turbidity, odor, bacterial contamination, and other issues pose a problem for use of SAS water. This is because the aquifers of the SAS are unconfined, near the land surface, and often overlain by sources of human and animal wastes. Near the coast, salt water intrusion is an important water-quality issue, especially in the Biscayne Aquifer.

Since there is no potential for karst development in the Sand and Gravel Aquifer, it is not discussed in the sections that follow. See Marsh (1966), Trapp (1975), Hayes and Barr (1983), and Fernald and Purdum (1998) for more information on the Sand and Gravel Aquifer of northwestern Florida.

## 4.3.2 Biscayne Aquifer

### 4.3.2.1 Location and Composition

The Biscayne Aquifer is developed in the Pleistocene Miami Limestone, Fort Thompson Formation, Anastasia Formation, and Key Largo Limestone in southeastern Florida (Fig. 4.6).

The Miami Limestone, which is calcitic with minor aragonite (Bock et al. 1969), includes two main facies: the oolitic facies (Figs. 2.42 and 4.11) that constitutes the Atlantic Coastal Ridge (Fig. 4.12a) and a bryozoan facies (Hoffmeister et al. 1967) that extends westward into the Everglades. The tidal, oolitic facies consists of a series of Pleistocene carbonate sand shoals separated by tidal passes

**Fig. 4.11** Exposure of the cross-bedded, oolitic facies of the Miami Limestone at Sunset Circle (25.7059°N, 80.2610°W) in Coral Gables, Dade County

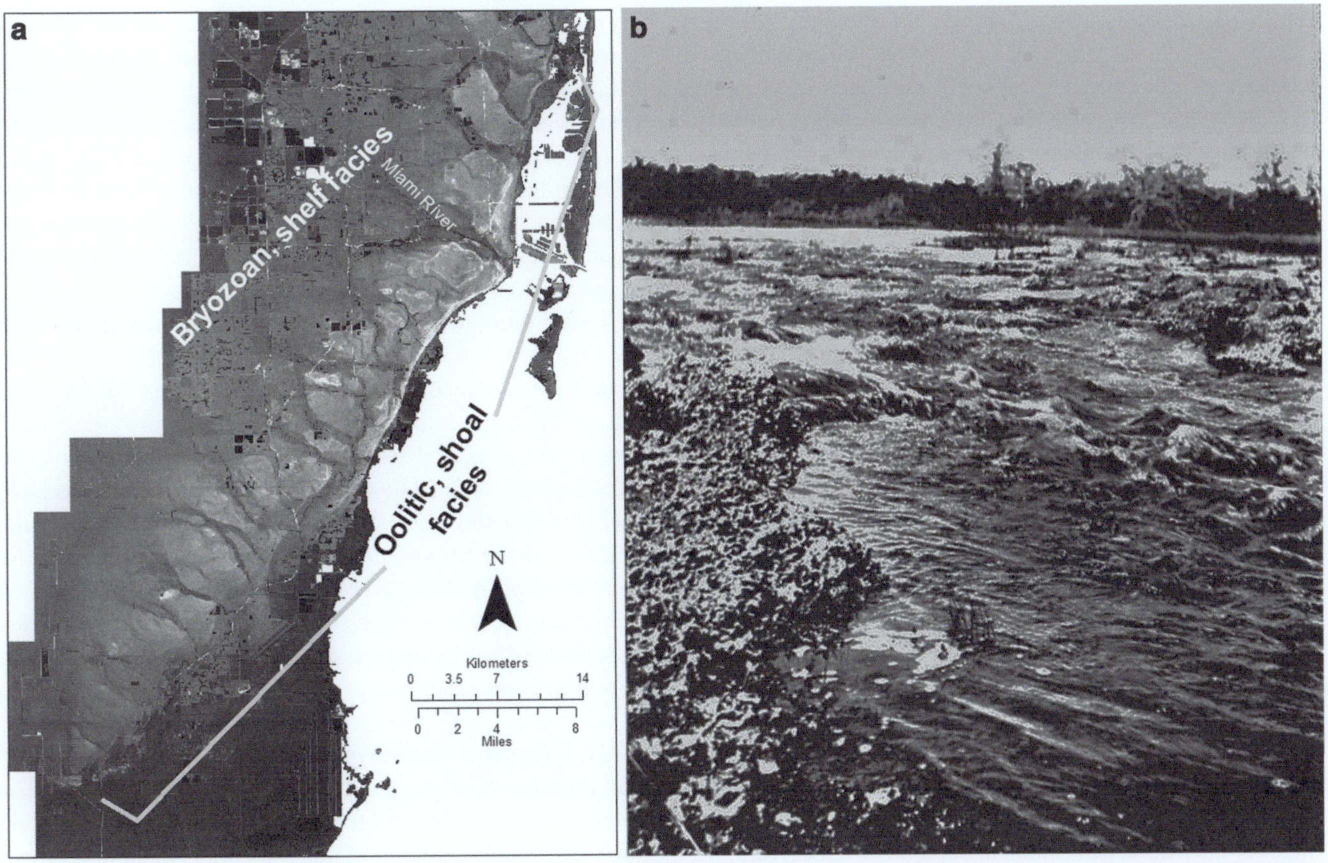


**Fig. 4.12** (**a**) Digital elevation map of coastal Dade and Broward counties showing the oolite shoals and tidal passes and portions of the bryozoan facies outcrop belt of the Pleistocene Miami Limestone. (**b**) Rapids on the Miami River, Dade County, in 1896. The limestone sill that caused these rapids retarded flow from the interior to the coast. (Image by an unknown photographer; from the Florida Memories Project, State Archives of Florida Image No38907)

(Fig. 4.12a) nearest the edge of the continental shelf. These shoals formed the high areas of the Atlantic Coastal Ridge. West of the oolite shoals, which are typically grainstones, the Miami Limestone transitions to a limestone facies consisting of abundant bryozoa in a packstone fabric. Within the framework of the two main facies defined by Hoffmeister et al. (1967), Cunningham et al. (2006) recognized 17 individual facies.

The underlying Fort Thompson is composed of interbedded sand, shell, and freshwater and marine limestone. The Key Largo consists of Pleistocene reef facies material, and the Anastasia consists of sand and shell beach-ridge and shallow shelf sediments. Of these, the Fort Thompson and Key Largo are the most permeable, with the Fort Thompson being the most utilized aquifer horizon (Miller 1990).

The oolitic shoals of the Miami Limestone form the Atlantic Coastal Ridge (Fig. 4.12a). This geomorphic feature is part of a beach-ridge and shoal complex that extends in a more-or-less continuous belt that parallels the Atlantic coast from southern Dade County almost to the Georgia border. In northern Broward County, the Miami Limestone interfingers with the Anastasia Formation, which consists of mixtures of quartz sand and shell with beds of cemented shell (coquinoid limestone lenses) and coquina, a sorted, clastic limestone composed of wave abraded and transported shell fragments. The coastal ridge complex underlain by the Anastasia Formation forms another geomorphic province (Williams et al. in prep.), the Atlantic Coastal Complex Province of the Barrier Island District (Fig. 3.15). The sand and shell beds of the Anastasia serve as an aquifer and are discussed in Sect. 4.3.3. In contrast to the shoals and tidal passes of the Atlantic Coastal Ridge, the Atlantic Coastal Complex consists of complex arrays of beach ridges and inter-ridge swales.

Because of relatively low land elevations and the presence of high hydraulic conductivities, the Biscayne Aquifer is more-or-less limited in extent to the oolitic limestone facies of the Miami Limestone and underlying Fort Thompson Formation under the Atlantic Coastal Ridge within Dade County and the southern part of Broward County (Fig. 4.6; Fish 1988; Fish and Stewart 1991).

Although heavily modified in recent years, the Everglades drainage system originally drained southward from Lake Okeechobee to Florida Bay (McVoy et al. 2011). Flow was generally as a thin sheet of water moving over the bryozoan facies of the Miami Limestone in the southeastern Everglades and Pliocene limestones in the southwestern Everglades. Recent peat and marl deposits formed *in situ* within the Everglades and created small channels and depressions within the drainage system.

The Atlantic Coastal Ridge served as a barrier to surface water flow to the east. Several small streams (i.e., the Miami River, Fig. 4.12b) carried overflow eastward from the Everglades to Biscayne Bay through the relic tidal channels between the shoals. These low areas and streams between the higher shoals (Fig. 4.12a) are called "transverse glades" because many contained the same marsh vegetation as the Everglades. In addition, seepage through the Miami Limestone as a result of recharge on the Atlantic Coastal Ridge and in the Everglades to the west caused small springs and general groundwater discharge to the east, on the coastline and in Biscayne Bay (Shaler 1890; Matson and Sanford 1913; Kohout and Kolipinski 1967; Cressler 1993; Langevin 2003). Dissolution of the calcite and minor aragonite by meteoritic water and by seepage from the Everglades caused karst features to develop on the Atlantic Coastal Ridge. These included a few large (greater than 5 m diameter) solution pipes (locally termed sinkholes), many small (less than 2 m diameter) solution pipes, springs, small caves (Cressler 1993), and small weathering features (karren) on the upper surface of the limestone of the ridge.

#### 4.3.2.2 Hydraulic Function of the Atlantic Coastal Ridge in Southeastern Florida

Prior to development and initiation of saltwater intrusion as a result of canalization and groundwater extraction, the Atlantic Coastal Ridge served as a barrier to salt-water intrusion because it retarded coastward discharge of water from within the Everglades, resulting in an elevated head in the oolitic ridge of the Biscayne Aquifer inland. The head difference between the Everglades and freshwater lens under the Atlantic Coastal Ridge caused the salt-water transition zone to remain near the coast.

Today, numerous drainage canals have been cut through the Atlantic Coastal Ridge (Fig. 4.13). The loss of hydraulic head in the mound of water under the Atlantic Coastal Ridge caused by these drainage projects and groundwater extraction on the ridge led to significant salt-water intrusion. To reduce and/or arrest salt-water intrusion, salinity dams have been constructed near the eastern termini of the canals. Because of the importance of the Biscayne Aquifer to the population of the Miami-Dade area, salt-water intrusion has been extensively investigated and the results of these investigations constitute a large body of our knowledge of saline-water intrusion processes.

#### 4.3.2.3 Aquifer Recharge

The limestones of the Biscayne Aquifer contain many dispersed recharge pathways through intergranular pore spaces, burrows, solution pipes, and other openings in the limestone. Over time, through karstification, these openings have been extensively enlarged and modified (Cunningham et al. 2008; Cunningham and Sukop 2011). The result is a highly productive aquifer that can recharge quickly and efficiently under natural (undeveloped) conditions. Today, however, extensive paved surfaces in developed parts of the Atlantic Coastal Ridge allow rainwater to run off instead of infiltrating therefore substantially reducing recharge to the aquifer.

**Fig. 4.13** Drainage canal excavated through the oolitic Miami Limestone of the Atlantic Coastal Ridge (a.k.a. Cutler Ridge) south of Miami, Dade County. Photograph courtesy of the Florida Geological Survey

Fernald and Purdum (1998) illustrated the potential recharge for coastal ridge portion of the Biscayne Aquifer. The less developed portions of the Atlantic Coastal Ridge (the southern and western parts of the area) have estimated potential recharge in excess of 107 cm/year, while the urbanized areas have potential recharge between 61 and 107 cm/year. When compared with rainfall and ET, the recharge potential is huge and is probably normally in excess of the available rainfall. Nevertheless, the high transmissivity of the Biscayne clearly indicates potential for through-put of water in the aquifer.

#### 4.3.2.4 Aquifer Potentials and Flow Systems

The transmissivities and yields to wells in the Biscayne Aquifer are highest in portions of the aquifer that are dominated by burrow porosity and karstic zones. Klein and Causaras (1982) reported yields to large wells as high as $3 \times 10^4$ L/min. in the karstic limestone of the Biscayne. Parker et al. (1955) reported that the transmissivities of the Biscayne ranged from $5 \times 10^4$ to $2 \times 10^5$ m$^2$/day from aquifer tests.

Klein and Hull (1978) discussed the spatial variations in transmissivity. To the north, where the calcium-carbonate facies of the Miami Limestone transitions to the sandy, shelly facies typical of the Anastasia Formation, the aquifer is thicker, but less transmissive as a result of the quartz sand content of the Anastasia facies. In southeast Broward County the transmissivity was reported to be about $4 \times 10^4$ m$^2$/day, while in northeast Broward County it declines to $4 \times 10^3$ m$^2$/day.

Because of the drainage canals and salinity dams, aquifer potentials in the Biscayne are managed to some extent (Miller 1990). In the past, the potentials of the water-table aquifer reflected local topography and the head differences between the Everglades and coastal waters. Ground- and surface-water flow was toward the coast with groundwater discharge to local streams that drained the Everglades, a few springs, and seepage. Shoals and rapids (Fig. 4.12b) in the streams held the groundwater levels above sea level and prevented salt-water intrusion. McVoy et al. (2011) discuss these features as controls on water levels in the Everglades (Fig. 4.12b).

### 4.3.3 Surficial Aquifer (Undifferentiated)

In addition to the larger, well-known, named aquifers, the SAS includes several water-producing aquifers of local to sub-regional extent (Fig. 4.6). These include strata that can effectively produce water to wells within the Atlantic Coastal Complex Province, a shallow limestone aquifer in Collier and southern Lee County, and water-bearing marine terrace sand deposits throughout the state. These aquifers range in age from Pliocene to Pleistocene.

Fig. 4.14 Sandy, shelly Anastasia Formation with partly cemented shell lens (arrow). Shell mine (29.1322°N, 81.0981°W), Volusia County. (Image courtesy of the Florida Geological Survey)

#### 4.3.3.1 Location and Composition

The undifferentiated SAS is locally exploited for water supply in several areas of the state. These include: (1) along the crest of the Atlantic Coastal Complex Province; (2) in southwest Florida, near Naples; and (3) locations throughout the state that support isolated domestic and irrigation wells.

**Surficial Aquifer System under the Atlantic Coastal Complex** The Atlantic Coastal Complex Province extends from northern Broward County north, nearly to the Georgia border (Fig. 3.15). It parallels the Atlantic coast line and generally extends from the coast 8–10 km westward.

Sediments within the Atlantic Coastal Complex include a ridge of quartz sand and shell beds (Fig. 4.14) that provides potable water to coastal municipalities such as St. Augustine in St. Johns County. These water-producing strata are part of the undifferentiated SAS. Other than SAS, the local aquifer in this area does not have a formal name, but is informally termed the "Atlantic Coastal Ridge aquifer" or the "Interstate aquifer" where it parallels I-95.

Sediments of the SAS in the coastal ridge consist of a complex of beach ridge, dune, and marine shoal sediments near the coast and crest of the ridge and estuarine/lagoonal and washover fan sediments on the landward (western) flank of the ridge. These Pleistocene sediments are predominantly sand with variable amounts of shell (Fig. 4.15) and nodules to small lenses of limestone (thoroughly cemented shell and sand).

Coquina is a notable deposit found at a number of locations in the Anastasia Formation (Figs. 4.15 and 4.16). Since the Anastasia coquina was deposited in a high energy, beach environment, there is often little or no matrix between the shell fragments. Where matrix is present, it consists of interstitial, fine quartz sand that is typical of the beaches on the coast today. The type section of the Anastasia Formation consists of this coquina deposit exposed in a small, seventeenth Century quarry on Anastasia Island, near St. Augustine at 29.879°N, 81.285°W.

The Castillo de San Marcos (29.898°N, 81.311°W) at St. Augustine was built of this coquina. The walls of the fort provide an interesting experiment in weathering of limestone in Florida. Construction of the fort began in 1672, so the walls of the fort have been exposed to Florida weather for over 300 years. Figure 4.17 illustrates a portion of the fort's exposed, curtain wall today. The blocks of coquina, which originally consisted of loosely cemented shell composed of low-Mg calcite and minor aragonite, have hardened over the years as they have been exposed to repeated wetting and drying. In the figure, the block of coquina has receded from the mortar approximately 1 cm, primarily by disaggregation and spalling of the shell with minor dissolution. As a result, the original bedding of the coquina has become emphasized (Figs. 4.15 and 4.17).

Locally, permeable shell beds are abundant in the undifferentiated surficial aquifer, and the shelly strata are sufficiently transmissive that good quality water can be produced to low-yield public supply wells. These shell beds have been developed for public water supply in numerous locations along the Atlantic Coastal Ridge, including supplies for St. Augustine (Spechler and Hampson 1984) and similar sized communities.

**Fig. 4.15** Cross bedded Anastasia Formation coquina exposed in Palm Beach County. The diagonal feature in the lower right is a paleosol. Location: 26.7414°N, 80.0410°W. Image courtesy of the Florida Geological Survey

**Fig. 4.16** Coquina from the Pleistocene Anastasia Formation near Washington Oaks State Park (29.6363°N, 81.1991°W), Flagler County

**Fig. 4.17** Detail of a block of coquina exposed in the exterior curtain wall of Castillo de San Marcos. The wall has receded from the mortar approximately 1 cm. Photograph courtesy of the Florida Geological Survey

**Tamiami Aquifer** The informal Tamiami aquifer of southwest Florida is a portion of the SAS that is used to supply water to communities in Collier County.

The Pliocene Tamiami Formation is composed of a complex mixture of interbedded sand, shelly sand, clay, and limestone (Figs. 4.18 and 4.19). It is found throughout southwestern Florida. The Tamiami Formation may form part of the SAS or IAS depending on the positions of clay-rich aquitards within the formation. Where it is near the land surface and unconfined, the Tamiami Formation is considered part of the SAS. Water production from the Tamiami is in shell beds and fractured limestone.

**Undifferentiated Surficial Aquifer System** While the Atlantic Coastal Ridge and Tamiami aquifers supply water to moderate-sized communities from carbonate and sandy shell strata, the remainder of the undifferentiated SAS typically does not. Where the SAS exists throughout the remainder of the state it typically consists of quartz sand with local beds of clay and shell (Fig. 4.6). Because the undifferentiated, surficial aquifer is predominantly composed of fine, quartz sand with ferric hydroxide and other interstitial materials, the aquifer is only capable of supplying small quantities of water to wells. The water is often slightly acidic, has high color and iron content, and often has unpleasant odors (Berndt and Katz 1992; Upchurch 1992). As a result, the sand aquifer is primarily used for low-yield domestic wells and irrigation.

**Fig. 4.18** Shelly limestone of the Pliocene Tamiami Formation in Collier County

Figure 4.20 illustrates a portion of the sandy SAS that has been dewatered for mining. Note the brown, ferric hydroxide soil zones, which represent former positions of the water table. The abundant iron staining in this stratum is typical of the sandy SAS.

#### 4.3.3.2 Aquifer Recharge

Estimated recharge to the Tamiami and Atlantic Coast Ridge aquifers in south Florida is local and dispersed throughout the extents of the aquifers. There is no appreciable inflow from outside of the areas where the water is utilized. Fernald and Purdum (1998) reported that recharge to the Tamiami aquifer ranged from less than 109 to over 142 cm/year. In St. Lucie and Martin counties, they reported that recharge was from less than 71 to over 112 cm/year. Statewide Adams et al. (undated) estimated annual recharge to the sandy, undifferentiated SAS to range from 18 to 23 cm/year.

#### 4.3.3.3 Aquifer Potentials and Flow Systems

As noted above, the SAS is generally characterized by local flow systems, meaning that groundwater typically flows short distances from the recharge site through the aquifer system to nearby discharge sites such as streams. Discharge from the SAS is either (1) by leakage into the underlying strata of the IAS/ICU and ultimately to the FAS or (2) to local lakes and streams by shallow flow systems as visualized by Tóth (1962, 1963). Note that this Tóthian flow is ultimately to the coasts, either by way of surface or groundwater.

**Fig. 4.19** Shelly, sandy facies of the Tamiami Formation near Naples, Collier County. The undercut area is a seepage face where a higher permeability stratum is allowing flow of groundwater to the mine cut

**Fig. 4.20** Exposure of the reworked, Pliocene Cypresshead Formation showing iron-rich soil zones caused by water-table fluctuations. Grandin Mine (29.2175°N, 81.9392°W), Putnam County

According to Miller (1990, 1997), transmissivities of the undifferentiated SAS are highly variable. Most reported transmissivities range from 93 to 930 $m^2$/day. However, transmissivities of 2300–4600 $m^2$/day have been reported where the aquifer is thick and permeable (Miller 1990, 1997). In contrast, the karstic Biscayne Aquifer has transmissivities greater than $1.5 \times 10^5$ $m^2$/day (Miller 1986).

Spechler and Hampson (1984) reported that the shell aquifer at the Tillman Ridge Wellfield, which supplies St. Augustine in St. Johns County, had transmissivities of 600–650 $m^2$/day in one aquifer performance test and 120–2400 $m^2$/day in another. The unusually high transmissivity was in a highly permeable shell deposit that extends 30 m below the land surface. Frazee and McClaugherty (1979) reported that the transmissivity of the sand and shell aquifer in northeastern Florida is 271 $m^2$/day, and the Anastasia (shell) aquifer was 101 $m^2$/day.

### 4.3.4 Karst Development in Strata of the Surficial Aquifer System

For most of the state, the only hydrogeologic involvement of the SAS in karst development is to deliver slightly acidic water to underlying, carbonate-rich strata, where karst development may occur. Where shell or limestone beds exist in the undifferentiated SAS, slow dissolution of the carbonate material may occur depending on the chemistry of the groundwater and hydraulic conductivity of the aquifer strata.

Some areas where shell and/or limestone are known to interact with surficial-aquifer water include the Biscayne, Atlantic Coastal Complex, and Tamiami aquifer areas and other regions where shallow limestone and shell beds are common. In these areas, the mildly acidic groundwater may cause localized, very slow dissolution of the strata to form shallow depressions and intricately sculpted rock surfaces. Small-scale (less than 1 m) karst features are common in limestone of the Biscayne Aquifer. Large, sinkhole-like depressions are rare, but solution pipes that are a meter or less in diameter and up to about 3 m deep are common. Where shell beds represent the host sediments for dissolution, the dissolution processes are: (1) slow, requiring thousands of years to form subsidence features; (2) limited by the degree of chemical saturation of the groundwater with respect to the calcite and aragonite of the shells; and (3) limited by depth below land surface. For the most part, shell beds only undergo dissolution when they are near the land surface, and dissolution is limited to the uppermost portion of the shell bed.

Saline-water transition zones may reflect exceptions to the limitations on development of dissolution in shell beds, however. Mixing of freshwater and saline water that is chemically equilibrated (chemically saturated) with respect to calcite or other carbonate mineral may create a zone where the water mixture is chemically undersaturated with respect to calcite. Where this water comes in contact with shell material, the shell or other calcium-carbonate material may undergo dissolution.

The materials of the SAS also constitute cover over underlying karst features developed in the carbonate-rich strata of the SAS, IAS/ICU, and FAS. Cover is the term used to characterize the overburden (sediments and soils that overly a karstic limestone or similar material). As such, the unconsolidated sediments of the SAS may migrate or settle into voids and/or depressions in the upper surfaces of carbonate-rich strata and void spaces within limestone or dolostone strata. When this happens, a new direct path for SAS water to enter the underlying carbonate sediment develops.

Direct entry of contaminated SAS water into the underlying carbonate aquifer may exacerbate dissolution and introduce contaminated water into the underlying carbonate aquifer (Chen Thomas and Upchurch 1995a, b; Chen Jones and Upchurch 1995). Note that collapse of cover sediment into a void in the underlying carbonate sediment or rock is a substantially different process from dissolution of shell material discussed in the previous paragraph.

## 4.4 Intermediate Aquifer System and Intermediate Confining Unit

The IAS/ICU is a complex system that remains controversial in terms of its potential to supply water to wells and its role in the development of Florida karst. The following is quoted directly from the Southeastern Geological Society (1986, p. 5) and illustrates the complexity of defining the roles of the unit as both a confining unit and source of water.

> *" .... intermediate aquifer system or intermediate confining unit – includes all rocks that lie between and collectively retard the exchange of water between the overlying surficial aquifer system and the underlying Floridan aquifer system. These rocks in general consist of fine grained clastic deposits interlayered with carbonate strata belonging to all or parts of the Miocene and younger Series. In places poorly-yielding to non-water-yielding strata mainly occur and there the term "intermediate confining unit" applies. In other places, one or more low to moderate-yielding aquifers may be interlayered with relatively impermeable confining beds; there the term "intermediate aquifer system" applies. The aquifers within this system contain water under confined conditions.*
> *The top of the intermediate aquifer system or the intermediate confining unit coincides with the base of the surficial aquifer system. The base of the intermediate aquifer is at the top of the vertically persistent permeable carbonate section that comprises the Floridan aquifer system, or, in other words, that place in the section where clastic layers of significant thickness are absent and permeable carbonate rocks are dominant. Where the upper layers of the persistent carbonate section are of low permeability, they are part of either the intermediate aquifer system or intermediate confining unit, as applicable to the area. The term "intermediate aquifer system" replaces previously used names such as "secondary artesian aquifer(s)" and "shallow artesian aquifer(s)."*

The IAS/ICU, therefore, consists, for the most part, of the Miocene Hawthorn Group. Where the carbonate rocks of the lower Hawthorn Group, including, for example, the Tampa Member of the Arcadia Formation, are in contact with the underlying carbonate rocks of the FAS without significant hydraulic separation, these Hawthorn Group carbonates are considered to be part of the FAS. Similarly, where sandy sediments at the top of the Hawthorn Group (e.g., the Bone Valley Member of the Peace River Formation) are in contact with and have hydraulic connection with the overlying Plio-Pleistocene sediments of the SAS, Hawthorn Group sediments are considered to be within the SAS.

For decades the IAS/ICU was ignored as a viable hydrogeologic unit capable of transporting water laterally and of supplying water to wells. The unit had been consistently treated as a barrier to vertical recharge of the underlying FAS in groundwater flow modeling. In fact, many quasi three-dimensional groundwater flow models in Florida have over-simplified the unit by assigning leakance values between SAS and FAS layers, essentially generalizing the conceptualizations of the IAS/ICU upon which these groundwater flow models were based.

Similarly, the clay strata of the IAS/ICU have been considered as barriers to sinkhole development because of:

1. Their cohesion and inability to disaggregate and ravel into underlying void spaces;
2. Their ability to span openings in the underlying limestone because of their cohesive properties; and
3. Their ability to prevent water and sediment movement from the sandy sediments overlying the clays into the underlying limestone.

Today, a different view of the IAS/ICU is emerging. Carbonate strata have been developed for public water supply in parts of southwestern Florida and to a limited extent northeast Florida (Duval County, Fig. 4.21); domestic and irrigation wells provide water elsewhere in the state; and we now understand that several of Florida's second and third magnitude springs discharge water from the IAS.

We also understand that the confining role of the IAS/ICU is highly variable. Many of the more recent groundwater flow models are fully three dimensional and include the IAS/ICU as at least one layer. We understand that the clays of the IAS/ICU vary considerably in thickness, ability to confine the aquifer, and prevent sinkhole development. Finally, we now understand that the carbonate-rich strata within the Miocene Hawthorn Group can develop interstratal karst (Wadsworth et al. 1983), play a significant role in the development of Florida karst escarpments (Upchurch 2002, 2005, 2014), and possibly contribute to the formation of sinkholes and other karst landforms.

As shown by data presented in Scott (1988), the relative difficulty of the Miocene Hawthorn Group to transmit groundwater vertically and horizontally is not just because the stratum is a clay bed. Rather, it is a highly complex system of interfingering beds of clay, sand, and carbonate sedi-

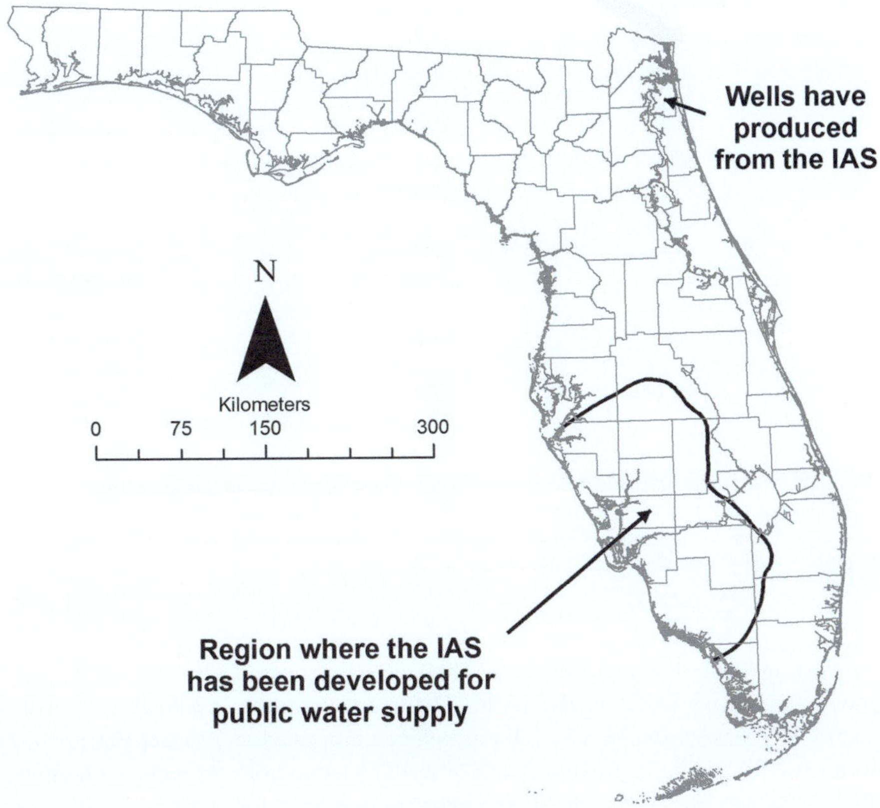


**Fig. 4.21** Locations of regions in Florida where strata within the intermediate aquifer system have been developed for public water supply

ments and rocks. While the clay beds are widespread and may attain some thickness, they are not all laterally extensive. As a result, much of the ability of the IAS/ICU to confine the FAS derives from the tortuous pathways that water must take to pass through the relatively permeable strata within the IAS/ICU.

Although the Miocene Hawthorn Group comprises a somewhat consistent group of lithosomes throughout its thickness, the relative proportions of these sediment types vary spatially within the group. It is this variation that provides a basis for breaking the Group into formations (Scott 1988). For example, two formations are recognized in central Florida. The upper formation, the Peace River Formation, contains interfingering sand, clay, and carbonate lithosomes, typically with abundant phosphatized clasts. The Peace River Formation, however, contains more sand and clay than carbonate lenses and layers. In contrast, the lower unit, the Arcadia Formation, contains more carbonate-rich strata, including dolostone and limestone, and the content of phosphatic sediments is typically lower. While the Arcadia still contains interfingering beds of the phosphatic sand, clay, and carbonate lithosomes, the increased importance of carbonate sediments makes it a location where karst features may develop. This dichotomy in carbonate content exists throughout much of the extent of the Miocene Hawthorn Group (Miller 1978, 1986, 1997; Scott 1988).

### 4.4.1 Aquifer Distribution

At present, southwestern Florida (Fig. 4.21) is the primary location where the IAS has been developed for public water supply. In the 1990s, the Southwest Florida Water Management District (SWFWMD), which manages water use in west-central Florida, recognized that saline-water intrusion was causing a loss of the freshwater resource in the highly confined, UFA in a large area of southwestern Florida (Fig. 4.21). The SWFWMD declared the area a "water-use caution area", which they termed the Southern Water Use Caution Area, or SWUCA. The District began rulemaking to limit and manage groundwater withdrawals from the UFA (Southwest Florida Water Management District 2006a, b). Within the most impacted area of the SWUCA, new withdrawals from the UFA have essentially been prohibited, unless there is a retirement of an existing water use permit (WUP), which results in a 10% net benefit (i.e., rebound) to the potentiometric surface within the MIA. In order to allow permitting of new water for growth and flexibility to accommodate changes in demand and increases in population growth, the SWFWMD also developed permitting incentives for use of alternative water sources, including tapping horizons capable of producing groundwater to wells (production zones) in the IAS/ICU (Table 4.2).

**Table 4.2** Groundwater production zones (PZ) in the IAS/ICU and UFA

| Series | Stratigraphic unit | | | Hydrogeologic unit | | Lithology |
|---|---|---|---|---|---|---|
| Holocene to Pliocene | Undifferentiated surficial deposits | | | Surficial Aquifer System | | Sand, silty sand, clayey sand. Peat and shell |
| Miocene | Hawthorn Group | Peace River Formation | | **UICU** | Intermediate Aquifer System | Predominantly sand with variable clay; limestone and dolostone; variably phosphatic |
| | | | | PZ 2 | | |
| | | Arcadia Formation | | **MICU** | | |
| | | | Tampa member | PZ 3 | | Limestone, gray to tan, sandy, soft, clayey, minor sand, phosphatic; chert found locally |
| | | | | **LICU** | | |
| Oligocene | Suwannee Limestone | | | UPZ | Upper Floridan Aquifer | Limestone, cream to tan, vuggy, fossiliferous |
| Eocene | Ocala Limestone | | | **SCU** | | Limestone, white to tan, friable to micritic, fine-grained, soft, abundant foraminifers |
| | Avon park Formation | | | LPZ<br>**Middle Semiconfining unit [of the FAS]** | | Limestone and dolostone. Limestone is tan, recrystallized; dolostone is brown, fractured, sucrosic, hard; peat found locally |

Modified from Basso (2002) and Southwest Florida Water Management District (2002)
*UICU* upper intermediate confining unit, *MICU* middle intermediate confining unit, *LICU* lower intermediate confining unit, *UPZ* upper production zone of the upper Floridan aquifer, *SCU* semi-confining unit, *LPZ* lower production zone of the upper Floridan aquifer. Row heights do not correspond to unit thicknesses

Note in Table 4.2 that the SWFWMD targeted several production zones (PZs) in the IAS/ICU for development. These PZs have been successfully developed for potable water use for public, agricultural, and private consumption. Water has also been developed from wells completed in the carbonate-rich strata of the Arcadia Formation above the Tampa Member.

South of the SWUCA, in Lee and Collier Counties where water is managed by the South Florida Water Management District (SFWMD), the Tamiami and Ft. Thompson formations and the Miocene Hawthorn Group may have strata that fall within the IAS/ICU. Here, the IAS/ICU has also been developed for potable use.

The local utility in Jacksonville (Duval County) previously used water from the IAS in one wellfield, and development of the IAS on a larger scale may happen in the future. Elsewhere in Florida, permeable zones within the IAS/ICU are known to exist and may serve as future sources of groundwater supply.

Because of its low transmissivities, the IAS/ICU tends to resist saline-water intrusion. However, where freshwater permeable zones intersect saline water, a mixing zone develops. If the permeable zone contains calcite, local dissolution and karstification may occur.

### 4.4.2 Location and Composition

Figure 3.5 illustrates the geologic map of Florida by Scott et al. (2001). The regions where the Miocene Hawthorn Group crops out or occurs in the subsurface constitute the general distribution of the IAS/ICU in Florida. The Miocene Hawthorn Group once covered the entire state (Scott 1981), but it has largely been removed by erosion near the crest of the Ocala Platform. Near the crest of the Ocala Platform, outliers of intact Hawthorn sediments and residual clays are widespread (Arthur et al. 2008), however. The clay deposits in these outliers locally form confining and aquifer-perching strata, reduce the probability of sinkhole development somewhat, and cause damage by shrink/swell behavior. The Miocene Hawthorn Group thickens and dips to the south, into the South Florida Basin, and to the east and north into the Jacksonville and South Georgia basins (Figs. 3.2, 3.3). Where the deeper, thicker Hawthorn Group sediments lie within the active or formerly active groundwater flow system, interstratal karst and sinkhole topography may develop (Wadsworth et al. 1983). Evidence of extensive interstratal karst development is limited in Florida to the highlands of the Okeechobee Basin District where they transition into the Cody Scarp (Upchurch 2002, 2005, 2014).

As noted above and in Chap. 3, the lithologic composition of IAS/ICU is complex. Individual lithosomes and strata are typically thin (less than 1 m) and they are complexly interfingered with each other. Beds of sand, sandy clay, clayey sand, calcareous clay (microcrystalline calcite and silt-sized dolomite), limestone, and dolostone are the primary lithosomes. All have varying amounts of phosphatic clasts and rock, and gravel-sized clasts of quartz and other rock types are not uncommon. Chert and porcelainite have replaced carbonate-rich sediments and cemented non-calcareous sediments in some locations. Gypsum crystals are found at several locations (Upchurch 1992). The depositional environments that caused the complex IAS/ICU include marine, fluvial, and coastal processes (Scott 1988; Strom and Upchurch 1985a, b).

### 4.4.3 Aquifer Recharge

Recharge to the permeable zones of the IAS/ICU is complex and limited by the leakance of the less permeable strata within the unit. Leakance is a function of the clay content of the strata and the highly tortuous routes that water must take in order to pass through the IAS/ICU system. Leakance through the IAS/ICU is often treated as a calibration variable when developing groundwater flow models. Other than modeling, there is little hard information about movement of water into and through the IAS/ICU. A reported leakance value for the IAS/ICU in the SWFWMD (2006a) was $2 \times 10^{-7}$ to 0.12/day.

### 4.4.4 Aquifer Potentials and Flow Systems

Miller (1990) reported well yields in southwestern Florida of up to 9800 L/min; however, most wells yield less than 760 L/min. Common transmissivities of the IAS in southwest Florida were reported to be 900 $m^2$/day or less.

Sandy strata within the Miocene Hawthorn Group have reported transmissivities of about 1200 $m^2$/day); and Hawthorn limestone and/or dolostone strata have reported yields of up to 4088 L/day (Miller 1986). Knochenmus (2006), Knochenmus and Bowman (1998), and Torres et al. (2001) provide detailed information of the hydraulic properties of the IAS/ICU in southwest and south-central Florida. Brown (1984) reported transmissivities of 9–900 $m^2$/day in northeast Florida and southeast Georgia.

Figure 4.22 illustrates the potentiometric surface of the IAS flow system in southwest Florida in 2001. The general pattern of flow is from potentiometric highs in south-central Polk County and the "Four Corners" area of Hillsborough, Polk, Manatee, and DeSoto counties, toward the coast. A known recharge area exists on the upper Peace River near the potentiometric high in south-central Polk County (Knochenmus 2004; Metz and Lewelling 2009), where the river discharges into a series of sinkholes and fractures that penetrate rocks of the Miocene Hawthorn Group. The potentiometric high in southwest Polk County is a location where several streams are incised into the Hawthorn Group sediments and phosphate mining is known to have breached portions of the upper Miocene Hawthorn Group (Peace River Formation). Flow from these potentiometric highs is generally southwest, toward the Gulf of Mexico. It is important to understand that the production zones in the IAS are typically discontinuous, so flow to the Gulf is not through a laterally continuous aquifer. Rather, flow is through lenses of elevated transmissivity connected by less permeable pathways through the semi-confining beds.

### 4.4.5 Role of the Intermediate Aquifer System in Karst Development

The carbonate strata in the IAS/ICU are known to have interstratal karst on the southwestern and western margins of the Okefenokee Basin District where the Cody Scarp is actively retreating (Wadsworth et al. 1983; Upchurch 2002, 2005, 2014). Given the predominance of dolostone and tortuous pathways of groundwater in the Hawthorn Group sediments, it is unclear whether the carbonate strata of the IAS in much of Florida have developed significant karst by simple dissolution. Sinkholes are known to have penetrated the strata of the IAS/ICU (Chap. 6), but the best known of these sinkholes are deep (greater than 60 m) and appear to have formed by collapse of Hawthorn Group sediments and rocks into void space developed in the underling Ocala Limestone or older strata. The Ocala Limestone is part of the FAS and is known to have a long and complex history of karstification. The extent of the karst in the IAS/ICU is clearly limited by the reduced solubility of dolostone, high degree of confinement, and retardation of flow in low hydraulic conductivity strata.

The development of karst features in the limestone and dolostone of the underlying UFA is highly influenced by the thickness of the IAS/ICU as cover (Miller 1990). The development of karst features in the IAS/ICU and FAS are coupled as a result of the processes required to strip the low hydraulic conductivity IAS/ICU sediments from the land surface and expose the underlying FAS. For example,

1. Where the Hawthorn Group clays have been largely removed by erosion (Fig. 4.23), karst development in the UFA has been extensive. Caverns, sinkholes, and other karst features are well developed and internal drainage characterizes the landscape.
2. Where the IAS/ICU is present, but less than 30 m in thickness, karst features vary in density and degree of development depending on the presence of breaches and thin areas within the Hawthorn. It is these areas where karst escarpments are likely to develop as the Hawthorn Group sediments are eroded away. The Cody Scarp is the best known escarpment in Florida. The carbonate rocks of the IAS/ICU in the Cody Scarp and contiguous portions of the Okeechobee Basin District have developed interstratal karst, and sinkholes caused by collapse of IAS/ICU sediments into voids in the underlying limestone have developed.
3. Where the IAS/ICU is over about 30 m thick (Fig. 4.23), sinkhole development at the land surface in the FAS is limited and karst is likely not developed within the carbonate units of the Miocene Hawthorn Group (Miller 1990; Upchurch 2002, 2005).

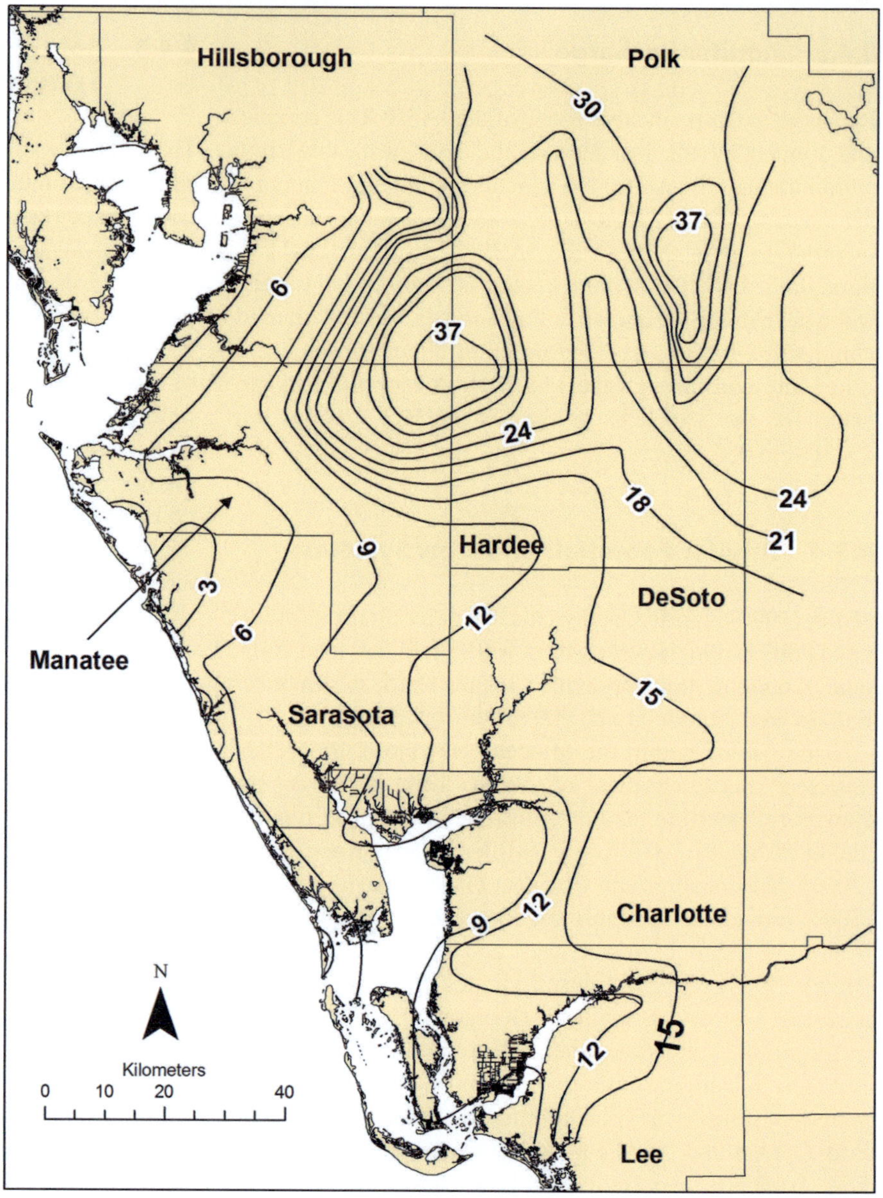


**Fig. 4.22** Potentiometric surface in meters NGVD of the intermediate aquifer system in 2001 in the Southwest Florida Water Management District. Data from the Southwest Florida Water Management District

## 4.5 Floridan Aquifer System

The Floridan Aquifer System (FAS) is subdivided into three hydrostratigraphic units:

1. Upper Floridan Aquifer (UFA);
2. Middle semi-confining units (MCUs); and
3. Lower Floridan Aquifer (LFA).

The FAS is present throughout Florida, and it extends into portions of Alabama, Georgia, and South Carolina. This discussion is, however, limited to those portions of the FAS that lie within Florida.

The FAS is one of the most productive aquifers in the world, and much of its ability to supply large quantities of water to wells is a result of a long and complex history of karstification. The FAS provides 59% of Florida's groundwater supply (Marella 2009). In coastal areas and in southern Florida, where the aquifer system is deeply buried (Fig. 4.24) and thick, water is saline, which limits use for potable supplies. Figure 1.8 illustrates those portions of the state where potable water supplies exist within the FAS.

As illustrated in Fig. 4.2, the FAS is exposed at the land surface on the crest of the Ocala Platform, and it dips to the south, where it is deeply buried. Near the state line with Alabama and Georgia, the FAS is less than 60 m in thickness. The FAS is thickest in southwestern Florida, within the South Florida Basin. Here, the aquifer system is in excess of 1000 m thick, and is entirely below sea level. The top of the FAS is about 440 m below sea level in southern Florida (Fig. 4.24).

Where the FAS lies below sea level and a freshwater lens has not developed, there appears to be an active saline-water

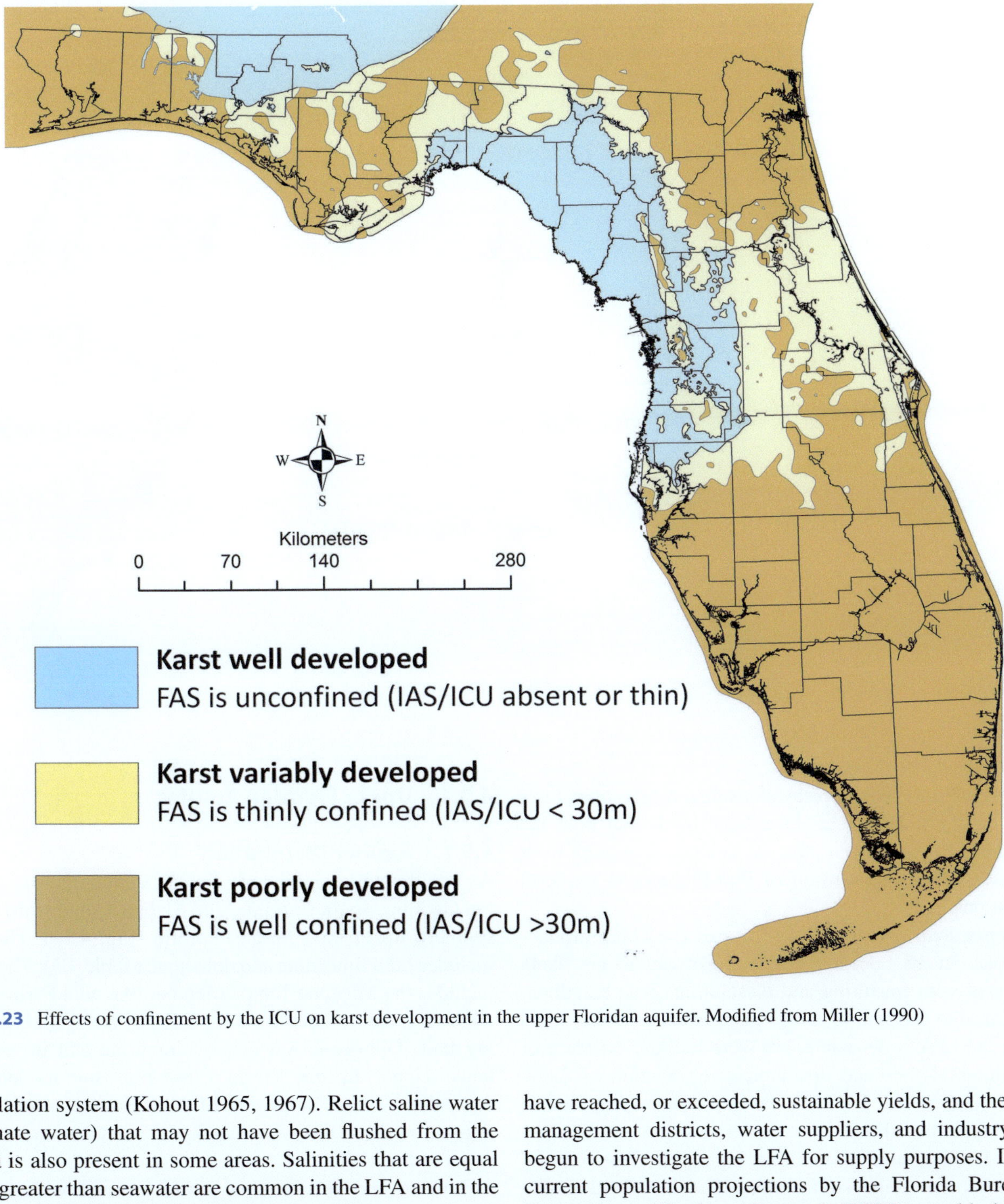


**Fig. 4.23** Effects of confinement by the ICU on karst development in the upper Floridan aquifer. Modified from Miller (1990)

circulation system (Kohout 1965, 1967). Relict saline water (connate water) that may not have been flushed from the strata is also present in some areas. Salinities that are equal to or greater than seawater are common in the LFA and in the deeper parts and coastal locations of the UFA.

Due to the relatively shallow depth and excellent water quality with limited need for treatment, the UFA has traditionally been utilized for potable supplies. Historically, the LFA was utilized for waste disposal and other uses where water quality was not considered a problem, because salinity and other water quality issues rendered the water non-potable and the MCU adequately separated UFA from the LFA. In the last two decades, however, water demands on the UFA have reached, or exceeded, sustainable yields, and the water management districts, water suppliers, and industry have begun to investigate the LFA for supply purposes. In fact, current population projections by the Florida Bureau of Economic and Business Research (BEBR) would place the majority of Florida into water-use caution areas by 2035. Portions of the LFA have turned out to contain water that can be used directly, after desalination or other treatment technologies, or by blending with UFA water. As a result, both the UFA and LFA are serving to supply water for public and private uses today.

In Florida, eogenetic karst occurs in much of the UFA with karst development taking place near the upper surface

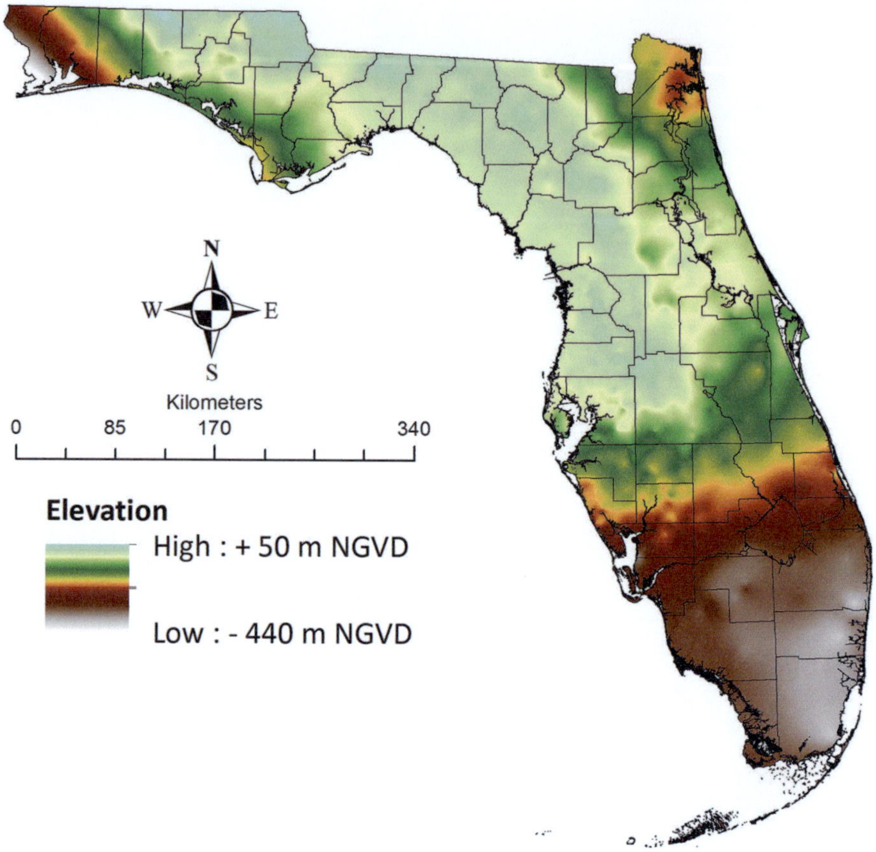


**Fig. 4.24** Depth to the top of the Floridan Aquifer System. (Based on spatial data from the Florida Geological Survey)

of the FAS and along freshwater/saline water transitions zones at depth. Arguments have been made that both the UFA and LFA have been affected to some degree by karst processes, largely as a result of Plio-Pleistocene sea-level fluctuations.

Deep-seated, potential karst features (i.e., high production zones in the Eocene Avon Park Formation) are likely associated with fracturing and dissolution. True karstification related to dissolution of carbonate or other soluble strata within the LFA is debatable, but there is some evidence of dissolution (Herbert and Upchurch 2016). Section 4.5.3 discusses the potential for karst development, as opposed to fracturing, in the LFA.

The MCUs consists of multiple, separate horizons of relatively low permeability strata. In central and southern Florida evaporite minerals reduce the vertical hydraulic conductivity of the MCU strata. As a result, the MCUs can serve as effective confining units where they are present and not penetrated by fractures or faults. The MCUs are not present throughout the state, however. They represent Paleocene to Early Eocene carbonate and evaporite sedimentation in the South Florida Basin, its margins, and adjacent shelf environments. In extreme south Florida, the MCUs are thick and consist of several tongues of evaporites.

### 4.5.1 Upper Floridan Aquifer

#### 4.5.1.1 Aquifer Distribution

As a hydrostratigraphic unit, the UFA extends from southern Georgia, South Carolina, and Alabama southward to the southern tip of the Florida Platform (Fig. 4.23). The unit includes both limestone and dolostone (Table 4.3). The UFA includes the Miocene Tampa Member (Arcadia Formation), where it is not isolated from the UFA by a clay-rich confining unit; Oligocene Suwannee Limestone and its equivalents; Upper Eocene Ocala Limestone and its laterally equivalent strata; and the Middle Eocene Avon Park Formation above the gypsum/anhydrite and dense carbonate strata of the MCUs.

Much of the UFA is a dual-porosity system with groundwater flow through interstitial pores between sediment grains and conduit flow through karst conduits and fractures in the rock (Vacher and Mylroie 2002). Often, the highest permeability zones are at unconformities between the formations where dissolution has produced relict, epigenetic karst. Flow systems are best developed in the Middle and Upper Eocene, Oligocene (where present) and lowest Miocene strata. Although the Ocala Limestone has zones of well-developed permeability, there are large sections of the Ocala that serve

**Table 4.3** Hydrostratigraphic relationships within Florida's aquifer systems in southern Florida

| Epoch | Formation | | Lithology | Hydrostratigraphic unit | Hydrostratigraphic unit thickness (m) |
|---|---|---|---|---|---|
| Holocene | Undifferentiated | | Quartz sand, silt, clay, shell | Surficial Aquifer System (SAS) | 0–100 |
| Pliocene | Tamiami Formation | | Silt, sandy clay, shell, limestone | | 0–30 |
| Miocene to late Oligocene | Hawthorn Group | Peace River Formation | Interbedded sand, silt, gravel, clay, limestone, dolostone, phosphatic sand | Intermediate Aquifer and Confining Unit (IAS/ICU) | 180–275 |
| | | Arcadia Formation | Sandy limestone, shell beds, dolostone, phosphatic sand, silt, clay | | |
| Early Oligocene | Suwannee Limestone | | Limestone | Upper Floridan Aquifer (UFA) | 210–370 |
| Late Eocene | Ocala Limestone | | Limestone | | |
| Middle Eocene | Avon Park Formation | | Limestone, dolostone | Middle Confining Unit (MCU) | 150–240 |
| | | | Dolostone, gypsum/anhydrite | | |
| Early Eocene | Oldsmar Formation | | Dolostone, gypsum/anhydrite | Lower Floridan Aquifer (LFA) | 430–550 Incl. Boulder zone(s) (120±) |
| Paleocene | Cedar Keys Formation | | Dolostone, dolomitic limestone | | |
| | | | Massive anhydrite | Sub-Floridan Confining Unit | 370± |

Modified from Herbert and Upchurch (2016, Table 1)

as a semi-confining unit (Table 4.2; Gaswirth et al. 2007) in southwestern Florida where fracturing and karst are not well developed and the limestone matrix has reduced bulk permeability because of its packstone to mudstone fabrics.

Where the MCUs are not present, the UFA is equivalent to the FAS. Where the IAS/ICU is not present and the UFA is a water-table aquifer; there is, by definition, no SAS, and the UFA is the water-table aquifer, even when the phreatic surface is within overlying sandy sediments (Southeastern Geological Society 1986).

### 4.5.1.2 Recharge and Discharge Patterns

Recharge to the UFA is controlled by the storage and vertical hydraulic conductivity of the SAS and IAS/ICU and the amount of confinement of the UFA. Recharge to the UFA is greatest where the UFA is unconfined near the crest of the Ocala Platform in north-central Florida (Fig. 4.25). As such, the largest recharge areas lie within the karst plains of the Ocala Karst District.

Recharge is often enhanced at locations where the ICU pinches out and karst escarpments separate karstic lowlands. Here, the UFA is more-or-less unconfined as opposed to the uplands, where the UFA is well confined. Such a karst escarpment, the Cody Scarp, extends from Alachua County in the south to Leon County in the north. This escarpment lies within substantial portions of the Alachua Hills, Madison Hills, and Tallahassee Hills geomorphic provinces. Streams drain off the highlands and go underground when they encounter sinkholes near the toe of the escarpment proximal to the lowlands. These stream-to-swallet systems represent surface-water drainage basins that focus recharge to the UFA within a small area. The Cody Scarp is discussed in detail in Chap. 8.

Note in Fig. 4.25 that recharge to the UFA is greatest at the crest of the Ocala Platform in west-central Florida (Stewart 1980; Miller 1990). These high recharge areas are identified by the presence of sinkholes and other geomorphic or hydrologic features and through groundwater flow modeling. When modeling groundwater flow, the actual amount of recharge is typically unknown, so it is treated as a "calibration variable" for model development. Note also in Fig. 4.25 that discharge from the UFA is spatially limited to coastal areas, offshore, and in reentrants that extend up many of the rivers in northern Florida.

The general discharge zone of the UFA extends around the coast of the state (Fig. 4.25). Greatest discharge is shown near the coast just southwest of the high recharge area corresponding to the unconfined portions of the UFA at the crest of the Ocala Platform. Figure 4.25 is not entirely accurate in its representation of discharge because the UFA discharge zones extend up many of Florida's rivers. For example, the lower and middle Suwannee and lower Santa Fe rivers are well known discharge areas with numerous springs (Fig. 4.3).

Note also that the discharge areas of the UFA (Fig. 4.25) are shown to extend offshore in some areas. Offshore springs and seeps indicate that discharge is not just confined to coastal areas of Florida. New offshore springs are being discovered, and known offshore springs that have either ceased to flow or have reduced flow suggest that prior to intensive extraction of water from the UFA, the freshwater lens extended offshore in many areas of Florida.

Extreme southern Florida is also shown as a minor discharge area. Figure 4.25 is somewhat misleading in this area as the discharge of water from the UFA to the SAS or local surface water is limited and near the zero end of the range in discharge values. Artesian, saline UFA water may move into

 

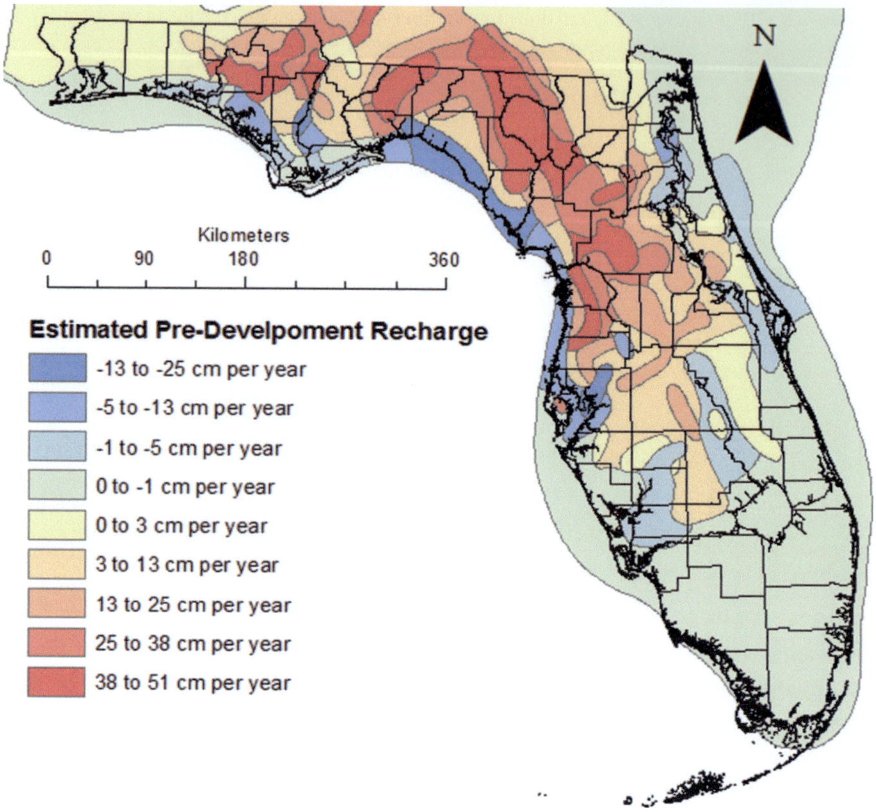


**Fig. 4.25** Recharge and discharge zones of the Upper Floridan Aquifer. Blue represents areas of groundwater discharge and reds represent areas of recharge to the UFA. (Data from Miller (1990))

the SAS or surface water bodies if pathways exist, thus contaminating any freshwater resources there. These pathways include improperly constructed or deteriorated wells and fractures (Sproul et al. 1972).

In the past, oil exploratory wells and water wells were often left without plugs or caps, metal casings deteriorated, and they commonly exhibited artesian flow of saline water from the UFA. This water contaminated overlying aquifers and, in some cases, surface waters (Boggess 1968; Boggess et al. 1977; Missimer and Scott 2001). For example, one site in Lee County experienced upwelling of artesian water from the saline UFA into the IAS. Water quality deteriorated from about 1000 mg/L chloride to as much as 150,000 mg/L chloride (Sproul et al. 1972). Absent pathways such as fractures or uncapped or deteriorated wells, the likelihood that significant discharge of saline UFA water to the land surface is negligible. The thickness of the ICU and depth of the UFA are just too great.

#### 4.5.1.3 Relationship of Potentiometric-Surface Highs and Recharge

The pre-development potentiometric-surface map (Fig. 4.26a) was developed by Bush (1982) by modeling the potentials of the UFA including estimated groundwater withdrawals in a steady-state groundwater flow model. By turning off the groundwater withdrawals in the simulation, it was thought that the resulting potentiometric surface would replicate potentials prior to the time when significant withdrawals began (prior to about 1950).

The UFA potentiometric surface is characterized by several potentiometric highs, from which groundwater flow is radial towards the large streams and the coast. There are four major potentiometric-surface highs identified in Fig. 4.26a. The high labeled "A" (Fig. 4.26a) is the southern extension of a large potentiometric surface high located in southern Alabama and southwestern Georgia. It extends into the Southern Pine Hills and western Dougherty Karst Plain provinces in areas where the UFA is confined. Potentiometric high "B" extends from southern Georgia as far south as the San Pedro Bay province in Taylor and Lafayette counties. The UFA is at least partly confined in this high, as well. "C" is a similar extension that is separated from the San Pedro Bay potentiometric-surface high by the Suwannee River, a major discharge corridor as a result of a multitude of springs. This potentiometric-high is also in an area where the UFA is largely confined. Finally, the distinctive potentiometric surface located at "D" on Fig. 4.26a, b is located in parts of Polk, Lake, and Pasco counties. It lies within the Lakes and Ocala Karst districts.

Figure 4.26b depicts the observed potentiometric surface in May, 2014. The May water-level monitoring period is considered to represent the end of peninsular Florida's dry season (October to May), so potentiometric surface levels are at a minimum. Comparison of the two maps (Fig. 4.26a, b)

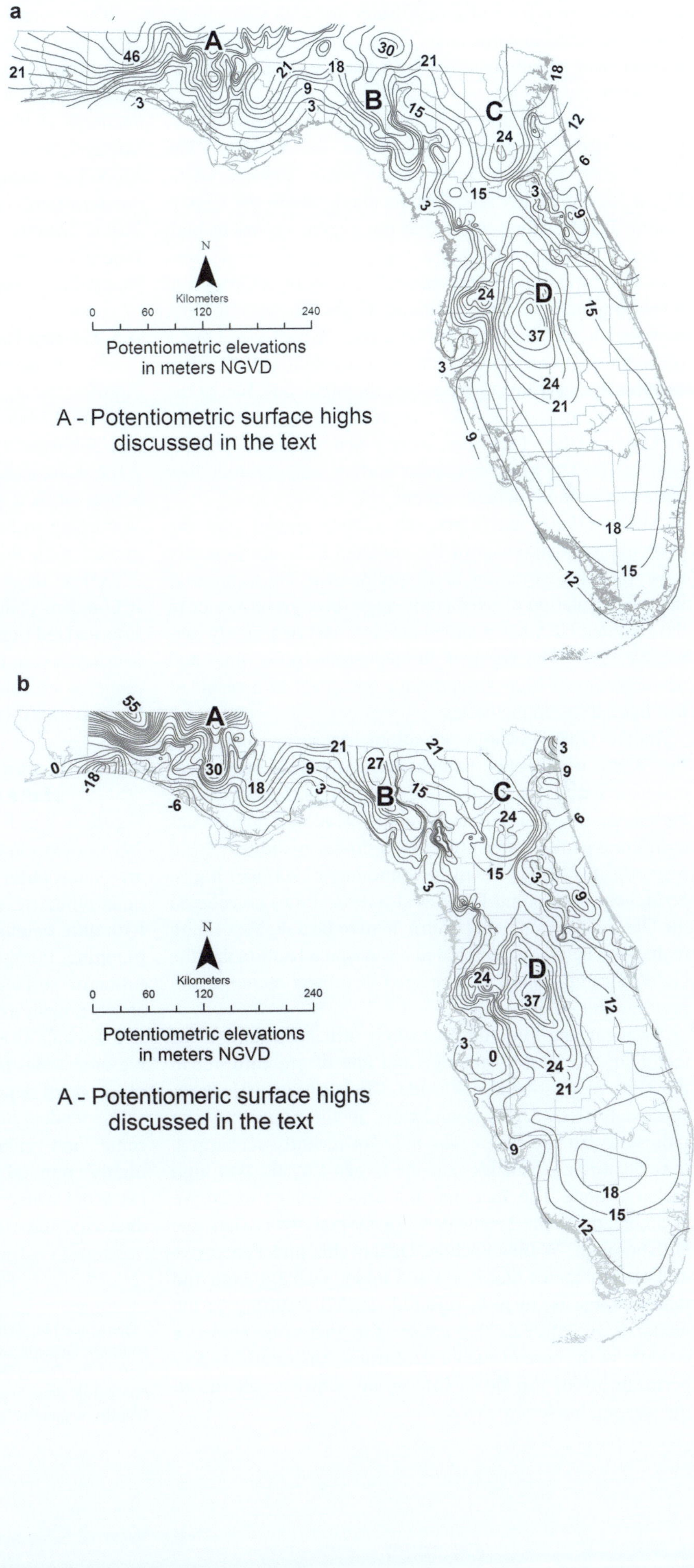


**Fig. 4.26** (**a**) "Pre-development" potentiometric surface of the Upper Floridan Aquifer. Elevations are in meters NGVD. (Geospatial data from the U.S. Geological Survey; based on Bush (1982)). (**b**) Potentiometric surface of the Upper Floridan Aquifer in May 2014. Elevations are in meters NGVD. (Geospatial data from the U.S. Geological Survey)

dramatizes the effects of groundwater use with water levels having declined in some areas more than 40 m. Clearly, the groundwater potentials of the UFA today do not reflect past conditions.

There is some confusion as to the relationship of the UFA potentiometric surface to recharge in Florida. The potentiometric-surface highs in northern Florida (A-C, Fig. 4.26) are clearly related to locations where the UFA is confined. Few would argue that these areas represent high recharge locations. They are topographic highland areas underlain by clay-rich sediments of the Hawthorn Group and other aquitards that limit recharge. Highs "B" and "C" are, however, associated with escarpments characterized by a transition from confined to unconfined conditions. At these escarpments, streams that flow off the highlands, where the UFA is confined, enter sinkholes (swallets) and recharge the underlying UFA. Therefore, these potentiometric highs are characterized by high recharge of surface water around their perimeters but not in their centers.

Miller (1990) and Upchurch (2002) argued that the hydraulic conductivities of the confined UFA are typically lower than in areas of low to no confinement. This is because limited circulation of chemically aggressive groundwater in the confined UFA has retarded development of a highly permeable UFA. The result is limited groundwater flow and development of high groundwater potentials as a result of low hydraulic conductivities.

In the Green Swamp potentiometric surface high (D, Fig. 4.26), it was once thought that greatest recharge was concentrated at locations over the potentiometric surface high in the UFA. The Green Swamp potentiometric high has a different geological setting, but a similar hydrogeological response to the northern potentiometric surface highs. Stringfield (1966) and Pride et al. (1961, 1966) considered the Green Swamp potentiometric high to be a region of high recharge. Their work resulted in a widespread notion that the Green Swamp should be managed to protect recharge and groundwater quality.

Swancar and Hutchinson (1995) utilized a number of techniques to identify the age and fate of groundwater in the UFA in west-central Florida. Their tritium ($^3$H) measurements from UFA groundwater in the Green Swamp indicated that the water was old, not recently recharged, thus disproving the idea that the Green Swamp is a high recharge area. The fact that the area contains extensive swamps and is the headwaters for several rivers (i.e., the Hillsborough, Withlacoochee, Ocklawaha, and Peace rivers) also indicates that it is not a major recharge area and that recharge is, in part, rejected rainfall resulting in the wetlands and streams that caused the area to be named a swamp. In the case of the Green Swamp, the relatively low permeability of the Ocala Limestone, which crops out in the swamp, hinders recharge.

The potentiometric surface highs in the UFA are important and should be protected, however. The land surface in each area is characterized by development of swamps and streams. As the streams flow out of the areas where recharge is limited, they encounter karstic limestone. The streams flow into sinkholes (swallets) and recharge the UFA. For example, Fig. 4.27 shows a location on the northwestern flank of the Green Swamp Potentiometric high in Pasco County where flow in the Withlacoochee River (south[2]) is captured by a sinkhole west of the river, resulting in significant recharge of the UFA.

#### 4.5.1.4 Aquifer Flow Systems and Hydraulic Properties

Aquifer testing and groundwater flow-models have been used to characterize transmissivities of the UFA (i.e., Bush and Johnston 1988; Miller 1990; Kuniansky and Bellino 2012; Kuniansky et al. 2012). Transmissivities ranged from a low of less than 900 m$^2$/day to over 90,000 m$^2$/day. According to Miller (1990), where the transmissivities are greater than 90,000 m$^2$/day and most of the areas from 23,000 m$^2$/day to 90,000 m$^2$/day are located where the IAS/ICU is less than 30 m thick. Kuniansky and Bellino (2012) summarized transmissivities from 1045 aquifer performance tests (APTs) in the UFA for the entire southeastern U.S. The range of transmissivities was $0.7 \times 10^5$–$8.6 \times 10^5$ m$^2$/day. Median transmissivity was $2.5 \times 10^3$ m$^2$/day.

#### 4.5.1.5 Declines in the Potentiometric Surface of the Upper Floridan Aquifer

Figure 4.26a, b compare the "pre-development" and May 2014 potentiometric surfaces of the UFA in Florida. There are substantial changes in the elevations(s) of the potentiometric surface. One way to test confinement and hydraulic conductivity of an aquifer is to draw it down by pumping. Pumping from any well usually causes the potentiometric surface around the well to be depressed in a cone shape known as a cone of depression. Large groundwater withdrawals from confined areas of the UFA have created regional cones of depression in Florida. The size and depth of a cone of depression caused by groundwater withdrawals are dependent on the ability of water to flow to the pumping center and replace the water withdrawn. If the aquifer is highly confined so surface water cannot readily replace the extracted water, or if the aquifer has a low hydraulic conductivity, the water has difficulty flowing laterally to the

[2]There are two Withlacoochee Rivers in Florida. One drains out of southern Georgia and joins the Suwannee River. The other drains a portion of the Green Swamp and flows into the Gulf of Mexico. To avoid confusion, these rivers are termed the Withlacoochee River (north) and (south), respectively.

**Fig. 4.27** Dobes Hole, a blind valley and swallet (sinkhole) that takes water from the Withlacoochee River in Pasco County during high river stages. River is in the distance, and the swallet is in the right foreground. Location: 28.4326°N, 82.1679°W

pumping center, and a regional cone of depression will develop.

Several large cones of depression have developed in the UFA, and they continue to expand as use of water drawn from the UFA increases. Bush and Johnston (1988) and Johnston and Bush (1988) examined several regional cones of depression that had developed in the UFA (Fig. 4.28). These cones of depression persist to the present, and most of them have continued to grow.

The map showing changes in the UFA potentiometric surface from about 1950 to 1980 (Fig. 4.28, Miller 1990) was developed by subtracting the results of a UFA 1980 potentiometric surface from the pre-development (c. 1950) potentiometric surface. Small-scale declines in the potentiometric surface have developed throughout the state. These have caused local problems, such as dehydration of wetlands, reduction in spring discharge, and changes in groundwater quality (Copeland et al. 2011). There are major areas, however, where large-scale withdrawals from the UFA have caused large cones of depression in the potentiometric surface (Fig. 4.28). All three areas have two factors in common:

1. The UFA is highly confined, which limits leakage of water through the ICU from significantly reducing the cones of depression; and
2. They are locations of heavy groundwater use.

The depression in the Florida panhandle is at Fort Walton Beach (Fig. 4.28), where public supply was withdrawing $5.7 \times 10^4$ m$^3$/day in 1980. The large depression centered on Savannah and Brunswick, Georgia, is in an area that was withdrawing $1.9 \times 10^6$ m$^3$/day for industrial and public supply in 1980. There was a second center of drawdown in the Brunswick/Savannah cone of depression at Fernandina, Florida. This was caused by a single industrial facility. The cone of depression near Fernandina has grown as the water demands in Nassau and Duval counties have increased. Finally, the drawdown in the UFA potentiometric surface in west-central Florida was a result of withdrawals of $4 \times 10^6$ m$^3$/day for mining, agricultural, and public supply in 1980. The UFA in all of these areas is well confined with little recharge and somewhat reduced horizontal conductivities. Steps are being taken to mitigate areas of significant UFA drawdown throughout the state. For example, in order to prevent saltwater intrusion, the SWFWMD has declared the area affected by the cone of depression in west-central Florida the SWUCA and developed water conservation and alternative sources requirements for permitting of withdrawals.

The regional cones of depression shown in Fig. 4.28 have interesting effects on karst features in their vicinity. Some springs have dried up (Scott et al. 2004) and sinkhole swarms have developed during heavy pumping episodes (Metcalf and Hall 1984; Bengtsson and Radin 2001; Upchurch et al. 2013), among other effects.

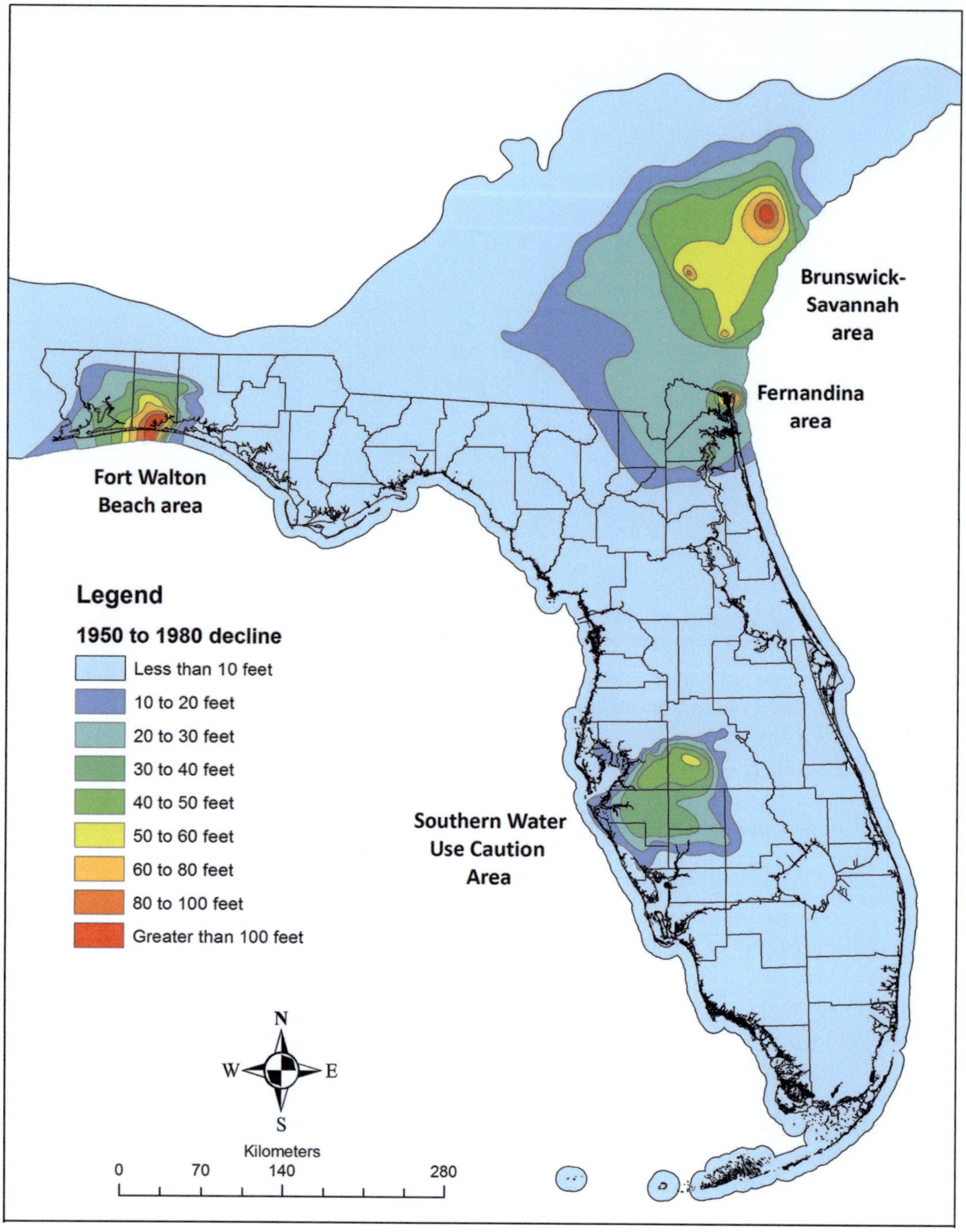


**Fig. 4.28** Difference in the pre-development (approx. 1950) upper Floridan aquifer potentiometric surface and the 1980 potentiometric surface (Bush and Johnston 1988)

#### 4.5.1.6 Saline-Water Transition Zones

The UFA has well-developed and well-documented saline-water transition zones along both coasts of Florida. Typically, the transition zones parallel and are located near the coasts. Offshore, freshwater springs (Brooks 1961; Kohout 1974; Hunn and Cherry 1969; Reich et al. 2001; Scott et al. 2004) and artesian wells near Florida's coasts testify to the fact that the UFA flow system maintains the saline-water transition zone near the modern coastline or even offshore in many areas. This is partly because the potentials in the UFA have not completely adjusted to the late Pleistocene and Holocene sea-level rise. As the freshwater lens adjusts to sea-level rise and as UFA water use increases (Marella 2009; Copeland et al. 2011), the position of the transition zone should slowly migrate landward throughout the state.

In areas where the UFA is well confined so freshwater recharge rates do not equal extraction rates, saline water has intruded well inland. Figure 4.28 illustrates areas in Florida where significant drawdown of the UFA potentiometric surface has developed. These drawdown areas are a result of extraction exceeding local recharge. Saline-water intrusion is an especially significant problem in Florida where these drawdown areas intersect the coast.

The distributions of salinity in the transition zones in the UFA are similar to the one shown for the Biscayne Aquifer (Fig. 4.8). The slope and amount of chemical dispersion across the transition zone are dependent on the hydraulic properties of the aquifer and rate of freshwater discharge to the sea. In southwestern Florida, confining beds create stacked flow systems, each with its own more-or-less independent transition zone, within the UFA.

The large cone of depression in the UFA in west-central Florida (Fig. 4.28) has an interesting saline-water transition zone at the base of the freshwater lens. This transition zone consists of a layer of high dissolved sulfate ($SO_4^{2-}$) concentrations (Fig. 4.29) formed by mixing of freshwater with water that has dissolved gypsum and anhydrite from the upper surface of the MCU (Sect. 4.5.2). Near the coast, water rich in sodium and chloride that reflects sea water lies below the layer of sulfate-rich water. As water is extracted from the UFA inland, the sulfate-rich saline water rises (upcones) into the freshwater lens more-or-less according to the Ghyben-Herzberg principle (Copeland et al. 2011). Near the coast, sodium-chloride-rich water may also intrude vertically.

While prehistoric positions of the transition zones are likely to have developed karstic zones as a result of groundwater mixing, none of the short-term, recent saline-water intrusion is likely to have affected dissolution-related karst formation in the UFA. The time frame of intrusion is too brief in comparison to the time necessary to dissolve significant volumes of carbonate sediment. The position of the predevelopment saline-water transition zone and the relationship of the transition zone to Plio-Pleistocene and Holocene sea-level fluctuations are of interest because these relationships have certainly affected development of karst features over geologic time.

#### 4.5.1.7 Role of the Upper Floridan Aquifer in Karst Development

The UFA has been heavily modified by karst, especially where it is unconfined. Given that late Tertiary and Quaternary sea-level fluctuations have affected the degree of exposure and weathering of the limestones and dolostones of the UFA and the position and flow of the groundwater system contained within the UFA, development of the karst has varied in intensity and position within the strata of the UFA. As a result, karst development has been multi-cyclic, and karst features developed on the freshwater/saltwater transition zone are likely "smeared" vertically as modifications of preexisting caverns (Upchurch 2017).

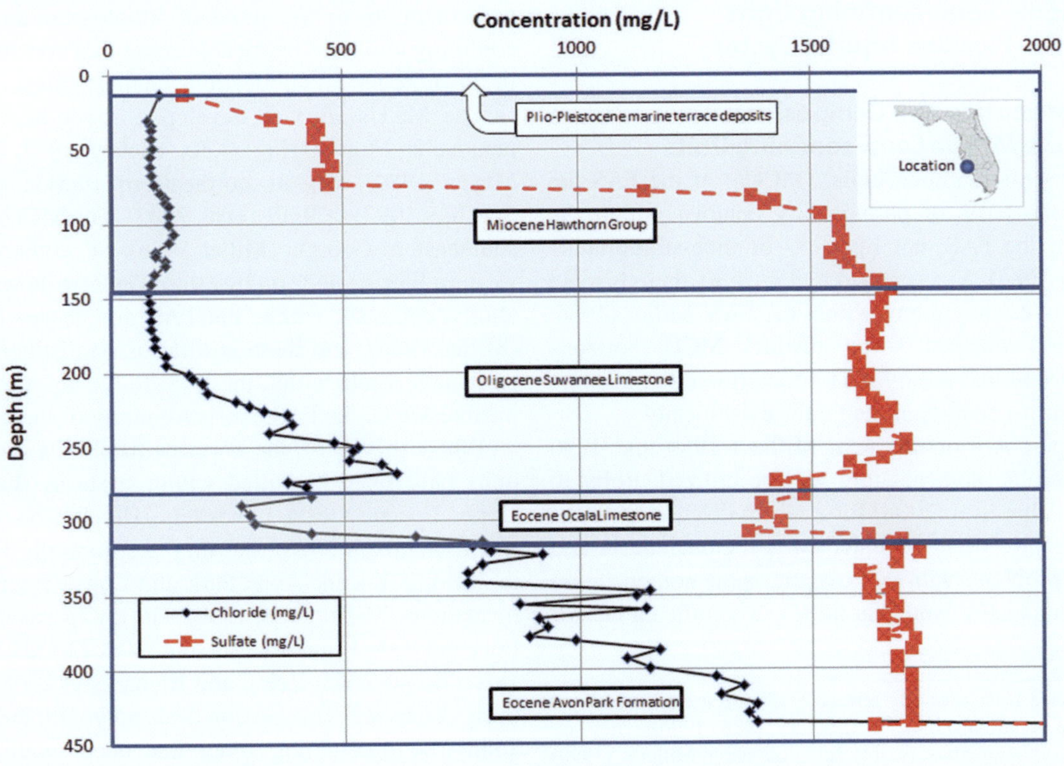


**Fig. 4.29** Geochemical profiles of upper Floridan aquifer water in the salt-water/freshwater transition zone near Osprey, Sarasota County. Data from the Southwest Florida Water Management District, ROMP 20 well

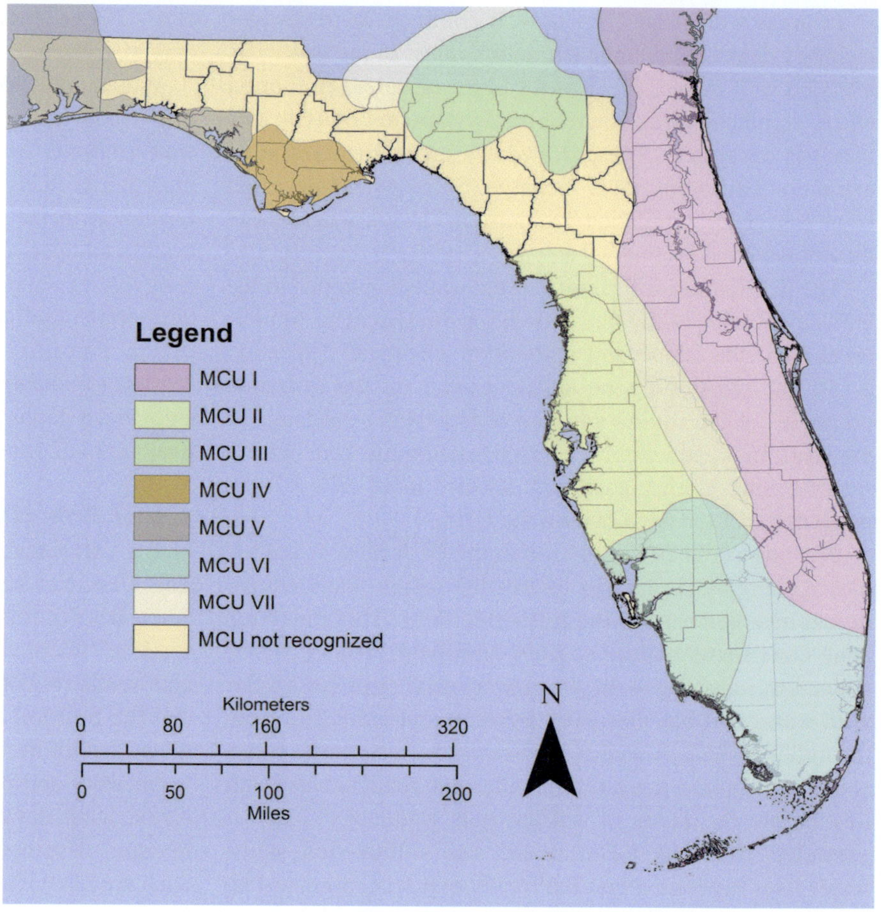


**Fig. 4.30** Distribution of Miller's (1986) seven middle semiconfining units in Florida. (Geospatial coverage from Bellino (2011))

## 4.5.2 Middle Semi-confining Units of the Floridan Aquifer System

### 4.5.2.1 Distribution and Composition of the Middle Semi-confining Units

The middle semi-confining units[3] (MCUs) of the FAS are defined on the basis of permeability contrast within the remainder of the FAS, not by rock- or time-stratigraphic units (Miller 1986). As suggested by Fig. 4.30, there is more than one semi-confining unit within the FAS. Miller (1986, pp. B57-B69) mapped seven distinct MCU horizons (Fig. 4.30). Note that these MCU horizons overlap in some areas, so multiple semi-confining units exist locally.

For the purposes of this text, Miller's 1986 and 1990 hydrostratigraphic nomenclature will be utilized. Refer to Reese and Richardson (2008) for a recent discussion of the hydrostratigraphy of these units on a regional basis and some of the problems with hydrostratigraphic nomenclature in the MCU and LFA. Note that there is a significant lack of agreement as to the number, lithologies, and depths of confining units and permeable zones that constitute the LFA in the more recent publications on these units.

The MCUs are well developed in southern and central peninsular Florida (Fig. 4.30, Hickey 1977, 1979, 1981; Meyer 1989) and in northeastern Florida (Phelps and Spechler 1997; O'Reilly et al. 2002). The MCUs extend into southeastern Georgia (Miller 1986). In northern and northwestern Florida and southwestern Georgia, however, permeability contrasts within the FAS are absent or indistinct (Miller 1986), and there is little or no distinction between permeable strata within the FAS. In the absence of a recognizable MCU, the FAS consists entirely of the UFA.

Where they exist, the lower permeability, semi-confining units have been identified within strata as shallow as the Upper Eocene Ocala Limestone (Hutchinson 1992; Reese and Richardson 2008), but they are normally found within the Middle Eocene Avon Park and Lower Eocene Oldsmar formations (Miller 1986, 1990). The low-permeability Upper Eocene Ocala Limestone in southwest Florida (Hutchinson 1992; Basso 2002; Reese and Richardson 2008) was previously discussed. This section deals only with the low permeability, semi-confining zones that occur within the lower Avon Park and Oldsmar formations.

[3]The MCUs have been variously termed confining units (i.e., Meyer 1989) and semi-confining units (i.e., Miller 1986, 1990). They are termed semi-confining units in this text following Miller's (1990) terminology.

The MCUs consist of fine-grained dolostones and limestones. Avon Park and Oldsmar MCU strata include significant interstitial gypsum and anhydrite included within low permeability limestone. Stewart (1966) and Navoy (1986) concluded that these evaporite minerals were primary and that they were present at time of carbonate deposition. They also noted some fracture and fossil filling by evaporite minerals. Stewart (1966) and Navoy (1986) also observed that evaporite nodules in the MCUs often have central cores of anhydrite and rims of gypsum, which may indicate that the evaporites were originally composed of anhydrite and that they have since been altered to gypsum by hydration. According to Hickey (1990), evaporite inclusions constitute about 10–30% of the host rock in central Florida. Hickey (1981, 1982, 1990), and Navoy (1986) concluded that the evaporite-rich strata are not laterally continuous.

#### 4.5.2.2 Confining Unit Properties

Where the MCUs exist, they generally serve as semi-confining horizons between the UFA and one or more LFA horizons. It is commonly recognized that the MCUs are not aquicludes and that slow leakage may exist between the contiguous strata. As a result of the ability of the MCUs to confine flow, the underlying LFA has been widely used for wastewater injection in Florida (Hickey and Vecchioli 1986). Where injection wells have been shown to have failed, contaminating overlying freshwaters, the failure has been a result of faulty well installation and the possible presence of faults or other breaches in the confining layer (i.e., Duncan et al. 1994). Concerns continue about the adequacy of confinement and, in the case of Class I injection wells,[4] migration of treated and injected wastewater to the continental margin abound (U.S. Environmental Protection Agency 2003).

One indicator that the MCU serves to confine circulation between the UFA and LFA is the development of a geothermal gradient across MCU II (Fig. 4.30) between the LFA and UFA. Hickey (1990), for example, found that the geothermal gradient in central Florida was 0.03 ° C/m. Hickey (1990) also noted a 91,000–110,000 kg/m$^2$ pressure differential across MCU II in central Florida. This gradient suggests a potential for upward flow from the LFA to the UFA across the MCU. However, the water-quality data indicate that upward flow across the MCUs is minimal in most areas. Kohout et al. (1977) noted similar geothermal gradients in deep petroleum exploratory wells in south Florida. These strong geothermal and pressure gradients testify to the general ability of the MCUs in southern and central Florida to isolate the UFA from the LFA.

[4]Class I injection wells are permitted for injection of wastes and hazardous materials into aquifers that are unsuitable for other uses and that are isolated from potable-water aquifers by appropriate confining units.

**Table 4.4** Summary of vertical hydraulic conductivity and transmissivity measurements in the middle semiconfining unit

| Hydraulic conductivity (m/s) or transmissivity (m$^2$/day) | Author(s) | Confining unit | Data source |
|---|---|---|---|
| $4.5 \times 10^{-11}$ to $7 \times 10^{-6}$ m/s | Stewart (1966) | MCU II | Laboratory estimates from cores |
| $2 \times 10^{-12}$ to $4.9 \times 10^{-6}$ m/s | Hickey (1981) | MCU II | Laboratory estimates from cores |
| $3 \times 10^{-4}$ m/day | Brown (1984) | MCU I | Packer tests |
| $8.5 \times 10^{-11}$ to $3 \times 10^{-6}$ m/s | Navoy (1986) | MCU II | Laboratory estimates from cores |
| $1 \times 10^{-9}$ to $1 \times 10^{-7}$ m/s | Haberfeld (1991) | MCU III | Packer tests |
| 65.6–274.2 m$^2$/day | Quiñones-Aponte et al. (1996) | MCU I | Packer tests |

Kuniansky and Bellino (2012a, b) summarized hydraulic properties of the FAS MCUs in the southeastern U.S. Fourteen aquifer performance tests (APTs) were evaluated. The reported transmissivities ranged from $9 \times 10^{-2}$ to $6 \times 10^4$ m$^2$/day. Median transmissivity was about $3 \times 10^3$ m$^2$/day. Only two APTs reported transmissivities less than 9 m$^2$/day, and eight reported transmissivities greater than $9 \times 10^3$ m$^2$/day. Table 4.4 summarizes the ranges of vertical hydraulic conductivities in MCU II and MCU III. Haberfeld (1991) estimated the transmissivities of the MCUs to be less than 0.6 m$^2$/day.

Where present, the MCUs serve as semi-confining units between the UFA and the LFA. As is shown below, the MCUs in some areas serve as the base of the fresh-water lens. In these areas, a second, different groundwater flow system driven by thermal convection appears to have developed in the LFA. Where the MCUs are absent or leaky, the LFA may be part of the gravity-driven, freshwater flow system of Florida.

### 4.5.3 Lower Floridan Aquifer

The LFA is defined by the presence of one or more MCUs that overlie the aquifer. Where the MCUs exist, the LFA exists. Where the MCUs are absent or ineffective as confining units, the LFA and the UFA cannot be differentiated and, by definition, the FAS consists entirely of the UFA.

Within the last decade, water managers in the central and northern parts of Florida have become interested in the LFA because fresh and/or treatable brackish water has been discovered in the LFA. Deep exploration wells and sub-regional studies are currently underway in central and northern Florida to determine if lower salinity water in the LFA constitutes a sustainable resource for water supply.

The hydrostratigraphy of the MCUs and LFA is confusing and has been defined differently by numerous authors. This confusion has largely been a result of (1) the somewhat limited spatial scopes of the studies, which only discuss subregional conditions of these units and (2), until recently, limited and local knowledge of the MCUs and their relationships to the LFA.

Our knowledge of the MCUs and LFA was limited for years because there was little incentive to explore for water in the deeper portions of the FAS owing to the abundant, high-quality water available in the UFA. What we did know about the MCU and LFA came from deep wells advanced for petroleum exploration. While there were many of these wells, their spatial distribution was not sufficient to accurately characterize the lower portions of the FAS.

In the last 30 years, however, the LFA has been locally developed for disposal of treated wastewater and saline-water concentrates from desalinization plants near the coast through Class I injection wells (Hickey and Vecchioli 1986; Muniz et al. 2003). All of Florida's 180 Class I injection wells (Florida Department of Environmental Protection 2014) are over 730 m deep, and are completed in the LFA. Injection of treated wastewater requires that the injection zone(s) in the LFA be well confined by the MCU (Florida Department of Environmental Protection 2014).

In addition to wastewater disposal, Florida has become a locus for wells installed for the purpose of aquifer storage and recovery (ASR, Florida Department of Environmental Protection 2014). ASR wells are allowed to inject water that meets Florida's drinking water quality standards into potable aquifers. The purpose of ASR is to store the water for future recovery and use. Therefore, some degree of confinement and assurance that the injectate will not migrate away from the recovery well is desirable. As a result, ASR wells tend to be deep, and they have provided considerable information concerning the deeper FAS. Deep wells are, therefore, becoming sufficiently abundant in some parts of Florida to allow for locally detailed characterization of the MCU and, where it is penetrated, the LFA.

#### 4.5.3.1 Aquifer Distribution

It is important to realize that the LFA and its relationships to the MCUs and confining beds within the LFA are complex. Figure 4.30 illustrates the distribution of the LFA based on the presence or absence of the MCUs. Wherever one or more MCUs exist, permeable strata below the MCU(s) are, by definition, termed the LFA.

There are two economically important, high permeability zones in the LFA. The so-called "Boulder Zone" in southern Florida (Fig. 4.31) is widely utilized for injection of wastewater via Class I injection wells. In northeastern Florida and southeastern Georgia, the Fernandina "permeable zone" (FPZ, Fig. 4.31) is also used for disposal of treated water through injection wells. The FPZ is a horizon of major concern as a location of saltwater intrusion.

#### 4.5.3.2 Aquifer Properties

Kuniansky and Bellino (2012) reported on the results of 64 APTs in the LFA in Florida, Georgia, South Carolina, and Alabama. The transmissivities ranged from 12 to 5 m$^2$/day with a median value of $4.4 \times 10^2$ m$^2$/day. The uppermost permeable zone of the LFA in northeast Florida and southeast Georgia was reported to have transmissivities that ranged from $4 \times 10^3$ to $6 \times 10^3$ m$^2$/day.

#### 4.5.3.3 Aquifer Potentials and Circulation Systems

Sproul (1977) characterized geothermal gradients in south Florida and identified anomalous areas of high and low heat flow. The high gradient anomalies were thought to reflect upwelling of warmer, normally to highly mineralized water. "Hot spots" identified by Sproul included:

- A region of upwelling mineralized, warm water that follows a north-south trend in Hardee and DeSoto counties (Kaufman and Dion 1967);
- A 93 km$^2$ township (T44S, R23E) area in western Lee County;
- A 5 km$^2$ area[5] in the McGregor Isles community of Lee County;
- A flowing, salt-water well that had been developed as a spa in the 1960s and 70s known as "Hot Springs of Florida"[6] in coastal southern Charlotte County (Kouout et al. 1977); and the
- Martin County "hot well", a 257 m deep well, whose water had low-salinity and temperature of 33 ° C, which is about 2.5 °C above ambient.

Warm springs are also found in southwest Florida (i.e., Warm Mineral and Little Salt springs in Sarasota County (Cockrell 1975; Clausen et al. 1979), and Mud Spring (Breland 1980; Fanning et al. 1979, 1981), which is located offshore of Ft. Myers, Lee County.

Sproul's (1977) "east coast cooling anomaly" was thought to reflect deep bottom water from the offshore Atlantic entering the LFA by "cyclic inland flow." This anomaly extends along the Atlantic coast from St. Lucie County, south into the upper Florida Keys of Monroe County (Sproul 1977). Lowest temperature reported in the anomaly was 15 °C in a 945 m deep well near Ft. Lauderdale.

[5] Sproul (1977) believed that upconing of saline water in McGregor Isles was along faults.

[6] This well, Humble Oil's Lowndes-Treadwell #1, was drilled in 1945. It was 610 m deep and had flowing water at 36 ° C and 18,700 mg/l chloride. The well has since been plugged and abandoned in order to protect the overlying aquifers.

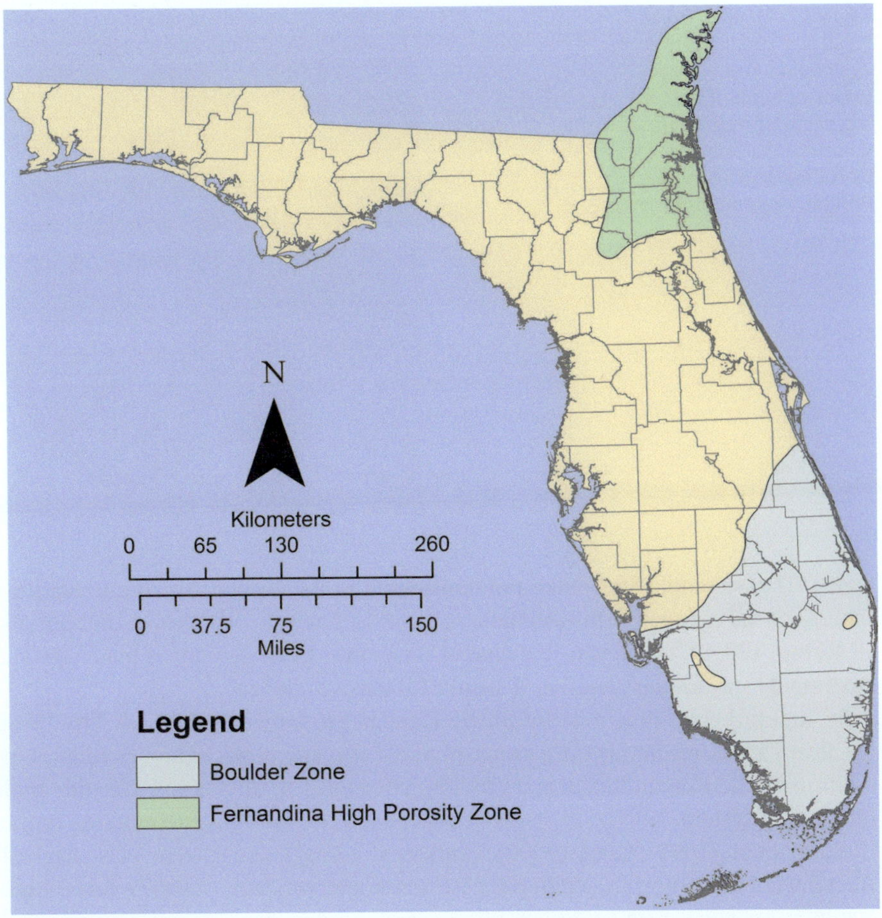


**Fig. 4.31** Extent of the Boulder Zone in south Florida and Fernandina permeable zone of northeastern Florida and southeastern Georgia. (From Miller (1990))

Two hypotheses have been proposed for this anomalous wedge of cool water that appears to be intruding into the LFA. Lichtler (1960), Stringfield (1966), and Vernon (1970) opined that the wedge of cool water was caused by conductive cooling through heat loss from the LFA to the cold ocean water off the edge of the continental shelf. Kohout (1967, 1975), Henry and Kohout (1972), and Kohout et al. (1977) proposed that the intrusion of cool water was a result of convective flow driven by the geothermal gradient within the FAS (Fig. 4.9).

While our knowledge of the MCUs and LFA has evolved considerably since the 1960s and 70s, the Kohout circulation model (Fig. 4.9) has weathered the test of time and is currently the favored groundwater circulation model (Meyer 1989). This model has been applied to the salt-water/freshwater interface in south Florida and the FAS elsewhere in Florida (Barlow 2003).

Note in Fig. 4.9 that Kohout included groundwater circulation in both the UFA (the "principal artesian zone") and the LFA (the Boulder Zone) as two, related and contiguous circulation cells that apparently share a salinity transition zone. The MCU is not indicated as being present, but it would occur along the dashed line that separates the principal artesian aquifer and freshwater lens from the geothermally driven circulation in the boulder zone, or LFA. The "Confining Beds" shown in the figure constitute the IAS/ICU. Under Kohout's model, relatively cold sea water (the upwelling, warm water and deep saline water are often near seawater composition and concentrations) is drawn into the aquifer system along the coast where it is heated and rises as a result of geothermal heat flow from below the FAS.

In southeast Georgia and northeast Florida (the area denoted by the UFA potentiometric-surface decline shown near Jacksonville and Savannah on Fig. 4.28), declines in UFA potentials combined with fractured and/or faulted MCU conditions and a strong upward hydraulic gradient in the LFA have resulted in upconing of LFA water into the UFA (Miller 1986; Bush and Johnston 1988; Spechler 1996; Phelps and Spechler 1997; O'Reilly et al. 2002; Kuniansky and Bellino 2012; Kuniansky et al. 2012), a major water-quality problem.

Assuming that there are one or more MCU horizons and high permeability zones, the Kohout model (Fig. 4.9) was a simple hypothesis that appears conceptually correct. There should be multiple circulation systems and salinity interfaces, and variably leaky strata between the permeable zones would control interchange of water between permeable

**Fig. 4.32** Boulder Zone "boulders" from a petroleum exploration well in Dade County, Florida. These are fragments of well-lithified dolostone that were rounded by rolling under a tri-cone drill bit. Keys provide scale

zones. These complexities were not considered in the early models of the convective flow model.

Rubin (2010), however, concluded that the LFA is "extremely cavernous" and, as a result, "intensive dispersion" and turbulent flow may invalidate Darcian groundwater flow. More recent investigations of the LFA are now confirming the Kohout model and providing details as to the circulation system.

Clark et al. (1997), Clark (2002), Clark et al. (2004), and Rectenwald (2010) examined noble gases, radiocarbon, and stable isotopes in LFA water in Florida and southeastern Georgia. They have consistently found evidence supporting the Kohout model, including helium concentrations and salinities consistent with Pleistocene seawater in southeast Georgia and post-Pleistocene water in southern Florida. Based on these findings, it was concluded by these authors that the saline water in the LFA and MCU intruded the aquifer system within the last 10,000 years. Rectenwald (2010) also noted evidence of post-Pleistocene seawater entering the MCU and LFA from the Gulf of Mexico in a well located in Collier County. From these data, he suggested that Kohout-type circulation enters the LFA from the Gulf of Mexico, as well as the Straits of Florida on the east coast.

Strong upward hydraulic gradients in the LFA create buoyancy but, owing to confinement, significant, focused upward flow is limited to fractures, possibly faults, and abandoned, deep test wells. Hydrostatic pressure generated by the Holocene sea level rise and by the geothermal gradient causes the gradient.

Sproull's (1977) "east coast cooling anomaly" and Miller's (1990) conclusion that, based on temperature and salinity in the boulder zone(s), these zones are connected to the sea on the continental shelf support the Kohout groundwater circulation model. The 150–300 m of low hydraulic conductivity sediments (the MCU) on top of the boulder zones may isolate injection zones from contaminating overlying aquifers. However, Duncan et al. (1994) have shown that not all Class I injection wells are vertically isolated from upward migration of injectate.

#### 4.5.3.4 The "Boulder Zone"

The "Boulder zone" is an informal hydrostratigraphic term used for the highly fractured, high permeability zone(s) encountered in the LFA in southern and central Florida. The term was originally coined by Kohout (1967) and Burke (1967) for a high-permeability zone encountered in deep petroleum test wells in southwestern Florida. In these wells, the drill rig would encounter highly fractured, hard dolostone, fragments of which would spall into the borehole. The rotation of the drill bits rounded the fragments, giving them the appearance of small boulders (Fig. 4.32), hence the term "Boulder Zone." There are no known horizons of boulder-sized clasts in the boulder zones.

The boulder zone(s) of south Florida consist of highly fractured dolostone (Kohout 1967; Maliva et al. 2001). The dolostone is diagenetic in origin and, owing to its high degree of lithification, easily fractured by tectonic forces and drilling processes.

There are multiple horizons of fractured dolostone. Maliva et al. (2001) concluded that the number of boulder zones in southwestern Florida decreased to the south and that the uppermost boulder zones were deeper in the southern part of their study area than in the northern part. The boulder zone described by Miller (1990) is fractured dolostone with an irregular top that ranges from −610 to −1040 m NGVD.

There is confusion relative to the importance of karst processes in the boulder zones. This confusion appears to be related to differences of the interpretation of such words as "cavernous." In some cases, contexts of early reports on the boulder zones suggest that the authors were communicating size of the cavities in the boulder zones, not origin.

Whatever the intent, words like karst, dissolution, and cavern are contained in discussions of high-permeability boulder zones in the literature.

There is no question that pore spaces in the boulder zones can be huge. For example, there was a 27 m drill bit drop in Sun Oil Company's Red Cattle 32-2 well in the Felda oil field in Hendry County (Kohout 1965, 1967; Burke 1967). This cavity extended from 768.7 to 796.1 m below land surface. Maliva et al. (2001) attributed this rod drop to a fracture in the dolostone.

Vernon (1970) attributed "cavernous" porosity in the boulder zones to calcite dissolution with concomitant dolomite precipitation. Puri and Winston (1974) described the boulder zones as an intricate network of vugs and large cavities. The boulders were thought to have dropped from the roofs of caverns. Miller (1990) thought that the boulder zones represent caverns that are interconnected by pipes or solution tubes. Winston (1995) argued that boulder-producing zone(s) contain cavities and wall-collapse zones.

More recently, Safko and Hickey (1991) and Duerr (1994) examined borehole video logs and geophysical logs in the LFA in detail. They concluded that the porosity of the boulder zone is a result of fracturing, not dissolution of the dolostone. Duerr (1994, p. 1) observed that:

> "*Fracture porosity was identified* [in 11 Class I injection wells and 3 test wells] *as the most common type of effective secondary porosity and was observed predominantly in dolomite and dolomitic limestone. Cavity porosity was the least common type of effective secondary porosity at the study sites. In fact, of the more than 17,500 feet of borehole studied, a total of only three cavities constituting effective secondary porosity were identified at only two sites. These cavities were detected in dolomite rocks. Most apparent cavities were caused by drilling-induced collapse of naturally fractured borehole walls. Also, fractures usually were observed above and below cavities. The majority of vugs observed in the television surveys did not constitute effective secondary porosity.*"

> "*No effective secondary porosity was evident in the limestone or dolomitic limestone in the 300-foot [91 m] interval* [of the LFA] *immediately above the injection interval at six sites on the southeastern coast of Florida. Injection wells commonly are cased through the 300-foot interval. Fractures or cavities that contribute to effective secondary porosity may be present in this interval, but were not detectable with the methods used. Widely dispersed, interconnected fractures or cavities can be present beyond the rock column intersected by the borehole and can provide local pathways for vertical migration of injected wastewater or the displaced saltwater. In the interval between the top of the Floridan aquifer system and a point 300 feet above the top of the injection interval, fractured rocks having effective secondary porosity were observed at five of six sites along the southeastern coast.*" (Duerr, 1994, p. 1)

Maliva et al. (2001) investigated Class I injection well sites in southwest Florida and concluded that the dolostones of the boulder zones are regional diagenetic features occurring at varying depths and stratigraphic positions. In many areas of Lee County, discontinuous anhydrite cement occludes fracture networks in the Oldsmar Formation. The presence of anhydrite as a cement in fractured dolostone suggests that the dolomitization and fracturing of the host strata preceded precipitation of the anhydrite.

The presence of the boulder zones is critical to the Kohout convection model. It is within the boulder zones that the groundwater circulation is best developed, and the groundwater mixing associated with the circulation system may be a source of porosity development. Clark et al. (2004) investigated the Kohout circulation system in the LFA using dissolved noble gases, stable isotopes of water, chlorinity, radiocarbon, and strontium isotopes. Based on geochemical data, they concluded that the circulation system was geologically rapid, having formed within the last 10,000 years as a result of sea-level rise, and that the sources of the saline water in the LFA are coastal.

Hughes et al. (2010) modeled Kohout convection in south Florida. As shown in Fig. 4.9, the Kohout convection model appears asymmetrical with primary inflow from the eastern continental margin. The Hughes et al. numerical model investigated this asymmetry and the influence of heat flow on circulation. They concluded that without geothermal heating of the system, the circulation system would still exist as a result of forced circulation in sympathy with the overlying, gravity-driven freshwater circulation system. However, the pattern of modeled circulation without geothermal heat flow was different. Addition of heat flow from below reduced the asymmetry of the Kohout circulation system and reduced seawater inflows from the Atlantic by 8% and from the Gulf of Mexico by half.

Cowart et al. (1978) used $^{234}U/^{233}U$ ratios to examine the flow system in the boulder zone of southeastern Florida. They concluded that the isotopic ratio in the boulder zone at −800 to −1000 m below sea level indicated a water source offshore in the Straits of Florida. These data were thought to support the Kohout LFA circulation model.

Meyer (1989) concluded that the saltwater entering the FAS from the Straits of Florida enters the aquifer by way of karst features (ancient sinkholes) exposed on the Miami Terrace near Fort Lauderdale. These karst features may reflect the intersection of the boulder zones with the continental margin.

As might be expected, aquifer testing of the boulder zones in south and central Florida has been limited. Table 4.5 presents some examples of the results of testing. Reported transmissivities range from $7.6 \times 10^3$ to $2.3 \times 10^6$ m²/day. Clearly, identification of boulder zones in south Florida is more a matter of permeability contrast than some critical value of transmissivity.

**Table 4.5** Transmissivities and hydraulic conductivities reported from boulder zone permeable zones in the Lower Floridan Aquifer in south Florida

| Hydraulic conductivity (m/s) or transmissivity ($m^2$/day) | Author(s) | Production zone |
|---|---|---|
| $4.0 \times 10^5$ to $2.3 \times 10^6$ $m^2$/day | Meyer (1974), Singh et al. (1983) | "Boulder zone" |
| $2.3 \times 10^6$ $m^2$/day | Myer (1989) | "Boulder zone" |
| $7.1 \times 10^4$ $m^2$/day | Quiñones-Aponte et al. (1996) | −377 to −509 m NGVD |
| $5.48 \times 10^{-2}$ m/s | | −389 to −398 m NGVD |
| $1.81 \times 10^{-2}$ m/s | | −419 to −424 m NGVD |
| $4.01 \times 10^{-2}$ m/s | | −456 to 462 m NGVD |
| $1.44 \times 10^{-2}$ m/s | | −472 to 476 m NGVD |
| $7.6 \times 10^3$ $m^2$/day<br>$>5 \times 10^4$ $m^2$/day | O'Reilly et al. (2002) | LFA, Northwest Orange County<br>LFA, south of Orlando |
| $5 \times 10^2$ to $7 \times 10^4$ $m^2$/day | Sepúlveda (2002) | LFA, model derived |
| $1.5 \times 10^5$ $m^2$/day<br>$<9.3 \times 10^3$ $m^2$/day | Reese and Richardson (2008) | Avon Park permeable zone, DeSoto County<br>Avon Park permeable zone, South Florida |

In conclusion, the boulder zones consist of fractured dolostone and, in some areas, dolomitic limestone or limestone. The Kohout circulation system, which includes the boulder zones, appears confirmed in southern Florida. Given the complexities of circulation in fractured-rock aquifers with multiple potential high permeability zones surrounded by reduced permeability zones, use of the boulder zones for wastewater disposal is a concern. The performance of Class I injection wells suggests general success in terms of containing vertical flow. The effects of lateral flow, dispersion and dilution along the Kohout circulation system flow paths, and confinement in areas distant from the injection wells remain unknown.

#### 4.5.3.5 The "Fernandina Permeable Zone"

A very high permeability zone at the base of the LFA was first discovered in 1945 in a 649 m deep well at Fernandina Beach, Nassau County (Krause and Randolph 1989). That highly permeable zone has since been known as the "Fernandina permeable zone" (FPZ). It is known to occur in extreme northeastern Florida and southeastern Georgia (Fig. 4.31). The FAS in this area consists of three productive zones: the UFA permeable zone and two permeable zones in the LFA (Bentley 1979; Brown 1984). The LFA permeable zones are termed the middle and lower permeable zones of the FAS. The uppermost permeable zone of the LFA (the middle permeable zone of the FAS) is used as a source of potable, fresh water. The lower, saline permeable zone (the FPZ) occurs at the base of the LFA (Brown 1984). According to Brown (1984), the uppermost production zone of the LFA occurs in the Middle Eocene lower Avon Park Formation[7] and the FPZ is within the Lower Eocene Oldsmar Limestone. The base of the LFA in the study area consists of evaporites and dense limestones of the Paleocene Cedar Keys Limestone (Brown 1984). The MCU that contains the middle permeable zone and overlies the FPZ is Miller's (1986) MCU 1.

The uppermost permeable zone of the LFA was found at 370–518 m below land surface in Nassau County, Florida. It is isolated by relatively impermeable limestone and dolostone from the overlying upper permeable zone, which is within the UFA, and the FPZ at the base of the LFA. The FPZ occurs at about 600–640 m below land surface. The thickness of the FPZ ranges from 30 m in the Jacksonville area to over 150 m in the Brunswick, Georgia area (Krause and Randolph 1989). The FPZ contains saline water with approximately 8000 mg/l $Cl^-$.

Both of the LFA permeable zones in northeast Florida are artesian. Potentials in the FPZ are higher than in the overlying permeable zone, so there is a potential for saline water derived from the FPZ to upcone into the overlying permeable zone along fractures, faults, and poorly installed wells. Fractures and possible faults that allow inter-aquifer flow and cause increases in the salinity of the potable, upper LFA are thought to occur in the area (Vernon 1951; Leve 1966, 1978; Krause and Randolph 1989; Spechler 1996, 2001; Phelps and Spechler 1997). However, none of these fractures and faults have been confirmed.

It is because of these inter-aquifer transfers of saline water that the FPZ is a concern. Because of natural head differences between the FPZ and the upper permeable zone of the LFA, saline water can migrate along fractures and cause salinity increases in the potable upper permeable zone of the LFA (Krause and Randolph 1989; Spechler 1996; Phelps and Spechler 1997; Spechler and Phelps 2001). Pumpage and resulting head losses in the UFA have, in recent years, caused saline water to migrate upward into the UFA, as well. Given that the UFA in the region is underlain by the FPZ and is highly confined by the ICU, drawdown and head loss in the UFA have been extreme, and upward migration of saline water has been greatly enhanced by development in northeast Florida and southeast Georgia.

It is important to understand that not all saline water intrusion into the FAS in southeast Georgia and northeast Florida is from the LFA permeable zones (Georgia Department of Natural Resources undated). Direct intrusion of seawater into poorly confined UFA has also been documented. However, upward migration from the LFA has been documented in the Brunswick area of Georgia (Georgia

[7]Brown (1984) used stratigraphic terminology that has since been changed. He placed the upper permeable zone of the LFA in the "Lake City Limestone", which is now considered to be part of the Avon Park Formation (Miller 1986; Scott 2001).

Department of Natural Resources undated; Krause and Clarke 2001; Jones 2001) and in Nassau and Duval counties in Florida (Krause and Randolph 1989; Spechler 1996; Phelps and Spechler 1997).

Hutchings and Tarbox (2003) utilized upward leakage from the LFA (undifferentiated) into the UFA as a calibration variable for a saltwater intrusion model in coastal Flagler County, Florida. They estimated that upward flux from the LFA into the UFA was 0.19 m/year. Sepúlveda and Spechler (2004) developed a groundwater flow model to evaluate use of freshwater injection wells to mitigate saltwater intrusion in the FPZ. Their simulated, steady-state flow budget suggested a net landward flow of saline water into the Fernandina permeable zone and then upward, into the upper zone of the LFA of 0.14 $m^3/s$. The modeled flux into the FPZ from the Atlantic was estimated to be 0.65 $m^3/s$ and out of the FPZ into the Atlantic was 0.51 $m^3/s$, which resulted in the calculated net inflow of 0.14 $m^3/s$. This circular flux suggests that there is a convective-type circulation system with losses by leakage through the confining strata that overlie the FPZ.

The circulation system in the FPZ is not fully understood. Fairchild and Bentley (1977) thought that the saline water was connate in origin, which would suggest a poorly developed circulation system. The model suggested by Krause and Randolpf (1989) and Spechler (1996) seems to suggest a more dynamic circulation system; however, it is uncertain at this time as to whether groundwater circulation is a result of Kohout geothermal mechanisms or some other causative mechanism. Both of the LFA aquifers in northeastern Florida and southeastern Georgia are highly transmissive (Table 4.6) and compare favorably with the boulder zones of southern Florida (Krause and Randolph 1989; Phelps and Spechler 1997).

As with the boulder zones of southern and central Florida, the high permeabilities of the upper LFA and FPZ have been attributed to karst processes (Miller 1986; Phelps and Spechler 1997; Spechler 1994, and others). Seismic subbottom profiling of the St. Johns River (Spechler 1994, 2001) and offshore (Popenoe et al. 1984, Snyder et al. 1989) has revealed down-warped strata and enhanced seismic penetrations that suggest karst features that extend to depths sufficient to intersect the FPZ. These seismic features, plus the high permeability and borehole video evidence of cavities in the LFA aquifers has led the authors of publications on the LFA in northeastern Florida and southeastern Georgia to conclude that the permeability is a result of karstification.

In order to explain karstification at depths greater than 600 m, Spechler (1994, 2001) and Krause and Clarke (2001) argued that a deep, ancient water table aquifer existed in the area, causing dissolution of the carbonate rocks in these aquifers. This argument was apparently a result of the assumption that karstification was epigenetic and occurred at, or near, the water table. It is also possible that the karst developed on a deeper freshwater/saline water transition zone and is hypogenetic in the sense of Palmer (1991). Unlike the boulder zones of south Florida, little or no consideration appears to have been given to fracturing as a cause of the porosity, Kohout circulation, or mixing-zone dissolution or dolomitization in the Eocene LFA rocks of northeast Florida. Data are sparse, and there is a need for a comparison of the origin of the rocks and porosity of the LFA in northeast Florida to the boulder zones of south and central Florida, continental margin tectonics, and modern concepts of mixing zone karstification.

**Table 4.6** Transmissivities reported from permeable zones in the lower Floridan aquifer in southeast Georgia and northeast Florida

| Transmissivity ($m^2$/day) | Author(s) | Production Zone |
|---|---|---|
| $4 \times 10^3$ to $6 \times 10^3$ $m^2$/day | Brown (1984) | "Middle water-bearing zone" of the FAS (uppermost permeable zone of the LFA) |
| $6.9 \times 10^3$ $m^2$/day | | "Lower water-bearing zone" of the FAS (Fernandina permeable zone); estimated from groundwater flow model |
| $4.5 \times 10^4$ to $6.5 \times 10^5$ $m^2$/day | Sepúlveda and Spechler (2004) | "Upper zone of the LFA", model developed |
| $5.6 \times 10^3$ $m^2$/day | | Fernandina permeable zone, calibration variable |

### 4.5.4 Sub-Floridan Confining Unit

The sub-Floridan confining unit (SFCU) consists of glauconitic, calcareous, argillaceous to arenaceous strata that range in age from Late Paleocene to Late Eocene, or massively bedded anhydrite in the Lower Paleocene strata (Miller 1986). In the Florida peninsula, strata of the SFCU include portions of the Lower Eocene Oldsmar Formation and the Paleocene Cedar Keys Formation (Southeastern Geological Society 1986). The SFCU forms the lower boundary of the Kohout circulation system and permeable zones of the LFA.

## 4.6 Summary of Florida's Karstic Groundwater Systems

This section summarizes the aquifer systems of Florida and their groundwater flow systems. The chapter began with a discussion of rainfall and recharge. It is important to understand that

- Rainfall and, therefore, recharge in Florida is highly variable with wet and dry seasons and numerous, multi-year droughts and wet episodes;

- Rainfall varies across the state with the northern portion receiving rainfall from continental derived weather patterns and tropical and sub-tropical weather;
- Rainfall in the Florida peninsula approximately south of the Suwannee River is typically characterized by seasonal rainfall often caused by convective thunderstorms and tropical storms and hurricanes;
- Evapotranspiration accounts for most of the rainfall, and the amount of water available for recharge is typically a small fraction of the available rainfall;
- Recharge is variable across the state with relatively high recharge in karstic areas where the carbonate aquifers are unconfined; and
- In areas where there is a shallow aquitard, generally the clay-rich strata of the Hawthorn Group, recharge is limited and surface-water runoff is enhanced.
- Florida has three regional aquifer systems: the Surficial, Intermediate, and Floridan Aquifer Systems.
  - Surficial Aquifer System (SAS)
    - The SAS is unconfined and, where present, includes the water table.
    - Where the UFA is unconfined and there is no confining unit to separate water in the UFA from the SAS, the unconfined, water-table aquifer is, by definition, considered to be part of the UFA.
    - The SAS includes three aquifers: the (1) Sand and Gravel Aquifer of the western Florida Panhandle (Santa Rosa and Escambia counties), (2) the Biscayne Aquifer of southeastern Florida (Dade and southern Broward counties); and (3) undifferentiated SAS in the remainder of the state.
    - The Sand and Gravel and Biscayne aquifers are the primary aquifers in their areas and supply much of the available potable water.
    - The Sand and Gravel Aquifer is composed of siliciclastic sediments and is not karstic.
    - The Biscayne Aquifer is composed of limestone and is karstic.
    - Because of development of karst features and high hydraulic conductivities, the Biscayne Aquifer is considered to be one of the most productive aquifers in the world.
    - The undifferentiated SAS is comprised of siliciclastic sediment with local shell and clay lenses.
    - The sandy and shelly strata locally provide water to domestic wells.
    - In the Atlantic Coastal Complex geomorphic province, shelly sands of the SAS provide water to a number of municipalities and industries.
    - In southwestern Florida, sandy, shelly, and limestone strata are also utilized for public and industrial supply.
    - Where shell beds or limestone lenses are shallow, they may develop karst features.
    - Strata of the SAS are largely Plio-Pleistocene in age.
    - SAS circulation systems are typically classical, gravity driven systems with recharge in the interiors of the shallow groundwater drainage basins and discharge along the coasts or regional streams and rivers. Tóthian flow (Tóth 1962, 1963) occurs within these small watersheds.
  - Intermediate Aquifer System and Confining Unit (IAS/ICU)
    - Strata of the IAS/ICU are typically Miocene in age and comprise the Hawthorn Group of sediments.
    - The IAS consists of one or more strata of sand and gravel or limestone/dolostone that are capable of supplying water to wells.
    - The IAS is considered a local aquifer that has been developed for public and industrial supply largely in southwestern Florida; it has been previously utilized in the Jacksonville area and is known to supply water to a few of Florida's springs.
    - The ICU comprises the silt- and clay-rich portions of the Hawthorn Group.
    - Individual lithosomes within the IAS/ICU range from laterally extensive layers of hundreds of hectares to lenses that have extents of less than a hectare, so lithologic variability is an important property.
    - The ICU is an aquitard and much of its ability to confine the underlying UFA is a result of a combination of flow retardation by clay and tortuous flow paths through the more permeable sediments interlayered with the clay-rich strata.
    - Flow in carbonate-rich strata, especially limestone, is known to develop small conduits, and some springs are known to discharge IAS water.
    - Karst features developed within the IAS/ICU are rare and limited to the carbonate-rich strata within the Hawthorn Group near karst escarpments where the Hawthorn Group has been penetrated by sinkholes and fractures. The IAS/ICU may be penetrated by collapse sinkholes developed in the underlying limestones of the FAS.
    - Lateral flow under gravity induced gradients occurs in the carbonate-rich (and siliciclastic sand and gravel beds) strata.
  - Floridan Aquifer System (FAS)
    - The FAS is subdivided into two aquifers by the Middle Confining Units (MCUs), a series of strata characterized by low permeability limestone and evaporites (gypsum and anhydrite).
    - In some parts of Florida there are more than one MCU; in other areas, the MCUs are absent.

- Where one or more MCUs are present, the FAS is subdivided into the UFA and LFA; where no MCU exists, the FAS consists of one aquifer.
- The UFA consists of limestone and dolostone of Eocene to Early Miocene age; any carbonate strata below the lowest clay-rich stratum of the Miocene Hawthorn Group is considered part of the UFA.
- The UFA contains well-developed karst landforms and features that have formed as a result of dissolution of limestone, and to a limited extent, dolostone by meteoric water derived from rainfall.
- The UFA may also have karst features that developed deeper in the aquifer along freshwater/saline-water mixing zones.
- The UFA includes numerous karstic horizons, including
  - Epigenetic karst developed at the current upper surface of limestone and within unconformities at formation contacts in the subsurface, and
  - Hypogenetic karst developed along saline-water transition zones.
- Sea-level fluctuations have apparently caused transition-zone features to develop at multiple levels within the FAS.
- The UFA is considered to be one of the most productive aquifers in the world and it is the primary source of potable water in most of Florida.
- The LFA is Paleocene to Eocene in age and consists of dolostone with minor limestone and evaporite mineral content.
- One or more boulder zones, horizons of very high hydraulic conductivity, are found within the LFA, especially in northeastern, central, and southern Florida.
- The LFA may contain deep karst developed along freshwater/saline-water mixing zones.
- The FAS is confined at its base by lower permeability, Paleocene and older strata.
- Flow in the UFA is typically simple, regional, gravity-driven flow with recharge where the aquifer is unconfined or poorly confined and discharge along the coasts and major rivers.
- Springs, sinkholes, and other karst features are common where confinement is not present or poorly developed; there is much less karst development in areas where the UFA is highly confined.
- The LFA has only been investigated in a few areas of the state; in southern Florida, circulation appears to be driven by geothermal heating in the interior of the state. Salty groundwater appears to enter the LFA at the continental margins and circulate to the interior where it is warmed, rises, and then flows back to the sea near the top of the LFA.

## References


Adams D North J Maddox G (undated) Conversion of DRASTIC maps from FastCad drawings to ArcInfo coverages. Fla Dept Environ Prot, accessed 2/6/2014 at http://proceedings.esri.com/library/userconf/proc98/proceed/to450/pap402/p402.htm

Arthur JD Fischler C Kromhout C Clayton J Kelley M Lee RA Li L O'Sullivan M Green R Werner C (2008) Hydrogeologic framework of the Southwest Florida Water Management District. Fla Geol Surv Bull 68

Ayers JF Vacher HL (1983) A numerical model describing unsteady flow in a fresh water lens. Water Res Bull, 19:785–792

Badon Ghyben W (1889) Nota in verband met de voorgenomen putboring nabij Amsterdam. Tijdschrift van het Koninklijk Instituut van Ingenieurs, The Hague, pp. 8–22

Barlow PM (2003) Ground water in freshwater-saltwater environments of the Atlantic coast. US Geol Surv Circ 1262

Basso RJ (2002) Hydrostratigraphic zones within the Eastern Tampa Bay Water Use Caution Area. Brooksville, Southwest Fla Water Mgt Dist

Bear J Cheng AH-D Sorek S Ouazar D Herrera I (eds.) (1999) Seawater intrusion in coastal squifers – Concepts, methods, and practices. Dordrecht Netherlands, Kluwer Academic Publishers

Bear J Todd DK (1960) The transition zone between fresh and salt waters in coastal aquifers. Water Res Ctr Cont No. 19, Hydraulic Laboratory, University of California

Beddows PA Smart PL Whitaker FF Smith SL (2007) Decoupled fresh–saline groundwater circulation of a coastal carbonate aquifer: Spatial patterns of temperature and specific electrical conductivity. Jour Hydrol 346:18–32

Bellino JC (2011) Digital surfaces and hydrogeologic data for the Floridan aquifer system in Florida and in parts of Georgia, Alabama, and South Carolina. US Geol Surv Data Series 584

Bengtsson T Radin H (2001) Water level elevation maps of the primary aquifers in the lower west coast of Florida. In: Missimer TM Scott TM (eds.), Geology and hydrology of Lee County, Florida - Durward H. Boggess Memorial Symposium, Fla Geol Surv Sp Pub 49, p. 139–149

Bentley CB (1979) Aquifer coefficients determined from multiple well effects, Fernandina Beach, Florida. Ground Water 17(6):525–531.

Berndt MP Katz BC (1992) Hydrochemistry of the surficial and intermediate aquifer systems in Florida. US Geol Surv Water-Res Inv Rept 91-4186

Bock WD Moore DR Neumann AC Supko PR (comp.) (1969, reprinted 1994) Late Pleistocene geology in an urban area. Miami Geological Society field trip guidebook, 31 p.

Boggess DH (1968) Water-supply problems in southwest Florida. Fla Div Geol Open-File Rept

Boggess DH Missimer TM O'Donnell TH (1977) Saline-water intrusion related to well construction in Lee County, Florida. US Geol Surv Water-Res Invest 77-1-33

Breland JA (1980) Chemical and physical characteristics of a saline geothermal submarine spring off Florida's southwestern coast. Master's thesis, Univ South Fla, St. Petersburg

Brooks H K (1961) The submarine springs off Crescent Beach, Florida. Fla Scient, 24:122–134

Brown DP (1984) Impact of development on availability and quality of ground water in eastern Nassau County, Florida, and southeastern Camden County, Georgia. US Geol Surv Water-Res Inv Rept 83-4190

Buddemeier RW Holladay H (1977) Atoll hydrology: Island groundwater characteristics and their relationship to diagenesis. In: Proc 3rd Int Coral Reef Symp, Miami, Univ Miami 2:167–174

Buddemeier RW Oberdorfer JA (1997) Hydrogeology of Enewetak Atoll. In: Vacher HL and Quinn TM (eds.), Geology and hydrogeology of carbonate islands. Amsterdam, Elsevier, Ch 23, pp. 667–692

Burke RG (1967) Sun whips Florida's boulder zone. Oil and Gas Jour 65:126–127

Bush PW (1982) Predevelopment flow in the Tertiary limestone aquifer, southeastern United States; A regional analysis from digital modeling. US Geol Surv Water-Res Inv Rept 82-905

Bush PW Johnston RH (1988) Ground-water hydraulics, regional flow, and ground-water development of the Floridan aquifer system in Florida and parts of Georgia, South Carolina, and Alabama. US Geol Surv Prof Pap 1403-C

Carr WJ Alverson DC (1959) Stratigraphy of Middle Tertiary rocks in part of west-central Florida. US Geol Surv Bull 1092

Chen J Jones IC Upchurch SB (1995) Mechanical impacts of air and water compression in karst conduits. Studia Carsologica 6, Czech Acad Sci, Inst. of Geonics, Bruno, Czech Republic, pp. 55–64

Chen J Thomas G Upchurch SB (1995a) Groundwater quality reflects karst development: A case study in central-west Florida. In: Beck BF (ed.), Karst Geohazards: Engineering and Environmental Problems in Karst Terrane, Rotterdam, A.A. Balkema, pp. 103–109

Chen J Thomas G Upchurch SB (1995b) A spatial-correlation model for assessment of large-area sinkhole and water quality data. In: Beck BF (ed.), Karst Geohazards: Engineering and Environmental Problems in Karst Terrane, Rotterdam, A.A. Balkema, pp. 111–116

Clark JF (2002) Isotopic measurements from the Floridan aquifer south of Lake Okeechobee. West Palm Beach, Florida, South Florida Water Management District

Clark JF Bennett MW Murrell MT Stute M Rademacher (2004) Time scale of seawater circulation and dolomitization within the lower Floridan aquifer. Geol Soc Amer Abst with Prog 36:327

Clark JF Stute M Schlosser P Drenkard S Bonani G (1997) A tracer study of the Floridan aquifer in Southeastern Georgia: Implications for groundwater flow and paleoclimate. Water Resour Res 33:281–289

Clausen CJ Cohen AD Emiliani C Holman JA Stipp JJ (1979) Little Salt Spring, Florida; a unique underwater site. Science 203: 609–614

Cockrell WA (1975) Warm Mineral Springs, 1975: A multi-disciplinary approach to a 10,000 BP archeological site (abs.). Fla Scientist 38(Suppl 1):19

Contractor DN (1983) Numerical modeling of saltwater intrusion in the Northern Guam Lens. Water Res Bull 19:745–751

Cooper HH Jr (1959) A hypothesis concerning the dynamic balance of fresh water and salt water in a coastal aquifer. Jour Geophys Res 64(4): 461–467

Cooper HH Jr (1964) A hypothesis concerning the dynamic balance of fresh water and salt water in a coastal aquifer. In: Cooper HH Jr Kohout FA Henry HR Glover RE (eds.), Sea water in coastal aquifers, US Geol Surv Water-Sup Paper 1613-C

Copeland R Dorian NA White AJ Upchurch SB (2011) Regional and statewide trends in Florida's spring and well groundwater quality (1991–2003). Fla Geol Surv Bulletin No. 69 (Revised)

Cowart JB Kaufman MI Osmond JK (1978) Uranium-isotope variations in groundwaters of the Floridan aquifer and Boulder Zone of south Florida. Jour Hydrol 36:161–172

Cressler A (1993) The caves of Dade County, Florida. Georgia Underground 30(3):9–16

Cunningham KJ Renken RA Wacker MA Zygnerski MR Robinson E Shapiro AM Wingard GL (2006) Application of carbonate cyclostratigraphy and borehole geophysics to delineate porosity and preferential flow in the karst limestone of the Biscayne aquifer, SE Florida. In: Harmon RS Wicks C (eds.), Perspectives on karst geomorphology, hydrology, and geochemistry—A tribute volume to Derek C. Ford and William B. White. Geol Soc Amer Sp Pap 404, pp. 191–208

Cunningham KJ Sukop MC Huang H Alvarez PF Curran HA Renken RA Dixon JF (2008) Prominence of ichnologically influenced macroporosity in the karst Biscayne aquifer: Stratiform "super-K" zones. Bull Geol Soc Amer 121:164–180

Delfino JJ Heaney JP (2004) Challenges to water resources sustainability in Florida. Conference on Allocating Water: Economics and the Environment, Universities Council on Water Resources, Portland, OR, July 20-22, 2004; accessed 8/20/2011 at http://warrington.ufl.edu/purc/docs/resources_ChallengesToWaterSustainability.pdf

Domenico PA Schwartz FW (1997) Physical and chemical hydrogeology. New York, John Wiley & Sons, 2nd Ed

Duerr AD (1994) Types of secondary porosity of carbonate rocks in injection and test wells in southern peninsular Florida. US Geol Surv Water-Res Inves Rept 94-4013

Duncan JG Evans WL III Taylor KL (1994) Geologic framework of the lower Floridan aquifer system, Brevard County, Florida. Fla Geol Surv Bull 64

Essaid HI (1986) A comparison of the coupled fresh water-salt water flow and the Ghyben-Herzberg sharp interface approaches to modeling of transient behavior in coastal aquifer systems. Jour Hydrol 86: 169–193

Fairchild RW Bentley CB (1977) Saline-water intrusion in the Floridan aquifer in the Fernandina Beach area, Nassau County, Florida. US Geol Surv Water-Res Invest Rept 77-32

Fanning KA Betzer PR Byrne RH Breland JA Jolley RR (1979) Characteristics of a submarine geothermal spring on the West Florida Shelf. Fla Scientist 42(Supl 1):21

Fanning KA Byrne RH Breland JA Betzer PR Moore WA Elsinger RJ Pyle TE (1981. Geothermal springs of the west Florida continental shelf: evidence for dolomitization and radionuclide enrichment. Earth and Planet Lett 52:345–354

Fernald EA Patton DJ (eds.) (1984) Water resources of Florida. Tallahassee, Inst Science and Public Affairs, Florida State University

Fernald EA Purdum ED (1998) Water resources atlas of Florida. Tallahassee, Inst Science and Public Affairs, Florida State University

Fetter CW Jr (1972) Position of the saline water interface beneath oceanic islands. Water Resour Res 8:1307–1314

Fetter CW Jr (2001) Applied hydrogeology. Upper Saddle River NJ, Prentice-Hall Publishers

Fish JE (1988) Hydrogeology, aquifer characteristics, and groundwater flow of the surficial aquifer system, Broward County, Florida. US Geol Sur Water-Res Invest Rept 87-4034

Fish JE Stewart MT (1991) Hydrogeology of the surficial aquifer system, Dade County, Florida. US Geol Surv Water-Res Invest Rept 90-4108

Florea LJ McGee D Wynn JG (2010) Stable isotopic and geochemical variability within shallow groundwater beneath a hardwood hammock and surface water in an adjoining slough (Everglades National Park, FL). 14th Symp Geol Bahamas, San Salvador, Bahamas

Florida Climate Center (2011a) Statewide average rainfall data, by year. Tallahassee, Florida State University, Center for Ocean-Atmospheric Prediction Studies, accessed 8/20/2011 at http://coaps.fsu.edu/climate_center/data/statewideavg_precip.shtml

Florida Climate Center (2011b) Statewide average temperature data, by year. Tallahassee, Florida State University, Center for Ocean-Atmospheric Prediction Studies, accessed 8/20/2011 at http://coaps.fsu.edu/climate_center/data/statewideavg_temp.shtml

Florida Department of Environmental Protection (2014) Underground injection control. Accessed 3-17-2014 at http://www.dep.state.fl.us/Water/uic/index.htm

Fratesi B (2013) Hydrology and geochemistry of the freshwater lens in coastal karst. In: Lace MJ Mylroie JE (eds.), Coastal karst landforms, Dordrecht, Springer Science, Coastal Research Library, pp. 59–76

Frazee, J.M., Jr., and D.R. McClaugherty, 1979. Investigation of Ground Water Resources and Salt Water Intrusion in the Coastal Areas of Northeast Florida. St. Johns River Water Management District, Technical Publication SJ 80-4, 248 p.

Freeze RA Cherry JA (1979) Groundwater. Upper Saddle River NJ, Prentice Hall Publishers

Gaswirth SB. Budd DA Farmer LG (2007) The role and impact of regionally extensive freshwater-seawater mixing zones in the maturation of regional dolomite bodies within the proto-Floridan aquifer. Sedimentology 54:1065–1092

Georgia Department of Natural Resources (n.d.)(undated) What have we learned? Georgia Dept Nat Res, accessed 5/12/2017 at www1.gadnr.org/cws/

Ghassemi F Chen TH Jakeman AJ Jacobson G (1993) Two and three-dimensional simulation of seawater intrusion: performances of the "SUTRA" and "HST3D" models. Jour Austral Geol Geophy 14:219–226

Ghassemi F Jakeman AJ Jacobson G Howard KWF (1996) Simulation of seawater intrusion with 2D and 3D models: Nauru Island case study. Hydrogeol Jour 4:4–22

Gingerich SB (1992) Numerical simulation of the freshwater lens on Roi-Namur Island, Kwajalein Atoll, Republic of the Marshall Islands, M.S. thesis, Honolulu, Dept. of Geol & Geophy, Univ Hawaii

Glover RE (1964) The pattern of fresh-water flow in a coastal aquifer. In: Cooper HH Jr Kohout FA Henry HR Glover RE (eds). Sea water in coastal aquifers, US Geol Surv Water-Sup Pap 1613-C: C32–C35

Gulley JD Florea LJ (2016) Caves as paleo-water table indicators in the unconfined Upper Floridan aquifer. Fla Scien, 79(4):239–256

Gulley JD Martin JB Moore PJ Murphy J (2013) Formation of phreatic caves in an eogenetic karst aquifer by $CO_2$ enrichment at lower water tables and subsequent flooding by sea level rise. Earth Surf Processes and Landforms, 38:1210–1224

Haberfeld JL (1991) Hydrogeology of effluent disposal zones, Floridan aquifer, south Florida. Ground Water 29(2):186–190

Halley RB Vacher HL Shinn EA (1997) Geology and hydrogeology of the Florida Keys. In: Vacher HL Quinn T (eds.), Geology and hydrogeology of carbonate islands. Elsevier Science, Developments in Sedimentology 54:217–248

Hanshaw BB Back W (1979) Major geochemical processes in the evolution of carbonate-aquifer systems. In: Back W Stephenson DA (eds.), Contemporary hydrogeology – The George Burke Maxey memorial volume. Jour Hydrol 43: 287–312

Harris WH (1967) Stratification of fresh and salt water on barrier islands as a result of differences in sediment permeability. Water Resour Res 3: 89–97

Hayes LR Barr DE (1983) Hydrology of the Sand and Gravel aquifer, southern Okaloosa and Walton counties, northwest Florida. US Geol Surv Water-Res Invest Rept 82-4110

Heath RC Conover CS (1981) Hydrologic almanac of Florida. US Geol Surv Open-File Rept 81-1107

Henry HR Kohout FA (1972) Circulation patterns of saline groundwater affected by geothermal heating - as related to waste disposal. In: Underground Waste Management and Environmental Implications, Amer Assoc Pet Geol, Memoir 18, pp. 202–221

Herbert TA Upchurch SB (2016) The potential role of hypogene speleogenesis in the lower Floridan aquifer and Sunniland Oil Trend, south Florida, U.S.A. In: Chavez T Reehling P (eds.), Carlsbad, NM, Nat Cave and Karst Res Inst, Symp 6, pp. 119–129.

Herzberg A (1901) Die Wasserversorgung einiger Nordseebder. Jour fur Gasbeluchtung und Wasserversorgung 44: 815–819, 45: 842–844

Hickey JJ (1977) Hydrogeologic data for the McKay Creek subsurface waste-injection test site, Pinellas County, Florida. US Geol Surv Open-File Rept 77-802

Hickey JJ (1979) Hydrogeologic data for the Bear Creek subsurface-injection test site, St. Petersburg, Florida. US Geol Surv Open-File Rept 78-853

Hickey JJ (1981) Hydrogeology, estimated impact, and regional well monitoring of effects of subsurface wastewater injection, Tampa Bay area, Florida. US Geol Surv Water-Res Invest Rept 80-118

Hickey JJ (1982) Hydrogeology and results of injection tests at waste-injection sites in Pinellas County, Florida. US Geol Surv Water-Sup Pap 2183

Hickey JJ (1984) Field testing the hypothesis of Darcian flow through a carbonate aquifer. Ground Water 22:544–547

Hickey JJ (1990) An Assessment of the flow of variable-salinity ground water in the middle confining unit of the Floridan aquifer system, west-central Florida. US Geol Surv Water-Res Invest Rept 89-4142u

Hickey JJ Vecchioli J (1986) Subsurface injection of liquid waste with emphasis on injection practices in Florida. US Geol Surv Water-Sup Pap 2281

Hine AC Chambers DP Clayton TD Hafen MR Mitchum GT (2016) Sea level rise in Florida: Science, impacts, and options. Gainesville. Univ Press Fla

Hoffmeister JE Stockman KW Multer HG (1967) Miami Limestone of Florida and its recent Bahamian counterpart. Geol Soc Am Bull 78:175–190

Hughes GH (n.d.) (undated) Runoff from hydrologic Units in Florida. Fla Bur Geol Map Ser 81

Hughes GH Hampton ER Tucker DF (1971) Annual and seasonal rainfall in Florida. Fla Bur Geol Map Ser 40

Hughes JD Vacher HL Sanford WE (2010) Three-dimensional flow in the Florida Platform: Theoretical analysis of Kohout convection at its type locality. Amer Assoc Petrol Geol Bull 94:889–917

Humphrey JD (1997) Geology and hydrogeology of Barbados. In: Vacher HL Quinn T (eds.), Geology and Hydrogeology of Carbonate Islands. Devel Sediment 54, Amsterdam, Elsevier, pp. 381–431

Hunn JD Cherry RN (1969) Remote sensing of offshore springs and spring discharge along the Gulf Coast of central Florida. In: Nat Aeron Admin, 2nd Ann Earth Res Aircraft Prog Status Rev, Hydrol and Oceanog 3:39–1 – 39–7

Hunt B (1979) An analysis of the groundwater resources of Tongatapu Island, Kingdom of Tonga. Jour Hydrol 49: 185–196

Hunt CD Peterson FL (1980) Groundwater resources of Kwajalein Island, Marshall Islands. Univ Hawaii, Water Resour Res Ctr Tech Rept 126

Hutchings WC Tarbox DL (2003) A model of seawater intrusion in surficial and confined aquifers of northeast Florida. Second Int Conf on Saltwater Intrusion and Coastal Aquifers – Monitoring, Modeling, and Management, Mérida, Yucatán, Mexico

Hutchinson CB (1992) Assessment of hydrogeologic conditions with emphasis on water quality and wastewater injection, southwest Sarasota and west Charlotte counties, Florida. US Geol Surv Water-Sup Pap 2371

Johnston RH Bush PW (1988) Summary of the hydrology of the Floridan aquifer system in Florida and parts of Georgia, South Carolina, and Alabama: Regional aquifer system analysis. US Geol Surv Prof Pap 1403-A

Jones LE (2001) Saltwater contamination in the upper Floridan aquifer at Brunswick, Georgia. Proc 2001 Georgia Water Res Conf, pp. 644–647

Kaufman MI Dion NP (1967) Chemical character of water in the Floridan aquifer in southern Peace River Basin, Florida. Fla Bur Geol Map Series No. 27

Kelly MH (2004) Florida river flow patterns and the Atlantic multidecadal oscillation. Brooksville, FL, Southwest Fla Water Mgt Dist,

Kelly MH Gore JA (2008) Florida river flow patterns and the Atlantic multidecadal oscillation. River Res Applic 24(5):598–616

Klein H Causaras CR (1982) Biscayne aquifer, south-east Florida. In: Franks BJ (ed.), Principal aquifers in Florida, US Geol Surv Water-Res Invest Open-File Rept

Klein H Hull JE (1978) Biscayne aquifer, southeast Florida. US Geol Surv Water Res Invest 78–107

Knochenmus LA (2004) Streamflow losses through karst features in the upper Peace River hydrologic area, Polk County, Florida, May 2002 to May 2003. US Geol Surv Fact Sheet 102-03

Knochenmus LA (2006) Regional evaluation of the hydrogeologic framework, hydraulic properties, and chemical characteristics of the Intermediate Aquifer System underlying southern west-central Florida. US Geol Surv Sci Invest Rept 2006-5013

Knochenmus LA Bowman G (1998) Transmissivity and water quality of water-producing zones in the Intermediate Aquifer System, Sarasota County, Florida. US Geol Surv Water-Resour Invest Rept 98-4091

Kohout FA (1960) Cyclic flow of saltwater in the Biscayne aquifer of southeastern Florida. Jour Geophy Res 65(7):2133–2141

Kohout FA (1964) The flow of fresh water and salt water in the Biscayne aquifer of the Miami area, Florida. US Geol Surv Water-Supply Pap 1613–C, pp. 12–32.

Kohout FA (1965) A hypothesis concerning cyclic flow of salt water related to geothermal heating in the Floridan aquifer. Trans New York Acad Science 28:249–271

Kohout FA (1967) Ground-water flow and the geothermal regime of the Floridan Plateau. Trans Gulf Coast Assoc Geol Soc 17:339–354

Kohout FA (1974) Hydrologic significance of a geothermal submarine spring off Florida west coast. Fla Scient, 37(Supplement 1):13

Kohout FA (1975) A hypothesis concerning cyclic flow of salt water related to geothermal heating in the Floridan aquifer. Trans New York Acad Science, 28:249–271

Kohout FA Kolipinski MC (1967) Biological zonation related to groundwater discharge along the shore of Biscayne Bay, Miami, Florida. Estuaries, Amer Assoc Adv Science, pp. 488–499

Kohout FA Henry HR Banks JE (1977) Hydrogeology related to geothermal conditions of the Floridan Plateau. In: Smith DL Griffin GM (eds.), The geothermal nature of the Floridan Plateau. Fla Bur Geol Sp Pub No. 21, pp. 1–41

Krause RE Clarke JS (2001) Saltwater contamination of ground water at Brunswick, Georgia and Hilton Head Island, South Carolina. Proc 2001 Georgia Water Res Conf, pp. 756–759

Krause RE Randolph RB (1989) Hydrology of the Floridan aquifer system in southeast Georgia and adjacent parts of Florida and South Carolina. US Geol Surv Prof Pap 1403–D

Kuniansky EL Bellino JC (2012) Tabulated transmissivity and storage properties of the Floridan aquifer system in Florida and parts of Georgia, South Carolina, and Alabama. US Geol Surv Data Ser 669

Kuniansky EL Bellino JC Dixon JF (2012) Transmissivity of the upper Floridan aquifer in Florida and parts of Georgia, South Carolina, and Alabama. US Geol Surv Sci Invest Map 3204

Langevin CD (2003) Simulation of submarine ground water discharge to a marine estuary: Biscayne Bay, Florida. Ground Water 41(6): 758–771

Langevin CD Stewart MT Beaudoin CM (1998) Effects of sea water canals on fresh water resources: An example from Big Pine Key, Florida. Ground Water 36:503–513

Leve GW (1966) Ground water in Duval and Nassau Counties, Florida. Fla Geol Surv Rept Invest No 43

Leve GW (1978) Altitude and configuration of the top of the Floridan aquifer, Duval County, Florida. US Geol Surv Water-Res Invest Rept No 77-114

Lichtler WF (1960) Geology and ground-water resources of Martin County, Florida. Fla Geol Surv Rept Invest No. 23

Maliva RG Walker CW Callahan EX (2001) Hydrogeology of the lower Floridan aquifer "boulder zone" of southwest Florida - Durward H. Boggess Memorial Symposium. In: Missimer TM Scott TM (eds.), Geology and hydrology of Lee County, Florida. Fla Geol Surv Sp Pub No. 49, pp. 167–182

Marella RL (2009) Water withdrawals, use, and trends in Florida, 2005. US Geol Surv Sci Invest Rept 2009-5125

Marsh OT (1966) Geology of Escambia and Santa Rosa counties, western Florida panhandle. Fla Geol Surv Bull 46

Matson GC Sanford S (1913) Geology and ground waters of Florida. US Geol Surv Water-Sup Pap 319

MacFadden BJ (1997) Fossil mammals of Florida. In: Randazzo AF Smith DL (eds.), Geology of Florida, Gainesville, Univ Press Fla, Ch 8, pp. 119–137

McVoy CW Said WP Obeysekera J VanArman JA Dreschel TW (2011) Landscapes and hydrology of the predrainage Everglades. Gainesville, Univ Press Fla

Metcalf SF Hall LE (1984) Sinkhole collapse induced by groundwater pumpage for freeze protection irrigation near Dover, Florida, January, 1997. In: Beck BF (ed.), Sinkholes: Their Geology, Engineering, and Environmental Impact. Rotterdam, A.A. Balkema Publ, pp. 3–10.

Metz PA Lewelling BR (2009) Hydrologic conditions that influence streamflow losses in a karst region of the upper Peace River, Polk County, Florida. US Geol Surv Sci Invest Rept 2009-5140

Meyer FW (1974) Evaluation of hydraulic characteristics of a deep artesian aquifer from natural water-level fluctuations, Miami, Florida. Fla Bur Geol Rept Invest No 75

Meyer FW (1989) Hydrogeology, ground-water movement, and subsurface storage in the Floridan aquifer system in southern Florida. US Geol Surv Prof Pap 1403-G

Miller JA (1978) Geologic and geophysical data from Osceola National Forest, Florida. US Geol Surv Open-File Rept 78-799

Miller JA (1986) Hydrogeologic framework of the Floridan aquifer system in Florida, and in parts of Georgia, Alabama and South Carolina. US Geol Surv Prof Pap 1403-B

Miller JA (1990) Ground water atlas of the United States: Alabama, Florida, Georgia, South Carolina. US Geol Surv Hydrol Invest Atlas 730-G, Segment 6

Miller JA (1997) Hydrogeology of Florida. In: Randazzo AF Jones DS (eds.), The geology of Florida, Gainesville, FL, Univ Presses Fla, pp. 69–88

Missimer TM Scott TM (2001) Geology and hydrology of Lee County, Florida – Durward H. Boggess Memorial Symposium. Fla Geol Surv Sp Pub No. 49

Moore YH Stoessell RK Easley DH (1992) Fresh-water/sea-water relationship within a ground-water flow system, northeastern coast of the Yucatan Peninsula. Ground Water 30:343–350

Muniz A Tobon M Bloetscher F (2003) Use of Class I injection wells provide an economical and environmentally sound approach for effluent and concentrate management. Int Symp Underground Injection Science and Technology, Lawrence Berkeley National Laboratory, Berkeley, California

Navoy AS (1986) Hydrogeologic data from a 2,000-foot deep core hole at Polk City, Green Swamp area, central Florida. US Geol Surv Water-Res Invest Rept 84-4257

Oberdorfer JA Hogan PJ Buddemeier RW (1990) Atoll island hydrogeology: flow and freshwater occurrence in a tidally dominated system. Jour Hydrol 120: 327–340

O'Reilly AM Spechler RM McGurk BE (2002) Hydrogeology and water-quality characteristics of the lower Floridan aquifer in east-central Florida. US Geol Surv Water-Res Invest Rept 02-4193

Palmer AN (1991) The origin and morphology of limestone caves. Geol Soc Amer Bull 103:1–21

Parker GG Ferguson GE Love SK (1955) Water resources of southeastern Florida. US Geol Surv Water Sup Pap 1255

Phelps GG Spechler RM (1997) The relation between hydrogeology and water quality of the lower Floridan aquifer in Duval County, Florida, and implications for monitoring movement of saline water. US Geol Surv Water-Res Invest Rept 96-4242

Popenoe P Kohout FA Manheim FT (1984) Seismic-reflection studies of sinkhole and limestone dissolution features on the northeastern Florida Shelf. In: Beck BF (ed.), Sinkholes: Their geology, engineering and environmental impact, Proc First Multidisc Conf Sinkholes, Rotterdam, A.A. Balkema, pp. 43–57.

Pride RW Meyer FW Cherry RN (1961) Interim report on the hydrologic features of the Green Swamp area in central Florida. Fla Geol Surv Inf Circ No. 26

Pride RW Meyer FW Cherry RN (1966) Hydrology of Green Swamp area in central Florida, Fla Geol Surv Rept Invest No. 42

Puri HS Winston GO (1974) Geologic framework of the high transmissivity zones in south Florida. Fla Bur Geol Sp Pub No. 20

Quiñones-Aponte V Kotun K Whitley JF (1996) Analysis of tests of subsurface injection, storage, and recovery of freshwater in the lower Floridan aquifer, Okeechobee County, Florida. US Geol Surv Open-File Rept 95-765

Rectenwald E (2010) Noble gases applied as paleotemperature indicators in the Floridan aquifer system. Proc 21st Salt Water Intrusion Meeting, Azores, Portugal, pp. 20–23

Reese RS Richardson E (2008) Synthesis of the hydrogeologic framework of the Floridan aquifer system and delineation of a major Avon Park permeable zone in central and southern Florida. US Geol Surv Sci Invest Rept 2007-5207

Reich CD Shinn EA Hickey TD Tihansky AB (2001) Tidal and meteorological influences on shallow marine groundwater in the upper Florida Keys. In: Porter JW Porter KG (eds.), The Everglades, Florida Bay, and coral reefs of the Florida Keys: An ecosystem sourcebook, Ch. 25, pp. 659–676

Renken RA Dixon J Koehmstedt J Lietz AC Marella RL Telis P Rogers J Memberg S (2005) Impact of anthropogenic development on coastal ground-water hydrology in southeastern Florida, 1900-2000. US Geol Surv Circ 1275

Rubin H (2010) Onset of thermohaline convection in a cavernous aquifer. Water Resour Res 12(2):141–147

Safko PS Hickey JJ (1991) A preliminary approach to the use of borehole data, including television surveys, for characterizing secondary porosity of carbonate rocks in the Floridan aquifer system. US Geol Surv Water-Res Invest Rept 91-4168

Scanlon BR Mace RE Barrett ME Smith B (2003) Can we simulate regional groundwater flow in a karst system using equivalent porous media models? Case study, Barton Springs, Edwards aquifer, USA. Jour Hydrol 276:137–158

Schmorak S Mercado A (1969) Upconing of fresh water-sea water interface below pumping wells, field study. Water Resour Res 6:1290–1311

Scott TM (1981) The paleoextent of the Miocene Hawthorn Formation in peninsular Florida (Abstract). Fla Sci, v 44, sup 1

Scott TM (1988) The lithostratigraphy of the Hawthorn Group (Miocene) of Florida. Fla Geol Surv Bull 59

Scott TM (2001) Text to accompany the geologic map of Florida. Fla Geol Surv Open File Rept 80

Scott TM Means GH Meegan RP Means RC Upchurch SB Copeland RE Jones J Roberts T Willet A (2004) Springs of Florida. Fla Geol Surv Bull 66

Scott TM Campbell KM Rupert FR Arthur JD Green RC Means GH Missimer TM Lloyd JM Yon JW Duncan JG (2001) Geologic map of the State of Florida. Fla Geol Surv Map Series 146

Sepúlveda N (2002) Simulation of ground-water flow in the intermediate and Floridan aquifer system in peninsular Florida. US Geol Surv Water-Res Invest Rept 02-4009

Sepúlveda N Spechler RM (2004) Evaluation of the feasibility of freshwater injection wells in mitigating ground-water quality degradation at selected well fields in Duval County, Florida. US Geol Surv Water-Res Invest Rept 03-4273

Shaler NS (1890) The topography of Florida. Bull Mus Comp Zoology 16:139–158

Singh UP Eichler GE Sproul CR Garcia-Bengochea JL (1983) Pump testing boulder zone aquifer, south Florida. Jour Hydraulic Eng 9(8):1152–1160

Snyder SW Evans ME Hine AC Compton JS (1989) Seismic expression of solution collapse features from the Floridan Platform. In: Beck BF (ed.), Engineering and environmental impacts of sinkholes and karst, Abington, England, Taylor & Francis, pp. 281–297

Southeastern Geological Society (1986) Hydrogeological units of Florida. Fla Geol Surv Sp Pub No 28

Southwest Florida Water Management District (2002) Saltwater intrusion and the minimum aquifer level in the Southern Water Use Caution Area. Brooksville, FL, Southwest Florida Water Management District, August 21, 2002 draft

Southwest Florida Water Management District (2006a) Aquifer characteristics within the Southwest Florida Water Management District. Brooksville, FL, Brooksville, Southwest Florida Water Management District, 4th Edition

Southwest Florida Water Management District (2006b) Southern Water Use Caution Area - Recovery strategy. Brooksville, FL, Southwest Florida Water Management District

Spechler RM (1994) Saltwater intrusion and quality of water in the Floridan aquifer system, northeastern Florida. US Geol Surv Water-Res Invest Rept 92-4174

Spechler RM (1996) Detection and quality of previously undetermined Floridan aquifer system discharge to the St. Johns River, Jacksonville to Green Cove Springs, northeastern Florida. US Geol Surv Water Res Invest Rept 95-4257

Spechler RM (2001) The relation between structure and saltwater intrusion in the Floridan aquifer system, northeastern Florida. In: Kuniansky EL (ed.), US Geol Surv Karst Int Grp Proc, Water-Res Invest Rept 01-4011

Spechler RM Hampson PS (1984) Ground-water resources of St. Johns County, Florida. US Geol Surv Water-Res Invest Rept 83-4187

Spechler RM Phelps GG (2001) Saltwater intrusion in the Floridan aquifer system, northeastern Florida. Proceedings of the 2001 Georgia Water Resources Conference, pp. 398–400.

Sproul CR (1977) Spatial distribution of ground water temperature in south Florida. In: Smith DL Griffin GM (eds.), The geothermal nature of the Floridan Plateau. Fla Bur Geol Sp Pub No 21

Sproul CR Boggess DH Woodard HJ (1972) Saline-water intrusion from deep artesian sources in the McGregor Isles area of Lee County, Florida. Fla Bur Geol Info Circ No75

Stewart HG Jr (1966) Ground-water resources of Polk County. Fla Geol Surv Rept Invest No. 44

Stewart JW (1980) Areas of natural recharge to the Floridan aquifer in Florida. Fla Bur Geol Map Series 98

Stringfield VT (1966) Artesian water in Tertiary limestone in the southeastern states. US Geol Surv Prof Pap 517

Strom RN Upchurch SB (1985a) Palygorskite distribution and silicification in the phosphatic sediments of central Florida. In: Snyder S Riggs S Partin B Mallette P Walker R (eds.), Eighth Inter Field Workshop and Symp (Southeastern United States), Guidebook, Inter Geol Correl Prog, Project 156 - Phosphorites, pp. 118–126.

Strom RN Upchurch SB (1985b) Palygorskite distribution and silicification in the phosphatic sediments of central Florida. In: Scott TM Cathcart JB (eds.), Florida Land-Pebble Phosphate District. Geol Soc Amer 1985 Ann Meeting Field Trip Guidebook, pp. 68–75.

Swancar A Hutchinson CB (1995) Chemical and isotopic composition and potential for contamination of water in the upper Floridan aquifer, west-central Florida, 1986-89. US Geol Surv Water-Sup Pap 2409

Todd DK (1980) Groundwater hydrology. New York, John Wiley & Sons

Torres AE Sacks LA Yobbi DK Knochenmus LA Katz BG (2001) Hydrogeologic framework and geochemistry of the intermediate aquifer system in parts of Charlotte, De Soto, and Sarasota counties, Florida. US Geol Surv Water-Res Invest Rept 01-4015

Tóth J (1962) A theory of groundwater motion in small drainage basins in central Alberta, Canada. Jour Geophy Res 67:4375–4387

Tóth J (1963) A theoretical analysis of groundwater flow in small drainage basins. Jour Geophy Res 68:4795–4812

Trapp H Jr (1975) Hydrology of the Sand-and-Gravel aquifer in central and southern Escambia County, Florida. US Geol Surv Open-File Rept 74-218

U.S. Environmental Protection Agency (2003) Relative risk assessment of management options for treated wastewater in south Florida. US Environ Prot Agency, EPA 816-R-03-010

Underwood MR Peterson FL Vos CI (1992) Groundwater lens dynamics of atoll islands. Water Resour Res 28:2889–2902

Upchurch SB (1992) Quality of waters in Florida's aquifers. In: Maddox GL Lloyd JM Scott TM Upchurch SB Copeland R (eds.), Florida ground water quality monitoring program – Volume 2, Background hydrogeochemistry. Fla Geol Surv Sp Pub No. 34 pp. 12–52, 64–84, 90–347

Upchurch SB (2002) Hydrogeochemistry of a karst escarpment. In: Martin JB Wicks CM Sasowsky ID (eds.), Hydrogeology and biology of post-Paleozoic carbonate aquifers. Charles Town, WV, Karst Waters Inst Sp Pub 7, pp. 73–75

Upchurch SB (2005) Geochemical and geomorphic evolution of a karst escarpment. In: Copeland R (comp.), Geomorphic influence of scarps in the Suwannee River basin. Tallahassee, Southeastern Geol Soc Field Trip Guidebook 44, pp. 38–42

Upchurch SB (2014) An introduction to the Cody escarpment, north-central Florida. In: Lawn A (comp.), Karst hydrogeology of the upper Suwannee River basin, Alapaha River area, Hamilton County, Florida. Tallahassee, Southeastern Geol Soc Field Trip Guidebook 63, pp. 13–28.

Upchurch SB (2017) Hypogene speleogenesis on the Floridan Platform, U.S.A. In: Klimchouk AB Palmer AN Waela JD Auler AS Audra P (eds.), Hypogene karst regions and caves of the world, Springer Int Pub, Ch. 49, pp.735–744.

Upchurch SB Dobecki TL Scott TM Meiggs SH Fratesi SE Alfieri MC (2013) Development of sinkholes in a thickly covered karst terrane. Proceedings, The Thirteenth Multidisciplinary Conference on Sinkholes and the Engineering and Environmental Impacts of Karst, Nat Cave and Karst Res Inst, Carlsbad, NM, pp. 273–277.

Vacher HL (1978) Hydrogeology of Bermuda – Significance of an across-the-island variation in permeability. Jour Hydrol 39:207–226

Vacher HL (1988a) Dupuit-Ghyben-Herzberg analysis of strip-island lenses. Geol Soc Amer Bull 100:580–591

Vacher HL (1988b) Ground water in barrier islands: Theoretical analysis and evaluation of the unequal sea level problem. Jour Coastal Res 4(1):139–148

Vacher HL (1997) Introduction: Varieties of carbonate islands and historical perspective. In: Vacher HL Quinn T (eds.), Geology and hydrogeology of carbonate islands. Develop Sedim 54, Amsterdam, Elsevier, pp. 1–34

Vacher HL Mylroie JE (2002) Eogenetic karst from the perspective of an equivalent porous medium. Carbonates and Evaporites 17:182–196

Vacher HL Wallis TN (1992) Comparative hydrogeology of freshwater lenses of Bermuda and Great Exuma Island, Bahamas. Ground Water 30:15–20

Vacher HL Wightman MJ Stewart MT (1992) Hydrology of meteoric diagenesis: Effects of Pleistocene stratigraphy on freshwater lenses of Big Pine Key, Florida. In: Fletcher CH III Wehmiller JF (eds.), Quaternary Coasts of the United States: Marine and Lacustrine Systems, Soc of Econ Paleont Mineral Sp Pub No. 48, pp. 213–219

Van der Veer P (1977) Analytical solution for steady interface flow in a coastal aquifer involving a phreatic surface with precipitation. Jour Hydrol 34:1–11

Vernon RO (1951) Geology of Citrus and Levy counties, Florida. Fla Geol Surv Bull 33

Vernon RO (1970) The beneficial uses of zones of high transmissivities in the Florida subsurface for waste storage and waste disposal. Fla Bur Geol Info Circ No. 70

Visher FN Hughes GH (1969) The difference between rainfall and potential evaporation in Florida. Fla Bur Geol Map Series 32

Wadsworth JR Brook GA Carver RE (1983) Surface expression of heavily mantled interstratal karst bordering Okefenokee Swamp, Georgia. Tech Paps, 49th Annual Meeting, Amer Soc of Photogrammetry, Washington DC, pp. 463–470

Watts WA Hansen BCS (1994) Pre-Holocene and Holocene pollen records of vegetation history from the Florida peninsula and their climatic implications. Palaeogeo Palaeoclim Palaeoecol 109(4):163–176

Watts WA Stuiver M (1980) Late Wisconsin climate of northern Florida and the origin of species-rich deciduous forest. Science 210(4467):325–327

Watts WA Hansen BCS Grimm EC (1992) Camel Lake: a 40 000-yr record of vegetational and forest history from northwest Florida. Ecology 73(3):1056–1066

Wentworth CK (1947) Factors in the behavior of ground water in a Ghyben-Herzberg system. Pacific Science 1:172–184

Wentworth CK (1948) Growth of the Ghyben-Herzberg transition zone under a rinsing hypothesis. Amer Geophy Union Trans 1:97–98

Wheatcraft SW Buddemeier RW (1981) Atoll island hydrology. Ground Water 19:311–320

Whitaker FF (1992) Hydrology, geochemistry, and diagenesis of modern carbonate platforms in the Bahamas. Ph.D. dissertation, University of Bristol

Whitaker FF Smart PL (1990) Active circulation of saline groundwaters in carbonate platforms: Evidence from the Great Bahama Bank. Geology 18:200–203

Whitaker FF Smart PL (1997) Hydrogeology of the Bahamian Archipelago. In: Vacher HL Quinn T (eds.), Geology and Hydrogeology of Carbonate Islands. Amsterdam, Elsevier, Develop in Sedimentol 54: 381–431

Williams CP Scott TM Upchurch SB Paul DT Hannon LM (in prep.) Florida geomorphology. Fla Geol Surv Bull 72

Williams LJ Kuniansky EL (2015) Revised hydrogeologic framework of the Floridan aquifer system in Florida and parts of Georgia, Alabama, and South Carolina 2015, US Geol Surv Prof Pap 1807

Winsberg MD (2003) Florida weather. 2nd Edition, Gainesville, Univ Press of Fla

Winston GO (1995) The boulder zone dolomites of Florida Volume 1: Paleogene & upper Cretaceous zones of the southeastern peninsula and the Keys: Miami Geol Soc, Accessed 2/4/2013 at http://sofia.usgs.gov/publications/reports/mgs_winstonv11995/.

World Health Organization (1971) European Standards for Potable Water, 2nd Ed., Geneva, World Health Org

World Health Organization (2010) Guidelines for drinking water quality. 3rd Ed., vol. 1. Geneva, World Health Org.

# Hydrogeochemistry of Florida Karst Waters

# 5

**Abstract**

The hydrogeochemistry of Florida's limestone aquifers is complex, dominated by carbonate equilibrium reactions and the influence of seawater.

Florida rainfall contains dilute $Na^+$ and $Cl^-$ from marine aerosols. Quartz-rich parts of the surficial aquifer system (SAS) contain Na-Cl water overprinted by human activity, whereas limestone and shell aquifers contain Ca-$HCO_3$ or Ca-Mg-$HCO_3$ water. The Floridan Aquifer System (FAS) also contains Na-Cl-rich residual marine waters from past sea-level fluctuations; Na-Cl-rich connate water; and $SO_4^{2-}$-rich waters from sulfide oxidation and dissolution of sulfate minerals.

Potential for dissolution of carbonate-rich strata is enhanced by mixing of groundwater such as at freshwater/saline-water interfaces, as well as chemical complexing, ion pairing, common ion effects, and high levels of dissolved organics.

Metastable minerals aragonite and high-magnesium calcite are present in Florida's youngest carbonate aquifers and sediments. These minerals are likely to be dissolved during early diagenesis and speleogenesis. High surface area relative to particle volume in fine-grained sediments causes water to equilibrate rapidly with carbonate constituents, resulting in preservation of shell and carbonate mud at depth.

There have been at least two episodes of diagenetic alteration of calcite and/or aragonitic sediments in the FAS. Dolomitization has occurred in Florida's aquifers, but the degree of subsequent involvement of the dolomite in diagenesis is unclear.

Groundwater-quality issues include arsenic released by aquifer storage and recovery, and $H_2S$ derived from reduction of $SO_4^{2-}$ and degradation of organics. Water quality is also impacted by evaporation and transpiration, biotic activity in soils and sediments, reduction/oxidation reactions, and generation of humic substances by plant decay. Nitrate from fertilizers and animal and human waste is a major contributor to eutrophication of Florida springs.

**Keywords**

Karst geochemistry · Diagenesis · Epigenetic speleogenesis · Hypogenetic speleogenesis · Groundwater mixing

## 5.1 Introduction

This chapter has several themes. The chapter begins with a brief introduction to inorganic and organic chemical processes that lead to carbonate mineral dissolution and precipitation, including acid-base reactions and reactions that involve mixing of two chemically different water masses, such as seawater and fresh, Ca-$HCO_3$ water.[1] The chapter then turns to characterizing inorganic chemical interactions in Florida and fresh-water/saline-water mixing-zone reactions. The latter part of the chapter includes a discussion of dolomitization and the reactions that involve natural organic compounds dissolved in surface water and groundwater.

## 5.2 Overview of Karst Geochemical Processes

### 5.2.1 Mechanisms of Carbonate Dissolution

Dissolution of calcite, aragonite, and, to a lesser extent, dolomite is generally due to action by an acid, most commonly carbonic acid ($H_2CO_3$) derived from atmospheric and soil carbon dioxide ($CO_2$). The following discussion is limited to the interaction of calcite and water. Similar reactions exist for the other carbonate minerals.

[1] When discussing hydrochemical facies (chemical characteristics of specific water masses), chemical symbols are utilized rather than spelling out chemical components. For example, calcium-bicarbonate water is noted as Ca-$HCO_3$ water and sodium-chloride water is noted as Na-Cl water.


S. Upchurch et al., *The Karst Systems of Florida*, Cave and Karst Systems of the World,
https://doi.org/10.1007/978-3-319-69635-5_5

#### 5.2.1.1 Chemical Equilibrium Concepts

Before we discuss the chemical processes of karst in Florida, it is important to introduce the concept of chemical equilibrium. This brief overview is based on discussions found in Garrels and Christ (1964); Palmer (1991, 2007); Ford and Williams (1989); Drever (1997); and Dreybrodt (2000). For more information on chemical equilibria and weathering, see Garrels and Christ (1965) or Drever (1985, 1997), for example.

Chemical equilibrium is the state of a chemical system where any reaction is complete and the products and reactants are balanced in the proportions dictated by the equilibrium constant ($K_{eq}$). For example, in the following chemical reaction

$$bB + cC \rightleftarrows \mathrm{dD} + \mathrm{eE} \quad (5.1)$$

$B$ and $C$ are reactants that are present in mole concentrations represented by $b$ and $c$, and $D$ and $E$ are products of the reaction represented by mole concentrations $d$ and $e$. Molarity (mole concentration, $M$) is expressed by

$$M = \frac{weight\ of\ solute \times 1000}{formula\ weight\ of\ solute \times volume\ of\ solution}. \quad (5.2)$$

In an aqueous chemical reaction, solutes $B$ and $C$ react to form $D$ and $E$ in proportions determined by the equilibrium constant

$$K_{eq} = \frac{a_D^d a_E^e}{a_B^b a_C^c} \quad (5.3)$$

where $a_X^x$ is the activity, or thermodynamic concentration, of the solute $X$. The symbol $\rightleftarrows$ in Eq. 5.1 represents the potential directions of the chemical reaction. In this example $D$ and $E$ can react to form $B$ and $C$, or $B$ and $C$ can react to form $D$ and $E$. The direction of the reaction depends on the equilibrium constant, which varies with temperature and other factors. When the proportions of the solutes reach the equilibrium ratio given in the equilibrium constant, the solutes are balanced according to the equilibrium constant, and no additional reaction occurs. This process is called chemical equilibration, and the state of the water with respect to equilibration is termed the equilibrium state of the water.

The state of chemical equilibration depends on several factors. Reaction kinetics dictate the rate of equilibration of the solutes. Some reactions are rapid while others may require thousands to millions of years (geologic time) to complete. The rate of progress of a chemical reaction involves temperature, complexity of the coordination of chemical constituents, and interferences by other solutes or solids in the system. For example, dissolution of calcite can be relatively rapid while precipitation of calcite from the fluid may be slow because of the need to arrange calcium and carbonate ions into a crystal lattice configuration.

When the activities of the reactants are greater than required by the equilibrium constant, then the reaction goes to the right and reaction products are produced until the equilibrium is obtained. Before this reaction occurs, the water is said to be under-saturated with respect to the reaction products. If there are excess products in the system, reaction 5.1 goes to the left and the products and reactants come into chemical equilibrium. At equilibrium, the water is saturated with respect to the reactants and products. If excess products exist, the water is said to be over-saturated with respect to the reactants. Collectively, the state of reaction 5.1 is called its equilibration state.

When the proportions of reactants and products are balanced, chemical equilibrium has been attained. For example, the following equilibrium reaction will occur if water is in contact with calcite ($CaCO_3$). The reactants are water ($H_2O$), carbon dioxide ($CO_2$), and calcite, and the products are dissolved calcium ($Ca^{2+}$) and bicarbonate ($HCO_3^-$). When dissolution of calcite occurs, the water is under-saturated with respect to calcite and the reaction moves to the right, creating dissolved calcium and bicarbonate. If the water is over-saturated with respect to calcite, the reaction moves to the left, calcite is precipitated and water and carbon dioxide are released.

$$H_2O + CO_2 + CaCO_3 \rightleftarrows Ca^{2+} + 2HCO_3^-. \quad (5.4)$$

Under equilibrium conditions, the concentration of $CO_2$ in water whose surface is in contact with air is expressed as a pressure equal to that exerted by $CO_2$ on the surface of the water itself. The concentration ("partial pressure") of $CO_2$ in the atmosphere at sea level is 38.5 pascals (Zhang et al. 1997). In the pore spaces of soils, the $CO_2$ concentration can be orders of magnitude higher (Enoch and Dasberg 1971; Palmer 2007; Ford and Williams 1989) because of microbial respiration which releases carbon dioxide. Meteoric waters are generally thought to gain much of their potential for dissolution of carbonate minerals by gaining $CO_2$ as they percolate through the soil column. This dissolution potential is further enhanced by mixing of vadose waters with phreatic waters and degassing of $CO_2$ at the water table (Gulley et al. 2012).

Freshwater with dissolved carbon dioxide contains $CO_2$, carbonic acid ($H_2CO_3$), bicarbonate ion ($HCO_3^-$), and carbonate ion ($CO_3^{2-}$). All of these species coexist within freshwaters under influence of the free atmosphere. They occur in proportion to several interconnected variables including temperature, pH, and partial pressure of $CO_2$ in the atmosphere. When the water comes in contact with calcite in limestone, dissolution according to Eq. 5.4 contributes calcium ion ($Ca^+$) as well as additional $CO_3^{2-}$ and $HCO_3^-$ to the reactions.

The combination of $H_2O$ and $CO_2$ (reaction Eq. 5.5) to form carbonic acid ($H_2CO_3$), a weak acid, involves several intermediate steps. The reaction that describes the formation of the $H_2CO_3$ is:

$$H_2O + CO_2 \rightleftarrows H_2CO_3. \quad (5.5)$$

The carbonic acid then dissociates according to:

$$H_2CO_3 \rightleftarrows H^+ + HCO_3^-, \quad (5.6)$$

It is the $H^+$ that reacts with calcite.[2] $HCO_3^-$ is a reaction product that forms during dissolution of calcite (Eq. 5.4) and also during dissociation of the carbonic acid (Eq. 5.6).

For reaction 5.4, the arrows indicate that the reaction involves chemical equilibria, and that the reaction can proceed in either direction. That is, the products of the reaction, $Ca^{2+}$ and $HCO_3^-$, are in equilibrium with the reactants, $H_2O$ and $CO_2$, which combine to form carbonic acid ($H_2CO^3$), and calcite. At any given temperature and water chemistry, the ratio of the products and reactants is a temperature (T)-dependent constant ($K_{eq., T}$), where:

$$K_{eq,\,T,calcite} = \frac{a_{Ca^{2+}} \cdot a^2_{HCO_3^-}}{a_{H_2CO_3} \cdot a_{CaCO_3}}. \quad (5.7)$$

In this equation, the calcite ($CaCO_3$), a solid, has an activity of 1 and can be ignored. Similarly, reaction 5.4 results in a very minor amount of $H_2CO_3$ in the solution. It, too, can be ignored for the purposes of this discussion. Therefore, Eq. 5.7 can be simplified as:

$$K_{eq,T,calcite} = a_{Ca^{2+}} \cdot a^2_{HCO_3^-} \quad (5.8)$$

When reaction 5.4 begins, $H_2CO_3$ dissolves calcite and produces $Ca^{2+}$ and $HCO^{3-}$. At equilibrium, this reaction ceases and it can be said that the water is at equilibrium with respect to calcite, which can no longer be dissolved or precipitated unless the equilibrium balance is disturbed. To determine how far dissolution or precipitation will proceed, we examine the degree of saturation of ions in a solution with respect to a mineral phase by comparing the activities of ions in the solution of interest to their theoretical activities in a saturated solution.

The equilibrium constant of a reaction describes the amount of a mineral required to dissolve in a solution (Eq. 5.8). If the water is not chemically equilibrated with respect to the mineral of interest, dissolution can occur and the activities of the dissolved ions are lower than in a solution that is saturated with respect to the mineral phase of interest:

$$K_{eq,T,calcite} > a_{Ca^{2+}} a^2_{HCO_3^-}. \quad (5.9)$$

In a supersaturated solution, the activities of the dissolved components are higher than in the saturated solution:

$$K_{eq,T,calcite} < a_{Ca^{2+}} a^2_{HCO_3^-}. \quad (5.10)$$

Reaction 5.8 can be rearranged to provide a convenient expression of the degree of saturation of a solution with respect to a mineral phase. The log ratio of the product activities to the equilibrium constant is called the *saturation index* (*SI*). For calcite, the saturation index is

$$SI_{calcite} = \log\left(\frac{a_{Ca^{2+}} a^2_{HCO_3^-}}{K_{eq.,calcite}}\right) \quad (5.11)$$

The equation for the saturation index is often written in a more general form that allows for more complex chemical equilibria. This equation is:

$$SI = \log\left(\frac{IAP}{K_T}\right), \quad (5.12)$$

where *IAP* is the ion activity product of the mineral-water reaction and $K_T$ is the equilibrium constant at a given water temperature. The logarithm is used to simplify comparisons of saturation states. For a saturated solution, SI = 0; for an under-saturated solution, SI is negative; for an over-saturated solution, SI is positive.

It is important to understand that the SI does not provide information as to the kinetics of a chemical reaction. Mineral phases may remain in contact with chemically undersaturated water for long periods of time without reaction or with slow equilibration rates. Other factors may also determine if a mineral will precipitate and grow, such as nucleation energy, surface poisoning, and mineral surface topography (Drever 1997). Because of uncertainties in measurement of activities and kinetics, the SI is evaluated allowing for a window of error. The water is generally considered to be saturated with respect to the mineral of interest if SI equal to $0 \pm 0.5$. If SI is less than $-0.5$, the water is considered undersaturated with respect to the mineral, and dissolution is probable if kinetics are appropriate. If SI) is greater than 0.5 and given suitable kinetics, precipitation of the mineral of interest is predicted.

Solubility of calcite in freshwater is influenced by several variables, including the partial pressure of carbon dioxide gas in the system ($P_{CO2}$), temperature, pressure, and solution pH. An open system, in which $P_{CO2}$ is held constant by transfer of $CO_2$ between an open-water surface (the water table or soil moisture) and the atmosphere will result in more potential for dissolution than in a closed system where reactions run to completion with only a fixed starting amount of $CO_2$.

[2]The reactions also apply to equilibrium reactions with aragonite and are similar for dolomite.

The role of temperature in these reactions is important. Gases, such as $CO_2$, are more soluble in cold water than in warm. Therefore, increases in water temperature decrease the solubility of $CO_2$ gas, decreasing the solubility of calcite. As a result of the effects of temperature on $CO_2$ solubility in water, calcite is more soluble in cold water than in warm water, an additive effect to the $CO_2$ solubility issue (White 1988; Palmer 2007).

The relationships between these variables are complicated by several chemical effects (Garrels and Christ 1965), including ion pairing and ionic strength effects, common ion effects, interactions with non-carbonate, inorganic acids, and interactions with organic acids. These effects are discussed in the following sections.

#### 5.2.1.2 Ion-Pairing and Ionic-Strength Effects

If dissolved ion concentrations are high enough to place the ions in close proximity to each other, ions with opposite charges tend to pair together, forming chemical complexes known as ion pairs. The constituents tied up in such ion complexes are unavailable for additional reactions, thus causing the activity of the chemical species involved in dissolution/precipitation reactions to be less than the total mass concentration of the ion. The amount of mineral that can be dissolved, therefore, increases as a result of sequestration of a proportion of the chemical in one or more ion pairs. For example, take the reaction in Eq. 5.4. If, as the calcite dissolves, some of the $Ca^{2+}$ ion forms a soluble complex with $SO_4^{2-}$ to form dissolved $CaSO_4^0$, a chemical complex formed by ion pairing, then additional calcium must be made available by dissolution of calcite in order to maintain equilibrium. This effect is important in solutions of high ionic strength, such as seawater. Large concentrations of ions, such as sodium ($Na^+$), potassium ($K^+$), and chloride ($Cl^-$), can decrease the activity of ions such as $Ca^{2+}$ and $HCO_4^-$ by ion pairing, thus increasing the solubility of carbonate minerals.

#### 5.2.1.3 Common Ion Effects

Calcite and dolomite release some of the same ions ($Ca^{2+}$, $HCO3^-$) into solution when they dissolve, such that the reaction products for dolomite impact the calcite dissolution process and vice versa. This chemical "interference" is known as the common ion effect. The potential of a solution to cause mineral dissolution or precipitation is governed by the kinetics of dissolution of calcite and dolomite. The amount of dissolution of both minerals is therefore less than if they were alone in contact with the solution.

#### 5.2.1.4 The Effects of Non-Carbonate, Inorganic Acids

The presence of inorganic acids, such as hydrochloric and sulfuric acid, provides additional dissolution potential in water in contact with limestone. These acids are often hypogenic, or derived from below rather than from atmospheric interactions. In Florida, for instance, deep groundwater circulation brings seawater into contact with gypsum and anhydrite deposits (Sect. 4.5.1), resulting in high groundwater sulfate concentrations that impact mixing-zone geochemistry in the area (Rye et al. 1981; Copeland et al. 2011) by formation of ion pairs and sulfuric acid ($H_2SO_4$). Consequently, sulfuric acid has the capability of developing karst.

The sulfide oxidation and dissolution reactions proposed by Hill (2000) for the formation of hypogene karst are:

$$H_2S + 2O_2 \underset{reduction}{\overset{oxidation}{\rightleftarrows}} 2H^+ + SO_4^{2-} \tag{5.13}$$

and

$$2H^+ + SO_4^- + CaCO_3 \rightleftarrows Ca^{2+} + SO_4^{-2} + H_2O + CO_2. \tag{5.14}$$

These reactions were first applied to karst in areas where hydrogen sulfide gas ($H_2S$) or sulfuric acid ($2H^+ + SO_4^{2-}$) welled upward from evaporite deposits consisting of gypsum ($CaSO_4.nH_2O$) or anhydrite ($CaSO_4$) in the western U.S. Herbert and Upchurch (2016) have suggested that similar hypogenic karst (speleogenesis) may have occurred in the deep strata of Florida. In the Upper Floridan Aquifer (UFA) $SO_4^{2-}$-rich water depicted in Fig.4.29, two opportunities for dissolution of limestone may exist. First, the dissolved $SO_4^{2-}$ may represent development of dilute sulfuric acid and dissolution. Second, mixing of the $SO_4^{2-}$-rich water with water of a different chemical composition may result in a new water mass that is undersaturated with respect to calcite.

#### 5.2.1.5 Organic Acid Effects

Organic molecules can generate acidity and increase the potential for mineral dissolution, including both carbonate and silicate minerals (Huang and Keller 1972a, b, c; Drever 1985). Very large organic molecules, known as humic substances, can also form chemical complexes with ions, isolating them from further reaction and thereby decreasing the activity of those ions (Upchurch et al. 1983). Divalent ions, such as calcium, are particularly susceptible to organic complexing.

#### 5.2.1.6 Consequences of Equilibrium Reactions Involving Carbonate

Because equilibrium reactions are governed by their equilibrium constants, important consequences may occur relative to the development of karst features. For example,

1. Since $H_2CO_3$ is formed by an equilibrium reaction with $H_2O$ and $CO_2$ gas (Eq. 5.5), evaporation of water from the system would cause the reaction shown in Eq. 5.4 to go to the left. As a result, $H_2CO_3$ concentrations in the system would decrease and $CO_2$ gas would be driven into the

atmosphere. With a decline in the concentration of $H_2CO_3$ in the system, the reaction shown in Eq. 5.4 would also be driven to the left, and calcite would precipitate from the solution. This consequence occurs in dry caves where water evaporation is significant, but the relative humidity of Florida's "dry" caves is often very high, and evaporation is unlikely to be a dominant process in many of Florida's caves.

2. If $CO_2$ degasses from groundwater at the water table and the $CO_2$ is allowed to leave the system, then calcite would precipitate, often as well developed crystals, according to the equilibrium reaction.
3. Conversely, if additional $CO_2$ is added to the system, reaction 5.4 would be driven to the right and additional $H_2CO_3$ would form and dissociate. The result would be a potential to dissolve additional calcite. There are at least two vectors by which $CO_2$ can be released into the water: (1) respiration by macrofauna and microbes in soil, guano, or other sources or (2) release of $CO_2$ by degassing from groundwater near the water table. This latter process is a new concept that is discussed in Sect. 7.2.5.
4. If a solution is at chemical equilibrium and nothing changes to disrupt the system, such as a change in temperature or $CO_2$ concentration, the water is no longer capable of dissolving or precipitating calcite. Water in a state of chemical equilibrium with respect to the calcium-carbonate mineral phases in the host sediment has no ability to dissolve more of the host sediment or precipitate cement and/or cave decorations. This is especially true in low permeability, fine-grained sediments, such as sandy clay, clayey sand, or clay, which blanket large areas of the Florida peninsula and somewhat moderate the karst development in those areas.
5. In active groundwater flow systems where chemically aggressive water is brought into contact with the calcite or aragonite along fractures or cavern walls, equilibrium may not develop for some distance down gradient. Therefore, in a karstic carbonate aquifer under active flow, there may be an active zone of dissolution that extends downgradient along fractures or karst conduits from the recharge location (Thrailkill 1965). At depth, where the water is at equilibrium, no additional dissolution occurs even though active flow continues.

It is the hydrogen ion ($H^+$) that reacts with the calcium-carbonate mineral phase. We measure the acidity of a solution by determining the activity of hydrogen ions using the pH scale.

$$pH = -log_{10}\left(a_{H^+}\right) = log_{10}\left(\frac{1}{a_{H^+}}\right). \qquad (5.15)$$

pH is a logarithmic concentration scale that ranges from near 0–14. Water with a pH of 7 is neutral, and pH values less than 7 are acidic, while greater than 7 are basic, or alkaline. Reactions involving pH are sometimes called acid-base reactions because changes in pH mark the progress of the reaction and equilibrium state of the solution.

In our example (Eq. 5.4), addition of $H^+$ to the system by the formation of additional $H_2CO_3$ makes the solution more acidic (pH decreases) and causes additional dissolution of the calcium-carbonate mineral phase. The pH of the solution is, therefore, a convenient and accurate way to measure the progress of dissolution. In many of the following examples, pH of the solution was used to determine the propensity for dissolution of carbonate sediments.

### 5.2.2 Geochemistry of the Freshwater/Saltwater Mixing Zone

In coastal aquifers, fresh groundwater flows towards the sea, floating over a body of more dense, saline groundwater (Sect. 4.2.7.3). Freshwater and saltwater are separated by the freshwater/saltwater transition zone, which is a location of mixing of saline and freshwater (Sect. 4.2.7.2). This mixing causes enhanced chemical reactivity and potential dissolution of carbonate rocks (e.g., Back et al. 1979, 1986; Budd 1988; Hanshaw and Back 1980; Mylroie and Carew 1990; Plummer 1975; Randazzo and Bloom 1985; Smart et al. 1988; Stoessell 1992; Wicks and Herman 1995). While it is reasonable to say that the interaction of freshwater and saline water dominates chemical processes within the saline-water transition zone in coastal karst systems, the end result of this interaction strongly depends upon the particulars of the coastal hydrogeological setting, including such factors as flow dynamics, chemical compositions of the waters and host sediment or rock, and other sources of chemical reactivity within the system. This section presents a theoretical overview of the major chemical processes that control karst development, namely dissolution of calcite (or dolomite) in the freshwater/saltwater mixing zone.

The processes to be considered are those in play where two waters of different chemical composition mix, one (seawater) with a much higher concentration of ions than the other (freshwater). It should be noted, however, that mixing and associated geochemical effects occur not only within the freshwater/saltwater transition zone, but also at the water table, where phreatic water mixes with vadose (meteoric) water.

#### 5.2.2.1 Effects of Mixing on Solubility

Depending on the chemistry of the end-member solutions and the proportions at which they are mixed, when two water masses with different compositions in a limestone aquifer mix, they may produce a solution that has a different set of chemical characteristics with respect to the variables that affect the solubility of the carbonate minerals. Because of the complexity of the interactions of these variables, the resulting solution may be undersaturated or supersaturated, and it may be difficult to predict which it will be (e.g. Bögli 1964; Plummer 1975; Wigley and Plummer 1976; Runnels 1969).

Several of the chemical interactions that influence carbonate solubility cause ionic activities to vary non-linearly with the mass concentrations of the ions in solution (Hanshaw and Back 1979; Wigley and Plummer 1976). These mixtures may be subject to changes in saturation state as a result of partial pressures of $CO_2$, ion pairing as a result of changes in ionic strength of the solution, changes in water temperature, and other processes.

- **PCO2 Effect**. The $P_{CO2}$ effect (Bögli 1964, 1980) is perhaps the most dominant of the mixing effects and, in most cases, causes two saturated solutions to become undersaturated when they mix. The relationship between $P_{CO2}$ and the solubility of calcite is nonlinear, exhibiting a convex upward slope (Fig. 5.1). In Fig. 5.1, the slope of the curve is steeper at lower $P_{CO2}$ values and lessens as $P_{CO2}$ increases. The relationship reflects the complex redistribution of carbonate species during mixing (Wigley and Plummer 1976).

  For example, when two calcite-saturated solutions are mixed, the composition of the resulting mixture is a weighted average of the compositions of the two end-member solutions. On a plot of concentration (expressed as dissolved $CaCO_3$) and carbon dioxide the mixture lies somewhere on the line connecting the two end members (Fig. 5.1; Palmer 1991). The new solution (C, Fig. 5.1) is below the saturation line, and the water becomes undersaturated with respect to calcite. This effect is most pronounced where the difference in $P_{CO2}$ between the two end members is high. The $P_{CO2}$ effect has been singled out as a potential mechanism for cave development (Back et al. 1986; Mylroie and Carew 1990; Palmer et al. 1977).

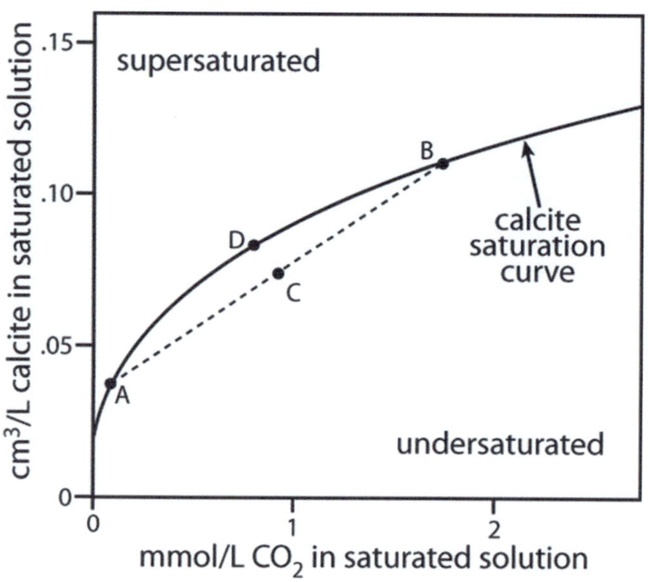


**Fig. 5.1** Effect of mixing two solutions (**a** and **b**) with differing partial pressures of $CO_2$. Line A-D-B represents the saturation line with respect to calcite. (From Fratesi (2013, Fig. 3.5))

- **Ionic-Strength Effect.** The addition of salts from saltwater with fresh groundwater also causes calcite and dolomite to become more soluble. The ionic-strength effect is only moderate in mixing of freshwaters, but may be considerable in freshwater-saltwater mixing zones due to the high ionic strength of seawater (Sect. 5.2.1.2; Table 5.1). The ionic-strength effect is a result of ion pairing. The ionic-strength effect is considered highly important in coastal mixing zones.
- **Algebraic Effect.** The algebraic effect describes a phenomenon wherein two saturated solutions tend to become oversaturated with respect to a mineral phase when they mix. As described above, the relative concentrations of species vary linearly during simple physical mixing. However, the saturation index depends upon the product of the ion activities (Eq. 5.16), which varies nonlinearly in these circumstances. As outlined in Wigley and Plummer (1976), it can be shown that:

$$SI_c = \log\left[\frac{(1+b^2)}{4b}\right], \quad (5.16)$$

  where, for calcite, $b$ is the ratio of $Ca^{2+}$ ion activities in the two end-member solutions. The part of Eq. 5.16 in brackets has a minimum value of 1 when the $Ca^{2+}$ activities are equal ($b = 1$) and increases as $b$ goes to zero or infinity. The algebraic effect is therefore strongest when two end-member solutions have greatly different activities of a particular ion of interest. This effect, however, can be overshadowed by other effects, such as the ionic-strength effect.
- **Temperature.** Mixing of waters of different temperatures produces a small undersaturation effect. The relationship of solubility to temperature traces a convex downward curve. The composition of a mixture lies somewhere along a line between the two end members, so that the concentration of any mixture of two saturated solutions lies above the saturation curve.

Figure 5.2 shows the combined effects of these mixing processes on the calcite saturation index for different sets of hypothetical waters, as calculated for demonstration

**Table 5.1** Distribution of major dissolved ions and chemical complexes in seawater at 19 ‰ chlorinity, 25 °C, pH = 8.15

| Ion | Total concentration in seawater (molality) | Percent free ion | Percent Me-$SO_4$ ion pair | Percent Me-$HCO_3$ ion pair | Percent Me-$CO_3$ ion pair |
|---|---|---|---|---|---|
| Sodium ($Na^+$) | 0.48 | 99 | 1+ | — | — |
| Potassium ($K^+$) | 0.010 | 99 | 1 | — | — |
| Magnesium ($Mg^{2+}$) | 0.054 | 87 | 11 | 1 | 0.3 |
| Calcium ($Ca^{2+}$) | 0.010 | 91 | 8 | 1 | 0.2 |

Modified from Garrels and Christ (1965, Table 4.4)

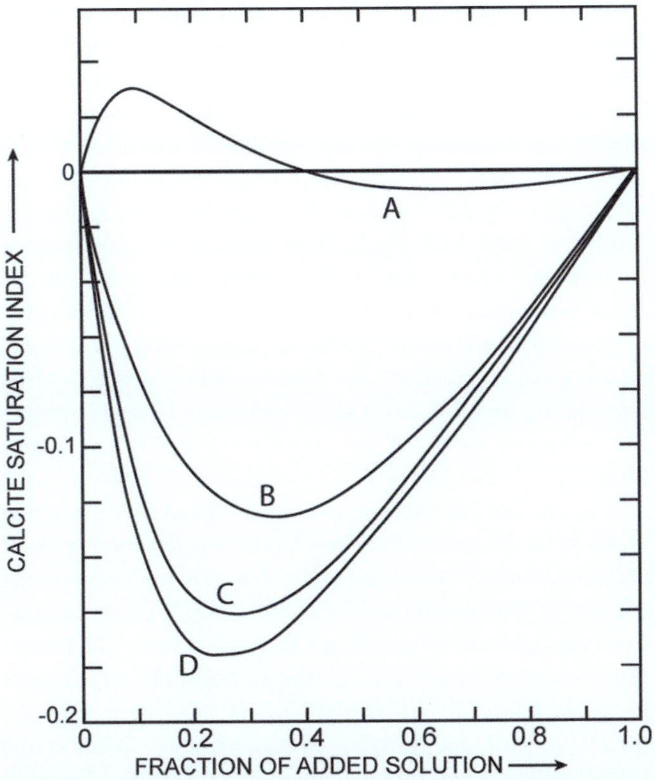


**Fig. 5.2** Effects of mixing of waters on the calcite saturation index. The different solutions (**a, b, c** and **d**) are based on hypothetical mixtures described in Wigley and Plummer (1976). (From Fratesi (2013, Fig. 3.6))

purposes by Wigley and Plummer (1976). Note that the end-member solutions in these scenarios were carefully chosen to demonstrate various degrees of impact for the different mixing effects and do not represent the most commonly encountered scenarios. For mixing of freshwater and seawater, the $P_{CO2}$ and ionic-strength effects will often dominate the resulting mixture.

The chemistry of coastal karst waters may be dominated by processes other than mixing dissolution, especially when mixing effects cause only a minor degree of undersaturation (Sanford 1987). In these cases, the deviation from these relationships gives insight into the processes operating in a particular system. Some important processes that occur concurrently with mixing are: (1) dissolution and outgassing of $CO_2$; (2) precipitation or dissolution of carbonate minerals; (3) reduction of sulfur compounds; and (4) kinetic (reaction-rate) effects.

Several field studies have found that the chemistry of coastal groundwater can be explained using some combination of these processes. For instance, Plummer et al. (1976) found that Bermuda groundwater chemistry is controlled by elevated $CO_2$ contents in soil, dissolution of aragonite, and mixing. The findings of Plummer et al. (1976) were inconsistent with simple mixing dissolution and indicate that the chemistry of Bermuda groundwater is dominated by dissolution and precipitation under non-equilibrium conditions influenced by dissolution and degassing of $CO_2$. In some Bermuda waters, outgassing of $CO_2$ overcomes the mixing dissolution effect and causes supersaturation with respect to aragonite or calcite.

Fanning et al. (1981) found calcium enrichment and magnesium depletion in waters emanating from geothermal springs on the continental shelf of southwestern Florida. There, anomalous calcium and magnesium concentrations were interpreted to be from dolomitization and gypsum dissolution caused by interaction with geothermal fluids derived from the Floridan Aquifer System (FAS).

Starks (1986) investigated mixing of UFA water in wells and springs in coastal Citrus County, Florida. Freshwater flowing toward the coastal discharge area was undersaturated with respect to both calcite and dolomite. Undersaturation was maintained until a mixture of about 12% local seawater was obtained. At 12% seawater, the mixture became oversaturated with respect to dolomite. The mixture remained undersaturated with respect to calcite until a mixture with 45% seawater was reached. Starks concluded that mixtures of fresh, UFA water with seawater are capable of causing dolomitization with concurrent dissolution of calcite. The dolomitization and karst formation were found to occur when the freshwater/seawater mixture is between 12 and 45% seawater. Mass-transfer calculations, however, indicated that calcite can dissolve about five times more rapidly than dolomite can precipitate. The result was thought to be a vuggy, dolomitized rock. Finally, assuming that the rates of calcite dissolution and dolomite precipitation remain constant, Starks estimated that 1 m of near-surface limestone would be removed every 15,000 years (a rate of 0.06 mm/yr.)

and that 20% of the volume of limestone removed would have been replaced by dolomite.

Wicks and Herman (1996) investigated the geochemistry of the UFA flow paths that originated near central Florida and flowed to the west coast of the peninsula. Groundwater flow and solute transport modeling were used to investigate the dynamics of the flow system. They found that between 8 and 15% of the freshwater and 30–31% of the saline groundwater penetrates nearly to the top of the Middle Confining Unit (MCU) where it interacts with the sulfate-rich water derived from dissolution of gypsum. When this Ca-$SO_4$-rich water returns to the near surface and mixes with fresh Ca-$HCO_3$ water, precipitation of calcite was predicted by the model.

In addition to mixing of freshwater and saltwater along the coast, mixing of fresh and saline waters near the base of the UFA is also known to be important. In both of these settings, the mixtures have been shown to have increased ability to react with the host limestone and dolostone. Hanshaw and Back (1971, 1972); Badiozomani (1973); and Plummer (1975), for example, showed that mixing of waters with different salinities in the freshwater/saline-water transition zone in Florida can cause dissolution of calcite and, under certain circumstances, precipitation of dolomite. This process was documented within the UFA in Citrus County by Starks (1986). The dissolution of calcite can lead to development of karstic features (caverns, enlarged fractures, etc.), formation of collapse breccias, and dolomite is thought to be precipitated in the space created by dissolution of calcite (Hanshaw and Back 1971, 1972; Folk and Land 1972; Badiozomani 1973; Plummer 1975).

#### 5.2.2.2 Rates of Porosity Development in Mixing Zones

As discussed above, Starks (1986) demonstrated that the rate of dissolution of coastal limestone in Citrus County, Florida was extremely slow at 0.06 mm/yr. There have been several studies that have approached porosity development in mixing zones using theoretical and empirical approaches. This section presents some of these studies.

Sanford and Konikow (1989) used a coupled reaction-transport model to study the rates of porosity development due to mixing dissolution. They found that porosity increases at rates proportional to three factors: the $CO_2$ content of the freshwater end member, the fluid flux through the system, and the rate of sea-level change. Rezaei et al. (2005) and Sainz-Garcia et al. (2011) extended this approach by reproducing Sanford and Konikow's model using reactive transport modeling. They found that the highest dissolution rates occur in areas where mixing is greatest.

The $CO_2$ content of the freshwater end member strongly impacts the dissolution potential of the mixed solution. In areas with typical groundwater velocities, kinetic effects in mixing dissolution are theoretically small (Sanford and Konikow 1989). Along the margin of the freshwater lens or wedge, groundwater velocities are fast enough that reaction rates become the limiting factor; kinetic effects thus become important in these areas (Sanford and Konikow 1989). Plummer et al. (1976) concluded that the rates of dissolution of aragonite and calcite are comparable to Bermuda groundwater residence times and that the kinetic effects are thus important.

In areas where kinetic effects are negligible, the rate of porosity increase varies linearly with specific discharge, which controls how fast the chemically aggressive water is brought in and dissolved constituents taken away. In coastal mixing environments, this implies that the coastline-to-area ratio of the land mass may be an important predictor of the rate of porosity development. All other things being equal, a large land mass discharging large amounts of water through a coastline mixing zone will undergo mixing dissolution much faster than a small circular island, where the available recharge is likely much smaller in proportion to the amount of coastline. It is partially for this reason that it is recognized that mixing dissolution on small carbonate islands is probably enhanced by decay of organics (Mylroie and Carew 1995).

The position of the mixing zone is linked to the position of sea level. If the coastline is migrating as a consequence of sea-level change, the mixing zone may have limited potential for dissolution at any one elevation before it migrates away from that position (Sanford and Konikow 1989). The formation of macroporosity (i.e., caves) is, therefore, much more likely at sea-level stillstands. This is consistent with the model of flank-margin cave development by Mylroie and Carew (1990). Flank-margin caves are consistently found to correspond with relative elevations of known high- and low-stands around the world (e.g., Carew and Mylroie 1987; Frank et al. 1998; Jenson et al. 2006; Kelley et al. 2006; Mylroie et al. 1995a, b; Mylroie and Carew 1995; Mylroie et al. 2001; Mylroie et al. 2005). Based on the duration of known high sea-level stands in the Bahamas, these caves are known to have formed in small freshwater lenses within a time span of 15,000 years or less (Carew and Mylroie 1995; Mylroie et al. 1991).

Hanshaw and Back (1980) found similarly high rates of dissolution when they calculated rates of limestone denudation in the Yucatan Peninsula and found that a block of limestone 1 $km^2$ by 3 m thick could have dissolved in 8,700 years. Specifically, they calculated that the incision of Xel Ha lagoon could have occurred as quickly as 3,000 years. Note that these denudation rates are high in comparison with those determined for the Floridan and Biscayne aquifers (Chap. 6).

#### 5.2.2.3 Reduction/Oxidation Reactions in the Mixing Zone

Within the mixing zone itself, the density contrast between saline and freshwater may trap organic matter and create stratified microenvironments in which sulfate reduction/oxidation (redox) reactions generate acidity (Socki et al. 2002). Bacterially mediated sulfate reduction produces elemental sulfur (Eq. 5.17) or sulfide (Eq. 5.18) near the base of the mixing zone. In a chemically reducing environment, organic substances, represented here by formaldehyde ($CH_2O$), undergo such reduction reactions as:

$$2CH_2O + SO_4^{2-} \xrightarrow{reduction} 2HCO_3^- + S^0 + H_2 \quad (5.17)$$

or

$$2CH_2O + SO_4^{2-} \xrightarrow{reduction} 2HCO_3^- + H_2S \quad (5.18)$$

during which $HCO_3^-$ and $S^0$ or $S^{2-}$ (represented by $H_2S$ in Eq. 5.18) are produced. The common ion effect may induce precipitation of $CaCO_3$ by driving the reaction in Eq. 5.4 to the left near the base of the mixing zone, or the sulfur species may be oxidized to form $H^+$ and $SO_4^{2-}$ in the upper mixing zone.

Oxidation of $H_2S$ and S in the upper part of the mixing zone generates acidity, such as indicated by the following equilibrium redox reactions. Oxidation drives the reactions to the right in each case.

$$FeS_2 + H_2O \underset{reduction}{\overset{oxidation}{\rightleftarrows}} Fe^{2+} + 2SO_4^{2-} + 2H^+. \quad (5.19)$$

or

$$H_2S + 2O_2 \underset{reduction}{\overset{oxidation}{\rightleftarrows}} SO_4^{2-} + 2H^+, \quad (5.20)$$

or

$$S^0 + \frac{3}{2}O_2 + H_2O \underset{reduction}{\overset{oxidation}{\rightleftarrows}} SO_4^{2-} + 2H^+. \quad (5.21)$$

The product of these oxidation reactions is $H_2SO_4$, which dissociates to form $2H^+$ and $SO_4^{2-}$. The $H^+$ ion is available to dissolve $CaCO_3$ (aragonite or calcite) by

$$2H^+ + CaCO_3 \rightleftarrows Ca^{2+} + 2HCO_3^-. \quad (5.22)$$

Field studies from the Bahamas (Bottrell et al. 1991, 1993; Smart et al. 1988), Isla de Mona, Puerto Rico (Wicks and Troester 1998) and the Yucatan Peninsula (Perry et al. 2002; Stoessell et al. 1993) have documented the contribution of sulfate redox reactions in systems open to a contribution of organic material from the surface (Wicks and Troester 1998). Similar reactions have been suggested in Florida, both at the freshwater/saltwater transition zone near the coast and inland, where sulfide-rich groundwater is exposed to oxidizing conditions elsewhere in the FAS. These sulfide/sulfate redox reactions have been investigated in the UFA by Rightmire et al. (1974); Rye et al. (1981); Sacks (1995); Sacks and Tihansky (1996); and Sacks et al. (2012). Garman and Garey (2005); Garman (2010); and Garman et al. (2011) have described anoxic and anchialine caves along the coast of Florida that have indications of these redox reactions.

#### 5.2.2.4 Mixing Zone Speleogenesis

Before we move on from the topic of mixing zone geochemistry and cave formation, it is important to note that the source of acidity in these sulfur-driven systems often derives from chemical reactions caused by the interactions of deep seated, saline or sulfate/sulfide-rich water with the overlying freshwater lens. This source of acidity qualifies as a form of hypogene speleogenesis according to Palmer (1991).

### 5.2.3 Textural Controls on Carbonate Mineral Stability

#### 5.2.3.1 Effect of Grain Size on Solubility

Most Florida sediments contain both coarse (sand- and gravel-sized) and fine (silt- and clay-sized) particles. Many, if not most, of the clay deposits of Florida have significant carbonate content as a result of deposition. The presence of silt- and clay-sized carbonate sediment (these particles can be aragonitic, calcitic, or dolomitic) has a profound effect on karst development by controlling where and when carbonate mineral dissolution occurs. Dissolution of fine-grained carbonate sediments results from a tradeoff between grain size of the particles and permeability of the sediment.

Chave and Schmalz (1966) and others in more recent years have shown that grain size greatly affects the solubility of carbonate particles. For example, Chave and Schmalz measured the relative solubilities of calcite in seawater by measuring the equilibrium pH of the sediment and seawater in the laboratory. They found that particles of calcite $10^{-6}$ cm (clay size) in diameter had activities (*a*; Eq. 5.4) more than eight times greater than particles 1 cm (pebble size) in diameter. This enhanced solubility is a result of the high surface area of a clay-sized particle relative to its volume. Because of their small size, the enhanced dissolution of calcite forces reaction Eq. 5.4 to the right, but it leaves excess calcium and bicarbonate in the water. Unless the calcium and bicarbonate are diluted to concentrations that approximate their equilibrium concentrations, the over-saturation will ultimately drive Eq. 5.4 to the left, and calcite will precipitate as cements in the sediment.

The consequence of the greater solubility of clay- and silt-sized calcite as compared to larger particles is that groundwater comes to equilibrium, and perhaps over saturation, with respect to calcite quickly and over short travel distances if small-sized calcite particles are present. This dissolution and chemical saturation of the water with respect to calcite and the other carbonate minerals, severely limits dissolution and karst development in underlying clayey sediments. As a result, it is uncommon to observe evidence of shell or fine particle dissolution in carbonate-rich, clayey sandy sediments.

Clay- and silt-rich sediments typically have low permeabilities. As a result, the soluble fractions of clay- or silt-rich sediment may be preserved for thousands of years owing to limitations of freely circulating water. Because of the high solubility of clay- and silt-sized calcite and aragonite, water that slowly passes through low permeability, clay/silt beds is likely to come to chemical equilibrium quickly and within only a few centimeters of penetration of the bed. After equilibration, this water is no longer able to cause dissolution, and the carbonate fraction of the sediment is preserved without significant modification. Such is the case in many of Florida's clay beds where the carbonate phases remain in sediments as old as Miocene with no evidence of dissolution. Both the silt- and clay-sized carbonate fractions and any shell materials may remain with little or no alteration.

Shell material that was originally aragonitic or high-Mg calcite may show evidence of diagenetic alterations and loss of shell integrity. As a result, the shells may appear "punky" and degraded. Also, the heterogeneity of clay-rich strata may provide pathways for chemically aggressive water deep into clay-rich strata. For example, the Hawthorn Group of clay-rich sediments consists of intricately interbedded layers of sand, carbonate, or clay mixed in varying proportions. Intense leaching of the uppermost Hawthorn Group sediments during the late Miocene-early Pliocene low sea stand and period of unconformity development resulted in deep alteration of smectitic clays (Upchurch et al. 1982) and opportunities for carbonate dissolution as water passed through the thin, interfingering permeable strata. As a result, shell fossils are uncommon in Hawthorn sediments near the land surface, and they become more abundant down dip to the south. The clay-rich sediments commonly contain soluble dolosilt and micrite throughout the extent of the Hawthorn because of the low permeability of the clay and inability of water to freely circulate. In contrast, permeable, sandy sediments may lose their fine-grained and even coarse carbonate fractions quickly where acidic water can freely circulate. This explains the absence of carbonate materials in the sandy sediments of the SAS.

#### 5.2.3.2 Effect of Grain Abrasion on Solubility

Chave and Schmalz (1966) also ground calcite in a mortar to determine the effects of mechanical disruption of the crystals on solubility. Since much of the silt- and clay-sized carbonate mineral fraction in marine sediment is derived from mechanical abrasion and breaking of shells and other hard parts of marine organisms, their experiment is also instructive. They found that the activity ($a$) of mechanically disrupted calcite is as much as seven times that of unground material of the same composition. In other words, mechanical abrasion of carbonate particles and surfaces enhances their solubility. The induced increases in grain surface area also increase grain solubility.

#### 5.2.3.3 Effect of Grain Shape and Surface Area on Solubility

The shape of a carbonate sediment particle also affects its solubility. This is especially true at locations on the particle where there is high surface area relative to local volume. For example, angular grains with points and sharp edges tend to undergo dissolution at the points and edges where there is much surface area but the nearby volume of the grain is low. For this reason, points and edges tend to be rounded and rough surfaces (on shells) smoothed. This is a useful property of carbonate sediment particles, because examination of the particles under the microscope can provide information as to whether dissolution of the grains has occurred or not. In most sandy and clayey, carbonate-rich sediments in Florida there is little evidence of rounding and smoothing of the grains by dissolution. Mechanical abrasion during transport and deposition may round and smooth grain surfaces, but the surfaces usually appear abraded in this circumstance.

#### 5.2.3.4 Application of Concepts of Solubility of Carbonate Sediments

These concepts are important for identification of evidence of dissolution in calcareous sediments and rocks in Florida. A common approach to identification of carbonate-mineral content in Florida sediments and rocks is to test a sample for the presence of a carbonate phase with 10 percent HCl (hydrochloric acid). Note that the presence of a carbonate phase in fine-grained sediments with this relatively strong acid does not indicate on-going or past dissolution and, therefore, potential for sinkhole development.

The processes discussed in the above paragraphs clearly demonstrate an approach for confirmation of dissolution in calcareous sediments. Clearly, there is a propensity for carbonate sediment particles to develop artifacts of dissolution as follows:

**Highest solubility** ⟵ **Lowest solubility**

Clay sizes > Silt sizes >> Sand sizes >> Corners and sharp edges on larger particles

Therefore, microscopic examination of the sediment and analysis of fine particles can present evidence of dissolution. It may be necessary to examine the fine-grained fractions of sediment separately to determine if they have undergone dissolution. Determination if dissolution is on-going is a geochemical, not petrographic, problem; physical evidence of dissolution does not indicate that the process is on-going. Searching for evidence of artifacts of dissolution only allows for determination of past dissolution events.

### 5.2.4 Measurements of Relative Solubility of Modern Sediments

Examination of the mineral content and saturation states of water and sediment in modern environments helps to understand the modes of formation and susceptibility to dissolution of ancient sediments.

Figure 5.3 compares the saturation states of sediment/water samples from Florida Bay and the Everglades to carbonate sediments from the Bahamas. A value of one on the horizontal axis of the graphs indicates that the water is chemically in equilibrium with aragonite. Recall that aragonite is slightly more soluble than low-Mg calcite, so Florida Bay water is saturated with respect to calcite for the most part. Note that a large proportion of water in contact with modern, biogenic sediment of southern Florida is chemically saturated with respect to low-Mg calcite and even aragonite. In comparison, sediment from the Bahamas is largely over-saturated with respect to calcite and aragonite, presumably because of its high-Mg calcite content.

The potential for dissolution of the biogenic phases (aragonite and calcite) during sedimentation may be low because the seawater is chemically saturated, or at equilibrium, with the sediment. This means that the fine-grain-sized calcite and aragonite is preserved in the sediment and available for reaction with chemically undersaturated groundwater at a later date. Unless the permeability of the sediment is too low for chemically aggressive water to interact with the carbonate sediment, there is a potential for eogenetic karst development upon exposure to chemically aggressive freshwater. Furthermore, the data represented in Fig. 5.3 clearly demonstrate that the potential for preservation and/or dissolution varies with location, composition and abundance of the biogenic phases in the sediment, and with grain size.

High-Mg calcite is likely the most reactive sediment component (Fig. 3.2, Table 5.2) while intermediate-Mg calcite and aragonite have somewhat lower potentials for preservation owing to their relatively high solubilities, or activities. Losses in sediment volume as a result of dissolution in meteoric and groundwater may affect these more soluble phases without involving the calcitic phases.

Small amounts of shell contained within coarse-grained siliciclastic sediments rarely reflect significant sediment volume loss as the shell fragments are dissolved away in Florida. This is because the volume loss is minor compared to the volume of grain-supported siliciclastic sediments intermixed with the carbonate material, and the internal friction between clasts can easily continue to support the sediment mass, even with loss of some of the shell component. In clay-rich sediments, water flow is restricted, and any biogenic sediment is usually isolated from dissolution by rapidly moving, chemically aggressive water. In shell beds, however, where the shells constitute the grain support of the stratum and water can freely circulate through the permeable sediments, there is a potential for dissolution and loss of sediment volume, thus causing minor subsidence.

An understanding of the stability of carbonate phases in modern through Quaternary-aged sediment helps understand how karst features may develop in such geologically young sediments. The karst features caused by dissolution of geologically young clay and shell beds are discussed in Chap. 7.

### 5.2.5 Karst Development as a Function of Sediment Mineralogy

Karst development is variably extensive in the shallow Quaternary and Tertiary limestones and unconsolidated carbonate sediments of Florida. The artifacts of karst processes range from subtle dissolution features in the youngest calcite- or aragonite-rich sediments to well-developed dissolution and depositional features in the older strata (Table 5.2). Dolomite rarely develops large-scale dissolution features and is not considered a major host to karst features in Florida.

For the purposes of this text, the importance of the three carbonate minerals to karst development lies in their relative solubilities in the waters of Florida. Table 5.2 summarizes the relative solubilities of the carbonate minerals and the role they play in eogenetic karst development in Florida.

As a consequence of the slight differences in solubility of the carbonate minerals, a few general principles for eogenetic karst development in Florida can be derived:

- Only the youngest (Pleistocene and Recent) sediments and rocks are likely to contain significant aragonite or high-Mg calcite. As a result, only modern carbonate sediments, some geologically young (Quaternary) shell beds and Pleistocene limestones are likely to contain dissolution features related to sediment mineralogy. These sediments are generally restricted to southern Florida and areas near the coasts.
- Most of the near-surface carbonate rocks in Florida consist of limestone containing low-Mg calcite or dolostone. It is in these sedimentary rocks that the karst of central and northern Florida is well developed. These rocks are Miocene and older in age and contain abundance evidence of karst processes.

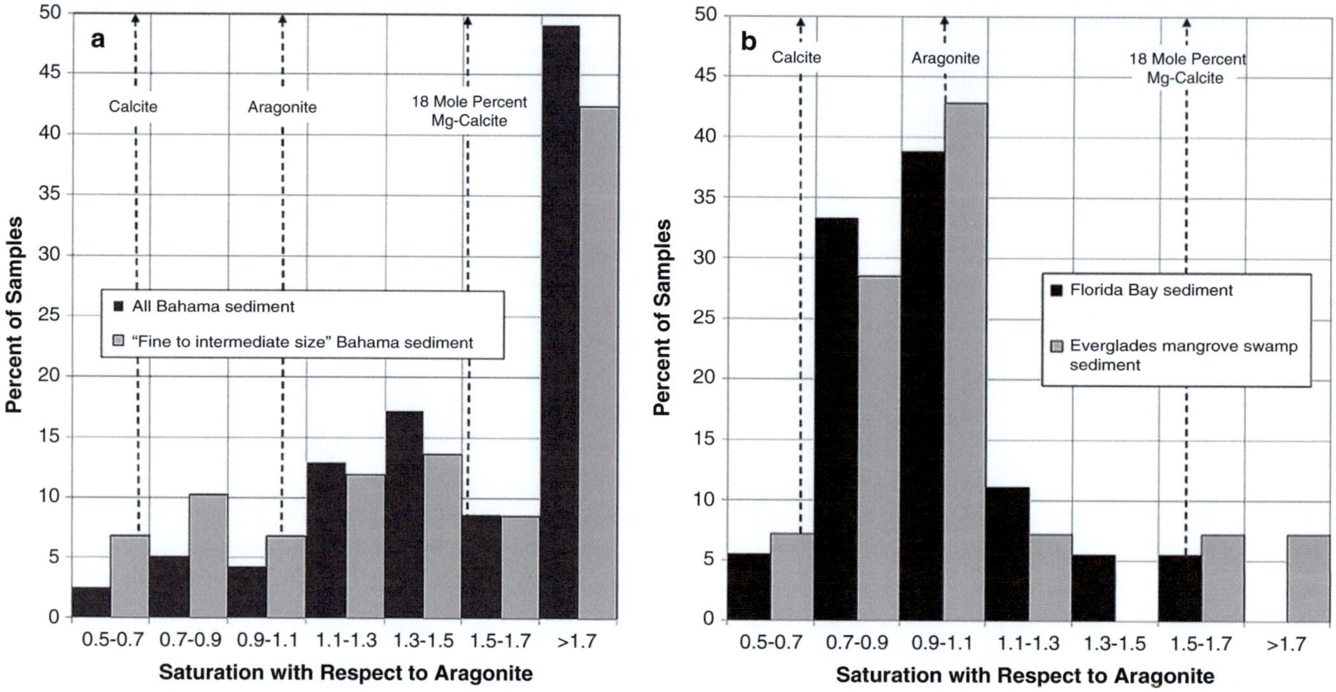


**Fig. 5.3** Histograms showing the range of relative solubilities of modern, shallow-water carbonate sediments from the Bahamas (**a**) and Florida (**b**) with respect to aragonite. From Morse et al. (1985). (Data from Florida Bay and the Everglades were recalculated by Morse and Mackenzie (1990) from data in Berner (1966))

**Table 5.2** Relative solubilities of the major carbonate minerals in Florida strata and the roles each mineral plays in eogenetic karst development. Compare with Fig. 2.2

| Predominant mineral | Importance to karst development |
|---|---|
| **Most soluble** | |
| ↑ High-Mg calcite<br>Aragonite | In geologically young (Pliocene through Recent) sediments; Development of moldic void space and some cover-subsidence sinkholes and sag features (sag ponds) |
| Intermediate- and Low-Mg calcite | Occurs in sediments of all ages; Predominant host of karst development In geologically older (Tertiary) sediments in Florida |
| Dolomite | Dominant sediment in some Tertiary deposits; Karst development is limited. Poorly ordered crystals are more soluble than well-ordered crystals. |
| **Least soluble** | |

- Since dolomite is less soluble than calcite under equivalent conditions, much of the dolostone in Florida has experienced substantially less karstification than the limestone. As rocks commonly formed by alteration of preexisting carbonate sediment or limestone, dolostones in Florida are often well lithified and brittle. Many features commonly attributed to karst in Florida dolostone deposits are fractures that may, or may not, have been modified by dissolution.

The mineral content of these rocks and the available time for karst development have affected the "styles" of karst development. These different "styles" of karst development are discussed in following chapters.

## 5.3 Chemistry of Florida Precipitation

The next several sections explore the specifics of the chemistry of Florida water with respect to water quality and karst development. The logical starting point for Florida's eogenetic karst processes is to consider the chemistry of Florida rainfall, the source of most of Florida's groundwater.

### 5.3.1 Composition of the Atmosphere

The earth's atmosphere is composed predominantly of nitrogen (78%), oxygen (21%), and argon (0.9%), a noble gas. Other gases are present in trace quantities. Of these, $CO_2$ is the most important to karst processes. $CO_2$ is approximately 0.09% of the atmosphere by volume and it represents about 94% of the trace gases present in the atmosphere. The reaction of $CO_2$ with moisture in the atmosphere forms $H_2CO_3$, the acid most responsible for dissolution of carbonate minerals in sediments and rocks.

The first step in formation of acidic precipitation and development of karst begins in the atmosphere where $CO_2$ reacts with $H_2O$ to form $H_2CO_3$ according to reaction 5.5. The $H_2CO_3$ then dissociates freeing $H^+$ and rendering the rainfall slightly acidic. Rainwater that is fully equilibrated with $CO_2$ gas at earth surface temperatures and pressures is about as acidic as a soda pop (pH ≈ 4.5 standard units [s.u.]).

Other acids form in the atmosphere as well. These include sulfuric acid ($H_2SO_4$) formed by the reaction of $SO_2$ and $H_2O$, and nitric acid ($NHO_3$) formed by the reaction of nitrogen-oxide compounds ($NO_X$) and $H_2O$. Geologically, the acids formed by these reactions are substantially less important than $H_2CO_3$ owing to the lower abundance of the source compounds in the pre-development atmosphere. In terms of karst processes, the interaction of $H_2CO_3$ with carbonate-rich sediments and rocks is of primary concern, both now and at most times in the geologic past. However, episodes of increased volcanic activity, which release large quantities of all of the acid-forming compounds into the atmosphere, may have altered the relative importance of these atmospheric acids at certain times in the past. Increases in the formation of these acids in modern rainfall are primarily caused by human impacts and have led to concerns about "acid rain" and acidification of soils, fresh-water systems, and the sea.

The National Atmospheric Deposition Program (2012; NADP) maintains a series of stations in Florida (Fig. 5.4) where rainfall is collected and the chemical content and pH of the rainwater are determined. These data are important because they assist in understanding the acidification of soils and water bodies that the earth is currently experiencing.

Figure 5.5 illustrates the median and range of pH in rainfall in Florida by NADP station. Statewide, the median pH of modern rainfall is 4.77 (average = 4.81), and there is only a small variation in pH from north to south. The variation is important, however, because it demonstrates the importance of the next subject, marine aerosols.

Given long periods of time, this mildly acidic rainfall can erode carbonate-rich strata, especially sediments rich in calcite. Figure 5.6 is an example of limestone (marble) erosion

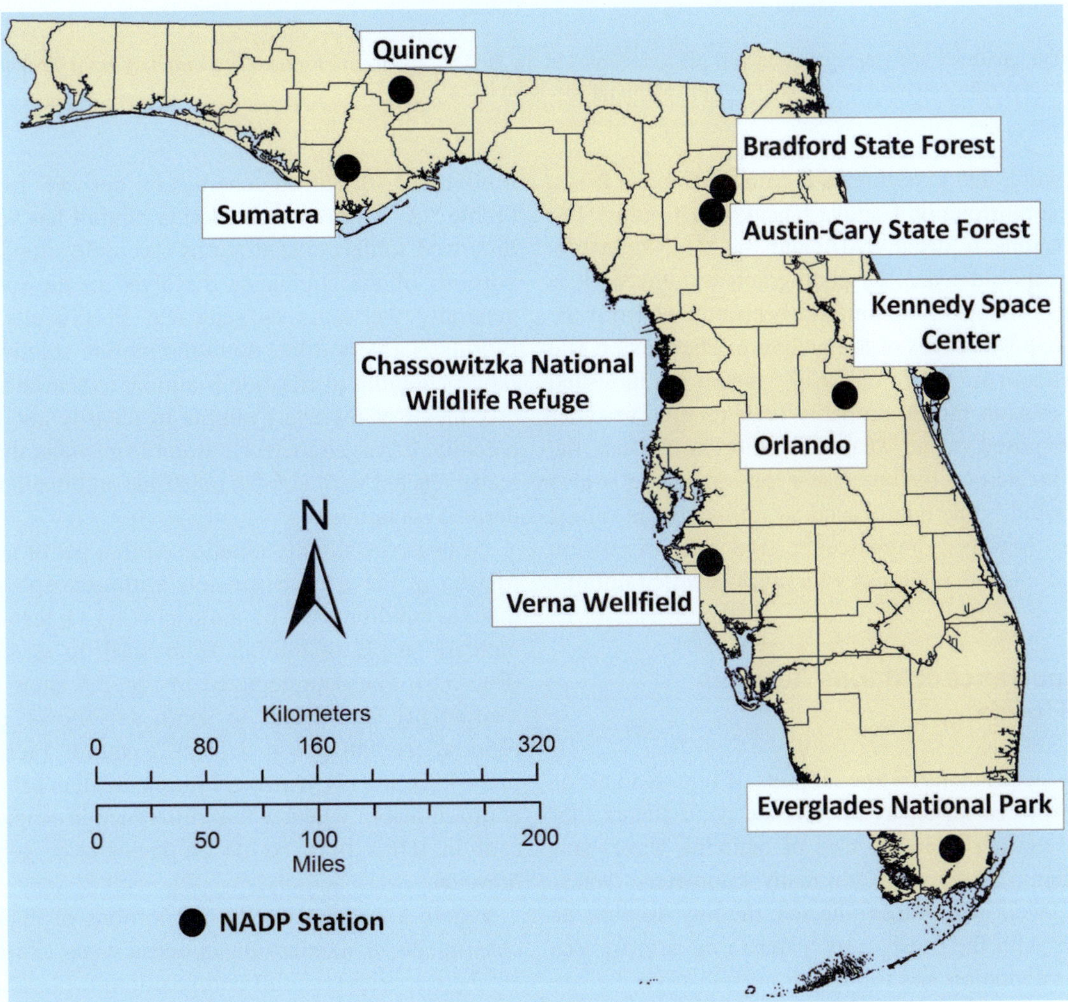


**Fig. 5.4** Locations of the National Atmospheric Deposition Program (2012) stations in Florida

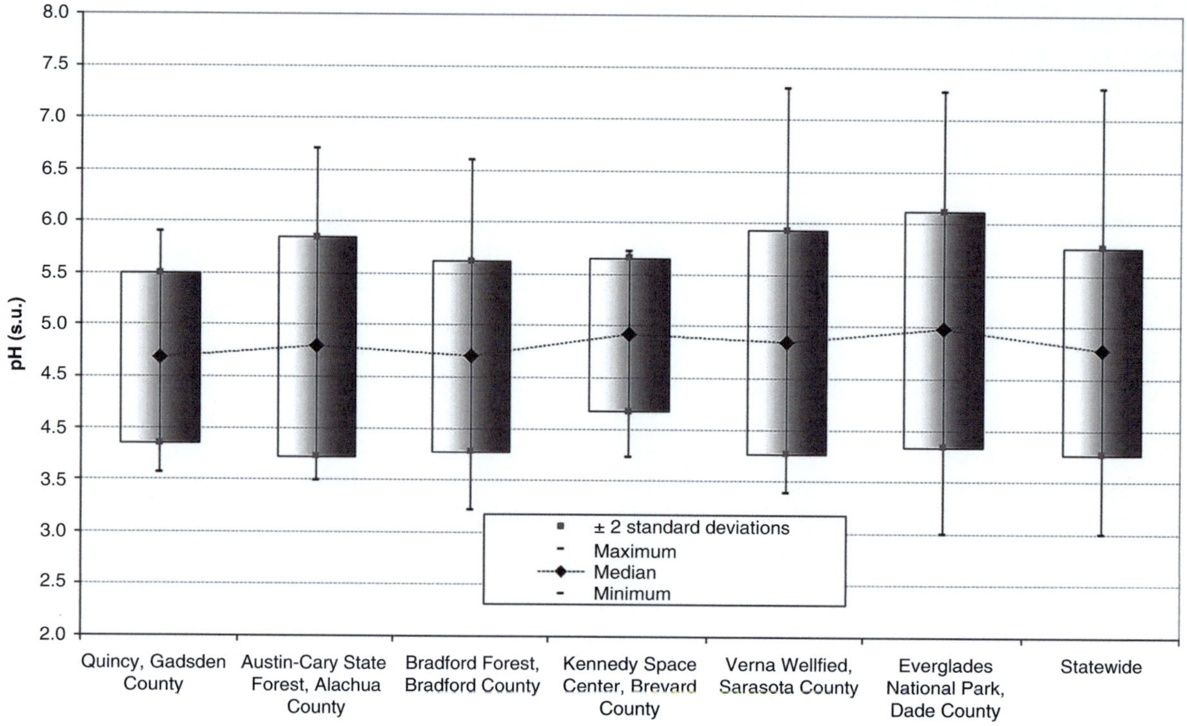


**Fig. 5.5** Median, quartiles, and extremes in rainfall pH as recorded at the NADP stations in Florida. Data summary from Upchurch (1992). On this graph, the stations are arranged in a general north-to-south pattern from left to right

by rainfall, wind, and exposure to thermal extremes. It is a portion of a tombstone on Captiva Island, Lee County. The marble tombstone is dated 1901, and the tombstone was about 110 years old when the photograph was taken. It is estimated that about 0.5–1 mm had been eroded from the corners and exposed edges of the sculpture. The erosion rate is, therefore about $4.5 \times 10^{-3}$ to $9 \times 10^{-3}$ meters per thousand years. This erosion rate is probably high relative to other limestones exposed to the weather in Florida because the cemetery is located on the beach and the tombstone is also abraded by wind-borne quartz sand as well as being rained upon. It does, however, dramatize the slow rates of erosion caused by bathing calcitic rocks with mildly acidic rainfall.

## 5.3.2 Importance of Marine Aerosols in Florida

Aerosols are suspensions of fine droplets of liquids in air or another gas. In coastal environments, the wind shears tiny droplets of seawater from the tops of breaking waves and transports them, as aerosols, for many kilometers. When aerosols are swept inland from the sea, the tiny droplets of seawater mix with the nearly pure water in the atmosphere derived by evaporation and transpiration (ET) from the sea and land surfaces. The mixing of the marine aerosols and water derived from ET, results in rainfall with small concentrations of dissolved chemicals derived from the sea (Table 5.3). Even though Florida rainfall has very low total dissolved solids concentrations and ionic strengths, the proportions of the chemicals dissolved in the rain water are generally the same as seawater. Specifically, rainfall in Florida is a very dilute sodium-chloride solution where the ratio of sodium to chloride is similar to seawater. This property makes it relatively simple to identify and differentiate recently recharged meteoric water from water that has chemically reacted with aquifer materials and that contains land-derived contaminants.

Seawater is slightly alkaline, with a pH of about 8.2 s.u. Mixing of the marine aerosols with atmospheric moisture that has equilibrated with atmospheric $CO_2$ results in a slight increase in pH of rainfall from north to south in Florida (Fig. 5.5). The data depicted in Fig. 5.5 show a change in median pH from north to south as follows: (1) northern Florida: median pH = 4.77; (2) central Florida: median pH = 4.80; and (3) southern Florida: median pH = 5.01 s.u. It is questionable whether this difference in acidity is significant in terms of karst development over geologic time, however.

Figure 5.7 compares the proportions of major chemical constituents in standard mean ocean water (SMOW; Drever 1997) to average rainfall composition in Florida (Upchurch 1992). Note that Florida precipitation is enriched in the proportion of sulfate compared to seawater. This is a result of

**Fig. 5.6** A 110-year old marble tombstone that has been exposed to mildly acidic rainfall, wind-borne sand, and insolation on the beach at Captiva Island, Lee County, Florida

**Table 5.3** Statewide median concentrations of major chemical constituents in Florida precipitation

| Constituent | Median Concentration (mg/L) or s.u. |
|---|---|
| Calcium | 0.12 |
| Magnesium | 0.05 |
| Potassium | 0.03 |
| Sodium | 0.38 |
| Ammonium | 0.09 |
| Chloride | 0.67 |
| Sulfate | 1.00 |
| Nitrate | 0.70 |
| pH | 4.78 s.u. |

Data from NADP (2012; Period of Record: Oct. 1978–Jan. 2012)

increased atmospheric sulfate (including $H_2SO_4$) because of modern air pollution. The elevated concentration of sulfuric acid increases the ability of modern rainfall to react with carbonate sediments slightly as compared to ancient, pre-industrialization precipitation. Even so, the proportions of sodium and chloride in Florida rainfall are similar to those in seawater, clearly indicating that the natural chemistry of Florida rainfall is controlled by marine aerosols.

It is instructive to examine the aerosol-derived sea-water component in Florida rainfall as a function of latitude. Recall that northern Florida climate is dominated by continental weather patterns that develop to the north of Florida. The climate of peninsular Florida, on the other hand, is significantly more maritime than extreme northern Florida. This pattern is reflected in the aerosol-derived component of Florida rainfall. Figure 5.8 compares the ratio and concentration range of $Na^+$ and $Cl^-$ in rainfall at Quincy (Gadsden County near the Georgia line) with rainfall at the Kennedy Space Center (Brevard County). The ratio is similar to seawater in both locations, demonstrating that the marine aerosol component is present in rainfall throughout the state. Data from all of the NADP sites in Florida have the same pattern of sodium relative to chloride in rainfall.

In Fig. 5.8, concentrations of the two constituents at each location reflect the north-to-south difference in rainfall composition and degree of influence of marine aerosols. At Quincy, median $Na^+$ concentration was 0.23 mg/L and $Cl^-$ was 0.39 mg/L (n = 1000 samples; Period of Record: 1984–2012). At the Kennedy Space Center, median $Na^+$ concentration was 0.70 mg/L and $Cl^-$ was 1.25 mg/L (n = 1096 samples; POR: 1984–2012). Fig. 5.8 also demonstrates that the ranges of sodium and chloride concentrations in marine aerosols are greater in peninsula Florida than in the northern tier of counties.

## 5.4 Processes That Affect Water Quality in Florida's Aquifers

Once rainfall enters the soils and sediments that compose the vadose and phreatic environments of Florida's aquifers, significant changes in water quality begin. Some of these processes are listed in Table 5.4 and discussed below.

### 5.4.1 Evaporation and Transpiration

Evaporation and transpiration from a mass of water increases the concentrations of chemical constituents that are left

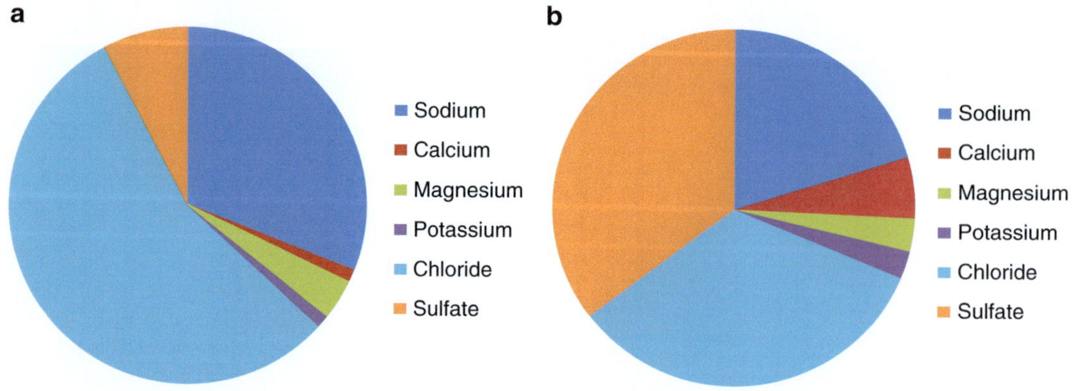


**Fig. 5.7** Proportions of major ions in (**a**) standard mean ocean water and (**b**) Florida average precipitation

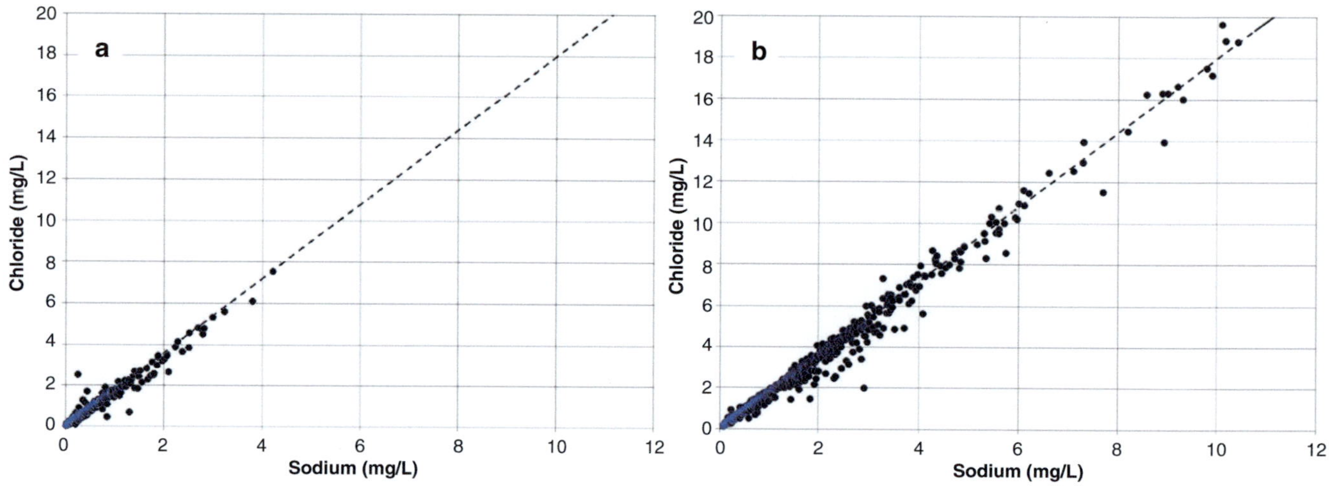


**Fig. 5.8** Comparison of sodium and chloride concentrations in rainfall at the NADP sites (Fig. 5.1) at Quincy (**a**) and the Kennedy Space Center (**b**). Dashed line represents the range of theoretical concentrations of sodium and chloride in diluted seawater

behind in the remaining water. We use the term evapotranspiration (ET) to describe the general transfer of moisture to the atmosphere from the earth surface.

Evaporation is the transfer of water from a free water surface in contact with the atmosphere. We tend to think of evaporation as being a process that is limited to open water bodies at the earth's surface, but evaporation is also an important process where moisture is in contact with atmosphere in the subsurface. That is why buildings are constructed with a vapor barrier under their floors at or near the ground surface to prevent moisture from entering the structure from the soil. Evaporation is also an important process in many caves and caverns in arid regions, where it can lead to precipitation of flowstone, stalactites, and other cave decorations. Formation of cave decorations by evaporation is apparently uncommon in Florida caves. Rather, degassing of $CO_2$ appears to dominate the processes of speleothem formation in Florida. Trees and other rooted plants extract moisture from both the vadose and phreatic zones. They metabolize nutrients and some of the moisture recovered from the moisture they take up and then pass the remaining moisture on to the atmosphere through transpiration. Plants can, therefore, not only remove substantial amounts of soil moisture and groundwater via transpiration, but also change the chemistry of the remaining water, depleting it of certain nutrients and leaving behind increased concentrations of other chemicals.

Tree roots can cause dehydration of shallow clay by transpiration (Bryant et al. 2001) which may cause clay shrinkage and damage to overlying structures. The soil dehydration can also result in ground penetrating radar anomalies that may be confused for potential soil disruption as a result of sinkhole development (Bryant et al. 2001).

### 5.4.2 Biotic Activity in Soils and Sediments

Soils and sediments are excellent media for the growth and sheltering of such organisms as microbes, fungi, and small, burrowing animals. Even the most sterile appearing soils and

**Table 5.4** Some processes that affect the chemical composition of groundwater in Florida

| Process | Unconfined or poorly confined aquifer | Confined aquifer |
|---|---|---|
| Evapotranspiration of soil moisture and from the surface and capillary zone of the water table | X | |
| Respiration by soil microbes and other biota | X | X |
| Creation of humic substances | X | [a] |
| Reduction – oxidation reactions induced by microbial activity | X | X |
| Nutrient uptake by plants and animals | X | |
| Sorption and ion exchange | X | X |
| Chemical equilibration with carbonate minerals in the soil/sediment | X | X |

[a]There is evidence that the organic sediments trapped in the Avon Park Formation are releasing organic carbon into Floridan Aquifer System groundwater, probably as a result of microbial action

**Table 5.5** Estimated biomass to a depth of 13 cm in a hectare of soil in a humid, temperate region

| Item | Examples | Estimated biomass (kg/ha) | Estimated no. individuals |
|---|---|---|---|
| Dead organic matter | Plant, animal debris and waste products | 105,400 | – |
| Living organisms | Bacteria, Protozoa, Fungi, higher plants and roots, etc | 14,600 | – |
| Animals, miscellaneous | Nematodes, snails and slugs, earthworms, etc. | 140 | $2.5 \times 10^9$ |
| Arthropods | Isopods, crayfish, etc. | 10 | $4 \times 10^{17}$ |
| Arachnids | Mites, spiders, centipedes, etc. | 50 | $4.8 \times 10^7$ |
| Hexapods | Springtails, ants, Diptera, etc. | 42 | $5.5 \times 10^7$ |
| Vertebrates | Mice, voles, gophers, etc. | 42 | $4 \times 10^5$ |
| **TOTAL** | | 120,000 | – |

Condensed from Buol et al. (1973)

sediments contain incredible populations of microorganisms and burrowing animals. Microbes and fungi can exist both in the vadose and phreatic environments as long as there is food and a hospitable environment. Air-breathing invertebrates and vertebrates are generally restricted to the vadose zone, above the water table.

In Florida, near-surface soils and sediments are typically moist and warm, and they contain abundant food supplies that are sources of carbon for microbially mediated chemical reactions. Table 5.5 presents estimates of the abundances of dead organic matter and living organisms in a humid temperate soil. While this soil is from a temperate region rather than Florida's sub-tropical soils, it does give a sense of the abundance of living organisms in a relatively warm and wet soil and for the huge amount of dead organic matter that is broken down as food by the living organisms. Living organisms range from microscopic fungi to microbes (bacteria and Protozoans) to large vertebrates.

These organisms, especially soil microbes, utilize the dead organic matter as food sources, converting dissolved and particulate organic materials (plant cellulose and animal waste material) to energy. They release waste compounds that include $CO_2$ or $CH_4$ (methane) gas; organic, humic substances dissolved or suspended in the water; and inorganic carbon, sulfur, and nitrogen compounds. These waste compounds, in turn, react with soil water to create dissolved organic acids, additional $H_2CO_3$, and sulfur and nitrogen compounds that are of great interest in Florida because of their relationships to karst (in the case of $H_2CO_3$) and groundwater contamination (in terms of the nitrogen and sulfur compounds). The $H_2CO_3$ dissociates as described by Eq. 5.6, which increases the acidity of water in the soil.

When microbes consume a food source, they tend to consume the lighter isotopes of the element or compound they are metabolizing. For example, if the microbes are metabolizing a carbon compound, they tend to select for $^{12}C$ rather than $^{13}C$.[3] The resulting metabolic waste products are, therefore, enriched in $^{13}C$ as compared to $^{12}C$. By comparing the ratio of the heavier to lighter isotope, one can learn much about the sources and forms of metabolism. For metabolism of carbon compounds the ratio of $^{13}C$ to $^{12}C$ is utilized to understand metabolic processes.

Rightmire and Hanshaw (1973) used carbon isotopic ratios to correlate decomposition of soil organic matter in aquifer water. They confirmed that the isotopic signature of UFA water was derived from the decomposition of soil organic matter. The ratio $^{13}C/^{12}C$, expressed as $\delta^{13}C_{PDB}$, in the carbon source materials they studied were

- Leaf litter: $\delta^{13}C_{PDB} = -18.2$ to $-29.5\ ‰$,
- Soil organic material: $\delta^{13}C_{PDB} = -22.4$ to $-24.1\ ‰$, and
- Soil $CO_2$: $\delta^{13}C_{PDB} = -14.7$ to $-21.3\ ‰$.

The dissolved carbonate in water from the UFA water at their sample site ranged from $-10.8$ to $-14.6\ ‰$. Rightmire and Hanshaw concluded that, at the time of recharge to the UFA, the soil $CO_2$ source was approximately $\delta^{13}C_{PDB} = -25 \pm 5\ ‰$, which corresponded to a leaf litter or soil organic material source. Their data suggested that micro-

[3]Carbon-12 ($^{12}C$) has 6 protons and 6 neutrons while carbon-13 ($^{13}C$) has 6 protons and 7 neutrons. Therefore, $^{13}C$ is the "heavier" isotope since it has a higher mass.

bial decay of these soil organics causes carbon isotopic fractionation, or separation by molecular weight, with the lighter carbon released to soil $CO_2$ gas generated by microbial respiration. The heavier $^{13}C$ remained in the water passing into the UFA.

Burrowing is another way in which biota affect the soil environment. For the purposes of this chapter suffice it to say that the invertebrates, such as earthworms and insects, and vertebrates, such as mice and voles, stir and mix the sediment, which brings new dead organic material into contact with the atmosphere and soil moisture and hastens microbial degradation of the organics.

### 5.4.3 Reduction/Oxidation Reactions: Microbial and Chemical Transformations

Soil microbes utilize the dead organic matter in soil and sediment as a source of carbon and nutrients. By doing so, they cause the organic matter to decay and release organic-rich wastes and dissolved chemicals, including nutrients that are utilized by other organisms. Microbial decay of organic-rich material is associated with many iron, sulfur, and nitrogen transformations. For example, chemical reaction Eq. 5.23 is an example of a microbially driven reaction in which organic carbon and $SO_4^{2-}$ are metabolized. This redox reaction, during which the sulfur in sulfate serves as an electron donor and is chemically reduced from $S^{6+}$ to $S^{2-}$, is commonly caused by soil microbes and is responsible for transformations of many compounds that contain iron, nitrogen, sulfur, and other polyvalent elements. In the example shown in reaction Eq. 5.23, microbes utilize the sulfate and organic carbon as energy sources and they release hydrogen sulfide ($H_2S$; the familiar gas with the "rotten egg" odor) and $HCO_3^-$ to the water.

$$SO_4^{-2} + 2C_{organic} + 2H_2O \xrightarrow[activity]{microbial} H_2S + 2HCO_3^- \quad (5.23)$$

Reaction 5.23 is important to Florida karst. As discussed in Chaps. 2 and 4, gypsum and anhydrite are important sedimentary components of the MCUs and the Lower Floridan Aquifer (LFA). These minerals are sources of dissolved $SO_4^{2-}$. If a carbon source is available, microbes in the aquifers metabolize the $SO_4^{2-}$ and release $H_2S$ according to reactions like Eq. 5.22 (Rightmire et al. 1974; Rye et al. 1981; Sacks 1995; Sacks and Tihansky 1996). Conversion of $SO_4^{2-}$ to $S^{2-}$ by aquifer microbes introduces both sulfate and sulfide to groundwater. In addition, the reaction indicates that bicarbonate is created. Bicarbonate, of course, is involved in the equilibrium reaction shown in Eq. 5.4. The common-ion effect (Sect. 5.2.1.3) is important here because consumption of $SO_4^{2-}$ and generation of $HCO_3^-$ affect other reactions involving these ions. Therefore, if bicarbonate in the groundwater is in equilibrium and microbial activity introduces additional $HCO_3^-$ to the water, then reaction Eq. 5.4 should go to the left and result in calcite precipitation if there is a sufficient source of $Ca^{2+}$.

Groundwater and soil microbes require a source of organic carbon, as well as the other nutrients. This organic carbon can be either dissolved in the water or occur as particulates in soil and the matrix of aquifer or cave sediment. The microbes decompose (degrade) the organic matter in successive steps, creating ever simpler compounds. These degraded organic compounds constitute humic substances if soluble in water and peat or muck particles if solid. Ultimately, the organic substances may be decomposed to the extent that simple gasses result. If the soil or water is chemically oxidizing, then $CO_2$ gas is a likely final degradation product, with carbon in the +4 oxidation state. If the environment is chemically reducing, methane ($CH_4$) gas generation is likely, with carbon in the −4 oxidation state.

Bacteria and other microbes often form mats on soil or aquifer surfaces (Canter 1997; Chapelle 2000). These mats are typically composed of a number of different species, which support each other in a complex community. In order for microbial mats to thrive, they typically require (1) a solid substrate for attachment and (2) a constant bath of water rich in organic carbon and other nutrients. In intergranular (sandy) aquifers, the networks of pore spaces provide abundant microbial substrate and mechanically filter particulate organic materials (particulate humic substances and dead microbes) from the water. In karstic flow systems, the caverns and karst conduits provide less substrate per unit volume of water, so there is likely less microbial activity within the large void spaces, as well as less mechanical filtration, and dissolved and particulate organic carbon may persist in the water. In karstic aquifers, therefore, dissolved and particulate organic carbon may travel some distance before suitable conditions for microbial decay or mechanical trapping can be realized. While they are produced in most soils, humic substances are best preserved in water-saturated, chemically reducing soils and aquifers where complete oxidation and aerobic microbial decay are inhibited.

Both $CO_2$ and $CH_4$ can pass out of the soil and aquifer water by degassing. $CO_2$ gas can participate in inorganic reactions as suggested in reaction 5.4. By this reaction, $CO_2$ released by microbes in the soil contributes to the production of $H_2CO_3$ in soil and aquifer waters. The partial pressure of $CO_2$ characteristically rises from about $10^{-3.4}$ in the open atmosphere to about $10^{-1.5}$ as a result of $CO_2$ production in soils (Garrels and Christ 1965; Drever 1997; Faimon et al. 2012). This results in a theoretical drop in water pH from around 5.5 to pH values of 4.0–4.5. Note in Fig. 5.5 that the mean pH of rainfall in Florida is approximately 4.8. This is a result of anthropogenic introduction of acids (acid rain;

Hendry and Brezonik 1984). In the geologic past, we assume that the pH of rainfall as a function of equilibration of water with $CO_2$ in the atmosphere was closer to 5.5. At any rate, rainfall is a mild acid and addition of soil $CO_2$ further lowers the pH and increases the acidity of the water and aggressiveness to carbonate minerals.

$CO_2$ degassing and generation in Florida caves were discussed by McGee (2010). She identified what appeared to be degassing of $CO_2$ into a cave in west-central Florida and elevated in a cave occupied by bats and guano. Gulley et al. (2013), Gulley et al. (2016), and Gulley and Polk (2017) showed that this $CO_2$ degassing is an important process for formation of caves near the water table in Florida's eogenetic karst. This process is discussed in Sect. 7.2.

### 5.4.4 Creation and Effect of Humic Substances

It is surprising that so little attention has been given to the role of humic substances in ground- and surface-water geochemistry, especially in Florida where so many water bodies are tea colored as a result of dissolved and colloidal organic substances. Humic substances constitute much of the dissolved organic fraction in natural waters. For example, Thurman et al. (1982) concluded that humic substances constitute 30–50% of the organic carbon in Florida surface water, and that they are a "principal component of aquatic organic matter." Humic substances form as a result of decomposition of plant debris (Thurman 1985; Qualls et al. 2003) and other organic materials in soils, streams and lakes, aquifers, and other environments. They may also be derived by decomposition of peat and muck sediments.

Because humic substances are chemically complex and contain a number of different chemical radicals (atoms, molecules, or ions that have unpaired valence electrons and are, therefore, chemically reactive) that can react with both dissolved and particulate material, they play a huge role in the chemical processes that occur in these environments.

Humic substances are complex molecules that can serve as chemical-complexing agents that can enhance or retard transport of metals (McKnight and Wershaw 1994; Murphy and Zachara 1995; Kar et al. 2011), mediate chemical transformations of nutrients, (Thorn and Mikita 2000; Kar et al. 2011) and inhibit or limit formation of carbonate sediments (Hoch et al. 2000; Reddy and Hoch 2000). They may flocculate (clump together) and form part of the sediment in certain environments, especially in estuaries and other locations where there is an increase in water salinity, or they may coat sediment grain surfaces (Wershaw 1999) and interfere with sediment-water reactions. They are sources of organic carbon for open water, soil, and aquifer microbes (Haitzer et al. 2003; Anesio et al. 2005). Humic substances are most abundant in surficial water and shallow groundwater, but they can occur in deep aquifers as well (Kovács et al. 2011). In general, humic substances rarely exceed 5 mg/L as dissolved organic carbon (Thurman and Malcolm 1981). However, levels of organic carbon (dissolved and total) in Florida water often exceed this concentration.

While dissolved and/or total organic carbon is reported in surface- and ground-water quality investigations in Florida, especially in studies sponsored by governmental agencies. Little has been done to evaluate the importance of humic substances in the development of karst and transport of metals as contaminants.

Most of the definitive work on the details of humic substances in Florida and Georgia relates to the Suwannee River (Fig. 5.9). Malcolm et al. (1994) described the origins of humic substances in the Suwannee River at its outfall from the Okefenokee Swamp in Georgia. They suggested the possibility that the humic substances being generated in the Okefenokee Swamp could be the result of decomposition of the well-developed peat deposits in the swamp. However, $^{14}C$ age dating of humic-substance extracts from the Suwannee River revealed ages of 0–25 years, while peat deposits in the Okefenokee have $^{14}C$ ages of 6000–7000 years. Clearly, the humic substances in the Suwannee are derived from decomposition of modern plant debris, not the peat deposits (Malcolm et al. 1994).

A large number of Florida's streams and lakes have dark water as a result of humic substances (Malcolm et al. 1994). These are called blackwater streams or lakes (Myers and Ewel 1990; Whitney et al. 2004). These dark-water bodies (Fig. 5.9) reflect high concentrations of humic substances. For example, wetland water in the Okefenokee/Suwannee River system usually has dissolved organic carbon concentrations of about 50 mg/L, of which about 75% consists of humic substances (Malcolm et al. 1994). Median concentrations of organic carbon in Florida's groundwater are 14 mg/L in the SAS, 4.8 mg/L in the Intermediate Aquifer System and Confining Unit (IAS/ICU), and 2.2 mg/L in the FAS (Upchurch 1992). So humic substances are abundant in Florida waters, and their concentrations are often in excess of concentrations reported elsewhere. In fact, humic substances have been commercially extracted from the Suwannee River in Georgia for use as laboratory standards (International Humic Substances Society 2014).

Humic substances have a wide range in molecular weights (Thurman et al. 1982; Thurman 1985). Thurman et al. (1982) found that their molecular weights ranged from approximately 500 to over 10,000 Daltons (unified atomic mass units). Since they are decomposition products, humic substances can include a number of different-molecular structures and functional groups, such as carboxyl, amino, and carbonyl groups. The proportions of the different humic substances and their compositions vary with environment and

**Fig. 5.9** The Suwannee River at Big Shoals, Hamilton County. Note the brown coloration of the water – a result of humic substances. Location: 30.3398° N, 82.6833° W

locality (Malcolm 1990). Humic-substance molecules have flexible structures, large sizes, and a diversity of functional groups, so they are effective chemical complexing agents, especially for metals. They are capable of conforming to clay surfaces and of flocculation as particulates in their own right. The outer radicals on dissolved humic substances are negatively charged (Thurman 1985), so humic substances are anionic. Because of the net negative charge, humic substances are able to coordinate with cations, especially dissolved metals. As such, humic substances are capable of complexing calcium and participating in the development of karst.

The wide range in molecular weight, diversity of molecular structures, and large number of functional groups make characterization of humic substances difficult. A common method of classification is by their response to the pH of surrounding water (Thurman 1985). The three basic fractions that comprise humic substances are as follows:

- Humic acids are humic substances that are soluble in basic and mildly acidic solutions and insoluble in acidic solutions with a pH less than 2 or in ethanol. Humic acids include large, complex molecules with molecular weights of 1,000 to more than 10,000 Daltons (Thurman et al. 1982; Thurman 1985). Aiken et al. (1994) determined that the median humic acid molecule from the Suwannee River in Georgia has a molecular weight of 1061 Daltons (median of two samples, pH = 9).
- Fulvic acids are less complex than humic acids (molecular weights of 500–2,000 Daltons; Thurman et al. 1982; Thurman 1985), and they are soluble under both acid and alkaline conditions. Aiken et al. (1994) determined that the median fulvic acid molecule from the Suwannee River in Georgia has a molecular weight of 711 Daltons (median of three samples, pH = 9). The smaller fulvic acid molecules do not color water, while humic acids may.
- Humins are insoluble in both acids and bases. The particulates and colloids that constitute the majority of organics in soils, especially organic hard pans, and sediments are humins.

Three processes affecting the mobility of organic acids (Thurman 1985) are

- Microbial decay, which converts metabolized organic molecules to $CO_2$ or $CH_4$ depending of the reduction/oxidation potential of the water and/or sediment;
- The pH of the host water, which may affect flocculation and chemical coordination with other dissolved constituents in water; and

**Fig. 5.10** Carbonized plant material in dolostone of the Eocene Avon Park Formation at the Gulf Hammock Mine, Levy County

- The salinity or total dissolved solids content of the host water, which causes flocculation of the organics in saline water.

Humic and fulvic acid molecules include numerous radicals (carbonyl, carboxyl, amino-, and similar sites) where chemical complexing can occur. These acids are capable of binding metals and inducing their transport. For this reason, water high in total organic carbon is usually high in iron and trace metals (Young and Comstock 1986). At contaminated sites, humic and fulvic acids may cause undesirable movement of metals.

Most organic material is derived from the land surface. Swamps and organic zones in soils are widespread in Florida and contribute significant organic carbon to groundwater. Thus, total organic carbon decreases in concentration with depth in most places in Florida aquifers. This is not to say that there are no other sources of organic carbon in Florida's aquifers. Clays in the Miocene Hawthorn Group contain organics that can be decomposed to produce soluble and/or particulate organic carbon. Limestone and dolostone in the Eocene Avon Park Formation (Fig. 5.10) contain widespread organic-rich zones that may also contribute organic carbon to FAS groundwater. Typically, Eocene Avon Park Formation water samples contain somewhat higher organic carbon concentrations than the overlying strata of the UFA.

### 5.4.5 Chemical Reactions with Soil and Sediment Minerals

The quality of ground and surface waters can be affected by chemical interactions involving humic substances that result in dissolution or precipitation of minerals. The reaction rates of these interactions are usually slow, occurring over geologic time, not human time. Given the right conditions, however, even silicate minerals can undergo dissolution. For example, organic acids, including humic substances, have been shown to be important weathering agents of silicate minerals (Huang and Keller 1971, 1972a, b, c; Drever 1985).

Chemical reactions of inorganic and organic acids with carbonates are the dominant processes in karst development. Equation 5.4 reflects the fundamental reaction for the interaction of $H_2CO_3$, an inorganic acid, with calcite. In this section, we address reactions involving organic acids and ligands.

#### 5.4.5.1 Organic Complexing

Consider the possible consequences of the equilibrium reaction shown in reaction Eq. 5.4. As an equilibrium reaction, this reaction is determined by the equilibrium constant (Eq. 5.3). If one adds $CO_2$ to the system described by the reaction in Eq. 5.4, additional $H_2CO_3$ has to be created to maintain the balance, and the added $H_2CO_3$ will dissociate to release $H^+$ and cause dissolution of the $CaCO_3$ in the equation. This new $CO_2$ is, of course, the likely consequence of respiration by microbes.

What if $H_2O$ or $CO_2$ were removed from the system? In order to maintain the equilibrium balance, $Ca^{2+}$ and $HCO_3^-$ would react (the equilibrium reaction Eq. 5.4 goes to the left) to replace the lost $CO_2$, and $CaCO_3$ would precipitate. Water is important because it affects concentrations and activities, and when evaporative removal causes the solution to become chemically oversaturated, calcite precipitates.

This equilibrium reaction is the basic inorganic process for karst formation. If we assume excess $CaCO_3$ as limestone, then the amounts of $CO_2$, $H_2O$, and other chemicals dissolved in the water control inorganic dissolution of the limestone. Conversely, if water is removed by evaporation or $CO_2$ is

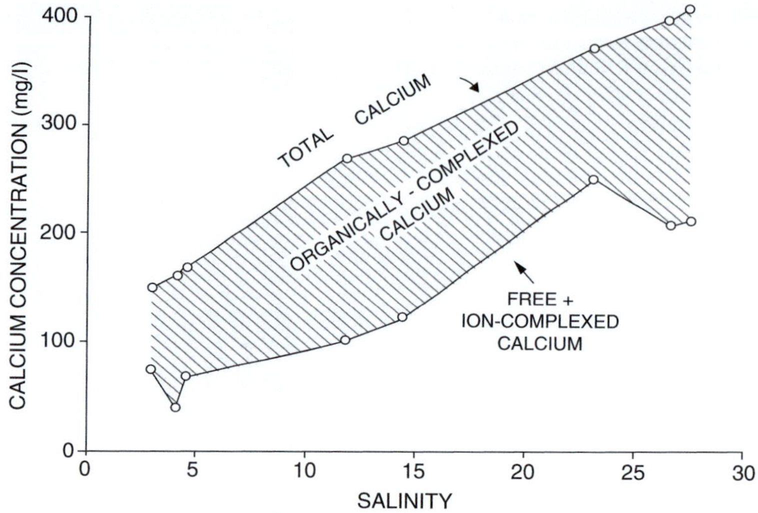


**Fig. 5.11** Partitioning of calcium in waters of the Blackwater River, Collier County, Florida. (From Upchurch et al. (1983). Salinity is reported in parts per thousand)

removed by degassing, $CaCO_3$ will precipitate to maintain equilibrium. The precipitated $CaCO_3$ constitutes cements and cave decorations that are so familiar to most of us.

Now consider the role of organic acids, such as humic and fulvic acids, in surface water and groundwater. While the chemical reactions are complex, and dissolved organic acids are very difficult to characterize thermodynamically, dissolution reactions are likely to occur. Figure 5.11 illustrates the partitioning of calcium dissolved in the Blackwater River, an organic-rich stream (hence its name) in coastal Collier County (Location: 2539724°N, 81.5931°W). The tidal stream drains portions of the Big Cypress Swamp and mangrove fringe along the Collier County coast, so it contains high concentrations of dissolved and particulate organic matter.

In order to determine the partitioning of calcium between the organics in the water and the water itself, Upchurch et al. (1983) compared the activities of calcium measured with an ion-sensitive electrode, which measures free and ion-complexed calcium, to total calcium in the water as measured by atomic absorption spectroscopy. The difference between these two measurements was assumed to reflect the calcium that was complexed with humic substances. On average, about half of the total calcium in the water was tied up as chemical complexes with ionic radicals on the dissolved humic substances. This complexing effectively removes that portion of the calcium from availability for participation in the equilibrium reaction in Eq. 5.4. Note that the total amount of calcium in the water increased with salinity of the water as it approached the Gulf of Mexico and that there was a dramatic increase in the proportion of calcium complexed by humic substances. This was a result of flocculation and concentration of organics near the open ocean. The important conclusion from the Upchurch et al. (1983) study is that humic substances are capable of creating karst features through chemical complexing of calcium in Florida's black- and brown-water environments.

As noted above, if the concentration of dissolved, free $Ca^{2+}$ decreases through the formation of chemical complexes, then Eq. 5.4 moves to the right, and more $CaCO_3$ can be dissolved. It is estimated that, in Florida's organic-rich waters, chemical complexing of calcium on humic substances essentially doubles the ability of the water to dissolve $CaCO_3$. Conversely, if the organics are destroyed by metabolic activity or other causes, then calcium would be freed to participate in the reactions once more, thus driving Eq. 5.4 to the left.

### 5.4.5.2 Inorganic Complexing

Inorganic chemical complexing is also a process of which we must be aware. At high concentrations, dissolved ionic metals coordinate with anionic constituents through ion pairing (Garrels and Christ 1965; Drever 1997). Ion pairing is a function of the ionic strength of the solution. Ionic strength is a measure of the charge concentrations in water as a result of the presence of ions.

Seawater and brines, for example, have high ionic strengths because of the abundant dissolved ions they contain. For example, in seawater with 19‰ chlorinity and at a pH of 8.5, calcium has a concentration of 400 mg/L, of which 91% is free ion ($Ca^{2+}$), 8% is paired with $SO_4^{2-}$ (dissolved $CaSO_4$), 1% is paired with $HCO_3^-$ (dissolved $Ca(HCO_3)_2$), and 0.2% is paired with $CO_3^{2-}$ (dissolved $CaCO_3$) (Garrels and Christ 1965). As long as the calcium is tied up as an ion pair with one of the anions, it is not available to participate in the reaction shown in Eq. 5.4. In dilute, freshwater systems, ion pairing is minimal and not an issue. However, in brines and high sulfate waters ion pairing contributes to the ability of the water to dissolve carbonate minerals. The ion-complexed calcium illustrated in Fig. 5.11 includes ion pairs that form as the river water mixes with seawater and the ionic strength of the solution increases.

#### 5.4.5.3 Sorption Reactions

Finally, sorption and ion exchange are important processes to consider in some Florida environments. Clay minerals have net negative charges on their outer surfaces. Positively charged cations, such as $Ca^{2+}$ and $Na^{+}$ are electrically attracted to the clay mineral surfaces where they may be loosely bound by weak electrical forces (van der Waals forces). In seawater which is sodium-rich, $Na^{+}$ may sorb[4] onto the clay mineral surface or onto interlayers within the clay structure, where it remains until the composition of the water changes. If the water in contact with the clay changes to $Ca^{2+}$-rich, the relatively high concentration of calcium in the water may replace the sodium sorbed on the clay through the process of ion exchange. When this ion exchange process occurs, there may be fundamental changes in the shrink and swell properties of the clay, and calcium may be removed from solution. Sorption of $Ca^{2+}$, a relatively small ion, by a clay that has been saturated by exchangeable $Na^{+}$, a relatively large ion, is likely to cause the clay to shrink in response to the smaller ions in the clay interlayer spaces and between clay particles. The sorbed $Ca^{2+}$ is removed from availability for equilibrium reactions and, in Eq. 5.4, the reaction would go to the left, dissolving additional $CaCO_3$.

The results of ion exchange have been noted to alter water quality in Florida's aquifers (Upchurch 1992). Na-$HCO_3$ and Ca-Cl waters detected in the UFA were thought by Upchurch (1992) to be the result of ion exchange. FAS water is normally rich in $Ca^{2+}$ and $HCO_3^-$ if $CaCO_3$ dissolution has occurred according to Eq. 5.4 and $Na^{+}$ and $Cl^-$ if seawater is a major component. The Na-$HCO_3$ and Ca-Cl water masses were created by exchange of $Ca^{2+}$ and $Na^{+}$ between aquifer water and clay-mineral surfaces.

### 5.4.6 Hydrochemical Facies

As groundwater moves along a flow path, it may encounter different rock types with different mineral assemblages and porosity/permeability configurations. The residence time of the water in contact with the rock varies with the nature of the porosity and permeability, flow velocity, tortuosity of the flow path, and hydraulic gradient. The water may also mix with seawater, connate water, or water that has a different chemistry from following a different flow path through the aquifer. All of these events affect the chemistry of the water. Recognition of changes in the water type requires that the water can be classified on the basis of its chemical composition.

Davis and DeWiest (1966) developed a groundwater classification system that allows differentiation of groundwater masses based on relative proportions of dissolved ions. The classification utilizes two standard trilinear diagrams - one for the dominant cations and the other for dominant anions in water. Each diagram is subdivided into fields, each of which represents a particular proportion of ions in the water (Fig. 5.12). For a particular sample, the proportion of each major cation ($Na^{+}$ + $K^{+}$, $Ca^{2+}$, and $Mg^{2+}$) is calculated in milliequivalents per liter with respect to the other major cations, and the result is plotted on the diagram. The sample is similarly plotted with respect to the major anions ($HCO_3^-$ + $CO_3^{2-}$, $SO_4^{2-}$, and $Cl^-$).

The arrangement of ions on the trilinear graphs is based on combinations of important, diagnostic ions in groundwater chemistry. For example, $Na^{+}$ and $Cl^-$ are placed in comparable positions on their respective trilinear diagrams because of their common association in marine aerosols and seawater, and $Ca^{2+}$ and $HCO_3^-$ are juxtaposed on the trilinear graphs because they are common weathering products of limestones and a number of other rock types.

The water type is designated by the dominant ions present. A water mass that contains predominantly calcium and magnesium in more-or-less equal proportions (plotting within area B on the trilinear cation diagram; Fig. 5.12) and bicarbonate as the dominant anion (plotting within area 1 on the anion trilinear diagram; Fig. 5.12) is said to be a Ca-Mg-$HCO_3$ water type. Based on the Davis and DeWiest (1966) classification, a Ca-Mg-$HCO_3$ water type would be assigned the symbol B1. After analyzing many samples, the predominant water type can be identified, and hydrochemical facies analysis can be applied to the system.

Back (1961) was the first to advocate use of hydrochemical facies analysis. The fundamental concept of hydrochemical facies analysis is that a water mass with more-or-less uniform chemical composition is thought to have a common history in terms of flow path, age, and geologic materials (water and rock) with which it had come in contact; therefore, the water mass develops a relatively homogenous chemical composition with the same types and proportions of ions throughout. The unique types and proportions of analytes in a water mass constitute its chemical "fingerprint", a characteristic chemical composition that conveys information about the hydrogeologic history of the water mass, and the spatial extent of water with a similar fingerprint represents a hydrogeochemical facies.

Several investigations have emphasized the distributions of hydrogeochemical facies in FAS groundwater. Hydrogeochemical facies analysis is especially useful where processes such as (1) mixing of groundwater types, (2) rock-water interactions, or (3) identification of connate or residual waters are important. Upchurch (1992) discussed the concept of hydrochemical analysis for Florida groundwater and presented maps showing the distribution of hydrochemical facies in Florida aquifers. He identified areas where ion

[4]The term sorb is used because both adsorption (uptake on outer surfaces) and absorption (uptake between the interlayers of the clay) are known to occur.

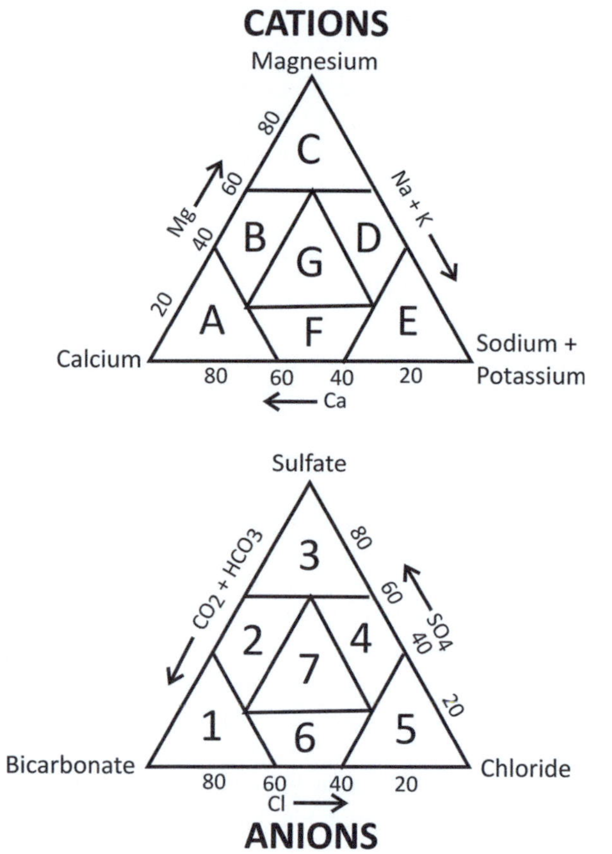


**Fig. 5.12** Trilinear diagrams representing groundwater types. (Modified from Davis and DeWiest (1966) and Upchurch (1992))

exchange had affected water quality and where seawater mixing and/or relict, connate waters were present in the aquifers. Mixing produces different hydrogeochemical facies, such as were noted by Frazee and McClaugherty (1979) and Upchurch (1992). Most of the time mixing is of little consequence. When, however, the mixing is between waters that differ significantly in chemical composition, important changes in composition and reactivity of the mixture may result (Runnels 1969, 1971).

Hydrochemical facies are interpretational, and use of the concept of facies analysis assists in assigning a common origin, history, or composition to a volume of water within an aquifer system. For example, if a large volume of water within the FAS has a common Ca-Mg-$HCO_3$ composition; one could interpret that water mass to reflect a hydrogeochemical facies controlled by dissolution of dolomite. In the following discussions, predominant groundwater types are attributed to hydrochemical facies where possible.

Some water types are highly unlikely and are not discussed in depth below. These include Mg-$HCO_3$, Mg-Cl, and Mg-$SO_4$ waters. Also, much of Florida's groundwater has complex histories (see, for example, Frazee and McClaugherty 1979) and consists of thoroughly mixed compositions. UFA water, for example, has been found to have a mixed composition (G7), where all major ions are present in subequal proportions. These samples are characteristically transitional between better defined water types, and they are not discussed below.

Upchurch (1992) used the Davis and DeWiest (1966) water-composition classification to delineate hydrochemical facies in Florida's aquifers. The facies he delineated and the fields on the Davis and DeWiest diagram (Fig. 5.12) where the water masses plot are listed below.

- Calcium-Bicarbonate Facies (Ca-HCO3; A1) – Ca-$HCO_3$ waters are among the most widespread in the carbonate-rich aquifers of Florida. They are derived from dissolution of calcite or aragonite in limestone, shell beds, and shelly siliciclastic sediments.
- Calcium-Magnesium-Bicarbonate Facies (Ca-Mg-HCO3, B1) – Ca-Mg-$HCO_3$, waters are either derived by (1) dissolution of dolomite in dolomitic limestones and dolostones, (2) by mixing of Mg-rich waters derived from weathering of magnesium-rich clay in the Miocene Hawthorn Group, and/or (3) by ion exchange. Determination of which of these processes influenced the chemistry of Ca-Mg-$HCO_3$ waters is largely based on which of the three processes is known to be active in the area from which the water was taken. If the water is from a dolomitic aquifer, the origin of the facies is attributed to dolomite dissolution. If it is near an area of active weathering of the Hawthorn Group sediments, which contain dolomite, or if the magnesium-rich clay beds of the intermediate confining zones are highly leaky, the facies is said to reflect weathering of the clays and/or dolostone.

  Examination of other cations and anions in the water often help differentiate the origin of these waters. For example, ion exchange and/or weathering of clay may also release dissolved potassium and silica (Lawrence and Upchurch 1976, 1982).
- Calcium-Sulfate (Ca-SO4; A3), Calcium-Magnesium-Sulfate (Mg-SO4; B3), and Calcium-Magnesium-Bicarbonate-Sulfate (Ca-Mg- HCO3-SO4; B2) Facies – Ca-$SO_4$, Ca-Mg-$SO_4$, and Ca-Mg-$HCO_3$-$SO_4$ waters are characteristically derived by interaction with the gypsum and anhydrite in the MCUs at the base of UFA and/or with LFA water. The mixed water facies result from mixing of calcium- or calcium-magnesium-bicarbonate waters with calcium-sulfate waters derived from dissolution of the gypsum or anhydrite.
- Sodium-Chloride Facies (Na-Cl; E5) – Na-Cl waters are found in two environments. Marine aerosol-dominated waters in the quartz-rich sediments of the SAS may have a dilute Na-Cl composition if little or no reaction with calcite, aragonite, or dolomite has occurred. Also, high salinity (high total dissolved solids, TDS) Na-Cl water masses are common in the saltwater/freshwater transition

zone. Occasionally, connate or relict Na-Cl water is found in the FAS. This water remains in the aquifers from previous high sea stands.

- Sodium-Bicarbonate (Na-HCO3; E1 or E6), Calcium-Chloride (Ca-Cl; A5 or A6), and Calcium-Magnesium-Chloride (Ca-Mg-Cl; B5 or B6) Facies – When cation exchange occurs, Na-$HCO_3$ and Ca-Cl waters may result. Na-$HCO_3$ waters are the more common of the two facies in Florida. Where Na-$HCO_3$ waters predominate, calcium has replaced sodium ion (Foster 1950; Upchurch 1992). This phenomenon develops when Na-saturated marine clays are bathed in Ca-$HCO_3$ waters. Ca-Cl or Ca-Mg-Cl waters are less common. They result from salt-water intrusion into aquifer systems that contain Ca- or Mg-saturated clays.
- Sodium-Sulfate Facies (Na-SO4; E4) – Na-$SO_4$ waters have been found in the SAS in central Florida (Hutchinson 1978). These waters are difficult to explain, but may result from addition of $SO_4$ through oxidation of organics or pyrite ($FeS_2$) to a Na-rich water. Upchurch et al. (1991) found similar water types near phosphogypsum waste disposal areas in Polk County.

Grouping the predominant water type data into aerially extensive hydrochemical facies allows for interpolation between data points and prediction of background water quality throughout the state. The following are just some of the benefits of hydrochemical facies analysis.

- Water History – Since the composition of the water reflects the sequence of rocks and sediments through which it has passed and any anthropogenic modifications that may have occurred, the water composition reflects, in a broad way, the history of the water. Understanding this history allows for deduction of flow paths, vulnerability of aquifer systems to contamination, and potentials for degradation by changing the flow paths, especially through upconing.
- Buffering Capacity – Buffering capacity is the ability of water to neutralize acids or bases. Within the FAS, buffering capacity involves interactions between groundwater and aquifer rock materials. Ca- and Ca-Mg-$HCO_3$ water masses have relatively high buffering capacities. In other words, application of acids and bases will result in some degree of neutralization as the water equilibrates with the added chemicals (Eq. 5.4). Na-Cl waters in siliciclastic aquifers have little buffering capacity and addition of acids or bases alters the facies of the aquifer. For example, consider the acid-rain problem. Lakes fed by buffered (Ca-$HCO_3$) groundwater have much higher tolerances for acidic precipitation than do lakes fed by Na-Cl water in siliciclastic soils (Hendry and Brezonik 1984).

### 5.4.7 Graphical Representations of Hydrochemical Properties and Facies

Hydrogeologists use two common graphical representations of water quality (Stiff and Piper diagrams) to depict sample compositions, hydrochemical facies, and chemical reaction pathways. These graphical methods are widely used to depict the chemical composition of individual samples and water masses.

#### 5.4.7.1 Stiff Diagrams

Stiff diagrams (Stiff 1951) are used to graphically represent the compositions of individual water samples. The method of representing water composition was developed to provide standard patterns that allow for visual pattern analysis. Figure 5.13 illustrates Stiff diagrams for (a) seawater (Na-Cl facies water) and (b) carbonate-aquifer water (Ca-$HCO_3$ facies water). Note that concentrations are represented in milliequivalents per liter (meq/L), and that Na and Cl and Ca and $HCO_3$ are plotted opposite each other to depict the common pairing of these analytes in natural waters. The results of these juxtapositions are characteristic patterns with Na-Cl (seawater) facies water presenting a "T-shaped" pattern (Fig. 5.13a) and Ca-$HCO_3$ facies water a diamond shape (Fig. 5.13b). The lengths of the arms of the Stiff diagram provide information as to the concentrations of each analyte in meq/L.

Figure 5.14 is an example of use of Stiff diagrams to demonstrate facies in a groundwater system. Figure 5.14a shows water quality in the Lake Tarpon surface-water drainage basin (Pinellas and Pasco counties) in the undifferentiated, sandy, SAS and Fig. 5.14b shows the hydrogeochemical facies in the UFA within the same basin. Note that both aquifers show a transition from Ca-$HCO_3$ facies inland to Na-Cl facies near the coast to the west. This transition reflects the saltwater transition zone along the Gulf of Mexico coastline.

Figure 5.14a, b illustrate the use of Stiff diagrams for hydrogeochemical analysis. The data used to develop these maps are based on water samples were taken from domestic and irrigation wells augmented by water samples taken using direct-push technology in the SAS near the coast. With the differences in water sampling methods in mind, we can derive the following conclusions from the map of the SAS (Fig. 5.14a).

- The water in the quartz sands of the SAS should have low dissolved solids contents with a highly dilute Na-Cl signature derived from rainfall. This pattern is rare within the basin.
- The dominant hydrogeochemical facies in the SAS is Ca-$HCO_3$ water. This type of water probably is a result of

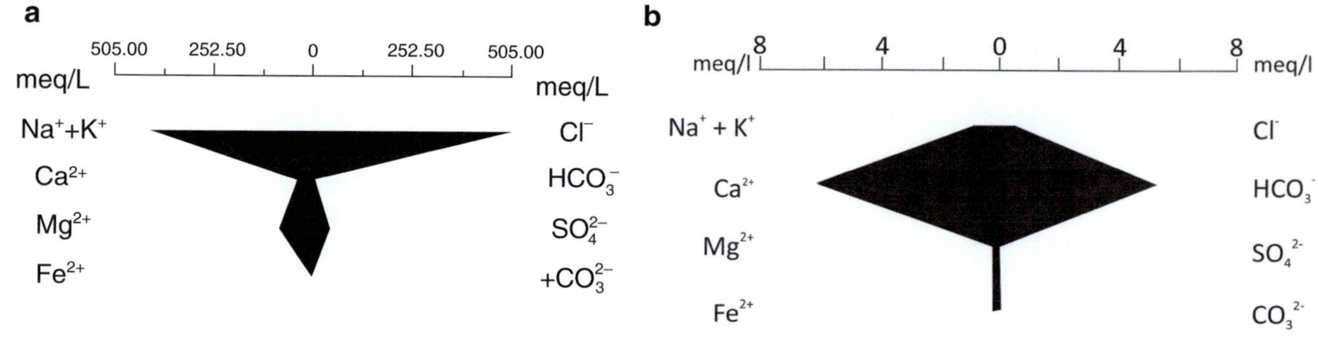


**Fig. 5.13** Sample Stiff diagrams of Floridan aquifer system groundwater. (**a**) Na-Cl water with a characteristic "T" shape. (**b**) Ca-$HCO_3$ water with a characteristic diamond shape. Sources: (**a**) – Southwest Florida Water Management District, ROMP 20 well 436 m below land surface, Osprey, FL; (**b**) – Frazee Jr. and McClaugherty (1979), Floridan aquifer well F-168

irrigation with Ca-$HCO_3$ water from the UFA. This pattern is common in the sandy, undifferentiated SAS in areas of Florida where irrigation is widespread.

- Near the coast and inland, the SAS in the Lake Tarpon drainage basin includes areas of upwelling of Na-Cl and Ca-Cl-$SO_4$ water from the FAS. These are indicated by the T-shaped pattern and weak C-shaped patterns, respectively.
- Variations in SAS water quality (Fig. 5.14a) reflect differences in well depth as well as local land use. For example, the sample with the Na-Cl signature in the SAS on the eastern side of the study area was from a domestic well and probably reflects nearby use of a salt (NaCl) based water softener.

The hydrogeochemical facies of the UFA (Fig. 5.14b) show a more predictable pattern.

- Inland, the UFA water belongs to the Ca-$HCO_3$ facies with diamond-shaped patterns being prevalent.
- The red outlier in the Ca-$HCO_3$ facies water on the eastern side of the basin may reflect a poorly cased or unusually deep well representing mixing.
- There is a well-developed salt-water transition zone indicated by Na-Ca-Mg-Cl-$HCO_3$ water. It is within this transition zone that speleogenesis as a result of groundwater mixing may occur.
- Finally, there is a coastal hydrochemical facies consisting of Na-Cl water. Well water from within this facies may be potable (total dissolved solids contents are relatively low as indicated by the moderate spread of the Stiff diagrams), but the dominant ions are derived from seawater. Thus, it appears that the coastal wells tap water from the upper part of the saltwater transition zone.
- Note that total dissolved solids contents of water from many of the wells in the transition zone facies are high as indicated by the large lateral spread of the arms of the Stiff diagrams. There are regional wellfields near these wells, and upconing of deeper, saltwater transition zone water appears to be occurring.

Herbert and Upchurch (2016) and Upchurch (2017) have suggested that hypogenetic speleogenesis may be occurring as a result of groundwater mixing in the saltwater/freshwater transition zone along the Gulf coast, including the area included in Fig. 5.14. The transition-zone mixing reflected by the hydrogeochemical facies in the UFA within the Lake Tarpon drainage basin provide insight as to the distribution of groundwater chemical facies and areas where modern-day speleogenesis may be occurring.

In Florida, water quality in the SAS often reflects land use rather than natural processes (Dalton and Upchurch 1978). As a result, deductions about pre-development water quality and speleogenesis may be problematic. Water quality in the FAS is less likely to have such confounding factors, and the Ca-$HCO_3$ facies, transition zone, and Na-Cl facies are usually easily identified. The confounding factors associated with sampling the UFA include variations in water quality related to the depth of the sample relative to the transition zone, upconing of deeper water as a result of pumpage, and coastal saltwater intrusion. Each of these confounding factors can affect interpretation of groundwater quality data in terms of karst processes.

#### 5.4.7.2 Piper Diagrams

Piper diagrams (Piper 1944) are intended to depict the relative proportions, as percentages, of ions in groups of water samples with different compositions. Figure 5.15 illustrates an example. Note that these triangular diagrams are the same as in Davis and DeWiest's (1966) facies classification diagram (Fig. 5.12). The trilinear diagram on the lower right represents the percentages, in meq/L, of anions with 100%

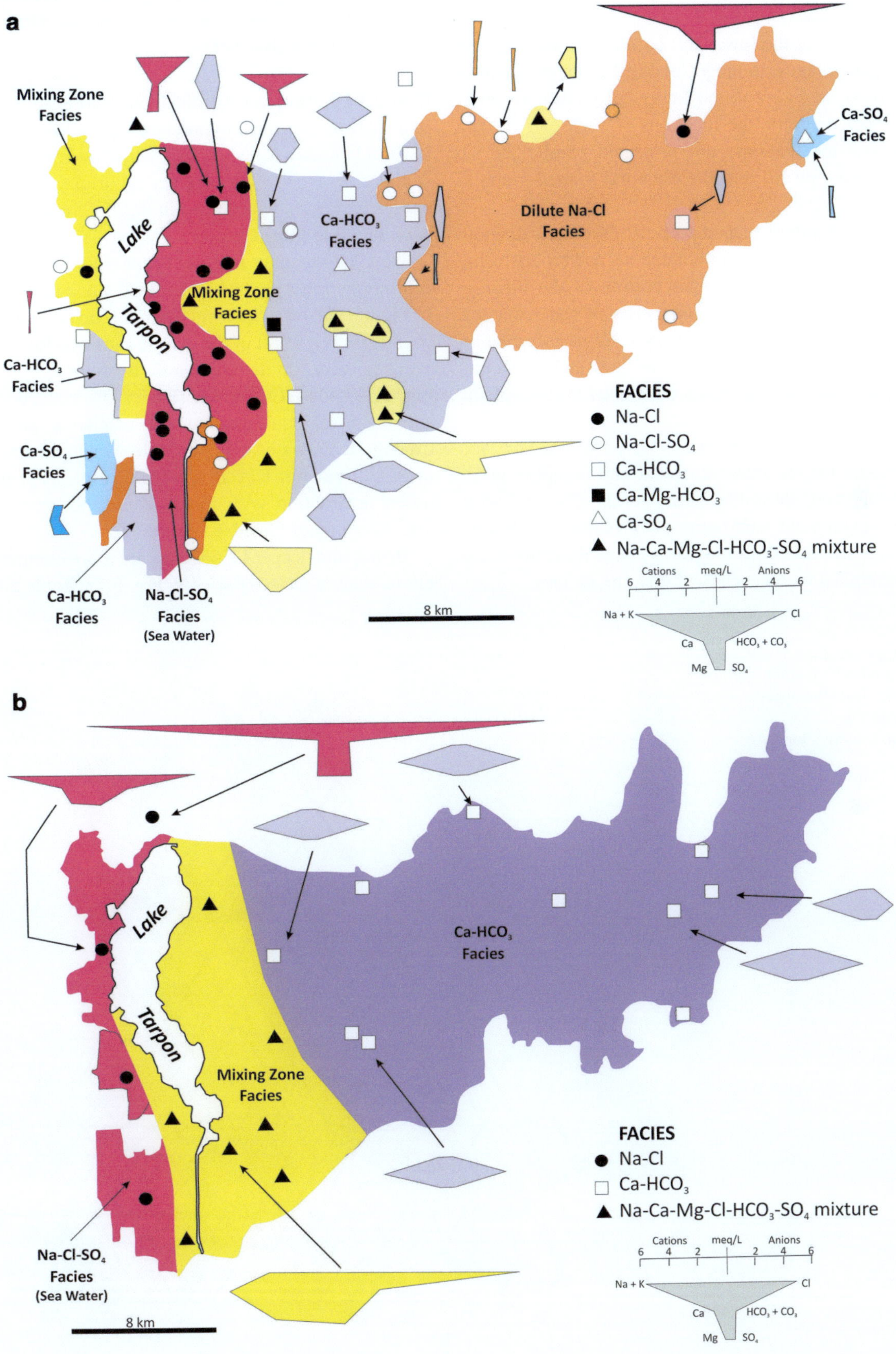


**Fig. 5.14** Hydrogeochemical facies present in the surficial aquifer system (**a**) and upper Floridan aquifer (**b**) in the Lake Tarpon drainage basin in Pinellas and Pasco counties. (From Upchurch (1998))

$SO_4^{2-}$ at the top apex, 100% $Cl^-$ on the lower right corner, and 100% $HCO_3^-$ at the lower left corner. The left triangle represents percentages, in meq/l, of $Mg^{2+}$ at the top, $Na^+$ at the lower right, and $Ca^{2+}$ at the lower left corner. To complete the Piper diagram, the locations of each pair of data points (anions and cations) are projected parallel to the appropriate axis upward into the diamond-shaped graph at the top.

The central, diamond-shaped portion of the Piper diagram is often all that is presented in publications and reports using Piper diagrams because the upper portion of the diagram depicts the anion plus cation constituency of water masses. Note that the graph depicts percentages. Some information content is lost by conversion to percentages and projecting the results into the central, diamond-shaped portion of the graph (Dalton and Upchurch 1978).

Frazee Jr. and McClaugherty (1979) compiled Stiff and Piper diagrams for the many different and complex water masses encountered in northeastern Florida. Figure 5.16 is one of their graphical summaries of groundwater quality showing both Stiff and Piper diagrams in one illustration and the hydrochemical facies each Stiff diagram defines.

### 5.4.8 Hydrochemical Facies of Florida Groundwater

Figure 5.17 illustrates hypothetical relationships of the common flow systems and hydrochemical facies of groundwater in Florida within the SAS, IAS/ICU, and FAS, and Table 5.6 summarizes each water type and its origin. All of the different water masses represented in Fig. 5.17 and Table 5.6 have been found in Florida.

Na-Cl water may represent seawater. The chemical concentrations must also be considered since rainwater (Facies I, Fig. 5.17) often presents similar proportions of major chemicals as a result of the marine aerosols. Rainwater, however, is very dilute compared to seawater (Facies VII, Fig. 5.17). Ca-$HCO_3$ water (Facies IIa) reflects dissolution and chemical equilibration with calcite or aragonite, and Ca-Mg-$HCO_3$ water with dolomite. Waters that are enriched in $SO_4^{2-}$, but not $Cl^-$, (Facies IIIa, Fig. 5.17) most likely reflect interaction with the gypsum and anhydrite of the MCUs or Lower Floridan Aquifer (LFA).

Brines and deeper LFA waters are more complex. The origin of shallow brine (Facies V, Fig. 5.17; Table 5.6) is prob-

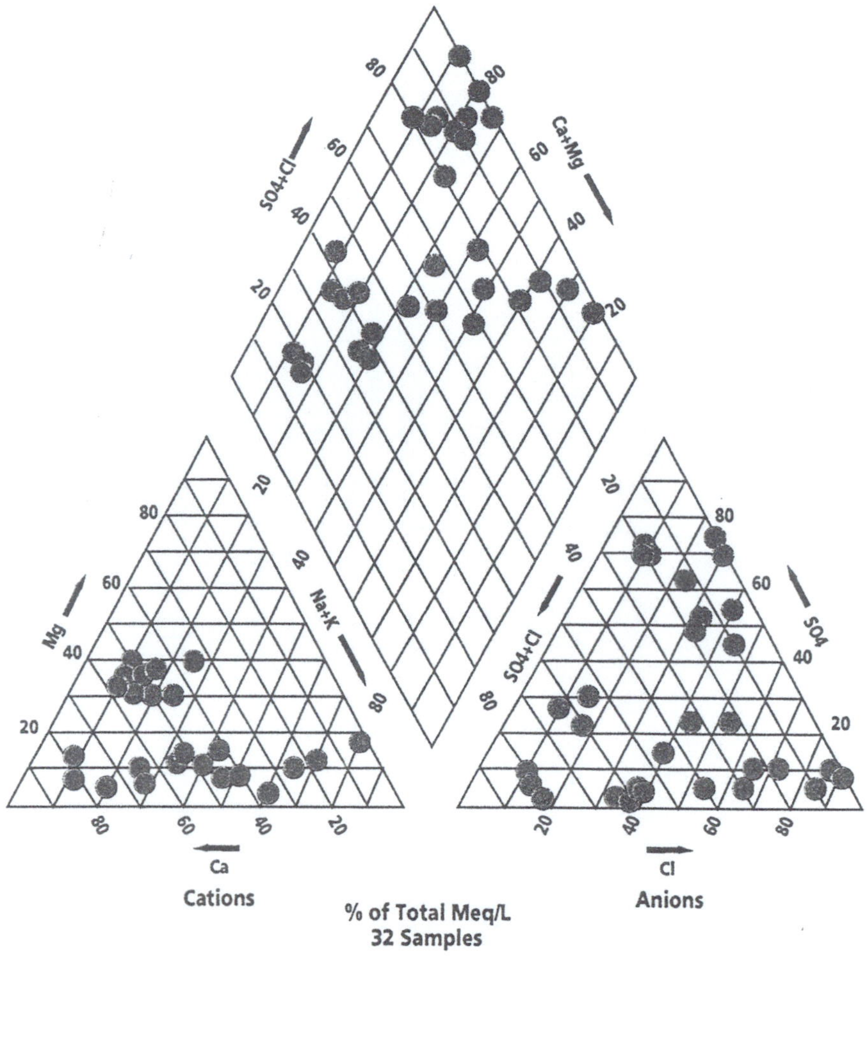


Fig. 5.15 Piper diagram showing groundwater quality for 32 samples from the upper Floridan aquifer in Hillsborough County, Florida

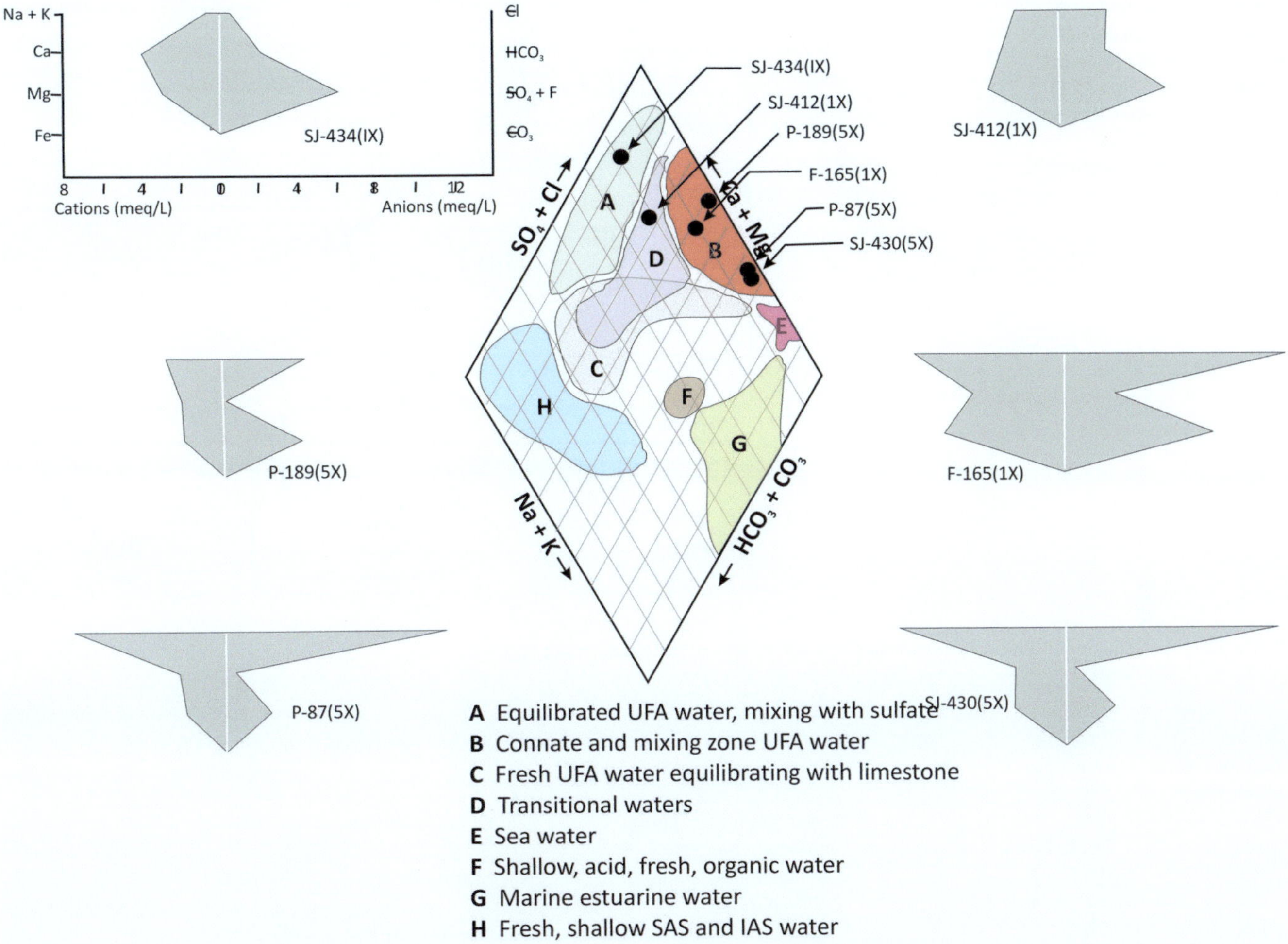


**Fig. 5.16** Piper and Stiff diagrams showing the relationships of the two types of diagrams to hydrochemical facies recognition. (Modified from Frazee and McClaugherty (1979, Fig. 31, p. 99))

lematic in the LFA, where clay content is minimal. Brines typically form as a result of interactions with clays that result in enrichment of the water in dissolved salts. They have higher concentrations of $Na^+$ and $Cl^-$ than seawater in Florida. For example, Peek (1958) reported on a water sample from the clay-rich Miocene Hawthorn Group at Anna Maria Island, Manatee County. The sample had chemical proportions of seawater but a $Cl^-$ concentration of 44,000 mg/L and total dissolved solids (TDS) concentration of 53,600 mg/L, which is considerably higher than normal seawater (Cl ≈ 19,000 mg/L, TDS ≈ 35,000 mg/L).

Connate and residual seawater[5] (Facies IV) may be indistinguishable from each other. The two facies were differentiated because of the likelihood of water with a seawater composition remaining in Florida's geologically young sediments either (1) from the time of deposition (Facies IVa) or (2) after saltwater intrusion during a later episode of high sea level (Facies IVb). Frazee and McClaugherty (1979) identified such water masses in northeastern Florida. The shallow, Holocene and latest Pleistocene strata of coastal Florida often contain pore water that is chemically identical to the water from which they were deposited. Finally, saline water (Facies VI) is defined as seawater that has mixed with type II and VII water in the saltwater transition zone.

The following sections discuss general groundwater water quality in regions of Florida defined by water management district areas (Fig. 5.17). While hydrogeochemical facies are discussed, it is important to understand that the facies designation only represents the dominant chemical components of the water mass. Groundwater facies are affected by numerous geochemical processes and the less abundant chemicals in the water tell a complex story of origins and interactions of the water with rainfall and aquifer materials.

For example, Lawrence and Upchurch (1982) investigated the origins of analytes in Ca-$HCO_3$ facies water of the UFA using principal components analysis (PCA). PCA

[5]Connate water is the water remaining from the time of sediment deposition. Residual seawater is water remaining in a sediment body after a high, post-depositional sea stand. If these water masses with somewhat different origins have similar chemical compositions, they cannot be differentiated.

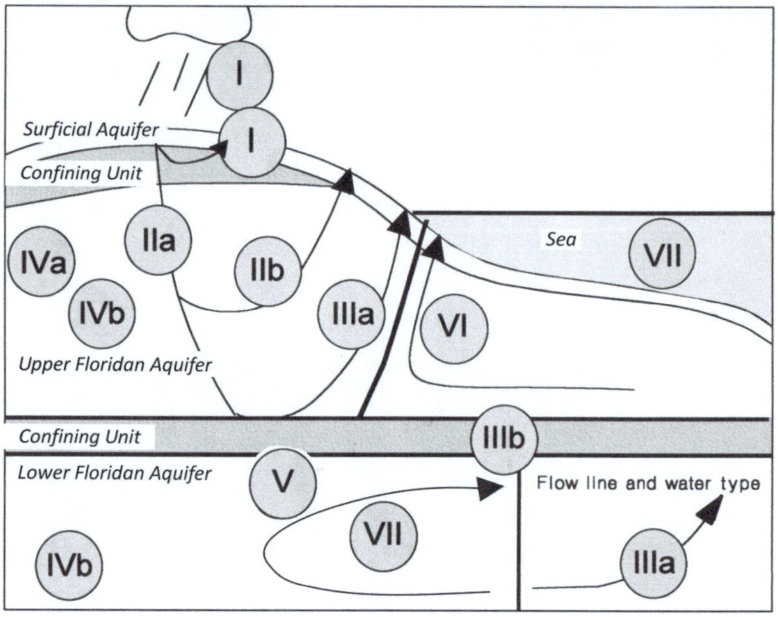


**Fig. 5.17** Hypothetical flow paths and hydrochemical facies in Florida's principal aquifer systems. Table 5.8 explains the compositions and origins of each facies. (Source: Upchurch lecture notes)

**Table 5.6** Classification of Florida groundwater by hydrochemical facies and water type

| Name | Facies and general water type (Facies, Fig. 5.17, and water type, Fig. 5.12) | Origin | Hydrochemical Facies | TDS Range (Approx. mg/L, based on FDEP monitoring data Upchurch (1992)) |
|---|---|---|---|---|
| Meteoric water (Rainwater) | I<br>E5 | Rainfall, soil moisture, and recent recharge | Na-Cl | 0–100 |
| Fresh groundwater | IIa<br>E6, E7, E8 | Interaction with aquifer materials | Na-Cl-$HCO_4$-($SO_4$) | 100–250 |
| | IIb<br>B2 | Interaction with aquifer materials | Ca-Mg-$SO_4$-$HCO_3$ | 250–1000 |
| Mineralized groundwater | IIIa<br>G2 | Interaction with evaporites at base of UFA | Ca-Mg-Na-$SO_4$-$HCO_3$ | 500–2000 |
| | IIIb<br>G4 | Interaction with evaporites within MCUs | Ca-Mg-Na-$SO_4$-Cl | 1000–30,000 |
| Connate groundwater | IVa<br>G7 | Original water of deposition | Na-Mg-Ca-Cl-$SO_4$-$HCO_3$ (seawater, poss. modified) | Rarely found as pure water type; often mixed with other waters |
| Residual groundwater | IVb<br>G7 | Water remaining from previous high sea stands | Na-Mg-Ca-Cl-$SO_4$-$HCO_3$ (seawater, poss. modified) | >2000–35,000 |
| Brine | V<br>E5-G7 | Water increased in salt content by mineral hydration, membrane filtration, etc. | Na-(Mg-Ca)-Cl-($SO_4$-$HCO_3$) in Florida | >35,000 |
| Saline groundwater | VI<br>G7 | Seawater mixed with type III or VII at saltwater/fresh-water transition zone | Na-Mg-Ca-Cl-$SO_4$-$HCO_3$ | >2000 |
| Seawater (marine ground water) | VII<br>E5-G7 | Holocene seawater in direct connection with the sea | Na-Mg-Ca-Cl-$SO_4$-$HCO_3$ (seawater) | Approx. 35,000 |

Based on data from Upchurch (1992a)

allows one to characterize the origins and spatial distributions of the dominant and minor chemicals in groundwater. Their investigation was based on UFA water quality from samples collected near Live Oak, Suwannee County. Live Oak is located on the Cody karst escarpment, and the 373 km$^2$ study area included regions of the UFA that are unconfined and confined. They found that a combination of $Ca^{2+}$, alkalinity, specific conductance, and pH represented the Ca-$HCO_3$ facies water but that other processes affected minor constituents of the water. $Mg^{2+}$, $F^-$, silica ($H_2SiO_4$), K+, $PO_4^{-3}$, and $SO_4^{2-}$ concentrations represented chemical weathering and ion exchange with minerals in the overlying Miocene

Hawthorn Group. These minerals include clays, pyrite, carbonate fluorapatite, dolomite, opal, and other sources of these chemicals. Other combinations of analytes were found to reflect chemical constituents derived from rainfall and from fertilization. The Lawrence and Upchurch (1982) paper emphasizes the sensitivity of UFA water quality to (1) dissolution of aquifer sediment, (2) proximity and exposure to minerals in the IAS/ICU, (3) recharge by precipitation in unconfined areas, and (4) land use. Clearly, the provenances of the different hydrogeochemical facies in Florida reflect a complex history of sources and interactions and that the facies designation only represents the dominant and perhaps most recent process that has affected the water.

## 5.5 Surficial Aquifer System Water Quality

### 5.5.1 Introduction

The text in this section is derived, in part, from Upchurch (1992). His 1992 paper was organized by water management district, and that organization has been retained in this section. See Fig. 5.18 for the areas of Florida that are managed by each district.

Water quality in the siliciclastic, undifferentiated SAS has a Na-Cl fingerprint derived from marine aerosols in rainfall until it has had an opportunity to interact with carbonate sediment or other water types. As a result, water quality in the SAS can be highly variable. In the Sand and Gravel Aquifer of the extreme western Panhandle of Florida, shell and other carbonate materials are limited in extent and the water retains its dilute Na-Cl fingerprint (Table 5.7) with a minor Ca-$HCO_3$ component where shell is present in the aquifer. In the Biscayne Aquifer, the chemical fingerprint is Ca-$HCO_3$ with a strong Na-Cl component, partly as a result of saltwater intrusion into the aquifer and partly because of evaporative concentration of water containing marine aerosols. Elsewhere, where undifferentiated sand with shell aquifer exists, the chemical fingerprint is mixed, with Na-Cl and Ca-$HCO_3$ components in various mixtures (Table 5.7). The following sections describe the chemical composition of groundwater in the SAS by location. Refer to Upchurch (1992) for maps showing the distributions of the different hydrochemical facies present in the aquifer system.

### 5.5.2 Marine Aerosol Components

Figure 5.19a and Table 5.7 show the medians and ranges of $Cl^-$ in the SAS (Upchurch 1992). At low concentrations, chloride is a chemically conservative ion, so it is not subject to reactions with aquifer materials or metabolic activity that might change its concentrations. Mixing by dilution and dispersion and concentration during evapotranspiration are the only significant processes that change the concentrations of $Cl^-$ in water. The $Cl^-$ data therefore provide insight as to evapotranspiration from the SAS.

The only major sources of $Cl^-$ in the SAS are rainfall, saline water along the coasts, and connate or residual water. When compared to the median concentration of $Cl^-$ in rainfall (Table 5.2; statewide median = 0.67 mg/L), the statewide median $Cl^-$ content of SAS water is over 40 times as high. Even the 25th percentile concentration (Table 5.7) is four times the median concentration in rainwater. While saline-water intrusion and relict water are probable causes of elevated $Cl^-$ in coastal areas, it is clear that there is an increase in $Cl^-$ and other analyte concentrations in the SAS as a result of evapotranspiration.

Figure 5.19 illustrates the median and range of pH (Fig. 5.19b), $Ca^{2+}$ (Fig. 5.19c) and $HCO_3^-$ (Fig. 5.19d) in the SAS by water management district[6] and aquifer. These box-and-whisker diagrams are included because they are excellent indicators of the nature of chemical reactions involving carbonate sediment and groundwater in the aquifer system.

Care must be taken in interpreting these data, however, because large areas in Florida where irrigation with FAS water is widespread have resulted in Ca-$HCO_3$ water types being found in siliciclastic sediments of the SAS (Dalton and Upchurch 1978). Also, well depth can skew data. Shallow wells in the SAS may have dilute Na-Cl water while nearby deeper wells may have Ca-$HCO_3$ water. It is likely that many of the outliers and anomalous water masses identified in the SAS by Upchurch (1992) reflect different sampling depths rather than spatially distributed facies.

There is a general change in predominant water type in the SAS from north to south. In the north, the SAS is largely siliciclastic, and shell content is limited to coastal areas, so Na-Cl water dominates. The origin of this water is mixed, with Na-Cl water near the coast being a result of the coastal transition zone and inland as a result of precipitation of marine aerosols. To the south, the carbonate content of the aquifer system increases and the water types become dominated by Ca-$HCO_3$ water. This trend is important because it suggests that water in contact with shell beds in central and southern Florida has equilibrated with the calcite and/or aragonite of the shell. As a result, rapid dissolution of shell is inhibited below the upper meter or less of sediment and shell preservation is typically excellent in the deeper shell beds.

[6]NWFWMD = Northwest Florida Water Management District, SRWMD = Suwannee River Water Management District, SJRWMD = St. Johns River Water Management District, SWFWMD = Southwest Florida Water Management District, and SFWMD = South Florida Water Management District.

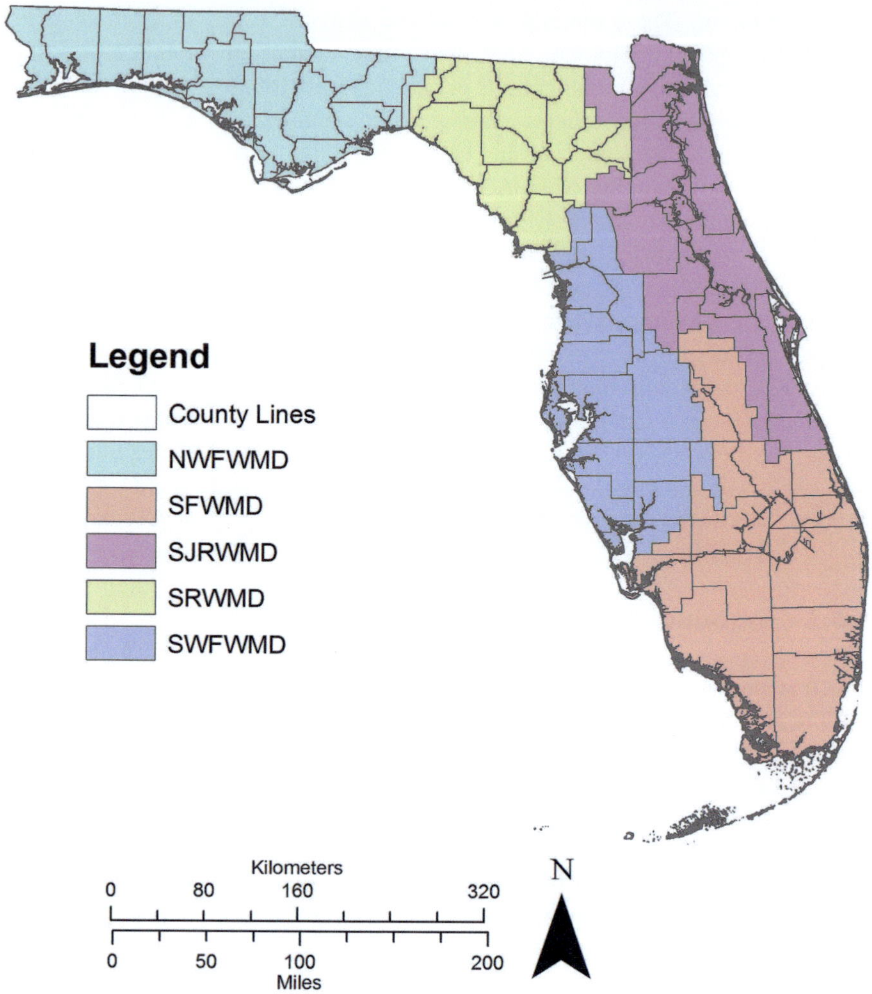


**Fig. 5.18** Locations of Florida's water management districts

**Table 5.7** Median and quartile concentrations (in parentheses) of major analytes in Florida's aquifer systems

| Analyte | SAS | | | | IAS statewide | UFA statewide |
|---|---|---|---|---|---|---|
| | Sand and Gravel Aquifer | Biscayne Aquifer | Undifferentiated Surficial Aquifer | Statewide | | |
| Sodium, mg/L | 5.0 (4.2–8.6) | 18.0 (11.1–31.0) | 19.4 (7.4–48.3) | 17.0 (7.0–39.0) | 41.0 (9.6–136.2) | 11.0 (4.5–84.6) |
| Calcium, mg/L | 4.6 (1.6–8.9) | 97.4 (79.4–125.0) | 85.6 (25.6–119.4) | 85.6 (27.7–118) | 58.0 (36.0–100.0) | 51.9 (28.5–84.1) |
| Magnesium, mg/L | 0.9 (0.6–2.0) | 4.9 (2.7–6.5) | 4.0 (1.4–8.8) | 4.1 (1.6–6.4) | 17.7 (9.5–40.9) | 14.6 (6.4–34.1) |
| Chloride, mg/L | 7.1 (5.0–11.0) | 58.0 (34.0–79.0) | 30.5 (14.0–74.7) | 40.5 (16.0–74.3) | 61.9 (18.0–334.5) | 21.0 (7.6–276.0) |
| Bicarbonate, mg/L | 5 (<1–13) | 263 (229–314) | 27 (8–101) | 138 (10–260) | 143 (95–200) | 146 (110–206) |
| Sulfate, mg/L | 4.6 (1.4–5.5) | 14.0 (<2.0–26.0) | 10.0 (<5.0–24.5) | 17.0 (7.0–39.0) | 5.4 (<1.0–65.5) | 5.4 (<1.0–54.0) |
| Total dissolved solids, mg/L | 74 (45–110) | 392 (316–468) | 339 (160–512) | 346 (181–474) | 390 (219–871) | 277 (176–715) |
| Total organic carbon, mg/L | 6.6 (4.1–8.4) | 14.3 (8.4–22.7) | 16.9 (7.8–36.0) | 14.0 (7.0–27.0) | 4.8 (<1.0–14.1) | 2.2 (<1.0–7.9) |
| pH | 4.9 (4.8–5.6) | 6.9 (6.6–7.2) | 6.7 (6.0–7.1) | 6.8 (6.4–7.1) | 7.3 (6.9–7.6) | 7.4 (7.0–7.8) |

Data from Upchurch (1992a). Where possible, sampling excluded areas of known saltwater intrusion

**Table 5.8** Hydrochemical facies of the Floridan Aquifer System as represented in Fig. 5.17

| Name | Water type (see Fig 5.17) | Origin | Hydrochemical facies (lesser or highly variable constituents are shown in parentheses) | Range of total dissolved solids content (mg/L) |
|---|---|---|---|---|
| Meteoric water | I | Soil water and recent recharge | Na-Cl | 0–100 |
| Fresh groundwater | IIa | Interaction with aquifer rocks during down flow | $Na\text{-}Cl\text{-}SO_4$ to $Ca\text{-}(Mg)\text{-}HCO_3$ | 100–250 |
| | IIb | Interaction with aquifer rocks during up flow | $Ca\text{-}Mg\text{-}HCO_3\text{-}(SO_4)$ | 250–1000 |
| Mineralized groundwater | IIIa | Interaction with evaporites of the Middle Confining Unit | $Ca\text{-}Mg\text{-}Na\text{-}SO_4\text{-}HCO_3$ | 500–2000 |
| | IIIb | Interaction with evaporites within the Middle Confining Unit | $Ca\text{-}Mg\text{-}Na\text{-}SO_4\text{-}Cl$ | 1000–30,000 |
| Connate groundwater[a] | IVa | Original water of deposition | $Na\text{-}Mg\text{-}Ca\text{-}Cl\text{-}SO_4\text{-}HCO_3$(sea water, poss. modified) | >2000–35,000 |
| Residual groundwater[a] | IVb | Water remaining from previous high sea stands | $Na\text{-}Mg\text{-}Ca\text{-}Cl\text{-}SO_4\text{-}HCO_3$(sea water, poss. modified) | >2000–35,000 |
| Brine | V | Water increased in salt content by mineral hydration, membrane filtration, etc. | $Na\text{-}(Mg\text{-}Ca)\text{-}Cl\text{-}(SO_4\text{-}HCO_3)$ in Florida | >35,000 |
| Saline groundwater[b] | VI | Sea water mixed with water type III or VII at transition zone | $Na\text{-}Mg\text{-}Ca\text{-}Cl\text{-}SO_4\text{-}HCO_3$ | 250–>2,000 |
| Sea water (marine groundwater)[c] | VII | Holocene sea water in direct connection with the sea | $Na\text{-}Mg\text{-}Ca\text{-}Cl\text{-}SO_4\text{-}HCO_3$(sea water) | Approx. 35,000 |

Based on data from Upchurch (1992)

[a]These water types are rarely found as pure hydrochemical facies; they are often mixed with other waters and, therefore, difficult to identify

[b]A geochemical gradient exists between fresh and salt water across this transition zone

[c]According to the Kohout circulation model, this facies mixes with water types IVb and V within the Lower Floridan Aquifer

### 5.5.3 Surficial Aquifer System Water Quality in Northern Florida

Water from the Sand and Gravel Aquifer in the western panhandle of Florida is predominantly Na-Cl in composition (see Fig. 5.19, specifically the columns labeled Northwest Florida Water Management District [NWFWMD] and Sand and Gravel Aquifer). Locally, there is $Ca\text{-}HCO_3$ and $Na\text{-}HCO_3$ water mixed with it, indicating a somewhat complicated history, different well depths, and minor carbonate sources. The Na-Cl water results from the coastal transition zone and marine aerosols.

There is insufficient data to speculate on predominant water type in central and coastal portions of the panhandle. However, Na-Cl and $Ca\text{-}SO_4$ waters do occur in places along the coast (Upchurch 1992), and there is a region centered on Jackson and Gadsden counties where $Ca\text{-}HCO_3$ water predominates. This region is characterized by exposures of limestones of the FAS at the land surface, and SAS waters undoubtedly reflect this influence.

The SAS is poorly developed in north central Florida (the area managed by Suwannee River Water Management [SRWMD]; Fig. 5.19). Much of the area managed by the SRWMD contains bare or thinly mantled karst, and the UFA is unconfined or poorly confined. Where the SAS exists, SAS waters are largely $Ca\text{-}HCO_3$ facies due to influences of FAS water. $Mg\text{-}HCO_3$ water reflects the magnesium-rich clays and dolostone deposits of the Miocene Hawthorn Group with which the SAS water interacts, and $Na\text{-}HCO_3$ waters reflect ion exchange related to Hawthorn Group clays (Upchurch 1992). In general, SAS waters in the district are $Ca\text{-}HCO_3$ facies derived from the influences of underlying IAS/ICU and FAS. Much of this influence may be a result of irrigation.

The undifferentiated surficial aquifer system in northeastern and east-central Florida (the area managed by St. Johns River Water Management District [SJRWMD], Fig. 5.18) includes some complex hydrochemical facies (Frazee and McClaugherty 1979), especially near the coast and the St. Johns River. In these regions of mixed water types, facies represent $Ca\text{-}HCO_3$, Na-Cl and other mixtures as a result of the presence of connate water, residual water, seawater intrusion, and an active, gravity-driven groundwater flow system. Figure 5.19 summarizes district-wide water quality within the SAS. There is a large area characterized by $Ca\text{-}HCO_3$ facies along the St. Johns River. This may reflect the upconing of FAS water described by Leve (1983) and Spechler (1996), irrigation or interaction with springs that discharge FAS water, or shelly horizons in the aquifer system. Elsewhere Na-Cl facies predominate along the river. There is a coastal Na-Cl facies, with large reentrants that Upchurch (1992) suggested might reflect connate seawater rather that active intrusion.

### 5.5.4 Surficial Aquifer System Water Quality in Central Florida

The undifferentiated SAS is poorly developed in much of the northern third of the Southwest Florida Water Management District (SWFWMD; Arthur et al. 2008). That area is underlain by poorly to unconfined UFA, and any water-saturated

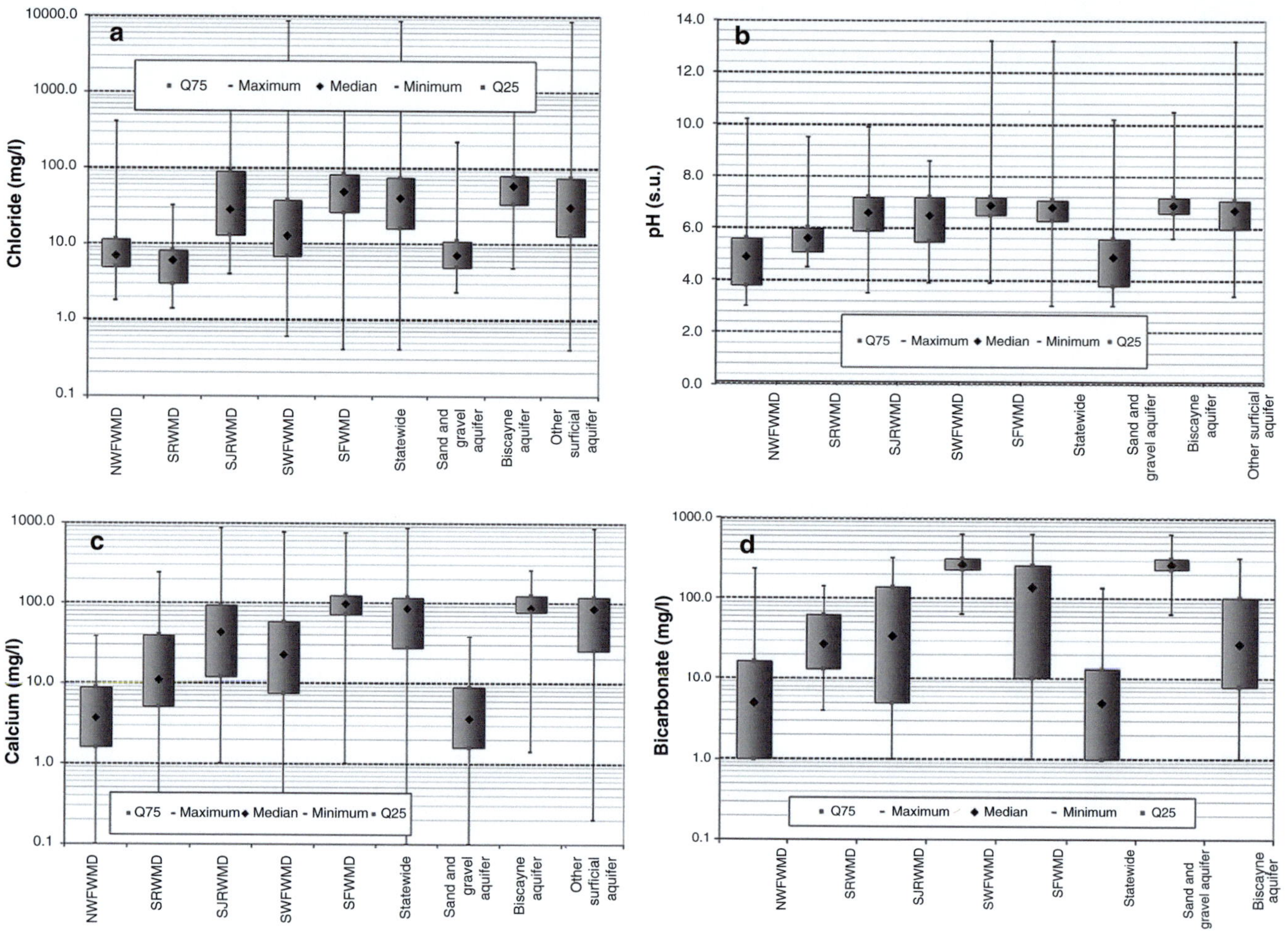


**Fig. 5.19** (**a**) Median and range of chloride in the Surficial Aquifer System, by water management district (Fig. 5.18) and aquifer. Data from Upchurch (1992) (**b**) Median and range of pH in the Surficial Aquifer System, by water management district (Fig. 5.18) and aquifer. Data from Upchurch (1992) (**c**) Median and range of calcium in the Surficial Aquifer System, by water management district (Fig. 5.18) and aquifer. Data from Upchurch (1992) (**d**) Median and range of bicarbonate in the Surficial Aquifer System, by water management district (Fig. 5.18) and aquifer. (Data from Upchurch (1992))

sand in the mantle over the limestone is considered to reflect the unconfined FAS (Southeastern Geological Society 1986).

The SAS varies in thickness, shell content, and isolation from UFA influences in the central third of the SWFWMD. Here, the SAS is largely quartz sand. It appears that any shell that was originally present in the older sands inland has long-since been removed by dissolution. This widespread dissolution appears to have occurred during the period of intense weathering in the Pliocene or early Pleistocene (cf. Hunt and Hunt 1957; Carr and Alverson 1959). Near the coast, shell is often well-preserved in Pliocene and Pleistocene sandy and argillaceous (clayey) sediments in the SAS. The water types present are mixed, with a slight predominance of Ca-$HCO_3$ over Ca-$SO_4$ and Na-Cl. Much of the Ca-$HCO_3$ water is UFA water introduced by irrigation. The Na-Cl water is a result of evaporative concentration of marine aerosols and saline-water intrusion near the coasts. The Ca-$SO_4$ water is problematical, in that an obvious source does not present itself (Upchurch 1992). Heavy pumpage is known to have induced up-coning of sulfate-rich water in the UFA in the well-confined southern part of the SWFWMD (2002, 2006), and some of that water may be present in the SAS as a result of irrigation. Sulfate from oxidation of organic sediments and/or pyrite may be another cause.

### 5.5.5 Surficial Aquifer System Water Quality in Southern Florida

The SAS becomes increasingly well developed to the south throughout the state. Near the coast and at lower elevations in the southern part of the state, shell is well-preserved in

Pliocene and Pleistocene sandy and argillaceous sediments and, of course, the Biscayne Aquifer and portions of the Tamiami Formation that are included in the SAS are predominantly limestone.

Waters in the southern third of the SWFWMD and South Florida Water Management District (SFWMD; Fig. 5.18) are predominantly Ca-Mg-$HCO_3$ facies inland and Na-Cl facies near the coast. The SAS in the southern third of the state is shelly, which results in Ca-$HCO_3$ waters. Upconing and irrigation withdrawals have resulted in the introduction of deeper, $SO_4$- and Cl-rich water.

Figure 5.19 reflects the SAS water types and facies in the SFWMD. The Biscayne Aquifer is predominantly composed of calcite with minor aragonite (Bock et al. 1969). The resulting facies is Ca-$HCO_3$ water. Salt-water intrusion near Miami (northern Dade County) has resulted in local Na-Cl enrichment (Upchurch 1992). On the west coast, the waters are derived from shelly sands and limestone and belong to the Ca-$HCO_3$ facies. Local Na-Cl water in Collier County may reflect wells intersecting the saltwater/freshwater transition zone or upconing of residual or connate water. The Kissimmee River corridor, located just north of Lake Okeechobee, is predominantly characterized by Ca-$HCO_3$ water. The water from wells in the Kissimmee corridor that exhibit a Na-Cl facies contains low ionic strengths and apparently reflects recharge by precipitation.

### 5.5.6 Organic Content of Surficial Aquifer System Water

Finally, dissolved and total organic carbon must be considered in the SAS because of karst features that may have been formed by organic acids. As noted above, near-surface soils and sediments host the microbial "factories" wherein plant and animal matter is converted to humic substances and $CO_2$ or $CH_4$. In addition, Florida has widespread organic (peat, muck) sediments (Davis 1946; Bond et al. 1986) and soils (histosols) which contribute humic substances to the SAS.

Figure 5.20 illustrates the total organic content of water from the SAS by water management district and aquifer. Note the interesting trend of increasing organic content from north to south. This trend reflects the transition from well-drained soils in the north to poorly drained soils in the south, as well as climatic influences (including an increase in plant productivity and crop turnover from north to south).

In terms of karst development, one can surmise that interaction of carbonate sediments with humic substances may be more important in the wet, organic-rich environments in the south than in the north.

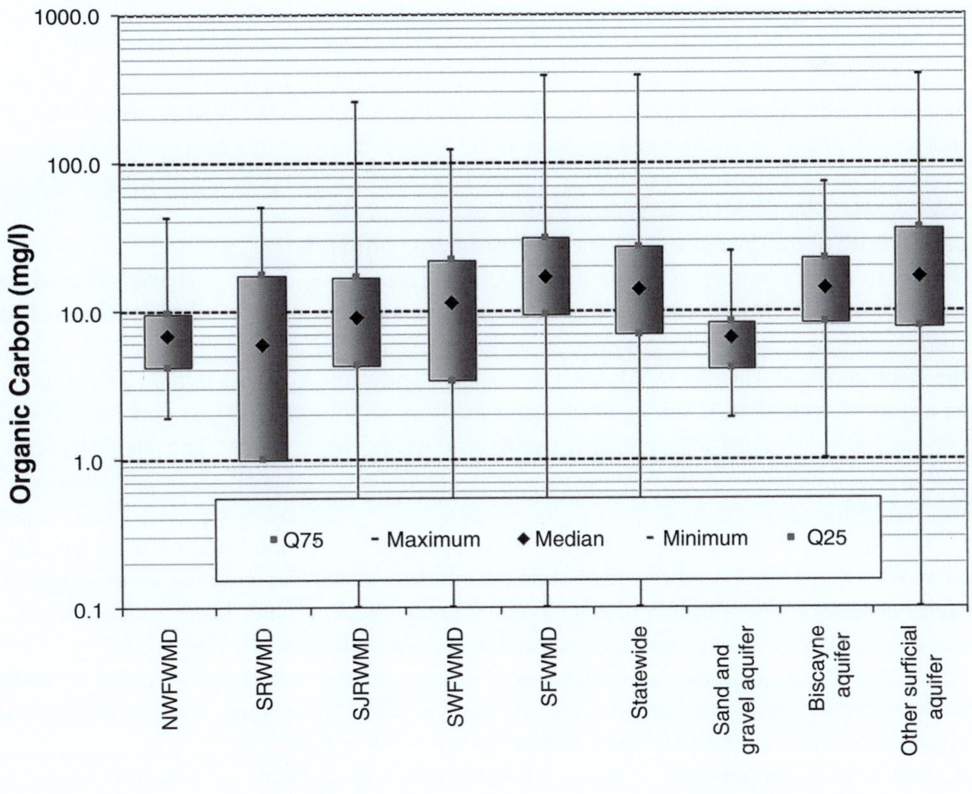


**Fig. 5.20** Distribution of total organic carbon in the Surficial Aquifer System of Florida. (Data from Upchurch (1992))

## 5.6 Intermediate Aquifer System Water Quality

### 5.6.1 Introduction

Keeping in mind that the IAS/ICU contains subregional permeable zones, the IAS data on water quality are biased by aquifer extent and location, water demand and use. While the IAS contains subregional permeable zones in many areas of the state, these potential aquifers have often been ignored in favor of the more productive UFA. The IAS is used as a source of potable water in southwestern Florida. Therefore, much of the data from the SWFWMD, for example, is from southwestern Florida, not from the eastern and central part of the water management district.

### 5.6.2 Regional Water Quality

With the exception of southwestern Florida (portions of the SWFWMD and SFWMD), the IAS is limited in extent. As a result, the number of samples of groundwater examined by Upchurch (1992) was limited.

Waters of the IAS in northwestern Florida (NWFWMD; Fig. 5.21) are predominantly Ca-$HCO_3$ type. This reflects the carbonates in the IAS. The absence of magnesium as a major cation in this water suggests that calcite dissolution, and not dolomite dissolution, controls water quality. Where it is present in the SRWMD, the predominant water type is either Ca-$HCO_3$ or Ca-Mg-$HCO_3$ facies (Fig. 5.21), which reflects the dolostone and limestone of the localized strata within the aquifer system. Where data exist in the SJRWMD (Fig. 5.21), there is an inland Ca-$HCO_3$ facies and a coastal Na-Cl facies.

The IAS is extensive and highly utilized in southern SWFWMD (Fig. 1.8). It is spotty to non-existent in the central third of the district and absent in the northern portion. Waters in the southern part of the district are characteristically Ca-Mg-$HCO_3$ facies due to the dolostone and magnesium-saturated clays of the Miocene Hawthorn Group. To the south and west, Ca-$HCO_4$-$SO_4$ and Ca-$SO_4$ facies develop as a result of addition of sulfate during upward flow on the coastal transition zone and, perhaps, pyrite oxidation. The outer coastal transition zone is characterized by Na-Cl facies derived from seawater.

The IAS is well developed on the west coast of the SFWMD (Fig. 1.8). Elsewhere there is no information as to facies or water types. The IAS in Lee County includes sandstone cemented by calcite and limestone and dolostone aquifers. There is considerable clay in the aquifer system. There is less $SO_4^{2-}$ than in the southern part of the SWFWMD (Upchurch 1992), so the $SO_4^{2-}$-rich belt at the lower part of the saltwater transition zone is absent. There is, however, a Na-$HCO_3$ facies belt derived by ion exchange. The clays of the aquifer system are Na-rich and, as Ca-rich water upwells along the inner transition zone, ion exchange releases the sodium into the water and fixes the calcium to the clays.

### 5.6.3 Organics in Intermediate Aquifer System Water

The variation of humic substances is relatively limited in the IAS (Fig. 5.22). With the exception of data from the SRWMD, the organic carbon concentrations reported in Fig. 5.22 are from unfiltered samples and represent total organic carbon. The data from the SRWMD represents filtered samples, which resulted in reporting of lower concentrations. However, when compared to the range in concentration of humic substances in groundwater reported by Thurman (1985), organic content is high.

## 5.7 Upper Floridan Aquifer Water Quality

### 5.7.1 Introduction

Water quality in the UFA is generally well understood. As the principal source of water for Florida's population, much attention is given to natural water quality and to contamination. In general, natural water quality reflects interactions with the limestone and dolostone of the aquifer. At the base of the aquifer, interaction with sulfate-rich water from interaction with the MCUs may create a sulfate-rich facies. This facies is especially well developed in the southern part of the SWFWMD, where the UFA is well confined and water extraction has caused sea-water intrusion along the coast and upconing of sulfate-rich water in inland areas.

### 5.7.2 Upper Floridan Aquifer Water Quality in Northern Florida

Waters in the UFA in the panhandle of Florida (NWFWMD, Fig. 5.23) can be subdivided into three facies (Upchurch 1992). The majority of the district is characterized by Ca-$HCO_3$ waters that reflect dissolution of aquifer calcite. Scattered Mg-$HCO_3$ water masses reflect dissolution of dolomite. Near the coastal transition zone, the predominant water types are mixed. Na-Cl and Na-$HCO_3$ waters predominate. The Na-Cl waters result from seawater at the transition zone. The Na-$HCO_3$ waters are apparently a result of ion exchange (Upchurch 1992). A mixed facies results where several water types exist because of ion exchange and local Ca-$HCO_3$ waters. Upchurch (1992) found Mg-$HCO_3$ facies water near Gulf County. This water is enriched in $Mg^{2+}$ from Mg-rich clays and/or dissolution of dolomite.

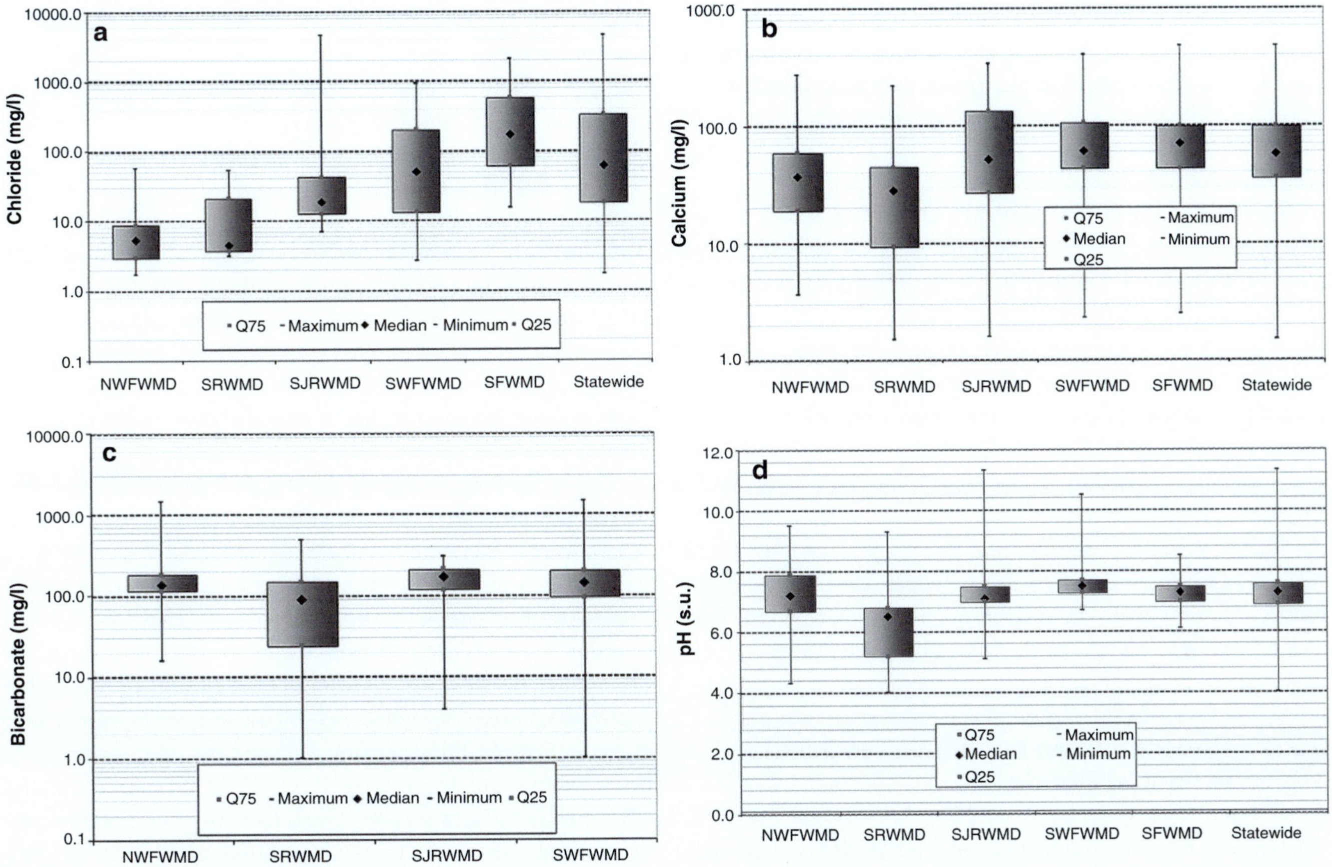


**Fig. 5.21** (**a**) Distribution of chloride of water of the Intermediate Aquifer System. Data from Upchurch (1992) (**b**) Distribution of calcium of water of the Intermediate Aquifer System. Data from Upchurch (1992) (**c**) Distribution of bicarbonate in water of the Intermediate Aquifer System. Data from Upchurch (1992) (**d**) Distribution of pH of water of the Intermediate Aquifer System. (Data from Upchurch (1992))

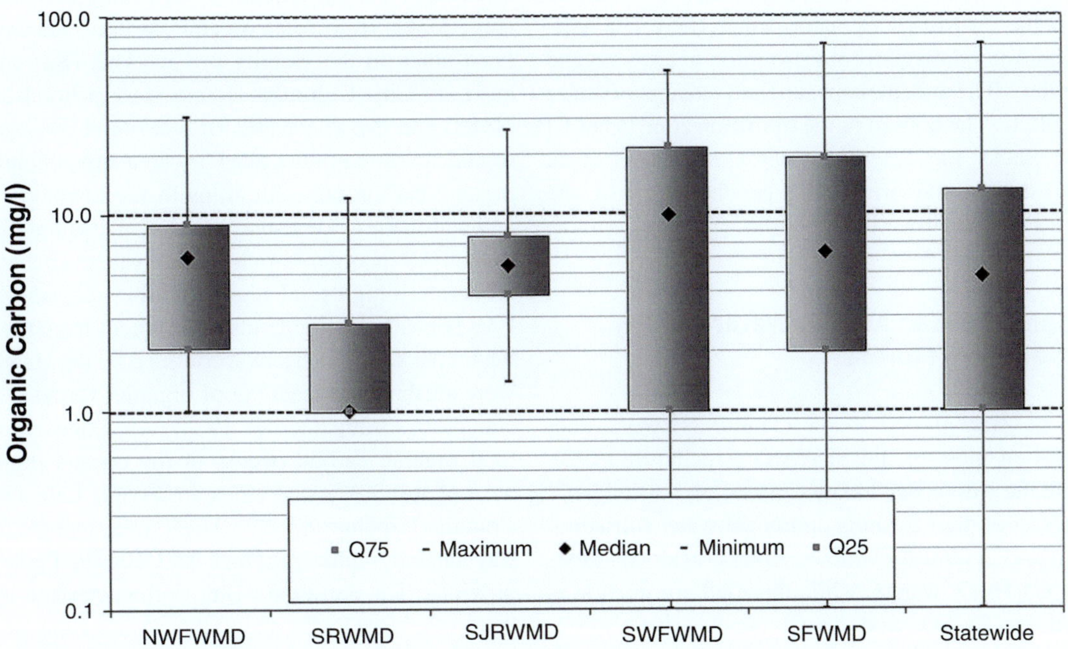


**Fig. 5.22** Distribution of organic carbon in water of the Intermediate Aquifer System. (Data from Upchurch (1992))

UFA water in the SRWMD can be subdivided into three facies. The UFA in most of the district is dominated by $Ca\text{-}HCO_3$ water. Local $Ca\text{-}Mg\text{-}HCO_3$ samples reflect dissolution of dolomite or possible additions of $Mg^{2+}$ from the overlying Hawthorn Group. Most of the UFA is characterized, however, by limestone dissolution.

Near the coast in the SRWMD, there is a Na-Cl facies, which reflects the coastal transition zone. In central Taylor County, there is a $Ca\text{-}Mg\text{-}HCO_3$ facies which reflects extensive dolomitization in the UFA.

Figure 5.23 illustrates water quality for water samples from the UFA in the SJRWMD. The dominant water type is $Ca\text{-}HCO_3$ facies. There are large reentrants of the Na-Cl facies near the coast. These reflect connate waters and modern, saltwater intrusion (Frazee and McClaugherty 1979; Spechler 1994; Spechler and Phelps 2001). Scattered $Na\text{-}HCO_3$ samples indicate ion exchange. The St. Johns River region includes two areas where the Na-Cl facies exists. These appear to reflect upconing or upward flow along fractures or faults as described by Leve (1983).

### 5.7.3 Upper Floridan Aquifer Water Quality in Central Florida

The UFA in the southern half of the SWFWMD is characterized by three facies. Inland, the hydrochemical facies consists of $Ca\text{-}Mg\text{-}HCO_3$ from dissolution of dolomite and calcite in the aquifer system and magnesium from the Miocene Hawthorn Group clays. Along the coast, the saltwater transition zone consists of three horizons. At the top is the freshwater lens, which consists of $Ca\text{-}Mg\text{-}HCO_3$ facies water. Below the $Ca\text{-}Mg\text{-}HCO_3$ facies water, there is a horizon of $Ca\text{-}SO_4$ water derived from dissolution of gypsum and anhydrite in the MCU and, perhaps, upward leakage of water through the MCU. The bottom of the transition zone is Na-Cl facies water. Sacks and Tihansky (1996) and Sacks et al. (2012) have studied the origin and geochemistry of the $SO_4^{2-}$in the UFA in southwestern Florida.

### 5.7.4 Upper Floridan Aquifer Water Quality in Southern Florida

There were no Background Water Quality Network data from the UFA in most of the SFWMD (Upchurch 1992). Samples from the west coast, in Lee County, suggest a Na-Cl facies as a result of poor flushing and/or saltwater intrusion. The northern and central Kissimmee River corridor is characterized by $Ca\text{-}HCO_3$ waters, while the southern portion is characterized by $Ca\text{-}SO_4$ and Na-Cl facies. The $Ca\text{-}SO_4$ facies is the inner half of the transition zone, while the Na-Cl facies is the outer portion. The Na-Cl facies persists into the Everglades because of little or no flushing of connate or residual waters.

### 5.7.5 Organic Content of Upper Floridan Aquifer Water

Total organic carbon in the UFA (Fig. 5.24) ranges from below detection limits to over 80 mg/L within the state (Upchurch 1992). While the median total organic carbon concentration is relatively low (2.2 mg/L) compared to other aquifer systems within the state, it is high compared to data from aquifer systems outside of Florida (Hem 1985; Thurman 1985).

The highest total organic carbon concentrations in northwestern Florida (NWFWMD column, Fig. 5.24) are in areas where the UFA is either unconfined or characterized by poor confinement in a well-developed karst terrain. Highest total organic carbon concentrations are in areas where the UFA is overlain by thin, organic-rich, sandy soils. These areas are dominated by freshwater swamps and poorly drained pine flatwoods, both of which are excellent sources of humic substances. Lowest total organic carbon concentrations are in areas where the UFA is well confined by the overlying Miocene Hawthorn Group. High total organic carbon concentrations also occur in the limestone outcrop belt (Suwannee Limestone and Ocala Limestone) in Walton, Holmes, Washington, and Jackson Counties. In this area UFA is unconfined and is characterized by internal drainage.

High total organic carbon concentrations are found in scattered wells along the karstic Cody Escarpment in the SRWMD (Fig. 5.24). High total organic carbon concentrations have been noted in the vicinity of Live Oak (Suwannee County) and Lake City (Columbia County) by Upchurch and Lawrence (1984) and Brown (1989). Brown found two sources of total organic carbon in the Lake City area and concluded that total organic carbon concentrations in the lowlands at the toe of and within the Cody Escarpment are a result of direct recharge to the FAS by organic-rich surface waters (Brown 1989). The moderately high total organic carbon concentrations in the FAS beneath the highlands up-gradient from the escarpment, where the aquifer system is confined by the Hawthorn Group, were attributed to leaching of organics found in the Miocene Hawthorn Group (Miller 1978). An extensive area of high total organic carbon occurs in the coastal drainage system west of the Suwannee River in Taylor, Lafayette and Dixie Counties (Upchurch 1992). These areas include the San Pedro Bay physiographic province and coastal Perry Karst Plain province. The San Pedro Bay is characterized by a thin clay layer (the "San Pedro clay") that overlies the UFA. The UFA is exposed at a number of sites in the San Pedro Bay province and the Perry Karst Plain. The San Pedro clay is penetrated by

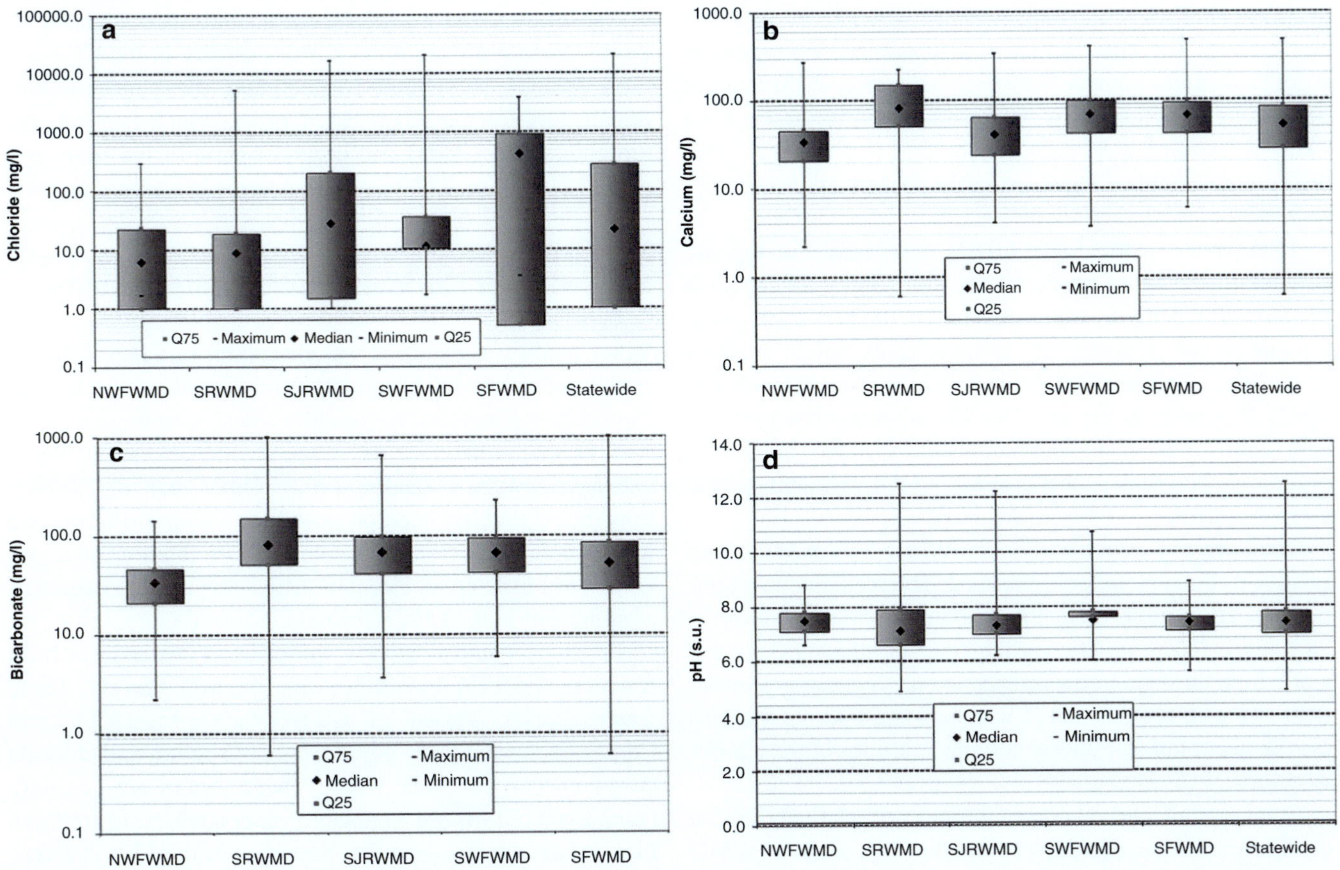


**Fig. 5.23** (**a**) Distribution of chloride in the Upper Floridan Aquifer, by water management district. Data from Upchurch (1992). (**b**) Distribution of calcium in the Upper Floridan Aquifer, by water management district. Data from Upchurch (1992) (**c**) Distribution of bicarbonate in the Upper Floridan Aquifer, by water management district. Data from Upchurch (1992) (**d**) Distribution of pH in the Upper Floridan Aquifer, by water management district. (Data from Upchurch (1992))

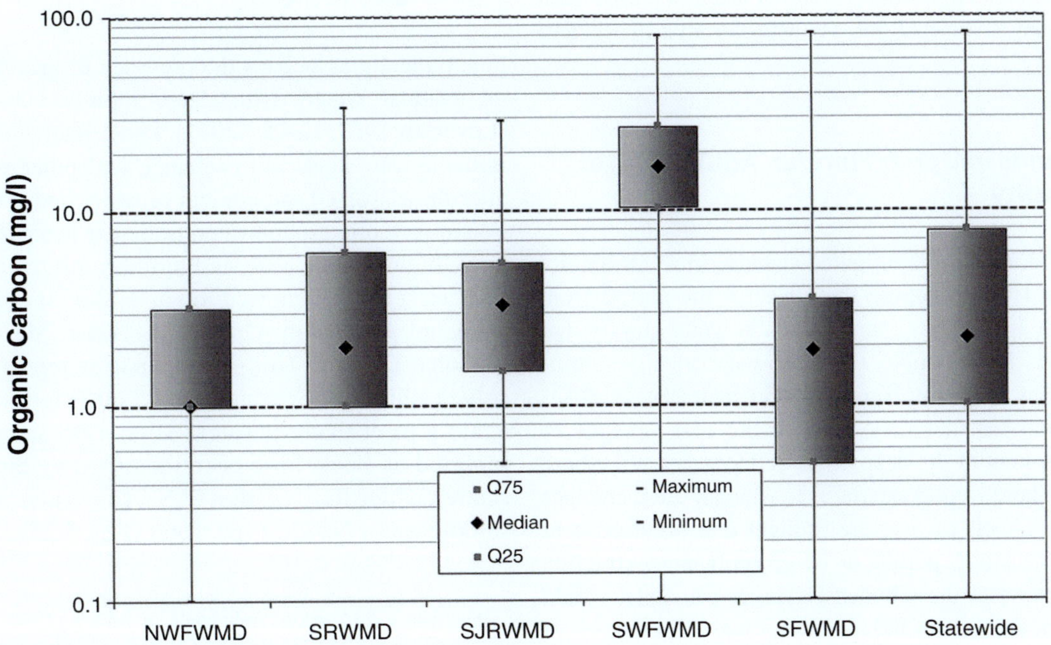


**Fig. 5.24** Distribution of total organic content of Upper Floridan Aquifer water, by water management district. (Data from Upchurch (1992))

scattered sinkholes, and sinkholes are abundant in the Perry Karst Plain. The land surface is characterized by extensive woodlands and swamps. Total organic carbon in these areas appears to be controlled by the local flow system and is most likely a result of local recharge through swamps and other areas with highly organic surface waters and organic-rich soils.

UFA water in the SJRWMD is comparable in organic content to the other northern districts (Fig. 5.24), with a median of 4.3 mg/L and a maximum of 29.0 mg/L total organic carbon. The majority of the samples high in organics came from a large area centered on Flagler County (Upchurch 1992). It appears from other data that this water has a significant connate or residual component, and the high total organic carbon may reflect organics in the sediments of the aquifer system.

Total organic carbon in the SWFWMD (Fig. 5.24) is high relative to the remainder of the state. The median concentration is 16.8 mg/L and the maximum concentration is 78.8 mg/L. Total organic carbon concentrations are consistently high throughout the district, but the highest concentrations occur in regions where wetlands and riverine systems overlie unconfined or poorly confined portions of the aquifer system.

Total organic carbon concentrations are relatively low in the limited number of samples from the UFA of the SFWMD (Fig. 5.24). Median concentration is 1.9 mg/L, but the maximum concentration is the highest recorded in the UFA at 80.6 mg/L. Subsequent sampling of this well yielded total organic carbon concentrations of 1.0 and 4.5 mg/L, indicating that the initial, high concentration may be anomalous (Upchurch 1992). High concentrations are associated with the Kissimmee River corridor and with an area in western Lee County.

### 5.7.6 Trends in Upper Floridan Aquifer Water Quality

Copeland et al. (2011) investigated changes in water quality of springs and wells tapping the UFA for the period from 1991 through 2004. They found trends in water quality that cause concern. For example, there were statistically significant reductions in spring discharge and increases in $SO_4^{2-}$ and other analytes that may indicate the first vestiges of upward movement of the deep, more saline water at the base of the UFA. There were concerns as to whether these changes represent the effects of a severe drought that occurred near the end of the study period or of groundwater extraction which, according to the Ghyben-Herzberg principle, would cause a rise in the base of the freshwater lens as potentials of the aquifer decline. The disturbing feature of the Copeland et al. study was that some areas that had not experienced the severe drought also showed these changes (Copeland et al. 2011). Note that even though water quality in the UFA and UFA springs is apparently declining the concentrations remain low for the most part. Nitrate, which is a concern throughout the central and northern parts of the state, is a result of land-use activities, not groundwater withdrawals.

## 5.8 Lower Floridan Aquifer Water Quality

Water quality in the LFA was discussed in Chap. 4 in the context of the Kohout circulation model. There are insufficient data to generate a statistically sound sampling program, such as was utilized in the UFA for the Background Water Quality Network (Maddox 1992). Other than the Kohout flow model that has been proposed for southern Florida and the upconing model of the Fernandina Production Zone, we know very little about circulation systems and the degree of confinement in the LFA elsewhere in Florida.

Freshwater and relatively low salinity water have been found in the LFA in northern and central Florida. These water masses consist of Ca-$HCO_3$, Ca-Mg-$HCO_3$, and Ca-$SO_4$ facies with varying amounts of sodium and chloride (Brown 1984; Spechler 1994). These waters are currently being explored for use as potable water sources. In southern Florida, the consensus is that the water in the LFA is essentially seawater (Na-Cl facies; Kohout 1965, 1967; Sproul 1977; Haberfeld 1991).

## 5.9 Connate Water, Residual Water, and Brine

One typically associates the presence of brines and connate or residual water[5] with deep, rapidly subsiding basins (Levorsen 1967; Leeder 2011) where interactions with clay minerals can cause ion exchange and other processes that alter the dissolved salt content of water and prevent circulation of meteoric groundwater. Little has been done to characterize highly saline water in Florida, yet it has been found.

Peek (1958) reported on a water sample from the Miocene Hawthorn Group on Anna Maria Island in Manatee County. This sample had a reported total dissolved solids (TDS) content of 53,600 mg/L, which is half again greater than seawater.[7] High TDS samples, such as reported by Peek, have been identified by others in other, scattered portions of the FAS. The Anna Maria Island sample was a Na-CL-type water (Fig. 5.25).

[7]There was a 232 meq/L imbalance in Peek's (1958) water analysis with equivalent concentrations of $Cl^-$ and $SO_4^{2-}$ higher than the corresponding cation milliequivalent concentrations. Therefore, one wonders if the analysis reported by Peek was in error.

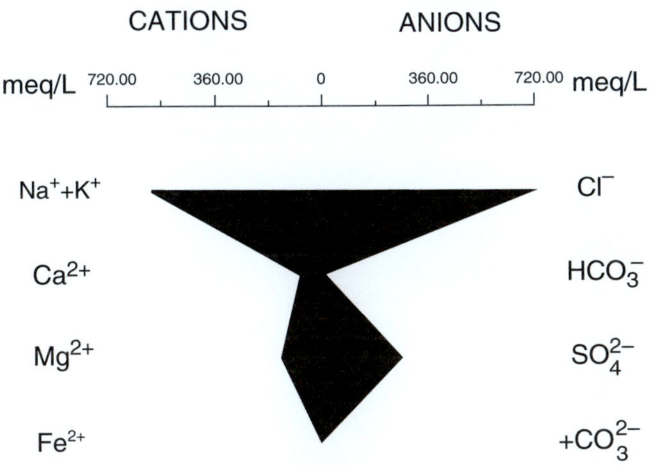


**Fig. 5.25** Stiff diagram illustrating the composition of a "brine" sample from the Miocene Hawthorn Group at Anna Maria Island, Manatee County. (From Peek (1958))

Frazee Jr. and McClaugherty (1979) identified connate and/or residual water in the FAS in portions of northeastern Florida. These waters ranged from Na-Ca-Mg-Cl-$SO_4$ to Na-Cl facies. Upchurch (1992) also reported water with elevated TDS concentrations and Na-Cl facies as being either connate or residual water.

## 5.10 Regional Studies of Saturation States of Floridan Aquifer Water

There have been a number of seminal studies of the chemical evolution of groundwater in Florida. They began with the work of Back and Hanshaw (1970, 1971) and Hanshaw and Back (1971, 1972). Back and Hanshaw identified groundwater flow paths in the UFA that began near the Green Swamp UFA potentiometric high in central Florida and radiate outward, toward the Atlantic Ocean and Gulf of Mexico in order to investigate changes in water chemistry as water flowed to the sea and the freshwater/saltwater transition zone. They produced a number of papers exploring these flow paths, including investigation of calcite and dolomite geochemical processes (Back and Hanshaw 1970, 1971; Hanshaw and Back 1979) and $^{14}C$ age dating of the water along the flow paths (Hanshaw et al. 1965). The wells they sampled had boreholes open across several formations and the water samples were, in a sense, integrals of multiple, sub-regional flow systems and formation-related geochemical processes. We now know that the different formations encountered along these flow paths have slightly different chemical signatures because the Eocene Ocala Limestone partially isolates flow systems in the Oligocene Suwannee Limestone from the Eocene Avon Park Formation. Even so, their work is excellent for understanding the regional behavior of UFA.

Back and Hanshaw sampled water in all directions from the potentiometric high. The Green Swamp potentiometric high was, at the time of their work, considered to be a major recharge area (Pride et al. 1961, 1966; Stringfield 1966). See the discussions in Chap. 4 concerning recharge around the Green Swamp UFA potentiometric surface high.

Back and Hanshaw (1970) and Hanshaw and Back (1971) showed that the area of "principal recharge" to the UFA does not coincide with the crest of the UFA potentiometric surface under the Green Swamp. Rather, it was located north and west of the potentiometric surface high. On the northwestern and western sides of the potentiometric surface high, Back and Hanshaw concluded that the areas of principal recharge extend to near the coast.

Back and Hanshaw (1970) and Hanshaw and Back (1971) also concluded that UFA water in the vicinity of the recharge area was undersaturated with respect to calcite (SI less than 0.0). Their work suggested that the groundwater becomes saturated with respect to calcite within a radius of roughly 50–75 km of the central recharge area. The area of low saturation indices is because of local, chemically aggressive recharge water mixing with the water flowing radially away from the Green Swamp potentiometric high.

As groundwater flow proceeds toward the coasts, it equilibrates with calcite, according to Back and Hanshaw's model. In the areas shown to be in equilibrium with respect to calcite (SI greater than 0.0) there should be no additional limestone dissolution. Near the coast, the saturation indices indicated that the groundwater became oversaturated (SI greater than +0.2) with respect to calcite. In the area where groundwater was thought to be oversaturated with respect to calcite, calcitic cements should be forming.

Plummer (1977) and Plummer and Back (1980) refined the saturation data proposed by Back and Hanshaw using a more comprehensive chemical equilibrium model. While the basic geochemical patterns remained the same, their mass balance approach improved their results in terms of degree of saturation of the groundwater and speciation as a result of ion pairing and thermodynamics. Many of the areas that were reported by Back and Hanshaw to contain UFA water that is oversaturated with respect to calcite were reduced in favor of simple saturation with respect to calcite. Budd and Vacher (2004) found little evidence of cement formation in the UFA where it is confined and down-gradient from the potentiometric-surface high by examination of aquifer rock samples.

Finally, through their numerous publications Back and Hanshaw (i.e., Back and Hanshaw 1970; Hanshaw and Back 1979) developed a conceptual model for the geochemical evolution and maturation of groundwater facies in the UFA. Figure 5.26 depicts a modified version of their Piper diagram showing the geochemical evolution of UFA groundwater. The shaded areas on the diagram represent the pattern

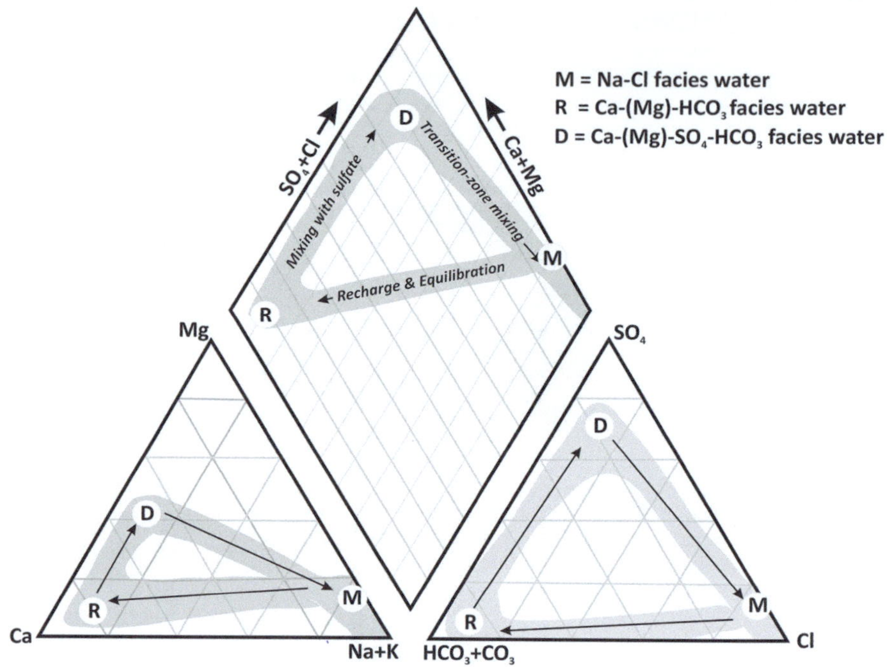


**Fig. 5.26** Pathways for the geochemical evolution of Upper Floridan Aquifer water. (Modified from Figs in Back and Hanshaw (1970) and Hanshaw and Back (1979))

of water-quality data from their work. They included other processes, such as brine development and saline-water intrusion in their model. Figure 5.26 simply examines the pathway that water takes from recharge as meteoric water through equilibration with limestone and/or dolostone to mixing with $SO_4^{2-}$ water and then seawater in the mixing zone.

Figure 5.26 is useful for understanding the geochemical behavior of groundwater in eogenetic karst systems, such as in Florida. Water-quality evolution begins with a dilute Na-Cl solution (position M; Fig. 5.26), which reflects marine aerosol-dominated precipitation. With dissolution of aragonite and calcite and inversion of high-Mg to low-Mg, water becomes a Ca-$HCO_3$ solution (position R, Fig. 5.26). If dolomite or Mg-rich clays are present in the system, then the water in the vicinity of R (Fig. 5.26) will also have a Mg component. This maturation of the groundwater includes flushing of residual and connate water from the system and increasing total dissolved solids (TDS) content as carbonate minerals are dissolved. As UFA water moves downward along its flow paths, it ultimately comes in contact with $SO_4$-rich water near the middle confining units (pathway R to D, Fig. 5.26) and the water evolves to a Ca-Mg-$SO_4$ facies with even higher TDS content through mixing. Finally, as mixing on the freshwater/saltwater transition zone continues, the water mixes with seawater and the proportions of Na and Cl increase, ultimately resulting in Na-Cl facies water, this time with TDS content near that of seawater.

Katz (1992) evaluated the composition of UFA water statewide. He utilized a large dataset of UFA water samples derived from the Florida Department of Environmental Protection's (FDEP) Background Water Quality Network (Maddox et al. 1992; Summers et al. 1998). Katz's results were in general agreement with Plummer's (1977). Figure 5.27 illustrates those areas of the state where UFA groundwater was under- (SI less than −0.5) and over-saturated (SI greater than +0.5) with respect to calcite.

There is little water-quality data from the UFA in southern Florida because (1) the aquifer is deeply buried and saline, so few wells were available to sample and (2) the wetlands of the Everglades system hinder access in many areas. Elsewhere in the state, data coverage is excellent. The Background Water-Quality Network (Maddox et al. 1992) intentionally avoided sampling areas known to have experienced saltwater intrusion and urbanized areas where contamination by anthropogenic sources was likely. Also, the Background Network utilized dedicated monitoring wells and existing, public and domestic, water-supply wells for sampling. There was some control on selection of the wells on the basis of depth, but the formations into which the wells penetrated and the intervals penetrated were not well controlled.

There are several important observations that can be made from the Katz (1992) data.

1. Much of the UFA water investigated by Katz (1992) was considered to be saturated with respect to calcite (SI equaling −0.5 to +0.5; colorless areas on Fig. 5.27).
2. In areas where the majority of water samples indicated over- or undersaturation with respect to calcite, water samples that were saturated were also found nearby, so

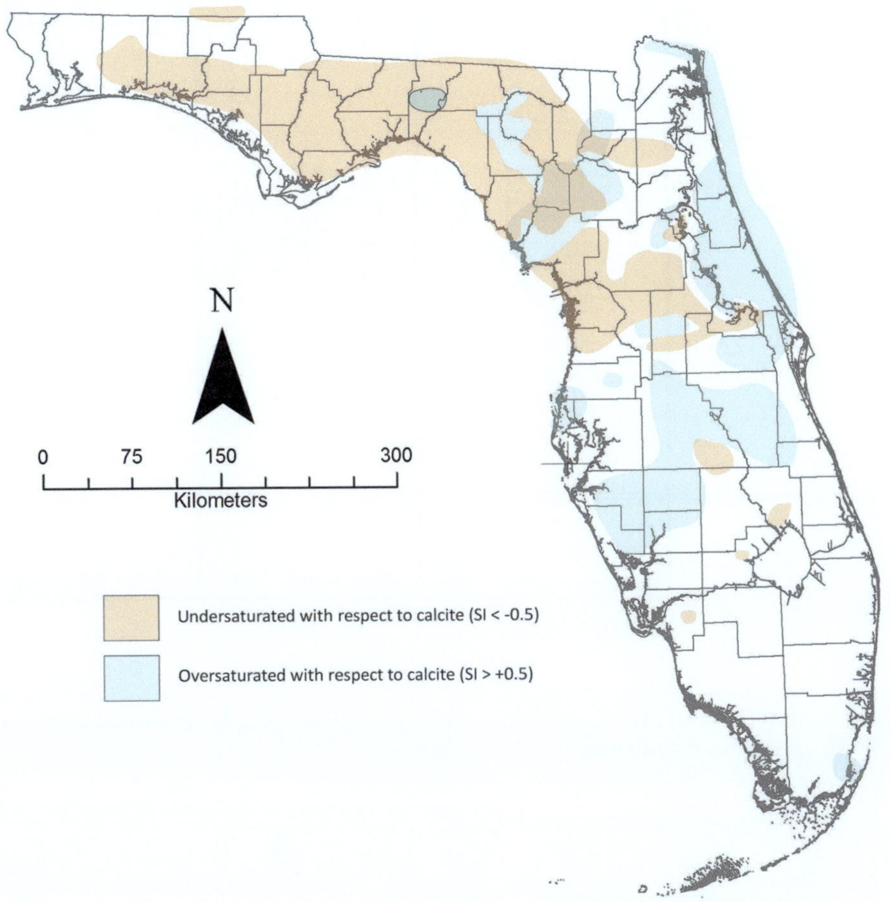


**Fig. 5.27** Distribution of the calcite saturation index for Upper Floridan Aquifer water in Florida. (Modified from Katz (1992))

well depth, conduit versus matrix flow, and other factors control the data with respect to degree of saturation of the water with respect to calcite.

Undersaturated water predominated in areas in northern and central Florida where the karstified limestone is bare or thinly mantled. This suggests that these areas are locations of aquifer recharge.

3. Areas where the UFA water was found to be oversaturated with respect to calcite were generally, but not always, located where confinement by the ICU is excellent. Deeper wells in the data set revealed water that was oversaturated at a number of locations.

Katz (1992) also summarized medians, quartiles, and ranges of saturation indices of UFA water with respect to calcite by region (Fig. 5.28). By far, the majority of water in Katz' sample set was considered to be saturated (i.e., −0.5 to +0.5, Fig. 5.27) with respect to calcite. A relatively small percentage of samples was found to be undersaturated with respect to calcite, and an even smaller percentage was oversaturated with respect to calcite.

Swancar and Hutchinson (1995) examined water quality in the UFA in the portion of the Back and Hanshaw study area that is west of the Green Swamp potentiometric surface high. They found that UFA groundwater was saturated to slightly undersaturated with respect to calcite over most of the northern half of the study area (Fig. 5.29). In the southern half of the study area, where the UFA is moderately to well confined, the groundwater was slightly oversaturated with respect to calcite.

Budd (2001) and Budd and Vacher (2004) investigated the evolution of porosity in the same strata that Back and Hanshaw (1970) and Swancar and Hutchinson (1995) studied in west-central Florida. Budd and Vacher concluded that the UFA in recharge areas consisted of a triple porosity system with intergranular, fracture, and karstic pore and void systems. Down gradient, the UFA was found to include intergranular and fracture porosity. There was little evidence of large-scale cementation, as was suggested by the work of Back and Hanshaw (1970) and Hanshaw and Back (1971). Both Budd (2001) and Budd and Vacher (2004) concluded that depth of burial of the limestone and dolostone in the area was not a major factor in determining porosity. In other words, compaction as a result of burial was much less important than the porosities resulting from depositional environments and early diagenesis.

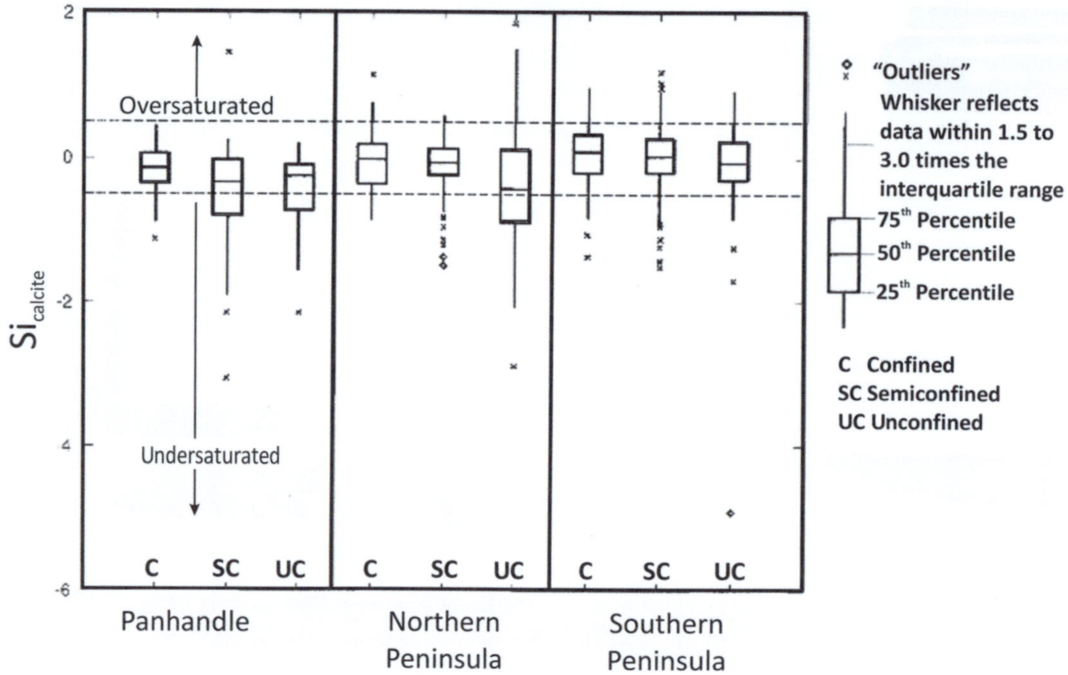


**Fig. 5.28** Box-and-whisker diagram illustrating the median and quartile distributions of the calcite saturation index for Upper Floridan Aquifer water, by region. (Modified from Katz (1992, Fig 12))

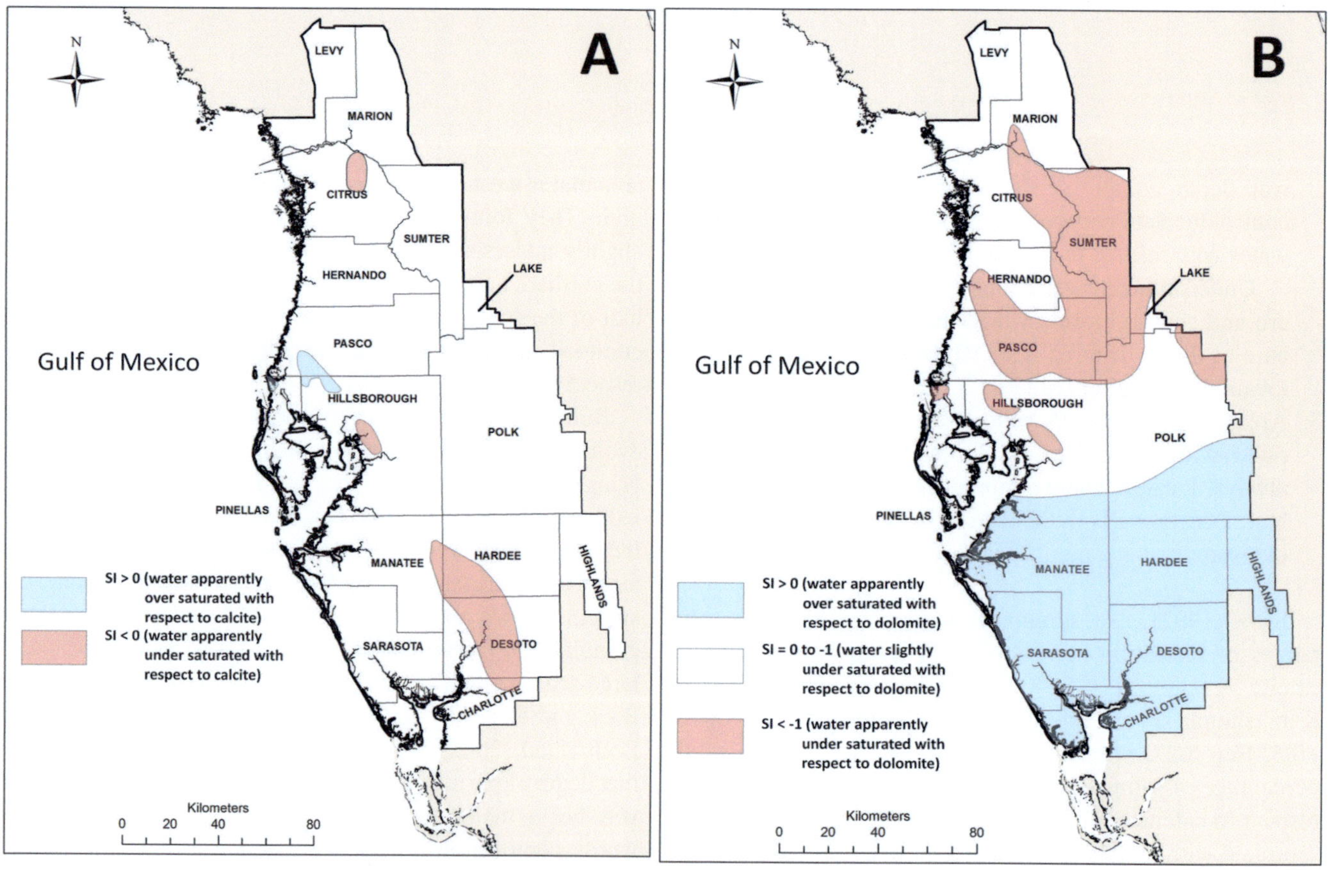


**Fig. 5.29** Saturation indices with respect to calcite (**a**) and dolomite (**b**) for Upper Floridan Aquifer groundwater in west-central Florida. (Modified from Swancar and Hutchinson (1995, Fig. 14))

Therefore, it appears that the shallow UFA water in recharge areas is undersaturated with respect to calcite, and that epigenetic karstification is predicted in the Budd (2001) and Budd and Vacher (2004) study locations. Deeper in the UFA in recharge areas and down gradient, toward the coastal discharge areas, the groundwater of the UFA is more-or-less saturated with respect to calcite, and karst development is limited to non-existent at the present time. Any karst features present in these areas is most likely a result of epigenetic dissolution during low sea stands or hypogenetic processes in saline-water mixing zones.

## 5.11 Age Dating of Upper Floridan Aquifer Water

Isotopes and $^{14}C$ dating began to be used to trace and date groundwater in the UFA in the early 1960's (Meyer 1989). $^{14}C$ dating was first used in central Florida by Hanshaw et al. (1965). Plummer (1977) examined the mass transfer of UFA water in the same area in relation to the earlier $^{14}C$ ages. He assigned new, slightly younger ages on the basis of mass-transfer reaction coefficients. The ages of groundwater in the UFA in west-central Florida ranged from essentially modern water to approximately 29,800 years in the deeper, down-gradient and confined UFA. Hanshaw et al. (1965) had concluded that the oldest water along the Green Swamp to the coast flow pathways was approximately 35,500 years. These dates indicate a slow rate of movement through the UFA, with flow from recharge area to discharge area generally taking tens of thousands of years. These slow flow rates indicate that there has been plenty of time for groundwater to chemically equilibrate with the host rocks.

Swancar and Hutchinson (1995) used tritium ($^3H$) to assess recharge of UFA water in the SWFWMD (Fig. 5.30). Their data agree with the recharge area delineated by Hanshaw and Back (1971). Their data also attest to the relative ages of UFA water. The oldest water is located in the southern half of their study area, where the UFA is well confined by the IAS/ICU. Note that the water sampled from the UFA Green Swamp potentiometric-surface high, which was formerly considered a recharge area, had low tritium activities and is, therefore, geologically old. The work of Swancar and Hutchinson (1995) disproved the notion that the Green Swamp potentiometric surface high is a region of high recharge to the UFA.

Age dating of UFA water is instructive because it indicates long residence times of the water in contact with the aquifer rocks. This long residence time results in chemical saturation of the water with respect to aquifer materials everywhere except within the vicinity of the shallow, recharge locations and within mixing zones at depth.

## 5.12 Dolomitization in Florida

In addition to karst conduit fractures, we must consider the porosity that can be developed by dolomitization. Dolomitization is the diagenetic process by which aragonite or calcite ($CaCO_3$) is altered to dolomite ($CaMg(CO_3)_2$) in sedimentary strata. During this process, porosity is created (Murray 1960; Mazzullo and Harris 1992; Lucia and Major 1994; Lucia 1995). Dolomitic sediments and rocks are widespread in Florida, and in many strata the literature is mixed as to whether porosity in these strata originated from karst-related dissolution or dolomitization. In fact, the origin of dolomite itself in Florida is highly debated. Therefore, it is important to discuss dolomitization in Florida and the conceptual models used to explain porosity in dolomitic strata.

Dolomitic facies of the sediments and rocks of the FAS contain numerous features that may appear to be karstic, and have in fact been interpreted as such, but were most likely formed by non-karstic processes connected to the formation of the dolomitic strata themselves. The origin of the Florida dolomites therefore has direct bearing on Florida karst and will be discussed in some detail here.

### 5.12.1 Dolomitization Chemistry

There are several models for the formation of dolomite in Florida. Dolomite is generally thought to form either by direct replacement of calcite or aragonite,

$$2CaCO_3 + Mg^{2+} \rightleftarrows CaMg\left(CO_3\right)_2 + Ca^{2+}, \quad (5.24)$$

or by dissolution of calcite or aragonite and precipitation of dolomite according to

$$Mg^{2+} + 2HCO_3^- + CaCO_3 \rightleftarrows CaMg\left(CO_3\right)_2 + CO_2 + H_2O. \quad (5.25)$$

Both reactions require elevated $Mg^{2+}/Ca^{2+}$ concentration ratios in the dolomitizing water.

Early diagenetic (penecontemporaneous) dolomite is most likely formed at the expense of metastable aragonite rather than calcite. Drever (1997) stated that the $Mg^{2+}/Ca^{2+}$ ratio had be about 6 for conversion of aragonite to dolomite. While there are many processes that can produce the high $Mg^{2+}/Ca^{2+}$ ratios necessary for dolomitization, two have been widely proposed for the formation of Florida dolomite. These are:

- Evaporation of seawater, which forms supratidal dolomite that is penecontemporaneous with sedimentation; and
- Saltwater/freshwater mixing, which occurs in coastal groundwater systems.

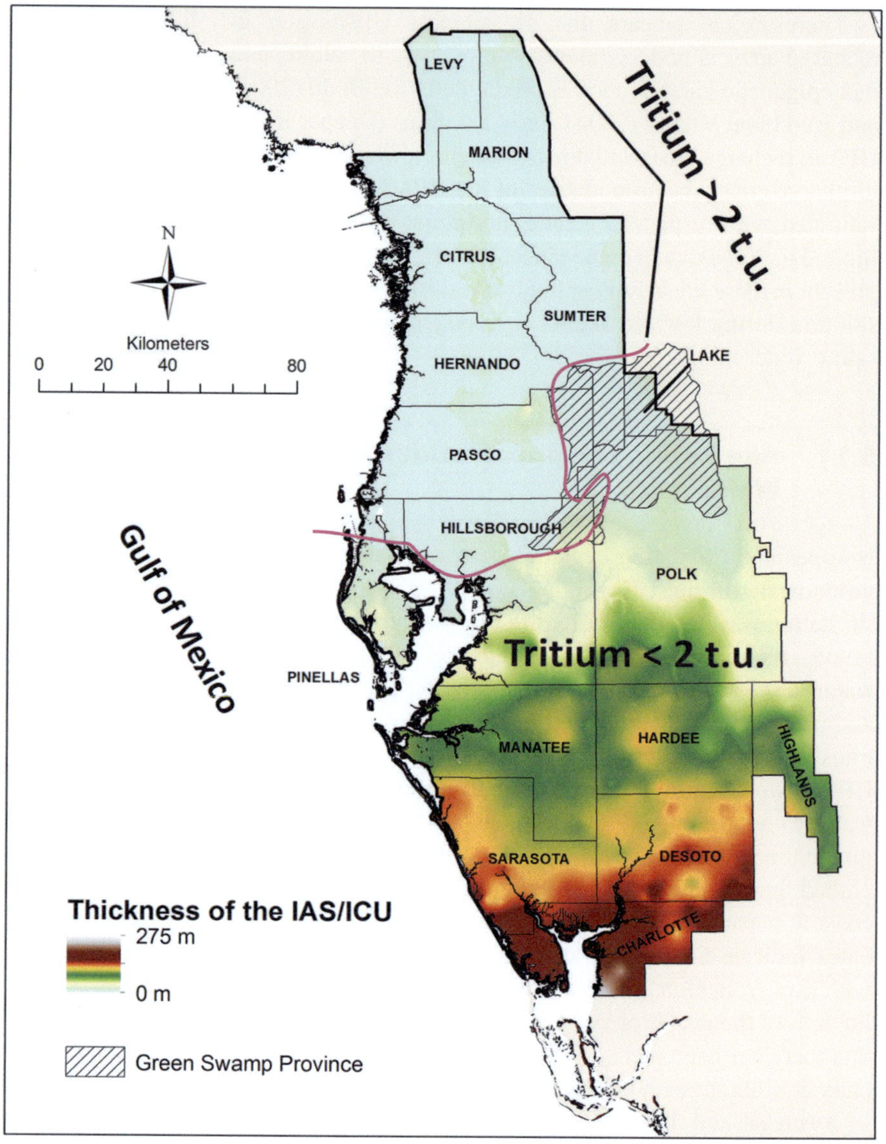


**Fig. 5.30** Comparison of tritium concentrations in the Upper Floridan Aquifer with Intermediate Confining Unit thickness in west-central Florida. (Modified from Swancar and Hutchinson (1995, Fig 18))

Florida's significant dolomite deposits occur in the Miocene Hawthorn Group and Eocene Avon Park and Oldsmar formations. In addition, penecontemporaneous dolomite that has been found in Recent sediments of southern Florida is a possible indicator of local environments within which dolomite can form.

## 5.12.2 Dolomite in Recent Florida Sediments

The presence of dolomite in modern sedimentary environments in southern Florida was first documented by Taft (1961); Shinn (1964a, b, 1968), and Taft and Harbaugh (1964). Small, silt-sized, limpid (clear), well-formed (euhedral) crystals were first noticed by Taft in Florida Bay. Many of these crystals had central cores of organic material, which was considered to represent nucleation sites (Taft 1961; Taft and Harbaugh 1964). They opined that the dolomite crystals were forming at the present time in Florida Bay. However, Deffeyes and Martin (1962) found that the Florida Bay dolomite contained no $^{14}C$, which indicated an age of the dolomite crystals that was greater than 35,000 years. In other words, the dolomite was detrital and had been derived from older strata.

Upchurch and Strom (1980) and Upchurch et al. (1983) described similar, silt-sized dolomite rhombohedra (Fig. 5.31) preserved in mangrove peats and tidal river sediments in the 10,000 Islands area of Collier and Monroe counties. They concluded that the dolomite crystals were detrital, but did not identify a point of origin for the dolomite crystals, which was presumably inland from the mangrove swamp.

Shinn and his colleagues first described dolomite formation in supratidal mud flats in the lower Florida Keys (Sugarloaf Key; Shinn 1964a) and Bahamas (Andros Island; Shinn 1964a, b). In both locations, the dolomite was poorly crystallized and had relatively high calcium content, such as the "protodolomites" described by Graf and Goldsmith (1956). Shinn (1968) concluded that the crusts of supratidal dolomite on Sugarloaf Key were forming in "brines" concentrated by evaporation on the supratidal flat and that they are penecontemporaneous with sedimentation. Carballo et al. (1987) obtained $^{14}C$ dates for the Sugarloaf dolomite crystals. The dates ranged approximately from 160 to 1420 years BP. They attributed dolomitization to tidal pumping of Florida Bay water. Budd (1997) has summarized the details of similar, penecontemporaneous dolomite occurrences worldwide.

### 5.12.3 Ancient, Syndepositional Dolomitization Processes

#### 5.12.3.1 Introduction

The Miocene Hawthorn Group and Eocene strata of the FAS are commonly partly or wholly dolomitized. Most of Florida's dolostone deposits are Middle Eocene (Avon Park Formation) and older.

Two depositional mechanisms have been proposed for the formation of these dolostones and dolomitic sediments – one mechanism is thought to be penecontemporaneous with sedimentation of the host carbonate sediments and the other mechanism is thought to be diagenetic, or post-depositional.

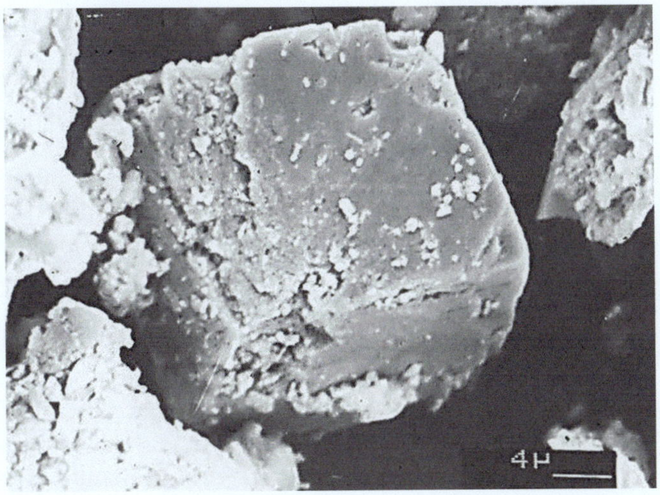


Fig. 5.31 Scanning electron microscope (SEM) micrograph of a silt-sized rhombohedron typical of the dolomite found in Recent southern Florida sediments. (Upchurch et al. (1983))

#### 5.12.3.2 Penecontemporaneous Dolomite in Post-Miocene Sediments

Dolomite can form at the land surface or at shallow sediment depths as sediments are deposited and undergo initial diagenesis in certain environments. In southern Florida, Shinn (1964a, b, 1968) and Carballo et al. (1987) showed that limited dolomitization is occurring today on supratidal, perimarine, carbonate mud flats as a result of evaporative concentration of seawater.

Elsewhere in the world, syndepositional dolomite deposits are forming as a result of evaporative concentration of seawater (i.e., Pray and Murray 1965; Blatt et al. 1972; Zenger and Mazzullo 1980; Boggs Jr 2011). In the Persian Gulf, large, highly evaporative, tidal, salt flats known as sabkhas, are characterized by deposition of evaporite minerals and dolomite (Illing et al. 1965). The chlorinity of Persian Gulf seawater is 22–24 g/L $Cl^-$. In the tidal flats, pore waters have been concentrated to 50–130 g/L $Cl^-$ through evaporation. As a result, aragonite, salt (NaCl), gypsum, dolomite, and traces of anhydrite and celestine ($SrSO_4$) are precipitating. At the edges of the sabkhas, pore waters have $Mg^{2+}/Ca^{2+}$ ratios as high as 10 (normal Gulf water has a $Mg^{2+}/Ca^{2+}$ ratio of about 3). In the interior of the sabkhas, the $Mg^{2+}/Ca^{2+}$ ratios fall gradually to below 5 as a result of dolomitization. According to Illing et al. (1965), the dolomite and gypsum form by replacement of aragonite in the pore waters.

Sabkha environments and similar evaporative systems have been documented to have existed in Florida in the Neogene Period, notably during the Miocene and Eocene Epochs. Depositional features of these evaporative systems are often confused with karst because they often contain unconsolidated beds with low penetration resistances, voids caused by volume reduction during dolomitization, and breccias caused by desiccation that resemble karst collapse features.

#### 5.12.3.3 Penecontemporaneous Dolomitization in the Hawthorn Group

The Miocene Hawthorn Group consists of a complex array of interfingering beds of phosphatic sand, clay, dolostone, and dolosilt (Scott 1988). These deposits interfinger, often over short distances as a result of deposition in perimarine, tidal, and shallow marine environments (Weaver and Beck 1977; Strom and Upchurch 1985a, 1985b; Scott 1988).

Dolosilt – Scott (1988) coined the term "dolosilt" to describe the unconsolidated, silt-sized dolomite particles that are commonly found in the Hawthorn Group sediments.

These crystals are much like the penecontemporaneous dolomite rhombohedra found in modern sediments of southern Florida. The crystals are translucent (limpid), often euhedral (well-formed rhombohedral crystals), and well crystallized. Figs. 2.31, 2.22 and 2.33 illustrate examples of dolosilt. Dolosilt deposits in the Miocene Hawthorn Group are normally mixtures of silt-sized dolomite and magnesium-rich smectite, palygorskite, and/or sepiolite (Fig. 2.33).

Dolostone – The dolostones are typically well lithified and light gray to tan colored rock. They commonly occur as thin beds and often contain black, tan, or white carbonate fluorapatite (phosphate) peloids. In places, the dolostone has either been cemented by calcite, dolosilt, or later-phase dolomite. The dolostone may be replaced by carbonate fluorapatite, in which case the replaced phosphatic dolostone may be dark brown, gray or black. The replaced dolostone beds were originally formed as hardgrounds on the sea floor. Storms ripped up these thin dolostone and dolostone/phosphate beds, so gravels, granules and coarse sand-sized clasts are commonly found in the sand and clayey sand strata of the Hawthorn. Care should be taken to not confuse these transported fragments of storm-generated carbonate and/or phosphate for the products of limestone dissolution.

Environments of Dolomite Formation – Much of the dolostone in the Hawthorn Group appears to have been deposited in shallow marine environments (Scott 1988). However, Weaver and Beck (1977) concluded that the dolomite and magnesian clay deposits, especially palygorskite and sepiolite, of the Hawthorn Group were formed in perimarine environments where evaporation elevated the $Mg^{2+}/Ca^{2+}$ ratios in pore water and shallow groundwater. Upchurch et al. (1982) and Strom and Upchurch (1985a, b) came to the same conclusion and described tidal flat sedimentary structures and paleo water-table deposits in Hawthorn Group sediments of central Florida. Rarely, gypsum crystals and chalcedony pseudomorphs after gypsum have also been found in the dolomitic silt and clay of the Hawthorn (Upchurch 1992), notably near New Port Richey (Pasco County), Gulfport (Pinellas County), and McKay Bay (Hillsborough County).

Prasad (1985) found that the dolosilt was associated with mud-supported, low permeability mudstone and wackestone fabric sediments. Silt-sized dolomite crystals were generally absent in calcite-rich, higher permeability, grain-supported grainstones. The dolosilt crystals contained about 45 mole percent magnesium as well-ordered, rhombic crystals. He concluded that the dolosilt formed "very early" (penecontemporaneous with aragonite (?) sedimentation).

Choquette and Hiatt (2008) concluded that sucrosic, dolomitic cement comprised 19–33% (mean = 25%) of Miocene Hawthorn Group dolostone. The dolomite was found to have formed in three phases that represent concentric layers within the crystals. Phase 1 is the initial formation of seed crystals, the phase II layer represents crystal growth that concentrically enlarged the core crystals, and the phase III layer is colorless and clear (limpid) dolomite filling the remaining free space. In other words, the dolostone of the Hawthorn began as penecontemporaneous dolomite that was cemented at a later time.

According to Prasad (1985), dolomitization of Hawthorn Group sediments was complete before freshwater diagenesis began. He concluded that dolomite formed directly from a metastable phase, probably aragonite. The dolomitization was attributed to mixing of water with different chemical compositions. The coexistence of mixtures of dolomitic and calcitic lithoclasts in many Hawthorn strata (Scott 1988) also suggests that dolomitization was early and exposed on the sea floor so that storms, waves, and currents could transport the clasts.

This description of the modes of formation of dolomite in the Miocene Hawthorn Group provides evidence that the dolomite is penecontemporaneous or diagenetic and that dolostone and dolosilt beds are not related to karst processes, a common misconception.

#### 5.12.3.4 Penecontemporaneous Dolomitization of Eocene Strata

There is little dolomite in the Ocala Limestone. Most of the Eocene dolostones are within the Avon Park Formation and appear to be penecontemporaneous. Dolomite of the Avon Park Formation consists of beds of limpid dolosilt crystals most likely formed in association with shallow water environments exposed near the crest of the Ocala Platform.

Choquette and Hiatt (2008) concluded that the dolostone of the lower Ocala Limestone (the former Inglis Formation of Puri (1957)) and Eocene Avon Park Formation exposed near the crest of the Ocala Platform were never deeply buried. The dolomite crystals were described as limpid with planar crystal faces and distinct zonation. In the shelf sediment of the Eocene Avon Park Formation, 50–70 percent of the dolostone had dolomitic cement.

Cander (1994) concluded that mixing-zone dolomite in the Eocene Avon Park Formation formed in a peritidal environment. This dolomite is fine grained and highly porous (20% mean porosity). He concluded that this early stage dolomite formed during the Eocene in "hypersaline" water.

Randazzo (1976); Randazzo and Saroop (1976); and Randazzo et al. (1977) showed that dolostone in the lower Ocala Limestone and the Avon Park formed as a result of multiple transgressive and regressive cycles in shallow water with intermittent subaerial exposure. The dolomite was thought to be penecontemporaneous and formed in Mg-rich groundwater. Figure 5.10 illustrates shallow-water seagrasses preserved in thinly bedded, shallow-water Avon Park dolostone and Fig. 5.32 shows the organics and dolostone in cross section.

**Fig. 5.32** Shallow-water, thinly bedded, well cemented dolostone with organics from the Eocene Avon Park Formation at the Gulf Hammock Mine, Gulf Hammock, Levy County

The origins of the dolomite discussed in this section have resulted in local, weathering and diagenetic features that developed during dolomitization and that may be confused with karst, including duricrusts and brecciation, which are commonly found in these strata.

#### 5.12.3.5 Soft Zones in Avon Park Formation Dolostone

"Soft zones" are frequently encountered in the Avon Park Formation dolostone in northern and central Florida. These soft, poorly consolidated zones have been erroneously attributed by some to karst-related voids. These horizons are termed soft zones because they are characterized by low penetration resistances, weight-of-rod or –hammer conditions, and, occasionally, lost drilling fluid circulation during standard penetration testing. The relationship of these zones to karst processes remains problematic, but a series of exposures of the Avon Park Formation along the Cross Florida Barge Canal in northern Citrus County has been instructive.

The geology of the Avon Park Formation was extensively explored during the excavation of the Cross Florida Barge Canal (Noll and Tegeder 2009) in the 1970s (Harrison 1969; Southeastern Geological Society 1970; Faulkner 1973). The early work noted excellent exposures of the Avon Park in Citrus County, including beds of dolosilt and calcite-cemented nodules. Figure 5.33a illustrates an exposure showing Ocala Limestone overlying Avon Park Formation dolostone.

Unlike the Miocene Hawthorn Group dolosilt deposits, the Avon Park dolosilt beds do not contain appreciable clay and they are often non- or poorly cohesive. The light tan beds consist of unconsolidated, fine sand- to silt-sized dolomite crystals. The soft zones are somewhat discontinuous and occur in a more-or-less horizontal plane (Fig. 5.33a). In Fig. 5.33b, the unconsolidated dolosilt has washed out of the exposure (a quarry face) leaving what appears to be a cavity. Remnants of the dolosilt deposit drape the floor of the washed-out zone (Fig. 5.33b). Probing of the washed out zones and observations by the senior authors during mining indicate that, away from the quarry face, open voids are minimal and the dolosilt appears to fill the available space.

These soft zones are located at, or near, the contact between dolomitic Ocala Limestone and the Avon Park Formation. This contact is unconformable and commonly contains zones of elevated permeability. It is unclear if the soft zones reflect paleosols or just poorly to uncemented horizons within the dolostone. The dolomite crystals are limpid (clear and translucent) euhedra (rhombic crystals), similar to the dolosilt of the Hawthorn Group (Fig. 2.31). Had they once been cemented, one would expect to observe eroded and embayed crystal faces, but none have been observed. It is clear that the soft zones are preferential locations for groundwater circulation, as suggested by the secondary mineralization found within the dolosilt-filled zones.

Note the concretion-like structures on the roof of the washed-out opening in the mine face (Fig. 5.33b). These nodules consist of dolosilt cemented by calcite and, rarely, dolomite. In some cases, it appears that the nodules of cemented dolostone are remnants of a stratum that was only partly cemented with late-stage dolomite or calcite. In others, the nodules consist of strings and layers of dolosilt cemented by calcite (Fig. 5.33). Many show little or no internal structure; they are simply masses of calcite-cemented dolosilt. The chains of nodules range from 0.1 to 2 m in length with individual nodules ranging from 1 cm to 10 cm.

### 5.12.4 Diagenetic Dolomite in the Floridan Aquifer System

As noted above, there are at least two generations of dolomite in many of the dolostone deposits of the FAS. In addition to the penecontemporaneous dolomite that formed at or near the time of deposition, there is clear evidence for later, post-deposition dolomitization. These later stage dolostones have been most extensively investigated in southern and central Florida where the Eocene Avon Park Formation is in the subsurface and comprises a highly transmissive part of the FAS. For the most part, the later phase dolomitization is attributed to mixing in the saltwater/freshwater transition zone within the aquifer as opposed to at the time of deposition.

Using geochemical data from FAS groundwater samples, Hanshaw and Back (1972) concluded that development of mixing zones caused dissolution of calcite and dolomitization in the deep, saline waters of the UFA. Cander (1994) reported on a 20 m interval in the Avon Park Formation that was found to have dolomitic cement that, in contrast to the penecontemporaneous dolomite, compared favorably with the findings of

**Fig. 5.33** (**a**) Wide-angle view of the working face of the Dolime Mine. Arrows point to locations of washed-out, dolosilt-filled horizons. N is a zone of well-developed nodules in a dolosilt horizon. Geologist provides scale. (**b**) View of a dolosilt horizon at the Dolime Mine showing calcite-cemented nodules (**A**), strings of crystals (**B**), and masses

Choquette and Hiatt (2008). This cement has created a "hard, Paleozoic-like" dolostone. The dolomite is not a replacement of early marine dolomite. Rather, it formed as overgrowths on the pre-existing, penecontemporaneous dolomite and formed after the Middle Miocene according to strontium isotope data. Cander (1994) concluded that this dolomite formed in a seawater-rich groundwater mixing zone.

Gaswirth et al. (2007) also found two dolostone textures in the Eocene Ocala and Oligocene Suwannee limestone formations in the subsurface southwest of the Green Swamp potentiometric surface high. The penecontemporaneous dolostone had a sucrosic texture and was permeable. The later cemented texture was less porous and relatively impermeable. Oxygen isotope data indicated a marine origin for the matrix dolomite and a freshwater/seawater mixing zone origin for the cements. Strontium isotope data indicated that the dolomitic cements developed in the Late Miocene to Pliocene in the Eocene Ocala Limestone and in the Pliocene in the Suwannee Limestone.

Maliva et al. (2002) and Melim et al. (2004) found similar results indicating multiple generations of dolomite in the FAS. They opined that the last phase of dolomitization was caused by mixing-zone diagenesis.

Note that these episodes of late stage, mixing-zone dolomitization correspond with the time of intense weathering and paleosol formation (the leached zone and post-Miocene soils superimposed on Miocene Hawthorn Group sediments) on the land surface according to Carr and Alverson (1959). Weathering of the Miocene Hawthorn Group sediments, which are magnesium rich, would supply additional magnesium to groundwater and, perhaps, increase the potential for dolomitization in the limestones of the transition zone.

Schijf and Byrne (2007) noted that saline-water springs in western Florida have seawater compositions with two notable exceptions. The water discharging from the springs is enriched in calcium and radium and depleted in magnesium. The authors hypothesized that this anomalous composition was a result of dolomitization in the FAS inland. Three offshore springs and one inland spring (Warm Mineral Springs in Sarasota County) were sampled for major and minor constituents, and it was discovered that calcium, strontium and barium were enriched by average factors of 1.6 and 3.8, respectively, in the offshore spring discharge. Warm Mineral Springs was enriched in strontium and barium by factors of 9 and 12, respectively. The enrichment in barium may provide an explanation for the occurrence of microscopic crystals of the mineral barite ($BaSO_4$) in silicified coral casts (Figs 5.34a, b) from Ballast Point, Hillsborough County. Schijf and Byrne (2007) also noted that magnesium depletion and strontium and barium enrichment increases in groundwater from south to north. Mass balance calculations indicated that these trends can be accounted for by dolomitization and possible dissolution of gypsum or anhydrite in the middle confining units. Mass balance calculations for water from Warm Mineral Springs suggested excess calcium after accounting for dolomitization and dissolution of the evaporite minerals. The authors suggested that this might reflect dissolution of limestone.

### 5.12.5 Relationship of Dolostone and Dolosilt to Karst Processes

After discussing the distribution and origin of dolomite in Florida's aquifers, it is important to place dolomitization in the context of karst development. Several conclusions can be drawn from reviewing the literature on Florida dolostone. These include:

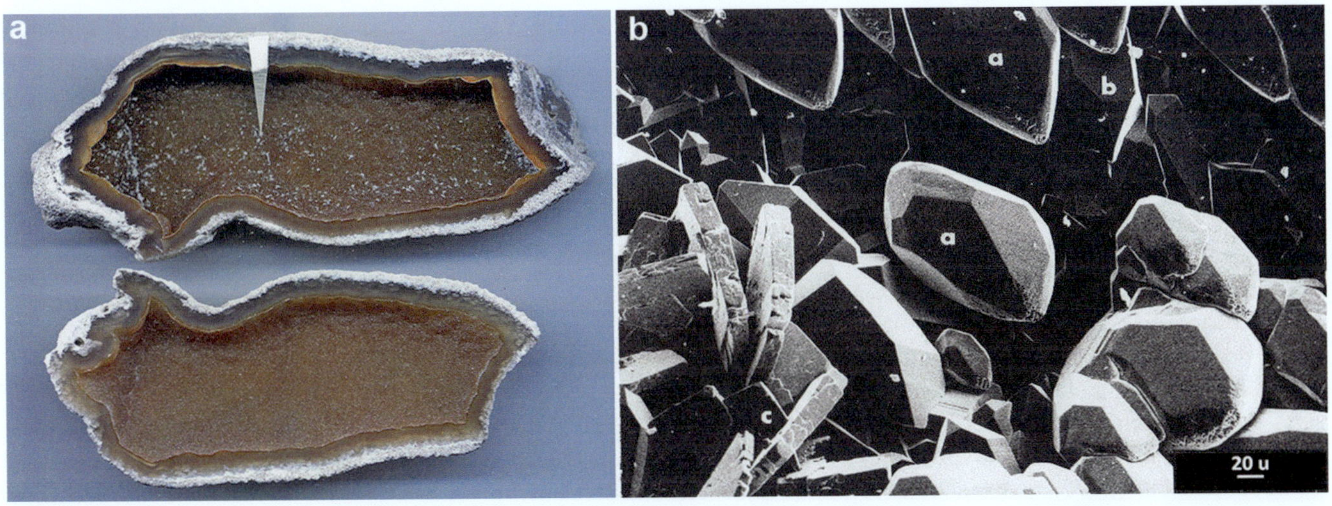


**Fig. 5.34** (**a**) Coral cast ("Tampa Bay geode") composed of chalcedony. White dots in top half are barite and calcite crystals. From Ballast Point (27.8889° N, 82.4800° W), Hillsborough County (**b**) Scanning electron micrograph of calcite (a), quartz (b), and barite (c) crystals in a chalcedony cast of a Miocene coral from Ballast Point, Hillsborough County

1. The models proposed for dolostone formation in Florida emphasize the prevalence of fine-grained dolomite in a matrix that was likely micrite prior to dolomitization. The resulting dolostones are typically fine grained and have low matrix permeability. An exception is the soft, dolosilt horizons found in the Avon Park Formation in northern Florida.
2. Many authors have suggested that the high porosity, high transmissivity zones of the dolomitic facies of the FAS are karstic. Borehole video logs of porosity development in these strata, however, reveal minimal evidence of dissolution. Fracturing appears to be the major cause of elevated permeability.
3. Studies of the dolomitic rocks in the FAS emphasize that there were at least two stages of dolomitization: a penecontemporaneous stage that produced porous dolosilt and dolostone and a late stage that cemented the dolosilt and greatly reduced porosity.
4. The first stage of dolomitization was penecontemporaneous with sedimentation in the Eocene and Miocene epochs. The second phase of dolomitization appears to have occurred near the end of the Miocene and into the Pliocene. This later phase of dolomitization appears to be contemporaneous with intense weathering and pedogenesis of the subaerially exposed Hawthorn Group and older limestone strata.
5. Borehole video logs of dolostone in the FAS have revealed abundant evidence of fracturing and minor evidence of dissolution. The thorough cementation of the dolostones during the second phase of dolomitization resulted in strata that are brittle and prone to fracturing and development of secondary porosity. Fracturing can be related to tectonic and tidal stresses.
6. Those who have studied mixing-zone dolomitization noted the potential for dissolution of calcite and precipitation of dolomite within the saltwater/freshwater transition zone. There is nothing that indicates that this dissolution/precipitation process must occur within the same volume of sediment or rock. While traces of fossils and sedimentary structures are common in the mixing-zone dolostones, it is entirely conceivable that calcite dissolution and dolomite precipitation might occur in different locations, resulting in pore or void space caused by calcite dissolution and cementation by dolomite elsewhere, probably down gradient.
7. Therefore, while karst development in dolomitic sediments and dolostones cannot be discounted, the preponderance of evidence suggests that development of karst in Florida dolomitic sediments and rocks is of substantially less importance than in calcitic sediments and rocks.

## 5.13 Other Karst-Related, Water-Quality Issues

In addition to development of karst through carbonate sediment and rock dissolution and possibly through dolomitization, there are several geochemical issues that affect human use of Florida's karstic aquifers. These are largely a result of groundwater quality issues and the "plumbing system" that develops in karstic aquifers.

### 5.13.1 Aquifer Storage and Recovery and the Arsenic Issue

Aquifer storage and recovery (ASR) of surplus water is a popular water-management strategy in Florida. Basically, highly treated wastewater and/or storm water is pumped into confined aquifer horizons for storage until needed. Storage must be within a horizon that is confined below aquifers utilized for potable water. The volume of stored water is conceptualized as a "bubble" even though it actually conforms to the distribution of pore and void space, including fractures, and hydraulic gradients in the injection zone (Vacher et al. 2006). When the need arises, the water is recovered from the "bubble", retreated, and distributed for potable use.

Upchurch et al. (1991) modeled the effects of an injection test of unbuffered, mildly acidic surface water in an ASR system in Manatee County. They demonstrated that the injected water dissolved calcite in the host aquifer, thus increasing porosity over time. However, the amount of void space enlargement over time was considered minimal and of little importance in terms of increased potential for development of karst-related problems. The dissolution was presumably on the surfaces of pre-existing void spaces. As it turns out, however, this small-scale dissolution results in an important water-quality issue.

Commonly, when ASR injectate is recovered in Florida there are elevated concentrations of arsenic in the water (Arthur et al. 2001, 2002, 2005, 2008; Pyne 2003, 2004; CH2M Hill 2007). Price and Pichler (2002, 2006) showed that the arsenic was incorporated in pyrite framboids (Fig. 2.39) that line many of the voids in the host aquifers. Injection of oxygenated water into these voids, which consist of fractures and karst openings (usually paleo-epikarst at or near unconformities), causes oxidation of the pyrite and release of arsenic to the injectate. Cyclic injection and recovery slowly exhausts the supply of pyrite framboids and arsenic, and many ASR wells continue to show long-term reductions in arsenic concentrations. Today, ASR facilities deal with the arsenic by chemical treatment, blending with low arsenic water, and reducing the oxidation potentials of injectate.

### 5.13.2 Sulfide and Sulfate Issues

Many areas around the world have problems with hydrogen sulfide ($H_2S$) in groundwater. Florida is no exception. The problems with hydrogen sulfide are (1) it forms a gas that imparts an unpleasant "rotten egg" odor to water, (2) it creates taste in water and water treatment issues, and (3) it may be toxic in high concentrations, especially as a gas. As discussed above, hydrogen sulfide in Florida aquifers has at least two sources:

1. Microbial sulfate reduction of the sulfate-rich waters in Florida's aquifers (i.e., Eq. 5.22); and
2. Oxidation of organic materials, humic substances and pyrite.

The processes of sulfate reduction and sulfide oxidation can cause karst development through creation of acidic waters that can dissolve carbonate minerals.

### 5.13.3 Nitrate Issues

Nitrate ($NO_4^-$) is not significantly involved in karst processes, but it is a major problem with Florida's most beloved karst features – springs. This section describes the geochemistry of nitrate and related nitrogen compounds in order to provide perspective on the problems with nitrate in springs.

Nuisance algae, odors, turbidity, and changes in the biota in springs have been linked to increases in the concentrations of $NO_3^{2-}$ in spring discharge. Sources of the $NO_3^{2-}$ include animal and human wastes and inorganic fertilizer use. The high $NO_3^{2-}$ in spring water is a result of surface-water runoff and flow to openings in karst aquifers, stream to swallet flow, and bank storage from major rivers in flood.

Nitrate is one member of a sequence of related nitrogen compounds that cycle through earth environments (Fig. 5.35). These other compounds include nitrogen gas ($N_2$), nitrogen dioxide gas ($NO_2$) and other oxides, ammonia and ammonium ($NH_3$, $NH_4^+$), nitrite ($NO_2$), a number of other inorganic compounds, and many organic compounds (Fig. 5.35). The gaseous phases exist in the atmosphere (Fig. 5.35) and in soil atmospheres, but are not of importance in the water-saturated zones of aquifers. Ammonia gas also escapes into the atmosphere. Ammonia is generally present in groundwater as the ammonium ion ($NH_4^+$) because of prevalent pH and reduction-oxidation potentials (the ammonium ion is a product of reducing environments). The complex organic compounds can occur as soluble organic molecules and as particulates.

Organic nitrogen, ammonium, nitrite, and nitrate are the compounds considered important in Florida groundwater systems. These compounds are related through a sequence of microbially driven reduction and oxidation reactions as indicated below:

$$\begin{gathered} Organic\ nitrogen \rightleftarrows NH_4 \rightleftarrows NO_2^- \rightleftarrows NO_3^- \\ ---Oxidation--\rightarrow \\ \leftarrow--Reduction--- \end{gathered} \tag{5.26}$$

with oxidation being the normal sequence in groundwater systems. The reduction/oxidation transformations between the compounds indicated in reaction Eq. 5.26 can be driven

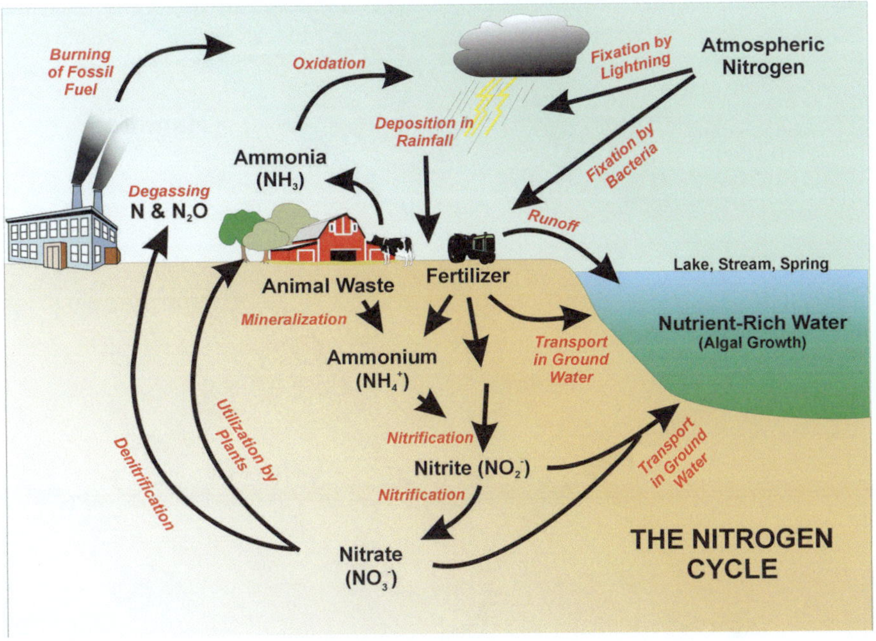


**Fig. 5.35** The nitrogen cycle. From Upchurch et al. (2007)

by inorganic processes, but the primary mechanisms for the reactions are microbial.

The largest reservoir of nitrogen is the atmosphere, which is 78% nitrogen, mostly as $N_2$ gas. $NH_3$ and $NO_2$ occur naturally in the atmosphere as a result of releases by terrestrial plants (Stallard and Edmond 1981). Atmospheric nitrogen is also converted to nitrogen oxides ($NO_X$) by lightning. As noted in Sect. 4.8, modern precipitation contains increased nitrogen oxides as a result of combustion of fossil and modern organic fuels. The oxides of nitrogen are then converted by oxidation and hydrolysis to nitric acid ($HNO_3$), which dissociates to $H^+$ and $NO_4^-$. Consequently, precipitation is a source of acidity and nitrate and ammonium derived from both natural and anthropogenic causes.

The concentration of nitrate in precipitation in Florida ranges from 0.00 to 10.32 mg/L (Upchurch 1992), and the statewide mean is 0.97 mg/L (Table 5.1). Ammonium ion ranges from 0.00 to 17.12 and the mean is 0.17 mg/L.

The nitrogen compounds in the atmosphere are transferred through precipitation to the groundwater system. Elevated nitrogen content of modern rainfall, however, cannot fully explain high nitrogen in most aquifers, a common problem in Florida's springs. This is because of the long time intervals involved in groundwater flow. Waters in surficial environments, including the SAS and shallow, unconfined portions of the UFA, are more likely to be affected by modern, high nitrogen precipitation. Deeper waters, in contrast, were recharged as meteoric waters before the advent of the industrial revolution. These older waters, which are potentially hundreds or thousands of years old, entered the aquifer system with some natural nitrogen content but at much lower concentrations than the present.

Certain microbes can fix nitrogen gas in soils. These microbes, in conjunction with plants such as the legumes, directly convert nitrogen into tissues and nitrogenous byproducts (Fig. 5.35). Plants require nitrate as a major nutrient, and they are responsible for removal of much of the nitrate that is taken from soils and groundwater. Average nitrogen content of living organisms is about 16 percent. These living tissues contain amino acids and other nitrogen compounds that can be released back into the environment upon death or waste elimination.

Animal wastes and decaying plant tissues release ammonia and ammonium, nitrite, nitrate, urea, and a number of nitrogenous organic molecules. Soil and aquifer microbes metabolize these according to the reduction-oxidation potentials of the soils and aquifers. Under reducing conditions, microbes convert these compounds to ammonium, and other reduced nitrogen species. Under oxidizing conditions, they are converted to nitrate, often with an intermediate nitrite step. Therefore, in reducing environments, such as water-saturated, reducing soils and aquifers, ammonium may persist and become a part of the ground-water system. Under these circumstances, ammonium can travel considerable distances before sorption, microbe metabolism, dilution, or dispersion reduce concentrations to below detection limits. Ammonium tends to sorb onto clays and soil particles, so some soil and aquifer materials mitigate ammonium migration. Septic tank systems, land-application waste-treatment systems, and feed-lot wastes can, under circumstances of overloading or failure of sorption systems, cause widespread ammonium contamination.

Oxidizing conditions are necessary for microbes to produce the complex reactions required to make the nitrogen

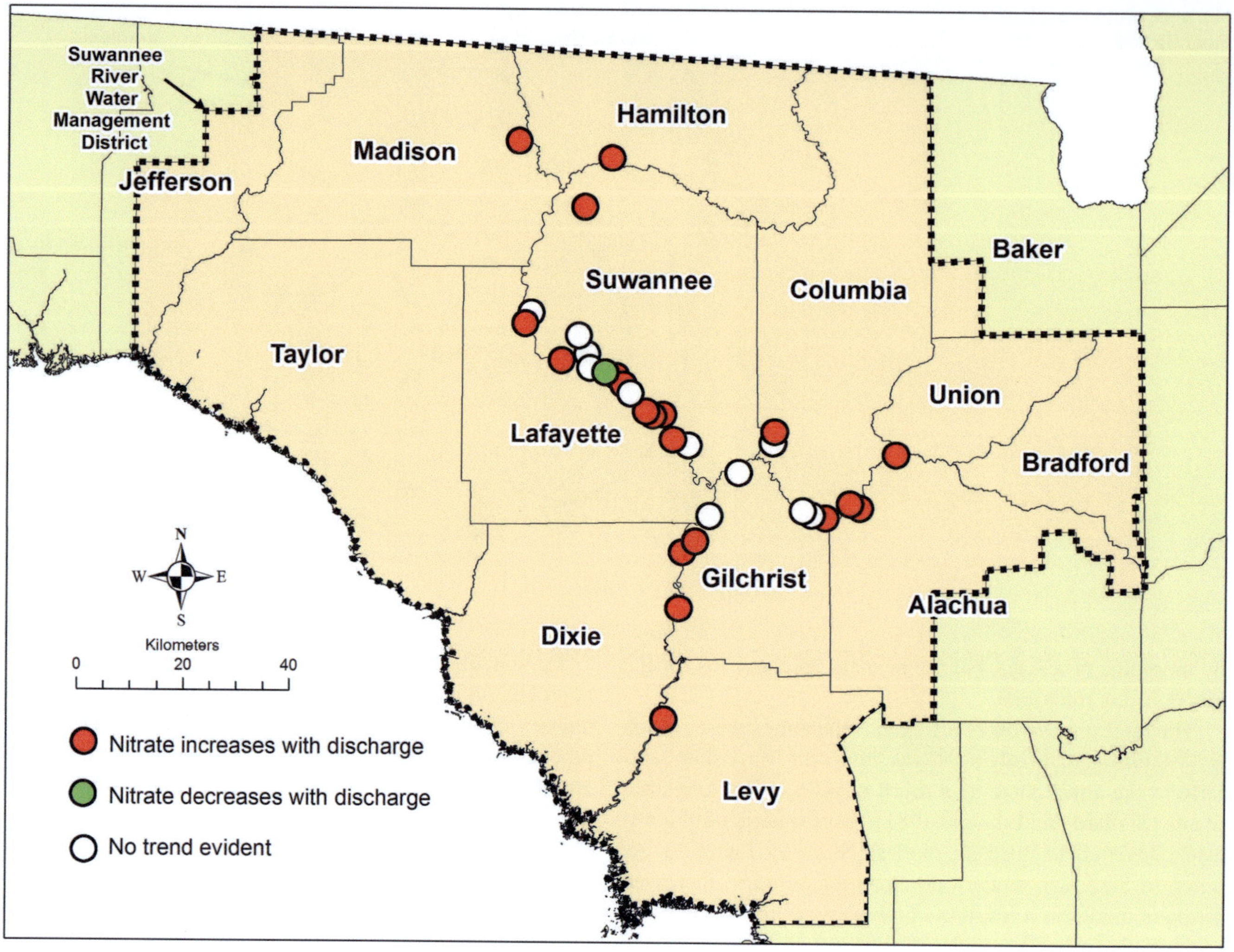


**Fig. 5.36** Relationships of spring discharge and nitrate concentrations in north Florida springs. (From Upchurch (2015))

useable for plants (Fig. 5.37). These aerobic microbes convert the ammonium and complex organic nitrogen molecules to nitrite and then nitrate. Ammonium and organic-nitrogen compound concentrations are low in most aerobic aquifers because oxidizing conditions are widespread near the land surface, where these nitrogen compounds are generated. Oxidizing conditions occur in oxygenated soils, vadose environments and in shallow, oxygenated portions of aquifers.

If nitrate is available in small amounts near the land surface, plants will utilize it. There are also microbes that denitrify soils by conversion of nitrate to nitrogen gas. If nitrate production from ammonium and more complex nitrogen compounds is not completed within the root zone, if the nitrate is unavailable to plants and denitrifying microbes, or if nitrate is produced in quantities too great for biological agents to fix, nitrate migrates with the groundwater. With the exception of plant and microbial activity, there are few mechanisms for nitrate removal in aquifers. Once nitrate enters the aquifer and is isolated from environments where denitrification and plant fixation occur, nitrate behaves more-or-less conservatively and can move long distances in aquifers (Fig. 5.35), and that is the problem.

Most of Florida's springs are showing patterns of increasing nitrate concentrations with time (Florida Springs Task Force 2000; Florida Department of Environmental Protection 2007; Copeland et al. 2011). The result has been uncontrolled algal growth, eutrophication, and other unwanted conditions in the springs (Mattson et al. 2007), accompanied by human health and economic impacts. In fact, several of the springs that discharge to the Suwannee River routinely exceed the federal water-quality standard of 10 mg/L, which is based on risk of methemoglobinemia, or "blue-baby syndrome."

For microbial decomposition of nitrogenous compounds to occur, there must be a source of organic carbon, and other nutrients. The role of nitrogen-utilizing microbes in deep aquifers has not been adequately evaluated. It appears that microbial transformations analogous to sulfate reduction (Eq. 5.22) may occur. Availability of organic carbon and

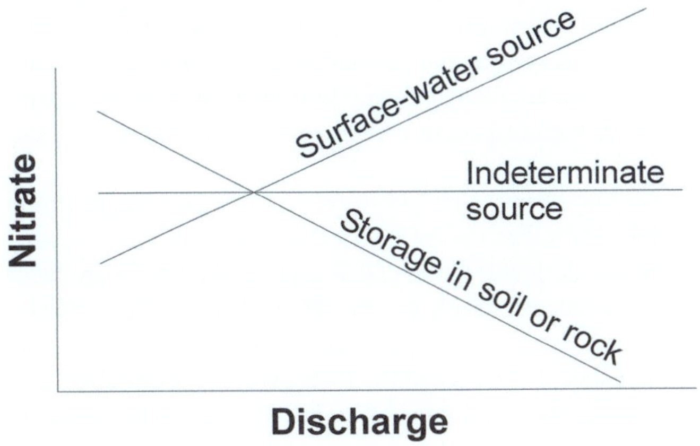


**Fig. 5.37** Conceptual model for interactions of nitrate and discharge in north Florida springs. Lines represent correlation trends. (From Upchurch (2015))

nitrogen compounds is limited in deeper portions of the FAS, so nitrogen-utilizing microbes are probably ineffective in the same way as are sulfate-reducing microbes. Our present concepts suggest that the majority of nitrogen fixation occurs in shallow, oxidizing aquifers and soils.

The presence of nitrate and the other nitrogenous compounds in groundwater is not a result of interaction of aquifer system water with surrounding rock materials. Nitrate in groundwater is a result of specific land uses associated with fertilization and waste disposal. If the land use is widespread, a body of nitrate-enriched water that is large enough to be contoured on a map may result. Otherwise, detection of elevated nitrate concentrations in groundwater is an isolated phenomenon.

Nitrate contamination of groundwater is a major concern in Florida. Numerous areas of the state have reported nitrate problems in groundwater. These problems are associated with areas of intense agriculture use and suburban housing. Some of the areas where nitrates are of concern include dairies and cattle ranches in the Suwannee and the Kissimmee River valleys, crop lands in the northern Everglades, and suburbs served by on-site waste treatment (septic tanks) throughout the state.

At present, the state and local agencies are dealing with the nitrate contamination problem by setting a maximum target concentration of nitrate in water of 0.35 mg/L (Harrington et al. 2010), developing total maximum daily loads (TMDLs) for karstic springsheds and other groundwater and surface water basins, and developing best management practices for agricultural and other industries. The FDEP first proposed the nitrogen threshold of 0.35 mg/L for nitrogen in clear-water streams, including spring vents. This target concentration is applicable to nitrate and nitrate plus + nitrite). Based on median nitrate plus nitrite concentrations, 36 of the 49 springs (73%) monitored by the FDEP exceed this proposed threshold. It will be interesting to see if this lofty goal can be met given that the statewide median nitrate concentration in rainfall was 0.70 mg/L for the period from 1978 to 2012 (Table 5.1).

Upchurch (2015) investigated the relationships between discharge, specific conductance, and nitrate in the discharge of 40 springs on the Suwannee and Santa Fe Rivers (Fig. 5.36) in northern Florida. He hypothesized that the relationships between spring discharge, specific conductance (a surrogate for total dissolved solids and an indicator of the amount of limestone dissolution), and nitrate (actually $NO_3^- + NO_2^-$) would indicate the modes of recharge and flow to the spring and the method of introduction of the nitrate.

Figure 5.37 depicts the hypothesized relationships of nitrate and spring discharge. If nitrate concentrations increase as discharge increases, the nitrate is contained in a surface-water source. A negative correlation suggests that the nitrate is stored in the soil or rock column and is diluted by conduit flow. If no correlation between discharge and nitrate exists, the origin of the nitrate cannot be identified.

Many of the springs in the Suwannee River basin are estavelles that take water when river levels are high. This form of "bank storage" complicates interpretation of the conduit-dominated models. If the spring discharges relatively elevated nitrate concentrations without high river levels, one can assume that the nitrate is derived from within the springshed and transported to the spring by conduit flow. However, if the elevated nitrate, low specific conductance discharge occurs in the waning phases or shortly after an episode of high river stage during which bank storage and estavelle action in the spring occurred, then the surface-water source may not be within the springshed.

Fifty percent of the springs show that nitrate concentrations increase as discharge from the spring increases (Fig. 5.36). Forty-five percent of the remaining springs showed no correlation between discharge and nitrate, and only 5% (2 springs with poor data) have relationships where high discharge was related to lower nitrate concentrations. Twenty percent of the springs had positive correlations of specific conductance with discharge, 37% showed no correlation, and 43% had negative correlations between specific conductance and discharge. Most important in terms of understanding the plumbing of the conduit systems, 40% of the springs had positive correlations between nitrate and specific conductance, 48% showed no correlation, and 12% had negative correlations.

Upchurch (2015) concluded that the positive correlation of discharge and nitrate concentrations indicated that the primary sources of nitrate in the springs is a result of surface water influx from runoff, sinking streams, and estavelle storage in springs marginal to the rivers. Therefore, while the nitrate problem is a major issue for Florida karst springs, it is an artifact of (1) conduiting and storage in the karst aquifer and (2) internal drainage and bank storage as a result of high rainfall events.

### 5.13.4 Ion Exchange Reactions

Ion exchange is a widespread phenomenon in aquifer systems deposited in coastal plain environments (Foster 1950). Clays, particularly smectites, have high ion-exchange capacities. The ion exchange reactions of sodium and calcium with a clay sorption (ion exchange) site can be characterized as follows:

$$\underset{\text{Sodium-saturated clay}}{\begin{matrix} Na \\ Na \end{matrix} > clay} + Ca^{2+}_{aq.} \rightleftarrows \underset{\text{Calcium-saturated clay}}{Ca = clay} + 2Na^{+}_{aq.} \qquad (5.27)$$

If deposited in seawater, clays with high ion-exchange capacities are initially saturated with sorbed sodium, which is loosely held on clay-mineral surfaces by electromagnetic forces (Fig. 2.40). With a higher positive charge and smaller ionic radii, calcium and magnesium have a higher sorption affinity for clay surfaces than sodium, so when calcium- or magnesium-rich groundwater bathes sodium-saturated clay, ion exchange is likely to occur. Calcium or magnesium exchange for sodium and the clay becomes calcium- and/or magnesium- saturated, while the water is enriched in sodium. The reverse reaction occurs upon salt-water intrusion into calcium or magnesium rich clay. Even though the sorption potential of sodium is low relative to calcium or magnesium, the high concentrations of sodium in seawater commonly cause exchange with a release of calcium and magnesium and a loss of sodium in the groundwater.

The exact effect of ion-exchange processes on karst development is unknown, but calcium enrichment or depletion in groundwater may affect chemical mass balance calculations and estimates of karst development rates and processes.

Upchurch (1992) identified several areas in the SAS, IAS, and FAS in Florida where Na-$HCO_3$ facies water was detected as part of the Florida Background Water-Quality Network (Maddox et al. 1992). Many of the areas where this relatively unusual hydrochemical facies was present were in proximity to the clay-rich Hawthorn Group.

## 5.14 Summary

The hydrogeochemistry of Florida's aquifer is complex. It is important to understand the following:

- Carbonate equilibrium reactions dominate water quality in limestone aquifers.
- Chemical complexing, ion pairing, common ion effects, and dissolved humic substances affect chemical equilibrium and are likely to result in somewhat more aggressive water with respect to carbonate minerals than simple carbonate equilibria may predict.
- Mixing of two different water types at groundwater mixing zones, such as on the freshwater/saline water mixing zones along the coasts and at depth in the FAS, may cause speleogenesis. With the Plio-Pleistocene sea-level fluctuations, artifacts of mixing-zone speleogenesis are likely distributed throughout the FAS. Artifacts of mixing-zone speleogenesis are most likely best developed as a result of sea-level still-stands.
- The high organic content of Florida surface water and groundwater contributes to the ability of the water to cause dissolution.
- Metastable minerals are present in Florida's youngest carbonate sediments and aquifers. These minerals include aragonite and high-magnesium calcite. Sediments and rocks with high contents of metastable minerals are likely to develop dissolution features during diagenesis and/or the earliest stages speleogenesis.
- Sediment grain size affects solubility of carbonates. Fine-grained particles are more soluble than coarse sediments because of their high surface area relative to particle volume. Carbonate mud (micrite composed of silt- and clay-sized particles) equilibrates with the surrounding water quickly.
- As a result, water contained in low hydraulic conductivity, clay-rich sediments is normally in chemical equilibrium with the carbonate phases present in the sediment. The rapid equilibration at shallow depths results in preservation of shell and carbonate mud at depths in clay-rich sediment bodies.
- Florida rainfall is dominated by inclusion of marine aerosols. As a result, the rainfall is a dilute solution of $Na^+$ and $Cl^-$. The ratios of $Na^+$ to $Cl^-$ in rainfall, meteoric water and the quartz-sand facies of the SAS are similar to the ratio in sea water.
- In the quartz sand-rich SAS, meteoric water and groundwater are composed of Na-Cl facies water. This fingerprint is commonly masked in the sandy SAS in Florida by introduction of Ca-$HCO_3$ facies water by irrigation with water from the FAS.
- Limestone and shell aquifers predominantly contain Ca-$HCO_3$ facies water. If dolostone is present, the facies is typically Ca-Mg-$HCO_3$.
- Sea-level fluctuations over the last few million years have resulted in residual waters in the FAS that are likely rich in sodium and chloride. Connate water remaining from the time of sediment deposition is also apparently present.

- Evaporation and transpiration, biotic activity in soils and sediments, reduction/oxidation reactions, and generation of humic substances by plant decay affect water quality in Florida streams and aquifers.
- Water in Florida aquifers frequently contains organic compounds (humic substances such as dissolved humic and fulvic acids and colloidal humins) that exceed published "norms." The dissolved organic fractions are capable of causing carbonate-mineral dissolution. Sequestering of $Ca^{2+}$ and $Mg^{2+}$ by chemical complexing with humic substances may double the ability of water to cause calcite dissolution.
- The FAS contains significant water masses rich in $SO_4^{2-}$ as a result of sulfide oxidation and dissolution of gypsum and/or anhydrite along the upper surface of the MCU.
- Mixing of Ca-$HCO_3$ and Ca-Mg-$HCO_3$ facies water with either Ca-$SO_4$-$HCO_3$ or Na-Cl facies water along freshwater/saline water transitions zones is likely to cause dissolution of the host carbonate aquifer.
- There has been considerable dolomitization within Florida's aquifers. Whether karst has affected the dolostone deposits is debatable. Most likely, dissolution is a minor process in dolomitic strata.
- There have been at least two episodes of diagenetic alteration of calcite and/or aragonitic sediments in the FAS. Geochemical evidence does not indicate significant dissolution of dolostone in the FAS; however, it has occurred. It appears that most of the high permeability zones in the LFA are a result of fracturing more than dissolution.
- Several groundwater-quality issues are of concern in Florida.
  - Aquifer storage and recovery of highly treated wastewater and of surface water are of concern because of arsenic released into the injectate plume at depth. When this water is recovered, the arsenic may result in additional treatment of the water to render it safe.
  - The potential for creation of secondary porosity by chemically aggressive injectate dissolving the carbonate host rock does not appear to be an issue. Dissolution of the host rock has been documented but is apparently volumetrically of little importance.
  - $H_2S$ dissolved in groundwater is widespread in the SAS, IAS, and FAS. It is derived from reduction of $SO_4^{2-}$ and degradation of organics. Because of the unpleasant odor, possible adverse reactions to consumption, and damage to plumbing and other infrastructure, $H_2S$ is a concern for potable use. "Sulphur springs", however, have been touted as being beneficial to your health in the past in Florida. $H_2S$ is dealt with by (1) avoiding use of the water and (2) treatment.
- Nitrate is a major concern relative to potential for eutrophication of Florida springs. Nuisance algae, odors, turbidity, and changes in the biota in springs have been linked to increases in the concentrations of $NO_3^{2-}$ in spring discharge. Sources of the $NO_3^{2-}$ include animal and human wastes and inorganic fertilizer use. The high $NO_3^{2-}$ in spring water is a result of surface-water runoff and flow to openings in karst aquifers, stream to swallet flow, and bank storage from major rivers in flood.
- Na-$HCO_3$ facies water is found in Florida aquifers as a result of ion exchange of $Ca^{2+}$ for $Na^+$ sorbed on clay surfaces. This water is not a hazard, but represents a relatively rare geochemical process.

## References


Aiken GR Brown PA Noyes TI Pinckney DJ (1994) Molecular size and weight of fulvic and humic acids from the Suwannee River. In: Averett RC Leenheer JA McKnight DM Thorn KA (eds.), Humic substances in the Suwannee River, Georgia: Interactions, properties, and proposed Structures, US Geol Surv Water-Sup Pap 2373, pp. 89–97

Anesio AM Graneli W Aiken GR Kiebler DJ Mopper K (2005) Effect of humic substance photodegradation on bacterial growth and respiration in lake water. App and Environ Microbiol 71:6267–6275

Arthur JD Cowart JB Dabous AA (2001) Florida aquifer storage and recovery geochemical study: Year three progress report. Fla Geol Surv Open File Rept 83

Arthur JD. Dabous AA Cowart JB (2002) Mobilization of arsenic and other trace elements during aquifer storage and recovery, southwest Florida. In: Aiken GR and Kuniansky EL (eds.), US Geol Surv Open-File Rept 02-89, pp. 47–50

Arthur JD Dabous AA Cowart JB (2005) Water-rock geochemical considerations for aquifer storage and recovery: Florida case studies. In: Tsang CF Apps JA (eds.), Underground injection science and technology, Develop Water Sci, Amsterdam, Elsevier, 52(24):327–339

Arthur JD Fischler C Dabous AA (2008) Bench-scale geochemical assessment of water-rock interactions: Orange County aquifer storage and recovery facility. Fla Geol Surv, contract report submitted to Barnes, Ferland and Associates, Inc., Orlando, FL

Back W (1961) Techniques for mapping of hydrochemical facies. US Geol Surv Short papers in the Geol and Hydro Sci, Pap 423, p. D-380–D382

Back W Hanshaw BB (1970) Comparison of chemical hydrogeology of the carbonate peninsulas of Florida and Yucatan. Jour Hydrol 10:330–368

Back W Hanshaw BB (1971) Rates of physical and chemical processes in a carbonate aquifer. In: Hem JD (ed.), Nonequilibrium systems in natural water chemistry, Amer Chem Soc, Advan Chem Ser No. 106, pp. 77–93

Back W Hanshaw BB Herman JS Van Driel JN (1986) Differential dissolution of a Pleistocene reef in the ground-water mixing zone of coastal Yucatan, Mexico. Geology 14:137–140

Back W Hanshaw BB Pyle TE Plummer LN Weidie AE (1979) Geochemical significance of groundwater discharge to the formation of Caleta Zel Ha, Quintana Roo, Mexico. Water Resour Res 15(6): 1521–1535

Badiozomani K (1973) The dorag dolomitization model—Application to the Middle Ordovician of Wisconsin. Jour Sed Pet 43:965–968

Berner RA (1966) Chemical diagenesis of some modern carbonate sediments. Amer Jour Sci 24:1–36

Blatt H Middleton G Murray R (1972) Origin of sedimentary rocks. Englewood Cliffs, NJ Prentice-Hall, Inc.

Bock WD Moore DR Neumann AC Supko PR (comp.) (1969). Late Pleistocene geology in an urban area. Miami Geol Soc field trip guidebook, accessed 3/20/2012 at http://sofia.usgs.gov/publications/reports/mgs_bock1969/mgs_bock1969.pdf

Boggs S Jr (2011) Principles of sedimentology and stratigraphy. Upper Saddle River, NJ, Prentice Hall, 5th Ed

Bögli A (1964) Érosion par Melange des Eaux. Inter Jour Speleol 1:61–70 (Translated as Mixed-water corrosion in Sweeting MM (1981) Karst geomorphology, Benchmark Papers in Geology, 59:90–99.)

Bögli A (1980) Karst hydrology and physical speleology. Berlin, Springer-Verlag

Bond P Campbell KM Scott TM (1986) An overview of peat in Florida and related issues. Fla Geol Surv Sp Pub 27

Bottrell SH Carew JL Mylroie JE (1993) Bacterial sulphate reduction in flank margin environments: Evidence from sulphur isotopes. In: White WB (ed.), 6th Symposium on the Geology of the Bahamas, Port Charlotte, FL, Proc Bahamian Field Station, pp. 17–21

Bottrell SH Smart PL Whitaker F Raiswell F (1991) Geochemistry and isotope systematics of sulphur in the mixing zone of Bahamian blue holes. App Geochem 6:99–103

Brown DE (1989) The role of natural organics in a groundwater system. M.S. thesis, Tampa, Univ South Fla

Brown DP (1984) Impact of development on availability and quality of ground water in eastern Nassau County, Florida, and southeastern Camden County, Georgia. US Geol Surv Water-Resour Invest Rept 83-4190

Bryant JT Morris DV Sweeney SP Gehrig MD Mathis JD (2001) Tree root influence on soil-structure interaction in expansive clay soils. In: Vipulanandan C Addison MB Hasen M (eds.), Expansive clay soils and vegetative influence on shallow foundations. Amer Soc Civil Eng, Geotech Sp Pub 115, pp. 110–131

Budd DA (1988) Aragonite to calcite transformation during freshwater diagenesis of carbonates: Insights from pore-water chemistry. Geol Soc Amer Bull 100:1260–1270

Budd DA (1997) Cenozoic dolomites of carbonate islands: their attributes and origin. Earth Sci Rev 42(1–2):1–47

Budd DA (2001) Permeability loss with depth in the Cenozoic carbonate platforms of west-central Florida. Amer Assoc Petrol Geo Bull 85:1253–2172

Budd DA Vacher HL (2004) Matrix permeability of the confined Floridan Aquifer, Florida, USA. Jour Hydro 12(5):531–549

Buol SW Hole FD McCracken RJ (1973) Soil genesis and classification. Ames, Univ Iowa Press

Cander HS (1994) An example of mixing zone dolomite, Middle Eocene Avon Park Formation, Florida aquifer system. Jour Sed Res A64:615–629

Canter LW (1997) Nitrates in groundwater. Boca Raton, FL, CRC Press

Carew JL Mylroie JE (1987) Submerged evidence of Pleistocene low sea levels on San Salvador, Bahamas. Symp Undersea Res, NOAA, 2(2):167–175.

Carew JL Mylroie JE (1995) Quaternary tectonic stability of the Bahamian Archipelago: evidence from fossil coral reefs and flank margin caves. Quat Sci Rev, 14:145–153

Carballo JD Land LS Miser DE (1987) Holocene dolomitization of supratidal sediments by active tidal pumping. Jour Sed Pet 57: 153–165

Carr WJ Alverson DC (1959) Stratigraphy of Middle Tertiary rocks in part of west-central Florida. US Geol Surv Bull 1092

CH2MHill (2007) Arsenic mobilization in two Suwannee Limestone aquifer storage recovery systems. Contract rept sub to the Southwest Florida Water Management District, Brooksville, Florida, accessed July 2, 2014 at https://www.swfwmd.state.fl.us/documents/reports/asr_arsenic_suwannee.pdf

Chapelle FH (2000) Ground-water microbiology and geochemistry. New York, John Wiley & Sons, 2nd Edition

Chave KE Schmalz RF (1966) Carbonate-seawater interactions. Geochim Cosmochim Acta 30:1037–1048

Choquette PW Hiatt EE (2008) Shallow-burial dolomite cement: A major component of many ancient sucrosic dolomites. Sediment 55:423–460

Copeland R Dorian NA White AJ Upchurch SB (2011) Regional and statewide trends in Florida's spring and well groundwater quality (1991–2003). Fla Geol Surv Bull 69 (Revised)

Dalton MG Upchurch SB (1978) Interpretation of hydrochemical facies by factor analysis. Ground Water 16:228–233

Davis JH Jr (1946) The peat deposits of Florida. Fla Geol Surv Bull 30

Davis SN DeWiest RJM (1966) Hydrogeology. New York, John Wiley & Sons

Deffeyes KS Martin EL (1962) Absence of carbon-14 activity in dolomite from Florida Bay. Science 136: 782

Drever JI (ed.) (1985) The chemistry of weathering. Dordrecht, D. Reidel Pub Co

Drever JI (1997) The geochemistry of natural waters. Englewood Cliffs, NJ, Prentice Hall, Inc.

Dreybrodt W (2000) Chemistry of speleogenetic processes. In: Klimchouk AB Ford DC Palmer AN Dreybrodt W (eds.), Speleogenesis: Evolution of karst aquifers. Huntsville, AL, Nat Speleol Soc

Enoch H Dasberg S (1971) The occurrence of high $CO_2$ concentrations in soil air. Geoderma 6:17–21

Faimon J Ličbinska M Zajíček P Sracek O (2012) Partial pressures of $CO_2$ in epikarstic zone deduced from hydrogeochemistry of permanent drips, the Moravian Karst, Czech Republic. Acta Carsologica 41(1):47–57

Fanning KA Byrne RH Breland JA Betzer PR Moore WS Elsinger RJ Pyle TE (1981) Geothermal springs of the west Florida continental shelf: Evidence for dolomitization and radionuclide enrichment. Earth Planet Let 52:345–354

Faulkner G L (1973) Geohydrology of the Cross Florida Barge Canal area with special reference to the Ocala vicinity. US Geol Surv Water Resour Invest 1-73

Florida Department of Environmental Protection (2007) Florida springs initiative: Program summary and recommendations. Tallahassee, Fla Dept Environ Prot

Florida Springs Task Force (2000) Florida's springs: Strategies for protection & restoration. Rept prep for Fla Dept Environ Prot

Folk RL Land LS (1972) Mg/Ca vs salinity: A frame of reference for crystallization of calcite, aragonite, and dolomite. Geol Soc Amer Abstracts with Prog 4(7):508

Ford DC Williams PW (1989) Karst geomorphology and hydrology. New York, John Wiley & Sons

Foster MD (1950). The origin of high sodium bicarbonate waters in the Atlantic and Gulf coastal plains. Geochim Cosmochim Acta 1:33–48

Frank EF Mylroie JE Troester J Alexander EC Jr Carew JL (1998) Karst development and speleogenesis, Isla de Mona, Puerto Rico. Jour Cave & Karst Stud 60(2):73–83

Fratesi B (2013) Hydrology and geochemistry of the freshwater lens in coastal karst. In: Lace MJ Mylroie JE (eds.), Coastal Karst Landforms, Dordrecht, Springer Science, Coastal Res Lib, Book 5, Ch 3, pp. 59–76

Frazee JM Jr McClaugherty DR (1979) Investigation of ground water resources and salt water intrusion in the coastal areas of northeast Florida. St. Johns River Water Mgt Dist Tech Pub SJ 80–4

Garman KM (2010) The biogeochemistry of submerged coastal karst features in west central Florida. Ph.D. dissert, Univ South Fla

Garman KM Garey JR (2005) The transition of a freshwater karst aquifer to an anoxic marine system. Estuaries 28(5):686–693

Garman KM Rubelmann H Karlen DJ Garey JR (2011) Comparison of an inactive submarine spring with an active nearshore anchialine spring in Florida. Hydrobiol 677:67–87

Garrels RM Christ CL (1965) Solutions, minerals, and equilibria. New York, Harper & Row

Gaswirth SB Budd DA Farmer LG (2007) The role and impact of regionally extensive freshwater-seawater mixing zones in the maturation of regional dolomite bodies within the proto-Floridan aquifer. Sedimentol 54:1065–1092

Graf DL Goldsmith JR (1956) Some hydrothermal synthesis of dolomite and protodolomite. Jour Geol 64:173–186

Gulley JD Martin JB Brown A (2016) Organic carbon inputs, common ions, and degassing: Rethinking mixing dissolution in coastal eogenetic carbonate aquifers. Earth Surf Process and Landforms, 41:2098–2110

Gulley JD Martin JB Moore PJ Murphy J (2013) Formation of phreatic caves in an eogenetic karst aquifer by $CO_2$ enrichment at lower water tables and subsequent flooding by sea level rise. Earth Surf Process and Landforms 38: 1210–1224

Gulley JD Martin JB Spellman P Moore PJ Screaton EJ (2012) Influence of partial confinement and Holocene river formation on groundwater flow and dissolution in the Florida carbonate platform. Hydrol Processes, 28:705–717

Gulley JD Polk JS (2017) Hypogene karst influences in the upper Floridan aquifer. In: Klimchouk AB Palmer AN Waela JD Auler AS Audra P (eds.), Hypogene karst regions and caves of the world, Springer Intern Publ, Ch. 50, pp. 745–755

Haberfeld JL (1991) Hydrogeology of effluent disposal zones, Floridan aquifer, south Florida. Ground Water 29(2):186–190

Haitzer M Aiken GR Ryan JN (2003) Binding of mercury(II) to aquatic humic substances: Influence of pH and source of humic substances. Environ Sci Tech 37(11):2436–2441

Hanshaw BB Back W (1971) A geochemical hypothesis for dolomitization by ground water. Econ Geol 66:710–724

Hanshaw BB Back W (1972) On the origin of dolomites in the Tertiary aquifer of Florida. In: Puri HS (ed.), Proc Seventh Forum on Geol Indust Minerals, Fla Bur Geol Sp Pub No. 17, pp. 139–153

Hanshaw BB Back W (1979) Major geochemical processes in the evolution of carbonate-aquifer systems. In: Back W Stephenson DA (eds.), Contemporary hydrogeology – The George Burke Maxey memorial volume. Jour Hydrol 43: 287–312

Hanshaw BB Back W (1980) Chemical mass-wasting of the northern Yucatan Peninsula by groundwater dissolution. Geology 8:222–224

Hanshaw BB Back W Rubin M (1965) Radiocarbon determinations for estimating groundwater flow velocities in central Florida. Science 148: 494–495

Harrington D Maddox G Hicks R (2010) Florida springs initiative monitoring network report and recognized sources of nitrate. Tallahassee, Fla Dept Environ Prot, Bur Watershed Restor, Ground Water Prot Sect

Harrison WB III (1969) Epigenetic growth of calcite-cemented nodules within a porous dolomite matrix – Avon Park Formation of central Florida. Masters thesis, Tampa, Univ South Fla

Hem JD (1985) Study and interpretation of the chemical characteristics of natural water. US Geol Surv Water-Sup Pap 2254, 3rd Edition

Hendry CD Brezonik PL (1984) Chemical composition of softwater Florida lakes and their sensitivity to acid precipitation. Water Resour Bull 20(1):75–86

Herbert TA Upchurch SB (2016) The potential role of hypogene speleogenesis in the lower Floridan aquifer and Sunniland Oil Trend, south Florida, U.S.A. In: Chavez T Reehling P (eds.), Proc DeepKarst 2016: Origins, resources, and management of hypogene karst, Carlsbad, NM, Nat Cave and Karst Res Inst, Symp 6, pp. 119–129

Hill CA (2000) Overview of the geologic history of cave development in the Guadalupe Mountains, New Mexico. Jour Cave and Karst Stud 62(2):60–71

Hoch AR Reddy MM Aiken GR (2000) Calcite crystal growth inhibition by humic substances with emphasis on hydrophobic acids from the Florida Everglades. Gechim Cosmochim Acta 64(1):61–72

Huang EH Keller WD (1971) Dissolution of clay minerals in dilute organic acids at room temperature. The Amer Mineral, 56:1082–1095

Huang EH Keller WD (1972a) Geochemical mechanics for the dissolution, transport, and deposition of aluminum in the zone of weathering. Clays and Clay Minerals 20:69–74

Huang EH Keller WD (1972b) Laboratory dissolution of plagioclase feldspars in water and organic acids at room temperature. Amer Mineral 57:1849–1859

Huang EH Keller WD (1972c) Organic acids as agents of chemical weathering of silicate minerals. Nature Phys Sci 239:149–151

Hunt CB Hunt AP (1957) Stratigraphy and archeology of some Florida soils. Geol Soc Amer Bull 68:797–806

Hutchinson CB (1978) Appraisal of shallow ground-water resources and management alternatives in the upper Peace and eastern Alafia River Basins, Florida. US Geol Surv Water-Res Invest 77-124

Illing LV Wells AJ Taylor JCM (1965) Penecontemporary dolomite in the Persian Gulf. In: Pray LC Murray RC (eds.), Dolomitization and limestone diagenesis: A symposium. Soc Econ Paleont Mineral, Sp Pub No. 13, pp. 89–111

International Humic Acids Society, 2014. Material safety data sheets for humic and fulvic acids. Accessed 4/15/2014 at https://ihss.humicsubstances.org/MSDS.html

Jenson JW Keel TM Mylroie JR Mylroie JE Stafford KW Taborosi D Wexel C (2006) Karst of the Mariana Islands: The interaction of tectonics, glacioeustasy and fresh-water/sea-water mixing in island carbonates. Geol Soc Amer Sp Pap 404, pp. 129–138

Kar S Maity JP Jean JS Liu CC Nath B Lee YC Bundschuh J Chen CY Li Z (2011) Role of organic matter and humic substances in the binding and mobility of arsenic in a Gangtic aquifer. Jour Environ Sci Health, Part A. Toxic/Hazard Subst and Environ Engin 46(11):1231–1238

Katz BG (1992) Hydrochemistry of the upper Floridan aquifer, Florida. US Geol Surv Water-Res Invest Rept 91-4196

Kelley K Mylroie JE Mylroie JR Moore C Moore PJ Collins L Ersek L Lascu I Roth M Passion R Shaw C (2006) Eolianites and karst development in the Mayan Riviera, Mexico. In: Davis RL Gamble DW (eds.), Proc 12th Symp on the Geol of the Bahamas and Other Carbonate Reg, Gerace Research Center, San Salvador, Bahamas, pp. 88–99

Kohout FA (1965) A hypothesis concerning cyclic flow of salt water related to geothermal heating in the Floridan aquifer. Trans New York Acad Sci 8:249–271

Kohout FA (1967) Ground-water flow and the geothermal regime of the Floridan Plateau. Trans Gulf Coast Assoc Geol Soc 17:339–354

Kovács K Gáspár A Sajgó CS Schmitt-Kopplin P Tombácz E (2011) Comparative study on humic substances isolated in thermal groundwater from deep aquifers below 700 m. Geochem Jour 46:211–224

Lawrence FW Upchurch SB (1976) Identification of geochemical patterns in ground water by numerical analysis. In Zaleem EA (ed.), Advances in groundwater hydrology, Amer Water Resour Assoc, p. 199–214

Lawrence FW Upchurch SB (1982) Identification of recharge areas using geochemical factor analysis. Ground Water 20:680–687

Leeder MR (2011) Sedimentology and sedimentary basins: From turbulence to tectonics. New York, Wiley-Blackwell, 2nd Ed

Leve GW (1983) Relation of concealed faults to water quality and the formation of solution features in the Floridan aquifer, northeastern Florida, U.S.A. Jour Hydrol 61:251–264

Levorsen AI (1967) Geology of petroleum. San Francisco, W.H. Freeman and Company, 2nd Ed

Lucia FJ (1995) Rock fabric/petrophysical classification of carbonate pore space for reservoir characterization. Amer Assoc Petrol Geolo Bull 79(9):1275–1300

Lucia FJ Major RP (1994) Porosity evolution through hypersaline reflux dolomitization. In: Purser BH Tucker ME Zenger DH (eds.), Dolomites, A Volume In Honour of Dolomieu, Inter Assoc Sediment Sp Pub No. 21, pp. 325–341

Maddox GL (1992) Data collection and management methods. In: Maddox GL Lloyd JM Scott TM Upchurch SB Copeland R (eds.), Florida ground water quality monitoring program - Vol 2, Background hydrogeochemistry, Fla Geol Surv Sp Pub No. 34, pp. 3–5

Maddox GL Lloyd JM Scott TM Upchurch SB Copeland R (1992) Florida's ground water quality monitoring program – Background hydrogeochemistry. Fla Geol Surv Spec Pub No. 34

Malcolm RL (1990) The uniqueness of humic substances in each of soil, stream, and maritime environments. Analy Chim Acta, 232:19–30

Malcolm RL McKnight DM Averett RC (1994) History and description of the Okefenokee Swamp – Origin of the Suwannee River. In Averett RC Leenheer JA McKnight DM Thorn KA (eds.), Humic substances in the Suwannee River, Georgia: Interactions, properties, and proposed Structures, US Geol Surv Water-Sup Pap 2373, Chapter A, pp. 1–12

Maliva RG Kennedy GP Martin WK Missimer TM Owosina ES Dickson JAD (2002) Dolomitization-induced aquifer heterogeneity: Evidence from the upper Floridan aquifer, southwest Florida. Geol Soc Amer Bull 114(4):419–427

Mattson RA Lehmensiek M Lowe EF (2007) Nitrate toxicity in Florida springs and spring-run streams: A review of the literature and its implications. St.Jjohns River Water Mgt Dist, Palatka, FL, Prof Pap SJ2007-PP1

McKnight DM Wershaw RL (1994) Complexation of copper by fulvic acid from the Suwannee River - Effect of counter-ion concentration. In: Averett RC Leenheer JA McKnight DM Thorn KA (eds.), Humic substances in the Suwannee River, Georgia: Interactions, properties, and proposed Structures, US Geol Surv Water-Sup Pap 2373, Chapter D, pp. 33–44

Mazzullo SJ Harris PM (1992) Mesogenetic dissolution: Its role in porosity development in carbonate reservoirs. Amer Assoc Petrol Geol Bull 76(5):607–620

McGee DK (2010) Microbial influences on karst dissolution: The geochemical perspective, with a chapter on assessment of the spreadsheets across the curriculum project. Ph.D. disser, Univ South Fla

Melim LA Swart PK Eberli GP (2004) Mixing-zone diagenesis in the subsurface of Floridan and the Bahamas. Jour Sed Res 74(6):904–913

Meyer FW (1989) Hydrogeology, ground-water movement, and subsurface storage in the Floridan aquifer system in southern Florida. US Geol Surv Prof Pap 1403-G

Miller JA (1978) Geologic and Geophysical Data from Osceola National Forest, Florida. US Geol Surv Open-File Rept 78-799

Morse JW Mackenzie FT (1990) Geochemistry of sedimentary carbonates. Amsterdam, Elsevier Sci Publ, Develop Sed, Vol. 48

Morse JW Zullig JJ Bernsteinm LD Millero FJ Milne P Mucci A Choppin GR (1985) Chemistry of calcium carbonate-rich shallow water sediments in the Bahamas. Amer Jour Sci, 285:147–185

Murphy EM Zachara JM (1995) The role of sorbed humic substances on the distribution of organic and inorganic contaminants in groundwater. Geoderma 67:103–124

Murray RC (1960) Origin of porosity in carbonate rocks. Jour Sed Pet 30(1):59–84

Myers RL Ewel JJ (eds.) (1990) Ecosystems of Florida. Orlando, Univ Central Fla Press

Mylroie JE Carew JL (1990) The flank margin model for dissolution cave development in carbonate platforms. Earth Surf Proc and Landforms 15:413–424

Mylroie JE Carew JL (1995) Karst development on carbonate islands. In: Budd DA Saller AH Harris PA (eds.), Unconformities in Carbonate Strata – Their Recognition and the Significance of Associated Porosity, Amer Assoc Petrol Geol Mem 63, pp. 55–76

Mylroie JE Carew JL Moore AI (1995a). Blue holes: Definition and genesis. Carbonates and Evaporites 10:225–233

Mylroie JE Carew JL Sealey NE Mylroie JR (1991) Cave development on New Providence Island and Long Island, Bahamas, Cave Science 18:139–151

Mylroie JE Carew JL Vacher HL (1995b) Karst development in the Bahamas and Bermuda. In: Curran HA White WB (eds.), Terrestrial and shallow marine geology of the Bahamas and Bermuda, Geol Soc Amer Sp Pap 300, pp. 251–267

Mylroie JE Jenson JW Taborosi D Jocson JMU Vann DT Wexel C (2001) Karst features of Guam in terms of a general model of carbonate island karst. Jour Cave and Karst Stud 63(1): 9–22

Mylroie JR Mylroie JE Jenson JW MacCracken RS (2005) Fresh-water lens anisotropy and flank margin cave development, Fais Island, FSM: Geol Soc Amer Abs with Progs, 37(7): 436

National Atmospheric Deposition Program (2012) Weekly precipitation chemistry data. Accessed 9/24/2012 at http://nadp.sws.uiuc.edu/data/ntndata.aspx

Noll S Tegeder D (2009) Ditch of dreams. Gainesville, FL, Univ Press of Fla

Palmer AN (1991) The origin and morphology of limestone caves. Geol Soc Amer Bull 103:1–21

Palmer AN (2007) Cave geology. Trenton NJ, Cave Books

Palmer AN Palmer MV Queen JM (1977) Geology and origin of the caves of Bermuda. In: Proc 7th Inter Speleol Cong, Sheffield England, pp. 336–339

Peek HH (1958) Ground-water resources of Manatee County, Florida. Fla Geol Surv Rept of Invest 18

Perry E Velazquez-Oliman G Marin L (2002) The hydrogeochemistry of the karst aquifer system of the northern Yucatan Peninsula, Mexico. Inter Geol Rev 44(5): 191 – 221

Piper AM (1944) A graphic procedure in the geochemical interpretation of water analyses. Amer Geophy Union Trans 25:914–923

Plummer LN (1975) Mixing of sea water with calcium carbonate ground water. In Whitten EHT (ed.), Quantitative Studies in Geological Sciences, Geol Soc Amer Mem 142, pp. 219–236

Plummer LN (1977) Defining reactions and mass transfer in part of the Floridan Aquifer. Water Resour Res 13:801–812

Plummer LN Back W (1980) The mass balance approach: Application to interpreting the chemical evolution of hydrologic systems. Amer Jour Sci 280:130–142

Plummer LN Vacher HL Mackenzie FT Bricker OP Land LS (1976) Hydrogeochemistry of Bermuda: A case history of ground-water diagenesis of biocalcarenites. Geol Soc Amer Bull 87: 1301–1316

Prasad S (1985) Microsucrosic dolomite from the Hawthorn Formation [Miocene] of Florida: Distribution and development. Masters thesis, Miami, FL, Univ Miami

Pray LC Murray RC (eds.) (1965) Dolomitization and limestone diagenesis: A symposium. Soc Econ Paleo Miner Sp Pub 13

Price RE Pichler T (2002) Naturally occurring arsenic in the upper Floridan aquifer, southwest Florida: Implications for aquifer storage recovery. Amer Water Resour Assoc, Proc AWRA Summer Specialty Conf, Ground Water/Surface Water Interactions, pp. 387–392

Price RE Pichler T (2006) Abundance and mineralogical association of arsenic in the Suwannee Limestone (Florida): Implication for arsenic release during water-rock interaction. Chem Geol 228:44–56

Pride RW Meyer FW Cherry RN (1961) Interim report on the hydrologic features of the Green Swamp area in central Florida. Fla Geol Surv Info Circ 26

Pride RW Meyer FW Cherry RN (1966) Hydrology of Green Swamp area in central Florida. Fla Geol Surv Rept Invest 42

Puri HS (1957) Stratigraphy and zonation of the Ocala Group. Fla Geol Surv Bull 38

Pyne RDG (2003) Water quality in aquifer storage recovery (ASR) wells. Amer Water Works Assoc (Florida Section), accessed 7/2/2014 at http://asrforum.com/fatestudy/Documents/ASRPoPa111503.pdf

Pyne RDG (2004) ASR dynamics, issues, and solutions. Amer Ground Water Trust IV, PowerPoint presentation accessed 7/2/2014 at http://www.dep.state.fl.us/geology/geologictopics/asr4/presentations/thursday_15/pyne_david.pdf

Qualls RG Takiyama A Wershaw RL (2003) Formation and loss of humic substances during decomposition in a pine forest floor. Soil Sci Amer Jour 67:899–909

Randazzo AF (1976) Petrographic and geohydrologic model of aquifer limestones in Florida. Gainesville, Univ Fla, Fla Water Resour Res Ctr Pub No. 35

Randazzo AF Bloom JI (1985) Mineralogical changes along the freshwater/saltwater interface of a modern aquifer. Sed Geol 43: 219–239

Randazzo AF Saroop HC (1976) Sedimentology and paleoecology of middle and upper Eocene carbonate shoreline sequences, Crystal River, Florida, U.S.A. Sed Geol 15:259–291

Randazzo AF Stone GC Saroop HC (1977) Diagenesis of middle and upper Eocene carbonate shoreline sequences, central Florida. Bull Amer Assoc Petrol Geol 61(4):492–503

Reddy MM Hoch AR (2000). Calcite crystal growth rate inhibition by aquatic humic substances. In: Amjad A (ed.), Adv Crystal Growth Inhibition Tech, New York, Kluwer Acad Pub, pp. 107–122

Rezaei M Sanz E Raeisi E Ayora C Vázquez-Sune E Carrera J (2005) Reactive transport modeling of calcite dissolution in the fresh–salt water mixing zone. Jour Hydrol 311(1–4): 282–298

Rightmire CT Hanshaw BB (1973) Relationship between carbon isotope composition of soil $CO_2$ and dissolved carbonate species in ground water. Water Resour Res 9:958–967

Rightmire CT Pearson FJ Jr Back W Rye RO Hanshaw BB (1974) Distribution of sulfur isotopes in ground waters from the principal artesian aquifer of Florida and the Edwards aquifer of Texas, United States of America. Isotope Tech Groundwater Hydrol, v. 2, Vienna, Inter Atomic Energy Agency, pp. 191–207.

Runnels DD (1969) Diagenesis, chemical sediments, and the mixing of natural waters. Jour Sed Pet 39:1188–1201

Runnels DD (1971) Chemical weathering of the Biscayne aquifer, Dade County, Florida. Jour Southeastern Geol 13:167–174

Rye RO Back W Hanshaw BB Rightmire CT Pearson FJ Jr (1981) The origin and isotopic composition of dissolved sulfide in groundwater from carbonate aquifers in Florida and Texas. Geochim Cosmochim Acta, 45(10): 1941 – 1950

Sacks LA (1995) Geochemical and isotopic composition of ground water with emphasis on sources of sulfate in the upper Floridan aquifer in parts of Marion, Sumter, and Citrus Counties, Florida. US Geol Surv Water-Resour Inves Rept 95-4251

Sacks LA Herman JS Kauffman SJ (2012) Controls on high sulfate concentrations in the upper Floridan aquifer in southwest Florida. Water Resour Res 31(10):2541–2551

Sacks LA Tihansky AB (1996) Geochemical and isotopic composition of ground water, with emphasis on sources of sulfate, in the upper Floridan aquifer and intermediate aquifer system in southwest Florida. US Geol Surv Water Resour Inves Rept 96-4146

Sainz Garcia AM Molinero Huguet JJ Saaltink MW (2011) Numerical modeling of coastal aquifer karst processes by means of coupled simulations of density-driven flow and reactive solute transport phenomena. Carbonates Evaporites 26:19–27

Sanford WE (1987) Assessing the potential for calcite dissolution in coastal salt-water mixing zones. Ph.D. thesis, University Park, Penn State Univ

Sanford WE Konikow LF (1989) Porosity development in coastal carbonate aquifers. Geology 17:249–252

Schijf J Byrne RH (2007) Progressive dolomitization of Florida limestone recorded by alkaline earth element concentrations in saline, geothermal, submarine springs. St. Petersburg, Univ So Fla Mar Sci Faculty Pub 5, accessed 5/25/2017 at http://scholarcommons.usf.edu/msc_facpub/5

Scott TM (1988) The lithostratigraphy of the Hawthorn Group (Miocene) of Florida. Fla Geol Surv Bull 59

Shinn EA (1964a) Formation of Recent dolomite in Florida and the Bahamas (abs). Amer Assoc Petrol Geol Bull 48:547

Shinn EA (1964b) Recent dolomite, Sugarloaf Key. In: Ginsburg RN (ed.), South Florida carbonate sediments, Geol Soc Amer Field Trip Guidebook 1, pp. 62–67

Shinn EA (1968) Selective dolomitization of Recent sedimentary structures. Jour Sed Pet 38:612–616

Smart PL Dawans JM Whitaker F (1988) Carbonate dissolution in a modern mixing zone. Nature 335:811–813

Socki RA Perry EC Romanek CS (2002) Stable isotope systematics of two cenotes from the northern Yucatan Peninsula, Mexico. Limnol Oceanog 47:1808–1818

Southeastern Geological Society (1970) Geology and geohydrology of the Cross-Florida Barge Canal area. Tallahassee, Southeastern Geol Soc 14th Field Conference Guidebook

Southeastern Geological Society (1986) Hydrogeological units of Florida. Fla Geol Surv Sp Pub 28

Southwest Florida Water Management District (2002) Saltwater Intrusion and the minimum aquifer level in the Southern Water Use Caution Area. Brooksville, FL, Southwest Fla Water Mgt Dist, Draft of August 21, 2002

Southwest Florida Water Management District (2006) Southern Water Use Caution Area - Recovery strategy. Brooksville, FL, Southwest Florida Water Mgt Dist

Spechler RM (1994) Saltwater intrusion and quality of water in the Floridan aquifer system, northeastern Florida. US Geol Surv Water-Resour Inves Rept 92–4174

Spechler RM (1996) Detection and quality of previously undetermined Floridan aquifer system discharge to the St. Johns River, Jacksonville to Green Cove Springs, northeastern Florida. US Geol Surv Water Resour Inves Rept 95-4257

Spechler RM Phelps GG (2001) Saltwater intrusion in the Floridan aquifer system, northeastern Florida. Proc 2001 Georgia Water Resour Conf, pp. 398–400

Sproul CR (1977) Spatial distribution of ground water temperature in south Florida. In: Smith DL Griffin GM (eds.), The geothermal nature of the Floridan Plateau. Fla Bur Geol Sp Pub 21, pp. 65–89

Stallard RF Edmond JM (1981) Geochemistry of the Amazon: 1. Precipitation chemistry and the marine contribution to the dissolved load at the time of peak discharge. Jour Geophys Res 86:9844–9858

Starks MJ (1986) Mixing-zone diagenesis in a carbonate aquifer, west-central Florida. Masters thesis, Tampa, Univ So Fla

Stiff HA Jr (1951) The interpretation of chemical water analysis by means of patterns. Jour Petrol Tech 3(10, sect. 1):15–16

Stoessell RK (1992) Effects of sulfate reduction on $CaCO_3$ dissolution and precipitation in mixing-zone fluids. Jour Sed Pet 62: 873–880

Stoessell RK Moore YH Coke JG (1993) The occurrence and effect of sulfate reduction and sulfide oxidation on coastal limestone dissolution in Yucatan cenotes, Ground Water 31(6): 566–575

Stringfield VT (1966) Artesian water in Tertiary limestone in the southeastern states. US Geol Surv Prof Pap 517

Strom RN Upchurch SB (1985a) Palygorskite distribution and silicification in the phosphatic sediments of central Florida. In: Snyder S Riggs S Partin B Mallette P Walker R (eds.), 8th Inter Field Workshop and Symp (Southeastern United States), Guidebook, Inter Geol Correl Prog, Project 156 - Phosphorites, pp. 118–126

Strom RN Upchurch SB (1985b) Palygorskite distribution and silicification in the phosphatic sediments of central Florida. In: Scott TM Cathcart JB (eds.), Florida Land-Pebble Phosphate District. Geol Soc Amer, 1985 Ann Meeting Field Trip Guidebook, pp. 68–75

Summers K Copeland R Singleton T Upchurch S Janicki A (1998) Designing a comprehensive, integrated water resources monitoring program for Florida. In: US Environ Prot Agency, Proc Natl Water-Quality Monitoring Council, Natl Conf on Monitoring: Critical Foundations to Protect our Waters, Reno, NV, p III-27–III-40

Swancar A Hutchinson CB (1995) Chemical and isotopic composition and potential for contamination of water in the upper Floridan aquifer, west-central Florida, 1986–89. US Geol Surv Water-Sup Pap 2409

Taft WH (1961) Authigenic dolomite in modern carbonate sediments along the southern coast of Florida. Science 134: 561–562

Taft WH Harbaugh JW (1964) Modern carbonate sediments of southern Florida, Bahamas, and Espíritu Santo Island, Baja California: A comparison of their mineralogy and chemistry. Stanford Univ Pubs, Geol Sci, Vol. VIII, No. 2

Thorn KA Mikita MA (2000) Nitrite fixation by humic substances: Nitrogen-15 nuclear magnetic resonance evidence for potential intermediates in chemodenitrification. Soil Sci Soc Amer Jour 64:568–582

Thrailkill JV (1965) Studies in the excavation of limestone caves and the deposition of speleothems. Ph.D. Dissertation, Princeton University

Thurman EM (1985) Organic geochemistry of natural waters. Dordrecht, Springer-Science+Business Media

Thurman EM Malcolm RL (1981) Preparative isolation of aquatic humic substances. *Environ Sci Tech* 15(4):463–466

Thurman EM Wershaw RL Malcolm RL Pinckney DJ (1982) Molecular size of aquatic humic substances. Organic Geochem 4:27–35

Upchurch SB (1992) Quality of waters in Florida's aquifers. In: Maddox GL Lloyd JM Scott TM Upchurch SB Copeland R (eds.), Florida ground water quality monitoring program – Vol 2, Background hydrogeochemistry, Fla Geol Surv Sp Pub 34, pp. 12–52, 64–84, 90–347

Upchurch SB (1998) Lake Tarpon ground-water nutrient study. Contract report prepared by Environmental Resources Management, Inc. for the Southwest Florida Water Mgt Dist, Brooksville, Florida

Upchurch SB (2015) Determination of the relationship of nitrate to discharge and flow systems in north Florida springs. In: Doctor DH Land L Stephenson JB (eds.), Sinkholes and the engineering and environmental impacts of karst, Proc 14th Multidisc Conf, Natl Cave and Karst Res Inst Symp 5, pp. 347–354

Upchurch SB (2017) Hypogene speleogenesis on the Floridan Platform, U.S.A. In: Klimchouk AB Palmer AN Waela JD Auler AS Audra P (eds.), Hypogene karst regions and caves of the world, Springer Int Pub, Ch. 49, pp. 735–744

Upchurch SB Chen J Cain CR (2007) Relationships of nitrate to flow in springs of the Suwannee River Water Management District. Live Oak, Florida, Suwannee River Water Magt Dist (Report available from the District at http://www.srwmd.state.fl.us)

Upchurch SB Lawrence FW (1984) Impact of ground-water chemistry on sinkhole development along a retreating scarp. In: Beck BF (ed.), Sinkholes: Their geology, engineering & environmental impact. Rotterdam, A.A. Balkema, pp. 23–28

Upchurch SB Strom RN (1980) Role of humic substances in transportation and deposition of calcium. In: Gleason PJ (ed.), Water, oil, and the geology of Collier, Lee, and Hendry counties, Miami Geol Soc, pp. 5–9.

Upchurch SB Strom RN Nuckels MG (1982) Silicification of Miocene rocks from central Florida. In: Scott TM Upchurch SB (eds.), Miocene of the southeastern United States, Fla Bur Geol Sp Pub 25, pp. 251–284

Upchurch SB Strom RN Williams MJ (1983) Preservation of dolomite in coastal peats of the Ten Thousand Islands area, Florida. In: Raymond R Jr Andrejko MJ (eds.), Mineral matter in peat: Its occurrence, form, and distribution, Los Alamos Natl Lab LA-9907-OBES, UC-11, pp. 215–224

Upchurch SB Taraszki MD Zimmerman J Pyne RDG (1991) Rock-water interactions during aquifer storage and recovery. In: Upchurch SB Moore JE Zaporozec A (eds.), Hydrology and hydrogeology in the '90s. Minneapolis, MN, Amer Inst Hydrol, pp. 309–319

Vacher HL Hutchings WC Budd DA (2006) Metaphors and models: The ASR bubble in the Floridan aquifer. Ground Water 44(2):144–154

Weaver CE Beck KC (1977) Miocene of the southeastern United Sates: A model for chemical sedimentation in a peri-marine environment. Amsterdam, Elsevier Sci Pub Co, Develop Sediment, Vol. 22 and Sed Geol 17:1–234

Wershaw RL (1999) Molecular aggregation of humic substances. Soil Sci 164:803–813

White WB (1988) Geomorphology and hydrology of karst terrains. New York, Oxford Univ Press

Whitney E Means DB Rudloe A (2004) Priceless Florida: Natural ecosystems and native species. Sarasota FL, Pineapple Press

Wicks CM Troester JW (1998) Groundwater geochemistry of Isla de Mona, Puerto Rico. Jour Cave and Karst Stud 60:107–114

Wicks CM Herman JS (1995) The effect of zones of high-porosity and permeability on the configuration of the saline fresh-water mixing zone, Ground Water 33(1–2): 733–740

Wicks CM Herman JS (1996) Regional hydrogeochemistry of a modern coastal mixing zone. Water Resour Res 32(2):401–407

Wigley TML Plummer LN (1976) Mixing of carbonate waters. Geochim Cosmochim Acta 40:989–995

Young TC Comstock WG (1986) Direct effects and interactions involving iron and humic acid during formation of colloidal phosphorus. In: Sly PG (ed.), Sediments and water Interactions. New York, Springer-Verlag, pp. 461–470

Zenger DH Mazzullo SJ (eds) (1980) Dolomitization, Benchmark Papers in Geology, Stroudsburg, PA, Hutchinson Ross Publ Co, Vol. 65

Zhang N Tans PP Thoning KW (1997) A high precision manometric system for absolute calibrations of CO2 in dry air. Jour Geophy Res 102(D5):5885–94

# 6 Controls on Karst Landforms in Florida

**Abstract**

Florida's eogenetic karst development is influenced by groundwater flow, sea level and water table position, sediment mineralogy and fabric, porosity, chemical reactions and geochemical saturation state of the water, residence time, and sediment surface area.

Primary porosity such as intergranular porosity, burrows and borings, intraparticle porosity, and shelter porosity can initially affect the pathways of water flow. Secondary porosity includes potentially extensive, interconnected, moldic porosity, and karstic secondary porosity includes caves and other karst conduits. Diagenetic processes, such as cementation of limestones and dolomitization, may destroy primary and secondary porosity.

Fractures or existing karst conduits provide fast flow routes along which karstification occurs. Photolineament analysis can, with ground truthing, confirm the existence of vertical fractures and faults in basement or overlying rocks. In Florida, photolineament sets occur in two main alignments that are consistent with earth tides.

Bedding planes are often poorly represented but can be dissolutionally enlarged. Epikarst consists of irregular, weathered surfaces exhibiting enhanced porosity and permeability, with cutters, pinnacles, and limestone fragments. Epikarst at unconformities may not relate to sinkhole development.

Reaction rates of groundwater with the carbonate rock are typically slow, and sudden collapse events are usually due to failure of cover materials rather than of the roofs of voids in limestone.

Florida's epigenetic caves form near the water table when the phreatic surface is stable. In most systems, vadose caves form concurrently with and up gradient from phreatic caves. Many caves reveal evidence of water-table fluctuations driven by climate and sea-level changes.

Some Florida caves have been attributed to mixing-zone dissolution, including caves with large rooms at depth. Evidence is relatively weak for sulfate dissolution-related hypogenetic karst. The dependence of epigenic and freshwater/saltwater mixing zones on sea level creates complex layering of caves with different positions and ages.

**Keywords**

Porosity · Permeability · Dissolution kinetics · Photolineaments · Fractures · Epigenetic karst · Hypogenetic karst · Rates of dissolution

## 6.1 Purpose of This and Subsequent Chapters

This chapter is an introduction to the controls on karst features, such as forms of porosity in carbonate rocks and sediments, effects of residence time of groundwater in contact with the host rock or sediments, recognition and effects of fracturing, effects of the phreatic and vadose zones on karst formation, timing of karst feature development, and weathering rates of carbonate sediments and rocks. The concepts of epigenetic and hypogenetic karst systems are introduced, as well.

Chapter 7 includes discussions of the mechanisms of karst development as a result of interactions of soil moisture, meteoric water, and groundwater. Chapter 7 concludes with a "catalog" of the many different small- and large-scale landforms and features that have resulted from dissolution and precipitation of carbonate minerals in the shell beds, limestone, and dolostone of Florida. Because there is much confusion in Florida about the mechanisms of sinkhole formation, emphasis is placed on sinkhole-development mechanics, including illustrations showing different authors' explanations of formation of the different types of sinkholes. Sinkhole risk is discussed in this chapter, as well.

Chapter 8 deals with the landforms and features developed as a result of interaction of the carbonate sediments and


S. Upchurch et al., *The Karst Systems of Florida*, Cave and Karst Systems of the World,
https://doi.org/10.1007/978-3-319-69635-5_6

rocks with surface water. This set of processes is important in Florida eogenetic karst where karst escarpments are forming.

## 6.2 Introduction

Karst occurs worldwide, from cold climates to the tropics, in arid and wet environments, in geologically young and old rocks, and in high mountains and low coastal areas. As noted in Chap. 1, however, most areas with well-known karst terrains that have been extensively researched differ from conditions in Florida in that they usually reflect telogenetic karst processes. Also, as discussed in Chap. 1, there are fundamental differences between eogenetic and telogenetic karst that reflect the basic, age- and burial-related differences in the host sediments and rocks and the ways that water interacts with these materials. These differences can be understood by considering the factors that affect karst development.

## 6.3 Factors that Affect the Distribution of Florida Karst

In Florida, as in most other karst landscapes, the dominant rock type with evidence of karstification is limestone. Near-surface shell beds and other, common carbonate-rich sediments and rocks (i.e., dolomitic sediment and dolostone) have also undergone karstification to a limited extent in Florida. Given enough time, these carbonate sediments and rocks are dissolved by mildly acidic precipitation, soil water, and/or groundwater. The dissolution process was discussed in detail in Chap. 5.

Development of karst is dependent on how this water interacts with the host carbonate sediment or rock. The development of karst landforms is dependent on six primary factors:

Hydrogeochemical Factors

1. Availability of water, including spatial and temporal variability of meteoric water and groundwater flow systems over geologic time;
2. The chemistry of the water, including acidity, dissolved constituent concentrations, and dissolved organic constituents in the water; and
3. Mineralogy of the carbonate sediment or rock.

Physical Factors

4. Distribution of porosity and permeability in the carbonate sediment or rock;
5. Residence time of the water in contact with the sediment or rock materials; and
6. The sediment or rock surface area exposed to the water.

The first three factors are hydrological and geochemical in nature and were discussed in Chaps. 4 and 5. The following sections discuss the physical factors that affect development of karst in Florida.

### 6.3.1 Porosity and Permeability

One of the most important differences between eogenetic and telogenetic karst is the distribution of porosity and permeability in the host limestone or dolostone. There is, of course, a wide variation in the sizes of openings in the host sediments or rocks. Eogenetic karst includes:

1. Interstitial pores in the rock matrix, many of which may have been enlarged by dissolution;
2. Fractures, including fractures that have been enlarged by dissolution;
3. Voids developed in the vicinity of the water table and along freshwater/saltwater transition zones;
4. Voids and sinkhole avens caused by collapse; and
5. Small to very large caves that have formed by dissolution along flow paths and that integrate the previously listed forms of secondary porosity.

The rock matrix in telogenetic karst typically has substantially less porosity. As a result, the porosity and permeability in telogenetic karst is usually limited to criteria 2 through 5 above.

Table 6.1 illustrates the different forms of porosity, or pore space, found in carbonate rocks. It is important to understand that not all pore or void space is created by dissolution of the carbonate sediment or rock and that not all openings are interconnected and capable of enhanced groundwater flow.

Most telogenetic karstic limestones have limited forms of matrix pore space, and, where present, the void spaces tend to be relatively large. Examples include caves, fractures in the rock, and enlarged sediment bedding planes. This is because the rocks are geologically old (tens to hundreds of millions of years old) and were previously deeply buried below the land surface (Chap. 1). As a result, the original pore spaces created during deposition and early stages of diagenesis of the sediments to limestones or dolostones have been destroyed by the heat and pressure of burial and by time. The matrix of these rocks, therefore, is usually relatively impermeable, and the primary way that water moves through these older rocks is through secondary fractures, bedding planes, conduits, and caves (Table 6.1).

In contrast, the eogenetic, karstic limestones of Florida have abundant forms of pore and void space, including primary porosity, such as intergranular pore space and molds, and secondary porosity, such as caves, fractures, and bedding plane separations. This is because the rocks are geologically

**Table 6.1** Types of void and pore space found in Florida's carbonate sediment and rock

| Void type | Origin | Relative size of pores | Relative permeability (pore or void space interconnection) | Example |
|---|---|---|---|---|
| **Primary porosity** | | | | |
| Intergranular pore space | Deposition of discrete sedimentary particles | Microns to millimeters | Low in clay-rich sediments<br>Moderate in sand-sized sediment<br>High in gravely sediments<br>Spaces interconnected unless filled with fine-grained particles or cement |  |
| Burrows and borings | Voids excavated by organisms in sediment or rock | Millimeters to fractions of a meter in diameter, may be many meters in length | Moderate to low (usually filled with later sediment or cement)<br>In granular sediments and rocks, burrows often have higher permeability that the matrix | 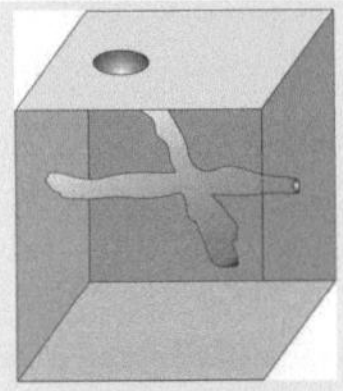 |
| Intraparticle porosity | Open spaces within fossils | Fractions of a millimeter to centimeters | Usually not interconnected |  |
| Shelter porosity | Voids created by tenting over or sheltering of open space by larger sedimentary particles, such as large mollusk shells | Millimeters to centimeters | Usually not interconnected in fine-grained sediments<br>May be interconnected in shell beds | 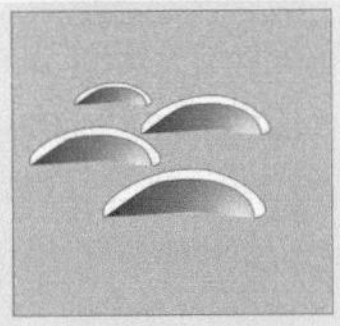 |
| **Secondary porosity** | | | | |
| Moldic porosity | Voids created by the dissolution of fossils and other particles within a sediment body. Only the outer shape (mold) of the object remains. | Millimeters to meters | Molds may be interconnected to form porous sediment or rock |  |
| Fracture porosity | More-or-less planar voids caused by sediment or rock shrinkage, tectonic forces, or tidal stresses. In carbonate rocks, fractures may be modified by dissolution | Centimeters to 100 s of meters | Unless filled with later sediment, fractures are highly transmissive | 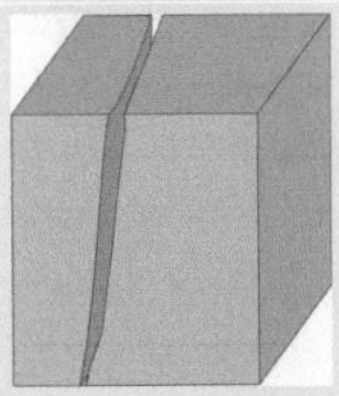 |
| Vugs, karst conduits, and caves or caverns | Formed by dissolution of matrix sediment or rock. May be filled with secondary cave decoration, such as flowstone, crystals, etc. | Vugs are centimeter-sized voids caused by the dissolution of carbonate sediment or rock.<br>Conduits are too small for human passage, so centimeters to meters<br>Caves or caverns are large enough for human passage, so meters to 10s of meters | Openings to vugs may be so small as to remain undetected. Low to moderately interconnections<br>Conduits are highly transmissive – a major path for water flow<br>Caves or caverns are highly transmissive – a major path for groundwater flow | 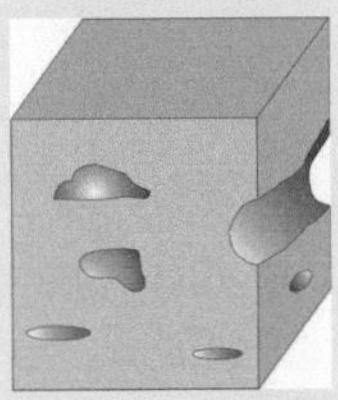 |
| Bedding planes and unconformities | Bedding planes form during deposition as a result of storms, interruptions in sedimentation, or abrupt changes in sediment texture. May be slowly enlarged by dissolution as a result of groundwater flow | Opening typically less than a centimeter across bedding plane<br>May extend for meters in the plane of sedimentation | Moderate to highly transmissive, especially if enlarged by rock dissolution | 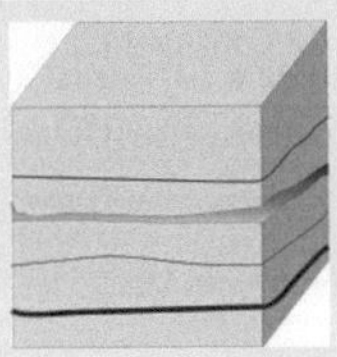 |

young (thousands to tens of millions of years old) and have never been deeply buried below the land surface. As a result, the original pore spaces created when the carbonate sediments accumulated and during early stages of diagenesis of the sediments to limestone and/or dolostone remain with cements that range from nonexistent to locally thorough cementation. The primary pore space has not been destroyed by the heat and pressure of burial and by time. The matrix of these eogenetic sediments and rocks, therefore, is relatively permeable, and the primary way that water moves through these young sediments and rocks is through both primary pore space and secondary fractures, bedding planes, conduits, and caves (Table 6.1).

The type and distribution of pore space affect karst development by (1) controlling flow paths of groundwater and (2) affecting growth of karst features. For example, small pore spaces and voids may not be active flow paths for groundwater and they may be volumetrically incapable of accepting cover material and/or collapsed host rock. Small voids are often detected and related to sinkhole development in Florida because they render the rock weak under penetration testing or they cause losses of drilling-fluid circulation. Because of the limitations on karst development by void space, including its origins and effects on the potential for sinkhole development, it is appropriate to discuss the variety of void spaces present in Florida's eogenetic limestones. For the pore space to participate in sinkhole development, the pore spaces must:

1. Be interconnected so that water and cover sediments and rock can easily and efficiently migrate into the void space; and
2. Be sufficiently large, either as a single void or as a complex of well-interconnected voids (touching-vug porosity; Lucia 1995) to accept the volume of cover sediments or collapsed rock that must migrate in to the void(s) and cause subsidence of the land surface.

Needless to say, collapse of cave roofs and sediments that bridge over large voids fit these requirements, but most of the smaller pore types are highly unlikely to cause sinkhole development.

### 6.3.1.1 Primary and Secondary Porosity

Carbonate sediments and limestones in the eogenetic karst limestones of Florida often have two major types of porosity: primary porosity (interstitial spaces between the grains of sediment that make up the rock itself; Figs. 6.1 and 6.2) and secondary porosity (pore space caused by fracturing, discontinuities in sediment accumulation, and dissolution of carbonate rock; Figs. 6.3 and 6.4).

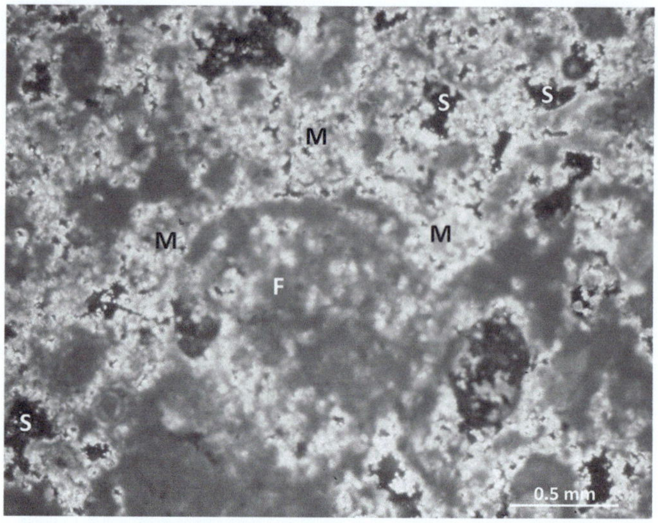


**Fig. 6.1** Micritic packstone with fossils (foraminifera, F), micrite (microcrystalline calcite, M) cement, and remnant interstitial pore space (S). From the Ocala Limestone near Belleview, Florida. The foraminifera have been partly micritized during diagenesis

### 6.3.1.2 Intergranular Porosity

As noted in Chap. 1, Florida's limestones are said to have dual porosity. That is, the matrix of the rock often has intergranular porosity (Table 6.1; Figs. 6.1 and 6.2) through which groundwater can be stored and slowly flow. Superimposed on this matrix permeability, secondary porosity in the form of fractures, bedding planes, and karst features provide additional and more efficient flow pathways (Figs. 6.3 and 6.4).

The dual porosity of eogenetic limestone is a critical property. Because of primary porosity, aquifers can store and transport large volumes of water. Because of secondary porosity, pathways of more efficient groundwater flow can occur and deliver water to wells, springs, and other flow destinations. The primary porosity presents great amounts of surface area at which water can quickly come to chemical equilibrium with the host sediment or rock. Thus, in recharge areas, water in intergranular porosity quickly loses its ability to dissolve rock and create karst. In contrast, secondary porosity may consist of large conduits where the surface area of the rock is limited relative to the volume of water moving through the large conduits. In this latter scenario, dissolution and conduit enlargement may extend some distance from the point of recharge.

Storage of water in the intergranular spaces in the matrix differentiates water flow behaviors to springs and wells in eogenetic versus telogenetic karst. In telogenetic karst, little water is stored in the matrix. As a result springs tend to be "flashy" with discharge peaks directly related to precipitation events. Between rainfall events, the springs may dry up

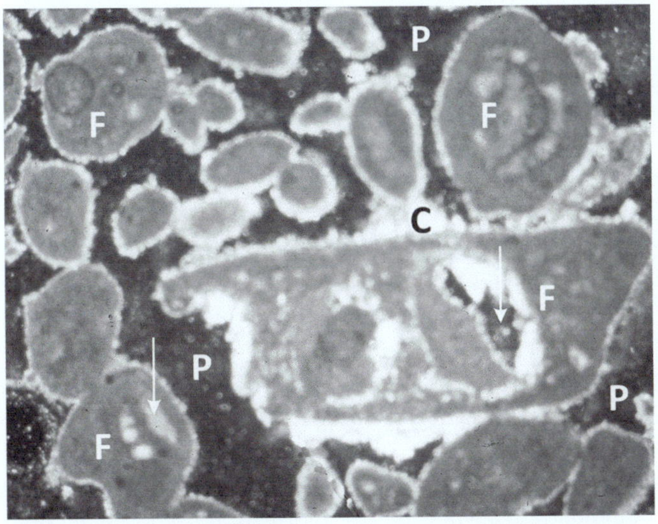


**Fig. 6.2** Primary, intergranular porosity in a typical sample of grainstone from the Ocala Limestone in the same core as the sample shown in Fig. 6.1. F = tests and fragments of fossil marine foraminifera that range in size from 0.5 to 1 mm in diameter. C = sparry calcite cement that is beginning to fill the pore space (P; dark gray areas) between the fossil. Arrows point to examples of intraparticle porosity, some of which is filled with sparry calcite

**Fig. 6.3** Secondary porosity in Florida limestone. Fractures in the Ocala Limestone at the Haile Quarry near Newberry, Florida. The fractures have been enlarged through dissolution of the limestone by groundwater, a karst process. Photograph courtesy of the Florida Geological Survey

**Fig. 6.4** A small vug with calcite crystals (C) and a stalactite (S) in the Suwannee Limestone near Lecanto, Citrus County, Florida. Note that the limestone is well cemented and very hard, as shown by the hammer bruising (B) around the vug. View is 25 cm (9.8 in.) across

or sustain minimal flow. In contrast, springs in eogenetic karst terrains can sustain flow between rainfall events by loss of water from storage in the host matrix. Spring discharge peaks during rainfall event are often somewhat muted because of water entering storage rather than flowing directly though karst conduits (Florea and Vacher 2007). Water wells are likely to behave in similar ways.

#### 6.3.1.3 Burrow- and Boring-Related Porosity

The remains of animal burrows and borings (Table 6.1) are common in Florida limestone and other carbonate and siliciclastic sediments. The process of disruption of sediment by the burrowing action of organisms is known as bioturbation. As illustrated in Figs. 6.5 and 6.6, burrows may be filled with sediment deposited after creation of the burrow. In Florida, heavily bioturbated sediment is usually less dense and more permeable than the surrounding sediment, which can lead to losses of drilling-fluid circulation during standard penetration testing, low relative strength sediment, and preferential groundwater flow. Highly bioturbated sediment and shallow rock in Florida is typically weaker than the unbioturbated equivalent owing to the increase of intergranular porosity caused by the burrowing and boring action of the organisms.

If near a recharge area where chemically aggressive water is introduced, bioturbated limestone may be preferentially subjected to dissolution. More deeply buried, heavily bioturbated sediments or rocks may produce false indicators of dissolution, such as low penetration resistances and losses of drilling fluid circulation.

The limestone shown in Fig. 6.5 is a highly burrowed example of the oolitic facies of the Pleistocene Miami Limestone. The matrix consists of lithified oolitic grainstone, and the less well cemented burrows are filled with tan, calcitic grainstone to packstone that washed into the burrows after they were created. The tan coloration is a result of preferential movement of dissolved iron in groundwater through the burrows. The bioturbated horizons in the Miami Limestone are stratiform facies that form groundwater flow zones (Cunningham et al. 2008; Cunningham and Sukop 2011).

Figure 6.6 illustrates the effects of bioturbation in siliciclastic, sandy sediment during pedogenesis. Animals and roots penetrate the sediment and, after passage of the animal or decay of the root, the burrow is filled with loose sediment. Note the dark orange staining around the burrows shown in Fig. 6.6. These stains were caused by leaching and reprecipitation of ferric hydroxide by groundwater preferentially moving through the burrows.

Boring is the process of tunneling in rock or highly consolidated sediment. Many marine organisms, most notably ship worms and pholads (rock-boring bivalve mollusks), have the ability to rasp or dissolve away hard substrates (wood, rock) in order to create shelters. This boring weakens the substrate and increases the surface area of the exposed rock. Evidence of boring is often seen on Florida's rocky coasts, in rocks at hardgrounds (Fig. 6.7) and on unconformities (see below), and in large fossil corals. From a karst perspective, holes created by boring that are preserved at ancient weathered upper horizons of limestone or dolostone and unconformities (1) weaken the rock, (2) represent entry points for clay and other siliciclastic materials that may be mistaken for raveled sediment, and (3) appear as fragmented rock when penetrated mechanically during subsurface testing.

**Fig. 6.5** Burrowed Miami Limestone, Greynolds Park (25.941°N, 80.154°W), Miami, Dade County, Florida

Figure 6.7 is an example of a hardground that is relatively common in the Miocene Hawthorn Group. The small circular depressions in the rock shown in Fig. 6.7 are pholad borings. The hardground was a surface that was exposed on the sea floor during the Miocene for a period long enough that the surficial sediment was cemented to form a hard rock. In this case, the sediment cement was the mineral carbonate fluorapatite (phosphate). In other situations, hardgrounds may reflect sediments cemented by calcite or aragonite. At any rate, because they were exposed on the sea floor, the hardgrounds provided shelter for boring marine organisms. Today, these highly cemented strata represent hard materials that may result in high penetration resistances during standard penetration testing or tip stress during cone penetrometer testing.

#### 6.3.1.4 Intraparticle Porosity

Intraparticle porosity (Table 6.1; Fig. 6.2) consists of open space within particles in sediment. Intraparticle pore spaces are small and usually not interconnected, so they are

**Fig. 6.6** Weathered and burrowed Citronelle Formation near New Hope, Washington County, Florida. Arrows indicate sediment-filled burrows

typically not involved with karst development. The small openings in the individual foraminifera tests shown in Fig. 6.2 represent intraparticle porosity.

#### 6.3.1.5 Shelter Porosity

Shelter porosity (Table 6.1) is typically formed in the open spaces that are inaccessible to sediment infiltration in larger fossils, such as mollusk shells, and when large, curved fossils are deposited with the concave portion of the shell facing down. The shell forms a tent-like shelter over the space, preventing sediment infiltration and preserving a small opening. Like intraparticle porosity, shelter porosity is typically isolated and not an important factor in groundwater flow or karst development.

**Fig. 6.7** Phosphatic hardground extensively bored by pholads, a common marine boring bivalve, from the Arcadia Formation (Miocene Hawthorn Group), Polk County, Florida. Boot toe gives scale. Photograph courtesy of the Florida Geological Survey

Figure 6.8a, b show examples of large clam shells with shelter porosity in their interiors. The clams (*Mercenaria* sp.) were buried in growth position in shelly sand that is partly cemented by calcite. Shell fragments in the sand are angular, indicating that dissolution of the shell in the bed has been insignificant.

Figure 6.9 shows a detail of one of the *Mercenaria* shells from the Ft. Drum fossil locality. The shell of the clam (S, Fig. 6.9) was originally aragonite, but during diagenesis the shell has been recrystallized into coarse calcite. The presence of the large calcite crystals in the interior space indicates that the shelter porosity prevented sediment, but not water, from washing into the void. The calcite cement in the sediment matrix and the calcite crystals in the opening inside the clam were derived from dissolution of shell higher up in the section and, possibly, recrystallization of aragonitic fossils during diagenesis. Calcium carbonate-rich groundwater migrated into the interiors of the clam shells, where calcite was deposited over a few hundred to thousands of years as large crystals. It is important to understand that these transformations (recrystallization and calcite deposition) are diagenetic processes, and that they are not considered to be karst processes, although diagenesis and karst development are closely related in Florida's karst systems. The calcium-carbonate-rich water was derived from weathering of shallow shells approximately 10 m above the shell bed shown. Dissolution of the shells at the top of the shell bed formed a shallow depression in the land surface. This depression is a form of slowly forming sinkhole and will be discussed in Chap. 7.

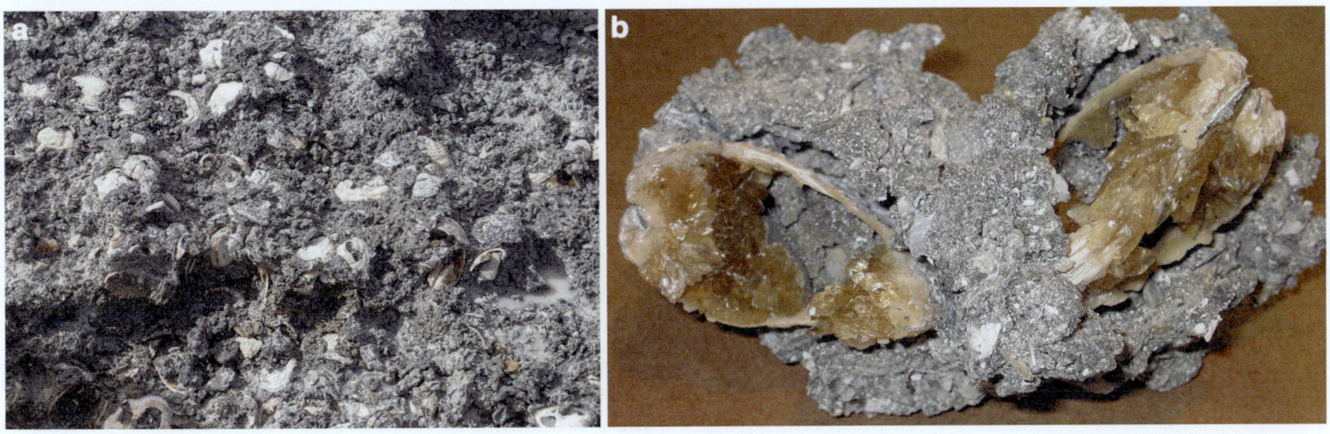


**Fig. 6.8** (**a**) Examples of shelter porosity. Note the open spaces within fossils of the mollusk *Mercenaria* sp. in clayey, calcareous, Pleistocene sand near Ft. Drum, Okeechobee County, Florida. Location: 27.512°N, 80.754°W. Openings within the mollusk shells at this locality were often lined with amber-colored calcite crystals. (**b**) Close up of *Mercenaria* fossils in growth position and with shelter porosity in cemented, calcareous, shelly sand from Ft. Drum, Okeechobee county, Florida. Pore space is lined with calcite crystals. Clams are each approximately 8 cm (3.1 in.) in maximum dimension

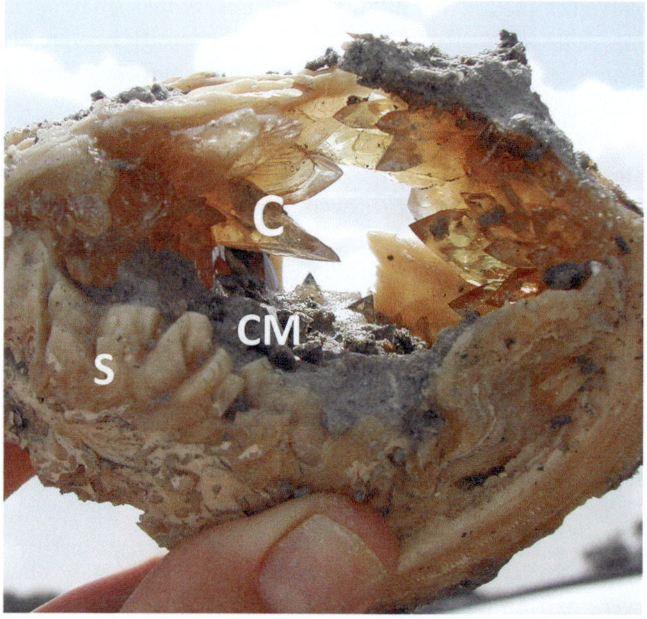


**Fig. 6.9** A *Mercenaria* shell with void-filling calcite crystals (C) from Ft. Drum, Okeechobee County. S is the mollusc shell, which has recrystallized to calcite. CM is a layer of sandy sediment that has been cemented by calcite during the crystal growth phase

#### 6.3.1.6 Moldic Porosity and the Formation of Casts

Many of the aragonitic fossils and other, highly soluble mineral phases incorporated in Florida clays and limestones have been completely dissolved away during diagenesis of the sediment. If the host sediment is firm enough to preserve the original external shape of the object, a mold preserving the shape and external details of the original material may remain as an imprint in the host sediment. The openings created by dissolution of a fossil constitute moldic porosity (Table 6.1).

Figure 6.10a, b depict some examples of molds preserved in Florida carbonate rocks. Figure 6.10a is a mold of a gastropod found in the Eocene Ocala Limestone. The gastropod shell was originally aragonite, which has been removed from the limestone matrix by dissolution during diagenesis. The details of the fossil are well preserved on the interior walls of the mold, and a plaster cast has been created to provide a replica of the original shell. Note that no interior details of the original shell have been preserved in the mold. Figure 6.10b also depicts molds of crystals of a soluble mineral – in this case gypsum crystals that grew in carbonate sediment that has been dolomitized. The gypsum has been dissolved away leaving an external mold in the shapes of the crystals.

While discussing moldic porosity, it is appropriate to contemplate the difference between early diagenesis and karst processes – a problem that is common in the youngest of eogenetic karst terrains. In this text, karst processes are considered to be driven by dissolution in meteoric water or groundwater that is chemically undersaturated with respect to the mineral that is being dissolved regardless of its thermodynamic stability. When carbonate sediments are deposited, they also undergo changes that include removal of chemically unstable phases, such as aragonite or high-Mg calcite. Selective dissolution of metastable phases such as aragonite or high-Mg calcite during diagenesis of the sediment is a result of mineral thermodynamics and is not considered a karst process. Dissolution of both metastable and stable phases (i.e., aragonite, high- and low-Mg calcite) by water that is chemically undersaturated with respect to the phases can be considered either diagenetic or karst-related depending on one's area of interest. For example, the ooliths that comprise the oolitic facies of the Pleistocene Miami Limestone were originally aragonitic. Today, they are calcite although some aragonite remains (Bock et al. 1969; Hester and Schmoker 1985). Many of the individual ooids are hollow today as a result of diagenetic removal of the aragonite in the center of the particles and strengthening of the outer shells by secondary calcification.

The time span between deposition and dissolution becomes the critical factor in determining whether the dissolution event was an artifact of early, thermodynamically driven diagenesis or a post-depositional weathering episode caused by exposure to chemically aggressive water. In most karst terrains, especially teleogenetic karst terrains, this distinction is moot because there is a long time interval between deposition and dissolution events. In eogenetic karst terrains, such as in Florida, the distinction is often difficult. There is no question that active, modern dissolution superimposed on Eocene limestone is karst related. However, what about voids developed in the Miami Limestone, which is roughly 125,000 years old and originally composed of aragonitic ooliths? Here, the distinction between diagenesis and karst-related dissolution is blurry. The presence and condition of oolite molds suggest diagenesis while numerous small caves (Cressler 1993) clearly reflect karstification.

If the molds are of particles or fossils that were originally composed of a metastable phase (aragonite or high Mg-calcite) but the particles and fossils that were originally low-Mg calcite remain intact, there is an indication that the dissolution event was early and may have been diagenetic rather than a later, karst-related event. For example, in the Miami Limestone the originally aragonitic ooliths are commonly moldic while the cements, which are low-Mg calcite, remain intact. This would suggest that the dissolution of the aragonitic ooliths was diagenetic since the calcite remains. Later dissolution events have affected the Miami Limestone in bulk, causing dissolution of all carbonate phases and creation of solution pits and small caves. The later dissolution events are considered karst-related.

Molds are common in most geologically young carbonate sediments. Normally, the cavities that represent the molds

**Fig. 6.10** (**a**) Mold (left) and plaster cast (right) of the Eocene gastropod *Eovasum* sp. from the Haile Quarry, near Newberry, Alachua County. Reproduced from Portell and Hulbert (2011) with permission. (**b**) Molds of gypsum crystals in the Avon Park Formation, Citrus County (Photograph courtesy of the Florida Geological Survey)

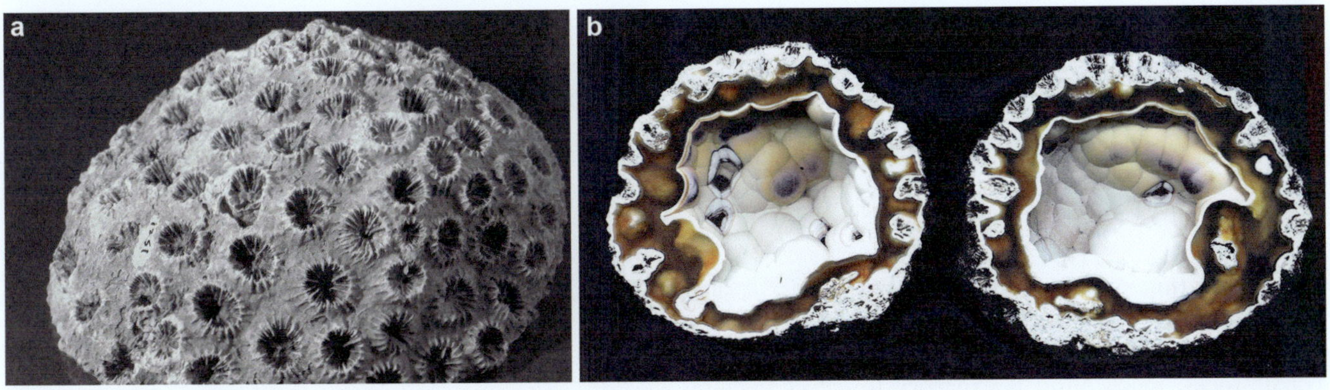


**Fig. 6.11** Exterior (left) and interior (right) of a Miocene coral (*Montastrea* sp.) cast composed of chalcedony. Cast is 9 cm in diameter. Sample is from a clay- and silicified-fossil-rich weathering mantle of the Miocene Tampa Member (Arcadia Formation) at Ballast Point, Tampa, Hillsborough County

are small and not significantly interconnected, as was the case of the molds shown in Fig. 6.10a, b. Where there are interconnections between molds (touching-vug porosity), groundwater may flow through the openings thus enhancing the hydraulic conductivity of the sediment and transmissivity of the unit.

Fossil molds and other voids in the host sediment may be filled at a later time to form casts, which retain the external pattern of the void. In a cast, however, the void space is partly to completely filled with a later sediment that preserves none of the original internal structure of the original object.

There are several localities in Florida where silica has filled or lined molds of mollusks and corals to form casts of either chalcedony or rock crystal (Upchurch et al. 1982; Fig. 6.11a, b). These are locally known as "Tampa Bay geodes." As discussed below, the presence of these and other quartz casts and replacements of limestone in Florida provides important information as to the timing of karst development.

#### 6.3.1.7 Fractures and Fracture Porosity

When fractures, or joints, in lithified materials open, fracture porosity is formed (Table 6.1). Fractures are formed in rocks by a variety of processes, including material shrinkage, tectonic movements, and earth tides (Hills 1972). They are important in karst terrains, because they form preferential pathways for water movement. As a result, fracture surfaces may undergo dissolution, creating enhanced groundwater

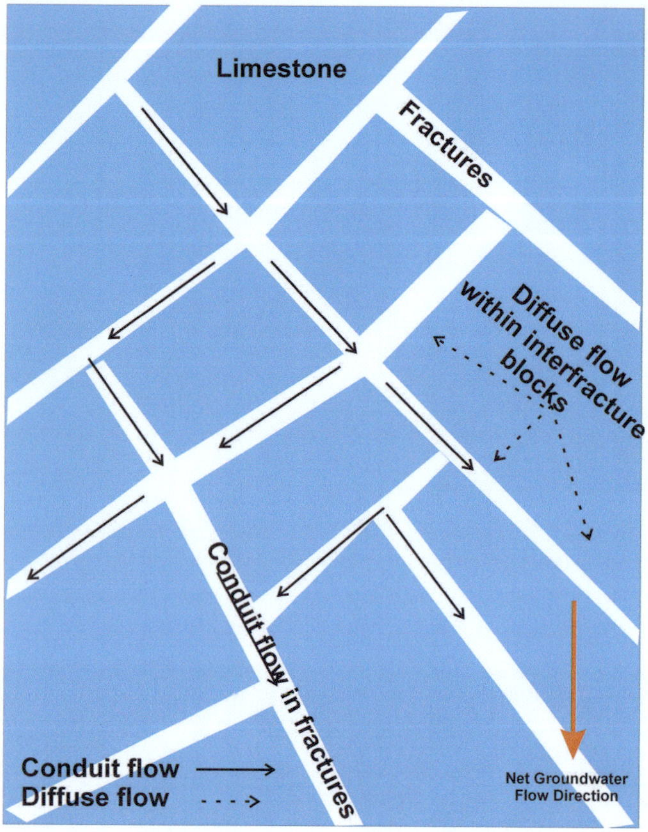


**Fig. 6.12** Potential flow through a fractured, doubly porous limestone

**Fig. 6.13** Cave diver in a water-filled cave at Hart Springs, Gilchrist County. (Photograph courtesy of Jill Heinerth)

flow paths (Fig. 6.12), with caves (Fig. 6.13), springs (Fig. 6.14), and sinkholes (Fig. 6.15).

Because of their importance to groundwater flow in carbonate rocks, much effort is expended to detect fractures through remote sensing using aerial photographs, satellite images, and topographic maps. Where present, vertical fractures may be visible as alignments of sinkholes, wetlands, streams, soil tonal differences, and vegetation at the land surface. Where they are exposed in the vertical plane, such as in quarries, fractures may be small with minor evidence of widening through dissolution (Fig. 6.16) or wide and clearly enlarged by dissolution (Fig. 6.17). Care should be taken, however, because not all alignments are a result of fracturing or jointing (Alfieri et al. 2018). For additional discussion on identification and verification of fractures and photolineaments (the alignments observed through remote sensing), see Sect. 6.4 on Fractures and Photolineaments.

#### 6.3.1.8 Karst Conduits and Caves

Caves and other karst conduits are openings caused by dissolution of limestone and other soluble strata (Table 6.1). Caves are karst conduits through which a person can move (Fig. 6.13). However, karst conduits range in diameter from a few centimeters (Fig. 6.16) to many meters in diameter and meters to kilometers in length. The term cave, or cavern, is based on anthropogenic properties, such as access and size relative to a human (Klimchouk 2004a).

Dissolution-related void space is extremely common in Florida limestone and dolostone. It is this void space, which consists of openings interconnected by karst conduits, caves, fractures, and other forms of overlapping pore space, that makes the Floridan and Biscayne aquifers such efficient and productive sources of groundwater. While void space is common and normal in Florida carbonate rocks, voids large enough and sufficiently connected to the overlying cover that raveling or collapse of near-surface sediments or rocks can occur and form a sinkhole are rare.

In Florida, the distribution of air-filled caves is limited owing to high water-table conditions that are so common throughout the state. Water-filled (phreatic) caves in Florida are famous as attractions to cave divers (Fig. 6.13; Florea 2008a). Several cave-diving sites (i.e., Ginnie Springs [Fig. 6.14], Gilchrist County; Blue Grotto and Devil's Den, Levy County) are operated commercially for cave diving. Commercial dry caves have been developed in Jackson County

**Fig. 6.14** The Devil's Eye Spring (29.8351°N, 82.6966°W), Gilchrist County. Photograph courtesy of the Florida Geological Survey

**Fig. 6.15** A linear sinkhole developed over a fracture located near Chiefland, Levy County. (Photograph by Kenneth Hill)

Fig. 6.16 Regularly spaced fractures in limestone in the wall of a quarry near Lecanto, Citrus County, Florida

Fig. 6.17 Dissolution-enlarged joints (fractures) in the Eocene Ocala Limestone exposed in a quarry near Lecanto, Citrus County, Florida. Note the rubble zone at the top of the exposure

(Florida Caverns State Park), and Marion County. In addition, there are numerous small, dry caves in the higher areas of Jackson, Alachua, Citrus, Marion, and Hernando counties (Florea et al. 2003; Florea and Vacher 2006a, b; Florea 2008a).

#### 6.3.1.9 Bedding Planes and Unconformities

Bedding planes are natural partings in sediment or rock caused by interruptions in sedimentation, erosion, or changes in sediment accumulation patterns or sediment types. Interruptions in sedimentation may be a result of short-term storm events or major time intervals where sedimentation is interrupted by tectonic or depositional conditions. Erosion can cause bedding planes during storms and when sea level drops exposing the sediment to subaerial weathering. Changes in sediment types, such as changes from accumulation of carbonate sediment to siliciclastic clay, can also result in bedding plane-like contacts. Bedding planes typically reflect planar surfaces that paralleled, more-or-less, the sediment surface at the time of accumulation. Burial and diagenesis often accentuate and distort bedding plane surfaces. Groundwater flow along the partings at bedding planes may further accentuate them by dissolution, as well.

Bedding planes in limestone may be poorly developed in the eogenetic limestones of Florida (Fig. 6.18) because (1) the rocks have not been deeply buried, (2) diagenesis related to compaction and lithification have not yet occurred, and (3) groundwater circulation has not enlarged them. The arrows in Fig. 6.18 delineate some of the more obvious bedding plane traces, which represent poorly developed partings caused by short-term interruptions in sedimentation, changes in limestone texture that developed during deposition in Oligocene time, or incipient dissolution at former water-table positions. Although these bedding planes have not been significantly enhanced by diagenesis or enlarged by dissolution, there are local openings that could cause losses of drilling fluid during standard penetration testing and that could produce increased water-well yields. In addition, the Suwannee Limestone shown in Fig. 6.18 is a grainstone to packstone through which small amounts of water may pass

**Fig. 6.18** Relict water-table positions or incipient bedding planes (arrows) in the Suwannee Limestone in a quarry face near Brooksville, Hernando County, Florida. Quarry face is approximately 6 m high

**Fig. 6.19** Cross-bedded (beds that dip down to the left) and planar, horizontal bed in the oolitic Miami Limestone, Dade County. Location: 25.7062°N, 80.2607°W

within the intergranular pore spaces, thus making this stratum a dual-porosity aquifer.

Where bedding planes are well developed in eogenetic limestones, they are either a result of distinctive breaks in deposition, such as occur between beds in the cross-bedded oolitic rocks of the Miami Limestone in southern Florida (Fig. 6.19) or at dissolution-enlarged unconformities (Fig. 6.20) or water-table positions. The exposure shown in Fig. 6.18 is about 10 m above the unconformity between the Eocene Ocala Limestone and the Oligocene Suwannee Limestone (Fig. 6.20) in the same quarry. The contrast in bedding plane development between the two exposures in

**Fig. 6.20** Dissolution-enlarged unconformity (black arrow) between the Eocene Ocala Limestone (To) and Oligocene Suwannee Limestone (Ts) and dissolution-enlarged bedding planes (white arrow) in a quarry near Brooksville, Hernando County

the same quarry is a result of the presence of an unconformity and possible paleo water-table positions in the stratigraphically lower section shown in Fig. 6.20.

When sediment or rock is exposed to weathering and erosion over geologically long periods of time and is then buried by another layer of sediment, the break between the two strata is termed an unconformity. The unconformity is typically a result of oscillations of sea level: a common occurrence in Florida's geological history. Unconformities in Florida are rarely simple planes between two strata of different ages. Ancient soils, small and large irregularities in the relative elevation of the eroded surface, thick rubble, and conduit-riddled strata also can characterize unconformities.

Figure 6.20 is an example of the unconformity between the Eocene Ocala Limestone and the Oligocene Suwannee Limestone in Hernando County, Florida. The unconformity between the Ocala and Suwannee represents a period of time during which sea level was low and the upper surface of the Ocala Limestone was exposed to weathering for several million years. The younger Suwannee Limestone was deposited later, when the sea advanced and again covered the area. Note, therefore, that the parting between the Ocala and Suwannee represents an interval of time and loss of sediment record between the two formations. This missing section represents a time gap of several million years, during which erosion was removing Eocene limestone from the section and, undoubtedly, karst was forming.

The erosional contact between the two formations is often highly porous and thick enough (it is an ancient, epikarst horizon) to serve as an excellent production zone for water flowing to wells. Permeability created by erosion at the unconformity between the two formations shown in Fig. 6.20 has enhanced dissolution and re-precipitation of the limestone above and below the actual unconformity. As a result, nodular or spherical deposits of calcite have accumulated at the unconformity, and bedding planes near the unconformity have been enlarged by dissolution coupled with calcite cement deposition in contiguous strata.

Paleosols (ancient soils; Valentine and Dalrymple 1976; Wright 1986) may develop and be preserved at unconformities; these are soils developed by weathering during the time of exposure. If paleosols remain at the unconformity, they serve as excellent markers of the unconformity during drilling and exploration.

Figure 6.21 shows an example of a paleosol developed in phosphatic, clayey sand in central Florida. Here, during the Late Miocene and Early Pliocene, sea level had retreated, exposing the phosphatic sediment of the Peace River Formation to intense chemical weathering. The phosphatic sediment in the upper part of the paleosol was removed by dissolution, resulting in "frothy", soft sediment (Puri and Vernon 1964) and re-precipitated as secondary aluminum-phosphate minerals (i.e., wavellite, crandallite, etc.) in the lower portions of the soil (Altschuler et al. 1951, 1956; Van Kauwenbergh et al. 1990). This soft horizon at the top of the Miocene Hawthorn Group sediments was termed the "leached zone" in the Central Florida Phosphate District (Altschuler et al. 1951; Cathcart 1963a, b, 1964, 1966). This unconformity and paleosol extends beyond the phosphate district, however, occurring throughout central and northern Florida, wherever the Miocene Hawthorn Group was in the shallow subsurface. Where the Mio-Pliocene weathered zone is superimposed on

**Fig. 6.21** The "leached zone" a paleosol developed at the unconformity overlying the upper-most Hawthorn Group sediments during the late Miocene and early Pliocene. LZ = leached zone paleosol, PP = Plio-Pleistocene marine terrace sand. Hookers Prairie Mine, Polk County

limestone as opposed to clayey sediments of the Hawthorn Group, the weathered surface is a karstic horizon characterized by residual clay and weathered limestone.

Where the weathered surface/paleosol is superimposed on limestone, weathered siliciclastic and/or carbonate clay, rubble, limestone pinnacles, and other indicators of weathering may occur. Voids caused by dissolution or rubble accumulation may cause losses of drilling fluid. In some cases, the voids may be large enough to accept movement or collapse of cover materials resulting in sinkhole development.

The unconformities between Tertiary Series (Eocene/Oligocene, Miocene/Pliocene, etc.) are often distinctive and easily identified. These unconformities typically represent major sea-level fluctuations largely caused by global tectonics. In the younger, Plio-Pleistocene sediments, sea-level fluctuations were caused by glacial events, and they occurred over much smaller time intervals than the Tertiary sea-level excursions.

In southern Florida, these geologically short-term, Pleistocene transgressions and regressions of the sea have left a complicated record. The unconformities that developed during low sea stands have been called "discontinuity surfaces" (Perkins 1977).

Since bedding planes and unconformities in limestone and other carbonate sediments and rocks often function as zones of preferential groundwater flow, a fractured rock with well-developed bedding planes or unconformities results in a three-dimensional network of flow paths. If the soil- or groundwater is chemically undersaturated with respect to the minerals in the sediment or rock, dissolution can enlarge both the fractures and bedding planes. The result is well-integrated, karstic complex of secondary porosity and an excellent aquifer.

### 6.3.2 Residence Time and Contact Surface Area

The last major factors affecting the development of karst features are (1) residence time of the water in contact with the carbonate sediment or rock and (2) the area of the rock or sediment surface in contact with the water. These are discussed below.

#### 6.3.2.1 Effect of Surface Area and Residence Time

The ability of water to chemically interact with minerals in soil or an aquifer depends on both (1) how much mineral, in terms of volume and surface area, is available to react with the water and (2) how long the water remains in contact with the mineral. To illustrate this concept, let us consider a glass of iced tea. If you dropped a large, single crystal of sugar in the tea and did not stir it, it might take many minutes for it to dissolve because the surface area of the crystal is small relative to the volume of tea in the glass and, with no stirring of the tea, the rate of dissolution is controlled by how quickly the dissolved sugar can diffuse into the tea and away from the crystal surface. On the other hand, if the sugar is granulated so that the surface area is greatly increased, there is significantly more surface area upon which dissolution can occur. If you stir the granulated sugar into the tea, mixing of the sugar with the tea is no longer dominated by slow diffusion. Advection, or turbulent mixing, makes it possible for the dissolved sugar to disperse in the tea and for dissolution of the crystalline sugar to repeatedly occur between fresh tea and sugar surfaces.

This analogy explains the effects of residence time and surface area, but it does not emphasize the much more dramatic contrasts in residence time and surface area that occur in groundwater systems. To emphasize these contracts, we must consider flow through a dual- porosity aquifer.

#### 6.3.2.2 Flow Systems in a Dual-Porosity Aquifer System

Groundwater flow through fractures and conduits is known as conduit or fracture flow (Fig. 6.12). Intergranular flow through the pores between particles in the limestone and through small-scale voids in the host rock is known as matrix flow. Groundwater flowing through both types of pore space can dissolve limestone and create karst landforms, especially in eogenetic karst systems where both forms of porosity coexist.

Intergranular, matrix flow is typically slow and laminar, with water slowly migrating past and chemically interacting with mineral surfaces. In contrast, water moving through conduits may be subject to laminar flow if the hydraulic gradient is low, turbulent flow under high hydraulic gradients, or variable flow with recharge events and changes in the hydraulic gradient. In either of these scenarios, the volume of water in contact with the walls of the conduit is low. Under laminar flow, diffusion may dominate over dispersion, and under turbulent flow dispersion dominates. Thus, there are wide ranges in the time and distance that water must travel in an eogenetic limestone aquifer before chemical equilibration reduces speleogenesis potential.

The rate and distance over which dissolution can occur is a function of the hydraulic conductivity of the medium through which the water flows and the hydraulic gradient. For systems where laminar flow exists, groundwater flow follows Darcy's Law (Freeze and Cherry 1979; Domenico and Schwartz 1997)

$$Q = AK\frac{\Delta H}{L}, \tag{6.1}$$

where Q is the flow rate (volume/time), A is the cross sectional area through which flow occurs, K is the hydraulic conductivity, ΔH is the hydraulic gradient (head), and L is the flow path length. If one considers the head (ΔH), flow distance (L), and cross-sectional area to be constant, then the rate of flow of the water (Q) through the medium is a function of the hydraulic conductivity (K). In granular media (say medium sand composed of calcite or aragonite) the horizontal hydraulic conductivity ($K_h$) is on the order of $10^{-7}$ to $10^{-4}$ m/s, whereas the horizontal hydraulic conductivity of a fractured limestone may be as high as $10^{-4}$ to $10^{4}$ m/s (Heath 1983). In other words, transfer of water through fracture and karst conduits is orders of magnitude more efficient in conduit flow than through the porous matrix. Chapter 4 discusses the ranges of hydraulic conductivities of the hydrostratigraphic units in Florida.

Within the blocks of rock between fractures or karst conduits (Fig. 6.12), hydraulic conductivities are relatively low and water transport through the rock matrix is along a gradient determined by the gradient field in the surrounding fractures and conduits. Within the fractures, hydraulic conductivities are high, so flow preferentially follows the larger fracture/conduit network with an overall gradient field in the direction of regional flow but with local, gradients determined by orientations of the conduits.

Owing to the relatively low hydraulic conductivities within the matrix materials between conduits, the movement of the water is slow, resulting in long residence times. The water is also in contact with the many small particles surrounding the intergranular pore spaces, which have high relative surface areas. For example, visualize water filling the pores in the grainstones shown in Fig. 6.2. The small particles in this limestone are the granulated sugar crystals and the slow-flowing water moving through the pore space is the tea, constantly, but slowly, passing over the crystals. Thus, one would expect chemical equilibration of water with the minerals of the host sediment over relatively short distances and times. After equilibration has occurred, the water can no longer dissolve the particles and it continues to flow through the matrix without effect.

In the fractures, transport is rapid and the specific surface areas of the walls of the conduits are low in comparison to the volume of water moving through the conduit. Chemical equilibration between the water and minerals of the wall rock is, therefore, limited to water in close proximity to the wall rock at any moment of time and the rate and turbulence of the water moving down gradient. It is this combination of slow and fast rates of flow, complex networking and interactions of matrix and conduit flow, variations in rates of equilibration with different local flow systems, and reactions with organic-rich water with matrices containing different carbonate minerals that characterize eogenetic karst in Florida. Given enough time, enlargement of conduit walls is far more effective that dissolution in the rock matrix, so growth and enlargement of caves tend to dominate during speleogenesis.

Because the exact locations and nature of aquifer heterogeneities, such as fractures and conduits, are rarely known, we can sometimes approximate their general effects rather than trying to model them explicitly. For example, in Fig. 6.12, the overall effect of the combination of flows within the conduits and matrix is the net groundwater flow direction. We approximate the flow in karst aquifers by assuming that they behave hydrologically as "equivalent porous media." The equivalent porous medium is the theoretical aquifer material that approximates the general combined effects of the conduits and matrix (Freeze and Cherry 1979; Hickey 1984; Huntoon 1994; Domenico and Schwartz 1997; Scanlon et al. 2003; Hill 2008). While we conceptually treat carbonate aquifers at a regional scale (square kilometers in area) as behaving like an isotropic and homogeneous medium with a porosity/permeability system similar to a simple porous sand medium, reality in most cases is far from such simplicity. See Vacher and Florea (2015) for an excellent discussion of the evolution and reality of the equivalent porous medium concept in karst aquifers.

Success in conceptualizing and modeling groundwater flow through a karstic aquifer is scale-dependent. In some cases, we may be able to treat an eogenetic karst aquifer as being more-or-less isotropic and homogeneous using concepts of an equivalent porous medium on a regional scale because the complex heterogeneities and interplay of matrix and conduit flow are spatially well integrated. At the local scale, integration of flow mechanisms is less valid and one must consider the effects of conduit versus matrix flow. At the local scale, consideration of processes and landform development must account for the complex interplay of the many variations in aquifer properties and groundwater flow that characterize an eogenetic karst system. The vast majority of aquifer performance tests completed within the state routinely fall on type curves for porous, permeable media and, whenever a fracture is intercepted, its signature is often very clear, so the effects of heterogeneities in the aquifer and flow system may be evident.

#### 6.3.2.3 Effects of Contact Surface Area on Karst Development

Table 6.2 compares the relative specific surface areas of a gram of sediment with different particle sizes and of a fracture in a block of rock of equivalent mass. Clearly, the surface area of a fracture is orders of magnitude less than in sediment made of silt- and clay-sized particles of calcium carbonate, or even the larger shell fragments of a coquina.

When one combines the permeability and surface area constraints of the medium through which the groundwater flows, two scenarios are evident (Table 6.3).

**Dominantly Matrix-Flow Scenario** The specific surface area (surface area per unit volume of rock) of the sediment affects how quickly chemical interactions can occur. The more surface of a reactive solid in contact with a volume of water, the more rapidly dissolution and chemical equilibration will occur. This is why water flowing through the intergranular pores of a limestone matrix comes to chemical equilibrium more quickly than water flowing through a conduit. Under matrix flow (Table 6.3) the high surface area of particles in a porous carbonate grainstone results in relatively rapid chemical equilibration driven by diffusion. After being in contact with the grain surfaces over a relatively short time, the water comes to chemical equilibrium and further dissolution is limited to locations where the $P_{CO2}$, temperature, or water volume change or where mixing occurs.

**Table 6.2** Orders of magnitude differences in total surface area in 1 g of close-packed sediment or rock as a function of particle size

| Particle type and size (cm) | Relative surface area ($cm^2/g$) |
|---|---|
| Single planar fracture in a block of rock 1 $cm^2$ on a side | $10^{-1}$ |
| Gravel (0.2) | 10 |
| Sand (.005) | $10^3$ |
| Silt (0.0002) | $10^4$ |
| Clay (0.00002) | $10^6$ |

**Table 6.3** Comparison of the relative consequences of groundwater/carbonate sediment interactions under conduit and matrix flow conditions

| Property | Conduit flow | Matrix flow |
|---|---|---|
| Relative local permeability | High | Low |
| Residence time of a parcel of water in contact with a specific volume of sediment or rock | Low | High |
| Relative surface area of water/sediment contact interface | Low | High |
| Probable nature of groundwater flow through media | May be turbulent | Laminar |
| Importance of chemical diffusion to dissolution of carbonate sediment | Low | High |
| Relative distance along a flow path over which water can chemically equilibrate with the sediment | Long | Short |
| Resulting dissolved solids content of groundwater | Likely relatively low | Likely relatively high |
| Response of spring discharge to a high rainfall event | Rapid | Slow |
| **Speleogenesis scenarios** | • Water comes to chemical equilibrium slowly and over long distances<br>• Dissolution increases hydraulic diameter quickly and efficiently<br>• Encourages development of extensive and integrated conduit networks | • Water comes to chemical equilibrium relatively quickly and over short distances<br>• Void space distributed between particles and initially small<br>• Conduit networks limited in extent |

Consider water recharging a poorly cemented grainstone at the upper limestone surface. Dissolution affects the upper meter or so of the limestone by dissolution of grain surfaces. If the grainstone still contains aragonite or high-Mg calcite, micritization of the particles may also occur. Void spaces developed by dissolution will be typically small, and evidence of karst may be limited to irregularities in the limestone surface, rubble and/or fine-grained carbonate debris, and the odd conduit.

Conduits can develop in matrix-flow scenarios if (1) there are fractures or depositional pathways that allow for some preferential groundwater flow or (2) random dissolution patterns result in locations of preferential water flow. A preferential flow path receives more water capable of dissolution than other flow paths and therefore becomes even more permeable, which results in more dissolution and so on. This feedback loop is critical to the development of karst conduit networks.

The more conventional possibility is that fracturing and development of preferential flow pathways begins when cementation of the carbonate sediment is initiated. As soon as the carbonate sediment begins to develop brittle characteristics, tidal and tectonic stresses (Sect. 6.4.4) will develop fractures and zones of enhanced permeability, which preferentially capture flow and initiate conduit development.

Even in the absence of fractures, water is able to find paths of least resistance through granular material. Recent investigations of the flow of water through unsaturated (dry) sand reveal the visible manifestations of slightly preferential flow paths through even a relatively homogenous granular medium. Glass et al. (1988, 1989) termed the pattern of wetting of an unsaturated porous medium “flow fingers” after the characteristic pattern of the wetting front as it penetrates the substrate. These flow fingers are thought to be randomly located and related to small-scale heterogeneities in the sandy sediment (Glass et al. 1988, 1989; Pendexter 1996; Pendexter and Furbish 1991). They consist of more-or-less cylindrical wetted zones that initially extend downward from the land surface during a precipitation event as a result of capillarity. Flow fingers developed at the laboratory scale are small, with depth and spacing on the order of centimeters. After development, they continue to function as locally enhanced, small-scale zones of increased hydraulic conductivity. De Rooj and Cho (1999) studied the properties of flow fingers and, among other things, considered the geometries of the flow fingers, including the fact that they tended to terminate at the water table, either by widening or dissipation.

The cylindrical pattern and vertical distribution of flow fingers and the fact that they seem to terminate at the water table are consistent with the patterns of solution pipes observed in Florida limestone (Fig. 6.22; Parker and Cooke 1944; Parker 1955; Perkins 1977). Only the scale differs. Development of the solution pipes may be a result of preferential dissolution within the flow fingers. Oncc pathways developed as a result of flow-finger development, they may have continued to grow by dissolution along the preferential pathways. Assuming that the upper surface of a carbonate sand or poorly cemented grainstone was unsaturated and in the vadose-zone at the time that the solution channels began to develop, the distribution and size of the solution pipes may have been controlled by the pattern of wetting of the unsaturated sand. See Sect. 7.3.2.2 for more on the origin of solution pipes.

**Fig. 6.22** Solution pipe developed in the epikarst at the top of the Oligocene Suwannee Limestone in the Vulcan Mine, near Brooksville (Hernando County)

Flow fingers and solution pipes are primarily near-surface phenomena; however, they both represent enhanced flow of water along slightly preferential flow paths that likely penetrate deeper into the aquifer and along which a conduit network may be formed.

**Dominantly Conduit-Flow Scenario** Once water finds a conduit through which to flow, the speleogenesis scenario changes dramatically. Under conduit flow, the water flow may be turbulent and the dissolution surface is limited to the walls of the conduit. Thus, the residence time of the water with respect to a specific location on the conduit wall is brief (Table 6.3). Under conduit flow, conduit expansion is favored as dissolution preferentially increases the diameter of the conduit. The consequence is that water that is undersaturated with respect to the minerals of the host rock can travel further and increase the extent of the karst conduit network. Because karst conduiting is typically within rock, fracturing can occur at any time in the history of the stratum as long as confining pressures are low.

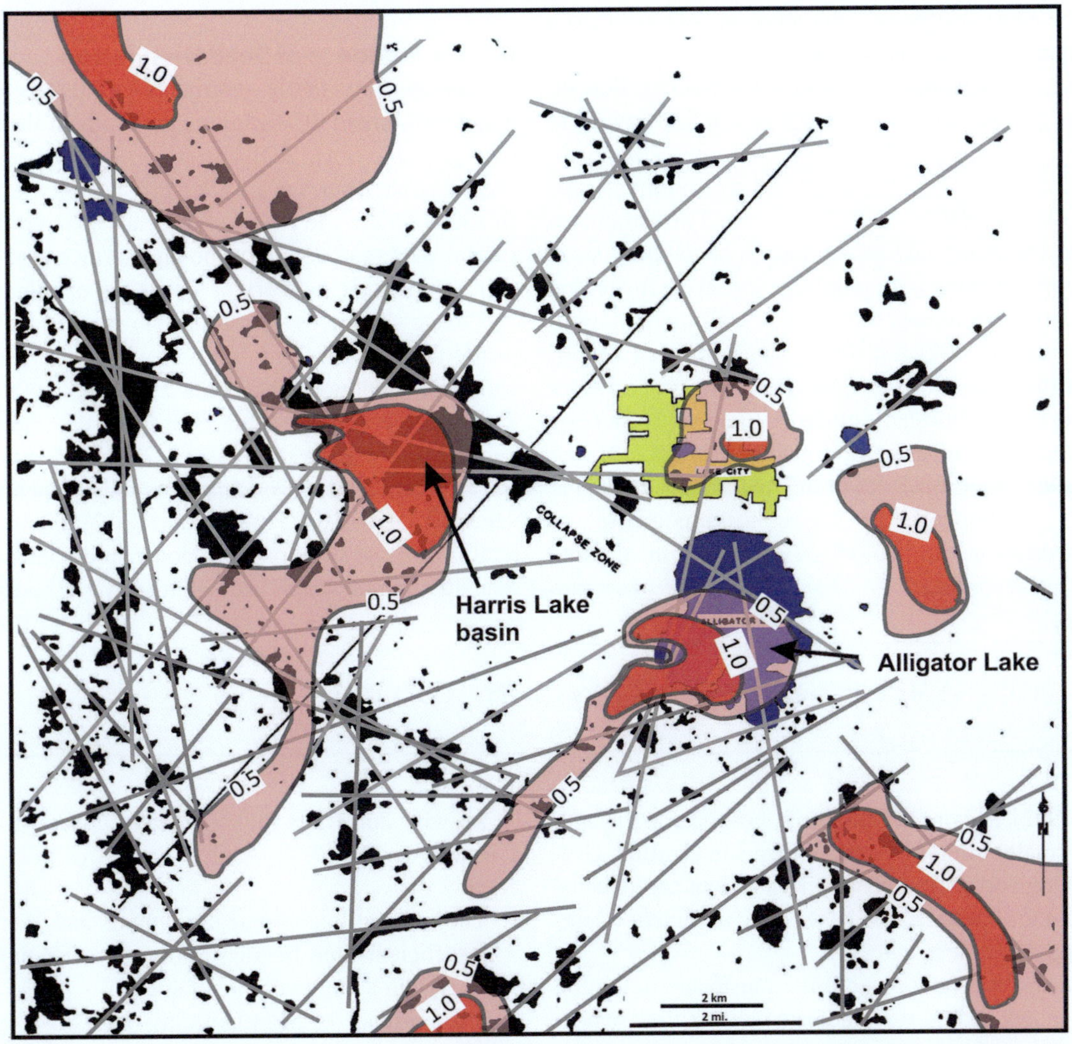


Fig. 6.23 Distribution of photolineaments and orthophosphate in Upper Floridan Aquifer water near Lake City, Columbia County, Florida in 1975. Alligator Lake is located at 30.170°N, 82.625°W. Gray lines are photolineaments, irregular black polygons are closed topographic depressions (paleosinkholes), and tinted areas delineate elevated orthophosphate-concentration contours (o-$PO_4^{3-}$ mg/L). The regional hydraulic gradient is to the southwest. (Modified from Lawrence and Upchurch (1976))

These two scenarios have been presented as different stages in the evolution of an eogenetic karst terrain; however, it is important to consider the impact that the initial permeability of the limestone has on the nature of the conduit network.

The contrast between the permeability of the matrix and the permeability of the preferential flow paths is critical in determining whether an extensive karst network is likely to form at all. In an eogenetic limestone with very high initial permeability, matrix flow can dominate the system because fracture permeability may not be so many orders of magnitude higher than matrix permeability. In a telogenetic limestone with very little matrix permeability, it is possible to start out with a fracture-dominated system that transforms into an extensive conduit system.

**Fate of Matrix and Conduit Flow Systems** As long as cementation has not occluded interconnected, intergranular pore spaces, matrix flow and rapid equilibration of the water with the carbonate sediment will continue. Equilibration under matrix flow is a self-limiting process and weathering at the upper surface of a carbonate sediment or rock will progress slowly.

Given the higher local permeabilities of conduits, more and more water will flow to the conduits and, over time, growth of the conduit network will continue. The result is a karstic stratum with extensive conduiting.

Figure 6.23 is an example of the effect of fracture flow on groundwater quality. Shown in the figure is a four township area (373 km$^2$) centered on Lake City, Columbia County, in north Florida. Lake City is named for a prominent lake (Alligator Lake; 30.169°N, 82.625°W) located just southeast of the town. This karst lake was prone to developing sinkholes and draining about once a decade. It is also situated on the Cody Scarp, a prominent fluviokarst escarpment in north Florida (Upchurch 2002). Lawrence and Upchurch (1976) investigated the effects of the Cody Scarp on groundwater quality in the Upper Floridan Aquifer (UFA) near Lake City. They detected two plumes of orthophosphate- ($PO_4^{-3}$) rich water migrating along fractures in the underlying limestone (Fig. 6.23). Orthophosphate tends to precipitate as carbonate-hydroxyapatite ($Ca_5(PO_4,CO_3)_3(OH)$) in alkaline groundwater according to the following reaction

$$5Ca^{2+}_{aq} + 1.5H_3PO_{4aq} + 1.5H_2CO_{3aq} + H_2O \rightarrow Ca_5\left(PO_4,CO_3\right)_3\left(OH\right)_{carbonate\ hydroxyapatite} + 8.5H^+ \quad (6.2)$$

(Upchurch 1992), so concentrations of phosphate in groundwater are normally limited to trace amounts in alkaline, UFA water. When high orthophosphate concentrations are detected in limestone, the above reaction has not had sufficient opportunity to precipitate the apatite.

The elevated concentrations of orthophosphate in the linear plumes shown in Fig. 6.23 indicate conduit flow. Under matrix flow, the orthophosphate would quickly be precipitated and there would be no indication of the plumes detected near Lake City. The plume originating at Alligator Lake (Fig. 6.23) was a result of Alligator Lake draining at the time of the Lawrence and Upchurch (1976) study. Water from some of the wells down gradient from Alligator Lake was discolored and contained sediment from the lake bottom (plant debris, fish scales). A second plume originating at Harris Lake, a small, ephemeral sinkhole lake that occupies a large, internally drained depression southwest of Lake City, was also identified.

These two plumes illustrate the distances groundwater can travel in karst conduits without chemical equilibration of the water with the surrounding limestone. Plumes as much as 4 or 5 km in length are not uncommon in conduits in Florida. The lack of chemical equilibration and resulting precipitation of carbonate-hydroxyapatite within the plumes is due to high water volumes in contact with limited limestone surface area. Within the plumes, water remains slightly acidic, discolored by dissolved and particulate humic substances, and enriched in surface-water-related chemical constituents, such as orthophosphate and nitrate.

#### 6.3.2.4 Evidence of Shell Dissolution

Unconsolidated, shelly sediments can undergo dissolution and development of shallow depressions in the land surface over geologic time. Shelly strata present a special case with respect to shell surface area and groundwater residence time. In shelly strata, the flow systems are through the shell and sand matrix, and contact time and surface area are initially controlled by the network of interstitial openings.

Coarse shell beds with little fine-grained matrix are subject to rapid flow of groundwater with short residence times and relatively low contact areas. Figure 6.24 is an example of a permeable, sandy, shell bed from the Plio-Pleistocene Tamiami Formation in Sarasota County. The pit where this picture was taken is world-famous for the quality of preservation of the fossils (Portell et al. 2012). Fossils commonly retain traces of their original color and details of the ornamentation on the shells are remarkably well preserved.

So why don't the shells show evidence of dissolution? The answers are:

1. They are still geologically young and there has been insufficient time for dissolution to cause any significant change in a large volume of shell;
2. The presence of abundant clay- and silt-sized carbonate fragments created by mechanical abrasion of the shell material at the time of deposition, causes groundwater to

**Fig. 6.24** Shell bed with sand and organics. Note the well preserved shells. From a shell pit in the Plio-Pleistocene Tamiami Formation near Sarasota, Sarasota County. (See Portell et al. (2012) for details of the geology of this pit)

come to chemical equilibrium with the carbonate sediment quickly, thus limiting additional dissolution of the coarser sediment deeper in the section; and
3. Residence time of the water in contact with the coarser sand and larger sized shell is very brief, which limits chemical reactions with large shells.

In clay-rich, shelly sediments, the fine-grained matrix slows groundwater flow, lengthening contact time and creating very large surface areas only limited by the amount of fine-grained carbonate in the stratum. Carbonate content and surface area in these shelly sandy clays to clayey sands is normally high because of the abundance of mechanically abraded, sand-sized shell fragments. In lower permeability, clay- or silt-rich sediments, chemical equilibration is rapid and restricted to the uppermost few centimeters of the stratum. As a result, shells in clayey sediments are typically very well preserved and show little or no evidence of dissolution (Fig. 6.25). Here,

1. The water residence time is high because of the low permeability of the sediment;
2. Chemical equilibration of the groundwater with any carbonate content is rapid; and
3. After equilibration, the water is no longer capable of dissolving shell.

This is the reason that, if cover-subsidence sinkholes (Chap. 7) develop, they develop at the top of a shelly bed where rainwater first penetrates the sediment and not by dissolution at depth within the bed.

To determine if shell material has undergone significant dissolution examine a suitable sample of whole and fragmental shells under the microscope or a strong loupe to determine if

1. Broken corners and edges are sharp and angular or rounded (Fig. 6.25); and
2. Decorations (ribs, whorls, etc.) on complete shells have been abraded or dissolved (abraded surface will be limited to exposed areas of the shell while shells that have undergone dissolution will likely show evidence of dissolution on decorations that are protected from abrasion).

## 6.4 Fractures and Photolineaments

As discussed above, fractures are a major area of study for karst scientists because they are locations of preferential flow of groundwater and, therefore, carbonate-rock dissolution and cave development. This section revisits fractures and presents a recommended technique used to search for them. This discussion begins with definitions and ends with methods and examples because the process of photolineament analysis is so often poorly done in Florida.

**Fig. 6.25** Shell fragments from Pleistocene clayey sand from near Ormond Beach, Volusia County. Scale is 1 cm

### 6.4.1 Definitions and Properties

Identification of fracture porosity is straightforward when seen in outcrops (Figs. 6.16 and 6.17). If outcrops are not available for study or if a large area is being investigated for the presence of rock fractures or fracture porosity, geologists often perform a photolineament analysis in order to identify potential fracture-trace locations.

There is considerable confusion about the terms "fracture trace" and "photolineament", which are very commonly and incorrectly assumed to be synonymous. A fracture trace is the pattern of a confirmed vertical fracture in the underlying rock that has sufficient extent as to be visible through remote sensing with aerial photographs, satellite images, or other forms of mapping. A photolineament is an unconfirmed alignment of features that may, or may not, reflect an underlying vertical fracture (Alfieri et al. 2018).

It is important to understand that the Plio-Pleistocene cover materials that overlie Florida's carbonate strata contribute to the uncertainty in attributing photolineaments to fracture traces. If the cover materials were absent and the soil, hydrologic, and topographic features observed in aerial photographs were closely associated with bedrock, the attribution of photolineaments to fracturing is much simpler and more reliable.

Photolineaments can be caused by spurious alignments of naturally occurring features, livestock or wildlife trails, human activities, and image processing. For this reason, it is recommended that several aerial photographs from varying years and seasons be reviewed for identification of photolineaments. It's often worth going back several decades. The authors have found aerial photographs taken shortly after World War II (1950s) had excellent photographic resolution due to the higher quality surplus military film used by the US Department of Agriculture. Until an underlying fracture has been confirmed, one should always use the term photolineament in lieu of fracture trace.

#### 6.4.1.1 Fractures

Fractures are cracks that occur in many different sediments and rock types. Fractures in bedrock have been studied for many years because of their relationships to earth movements, groundwater flow, and economic mineral deposits.

Fractures typically have the following properties:

- They normally form linear patterns when viewed from above in aerial photographs and satellite images;
- They represent planes that are nearly vertical in orientation (Figs. 6.16 and 6.17); and
- They usually occur in statistically similar compass orientations.

When the joints are in limestone or other soluble rock, they may be enlarged by dissolution of the rock through percolating groundwater. Figure 6.17 shows a set of fractures that is spaced about a meter apart. The limestone has been subjected to dissolution, and the fracture surfaces are irregular and coated with clay, ferric hydroxide, and other materials washed down from above. The fractures shown are just below a rubble zone in a variably cemented limestone.

In contrast, the fractures shown in Fig. 6.16, which are also about a meter apart and stratigraphically below the horizon shown in Fig. 6.17, have not been substantially enlarged. Here, the Ocala Limestone has been fractured and water has passed through the fractures with minimal dissolution. In fact, note the white, approximately 2 cm rim on the left-hand fracture. The dark gray color of the rock is black mold that commonly grows on long-exposed, porous grainstone where moisture can be trapped in interstitial pores. The white zone is well cemented by calcite from water passing through the fracture. The lower porosity of the cemented zone limits growth of the black mold. Such features are commonly observed in eogenetic limestones in humid, subtropical climates.

A group of parallel fractures is termed a set. The parallel fractures shown in Figs. 6.16 and 6.17 constitute a set of fractures. The orientations and spacing of such fractures persisted through much of the quarry where these exposures were photographed.

Two or more sets that intersect at more-or-less consistent angles are termed a fracture system. Fractures that form sets or systems with a repeatable geometric pattern are systematic fractures. Fracture systems where the two or more orientations appear to have formed simultaneously are termed a conjugate fracture system (Hills 1972). Figure 6.26 shows a rose diagram for photolineaments, many of which have been confirmed as fractures by geophysical techniques, at a wellfield north of Tampa. There are two prominent sets of photolineaments/ fractures, one from northeast to southwest and the other from northwest to southeast. This pattern of photolineaments/fractures is common in Neogene limestones and dolostones throughout Florida.

Vertical fractures can extend to great depths. For example, Lattman and Matzke (1961) were able to determine that fractures extended to depths of over 915 m at a site in Arizona, and, using microgravity techniques, Culbreth (1988) was able to show that several large-scale vertical fractures con-

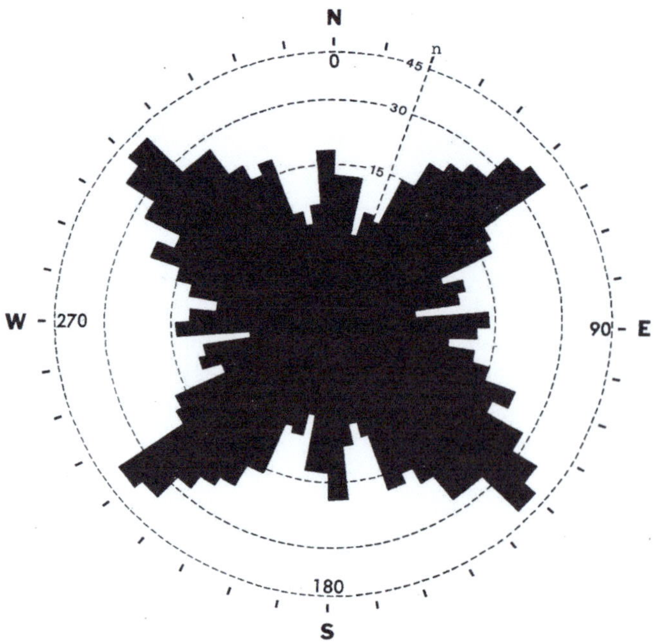


Fig. 6.26 Rose diagram illustrating the compass orientations of photolineaments at the Cross Bar Ranch Wellfield, near Loyce, Pasco County. Diagram from Williams (1985)

**Fig. 6.27** The Plant City (Florida) Lineament. At this location, the alignment is of stream and wetland vegetation. Center of image is located at 27.8247°N, 82.1171°W. Aerial photograph from Google Earth

**Fig. 6.28** (**a**) Example of a photolineament identified by the dark, linear soil tone pattern. Photograph from a site in eastern Gilchrist County. (**b**) Soil tone photolineaments near Lee, Madison County

firmed by microgravity measurements in central and south Florida extend to depths over 600 m. For example, the "Plant City lineament" (Fig. 6.27) is a straight photolineament that includes aligned streams, wetlands and swamps, and soil tones. Culbreth (1988) determined from satellite imagery that the photolineament is at least 37 km long. His microgravity data showed that the feature is a fracture trace that extends into faulted, pre-Tertiary basement rocks over 5 km below land surface.

Fractures occur at many different scales from cracks that are only visible in thin section or hand sample to large-scale features that can be seen in satellite images.

#### 6.4.1.2 Photolineaments

The features one typically utilizes to delineate photolineaments in an aerial photograph or satellite image include:

1. Alignments and elongations of three or more closed depressions[1] (potential sinkholes and sinkhole lakes);
2. Alignments of soil tones[2] (Fig. 6.28a, b);
3. Alignments of vegetation patterns (Fig. 6.27); and
4. Alignments of wetlands, lakes, stream segments and stream valleys (Fig. 6.27).

Photolineaments should be field verified prior to concluding that they represent fracture traces (Alfieri et al. 2018).

While they are not true photolineaments, alignments of closed depressions as depicted on topographic maps are considered to be "photolineaments", as well. Figure 6.29 illustrates an analysis of an area near Rainbow Springs in Marion County that was constructed using only topographic depressions.

It is important to understand that elongation of closed depressions may not reflect the presence of a fracture trace. Confidence (a qualitative estimate of the reliability of photolineament identification based on the quality and abundance of indicators) of the photolineament identification should reflect the (1) number of depressions, strength of soil tone indicators, and number of aligned stream or drainage segments along the alignment; (2) the degree of linearity within the individual depressions, stream segments, and soil tonal indicators; and (3) absence of indications of anthropogenic modifications.

### 6.4.2 Importance of Fracture Trace Identification

Identification of fractures in bedrock is important because of (1) the information their orientations may provide information as to the tectonic history of an area and (2) the economic importance of the fractures to mineral deposits, groundwater flow, and karst development. In Florida, we are mainly concerned with groundwater flow and the development or enhancement of karst features. Identification of fractures on a regional scale is important because the flow of groundwater tends to preferentially occupy the fractures owing to higher permeabilities within the fractures as compared to the bulk host rock.

[1] It is poor practice to assume that single closed depressions that are elongated have obtained their orientations by fractures since sinkholes tend to develop multiple, nearby depressions and cover materials that are washed into depressions may be affected by drainage, land slope, or other factors.

[2] Soil tones are anomalous color or darkness variations observed on the ground or in aerial photography. Often the soil tones are darker than surrounding areas because the soils have higher moisture content. Photolineament indicators are linear strips or aligned spots of darker (or lighter) soil.

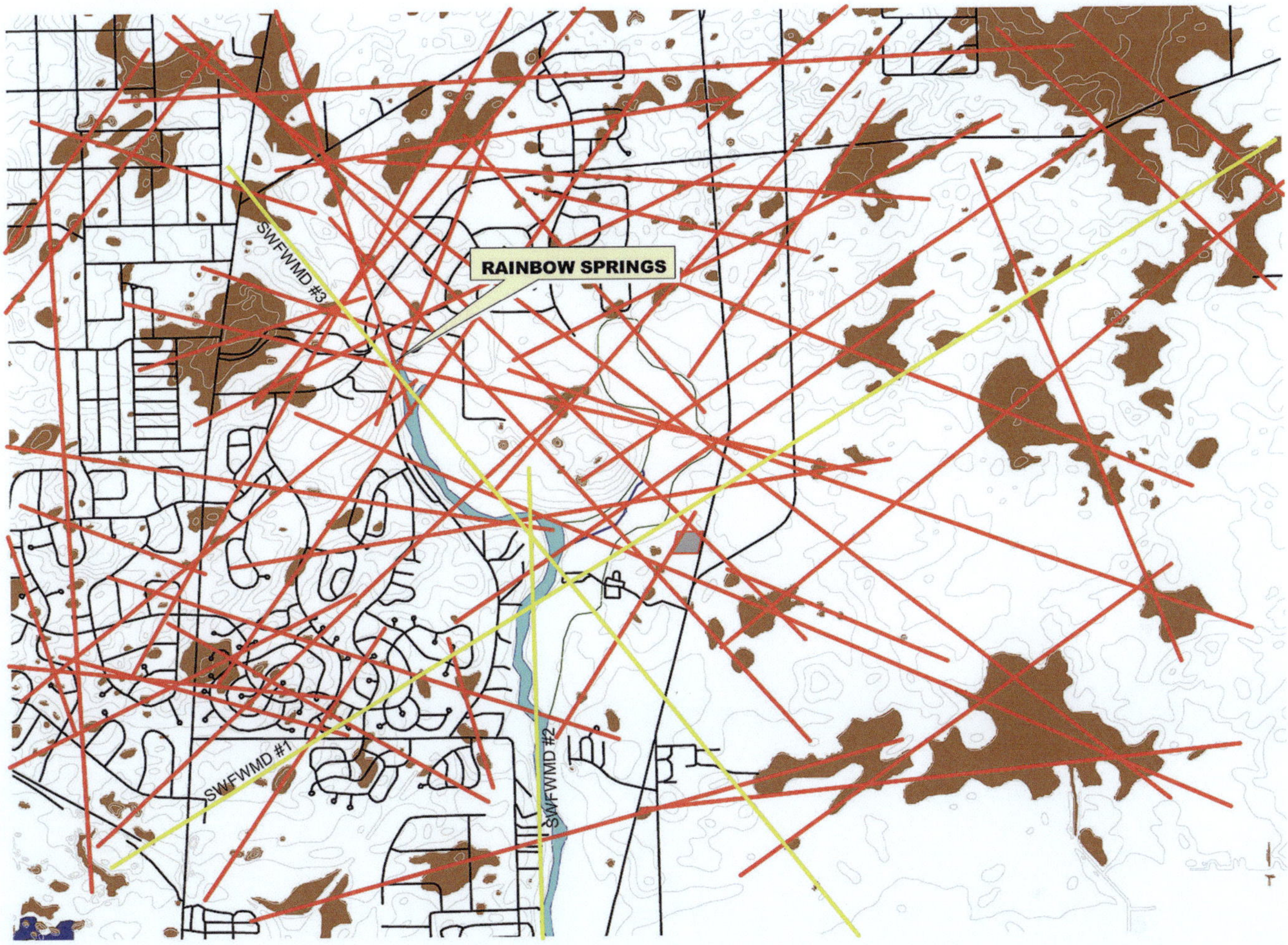


**Fig. 6.29** A photolineament analysis based on alignments of closed depressions (shaded areas) near Rainbow Springs (29.102°N, 82.438°W), Marion County. Colors of the lines indicate relative confidence in the photolineament designation (yellow = low confidence, red = high)

Lattman and Parizek (1964) showed that exploration for groundwater in fractured rock could be greatly enhanced by identification of fractures, especially intersections of fractures. Identification of fracture trace intersections has since become a standard investigation procedure for locating water-supply wells. Similarly, Littlefield et al. (1984) showed that sinkholes tend to preferentially occur on photolineaments and at photolineament intersections in Hillsborough County.

### 6.4.3 Importance of "Ground Truthing" Photolineaments

Ground truth is the key to the effectiveness of photolineament analysis for identifying vertical fracture traces. Ground truthing of the photolineaments should be utilized to confirm that the identified features represent actual fractures, but that is seldom done. More often than not, a photolineament analysis is conducted and the assumption is made that linear features reflect fracture traces. In Florida, the weakness of photolinear analysis lies in masking of bedrock features by the cover materials. When the land surface consists of bare rock or thin, residual soils, confidence is much higher that photolinears represent fractures or joints. When there is cover that ranges to over 30 m in thickness and that was deposited thousands to millions of years after the bedrock and over several possible episodes of marine transgression and regression, one simply cannot make the assumption that photolinears reflect vertical fractures with confidence.

This is not to say that photolineament analysis has not been useful and successful in Florida. The analysis has often been used to locate production wells in wellfields and injection and monitoring wells at waste disposal facilities. Most of these wells have functioned as expected. The problem with most of these analyses is that testing has not been done to prove that the photolineaments are truly fractures in the bedrock and that they differ from the surrounding areas in lithology, structure, hydraulic conductivity, or some other property that suggests a linear feature in the bedrock.

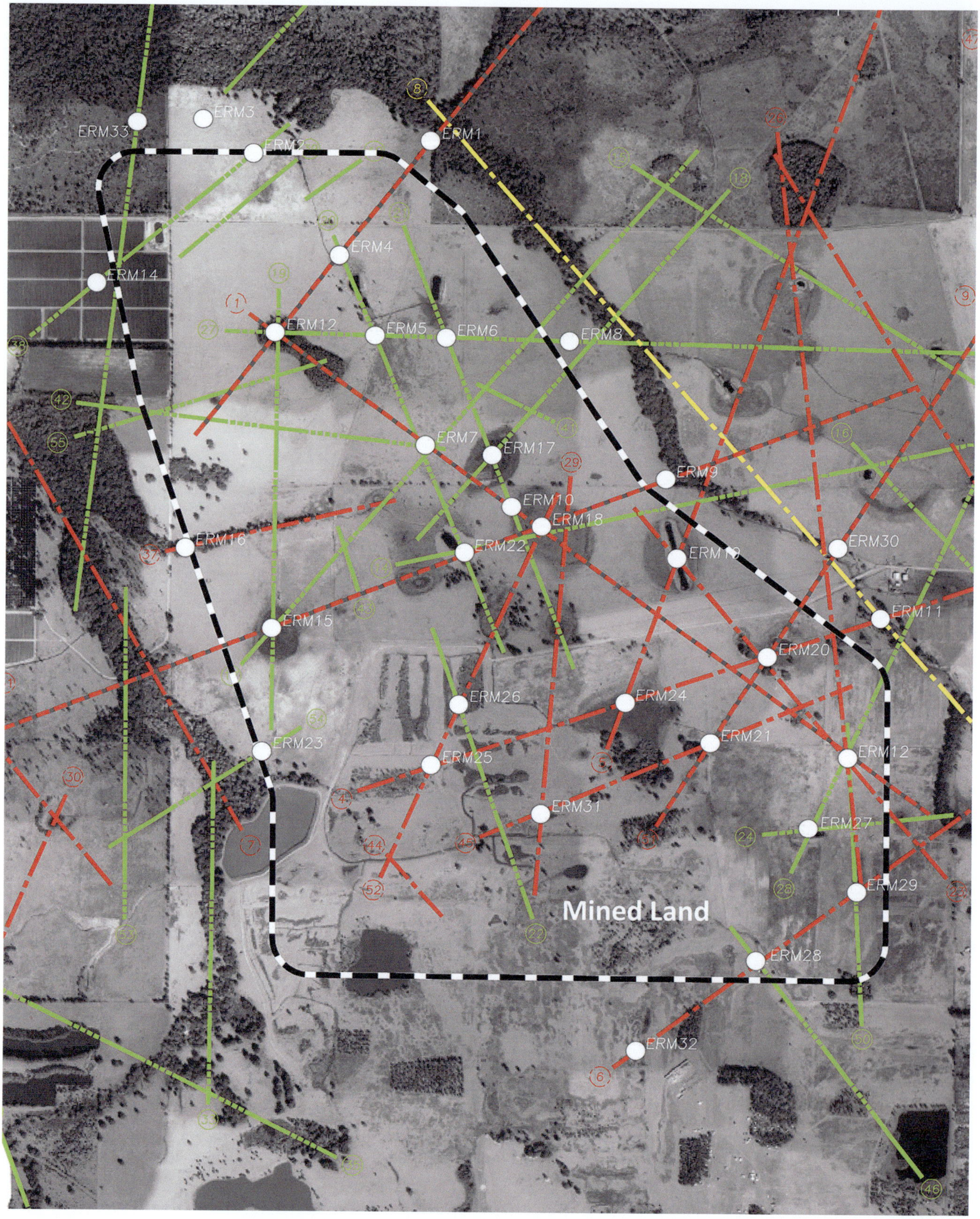


**Fig. 6.30** Pre-development aerial photograph of the C.W. Bill Young Reservoir site near Tampa, Hillsborough County showing photolineaments ranked by confidence. Dashed line is the center line of the proposed 91 m wide reservoir berm. Red = high, green = medium, yellow = low. White dots are the subsurface testing locations. Location: 27.8008 N, 82.1782 W

It may be possible to verify a system of photolineaments in Florida that consists of two sets oriented in northeast to southwest and northwest to southeast (Fig. 6.26) as representing a fracture system, but attribution of any one photolinear to a fracture in the bedrock is ill advised.

The authors are only aware of one large-scale photolineament analysis in Florida wherein the photolinears were specifically and comprehensively ground truthed. The investigation (Upchurch et al. 1999, Dobecki and Upchurch 2010) was an analysis of photolineaments at the C.W. Bill Young (Tampa Bay Regional) Reservoir site (Fig. 6.30) in Hillsborough County. The site is located on the Hardee Upland (Fig. 3.16) in an area characterized by over 120 m of Miocene Hawthorn Group sediments (clay, sand, dolostone). Prior to construction of the 445 ha reservoir, several circular features that might reflect sinkholes, as well other photolinear features of interest, including alignments of soil tones and depressions (Fig. 6.30) had been observed. Because of concerns for the safety of the reservoir and its berm, Upchurch et al. (1999) conducted a photolineament analysis and ranked the photolinears according to confidence. In Fig. 6.30, color and patterns of the lines indicate the level of confidence. Confidence was based on the number and extent of alignments of soil tones, streams, depressions, and other evidence.

The photolineaments were then extensively ground truthed using ground penetrating radar, refraction and reflection seismic profiling, standard penetration testing, and cone penetrometer testing (Dobecki and Upchurch 2010). Dots on Fig. 6.30 are the subsurface testing locations targeted specifically at the photolineaments. In addition to the standard penetration test borings used to ground truth the interior photolineaments, standard penetration tests and cone penetrometer soundings were advanced every 30 m along the center line of the proposed berm.

Figure 6.30 shows the photolineament analysis with photolineaments ranked according to relative confidence. Following geophysical and subsurface testing, Upchurch et al. (1999) were able to evaluate the accuracy of the photolineament analysis using standard methods. In 2018, Alfieri et al. conducted a *post hoc* analysis of the ground truthing exercise at the reservoir. The unlined reservoir has successfully held water with less than predicted leakage for over a decade.

**Table 6.4** Results of ground truthing the photolineament analysis at the C.W. Bill Young Reservoir construction site (Upchurch et al. 1999)

| Outcome | Percentage |
|---|---|
| Photolineaments confirmed as fracture traces, no modifications of delineation required | 5.8% |
| Photolineament confirmed as a fracture trace, length was shorter than predicted | 32.7% |
| Photolineament confirmed as a fracture trace, length was longer than predicted | 9.6% |
| Photolineament was not confirmed to represent a fracture trace | 57.7% |

Table 6.4 summarizes the outcomes of ground truthing the photolinears. It is important to note that even photolinears with high levels of relative confidence (that is, they had multiple, strong visual indicators of linear subsurface features) were found to have no or very weak subsurface indications of the causes of the photolinears over half the time. The Alfieri et al. (2018) study clearly demonstrates that there is considerable risk to the assumption that any photolineament represents a fracture trace. Over half of the 57 photolineaments identified at the reservoir (Fig. 6.30) were not found to have subsurface evidence related to karst or fracturing (Table 6.4) in bedrock. In addition, 42% of the photolinears were either shortened or lengthened after they were found to represent subsurface features. Only about 6% of the photolineaments were confirmed as fracture traces without any modification.

### 6.4.4 Causes of Fractures

Fractures are caused by many different processes, including stress release as a result of tectonic movement and removal of overburden materials, tidal stresses, tectonic movements, and other processes. Folding and faulting (tectonic movement) are not of great concern in Florida, but tidal stresses are. Most of Florida's photolineaments form sets that are oriented in a conjugate system with primary orientations in the northeast-southwest and northwest-southeast orientations (Fig. 6.26) suggesting that they were created by tidal stresses (Hills 1972). See Merritt (2004) for a discussion of earth tides in Florida.

### 6.4.5 Regional Photolineament Analyses

There have been several regional photolineament analyses in Florida. The first was by Vernon (1951). He produced a map (Fig. 6.31) showing proposed lineaments and faults in central Florida. This map is controversial because many of the proposed faults have yet to be confirmed by ground truthing investigations. Note that the predominant orientation of Vernon's photolinears was northwest to southeast.

In 1988, Culbreth prepared a photolineament map (Fig. 6.32) based on satellite image data. The use of satellite images limited the photolinears to features large enough to be observed by the satellite. Culbreth evaluated a number of the major photolineaments using microgravity techniques. He determined that the photolineaments he investigated all had basement-related tectonic features (see the discussion of the Plant City lineament above).

In 1973, the Florida Department of Transportation prepared a photolineament map that encompassed much of the

**Fig. 6.31** Photolineament map of Vernon (1951)

state (Fig. 6.33). Finally, the U.S. Army Corps of Engineers (2004) completed a draft regional photolineament map for south Florida (Fig. 6.34).

It is important to note that all four of these regional maps reveal dominant northeast-southwest and northwest-southeast sets of photolinears. One prominent orientation is a northwest-southeast set that appears to parallel basement faults proposed by Barnett (1975), Smith and Lord (1997), and Reese (2004).

## 6.5 Development of Epigenetic and Hypogenetic Karst Systems

### 6.5.1 Epigenetic Versus Hypogenetic Karst Systems

In the late 1980s and early 1990s, researchers in the western United States and elsewhere discovered that not all caves were formed by acidic groundwater derived from rainfall. One of the locations where these revelations occurred was at

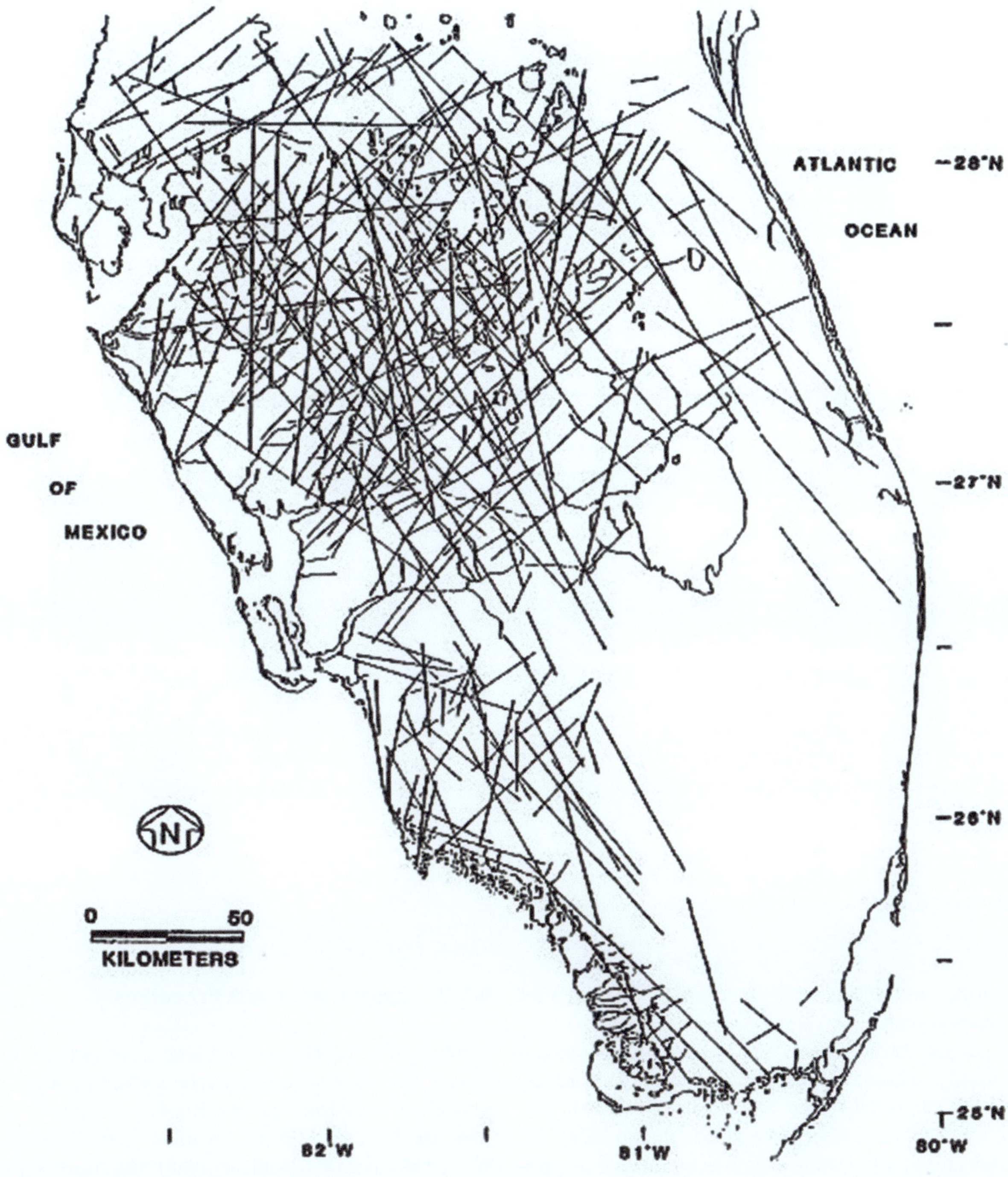


Fig. 6.32 Photolineament map of Culbreth (1988)

Carlsbad Caverns in southern New Mexico. It was discovered that the caverns were formed by acidic water derived from below (Hill 1981, 1990, 1995, 2000). Since then, karst scientists worldwide have discovered many cave systems that had similar origins (Klimchouk et al. 2017). This discovery has resulted in differentiating cave systems on the basis of the source water that formed them.

Palmer (1991) differentiated cave systems and their origins into two categories based on the origins of the chemically aggressive water responsible for speleogenesis. Epigenetic cave systems were defined as cave systems formed by dissolution of the host rock by acidic water derived from infiltrating rainwater and shallow, fresh groundwater. This epigenetic mode of cave development is the

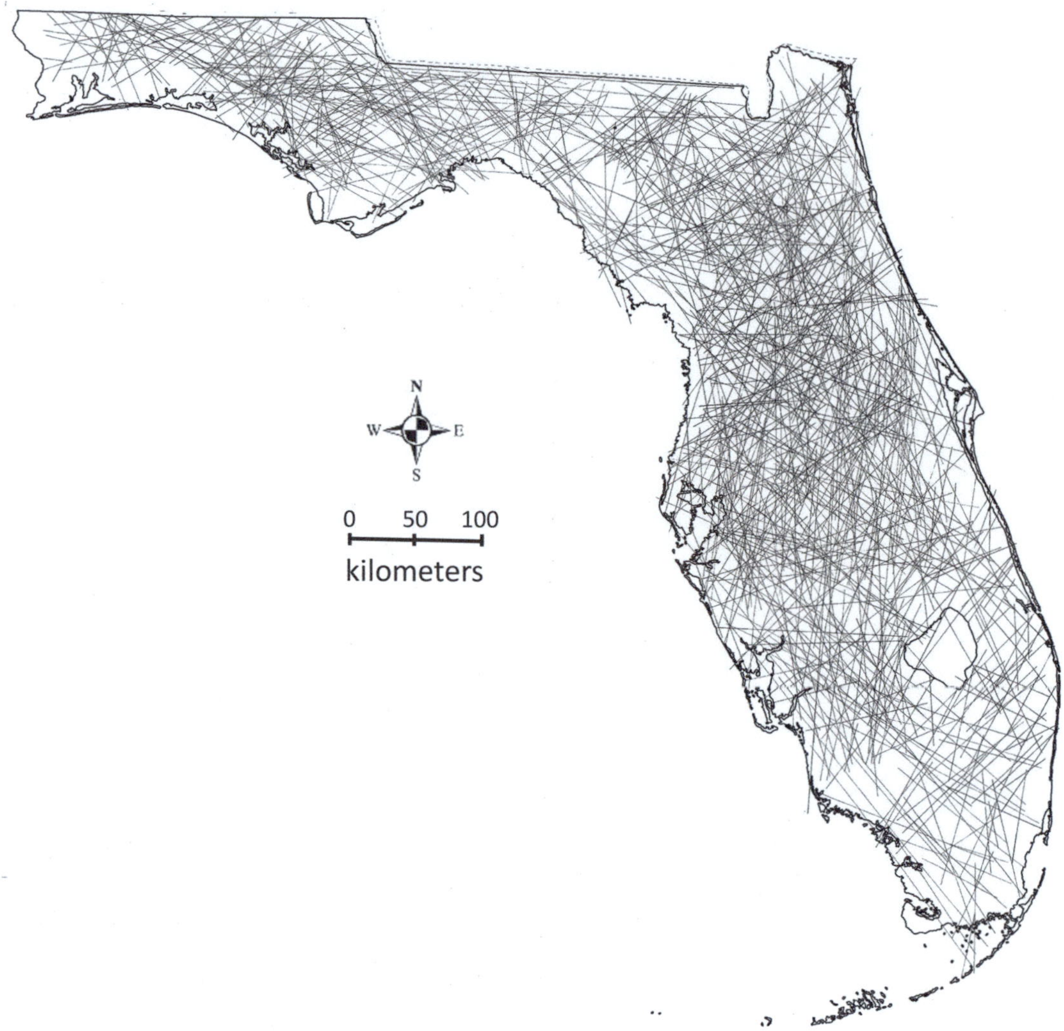


**Fig. 6.33** Photolineament map of Florida Dept. Transportation (1973)

classic process of speleogenesis observed and described for over a hundred years.

Hypogenetic karst systems, in contrast, were formed by chemically aggressive water derived in the subsurface. Palmer (1991) argued that there are two major causes of dissolution by deep, subsurface waters. Dissolution along the freshwater/saltwater transition zone was considered a hypogenetic process because the aggressiveness of the water originated at depth by groundwater mixing. The systems that Hill and her colleagues investigated were formed by the action of sulfuric acid on limestone. The sulfuric acid ($H_2SO_4$) formed by (1) oxidation of $H_2S$ gas that migrated from petroleum or evaporite deposits at depth or (2) direct dissolution of gypsum or anhydrite. The following sections are based on Palmer's (1991) classification, starting with epigenic karst systems.

### 6.5.2 Epigenetic Karst Processes

In this section, we discuss epigenetic karst, or karst formed at or near the water table or phreatic surface by water derived from above the phreatic surface. Karst development in recharge areas has been investigated for many years (Jennings 1985; White 1988; Ford and Williams 1989). The chemical processes of speleogenesis at the phreatic surface were discussed in Chap. 5. Carbon dioxide combines with water to form carbonic acid, which causes dissolution of carbonate minerals in the aquifer. Other ligands, such as humic substances, may also cause dissolution of the host carbonate rock. The key to epigenetic karst is that the chemically aggressive water began to increase in acidity at or near the land surface and incorporation of $CO_2$ and formation of carbonic acid continues through the vadose zone (White 1988; Ford and Williams 1989).

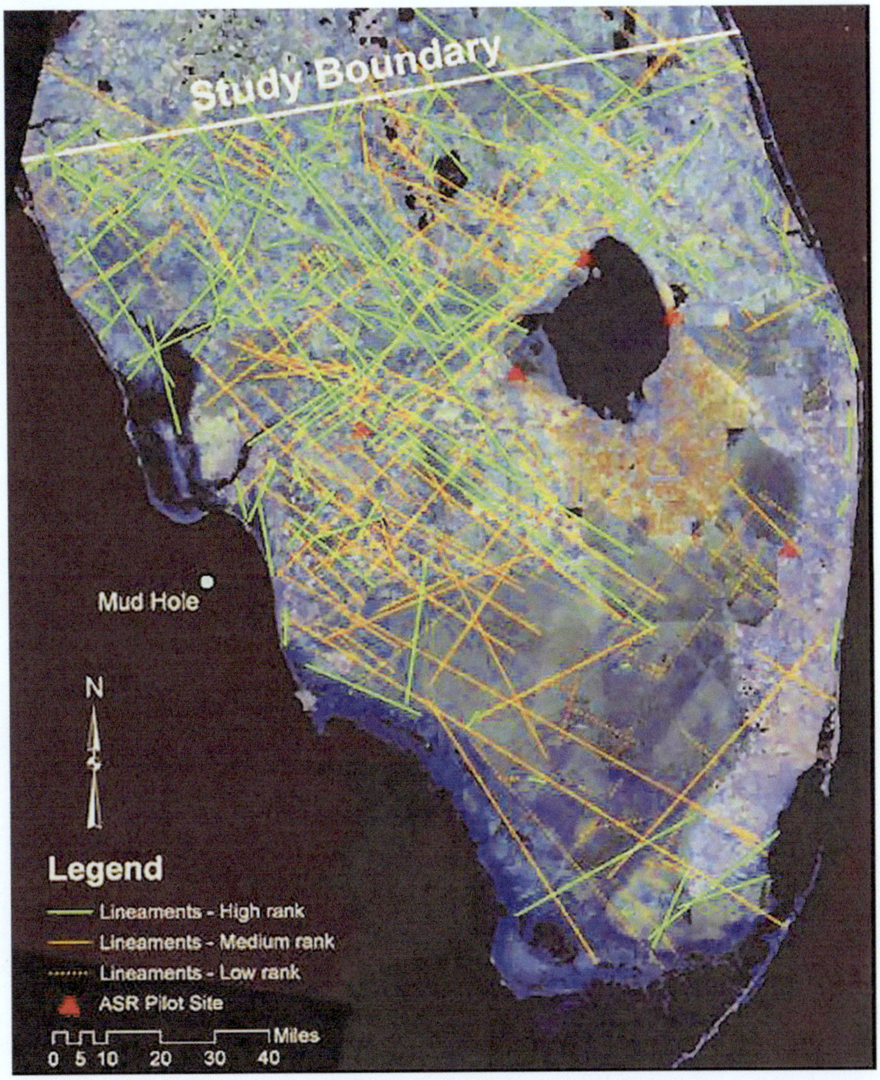


**Fig. 6.34** U.S. Army Corps of Engineers (2004) draft photolineament map of south Florida

Recently, studies of $CO_2$ in caves and the geometries of caves and cave decorations in Florida have resulted in reconsideration of the sources of $CO_2$ to form carbonic acid in near the water table. These processes are discussed in Sect. 6.5.2.5. For the present, we assume that the infiltrating groundwater becomes acidic as it passes through the vadose zone and enters the phreatic zone at the water table.

As meteoric water enters the sediments of Florida, it may react with shell within the cover materials. If this is the case, dissolution of the underlying limestone may be limited because the water has equilibrated chemically with the shell. If, on the other hand, the water infiltrates a thick, permeable, siliciclastic sand mantle with little or no shell, the water gains $CO_2$ from the atmosphere and then from soil microbial activity. At the water table, incorporation of additional $CO_2$ may lead to additional acidity and dissolution (Gulley et al. 2013; 2015; 2016; Gulley and Florea 2016; Gulley and Polk 2017). When this water encounters limestone, it creates karst features, such as caves, pinnacles, and epikarst within the shallow limestone. These processes are slow, requiring hundreds to thousands of years to develop (Sect. 6.6).

Epigenic karst is greatly dependent on the position of the water table in the aquifer. At this point, it is important to discuss the different conditions in the vicinity of the water table that are important to karst development in Florida.

#### 6.5.2.1 The Vadose Zone

The vadose zone is that volume of soil, sediment, or rock that lies above the water table. The interstitial pore spaces and voids in vadose zone sediments or rocks are filled with atmosphere that is most likely enriched in soil gases, such as $CO_2$, $CH_4$, and other compounds produced by microbial activity and decay. In soils and sediments of the vadose zone, thin films of water may coat grain and rock surfaces and interstitial pore space may be temporarily filled with water during recharge events. With time, this moisture (1) evaporates, (2) is utilized by plants and animals, (3) hydrates minerals, or (4) drains downward to the water table and into the phreatic zone.

It is within the vadose zone where the acidity of the moisture is increased by (1) addition of $CO_2$ to form carbonic acid and (2) creation of humic substances with acidic radicals. It is within the vadose zone that pH declines and the moisture increases in aggressiveness with respect to carbonate minerals (White 1988; Ford and Williams 1989; Upchurch 1992). Because of the abundance of carbonate and other acid-reactive minerals in the surficial sediments in much of Florida, the chemical aggressiveness of the percolating water does not necessarily persist as the water moves through the vadose zone and into the underlying aquifer. Rather, it begins to be buffered by interaction with sediment minerals. Chapter 5 presents pH and other geochemical data for the Surficial Aquifer System (SAS). If carbonates are present in the vadose zone and SAS, buffering begins quickly and the pH of the soil moisture or groundwater rises.

Statewide, the median pH of water in the SAS in Florida is 6.8, and pH ranges from 6.4 to 7.1 (Table 5.7). The Sand and Gravel Aquifer contains the least reactive mineral content of the SAS aquifers with a median pH of 4.9 and the pH ranges from 4.8 to 5.6, which is about the same as rainfall (Sect. 4.8; Table 5.3). Contrast the pH of water in the Sand and Gravel Aquifer to the other aquifers of the SAS. The median pH of the Biscayne Aquifer is 6.9 (range: 6.6–7.2; Table 5.7) and the undifferentiated SAS has a median pH of 6.7 and range of 6.0 to 7.1. The difference in acidity of the Sand and Gravel Aquifer as compared to the remainder of the SAS is a result of the onset of dissolution reactions in the vadose zone in carbonate-rich sediments and rocks and below the water table as epigenic karst development begins in carbonate-rich aquifers.

The vadose zone is an important horizon for formation of karst features in Florida. When water enters the pore space or forms thin films on grain surfaces, minor dissolution of carbonate minerals may occur. The rate of dissolution is slow, and it is controlled by the frequency of precipitation events, rate of interaction of soil gas and moisture, rate of migration of moisture through the vadose zone, and kinetics of the dissolution reactions. Offsetting dissolution is the possibility of precipitation of calcium carbonate as soil moisture evaporates or is removed by plant roots. As the dissolution reaction proceeds, the water is enriched in calcium, magnesium, bicarbonate, and other dissolution products. If it equilibrates with the carbonate minerals in the vadose zone, the ability of the same water to produce additional dissolution in the underlying, water-saturated horizons may be limited.

Once moisture encounters carbonate minerals in the vadose zone, it tends to erode downward along fractures or joints or in vertical pathways between carbonate sediment grains or in void spaces. Over time, this vertical "downcutting" produces solution pits, vertical shafts, and vertically oriented caves. The origin of solution pits, shafts, and vertical caves is further discussed in Chap. 7.

**Fig. 6.35** Example of the capillary zone exposed on the banks of Washington Blue Spring (30.513°N, 85.847°W) in Washington County

Given Florida's humid, semi-tropical climate and siliciclastic cover materials, slow dissolution in the vadose zone is an important process. The result is many areas where the SAS is wholly siliciclastic and relatively free of remaining carbonate minerals. Much of the process of carbonate dissolution in the vadose zone occurred during the Plio-Pleistocene low sea stand (Hunt and Hunt 1957; Carr and Alverson 1959).

In Florida, vadose-zone moisture has a high probability of evaporation or removal by transpiration if it enters highly porous and permeable sediments. The removal of water increases the saturation state of remaining dissolved constituents. As a result, it is not uncommon to encounter calcium carbonate cements where limestone is present in the vadose zone. This is especially true on the upper surface of the Biscayne Aquifer in southern Florida and in the Florida Keys. Artifacts of precipitation of calcite in Florida soils include pisolites, cemented rubble, and caprock. Cemented rubble with pisolites been termed caliche in southern Florida (Coniglio and Harrison 1983).

#### 6.5.2.2 The Capillary Zone

Capillary action is the ability of a liquid to move upward between solid surfaces as a result of the intermolecular forces (surface tension) between the fluid and solid surfaces. In granular sediments, the close proximity of sediment grains creates an ideal environment for water to "wick" upward, opposite to gravitational forces, between the grains.

The capillary zone is the transition from vadose conditions to phreatic, or water-saturated, conditions. The capillary zone is considered to be part of the vadose zone because larger interstitial pore space may be too large to draw moisture into some of the voids and they remain air filled.

Figure 6.35 illustrates an example of the capillary zone in quartz sand. The surface of the spring pool in the foreground is the water table. Capillary forces have drawn moisture upward into the sand from the top of the water table. As a result, the sediment is moist, but not water saturated. The moisture content gives the sand a darker tone in the image. The brown coloration is a result of precipitation of ferric hydroxide in the capillary zone.

Residence time of moisture in the capillary zone is typically longer than in the upper part of the vadose zone. As a result, dissolution near, but just above, the water table may be enhanced. Conversely, concentration of the chemicals as a result of evapotranspiration may also result in precipitation of calcium-carbonate and ferric hydroxide nodules, root casts, and cements in the capillary zone.

#### 6.5.2.3 The Phreatic Zone

In the phreatic zone, the pore space and voids are filled with water. It is for this reason that the phreatic zone is sometimes called the "saturated zone" as opposed to the unsaturated, vadose zone. Use of the terms vadose and phreatic is preferred to avoid confusion with the geochemical terms saturated and unsaturated, which relate to the state of chemical equilibration of the water with host rock materials.

The phreatic zone consists of the volume of aquifer below the water table[3] or potentiometric surface, within which the interstitial pore space and voids are filled with groundwater. It is within the phreatic zone that lateral, as opposed to vertical, groundwater flow generally occurs. Depending on the geochemical history of moisture in the vadose zone, groundwater may be chemically undersaturated with respect to carbonate minerals, and active dissolution of the host carbonate minerals may occur. Thrailkill (1965) argued that much of the water movement within the phreatic zone occurs within the upper 100 m of a limestone aquifer, and that most of that flow is within the upper 10 m. This is where chemically aggressive water is likely to remain and development of epigenetic karst conduits and voids below the water table is most likely to occur.

[3]The water table is also termed the phreatic surface.

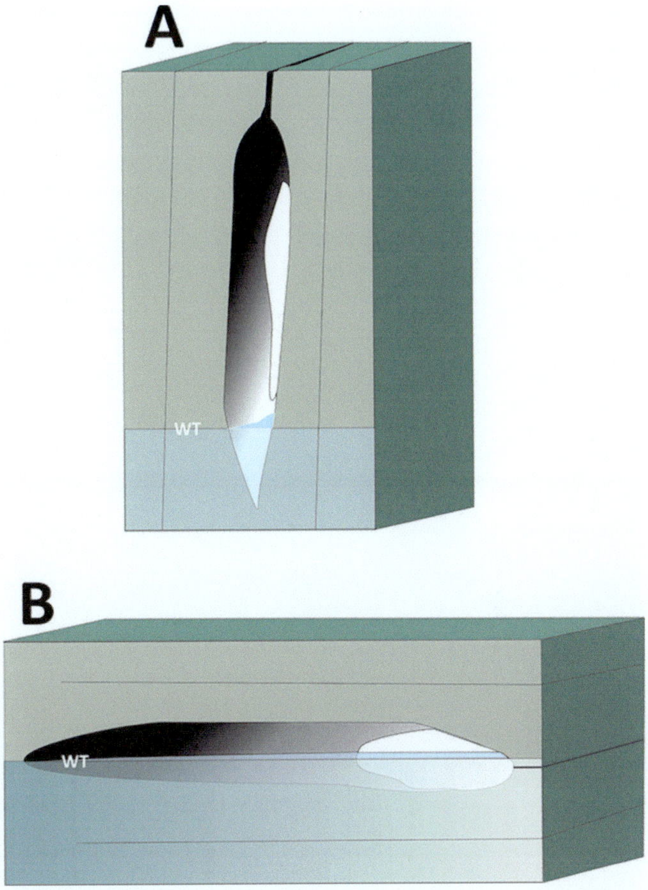


Fig. 6.36 Possible cave morphologies in vadose (**a**) and phreatic (**b**) environments. WT indicates the water table, the contact between the vadose and phreatic zones

When water capable of dissolving carbonate sediments and rocks enters the phreatic zone, dissolution continues down the hydraulic gradient until chemical equilibration of the water with the host sediment has occurred. Much of cave development occurs in the lower vadose zone and upper phreatic zone, and along shallow phreatic flow paths beginning at the water table. After the water has equilibrated with the host sediment, no further dissolution can occur unless the water mixes with water of different chemistry or unless additional sources of $CO_2$, $SO_4$, or humic substances are introduced to the groundwater.

Groundwater flow in the phreatic zone has a significant horizontal component, so the normal tendency of the groundwater to create vertical shafts and caves is reduced, and development of horizontal openings is enhanced. Figure 6.36 illustrates the geometries of caves developed near the water table in limestone. If the water table is dropping or if dissolution is within the vadose zone, the caves developed in limestone tend to be vertically oriented along joints or fractures (Fig. 6.36a). Most often, horizontal openings develop (Fig. 6.36b) in the shallow phreatic zone, especially near the water table, where groundwater is

Fig. 6.37 Diver exploring a passage in the Hart Springs cave system. (Photograph courtesy of Jill Heinerth)

undersaturated with respect to the minerals in the host aquifer and flow is has a horizontal component. Gulley and Florea (2016) demonstrated the importance of dissolution at the water table.

#### 6.5.2.4 Cave Cross-Section Morphology in Florida

While Florida's vadose and phreatic caves often have cross sections similar to the schematics shown in Fig. 6.36, Florea and Vacher (2006b) note that many Florida caves exhibit a "plus sign" cross section (Fig. 6.37). This suggests that the conduits have been enlarged by dissolution along both fractures and the mean position of the water table. In telogenetic karst, bedding planes in the limestone often represent sites of lateral speleogenesis. In Florida's geologically young limestones, bedding planes are usually poorly developed and indistinct. Where they exist and provide locations of preferential flow, voids may develop, as well. It may also be possible that the vertical movement of the water table as a result of sea-level (or base-level) fluctuations has contributed to superposition of vadose and phreatic cave morphologies.

Figure 6.37 is an excellent example of the artifacts of water-table fluctuations during speleogenesis. Several positions of the water table and of karst epigenic speleogenesis are documented in the image. The water-table fluctuations may be a result of long-term changes in climate or sea level, which ultimately controls the position of the phreatic zone in Florida. The enlarged vertical shaft is apparently fracture controlled and represents down-cutting by dissolution of the Ocala Limestone. The horizontal opening is a former position of the water table. Today, the cave is completely flooded, so the photograph provides evidence of (1) either a declining or low water-table position, (2) a still-stand of the water table, and finally, (3) flooding of the entire cave system. Note the bifurcations of the passageway in the background and the minimal amount of breakdown debris on the cave floor. Clearly, dissolution has been the dominant process of cave enlargement at this location. Note also the scalloped surfaces of the cave walls. The scallops are a result of erosion during turbulent flow (Curl 1974).

#### 6.5.2.5 The Role of $CO_2$ Degassing at the Water Table

The traditional concept of incorporation of $CO_2$ in water to form carbonic acid in karst systems includes interactions of water and $CO_2$ in the atmosphere and with microbe-generated $CO_2$ in the vadose zone. In recent years, Florida has been the location of important research on the dynamics of $CO_2$ near the water table and in caves. This research has indicated that $CO_2$ degassing, respiration by bats and other fauna, and other processes are likely to result in renewed chemical aggressiveness of water in the vicinity of the water table.

McGee (2010, summarized in Florea et al. 2011) studied the distribution of $CO_2$ in the atmosphere and pools of a cave in west-central Florida. $CO_2$ concentrations in cave atmosphere ranged from 452 to 573 parts per million (ppm) in the open atmosphere and 982 ppm in a portion of the cave occupied by a colony of bats. Isotopically, this $CO_2$ was "light" ($\delta^{13}C_{CO2} = -10.9$ to $-16.6$ parts per thousand [‰]) indicating depletion of the heavier $C^{13}$ isotope. Isotopic depletion indicates either an evaporative or respiratory origin of the $CO_2$. The lowest isotopic ratio was found in the bat wing of the cave. Pools, which indicate positions of the water table, within the cave had $CO_2$ concentrations that ranged from $1.5 \times 10^{-3}$ to $3.6 \times 10^{-3}$ atm. Water in a pool in the bat wing of the cave had the lowest concentration. Isotopic ratios of water samples were also isotopically light ($\delta^{13}C_{CO2} = -4.6$ to $-5.3$ ‰). McGee (2010) also measured $CO_2$ production in the cave as follows:

- Dry cave rock = 0.18 μmol $m^{-3}$ $s^{-1}$;
- Wet cave rock = 0.23 μmol $m^{-3}$ $s^{-1}$;
- Wet cave sediment = 0.15 μmol $m^{-3}$ $s^{-1}$; and
- Surface soils outside of the cave = 0.33 μmol $m^{-3}$ $s^{-1}$.

$CO_2$ emanating from these substrates was also isotopically light ($\delta^{13}C_{CO2} = -13.2$ to $-23.1$ ‰). Clearly, $CO_2$ was emanating from the water table, rock and sediment, and bat population.

Recent work on $CO_2$ partitioning and sources in Florida caves has resulted in an important modification of the fate and transport of $CO_2$ in Florida caves and at the water table (Gulley et al. 2013, 2015, 2016; Gulley and Florea 2016; Gulley and Polk (2017). They argued that $CO_2$ emanating from limestone with intergranular porosity to form common, water-table-related voids requires a source other than meteoric water percolating downward from the vadose zone. They argued that the "spongey" nature of Florida's eogenetic limestones disperses groundwater flow, reducing the dominance of fracture flow and dissolution related to fracture conduiting. Many of the apparent water-table-related cave systems (Florea and Vacher 2006a, b) do not appear to be related to fracture flow, and a dispersed source of carbonic acid is required to explain these caves.

The argument that $CO_2$ "exhaled" (Gulley 2013) by tree roots and microbes in the shallow, spongey aquifer migrates to the water table, as suggested by McGee's (2010) data, makes sense. The $CO_2$ produces carbonic acid and dissolution ensues. $CO_2$ emanations near fractures should migrate to the fractures and cause dissolution focused on the intersection of the water table and fracture. This explains the plus-sign morphology (Sect. 6.5.2.4; Fig. 6.37) that is so common in Florida caves. Gulley and Polk (2017) attributed this process to hypogenetic speleogenesis because the source of the $CO_2$ is in the subsurface and not of meteorological origins.

### 6.5.3 Development of Hypogenetic Karst

As discussed above, hypogenetic speleogenesis is a result of dissolution by waters that gain their chemical aggressiveness from subsurface conditions.

Speleogenesis as a result of the formation of sulfuric-acid solutions was first recognized in the western United States. Hill and her colleagues (i.e., Hill 1981, 1990, 1995, 2000, 2008; DuChene and Hill 2000) championed this mode of cave formation and, as a result, many cave systems throughout the world are now recognized as having formed by interactions with sulfuric acid solutions derived from evaporite strata (Klimchouk et al. 2017).

Cave systems formed at groundwater mixing zones have also been recognized as being hypogenic (Palmer 1991). Mixing-zone karst features are of special interest because of the proximity of the sea and presence of freshwater/saltwater mixing zones around the coastline.

In this section, we discuss both forms of hypogene karst and their potential development in Florida. More on hypogene karst, including examples, is presented in Chap. 7.

#### 6.5.3.1 Mixing-Zone Hypogene Speleogenesis

Because of the extensive groundwater geochemical data developed by studies in the 1970s and 1980s (i.e., Back and Hanshaw 1970; Herman et al. 1985), Florida data have been used to demonstrate how groundwater mixing at freshwater/saltwater transition zones can cause dissolution of limestone (e.g. Plummer 1975; Plummer and Back 1980). Water-quality data from shallow (<200 m) UFA groundwater-mixing zones in Florida theoretically demonstrate the presence of waters that are chemically undersaturated with respect to calcite (Plummer 1975). Where coastal groundwater water is stratified with a lower $SO_4$-rich facies and, below that, a Na-Cl facies, there may be two mixing-zone contacts along which speleogenesis may occur.

A limited number of vadose and phreatic caves have been attributed to groundwater mixing in Florida (Polk and Brinkman 2013). Where cave origin has been discussed, most caves are considered epigenetic (i.e., Florea and Vacher 2006a, b; Florea et al. 2003, 2007). Mixing-zone origins of some caves have been suggested as a result of (1) proximity to modern or ancient coastlines, (2) modern water-quality conditions, and/or (3) rare erosional patterns (e.g. mazework caves and cupolas, Klimchouk 2009) in vadose caves (Polk and Brinkman 2013).

Given the many Plio-Pleistocene sea-level oscillations, it is probable that epigenetic caves have been modified by later hypogene processes multiple times. While these modifications are likely to be overlooked during cave exploration, their presence seems likely. Garman and Garey (2005) and Garman (2010) noted, for example, the potential for limestone dissolution in coastal anchialine caves (caves that are isolated from the sea but contain a freshwater/saltwater transition zone in the subsurface) in west-central Florida (A, Fig. 6.38). At the transition zone, $H_2SO_4$-rich water is reduced to $H_2S$ in the presence of organics swept into the cave from above. The pH in the transition zone is reduced as this $H_2S$ is oxidized to $H_2SO_4$ at shallower depths, and a local "notch" is likely to develop on the walls of the epigenetic cave. If sea stands persist and the transition zone remains for a long period, extensive modification of pre-existing caves may be possible. Gulley (2013) argued that $CO_2$ migration to the transition zone could also cause mixing-zone speleogenesis.

There are several unusually deep, coastal cave systems on the west coast of Florida that are currently being explored by deep, mixed-gas diving. The best known of these deep caves are Diepolder II and Weeki Wachee Springs (B, Fig. 6.38) in Hernando County. These cave systems have been touted as the deepest explored, phreatic caves in the United States. The Weeki Wachee Springs cave system has been explored to a depth of at least 124 m in a cave that has been explored to 1400 m in length (Neill and Anderson 2009). The Diepolder II cave has been explored

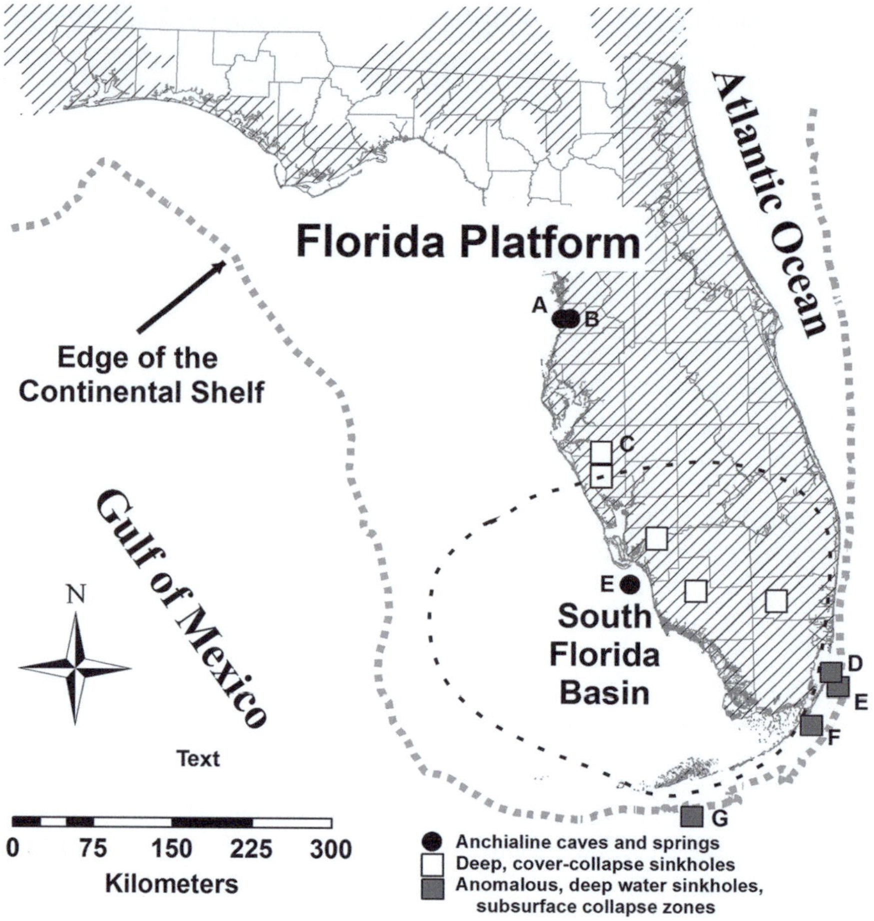


**Fig. 6.38** The Florida Platform and South Florida Basin. Shaded areas are areas where middle confining units of the Floridan Aquifer System were mapped by Miller (1986, 1990). Symbols are sites discussed in text (From Upchurch 2017)

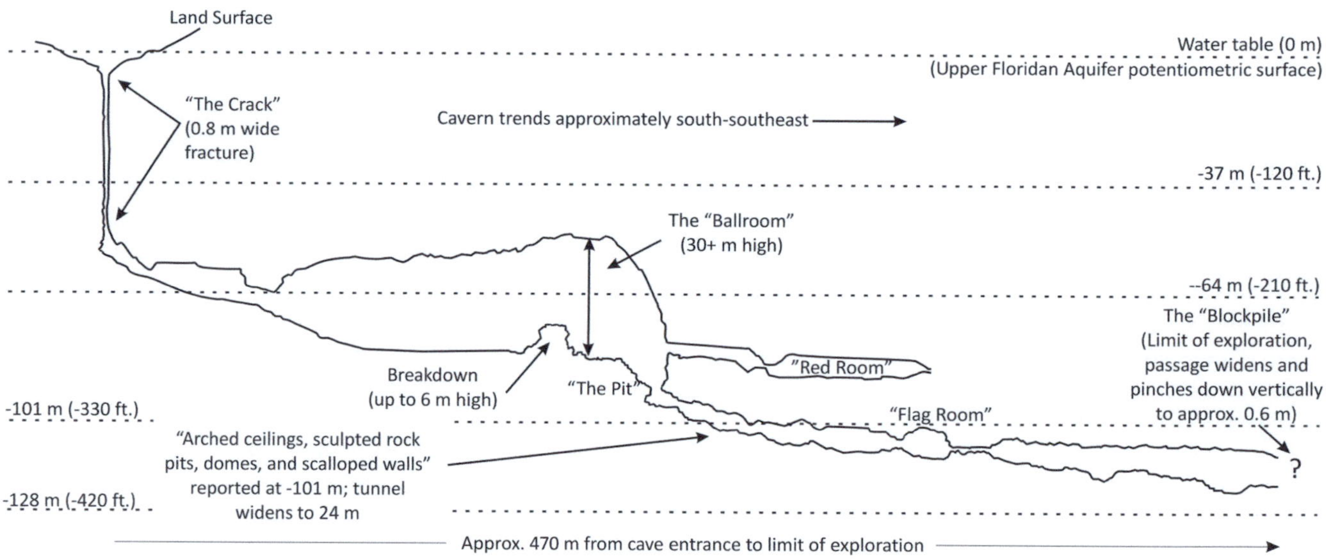


**Fig. 6.39** Diepolder II cave profile. Vertical datum is the phreatic surface. (From Upchurch (2017))

**Fig. 6.40** Divers explore the Ballroom in Diepolder II. (Photograph courtesy of Mark Long)

to a depth of 110 m at 470 m from the base of the entrance sinkhole (Clesi and King 1991; Upchurch 2017). These coastal cave systems are unique because of their depth and the large rooms that exist at depth (Figs. 6.39 and 6.40). The profiles of the caves are approximately parallel to, but currently above, the groundwater mixing zone. Little else is known about their architecture. These coastal cave systems appear to be the best candidates for hypogenic speleogenesis by groundwater mixing in Florida at the present time (Upchurch 2017).

#### 6.5.3.2 Sulfate-Driven Speleogenesis

The presence of sulfate-driven speleogenesis is problematic in Florida. Mechanisms that have been cited for sulfate-driven speleogenesis include hydration and oxidation of $H_2S$ gases emanating from $H_2SO_4$-rich groundwater, petroleum, or evaporite strata at depth. Given the abundant $SO_4^{2-}$-rich water in the UFA, Lower Floridan Aquifer (LFA), and the Cretaceous strata that lie below the Floridan Aquifer System (FAS), hypogenetic speleogenesis cannot be ruled out. Chapter 4 discussed the presence of evaporite beds that contain gypsum or anhydrite in Florida aquifer systems and Chap. 5 discusses the presence and circulation of $SO_4^{2-}$-rich groundwater.

Recently, researchers (i.e., Randazzo 1997; Cunningham and Walker 2009; Herbert and Upchurch 2016; Upchurch 2017) have suggested that sulfate-driven, hypogene speleogenesis may have affected Florida karst systems, especially the LFA and Cretaceous Sunniland Formation.

In the LFA, mixing of groundwater and the presence of $SO_4^{2-}$ water circulating under the Kohout circulation model (Fig. 4.9; Sect. 4.5.3) provide opportunities for speleogenesis by groundwater mixing and by interaction with waters containing sulfate. Rocks of the LFA have long been described as having caves and other dissolution-related features (see Sect. 4.5.3 for a discussion of these features). Randazzo (1997) was the first to suggest that, based on the presence of sulfate-rich groundwater, sulfate-driven speleogenesis might exist in the LFA. Cunningham and Walker (2009) observed what appear to be deep collapse features in Biscayne Bay and the continental shelf near Miami (D and E, Fig. 6.38. They attributed these collapse features, which penetrate younger strata to near the seafloor, to hypogene processes in the LFA or deep UFA. In 2017, Cunningham et al. reported on deep seismic exploration in Broward County. They described numerous collapse zones that penetrate to the LFA. Herbert and Upchurch (2016) and Upchurch (2017) summarized the earlier authors' findings and hypothesized that deep cenote (water-filled sinkholes; C, Fig. 6.38) and other features on the mainland and saline springs offshore (E, Fig. 6.38) may reflect LFA-related speleogenesis. These deep, rock-collapse sinkholes likely correspond to collapse zones such as were observed by Cunningham and Walker (2009) and Cunningham et al. (2017).

The Cretaceous strata of the South Florida Basin (Fig. 6.38) have long produced sour, $H_2S$-rich oil and saline water from oil fields in a trend that extends from Lee County in the northwest to Dade County to the southeast. The oil fields are small, have unusually well-developed permeabilities, and have locally produced millions of liters of crude oil. Herbert and Upchurch (2016) and Upchurch (2017) summarized the features of this oil-producing trend, the Sunniland Trend, and suggested that there may be evidence of sulfate-related hypogene speleogenesis. Evidence includes the fact that the trend is bounded above and below by anhydrite beds and there is abundant evidence of brecciation and, possibly dissolution. If the Cretaceous strata of the Sunniland Trend do contain artifacts of hypogene speleogenesis, the timing of the karst is problematic. Today, the oil-producing strata are about 3300–3400 m below the land surface. It does not seem reasonable that karst is developing at the present time. Herbert and Upchurch (2016) and Upchurch (2017) suggested that the karst, if present, formed prior to subsidence and at much shallower depths.

## 6.5.4 Speleogenesis and Cave Morphology

### 6.5.4.1 Introduction

So far in this chapter, we have discussed the artifacts of dissolution processes, including the effects of recharge, dissolution at the water table, and epigene versus hypogene speleogenesis. This section presents Palmer's (1991) models for the patterns of cave development as a result of types of recharge and dominant types of porosity.

Recall that the nature of karst terrains and cave morphology can differ greatly depending on boundary conditions, including initial porosity distribution (i.e., intergranular porosity, fracturing, etc.), cover (i.e., permeability, thickness), and stability and position of the phreatic surface. Since Florida karst is eogenetic and geologically young, (1) primary and secondary porosity are often present, (2) cover thickness and permeability are variable, and (3) the phreatic surface is currently near land surface, but has varied significantly over the last few millions of years.

Palmer (1984, 1991, 2007) related cave morphology to recharge, the nature of the cover materials and dominant porosity. The porosity of Florida carbonate sediments and rocks is discussed in detail in Chap. 5. Suffice it to say that sandy shell beds and poorly cemented limestone are common in Florida, both having extensive intergranular pore networks. As a result, all three of the types of porosity (fractures, bedding planes, and intergranular porosity) shown in Palmer's diagram are commonly encountered in Florida. Similarly, recharge through sinkholes and diffuse recharge are common in Florida. Florea (2006a, b, 2008a, b) discussed many of the forms of caves that have developed in Florida as a result of epigenetic processes.

Hypogenic karst is problematic in Florida, largely because exploration of caves that have developed on the freshwater/saltwater transition zone or by sulfuric-acid derived from deeper in Florida aquifers remain largely unobserved because they are in the deep, phreatic environment.

### 6.5.4.2 Role of Sea Level in Cave Development

As we continue our discussion of epigenetic and hypogenetic speleogenesis, it is important to reiterate that both the positions of the phreatic surfaces of the FAS and SAS and the configuration of the freshwater/saltwater transition zone depend upon the position of sea level among other factors. Speleogenesis requires time, so still-stands of sea level are most likely to result in development of caves. Sea level is known to have made wide excursions during the Plio-Pleistocene, and still-stands that are higher than the most current one are well documented (e.g., MacNeil 1950; Alt and Brooks 1965; Pirkle et al. 1970; Healy 1975; Vail and Mitchum 1979; Tanner et al. 1989; Tanner 2000; Hine et al. 2016, and others). Sea stands below modern sea level are less well known, but have been documented (Bloom 1983; Miller 1998; Davis 2011; Hine et al. 2016). The development of epigenetic karst is related to sea level (Florea and Vacher 2006a; Florea et al. 2007) in Florida, and the position of the freshwater/salt water transition is related to sea level (Herzberg 1901; Badon Ghyben 1889; Wentworth 1947, 1948), so mixing-zone speleogenesis must be as well.

The pattern of speleogenesis in Florida, therefore, must be complex and polygenetic (Fig. 6.41). In Florida, one should expect cavernous zones to represent tiers of epigenetic features as a result of ancient sea-level stands with hypogenetic modifications superimposed, assuming that mixing-zone speleogenesis is represented in the karst record. This complexity (Fig. 6.41) is likely to add complications to the attributes of relict karst, especially cavernous zones at depth in the UFA.

### 6.5.4.3 Epigenetic Speleogenesis

Focused recharge via karst depressions and sinking streams is well documented and pervasive in Florida karst terrains. Considering Palmer's (1991) well-known arguments showing patterns of cave passages as a function of different recharge mechanisms and dominant types of porosity, it is instructive to think about recharge mechanisms with respect to caves known or suspected to exist in Florida. Palmer classified the expected forms of karst development based on patterns of recharge (i.e., via karst depressions, diffuse recharge, and hypogenic origins) and dominant types of porosity (i.e., fractures, bedding planes, and intergranular). The following discusses each of Palmer's types of karst (cave) development as a function of recharge and porosity in Florida limestones.

**Recharge via karst depressions** All six of the cave types posited by Palmer (1991) as having developed as a result of recharge through karst depressions are well documented in Florida. Two categories of recharge via karst depressions were identified by Palmer. These are (1) meteoric water that passes into the limestone by way of sinkholes that do not

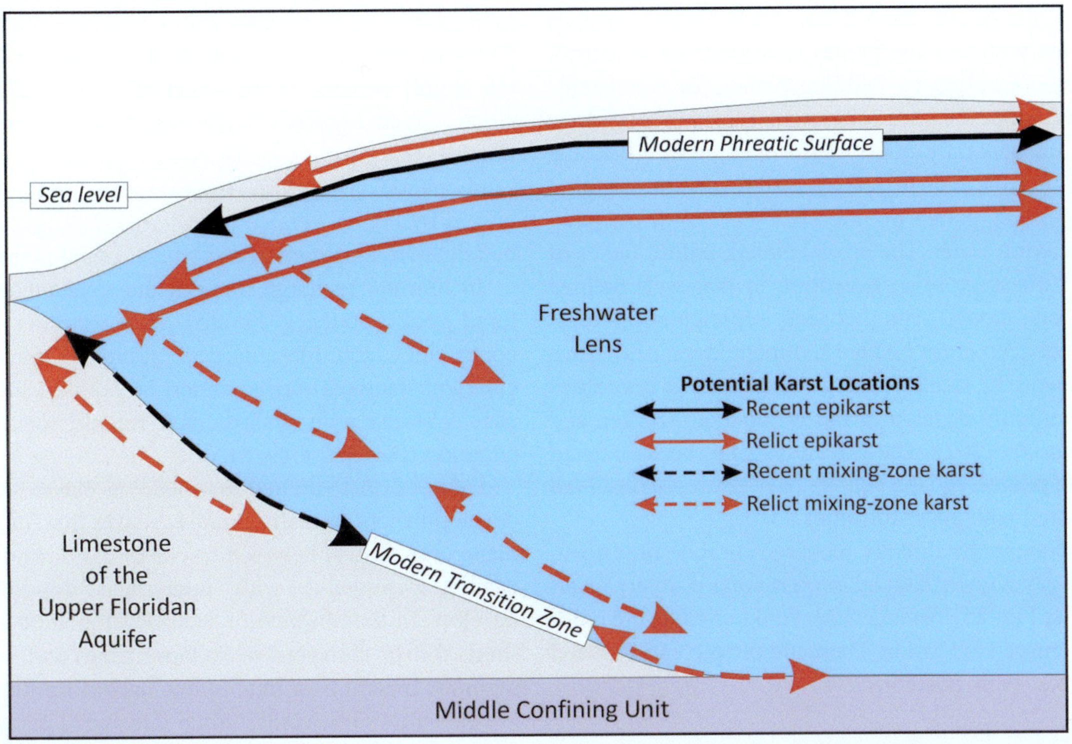


**Fig. 6.41** Superposition of epigenetic and hypogenetic karst features after several excursions of sea level and development of paleo-karst horizons

capture stream flow and (2) water that recharges the limestone aquifer by way of sinkholes that capture stream flow.

**Recharge Through Sinkholes** Water that recharges as local runoff into sinkholes were thought to result in branchwork (branching networks of caves), usually with several levels or with single passages. Water that enters fractured rock develops angular passages with trellis-drainage-like cave patterns. Fracture controlled cave systems are common in Florida (Florea 2008a, b).

Water that recharges limestone with well-developed bedding planes was thought by Palmer to develop curvilinear passages that may appear as dendritic or sinuous cave systems. Few locations in the eogenetic limestone of Florida have well-developed bedding planes, so this form of cave development is apparently limited.

Water that passes from the sinkholes directly into porous limestone was thought to develop rudimentary branchwork caves. Such cave patterns are relatively common in Florida. Given (1) the relatively high porosity of much of Florida eogenetic limestone and (2) the presence of reactive mineral phases (aragonite, high Mg-calcite) in the youngest carbonate sediments and limestones, this form of cave development appears to have been an early phase in the development of many of Florida's better developed branchwork caves.

**Recharge Through Swallets** Sinkholes that capture surface-water streams are known as swallets. In Florida, swallets are common, especially near karst escarpments where streams originate in areas with limited recharge as a result of shallow, clay-rich sediments. As the streams flow off the area of limited recharge onto poorly confined limestone, swallets develop. This form of recharge introduces high volumes of water to the underlying aquifer and substantially reduces loss of water to evapotranspiration. The slightly acidic water contains carbonic acid and humic substances, both of which encourage speleogenesis. As a result of introduction of relatively high volumes of chemically aggressive water, single passages and crude branchwork caves develop. As will be discussed in later chapters, many of Florida's swallets are involved in fluviokarst escarpment retreat and development of long cavern systems with numerous collapse sinkholes (karst windows). The following patterns were shown by Palmer (1991) to result from recharge through swallets.

If water entering the limestone aquifer via swallets passes into fractured limestone, Palmer showed that fissures and irregular networks controlled by fractures develop. It is important to note that many of the caves created in fractured limestone and fed by swallets in Florida remain as well-defined, single caves or cave networks with one dominant passage (e.g., the Aucilla trace, Chap. 8). If the water passes into limestone with bedding planes, anastomoses or anastomotic mazes controlled by bedding planes [or water-table positions] develop. Bedding planes in Florida's eogenetic limestones are generally poorly developed. While bedding planes can often be recognized by textural, color, or other indicators, bedding planes do not often serve as direct pathways for meteoric water. The cross-bedded, oolitic facies of the Miami Limestone is an exception in that well defined breaks are commonly seen between sedimentation units within and between cross-bedded sediment layers. Horizons of what appear to be small, anastomosing caves or cavernous porosity commonly develop at the water table (Florea and Vacher 2006a, b, 2007). These features may be associated with bedding planes, and the authors have certainly observed them associated with unconformities.

Finally, if the water directly recharges porous limestone, spongework caves controlled by intergranular porosity result (Palmer 1991; Fig. 25). Spongework void networks are commonly encountered in Florida. These are often clearly related to former water-table positions.

**Diffuse recharge** The effects of diffuse recharge are especially interesting in Florida's eogenetic karst. Palmer (1991) considered two types of diffuse recharge: through sandstone and directly into porous, soluble rock. The process of diffuse recharge through sandstone is applicable to Florida limestone that is overlain by a cover of marine-terrace sand.[4] Here, the additional thickness of unconsolidated sand [or, rarely, sandstone] allows for microbial decomposition of plant material and creation of carbonic acid and humic substances, thus increasing the acidity of meteoric water that has passed through the sandy sediments of the vadose zone.

**Water Passing Through a Sand Mantle** According to Palmer (1991), water that passes through sandstone constitutes a major category of recharge. The examples provided by Palmer are related to telogenetic karst systems where sandstone strata overlie limestone where it forms an erosion-resistant cap. His examples typically show caves forming just under the sandstone and, in many cases, extending down gradient as complex cave forms.

In Florida, sandstone in Neogene and Quaternary strata is rare and is typically not erosion resistant. Rather, Florida limestone is often overlain by unconsolidated or poorly lithified sand. Geochemically, this sand mantle plays a role that is somewhat similar to a porous, sandstone cap. As meteoric water passes through the sand or sandstone, it is likely to gain carbonic acid by interaction with microbially generated $CO_2$ and, possibly, organic acids, both of which may enhance the aggressiveness of the water when it reaches the limestone. Given Florida's warm and humid climate and high rainfall, this process is likely to be intense in the shallow sand mantle. Therefore, by extension, the processes discussed by Palmer related to recharge through sandstone can be related to epigenetic karst where a sand mantle exists.

In Florida, recharge through the sand mantle into fractured rock produces fissures and networks of fracture-controlled, solution-enlarged joints. Grikes (solution enlarged fractures) are common in exposed limestone, and cave systems with well-defined fracture control are also common (Florea 2008a, b).

Palmer (1991) argued that bedding planes would control development of laterally extensive caves that follow bedding-plane orientations between fractures. As a result, shafts and canyon complexes with interstratal dissolution would develop. In the absence of well-developed bedding planes, shafts follow the avens of collapse zones and fracture intersections. Lateral flow may follow unconformities and galleries of water-table voids. Voids developed at ancient water tables form more-or-less planar horizons (Florea and Vacher 2006a, b) that may be confused with, or follow, incipient bedding planes. These arrays of voids are, however, a result of dissolution in strata with intergranular porosity.

Water that passes through the sand mantle into porous limestone creates poorly developed spongework; a common form of karst in Florida.

**Water Directly Recharging Porous, Soluble Rock** If meteoric water directly passes into porous carbonate sediment by diffuse recharge without contact with a significant sand mantle, dissolution begins with enlargement of intergranular pore spaces. As dissolution continues, pore spaces become enlarged and networks of enhanced permeability evolve. The pattern of cave development known as spongework appears to be an end result of this process. Depending on hydraulic gradient, fracturing, and mineralogy (low Mg-calcite versus high Mg-calcite or aragonite), cave patterns can be complex. Palmer (1991) argued that caves would be formed "at depth" by groundwater mixing. His predicted form of karst assumes telogenetic karst with pre-determined pathways for dissolution. In dual porosity, eogenetic limestone direct recharge to the intergranular porosity of the limestone is clearly a first step in the development of karst. Fractures may be present in the older limestones, but absent in younger ones. For example, fractures are well developed in Eocene limestone and poorly developed in the late Pleistocene Miami Limestone. With the exception of the Miami Limestone, bedding planes are typically poorly developed in most Florida limestone. Therefore, flow through intergranular porosity is a likely first step in karstification.

[4]Marine terraces are the accumulations of sediment that resulted from a Plio-Pleistocene high sea stand. In Florida, the terrace deposits are typically quartz sand. Wave-cut scarps, relict aeolian dunes, and relict sand bars often mark the shorelines that formed during these high sea stands.

If the rock is fractured, isolated, dissolution-enlarged fissures and rudimentary networks result according to Palmer (1991). If the water follows bedding planes (remember that bedding planes are often poorly developed in much of Florida's limestone) or a lithologic contact along which water can move, the meteoric water is likely to develop spongework. Similarly, water that directly recharges porous limestone is likely to develop spongework. Galleries of voids associated with the water table (Florea and Vacher 2006a, b) may result.

The drainage networks described by Palmer (1991) begin where chemically aggressive water enters openings in the soluble rock. Over geologic time, these openings are widened and extended by dissolution (Fig. 6.42). Holding the drainage into the network constant, over time the chemically aggressive water is spread over a larger and larger void space as the network grows. Surface areas of rock with which the water is in contact become less. This means that the acidity of the water is initially neutralized and comes to chemical equilibrium near the point where it enters the carbonate rock. Over time, however, the growth of the void network means that the water can travel further before it is neutralized.

While Palmer's (1991) paper dealt with the patterns developed in entire cave systems, many of his concepts scale down to the outcrop scale. Figures 6.42 and 6.43 illustrate the effects of exposure to chemically aggressive water in a recharge area. The host rock is the Oligocene Suwannee Limestone, which often has well developed intergranular porosity. In Fig. 6.42, chemically aggressive water has developed voids at relict water-table positions or along incipient bedding planes, dissolving passages along the horizontal planes and partly cementing adjacent strata. One bed (S, Fig. 6.42) includes some spongework created by dissolution within intergranular porosity. This spongework is probably not hydraulically well connected to the larger openings. The exposure is located on the Suwannee River, whose water is rich in humic substances, so dissolution on the rock face and extending into the rock as a result of bank storage may have contributed to the complex array of voids and differentially eroded limestone strata. In Fig. 6.43, the upper meter of limestone has been thoroughly cemented by calcite, apparently as a result of repeated wetting and drying of the limestone during episodic recharge and/or flood events. In thin section, the cement ranges from microscopic, sparry crystals

**Fig. 6.42** Well-developed solution channels along bedding planes and/or former water-table positions (B) in the Oligocene Suwannee Limestone exposed on the bank of the Suwannee River near Ellaville Spring, Suwannee County. Deep cutters (C) on each side of this pinnacle of limestone are solution-enlarged fractures. Poorly developed spongework is preserved within the stratum marked (S). Location: 30.385°N, 83.172°W

**Fig. 6.43** Thoroughly cemented "caprock" over spongework in Oligocene Suwannee Limestone exposed on the bank of the Suwannee River near Dowling Park, Suwannee County. The object growing out of the small cave is a tree root

**Table 6.5** Some possible effects of recharge water in vadose and phreatic environments on shell and limestone deposits

| Relation to water table | Bare Limestone | Sand mantle over Limestone | Shelly sand over Limestone | Shelly clay over Limestone | Clay confining unit over Limestone |
|---|---|---|---|---|---|
| Vadose zone | Development of spongework, dissolution at top of limestone and in shallow fractures, on water-table positions, or bedding planes; vertical shafts and passageways predominate; epikarst development | Enhanced dissolution near top of limestone in spongework and in shallow fractures, water table positions, or bedding planes; vertical shafts and passageways predominate; epikarst development | Shell dissolution may buffer water before reaching limestone; groundwater aggressiveness reduced; spongework and epikarst development possible | High clay content is likely to prevent dissolution of shell or uppermost limestone; epikarst unlikely to develop, pre-existing epikarst may be present | Clay capable of aquifer confinement may prevent chemically aggressive water from reaching limestone |
| Shallow phreatic zone | Dissolution at top of limestone, in spongework, and in shallow fractures, water-table positions, or bedding planes; horizontal passageways begin to develop | Dissolution at top of limestone and in spongework, shallow fractures, water-table positions, or bedding planes; horizontal passageways likely to develop | Shell dissolution may buffer water before reaching limestone; groundwater aggressiveness reduced; horizontal passageways likely to develop | High clay content is likely to prevent dissolution of uppermost limestone; pre-existing epikarst may be present | Clay capable of aquifer confinement may prevent chemically aggressive water from reaching limestone; pre-existing epikarst may be present |
| Deep phreatic zone | Water is buffered and no longer aggressive; no dissolution until mixing zone is encountered | Water is buffered and no longer aggressive; no dissolution until mixing zone is encountered | Water is buffered and no longer aggressive; no dissolution until mixing zone is encountered | Water is buffered and no longer aggressive; no dissolution until mixing zone is encountered | Water is buffered and no longer aggressive; no dissolution until mixing zone is encountered |

to silt- and clay-sized micrite. Below the cemented horizon, which forms a caprock-like stratum, is a layer of spongework and a maze of voids has developed. The spongework has been enlarged as recharge water slowly dissolved away the stratum. At the contact between the spongework and "caprock", larger, cavernous porosity has begun to develop. Note also that the "caprock" is much more resistant to dissolution because there is no longer significant intergranular porosity within the stratum. Chemically aggressive water is only in contact with this stratum at surfaces such as the walls of the small cave and the surface exposed to the river. In contrast, within the spongework there are many small surfaces, including the intergranular porosity and walls of spongework openings.

The purpose of presenting Figs. 6.42 and 6.43 is to demonstrate the karst-related complexity of Florida limestones that initially contain intergranular porosity. These limestones include the Pleistocene Miami and Key Largo limestones of southeastern Florida, portions of the Pliocene Tamiami Formation in southwestern Florida, and the Oligocene Suwannee and Miocene Tampa Member limestones of central and north Florida. The Eocene Ocala Limestone ranges from having pore space filled with carbonate mud to unfilled, intergranular porosity. Because of the lower initial porosities of the Eocene Ocala Limestone, the style of karst development is somewhat different, with much of the karst developed along fractures, water-table positions, and incipient bedding planes. The common lack of well-developed spongework and intergranular porosity is the reason that the Eocene Ocala Limestone is considered in some areas of Florida as a semi-confining unit (Table 6.5).

In Florida's eogenetic limestone, therefore, there appears to be a progression from diffuse recharge and formation of spongework to development of larger openings along bedding planes and fractures; this pattern extends to larger openings such as caves. Cementation of strata contiguous to these openings is common at shallow depths where the episodic influx of chemically aggressive water has resulted in localized dissolution followed by reprecipitation of $CaCO_3$ as cement. Where this has happened, there is a loss of intergranular porosity and dissolution is confined to the surfaces of bedding planes and fractures.

Table 6.5 synthesizes the relationships of the vadose and phreatic zones on karst development in the context of buffering of the acidity in recharge water. Based on the degree of contact with carbonate minerals, the chemical aggressiveness of recharge water diminishes with distance from the location of initial contact with carbonate minerals. This results in the important concept that buffering of the acidic recharge water occurs quickly and at shallow depths in young limestones and that void development at depth in Florida is most likely a result of development (1) at unconformities or long-term water-table positions developed during previous low sea stands or (2) along a freshwater/saltwater mixing zones.

#### 6.5.4.4 Hypogenetic Speleogenesis

Of the three forms of recharge described by Palmer (1991), the possibility of hypogenetic "recharge" is most problematic in Florida. Hypogenetic karst forms when (1) deep sources of acidic water or (2) mixing along a freshwater/saline water transition zone exist. The formation of caves by sulfuric acid derived from subsurface gypsum or anhydrite beds or petroleum elsewhere in the world is well known (Hill 1981, 2000; Klimchouk 2007; Palmer 2006). These hypogenetic cave systems form as a result of sulfuric acid developed during sulfide oxidation.

Hypogenic karst that has resulted from formation of sulfuric acid ($H_2SO_4$) from sulfide-rich water or evaporite minerals deep in the subsurface may be occurring in Florida (Herbert and Upchurch 2016; Upchurch 2017). One such possibility is that hypogenetic karst forms as a result of geothermal processes related to the Kohout's geothermal circulation model for the LFA (Fig. 4.9; Sect. 4.5.3.3) or the formation of sulfuric acid through interaction with the sulfate-rich evaporite beds of the middle confining units and deeper Mesozoic strata (Herbert and Upchurch 2016). Stoessell et al. (1993) investigated the possibility of formation of Yucatan cenote (deep, water-filled, rock collapse sinkholes). They concluded that some Yucatan cenote were undergoing calcite dissolution as a result of sulfide oxidation and formation of sulfuric acid. Similar processes may be in action in the boulder zones and vicinity of the deep cenote of Florida (e.g. Warm Mineral Springs and Little Salt Springs in Sarasota County; C, Fig. 6.38).

Water-quality data from shallow (<200 m) UFA groundwater systems in Florida have long been used to demonstrate the mechanisms of mixing-zone speleogenesis at transition zones (Plummer 1975, 1977; Wicks and Herman 1995). These water masses mix, resulting in waters that may be chemically undersaturated with respect to calcite. Where coastal UFA water is stratified with a lower $SO_4$-rich facies and, below that, a Na-Cl facies, there may be two mixing-zone contacts along which speleogenesis may occur.

A limited number of vadose and phreatic caves have been attributed to groundwater mixing in Florida (Polk and Brinkman 2013). Given the many Plio-Pleistocene sea-level oscillations, it is likely that epigenetic caves have been modified by hypogene processes.

The unusually deep coastal cave systems (A, B, Fig. 6.38) on the west coast of Florida are the best candidates for mixing-zone speleogenesis (Upchurch 2017). Most of these water-filled caves are entirely filled with freshwater today. However, their great depth, proximity to the coast, and large rooms at depth suggest a prior mixing zone origin.

#### 6.5.4.5 Comparison of Epigenetic and Hypogenetic Attributes of Florida Caves

In spite of the fact that hypogenetic cave systems in Florida are poorly understood, we can begin to compare and contrast the suspected hypogenetic cave systems to the much better known epigenetic systems (Table 6.6).

**Epigenetic Cave Systems** In Florida, the epigenetic cave systems are closely related to the modern or ancient water table positions, or paleo-karstic unconformities in the subsurface. While most are water filled, dry caves can be found in some ridges and other areas where there is a relatively

**Table 6.6** Comparison of epigene and hypogene attributes of Florida Cave Systems

| | Epigenetic Karst Systems | Hypogenetic Karst Systems | |
|---|---|---|---|
| | Dissolution near the phreatic surface | Dissolution on the mixing zone | Dissolution by sulfuric acid from below |
| Location | Near the land surface or an unconformity | More-or-less planar surface that dips inland from modern or ancient sea-level still stands | Variable, a sulfur source (evaporite beds) nearby |
| Source of aggressive water | Meteoric water (recharge from rainfall) | Freshwater/saltwater transition zone | Upward migrating gases ($H_2S$) or $SO_4^{2-}$-rich water |
| Depth of cavern penetration | Shallow, in Florida approx. 30–60 m | Shallow to deep, in Florida approx. 30–150 m | Shallow to deep, approx. 30–180+ m |
| Occurrence of vadose caverns | Common in some ridge areas; otherwise most are currently phreatic | Phreatic caverns predominate | Phreatic caverns predominate;<br>No known existing vadose caves other than shallow, anchialine caves |
| General orientations of cavern systems | Often elongated in the direction of local groundwater flow or fractures<br>If porosity is intergranular, spongework may be directionless<br>Pathway from swallet or sinkhole recharge location may control | Poorly understood in Florida<br>May be anastomosing if flow within the transition zone is weak or oriented perpendicular to the coast if flow across the interface is strong;<br>Fractures may strongly affect directionality;<br>May be expressed as modifications of pre-existing epigenetic caverns | In Florida, largely unknown;<br>In upper Floridan aquifer, anchialine caves may have localized dissolution at saltwater transition zone;<br>In the lower Floridan aquifer, karst, if present, may be (1) widespread and stratigraphically bound in fractured dolostone and (2) related to Kohout's geothermally driven circulation |
| Chamber size and distribution in cavern system | Shallow, large chambers are common;<br>Chambers and conduits usually become smaller with depth and with branching | Chambers in caverns on west coast of Florida get larger with depth to approx. 120 m;<br>Unknown below 120 m | Unknown for deep systems;<br>"Rod drops" in oil wells suggest that large chambers may exist;<br>In anchialine caves, minor enlargement of pre-existing caverns |

thick vadose zone. Typically, where accessible, these dry cave systems may have relatively large rooms near the downgradient end of the cave system and the conduits tend to branch and become smaller upgradient. They may consist of a single master conduit with minor branches or they may represent networks, many of which are fracture controlled.

**Hypogenetic Cave Systems** Few, if any, hypogenetic cave systems that formed by dissolution at the freshwater/saltwater transition zone have been documented in Florida. In contrast to many epigenetic caves, those cave systems that have been tentatively attributed to transition-zone mixing (Upchurch 2017)

- Are located near the modern coastline, mostly on the Gulf Coast;
- Tend to dip downward away from the coastline;
- Have unusual, explorable depths up to 122 m; and
- Large rooms occur at depth (Table 6.6).

Since sea level has varied over 100 m several times since Pliocene times, it is appropriate to assume that epigenetic cave systems have been modified by mixing-zone dissolution in the past and vice versa (Fig. 6.41).

Hypogenetic cave systems formed by dissolution with sulfuric acid are the least understood in Florida and, if they exist, they are all phreatic and deep. Where we suspect they exist (Herbert and Upchurch 2016; Upchurch 2017), they can be subdivided into three settings (Table 6.6). In the LFA, the boulder zones may have dissolution-related porosity superimposed on fractured limestone or dolostone. The boulder zones are bounded by gypsum or anhydrite beds and the Kohout circulation system sets up opportunities for dissolution by sulfuric acid or by mixing. Also, limestones within strata composed of anhydrite in the Cretaceous have breccias and dissolution-related features that may reflect an ancient sulfate-driven karst system, possibly be $H_2S$ derived from petroleum or evaporate minerals.

In the UFA, $SO_4^{2-}$-rich water is present at the base of the aquifer. This water, derived from dissolution of gypsum and/or anhydrite at the base of the aquifer or by oxidation of $H_2S$, may be responsible for karst within the aquifer. The deep sinkholes scattered around central and southern Florida may reflect collapse cenote created by dissolution. Geophysical data from the Miami area (Cunningham and Walker 2009) suggest such collapse features, as well.

Finally, Garman and Garey (2005) and Garman (2010) have suggested that oxidation of $H_2S$ at the saltwater transition zone in anchialine caves may be locally enlarging epigenetic caves.

## 6.6 Limestone Dissolution and Denudation Estimates

One of the great public concerns in Florida is of collapsing sinkholes, which cause millions of dollars in damage each year. These collapses are normally a result of failures of the cover materials and rapid raveling into pre-existing voids in the underlying limestone. Rarely, the limestone roof of a cave will collapse, but the roof failure is not a result of rapid dissolution of the limestone. The cause is mechanical and deals with the strengths and loading of the rock and overlying cover. Because of the concern that limestone is "dissolving away below our feet", it is appropriate to discuss our knowledge of the rates of dissolution in Florida strata.

Karst scientists agree that, as in other karst areas of the world, the rate of dissolution of limestone in Florida is slow (Sellards 1909; Brooks 1967; Opdyke et al. 1984; Beck and Sayed 1991; Upchurch and Randazzo 1997; Adams et al. 2010; Adams 2013), requiring many thousands of years to dissolve a physically measurable amount of limestone. As a result of this slow dissolution rate, the rate of denudation and/or void development is not a critical human risk issue in the development of karst landforms, such as sinkholes, in the eogenetic karst of Florida.

There are few ways to quantify the site-specific rate of dissolution of limestone in the field. To do so requires a geochemical investigation. See Ford and Williams (2007) for a discussion of methods for determining the rates of dissolution of limestone.

One example of use of geochemistry to characterize the site-specific rate of dissolution of Florida limestone was by Upchurch et al. (1991). Upchurch et al. used the chemical differences between injected and recovered water in an aquifer storage and recovery (ASR) well in Manatee County to determine how much, if any, limestone was dissolved by the injection of the treated but not buffered (and therefore chemically undersaturated with respect to calcite) surface water during each injection and recovery cycle. Injection is much more rapid than natural recharge, creating a forced system. Water in the UFA had a pH of 7.7 and the surface water injected for storage had a pH of 6.8. The injection and storage cycle was 3 months. Upchurch et al. found that injection of $3.10 \times 10^6$ $m^3$ of water resulted in dissolution of 4.08 $m^3$ of limestone and dolostone. Spread over an assumed spherical "bubble" of injected water, 0.0013% of the rock volume was removed. The actual distribution of injectate in an ASR well does not have a spherical bubble configuration; rather, it occupies a complex space determined by variability in permeability and head in the host aquifer (Vacher et al. 2006). Upchurch et al. (1991) concluded that this forced system of injecting chemically undersaturated water into the UFA would only result of a 1% change in rock volume every 7600 years.

There have been a few estimates of the regional rate of dissolution of the limestone in Florida. These typically utilize measurements of the amount of dissolved calcium being carried each year by the groundwater, rivers, and springs of Florida. Needless to say, these estimates may be in error because they do not reflect dissolution (and reprecipitation) in the deeper portions of the aquifer systems and the emergence of undocumented seepage along coasts and the continental margins.

### 6.6.1 Denudation Rate Estimates for the Limestones of the Floridan Aquifer

Denudation is sum of all processes that wear away the earth's surface. In Florida, denudation is largely a result of dissolution of carbonate sediments and rocks. Streams carry particulate sediments to the sea as well, but in Florida the rate of sediment erosion is low in comparison to chemical weathering. The denudation rates discussed in this section are presented in the equivalent metric units and dimensions presented by the original authors.

For comparison of dissolution-related denudation rates in Florida, the chemical denudation rates for the south Atlantic and eastern Gulf area of the United States have been summarized by Judson and Ritter (1964). They estimated the regional denudation rate to be $28–39 \times 10^3$ $kg/km^2$ per year.

The first estimate of the denudation rate (void space creation or equivalent surface lowering rate) of Florida limestone was by Sellards (1909) who examined water quality in a small number of springs in north Florida and estimated that the "mineral solids" removed annually was a little more than $6 \times 10^4$ $kg/km^2$.

Using data from the Suwannee River watershed, Brooks (1967) concluded that the overall, regional dissolution rate of limestone in the Suwannee Basin was 3.8 cm/1000 years. In the Okefenokee Basin District, where the Floridan aquifer is confined, Brooks concluded that the denudation rate of the underlying limestone was insignificant. In areas where stream runoff from the Okefenokee Basin contributed to the UFA by drainage into swallets at the western margin of the district, Brooks estimated a denudation/dissolution rate of 5 cm/1000 years. The high dissolution rate by surface water entering the UFA along the escarpment at the margin of the Okefenokee Basin District is a result of the focused recharge of acidic, organic-rich surface water into swallets. In the Ocala Karst District, which is characterized by thinly covered limestone with numerous, dispersed sinkholes, the UFA is unconfined. Brooks estimated the dissolution rate in the karst plain to be 3 cm/1000 years.

Opdyke et al. (1984) used spring discharge to estimate a denudation rate for northern Florida. Based on data in Rosenau and Faulkner (1975), Rosenau et al. (1977), and Slack and Rosenau (1979), Opdyke et al. estimated that the minimum denudation rate was 12 × $10^6$ $m^3$/year, or about 1 m of limestone dissolution every 38,000 years. Opdyke et al. utilized this loss of mass of limestone to calculate the maximum amount of isostatic uplift of northern peninsular Florida to be at least 36 m during Pleistocene and Holocene time. This isostatic uplift was used to demonstrate why Pleistocene and Holocene fossils are found in northern Florida at elevations above those predicted by Plio-Pleistocene shoreline elevation estimates (Healy 1975).

Screaton et al. (2004) investigated the interaction of stream-derived water in the cave between the Santa Fe River Sink, a swallet where the Santa Fe River goes underground, and Rise, a resurgence where the Santa Fe River rises to the land surface again, in Alachua and Columbia counties. They concluded that organic-rich flood waters were lost from the conduit through infiltration into the conduit walls. This water was undersaturated with respect to calcite, so limestone was dissolved during and following each flood event. They estimated the erosion rate of the conduit between the swallet and resurgence on the river assuming the conduit is approximately 6 km in length and the width of the area affected by dissolution is approximately 5 km. The estimated dissolution rate was 5.6 × $10^{-4}$ m/year, which is an order of magnitude greater than the dissolution rates of Brooks (1967), Runnels (1971), and Opdyke et al. (1984).

Adams et al. (2010) continued the work of Opdyke et al. (1984) and modeled and revised the denudation rate estimates for north-central Florida. They concluded that the denudation rate (1 m/11,200 years) was 3.5 times the previous estimate of 1 m/38,000 years and that isostatic uplift was, therefore, approximately twice the previous estimate. Modeled uplift was estimated to be 0.047 mm/year as opposed to 0.024 mm/year as given in Opdyke et al. (1984). Modeling the uplift resulted in estimated ages of 1,440,000 years for Trail Ridge (a prominent Pleistocene relict barrier island complex in the Lakes District of northeast Florida), 408,000 years for the Penholoway Terrace, and 120,000 years for the Talbot Terrace.

The crustal movement issue is especially interesting, because it appears to be related to karstification. Hoyt (1969) and Winker and Howard (1977) observed that the Plio-Pleistocene shorelines of Florida and elsewhere on the coastal plain were not level, and that they had been deformed by crustal movement at a sub-regional scale. They argued that Pleistocene marine fossils found in Florida and Georgia at elevations of 42–49 m MSL are too high for conventional elevations of Pleistocene high sea stands. These high elevations were thought to be a result of uplift caused by unloading of the crust as limestone was being dissolved.

Willet (2006) investigated the effects of loss of limestone by dissolution using data from springs in central and north Florida. His estimate of dissolution from the UFA was 5.8 × $10^5$ $m^3$/year, or 1 m of limestone every 160,000 years. This loss was estimated to result in at least 9 to 58 m of sub-regional uplift since the beginning of the Quaternary.

In 2012, Ezell et al. presented a paper describing carbonate dissolution in estavelles (springs that periodically reverse and recharge the aquifer when the stream into which the spring normally discharges in at flood) in north-central Florida. They studied the effects of organic-rich, riverine flood waters on dissolution in the caves associated with the springs during flood events. While the rate of dissolution varied with flood magnitude and duration, they concluded that erosion rates on the cave walls were substantially higher than previous regional estimates. In one, two-month long flood in 2009, it was estimated that the cave underwent a 0.16 mm increase in diameter.

### 6.6.2 Denudation Rates in the Biscayne Aquifer

Halley and Evans (1983) estimated the rate of dissolution of the Miami Limestone in southeastern Florida based on estimates of precipitation, evapotranspiration, and groundwater recharge. Net recharge, based on current climatic conditions, was estimated to be 16 $cm^3$ of meteoric water recharge through every square centimeter of limestone each year. Over the 120,000 years that the Miami Limestone has existed, this amount of recharge accounted for 1980 L/$cm^2$ of water available to react with the limestone. At 80 mg/L $Ca^{2+}$ (Parker et al. 1955), the groundwater was assumed to have dissolved about 200 mg $CaCO_3$ for each liter of water. Halley and Evans (1983) estimated that the net dissolution rate of the Miami Limestone was 264 cm over the 120,000 years, assuming that the dissolution was entirely at the top of the limestone. This estimate equates to 0.0022 cm/year. Assuming that half of the dissolution occurred within the body of the limestone aquifer, the total erosion at the upper surface of the limestone was estimated to be 132 cm, for a rate of 0.0011 cm/year.

One can get a sense of the amount of dissolution of exposed limestone surfaces (Fig. 6.44) in southern Florida by examining exposures of the upper surface of the oolitic Miami Limestone at Silver Bluff in Coconut Grove, Dade County. These surfaces are:

- Less than about 125,000 years old – the uppermost limestone was deposited between 127,000 and 116,000 years

**Fig. 6.44** Exposure of the Silver Bluff wave-cut scarp, which has developed on the Miami Limestone. Note the partially obscured cross bedding and thin duricrust that occludes the bedding, Location: 25.7375°N, 80.2267°W

ago[5] (marine isotopic stage 5e; Broecker et al. 1968; Neuman and Moore 1975);

- Modified by marine erosion during the Silver Bluff[6] sea stand, either
  1. During a still-stand at about +2.4 m NAVD toward the end of marine isotopic stage 5e[7] as the Sangamon-aged sea retreated (Hine 2013); or, possibly,
  2. During a possible sea stand that rose to about +3 m NAVD about 2100 years before present (Missimer 1973; Mitchell-Tapping et al. 1998[8]);
- Developed on strata that were once composed of aragonitic ooliths and had been more-or-less lithified prior to much of the erosion evidenced in exposures today; and
- Partly coated by a thin calcitic duricrust, suggesting that significant dissolution and reprecipitation of calcite has occurred after the exposure was created.

In the exposure (Fig. 6.44) of the Miami Limestone exposed on the Silver Bluff in Coconut Grove, traces of the original cross bedded oolite remain, but the upper surface appears rounded and weathered as a result of subaerial exposure and incipient duricrust development. Part of the erosion that created the exposure was a result of marine erosion during the Silver Bluff sea stand. The remainder is a result of post-marine dissolution and reprecipitation. Note the cylindrical pit exposed on the left-hand side of Fig. 6.44. This is a remnant of a solution pipe, a karst feature developed by dissolution (Chap. 8). The fact that the edge of the solution pit appears to have been breached, perhaps by marine erosion, suggests that substantial karstification of the Miami Limestone had occurred prior to development of the Silver Bluff. If so, the younger age (approx. 2100 years) of the Silver Bluff appears reasonable, and the final rounding and development of the duricrust is only a few thousand years old. Using Halley and Evans' (1983) denudation rate and about 2000 years of exposure, about 4 cm of limestone

[5]The ages of corals and other fossils in the Miami Limestone and equivalent strata have been shown to be 127,000 to 116,000 years before present. We generalize the age of the Miami Limestone, therefore, at about 125,000 years. This age is equivalent to marine isotopic stage 5e based on oxygen isotope data.

[6]The age and relationship of the Silver Bluff wave-cut exposures to Pleistocene/Holocene sea levels in southern Miami and Coconut Grove (Dade County) is poorly known.

[7]Since the age of the Miami oolite has been dated to isotopic marine stage 5e, this date appears too old for the Silver Bluff erosional episode. The "bluff" is clearly younger than the oolitic facies of the Miami Limestone and early karstification, and it must have developed after substantial lithification of the oolite had occurred.

[8]The investigations by Missimer (1973) and Mitchell-Tapping et al. (1998) are based on sediments observed in southwestern Florida where the post-Pleistocene sea-level record is better preserved in the stratigraphic record.

should have been removed by dissolution. This estimate appears high given the preservation of the cross bedding in the exposed surface.

### 6.6.3 Overview of the Rate of Carbonate Rock Dissolution

These estimates of dissolution rates suggest that the overall dissolution of carbonate sediment is very slow in terms of a human's experience. If one assumes that the estimated loss of limestone is limited to the upper few meters of carbonate rock (epigenetic karst), the estimated losses range from a low of approximately 0.0006 mm/year to a high of approximately 0.089 mm/year with an average of approximately 0.041 mm/year. In other words, using the average estimate, this model suggests that the upper limestone surface is lowered approximately 5.1 mm a century.

It is important to understand that these loss estimates are not just from the upper surface of the limestone. They represent estimates of total limestone loss through dissolution within the three-dimensional, shallow groundwater-flow system. Clearly, these dissolution rates are too slow for karst features to develop entirely during a human lifetime, let alone a few years. Therefore, the argument that a void or lowered top-of-rock surface in carbonate strata develops within a few years and suddenly forms a sinkhole is specious.

Clearly, estimates of natural erosion rates are slow relative to a human life span, but they are surprisingly rapid in geologic time. Comparison of the results of these studies is instructive. The regional studies (Brooks 1967; Opdyke et al. 1984; Adams et al. 2010) suggest a relatively slow regional dissolution rate, while site specific studies (Screaton et al. 2004; Ezell et al. 2012) and examination of the exposure shown in Fig. 6.44 indicate more rapid, local dissolution rates, including accelerated dissolution during extreme events, such as floods. Clearly, the differences in the results of these studies indicate that the regional denudation rates do not uniformly reflect distributed dissolution. Rather, these studies and logic dictate that actual dissolution sites are discrete and site-specific, related to pre-existing fractures, water-table positions, bedding planes, and karst conduits, surface drainage features, water residence time, and, in some cases, extreme hydrologic events.

## 6.7 Timing of Karst Development

There is some interesting and relevant information to assist in identification of periods when dissolution and karst development were active, as opposed to periods when karstification was less intense. Clearly, epigenetic karstification was inhibited during episodes of high sea level. During episodes of low sea level, active erosion created the majority of the karst terrain observed today. This karst is reflected by unconformities, paleosols, the fossil record, relative maturity of karst terrains, and certain diagenetic modifications of sediment.

Fig. 6.45 An epikarst horizon developed on the top of the Oligocene Suwannee Limestone on the east shore of the Suwannee River at Rock Bluff Landing, Gilchrist County. Note the pinnacles and soil-filled depressions in the limestone. Cover thickness is less than 1 m (3 ft.). Location: 29.795°N, 82.919°W

## 6.7.1 Unconformities and Paleosols

### 6.7.1.1 Introduction

Development of unconformities can usually be identified by the fossils in the sediments above and below the unconformity, associated paleosols, and discontinuities in sedimentation. The time interval missing from the stratigraphic record can be assumed to include the period during which any karstification developed if carbonate rocks are present below the ancient erosion surface.

The relationship of karstification to unconformities is an important concept because the karst features that developed in carbonate rocks below an unconformity can be economically important. In Florida, the carbonate rock below an unconformity associated with karstification usually includes an ancient epikarst horizon, or erosion surface.

### 6.7.1.2 Epikarst

Epikarst is the uppermost part of a carbonate rock stratum that has been subjected to weathering and possibly soil formation (pedogenesis). Epikarst horizons are typically characterized by:

- Enhanced permeability;
- Variability in the amount of dissolution and/or precipitation and recrystallization of the carbonate materials;
- Variability in the thickness of the weathered zone;
- Variability of the elevation of the upper surface of the carbonate rock, including pinnacles, depressions (Fig. 6.45); and
- Mixing of the underlying carbonate material with the sediments that overlie the epikarst if it is covered (Klimchouk 2004b).

While the epikarst horizon does not represent sinkhole formation unless a depression has formed at the land surface in bare epikarst or there is evidence of cover materials migrating into a void within the epikarst in covered epikarst, the epikarst is considered to be a primary major location of weathering and dissolution.

The uppermost epikarst horizon in Florida is usually covered by a mantle of younger marine sediments and is considered covered epikarst (Klimchouk 2004b). Figure 6.46 illustrates an example of covered epikarst exposed under about 4 m of Plio-Pleistocene marine terrace sand. The epikarst is developed in carbonate rocks of the Miocene Arcadia Formation (Hawthorn Group). Covered epikarst is manifested in test borings by very soft and/or refusal strength limestone, mixing of the limestone with the overlying sediments, losses of drilling fluid circulation, and other artifacts of the ancient weathering that created the epikarst.

**Fig. 6.46** Mantled (covered) epikarst horizon exposed in Gator Sink (27.873°N, 81.805°W), near Bartow, Polk County, Florida. White arrow indicates cover, and the epikarst layer is indicated by the black arrow. Cover is well-sorted sand approximately 5 m (16.4 ft.) in thickness

**Table 6.7** Elevations above sea level of prominent Plio-Pleistocene marine terraces in Florida

| Terrace | Terrace elevations | |
|---|---|---|
| | Cooke (1939, 1945) | Healy (1975; see Fig. 6.26) |
| Hazlehurst | 65–82 m | 66–98 m |
| Coharie | 52–65 m | 52–66 m |
| Sunderland | 30–52 m | 30–52 m[a] |
| Wicomico | 21–30 m | 21–30 m |
| Penholoway | 13–21 m | 13–21 m |
| Talbot | 8–13 m | 8–13 m |
| Pamlico | 2–8 m | 2–8 m |
| Silver bluff | 0–3 m | 0.3–3 m |

[a]Includes the Sunderland and Okefenokee terraces, as recognized by Healy (1975).

Ancient epikarst horizons are also common in Florida. If these paleo-epikarst horizons are at erosional unconformities between limestone or dolostone, the epikarst horizon may be modified by many years of groundwater flow through the enhanced porosity within the epikarst horizon. When investigating water production in wells, it is common to find that the most productive horizons in the well are at formation contacts, where paleo-epikarst exists. It is also common to encounter cavernous porosity, ancient water-table positions and bedding planes enlarged by dissolution, and post-depositional deposits of carbonate and siliciclastic materials, such as nodules of calcite or dolomite (Fig. 6.20) and deposits of sand and clay.

In summary, paleosols and unconformities, which can be dated by their fossil content and stratigraphic relationships to adjacent strata, allow us to understand approximately when episodes of karstification occurred.

### 6.7.2 Fossils

Many of Florida's vertebrate terrestrial fossils have been found in sediment deposits in ancient sinkholes (McFadden 1997). These animal remains entered the sinkholes at the time of, or after, formation of the sink. The oldest terrestrial vertebrate fossils found in Florida sinkhole deposits are Oligocene in age. A large number of occurrences of vertebrate fossils can be found in Miocene sinkholes (Upchurch and Randazzo 1997). From these occurrences, we can conclude that near-surface karstification in Florida occurred at least as early as the Oligocene, and that there was a major episode of karstification, apparently near the end of the Miocene Epoch, during which there was a major low sea stand.

### 6.7.3 Relationship of Karst Landforms to Sea-Level Fluctuations

There have been two studies that attempted to relate karst landforms to ancient sea levels in Florida. These studies related the degree of development of sinkholes based on size, shape, complexity, and spatial density to presumed elevations of marine terrace deposits (Table 6.7). Their hypotheses hinged on the assumption that the depressions that reflect sinkhole development must be younger than the terrace deposits within which they occur.

Bahtijarevic (1996) investigated the distribution and morphology of closed depressions within a set of random areas in the Ocala Karst District. Each quadrat area had to include a geological borehole from which data concerning the local stratigraphy could be obtained and related to depression development. She sampled a total of 148.8 km$^2$ and characterized 774 closed topographic depressions within that space. Her conclusions included the following.

1. The number, average area, range of depression areas, and percent coverage by closed depressions per unit area in covered, unconfined karst increased with an increase in elevation. Assuming that the marine terraces are older at higher inland elevations, this conclusion is consistent with establishment of relative age of the karst features. Her data suggested that the inland karst areas are older and the karst is more "mature" than settings nearest the coast.
2. The degree of confinement by Miocene Hawthorn Group sediments that underlie the terrace sands affects the morphological variability and depression area and depth. In other words, where covered karst is superimposed by strata that limit recharge to the underlying UFA, the relative geomorphic age of the karst depressions is less certain.
3. The number of closed depressions in confined karst decreases with increasing elevation and/or thickness of the confining unit.
4. Multicyclicity of the karst process and sea-level change disrupts the patterns of karst landforms and determination of their relative ages.

Denizman and Randazzo (2000) investigated approximately 25,000 closed depressions in the lower Suwannee River basin. They concluded that the karst development was controlled by lower sea-level stands of the Pleistocene and by development of the Suwannee River. Average density of karstic depressions in their study area was 6.07/km$^2$.

Denizman and Randazzo (2000) also examined the morphometry of closed depressions as a function of elevation, as related to marine terrace elevations. For reference, the terraces range from the Pamlico terrace (Table 6.7), the lowest widespread terrace at approximately 2–8 m elevation above sea level to the highest terrace in their study area, the Sunderland (Table 6.7) at approximately 100–170 m above sea level. Ages of the terraces are controversial, but a good working hypothesis is that the highest terraces are the oldest, possibly Pliocene in age (Tanner 1982).

The mean depression area increases with assumed age of the marine terraces (Denizman and Randazzo 2000). On the Pamlico marine terrace, the mean depression area was nearly 10,000 $m^2$, while the mean area on the higher and presumed older Sunderland terrace was over 20,000 $m^2$. They also showed that the younger sinkholes associated with the Pamlico terrace are more likely to be circular, while sinkhole features associated with the Sunderland terrace are substantially less circular.

Both investigations, therefore, suggest that there is a relationship of relative sinkhole shape and size to elevation and age. Higher, inland areas with older terrace deposits include classical, rounded sinkhole depressions, and a complex array of large, coalesced and complex-shaped depressions, while coastal and low elevation areas are more typically characterized by simple, nearly circular, cone-shaped sinkholes.

Florea and Vacher (2006a) and Florea et al. (2007) examined the relationship of cave levels in west-central Florida to sea levels, marine terraces, and paleoshorelines. They found correlations between the elevations of caves in the subsurface and sea-level fluctuations, which suggested that groundwater levels and resulting cave development respond to sea-level fluctuations.

While these studies allow for estimation of the relative maturity of karst features, the data clearly suggest that karstification has been episodic and is related to sea-level fluctuations (Florea and Vacher 2006a).

### 6.7.4 The Mio-Pliocene Unconformity: Florida's most Intense Episode of Karstification

The interval from the Late Miocene through Early Pliocene was an episode during which global sea level was extremely low (Scott 1997; Hanada 2010) and the climate was conducive to chemical weathering of carbonate, phosphate, and even silicate minerals (Carr and Alverson 1959). This episode of intense weathering resulted in an unconformity and intense weathering of the underlying Miocene and older strata. This weathering resulted in distinctive features that assist in understanding the timing of some of the early karst development in Florida. It is within this episode of weathering that the paleosol locally known as the "leached zone" (Fig. 6.21) was created in what is now the Central Florida Phosphate District. Development of the leached zone paleosol and silicification of the strata that lie under the paleosol provide evidence of the timing of the Mio-Pliocene karst event. Weathering during the Late Miocene and Early Pliocene resulted in well-developed karst features throughout central and northern Florida.

Two artifacts of this period of intense weathering are shown in Fig. 6.47. These are silicified karst features that assist in determining the age of what appears to have been the most intense period of epigenetic karst development in Florida's history. The object shown in Fig. 6.47a is a silicified, duricrust rim of a solution channel, such as are shown in Fig. 6.22. Figure 6.47b is a silicified corner from an intersection of two fractures. Both are karst-related features that developed in the limestone of the Tampa Member of the Arcadia Formation (Miocene Hawthorn Group) according to the following history:

1. The Tampa Member, which lies at the base of the Hawthorn Group in west-central Florida, was deposited in Late Oligocene or Early Miocene time.
2. After deposition, the limestone was lithified and other Miocene Hawthorn sediments were deposited above the Tampa limestone.

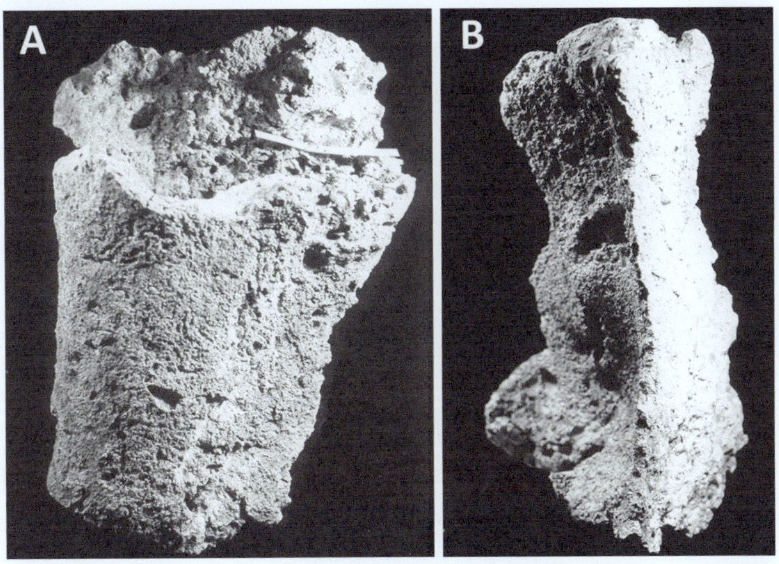


**Fig. 6.47** Casts of relict karst features that have been replaced with chalcedony. A is a silicified duricrust that once lined a solution pipe and B is a silicified fracture intersection. Both specimens, which are from offshore of Ballast Point, Tampa, Hillsborough County, are approximately 60 cm in height

3. At some point in this history, the limestone was fractured (Fig. 6.47b).
4. Toward the end of the Miocene and extending into the Pliocene Epoch, intense weathering occurred, the sediments overlying the Tampa Member limestone were removed, and the Tampa Member was subjected to epigenetic karst development as illustrated by the silicified solution pit (Fig. 6.49a).
5. With the intense weathering, clays of the Hawthorn Group were altered or destroyed, releasing silica to the groundwater (Upchurch et al. 1982). This silica replaced and/or lined the surfaces of the intersection of the two fractures and the solution pit.
6. The limestone supporting the silicified fracture and solution pit walls was removed by subsequent erosion, leaving the two artifacts in the residual, paleosol at the top of the remaining Tampa Member limestone. This later erosional episode was probably Pleistocene in age.

The widespread paleosols, occurrences of Mio-Pliocene vertebrate fossils in sinkholes, loss of stratigraphic record, and other consequences of this major weathering event suggest that the Mio-Pliocene weathering episode was the most intense erosion event in the last 20 million years. Where limestone was exposed to this erosion, development of epigenetic karst proliferated.

## 6.8 Caves and Void Spaces

### 6.8.1 Introduction

When many of us think of karst landforms, caves come to mind. The origin and properties of Florida caves are discussed in the next chapter. The purpose of this section is to address definitions of pore and void spaces and introduce their properties as a factor that affects karst development.

The openings in Florida carbonate sediment and rock range in origin, age, and size (Table 6.1). The openings in Florida limestones range from the pore space between sediment particles (intergranular spaces; Fig. 6.2), to molds caused by the dissolution of fossils, to large caves. The limestones of the Biscayne, Intermediate, and Floridan aquifers are highly productive because they contain this broad spectrum of openings which make the aquifers highly porous and permeable.

### 6.8.2 Pore Versus Void Space

Voids, as used in this text, are openings large enough to be involved in sinkhole development. A void would likely greatly exceed 1 $m^3$ in volume. Openings as large as a football are simply too small to accept sufficient material from above to create a destructive sinkhole.

The term pore space is restricted to openings too small to accept sediment and cause a sinkhole to develop. In this context, pore space includes interstitial space between sediment grains, typical moldic porosity, and small dissolution features that are typically less than 1 $m^3$ in volume.

While the void space in Florida's limestones and dolostones may be related to past or even present-day karst processes, they may also be a result of diagenesis (i.e., dissolution of aragonitic shell material), fracturing, or other processes.

### 6.8.3 Distribution of Void Space

No one can predict where voids occur in Florida's carbonate rocks. We know that they are extremely abundant, because of the ability of the limestones and dolostones to produce water to typically high yielding wells.

Evaluation of *apparent* voids[9] reported in well logs was used by Stewart (1966) and Wilson (1994) to classify voids according to evidence and formation. Finally, there have been many studies that have combined geophysical and borehole data to characterize void space (Schmoker et al. 1979; Schmoker and Halley 1982; Navoy 1986; Hester and Schmoker 1985; Safko and Hickey 1991; Duerr 1994; Cunningham et al. 2006; Cunningham and Sukop 2011). All of these studies have produced dramatic and compelling evidence that there is abundant pore and void space in Florida carbonate rocks and that the void space has many origins, including fracturing, dissolution, and primary depositional processes.

Table 6.8 summarizes the properties of void-like "solution features" identified by Stewart (1966) in 191 deep boreholes in Polk County. The data ranged from drillers' logs to detailed borehole logs prepared by qualified geologists, so care should be taken to avoid over interpretation of the data. Stewart identified four categories of apparent "void spaces." These included (1) cavities or open voids, (2) sediment-filled cavities, (3) porous zones (lost drilling fluid circulation zones, wherein drill cuttings were lost), and (4) honeycomb zones. Stewart (1966) does not make the properties of "honeycomb zones" clear in his report. There is no distinction between voids and/or sediment-filled voids, caused by

[9]Note that losses of drilling fluid circulation during drilling reflect openings in the host sediment large enough to accept a somewhat viscous fluid, but they may not be large enough to accept a volume of raveled overburden sediment or rock collapse material. Studies that report losses of drilling fluid circulation as representing voids, make no distinction as to whether the opening is a void that might participate in sinkhole development or a pore system too small to accept sediment and cause sinkhole development. Such usage is problematic because it is likely to exaggerate the importance and size of the pore or void space that represents the loss of drilling fluid circulation.

**Table 6.8** Proportion of cavity types arranged by formation in Polk County, Florida

| Formation | Cavities (percent within formation) | Sediment-filled cavities (percent within formation) | Porous zones (percent within formation) | Honeycomb zones (percent within formation) |
|---|---|---|---|---|
| Miocene hawthorn group (excl. Tampa member) | 62 | 0 | 29 | 9 |
| Miocene Tampa member | 43 | 14 | 43 | 0 |
| Oligocene Suwannee limestone | 46 | 17 | 29 | 8 |
| Eocene Ocala limestone | 45 | 37 | 9 | 9 |
| Eocene Avon Park formation | 48 | 20 | 26 | 6 |

Data are from Stewart (1966)

dissolution, as opposed to fractures and voids with other origins.

The proportions of different types of voids and in-filled voids presented in Table 6.8 are informative. With the exception of the Miocene Hawthorn Group (excluding the Tampa Member) sediments, the proportion of void space identified as being cavities is relatively uniform across all formations tested, and the dominant form of void space in all formations. The Upper Eocene Ocala Limestone has a well-developed epikarst with common in-filled cavities in west-central Florida. Stewart's data appear to bear this observation out with almost twice as many in-filled cavities as the other strata.

## 6.9 Summary and Conclusions

Chapter 6 has introduced a number of issues that affect karst in Florida's eogenetic karst. The most important topics covered in the chapter are listed below.

- With respect to the subjects covered in this and previous chapters, it is important to understand the many factors that affect eogenetic karst development geologically young carbonate sediments and rocks. These include
  - Mineralogy of the sediments, including metastable minerals such as aragonite and high-Mg calcite;
  - Fabric and texture of the sediments and any diagenetic modifications that have occurred before karst development begins;
  - Diagenesis and karstification may be reflected by the same processes and may coincide or overlap;
  - Pathways of water migrating over and through the carbonate sediments or rocks;
  - Effects of sea-level oscillations and changes in water-table position;
  - Causes of porosity in the sediments and relationships of porosity to water flow rates and chemical reactions;
  - Residence time of the water in contact with carbonate sediment particles and/or surfaces;
  - Geochemical saturation state of the water with respect to the minerals present in the carbonate strata; and
  - Relative surface area in contact with chemically aggressive water.
- One of the important factors covered in this chapter is porosity of the carbonate sediment and rock. Porosity can be primary or secondary.
  - Primary porosity is developed in the sediment by depositional processes and packing of the initial sediment.
  - Secondary porosity is developed in the sediment after deposition, either by diagenesis, fracturing, or karst-related dissolution.
- Intergranular porosity in carbonate strata is essentially unique to eogenetic karst systems.
- Other forms of primary porosity that are common in Florida carbonate sediments include burrows in soft sediment and borings in lithified sediments, intraparticle porosity, and shelter porosity.
  - These forms of porosity initially affect the pathways of water flow through the sediment or rock;
  - If the void and pore spaces of primary porosity are interconnected, the efficiency of water flow through the strata is enhanced and drilling fluid circulation may be lost during penetration testing; and
  - When abundant in the sediment or rock, these forms of primary porosity are likely to affect penetration resistances when drilling to explore for evidence of speleogenesis.
- Secondary porosity as a result of diagenetic alterations includes formation of molds as a result of dissolution of aragonitic or high-Mg calcite objects in the sediment. If moldic porosity is widespread and the molds are interconnected, permeability of the stratum in enhanced and low penetration resistances and losses of drilling fluid circulation may result.
- Diagenetic processes, such as cementation of limestones and dolomitization, are likely to destroy or occlude some primary and many forms of secondary porosity.
- In eogenetic karst systems, primary and secondary porosity, including growth of voids as a result of dissolution may result in touching-vug porosity. Touching-vug porosity results in horizons with high hydraulic conductivities.

- Fracturing of brittle strata is common in Florida carbonate sediments.
  - Fractures in Florida commonly form the dominant pathways of groundwater flow in aquifers;
  - Infiltration and transport of chemically aggressive, meteoric water through fractures result in development of many common karst landforms, including sinkholes and sinkhole lakes, linear stream and vegetation patterns, and water-saturated soils or cover sediments; and
  - Much of the porosity in Florida dolostones is a result of fracturing.
- Photolineaments are linear patterns observed in aerial photographs, satellite images, and topographic maps.
  - Photolineaments are identified by alignments of three or more closed depressions, aligned stream or wetland segments, and soil tone patterns.
  - In Florida, photolineaments commonly occur as two conjugate sets of alignments – northeast to southwest and northwest to southeast. These alignments are consistent with fracturing as a result of stresses imposed by earth tides.
  - While the aggregate pattern of photolineaments in an area appears to reflect vertical fracturing in the underlying carbonate strata, one cannot assume that any one photolineament accurately reflects the presence of a fracture. Ground truthing is required to confirm the presence of a fracture represented by a photolineament.
  - Some of the major photolineaments that can be observed in satellite images have been shown to represent faulting in the basement rocks of Florida.
- Karst-related secondary porosity includes caves and other karst conduits. Fracturing and other zones of enhanced permeability often control the development of these conduits.
- Bedding planes are often poorly represented in the geologically young carbonate sediments and rocks of Florida. If they represent horizons of enhanced permeability, they may be enhanced or modified by dissolution.
- More-or-less horizontal galleries of voids are common in Florida limestones. Rather that reflecting dissolution on bedding planes, these galleries appear to represent dissolution horizons located on ancient water-table positions.
- Epikarst is the uppermost karstic layer of a limestone stratum. Epikarst consists of the irregular, weathered surface characterized by
  - Cutters and pinnacles (cutters are the shallow depressions between pinnacles);
  - Coarse rubble and fine-grained, disaggregated limestone fragments;
  - Porosity through which water and drilling fluid freely move; and
  - Mixtures of the disaggregated limestone with any overburden sediments.
- Unconformities between and within formations affected by sea-level fluctuations often have enhanced permeability as a result of weathering and karstification at the time that they were subaerially exposed. Today, these paleo- or relict-epikarst horizons are likely locations of enhanced groundwater flow.
- In Florida, the common occurrence of both primary and secondary porosity in limestone results in dual-porosity aquifers. Because of the abundance of interconnected, primary and secondary porosity, the UFA and Biscayne aquifers are considered by many to be amongst the most productive aquifers in the world.
- Intergranular pore spaces allow for storage and slow drainage of the aquifers while fractures and cavernous systems allow for rapid drainage after precipitation events.
- Epigenetic karst forms by the interaction of meteoric water with carbonate sediments in the vadose and phreatic zones, near the water table.
- Carbon dioxide interacts with water to form carbonic acid both within the vadose zone as meteoric water percolates toward the water table and at the water table by degassing and other processes.
- Water that recharges the limestone directly, through cover materials and into small-scale porosity may result in disaggregated particles of sand to clay size. Chemical equilibration with the water with respect to calcite occurs within short distances after contact with carbonate minerals. If the carbonate contains metastable minerals, dissolution is enhanced and micritization of carbonate particles is likely.
- Epikarst does not constitute sinkhole development, although unusually large voids within the epikarst can result in sinkhole development.
- Recharge by way of fractures or existing karst conduits will contribute to epikarst development but, more importantly, equilibration with the carbonate host rock is delayed because of short residence times and low surface areas in contact with the water. As a result, the conduits are enlarged and lengthened over time.
- Reaction rates of groundwater with the carbonate rock are typically slow.
  - The time required to create a void large enough to accept cover or rock materials and create a sinkhole is thousands of years.
  - Sudden collapse events that form sinkholes are rarely a result of collapse of a rocky cave roof. More often the collapse event is a result of failure of cover materials that overlie a pre-existing void.

- Florida caves form parallel to the water table when the phreatic surface is stable and perpendicular to the water table if water levels are low or dropping. Many Florida caves reveal evidence of water-table fluctuations that are driven by climate and sea-level changes. Dissolution as a result of $CO_2$ degassing at the water table is often evident.
- Hypogenetic karst is poorly documented in Florida.
  - Hypogenetic speleogenesis is the process of cave formation by chemically aggressive water that originates at depth rather than near the land surface
  - Mixing-zone speleogenesis along freshwater/saltwater transition horizons has been predicted from groundwater geochemical modeling.
  - A small number of vadose and phreatic caves have been attributed to mixing zone dissolution. These attributions are usually related to proximity to modern or ancient shorelines or cave morphologies.
  - Caves that are open to the land surface by water-filled sinkholes and that extend to unusual depths on the west coast of Florida are likely candidates for mixing-zone origins. These caves have unusually large rooms at depth, which is contrary to the more normal pattern in epigenetic caves.
- Hypogenetic cave systems created by sulfate reactions related to the gypsum/anhydrite beds are problematic in Florida.
  - Florida has evaporite beds that may be sourcing sulfate/sulfide reactions in (1) the UFA from MCU evaporite sources, (2) the LFA from sub-FAS evaporite sources, and (3) the Cretaceous from anhydrite sources and sulfide-rich petroleum.
  - Dissolution related to the sulfate-rich zone at the base of the UFA and created by dissolution and/or upward leakage from the MCU may be creating or enlarging karst conduits, but evidence is weak.
  - Evidence of sulfate-driven hypogenesis in the UFA consists of formation of deep collapse features, such as some salt springs, cenote and collapse features observed by geophysical means.
  - The Kohout circulation system that prevails in the LFA may also be developing karst by dissolution through groundwater mixing or sulfate reduction/oxidation processes.
  - There is also some evidence of a former hypogene karst system in the Cretaceous, oil-bearing strata of southwestern Florida. This evidence consists of confining beds of anhydrite, collapse and breccia zones, and springs on the continental margin that appear to be involved in sulfur reactions.
  - Anchialine caves on the west coast of Florida have small-scale sulfate to sulfide to sulfate systems at the chemocline and groundwater mixing horizon. While these are extremely local, sea-level fluctuations are likely to have "smeared" the dissolution features throughout otherwise epigenetic cavers systems.
- Sea-level fluctuations beginning in the Late Miocene and continuing to the present are clearly causing significant variations in the elevation of the phreatic surface, freshwater/saltwater transition zone, and ages of cavernous systems.
- Apparently, the episode of most intense weathering, pedogenesis, and karstification was during the Mio-Pliocene low sea stand.
- Unconformities in the Neogene carbonate strata contain relict epikarst with elevated permeabilities. These pale-epikarst systems constitute significant water-producing horizons in the UFA.

## References


Adams PN (2013) A plausible explanation for raised coastal ridges and terraces along the axis of peninsular Florida. Southeastern Geol Soc Guidebook 50

Adams PN Opdyke ND Jaeger M (2010) Isostatic uplift driven by karstification and sea-level oscillation: Modeling landscape evolution in north Florida. Geology, 38:531–534

Alfieri MC Upchurch SB Dobecki TL (2018, in press). Photolinears, fractures, and fallacies: A *post hoc* study of photolineaments, Hillsborough County, Florida. Proceedings, 15th Multidisciplinary Conference on Sinkholes and the Engineering and Environmental Impacts of Karst, Shepherdstown, West Virginia.

Alt D Brooks HK (1965) Age of Florida marine terraces. Jour Geol, 73:406–411.

Altschuler ZS Jaffe EB Cuttitta F (1956) The aluminum phosphate zone of the Bone Valley Formation, Florida, and its uranium deposits. US Geol Surv Prof Pap 300, pp. 495–504

Altschuler ZS Jaffe EB Dwornik E (1951) The stratigraphy of the upper part of the Bone Valley Formation and Its relation to the leached zone. US Geol Surv Trace Elements Memo Repo 237

Back W Hanshaw BB (1970) Comparison of chemical hydrogeology of the carbonate peninsulas of Florida and Yucatan. Jour Hydro 10:330–368

Badon Ghyben W (1889) Nota in verband met de voorgenomen putboring nabij Amsterdam. Tijdschrift van het Koninklijk Instituut van Ingenieurs, The Hague, pp. 8–22.

Bahtijarevic A (1996) Karst landforms in Florida, geomorphological analysis. Master's thesis, Tampa, Univ South Fla

Barnett RS (1975) Basement structure of Florida and its tectonic implications. Trans Gulf Coast Assoc Geol Soc 25:122–142

Beck BF Sayed S (1991) The sinkhole hazard in Pinellas County: A geologic summary for planning purposes. Fla Sinkhole Res Inst Report 90-91-1

Bloom AL (1983) Sea level and coastal morphology of the United States through the Late Wisconsin glacial maximum. In: Porter SC (ed.), Late Quaternary environments of the United States, Volume 1 - The late Pleistocene, pp. 215–229

Bock WD Moore DR Neumann AC Supko PR (compilers) (1969) (reprinted 1994) Late Pleistocene geology in an urban area. Miami Geol Soc, field trip guidebook, accessed at http://sofia.usgs.gov/publications/reports/mgs_bock1969/mgs_bock1969.pdf on 3/20/2012.

Broecker WS Thurber DL Goddard J Ku TL Mesolella KJ (1968) Milankovitch hypothesis supported by precise dating of coral reefs and deep-sea sediments. Science, 159:297–300

Brooks HK (1967) Rate of solution of limestone in the karst terrane of Florida. Fla Water Resour Res Ctr Pub 6

Carr WJ Alverson DC (1959) Stratigraphy of middle Tertiary rocks in part of west-central Florida. US Geol Surv Bull 1092

Cathcart JB (1963a) Economic geology of the Chicora Quadrangle, Florida. US Geol Surv Bull 1962-A

Cathcart JB (1963b) Economic geology of the Keysville Quadrangle, Florida. US Geol Surv Bull 1128

Cathcart JB (1964) Economic geology of the Lakeland Quadrangle, Florida. US Geol Surv Bull 1962-G

Cathcart JB (1966) Economic geology of the Fort Meade Quadrangle, Florida. US Geol Surv Bull 1207

Clesi D King J (1991) Team Diepolder '91: Our story. Underwater Speleology, 18(3):7–12

Coniglio M Harrison RS (1983) Holocene and Pleistocene caliche from Big Pine Key, Florida. Bull. Can. Petrol. Geol., 31:3–13.

Cooke CW (1939) Scenery in Florida interpreted by a geologist. Fla Geol Surv Bull 17

Cooke CW (1945) Geology of Florida. Fla Geol Surv Bull 29

Cressler A (1993) The caves of Dade County, Florida. Georgia Underground, 30(3):9–16

Culbreth MA (1988) Geophysical investigation of lineaments in south Florida. Master's thesis, Tampa, Univ South Fla

Cunningham KJ Kluesner JW Westcott RL Robinson E Walker C Khan SA (2017) Sequence stratigraphy, seismic stratigraphy, and seismic structures of the lower intermediate confining unit and most of the Floridan aquifer system, Broward County, Florida. U.S. Geol Surv Sci Invest Rept 2017-5109

Cunningham KJ Renken RA Wacker MA Zygnerski MR Robinson E Shapiro AM Wingard GL (2006) Application of carbonate cyclostratigraphy and borehole geophysics to delineate porosity and preferential flow in the karst limestone of the Biscayne aquifer, SE Florida, In: Harmon RS Wicks C (eds.), Perspectives on karst geomorphology, hydrology, and geochemistry—A tribute volume to Derek C. Ford and William B. White, Geol Soc Amer Sp Pap 404, pp. 191–208

Cunningham KJ Sukop MC (2011) Multiple technologies applied to characterization of the porosity and permeability of the Biscayne aquifer, Florida. US Geol Surv Open-File Rept 2011–1037

Cunningham KJ Sukop MC Huang H Alvarez PF Curran HA Renken RA Dixon JF (2008) Prominence of ichnologically influenced macroporosity in the karst Biscayne aquifer: Stratiform "super-K" zones. Bull Geol Soc Amer, 121:164–180

Cunningham KJ Walker C (2009) Seismic-sag structural systems in Tertiary carbonate rocks beneath southeastern Florida, USA: Evidence for hypogenic speleogenesis? In: Klimchouk A Ford D (eds.), Hypogene speleogenesis and karst hydrogeology of artesian basins, Ukrainian Inst of Speleo and Karst Special Paper 1, pp. 151–158

Curl RL (1974) Deducing flow velocity in cave conduits from scallops. Bull Nat Speleol Soc, 36:1–5

Davis RA Jr. (2011) Sea-level change in the Gulf of Mexico. College Station, Texas A&M Univer Press

Denizman C Randazzo AF (2000) Post-Miocene subtropical karst evolution, lower Suwannee River basin, Florida. Geol Soc Amer Bull 112(12):1804–1813

De Rooj GH Cho H (1999) Modeling solute leaching during fingered flow by integrating and expanding various theoretical and empirical concepts. Hydrol Sci – Jour des Sciences Hydrologiques 44(3):447–465

Dobecki TL Upchurch SB (2010) A multi-level approach to site characterization - C.W. Bill Young Regional Reservoir, Hillsborough County, Florida. In: LoCoco (ed.), Symposium on the Application of Geophysics to Engineering and Environmental Problems, Denver, CO, Environ and Engineer Geophysical Soc, pp. 508–516

Domenico PA Schwartz FW (1997) Physical and chemical hydrogeology. 2nd Ed, New York, John Wiley & Sons

DuChene HR Hill CA (eds.) (2000) Caves of the Guadalupe Mountains. Jour Cave and Karst Stud 62(2):91–108

Duerr AD (1994) Types of secondary porosity in the injection and test wells in southern peninsular Florida. US Geol Surv Water-Resour Invest Rept 94-4013

Ezell J Martin JB Screaton E Brown A Gulley J Spellman P (2012) The effect of flood magnitude on carbonate dissolution rates during spring reversals (abs.). Geol Soc Amer Abst with Prog 44(7):412

Florea LJ (2006a) Architecture of air-filled caves within the karst of the Brooksville Ridge, west-central Florida. Jour Cave and Karst Stud 68(2):64–75

Florea LJ (2006b) The karst of west-central Florida. Ph.D. Dissert, Univ South Fla

Florea LJ (ed.) (2008a) Caves and karst of Florida: A guidebook for the 2008 National Convention of the National Speleological Society. Huntsville, AL, Nat Speleo Soc

Florea LJ (2008b) Geology and hydrology of karst in west-central Florida and north-central Florida. In: Florea LJ (ed.), Caves and karst of Florida: A Guidebook for the 2008 National Convention of the National Speleological Society. Huntsville, AL, Nat Speleo Soc, pp. 225–239

Florea LJ Noe-Stinson CL Brewer J Fowler R Kearns JB Greco AM (2011) Iron oxide and calcite associated with *Leptothrix* sp. biofilms within an estavelle in the upper Floridan aquifer. International Journal of Speleology, 40(2):205–219

Florea LJ Vacher HL (2006a) Cave levels, marine terraces, paleoshorelines, and the water table in peninsular Florida. In: Onac BP Tamas T Constantin S Persoiu A (eds.), Archives of Climate Change, Karst Waters Inst Sp Pub 10, pp. 188–192

Florea LJ Vacher HL (2006b) Morphologic features of conduits and aquifer response in the unconfined Floridan aquifer system, west central Florida. The 12th Symp Geol Bahamas and other Carb Regions, pp. 32–44

Florea LJ Vacher HL (2007) Eogenetic karst hydrology: Insights from the 2004 hurricanes, peninsular Florida. Ground Water, 45(4):439–446

Florea LJ Vacher HL Donahue B Naar D (2007) Quaternary cave levels in peninsular Florida. Quaternary Sci Rev 26:1344–1363

Florea LJ Vacher HL Oches EA (2003) Karst studies in west central Florida: USF seminar in karst environments. Brooksville, Southwest Florida Water Management District, accessed 9/4/2013 at http://uweb.cas.usf.edu/geokarst/Publications/SWFWMD2003/intro.pdf

Ford DC Williams PW (1989) Karst geomorphology and hydrology. New York, John Wiley & Sons

Ford DC Williams PW (2007) Karst hydrogeology and geomorphology. New York, John Wiley & Sons

Freeze RA Cherry JA (1979) Groundwater. Englewood Cliffs, NJ, Prentice Hall Pub

Garman KM (2010) The biogeochemistry of submerged coastal karst features in west central Florida. Ph.D. diss, Univ South Fla

Garman KM Garey JR (2005) The transition of a freshwater karst aquifer to an anoxic marine system. Estuaries 28(5):686–693

Glass RJ Steenhuis TS Parlange JY (1988) Wetting front instability as a rapid and far-reaching hydrologic process in the vadose zone. Jour Contam Hydrol 3:207–226

Glass RJ Steenhuis TS Parlange JY (1989) Wetting front instability: 2. Experimental determination of relationships between system parameters and two-dimensional unstable flow field behavior in initially dry porous media. Water Resour Res 25:1195–1207

Gulley J (2013) On the origin of Florida's underwater caves. Underwater Speleology, 40(4):10–11, 34

Gulley JD Florea LJ (2016) Caves as paleo-water table indicators in the unconfined Upper Floridan aquifer. Florida Scientist, 79(4):239–256

Gulley JD Martin JB Brown A (2016) Organic carbon inputs, common ions, and degassing: Rethinking mixing dissolution in coastal eogenetic carbonate aquifers. Earth Surface Processes and Landforms, 41:2098–2110

Gulley JD Martin JB Moore PJ Brown A Spellman PD Ezell J (2015) Heterogeneous distributions of $CO_2$ may be more important for dissolution and karstification than mixing dissolution. Earth Surface Processes and Landforms, 40:1057–1071

Gulley JD Martin JB Moore PJ Murphy J (2013) Formation of phreatic caves in an eogenetic karst aquifer by $CO_2$ enrichment at lower water tables and subsequent flooding by sea level rise. Earth Surface Processes and Landforms, 38:1210–1224.

Gulley JD Polk JS (2017) Hypogene karst influences in the upper Floridan aquifer. In: Klimchouk AB Palmer AN Waela JD Auler AS Audra P (eds.), Hypogene karst regions and caves of the world, Springer Intern Publ, Ch. 50, pp. 745–755

Halley RB Evans CC (1983) The Miami Limestone: A guide to selected outcrops and their interpretation. Miami Geol Soc Fieldtrip Guidebook, accessed on 3/20/2012 at http://sofia.usgs.gov/publications/reports/mgs_halley1983/index.html

Hanada M (2010) Global sea level in the Messinian Stage of the late Miocene. Inst Oceanic Research & Develop, Tokai Univ, Bulletin 31:65–77

Haslett RC (1989) A study of lineaments and fracture traces mapped on the Crescent City Ridge, Putnam and Volusia Counties, Florida. Masters thesis, Gainesville, Univ Fla

Healy HG (1975) Terraces and shorelines of Florida. Fla Bur Geol Map Series 71

Heath RC (1983) Basic ground-water hydrology. US Geol Surv Water-Sup Pap 2220

Hester TC Schmoker JW (1985) Porosity, grain-density, and inferred aragonite-content data from the Miami Limestone, Miami area and lower Florida Keys. U.S. Geol Surv, Open-File Rept 85-21

Herbert TA Upchurch SB (2016) The potential role of hypogene speleogenesis in the lower Floridan aquifer and Sunniland Oil Trend, south Florida, U.S.A. In: Chavez T Reehling P (eds.), Proceedings of DeepKarst 2016: Origins, Resources, and Management of Hypogene Karst, Carlsbad, NM, Nat Cave and Karst Res Inst, Symposium 6, pp. 119–129

Herman JS Back W Pomar L (1985) Geochemistry of groundwater in the mixing zone along the east coast of Mallorca, Spain. Karst Water Resour, IAHS Pub 161, pp. 467–479

Herzberg A (1901) Die Wasserversorgung einiger Nordseebder. Jour fur Gasbeluchtung und Wasserversorgung, 44: 815–819, 45: 842–844

Hester TC Schmoker JW (1985b) Porosity, grain-density, and inferred aragonite-content data from the Miami Limestone, Miami area and lower Florida Keys. US Geol Surv, Open-File Rept 85-21

Hickey JJ (1984) Field testing the hypothesis of Darcian flow through a carbonate aquifer. Ground Water 22:544–547

Hill CA (1981) Speleogenesis of Carlsbad Caverns and other caves of the Guadalupe Mountains, New Mexico and Texas. New Mexico Bur Mines and Mineral Resour Bull 117

Hill CA (1990) Surfuric acid speleogenesis of Carlsbad Cavern and its relationship to hydrocarbons, Delaware Basin, New Mexico and Texas. Amer Assoc Petrol Geol Bulletin 74:1685–1694

Hill CA (1995) $H_2S$-related porosity and sulfuric acid oil-field karst. In: Budd DA, Sailer AH, Harris PM (eds.), Unconformities and porosity in carbonate strata. Amer Assoc Petrol Geols Mem 63, p. 301–306

Hill CA (2000) Overview of the geologic history of cave development in the Guadalupe Mountains, New Mexico. Jour Cave and Karst Stud 62(2):60–71

Hill ME (2008) An evaluation of conduit conceptualizations and model performance. Ph.D. diss, Tampa, Univ South Fla

Hills ES (1972) Elements of structural geology. New York, Springer Verlag, 2nd Ed

Hine AC (2013) Geologic history of Florida: Major events that formed the sunshine state. Gainesville, Univ Press Fla

Hine AC Chambers DP Clayton TD Hafen MR Mitchum GT (2016) Sea level rise in Florida: Science, impacts, and options. Gainesville, Univ Press Fla

Hoyt JH (1969) Late Cenozoic structural movements, northern Florida. Gulf Coast Assoc Geol Soc Trans 19:1–9

Hunt CB Hunt AP (1957) Stratigraphy and archeology of some Florida soils. Geol Soc Amer Bull 68:797–806

Huntoon PW (1994) Is it appropriate to apply porous media groundwater circulation models to karstic aquifers? In: El-Kadi AI (ed.), Groundwater models for resources analysis and management, Pacific Northwest/Oceana Conf, Honolulu, HA, pp. 339–358

Jennings JN (1985) Karst geomorphology. Oxford, Blackwell Publ, 2nd Ed

Judson S Ritter DF (1964) Rates of regional denudation in the United States. Jour Geophy Res 69:3395–3401

Klimchouk A (2004a) Caves. In: Gunn J (ed.), Encyclopedia of caves and karst science. New York, Fitzroy Dearborn, pp. 203–205

Klimchouk A (2004b) Towards defining, delimiting and classifying epikarst: Its origin, processes and variants of geomorphic evolution. Speleo Evol Karst Aquifers 2(1):1–13

Klimchouk A (2007) Hypogene speleogenesis: Hydrogeological and morphogenetic perspective. Carlsbad, NM., Nat Cave and Karst Res Inst

Klimchouk A (2009) Principal characteristics of hypogene speleogenesis. In: Stafford KW Land L Veni G (eds.), NCKRI Symposium 1 Adv Hypogene Karst Stud, Carlsbad, NM, Nat Cave Karst Res Inst, pp. 1–11

Klimchouk AB Palmer AN Waela JD Auler AS Audra P (eds.) (2017), Hypogene karst regions and caves of the world, New York, Springer Sci Pub

Lattman LH Matzke RH (1961) Geological significance of fracture traces. Photogram Eng 27:435–438

Lattman LH Parizek RR (1964) Relationship between fracture traces and the occurrence of groundwater in carbonate rocks. Jour Hydrol 2:73–91

Lawrence FW Upchurch SB (1976) Identification of geochemical patterns in ground water by numerical analysis. In: Zaleem EA (ed.), Advances in Groundwater Hydrology, Amer Water Resour Assoc, pp. 199–214

Littlefield JR Culbreth MA Upchurch SB Stewart MT (1984) Relationship of modern sinkhole development to large-scale photolinear features. In: Beck BF (ed.), Sinkholes: Their geology, engineering & environmental impact, Rotterdam, B.A. Balkema, pp. 189–195

Lucia FJ (1995) Rock fabric/petrophysical classification of carbonate pore space for reservoir characterization. Bull Amer Assoc Petrol Geologists 79(9):1275–1300

MacNeil FS (1950) Pleistocene shore lines in Florida and Georgia. U.S. Geol Surv Prof Pap 221-F

McGee DK (2010) Microbial influences on karst dissolution: The geochemical perspective, with a chapter on assessment of the spreadsheets across the curriculum project. Ph.D. dissert, Tampa, Univ South Fla

Miller JA (1986) Hydrogeologic framework of the Floridan Aquifer System in Florida, and in parts of Georgia, Alabama and South Carolina. US Geol Surv Prof Pap 1403-B

Miller JA (1990) Ground water atlas of the United States: Alabama, Florida, Georgia, South Carolina. US Geol Surv Hydrol Invest Atlas 730-G, Segment 6

Miller JJ (1998) An environmental history of northeast Florida. Gainesville, Univ Press Fla

Mitchell-Tapping HJ Mitchell-Tapping AM Lee TJ Williams CR (1998) Core evidence of sea level high stands in southwestern Florida during the last 5,000 years. In: Treat S (ed.), Proc Charlotte Harbor Public Conf and Tech Symp, Palm Beach, Fla, South Fla Water Mgt Dist, pp. 47–53

Merritt ML (2004) Estimating hydraulic properties of the Floridan aquifer system by analysis of earth-tide, ocean-tide, and barometric effects, Collier and Hendry Counties, Florida. US Geol Surv Water-Resour Invest Rept 03-4267

Missimer TM (1973) Growth rates of beach ridges on Sanibel Island, Florida. Gulf Coast Assoc Geol Soc Trans, 23:383–388

Missimer TM (2002) Late Oligocene to Pliocene evolution of the central portion of the south Florida Platform: Mixing of siliciclastic and carbonate Sediments. Fla Geol Surv Bull 65

Navoy AS (1986) Hydrogeologic data from a 2,000-foot deep core hole at Polk City, Green Swamp area, central Florida. US Geol Surv Water-Resour Invest Rept 84-4257

Neill L Anderson J (2009) Cave divers explore deepest parts of Weeki Wachee Springs. Tampa Bay Times, April 19, 2009 (http://www.tampabay.com/news/humaninterest/cave-divers-explore-deepest-parts-of-weeki-wachee springs)

Neumann AC Moore WS (1975) Sea level events and Pleistocene coral ages in the northern Bahamas. Quaternary Res, 5:215–224

Opdyke ND Spangler DP Smith DL Jones DS Lindquist RC (1984) Origin of the epeirogenic uplift of Pliocene-Pleistocene beach ridges in Florida and development of the Florida karst. Geology 12(4)226–228

Palmer AN (1984) Geomorphic interpretation of karst features. In: LaFleur RG (ed.), Groundwater as a geomorphic agent, Boston, Allen & Unwin, Ch. 8, pp. 173–209

Palmer AN (1991) The origin and morphology of limestone caves. Geol Soc Amer Bull 103:1–21

Palmer AN (2006) Support for a sulfuric acid origin for caves in the Guadalupe Mountains, New Mexico. New Mexico Geol Soc Guidebook, 57th Field Conf, Caves and karst of southeastern New Mexico, pp. 195–202, accessed 4-28-2014 at http://nmgs.nmt.edu/publications/guidebooks/downloads/57/57_p0195_p0202.pdf.

Palmer AN (2007) Cave geology. Trenton NJ, Cave Books

Parker GG (1955) Geomorphology. In: Parker GG Ferguson GE Love SK and others. Water resources of southeastern Florida, U. S. Geol Surv Water Sup Pap 1255, pp. 127–155

Parker GG Cooke CW (1944) Late Cenozoic geology of southern Florida, with a discussion of the ground water. Fla Geol Surv Bull 27

Parker GG Ferguson GE Love SK and others (1955) Water resources of southeastern Florida, US Geol Surv Water Sup Pap 1255

Pendexter WSS (1996) The influence of media heterogeneities on the development of flow fingers. Ph.D. Diss, Tallahassee, Fla State Univ

Pendexter WS Furbish DJ (1991) Development of a heterogeneous moisture distribution and its influence on the evolution of preferred pathways of flow in an unsaturated sand soil. Proc Nat Symp Preferential Flow, Chicago, pp. 104–112

Perkins RD (1977) Part II, Depositional framework of Pleistocene rocks in south Florida. In: Enos P Perkins RD (eds.), Quaternary sedimentation in south Florida, Geol Soc Amer Memoir 147, p. 131–198

Pirkle EC Yoho WH Hendry CW Jr. (1970) Ancient sea level stands in Florida. Fla Bur Geol, Geol Bull 52

Plummer LN (1975) Mixing of sea water with calcium carbonate ground water. In: Whitten EHT (ed.), Quantitative studies in geological sciences, Geol Soc Amer Memoir 142, pp. 219–236

Plummer LN (1977) Defining reactions and mass transfer in part of the Floridan Aquifer. Water Resour Res, 13:801–812

Plummer LN Back W (1980) The mass balance approach: application to interpreting the chemical evolution of hydrologic systems. Amer Jour Sci 280:130–142

Polk J Brinkman R (2013) Climatic influences on coastal cave and karst development in Florida. In: Lace MJ Mylroie JE (eds.), Coastal karst landforms, Dordrecht, Springer Science, Coastal Res Lib, Book 5, Chapter 15, pp. 317–345

Portell RW Hulbert RC (2011) Haile Quarries field guide, Newberry, Florida. Southeastern Geol Soc Guidebook No. 53

Portell RW Means GH Hulbert RC Jr. (2012) SMR Aggregates, Inc., Sarasota, Florida. Southeast Geol Soc Fieldtrip Guidebook 56

Puri HS Vernon RO (1964) Summary of the geology of Florida and a guidebook to the classic exposures. Fla Geol Surv Sp Pub 5 (revised)

Randazzo A F (1997) The sedimentary platform of Florida: Mesozoic to Cenozoic. In: Randazzo AF Jones DS (eds.), The geology of Florida, Gainesville, Univ Press Fla, pp. 39–56

Reese RS (2004) Hydrogeology, water quality, and distribution and sources of salinity in the Floridan aquifer system, Martin and St. Lucie counties, Florida. US Geol Surv Water-Resour Invest Rept 03-4242

Rosenau JC Faulkner GL (1975) An index to springs of Florida. Fla Bur Geol Map Series 63 (Revised)

Rosenau JC Faulkner GL Hendry CW Jr. Hull RW (1977) Springs of Florida. Fla Bur Geol Bulletin 31 (Revised)

Runnels DD (1971) Chemical weathering of the Biscayne aquifer, Dade County, Florida. Jour Southeast Geol 13:167–174

Safko PS Hickey JJ (1991) A preliminary approach to the use of borehole data, including television surveys, for characterizing secondary porosity of carbonate rocks in the Floridan aquifer system. US Geol Surv, Water-Resour Invest Rept 91-4168

Scanlon BR Mace RE Barrett ME Smith B (2003) Can we simulate regional groundwater flow in a karst system using equivalent porous media models? Case study, Barton Springs, Edwards aquifer, USA. Jour Hydrol 276:137–158

Schmoker JW Halley RB (1982) Carbonate porosity versus depth: a predictable relation for south Florida. Amer Assoc Petrol Geol Bull 66:2561–2570

Schmoker JW Halley RB Meyer FW Robbins SL (1979) Preliminary porosity estimates of south-Florida Cenozoic carbonate rocks based on borehole gravity measurements. US Geol Surv Open-File Rept 79-1652

Scott TM (1997) Miocene to Holocene history of Florida. In: Randazzo AF Jones DS (eds.), The geology of Florida, Gainesville, Univ Press Fla, pp. 57–67

Screaton E Martin JB Ginn B Smith L (2004) Conduit properties and karstification in the unconfined Floridan aquifer. Ground Water 42(3):338–346

Sellards EH (1909) Some lakes and lake basins. Fla Geol Surv 3rd Ann Rept, pp. 43–76

Slack IJ Rosenau JC (1979) Water quality of Florida springs. Fla Bur Geol Map Series 96

Smith DL Lord KM (1997) Tectonic evolution and geophysics of the Florida basement. In: Randazzo AF Jones DS (eds.), The geology of Florida. Gainesville, Univ Presses Fla, pp. 13–26

Stewart HG Jr (1966) Ground-water resources of Polk County. Fla Geol Surv Rept Invest 44

Stoessell RK Moore YH Coke JG (1993) The occurrence and effect of sulfate reduction and sulfide oxidation on coastal limestone dissolution in Yucatan cenotes, Ground Water 31(6): 566–575

Tanner WF (1982) High marine terraces of Mio-Pliocene age, Florida panhandle. In: Scott TM Upchurch SB (eds.), Miocene of the southeastern United States, Fla Bur Geol Sp Pub 25, pp. 200–209

Tanner WF (2000) Beach ridge history, sea level change and the A.D. 536 event. In: Gunn JD (ed.), The years without summer: Tracing A.D. 536 and its aftermath, British Archaeol Repts, Intern Ser 872, p. 89–97

Tanner WF Demirpolat S Stapor FW Alverez L (1989) The "Gulf of Mexico" late Holocene sea level curve. Gulf Coast Assoc Geol Socs, Trans 39:553–562

Thrailkill JV (1965) Studies in the excavation of limestone caves and the deposition of speleothems. Ph.D. Diss, Trenton, Princeton Univ

US Army Corps of Engineers (2004) Lineament analysis: South Florida Region. US Army Corps of Eng, Jacksonville Dist, Draft Tech Mem, Central and Southern Florida Proj, Comprehensive Everglades Restoration Plan, Aquifer Storage and Recovery Regional Study

Upchurch SB (1992) Quality of waters in Florida's aquifers. In: Maddox GL Lloyd JM Scott TM Upchurch SB Copeland R (eds.), Florida ground water quality monitoring program -- Volume 2, Background Hydrogeochemistry, Fla Geol Surv Sp Pub 34, Ch. IV, pp. 12–52, 64–84, 90–347

Upchurch SB (2002) Hydrogeochemistry of a karst escarpment. In: Martin JB Wicks CM Sasowsky ID (eds.), Hydrogeology and biology of post-Paleozoic carbonate aquifers. Charles Town, WV, Karst Waters Inst Sp Pub 7, pp. 73–75

Upchurch SB (2017) Hypogene speleogenesis on the Floridan Platform, U.S.A. In: Klimchouk AB Palmer AN Waela JD Auler AS Audra P (eds.), Hypogene karst regions and caves of the world, New York, Springer Int Pub, Ch. 49, pp.735–744

Upchurch SB Dobecki TL Daigle DM (1999) Geological, hydrogeological, and geophysical investigation. In: Law Eng Serv and others, Geotechnical Site Characterization Report – Tampa Bay Regional Reservoir, Volume I, Section 3

Upchurch SB Randazzo AF (1997) Environmental geology of Florida. In: Randazzo AF Smith DL (eds.), Geology of Florida, Gainesville, Univ Press Fla, Ch 13, pp. 217–249

Upchurch SB Strom RN Nuckels MG (1982) Silicification of Miocene rocks from central Florida. In: Scott TM Upchurch SB (eds.), Miocene of the southeastern United States, Fla Bur Geol Sp Pub 25, pp. 251–284

Upchurch SB Taraszki MD Zimmerman J Pyne RDG (1991) Rock-water interactions during aquifer storage and recovery. In: Upchurch SB Moore JE Zaporozec A (eds.), Hydrology and hydrogeology in the '90s. Minneapolis, MN., Amer Inst Hydrol pp. 309–319

Vacher HL Florea LJ (2015) Quantitative hermeneutics: Counting forestructures on a path from W.M. Davis to the concept of multiple permeability karst aquifers. Intern Jour Speleol, 44(3):207–230

Vacher HL Hutchings WC Budd DA (2006) Metaphors and models: The ASR bubble in the Floridan aquifer. Ground Water, 44(2):144–154

Vail PR Mitchum RM Jr. (1979) Global cycles of relative changes of sea level from seismic stratigraphy. In: Watkins JS Montadert L Dickerson PW (eds.), Geological and Geophysical Investigations of Continental Margins, Amer Assoc Petrol Geol, Memoir 29, pp. 469–472

Valentine KWG Dalrymple JB (1976) Quaternary buried paleosols: A critical review. Quat Res 6:209–220

Van Kauwenbergh SJ Cathcart JB McClellan GH (1990) Mineralogy and alteration of the phosphate deposits of Florida. US Geol Surv Bull 1914

Vernon R O (1951) Geology of Citrus and Levy counties, Florida. Fla Geol Surv Bull 33

Wentworth CK (1947) Factors in the behavior of ground water in a Ghyben-Herzberg system. Pacific Sci, 1:172–184.

Wentworth CK (1948) Growth of the Ghyben-Herzberg transition zone under a rinsing hypothesis. Amer Geophy Union Trans, 1:97–98

White WB (1988) Geomorphology and hydrology of karst terrains. New York, Oxford Univ Press

Wicks CM Herman JS (1995) The effect of zones of high-porosity and permeability on the configuration of the saline fresh-water mixing zone, Ground Water, 33(1–2): 733–740.

Williams SR (1985) Relationship of ground water chemistry to photolineaments in a karst aquifer. Masters thesis, Tampa, Univ South Fla

Willet MA (2006) Effect of dissolution of the Florida carbonate platform on isostatic uplift and relative sea level. Ph.D. Diss, Tallahassee, Fla State Univ

Wilson WL (1994) Cave occurrence and ground-water velocities in caves of the Floridan aquifer, Polk County, Florida. Winter Springs, FL, Subsurface Evaluations, Inc.

Winker CC Howard JD (1977) Correlation of tectonically deformed shorelines on the southern Atlantic coastal plain. Geology 5:124–127

Wright VP (1986) Introduction. In: Wright VP (ed.), Paleosols, Trenton, Princeton Univ Press, pp. x–xiv

# Caves and Sinkholes in Florida

# 7

**Abstract**

Types of sinkholes in Florida include (1) solution sinkholes formed by dissolution of shallow limestone; (2) suffosion sinkholes where non-cohesive sediments ravel directly into limestone voids; (3) rock-collapse sinkholes formed by collapse of cave roofs; (4) cenote, deep water-filled rock-collapse sinkholes; (5) cover-collapse sinkholes formed by the rapid failure of sand and clay over a limestone void; (6) cover-subsidence sinkholes formed by slow dissolution of the upper surface of a carbonate-rich sediment; and (7) alluvial, or relict, sinkholes.

Other surface depressions include those similar to "Carolina bays" with hybrid karst and aeolian origins, and pseudokarst depressions formed by differential compaction in marine terrace sands, pedogenesis, and other non-karst processes. Karst windows are sinkholes formed by rock or cover collapse and that expose the water-table aquifer. Uvalas are coalescent sinkholes with complex outlines and multiple sinkholes. Although they do not have steep walls, Florida's poljes are uvalas that have been filled by lacustrine or paludal sediments. Collapsed caves occur in traces as a string of karst windows or a wet linear depression.

Sinkhole risk has been evaluated several times in Florida; while sources of elevated risk are known, there is no way to predict where or when sinkholes will develop. Risks include injury, structural damage, flooding, and aquifer contamination.

Both vadose and phreatic, epigenetic caves are associated with current and former water table positions. Because of the high water table or potentiometric surface of the Upper Floridan Aquifer (UFA), air-filled, vadose caves are mostly found on ridges and topographically high areas. Deep phreatic caves near the coasts may be hypogenetic, caused by mixing in freshwater/saltwater transition zones.

Vadose caves are commonly decorated with secondary speleothems, such as stalactites, flowstone, soda straws, and calcite crystals. Phreatic caves rarely contain secondary cave formations. Deposition of speleothems can be caused by evaporation, $CO_2$ degassing, and chemical oversaturation of water with respect to calcite. Analyses of stalactites from caves in central Florida reveal a record of deposition correlated to major climatic patterns extending to the late Pleistocene.

**Keywords**

Sinkholes · Caves · Cover-collapse sinkholes · Cover-subsidence sinkholes · Vadose speleogenesis · Phreatic speleogenesis

## 7.1 Purpose

This chapter is an introduction to the cave and sinkhole features in Florida's eogenetic karst. There is considerable literature discussing the origins and risks of sinkhole development. It is because of the need for a clear understanding of how sinkholes develop, their rates of development, and their risks, that this chapter dwells on sinkholes. In addition to the risks related to sinkholes, cave exploration, both air-filled and water-filled, can be a dangerous endeavor and require significant planning, experience, and training for safety. Many caves and sinkholes are on privately owned lands and, for obvious legal reasons, may require the owner's permission to enter. In order to protect them from trespass, geographic coordinates for caves and sinkholes on private property are not included in this chapter and maps showing


S. Upchurch et al., *The Karst Systems of Florida*, Cave and Karst Systems of the World,
https://doi.org/10.1007/978-3-319-69635-5_7

Fig. 7.1 Cave divers exploring the Diepolder II cave in Hernando County. This cavern is unusually deep and requires use of mixed gases and deep-diving techniques by these highly experienced divers. Note the line showing in the light of the diver on the right. These lines are used to navigate in the caves and prevent getting lost. (Photograph courtesy of Mark Long (©marklongphotography.net))

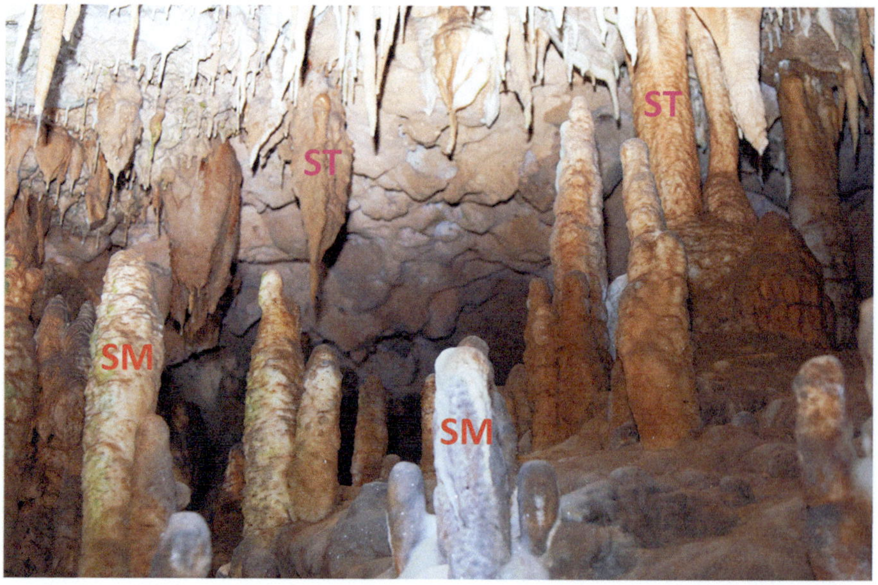


Fig. 7.2 Cave decorations in Florida Caverns State Park, Mariana, Jackson County. ST indicates stalactites and SM represents stalagmites

cave locations are designed to provide geological understanding without revealing specific cave locations.

We begin with the origin and examples of caves in Florida, because they are intimately involved in sinkhole development.

## 7.2 Cave Development

To most people, caves are symbols of karst terrains. The hydrogeochemical origin and properties of caves have been previously discussed. Suffice it to say that Florida limestones

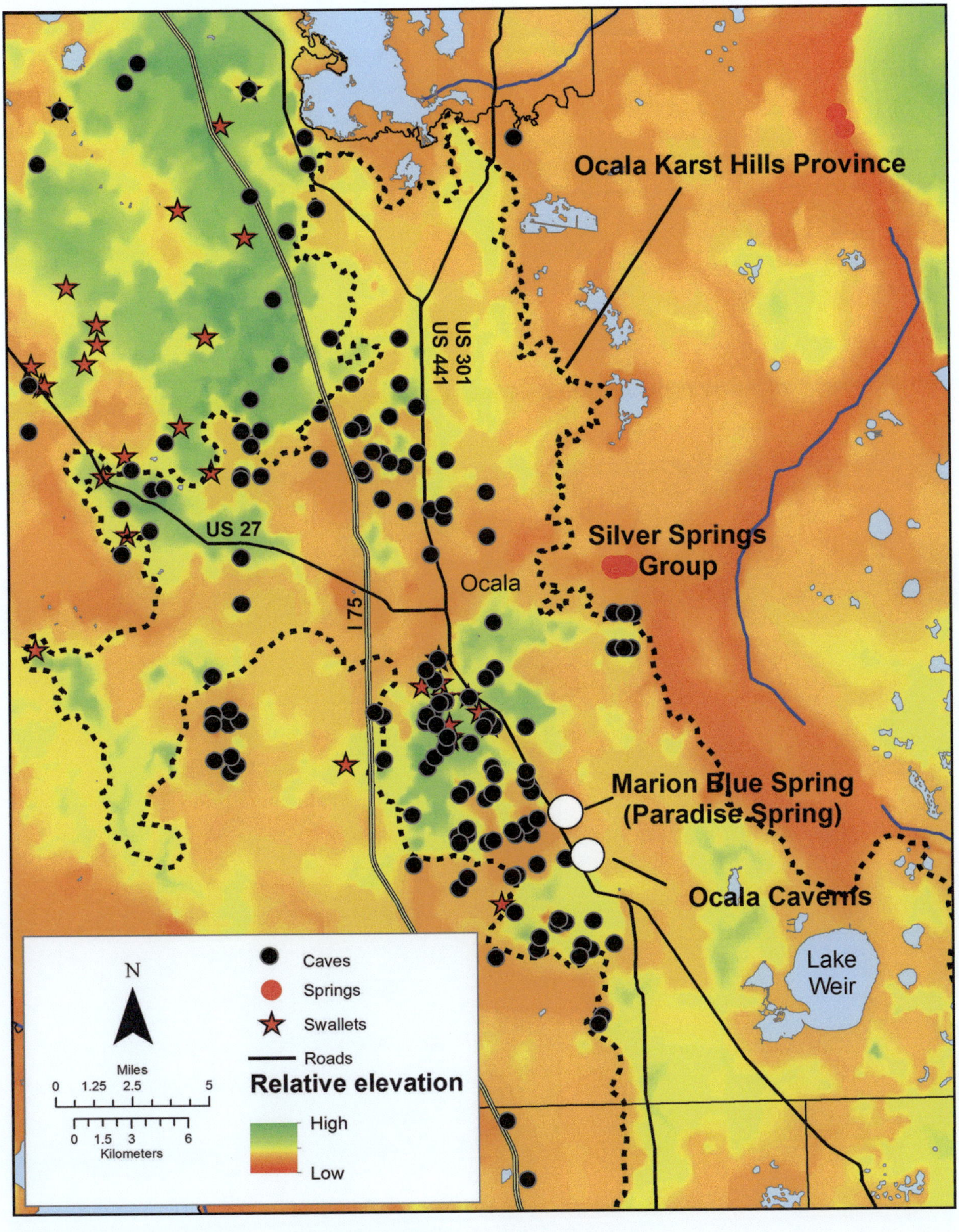


**Fig. 7.3** General locations of known caves in the Ocala Karst Hills

are riddled with karst conduits that range in size from small networks of anastomosing and branching conduits to very large caverns. Most are below the water table and are water-filled (Fig. 7.1), others are dry (Fig. 7.2).

### 7.2.1 Phreatic Caves

Cave detection and exploration is problematic in many areas of Florida because the water table is so high. Several of the

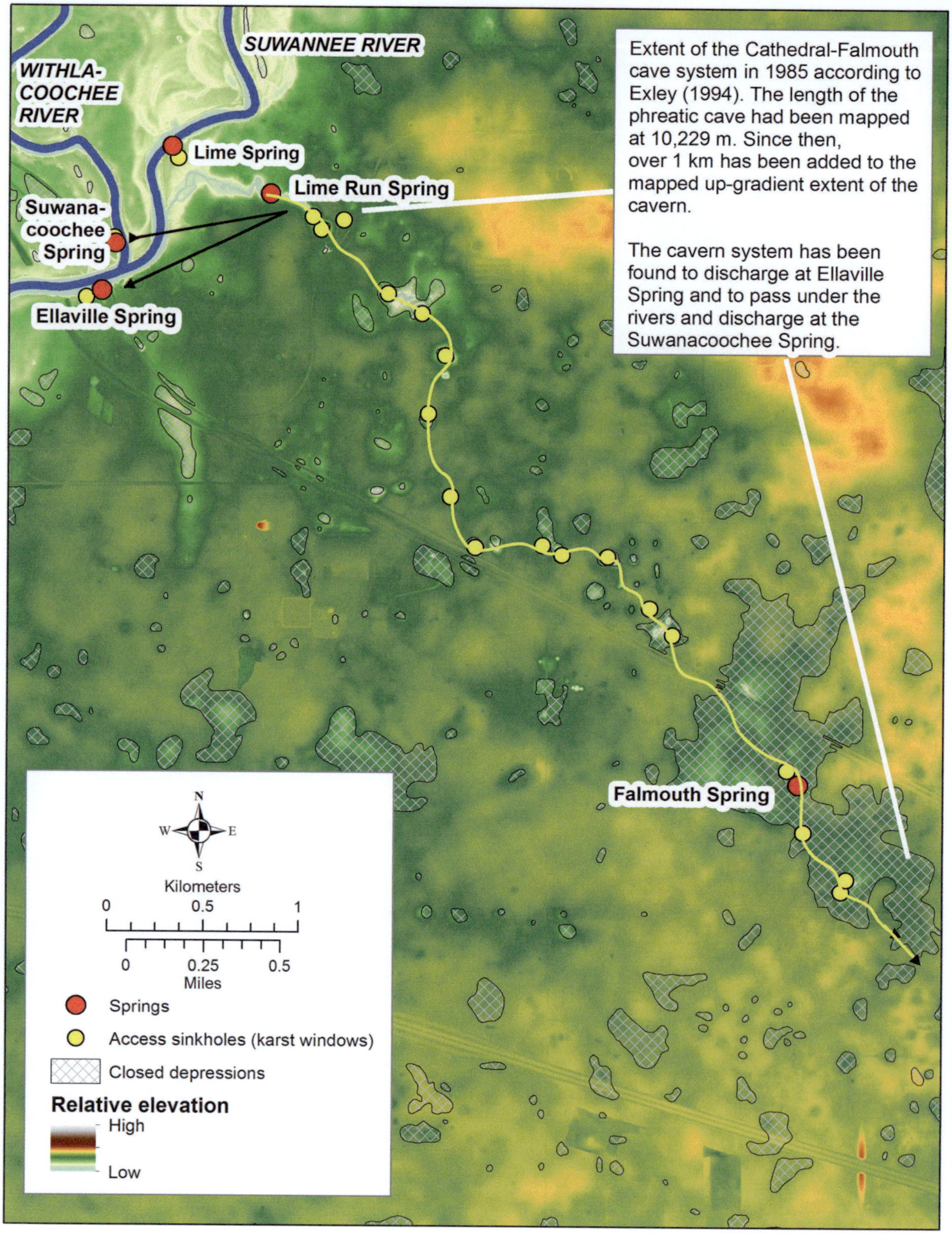


**Fig. 7.4** LIDAR map showing (1) topography, closed depressions, and karst windows (cave access points) on the land surface and (2) the general location of the Cathedral Canyon-Falmouth cavern system, Suwannee County. Cavern location is based on a map by Exley (1994)

**Fig. 7.5** Falmouth Spring, a karst window on the Cathedral-Falmouth cavern system in Suwannee County. Image courtesy of the Florida Geological Survey

prominent ridges (i.e., Brooksville Ridge and Ocala Karst Hills) and areas with higher elevations relative to water-table elevations (i.e., areas on the Chattahoochee Arch [Fig. 3.3]) have dry caves. Figure 7.3, for example, illustrates the locations of known caves in the Ocala Karst Hills. Marion Blue Spring (Paradise Spring) is a karst window that is currently being operated as a dive center and Ocala Caverns was once a commercial cave but is now on private property.

Many of Florida's caves follow fractures, particularly with the northeast to southwest and northwest to southeast orientations. Figure 7.4 is a map showing the Cathedral-Falmouth cave system, one of the longest known caves in Florida, in relation to surface depressions and karst windows. Since the map was made in the 1980s (Exley 1994), the known length of the cave has been extended upgradient to the southeast by cave divers for several kilometers. It can be shown by dye tracing and exploration by divers that the cave connects to the Suwannee River at Ellaville Spring (30.3845°N, 83.1724°W). Recent dye tracing also shows connections of the Cathedral-Falmouth system with springs across the Suwannee River, demonstrating that the cave network passes under the river. Section 7.2.5 offers an explanation for caves that pass under modern streams.

Falmouth Spring (30.3611°N, 83.1350°W; Figs. 7.4 and 7.5) is a karst window (a sinkhole caused by a collapse of the cavern) through which the water flowing in the cave emerges on the up-gradient (southeast) side and disappears into the cavern again on the northwest side. When the Suwannee River is at flood stage, flow ceases and brown water may emerge from the northwest opening into the cavern.

Phreatic caves in Florida rarely have well-developed cave decorations, such as stalactites and stalagmites (Sect. 7.2.7.3). There is evidence of breakdown caused by roof collapse and erosion by dissolution above and below the water table (Figs. 6.36 and 6.37; Sect. 6.5.2).

### 7.2.2 Vadose Caves

Briar Cave (Fig. 7.6), a small, mostly dry cave near Ocala, Marion County, is a popular[1] cave for amateur and professional cave explorers. This cave consists of multiple passages, with the primary passages aligned in a northeast-southwest orientation, parallel to the dominant regional fracture orientation. Some portions of the cave are permanently or seasonally flooded.

Briar Cave is the third longest vadose cave in Florida according to Krause (2008b) at 1747 m. The caves that rank first and second are Warren Cave in Alachua County at 6858 m and The Catacombs in Marion County at 1980 m. Clearly, Florida's vadose-zone caves are significantly shorter and do not penetrate the limestone as deeply as do the record-setting vadose caverns elsewhere in the world.

Vadose caverns are often well decorated with flowstone, draperies, stalactites, and stalagmites (Figs. 7.2 and 7.7). Breakdown from roof failures is common and the caves commonly have red clay deposits as a result of surface water flowing into the cave.

As suggested by Fig. 7.6, the orientations of many vadose caves are controlled by fracturing, usually in the northwest-southeast or northeast-southwest orientations.

[1] Access is by permission only.

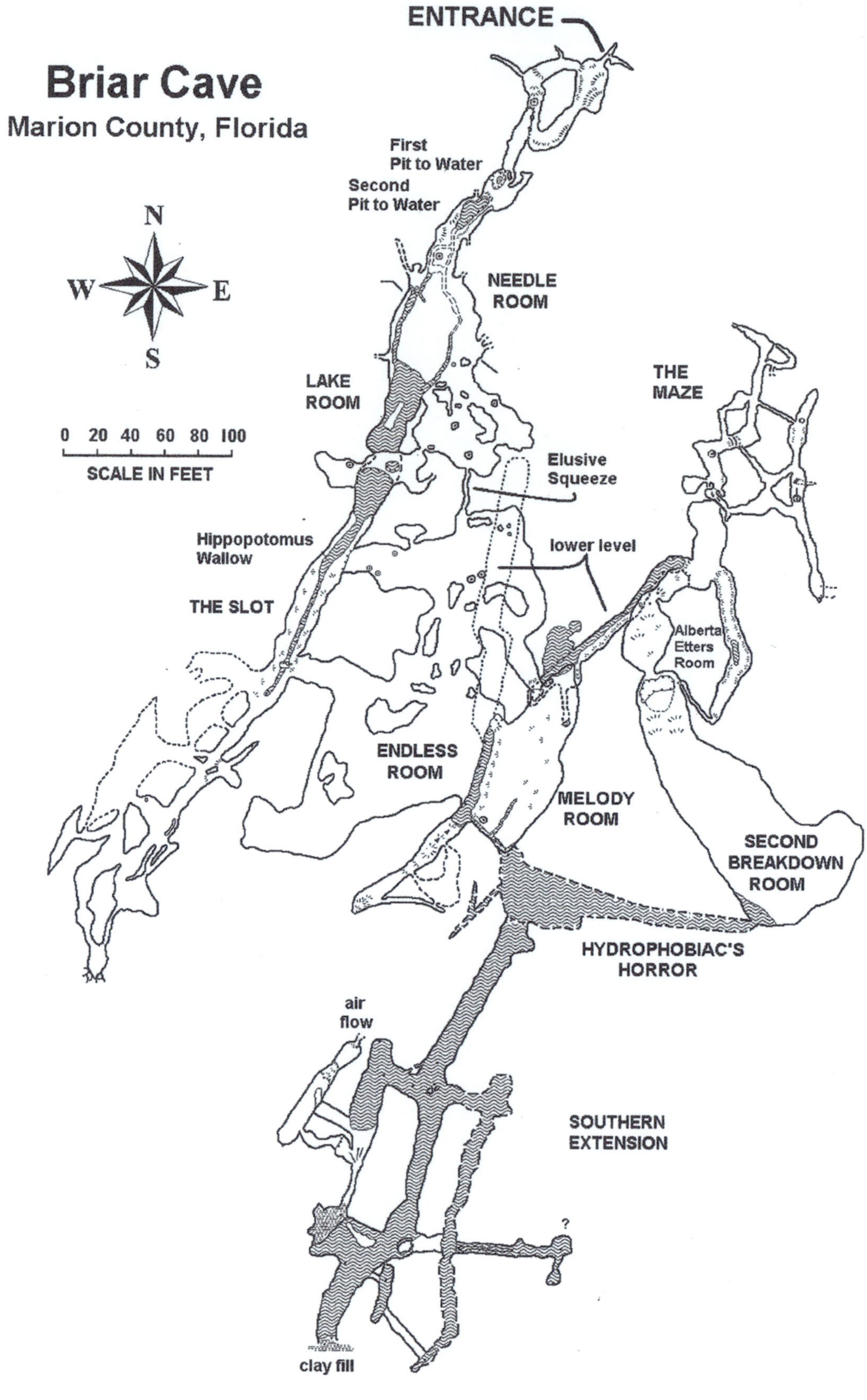


**Fig. 7.6** Map of Briar Cave, near Ocala, Marion County. Shaded passages are either water filled or have the phreatic surface exposed. Map prepared by Krause (2008a)

**Fig. 7.7** Stalactites and helectites in a cave exposed in a limestone quarry in Washington County. (Photograph courtesy of the Florida Geological Survey)

**Table 7.1** Top ten longest mapped caves in Florida

| World ranking | Cave name | County | Length (meters) | Depth (meters) | Geomorphic Province |
|---|---|---|---|---|---|
| 6 | Wakulla-Leon Sinks Cave System | Leon and Wakulla | 51,483 | 106.7 | Woodville Karst Plain |
| 11 | Beacon Woods – Wayne's World System | Pasco | 18,289 | 48.8 | Land O'Lakes Karst Plain |
| 13 | Cathedral-Falmouth System | Suwannee | 14,386 | 57.9 | Branford Karst Plain |
| 19 | Old Bellamy (Derekson) (Santa Fe River Rise) Cave Sys. | Alachua | 12,657 | 48.8 | Alachua Karst Hills |
| 21 | Peacock Cave I-II System | Suwannee | 11,587 | 64.0 | Branford Karst Plain |
| 31 | Hole in the Wall (Rock Spring) Cave | Jackson | 8,452 | 31.1 | Marianna Karst Plain |
| 34 | Madison Blue Spring | Madison | 7,775 | 36.6 | Branford Karst Plain |
| 35 | Telford Springs | Suwannee | 7,620 | 21.9 | Branford Karst Plain |
| 36 | School Sink | Pasco | 7,400 | Unknown | Land O'Lakes Karst Plain |
| 37 | Devils Eye Cave System | Gilchrist | 7,174 | 48.8 | Williston and Branford Karst Plains |

Gulden and Coke (2015)

### 7.2.3 Lengths and Depths of Florida Caves

Vadose-zone caves are typically short and terminate at shallow depths or the water table because they are epigenetic and limited by the elevations of the water table at the time of formation. In contrast, many of Florida's phreatic caves have been found to be kilometers in length and over 100 m in depth. The presence of these caves is a testament to fluctuations in past sea levels and water-table elevations.

The Cathedral-Falmouth cave system (Fig. 7.4) is one of the longest known caves in Florida – and its exploration is completely by cave divers! Cave diving has become a challenging and popular sport in Florida, and many of the longest and largest underwater caves occur in Florida (Table 7.1). For example, the Cathedral Canyon – Falmouth underwater cavern had been mapped as a single, linear cavern for over 10,229 m in 1985 when the late Sheck Exley (2009), a well-known cave explorer and leader in the cave-diving sport,

Table 7.2 Unusually deep caves in Florida

| Cave Name | County | Location | Mapped depth (meters) |
|---|---|---|---|
| Diepolder II | Hernando | 28.5230°N, 82.5289°W | 128 |
| Weeki Wachee | Hernando | 28.5173°N, 82.5734°W | 124 |
| Wakulla-Leon Sinks | Leon and Wakulla | 30.2652°N, 84.3026°W | 107 |
| Eagles Nest | Hernando | 28.3555°N, 82.6092°W | 95 |

Data from Gulden and Coke (2015) and Upchurch (2017)

published the cave map from which Fig. 7.4 was generated. By 2015, the explored extent of the cave system was 14,386 m, an increase of over 4 km (Table 7.1). The length and depth statistics of Florida caves routinely change because of the active exploration being conducted by cave divers.[2]

Determining cave depth is especially challenging. Most of Florida's phreatic caves extend to depths of less than 50 m below the phreatic surface (depths are measured by divers measuring water pressures in the cave). There are some extraordinarily deep caves in Florida, however. The recorded depths of these caves are limited by dive and decompression times, available gas mixtures, and other constraints on safe diving. The caves may extend to greater depths. Table 7.2 lists the four known caves that are deeper than 100 m. Note that all four are located near the Gulf of Mexico coast. Upchurch (2017) suggested that Weeki Wachee and Diepolder II may have formed as hypogenetic karst in the freshwater/saltwater transition zone rather than as epigenetic caves.

### 7.2.4 Relationship of Cave Elevations to Ancient Water-Table Elevations

Florea (2006, 2008a, b, 2009), Florea and Vacher (2006a, b), and Florea et al. (2007) present data on elevations of caverns in central Florida, including the Brooksville Ridge and Ocala Karst Hills. They have shown that the horizontal components of the caves have common elevations, which suggests positions of ancient, regional water-table elevations that have varied as a result of sea-level and, perhaps, climate changes.

Figure 7.8 shows the elevations and proposed relationships of some central Florida caves to Plio-Pleistocene marine terrace elevations. These data by Florea and his colleagues strongly suggest that caves in central Florida were developed during Plio-Pleistocene high and low sea stands, and that the karst is a result of dissolution at the water table. Note that the water tables referenced in these studies are clearly the phreatic surfaces of the UFA over time.

Conventional thinking on the generation of carbonic acid involves interactions with atmospheric and microbe-generated $CO_2$ in the atmosphere and vadose zone. This pattern of creation of acidic water would result in vertical components of cave systems and a lack or weak correlation of cave elevations from site to site. While caves generated by meteoric waters migrating to and through the limestone surface are important and common, the persistence of horizontal caves at common elevations suggests the dissolution of the limestone at, or near, the water table is also a highly significant process. At the water table, incorporation of additional $CO_2$ may lead to additional acidity and selective dissolution in the immediate vicinity of the water table (Gulley 2013; Gulley et al. 2013, 2015, 2016; Gulley and Florea 2016; Gulley and Polk 2017). Section 6.5.2 discusses this process.

Caves and caverns in Florida include the morphologies common to caves worldwide. Their cross-sectional outlines commonly indicate down cutting in the vadose zone and lateral expansion at the water table (White 1988; Ford and Williams 1989; Fig. 6.37). The combination of (1) patterns of cave development that are similar to caves in telogenetic karst systems and (2) cave development in the zone of water-table fluctuations within porous limestone appears to be more-or-less unique to eogenetic limestones (Gulley 2013).

Where bare walls are exposed, and water velocities are sufficient to cause advective groundwater circulation, caves frequently display scalloped or fluted surfaces which indicate more-or-less turbulent flow of the water within the cavern conduits (Curl 1974; White 1988). Scalloped walls are most common near springs or at restrictions in groundwater flow where laminar flow gives way to turbulent flow.

### 7.2.5 Relationships of Caves to Modern Stream Systems

#### 7.2.5.1 Caves That Pass Under Major Streams or Rivers

While many of Florida's caves are graded to modern streams and discharge at springs near the rivers, one of the interesting puzzles in Florida karst is why some caves pass under rivers that are incised in limestone without apparent, significant interactions with the river. For example, dye tracing in the Cathedral-Falmouth cave system (Fig. 7.4) indicates that the cavern system discharges at springs with several different functions. Lime Run Spring (Fig. 7.4) appears to be a "pop-off" spring that only flows when water levels in the UFA are high. Lime Spring (Fig. 7.4) is a small, perennial spring that discharges year-round. At least one spring,

[2]Cave diving is extremely dangerous and should not be attempted by persons who have not earned cave-diving certifications. Each year, we lose divers who become lost, run out of air, or otherwise fail to follow the rules of the cave diving community.

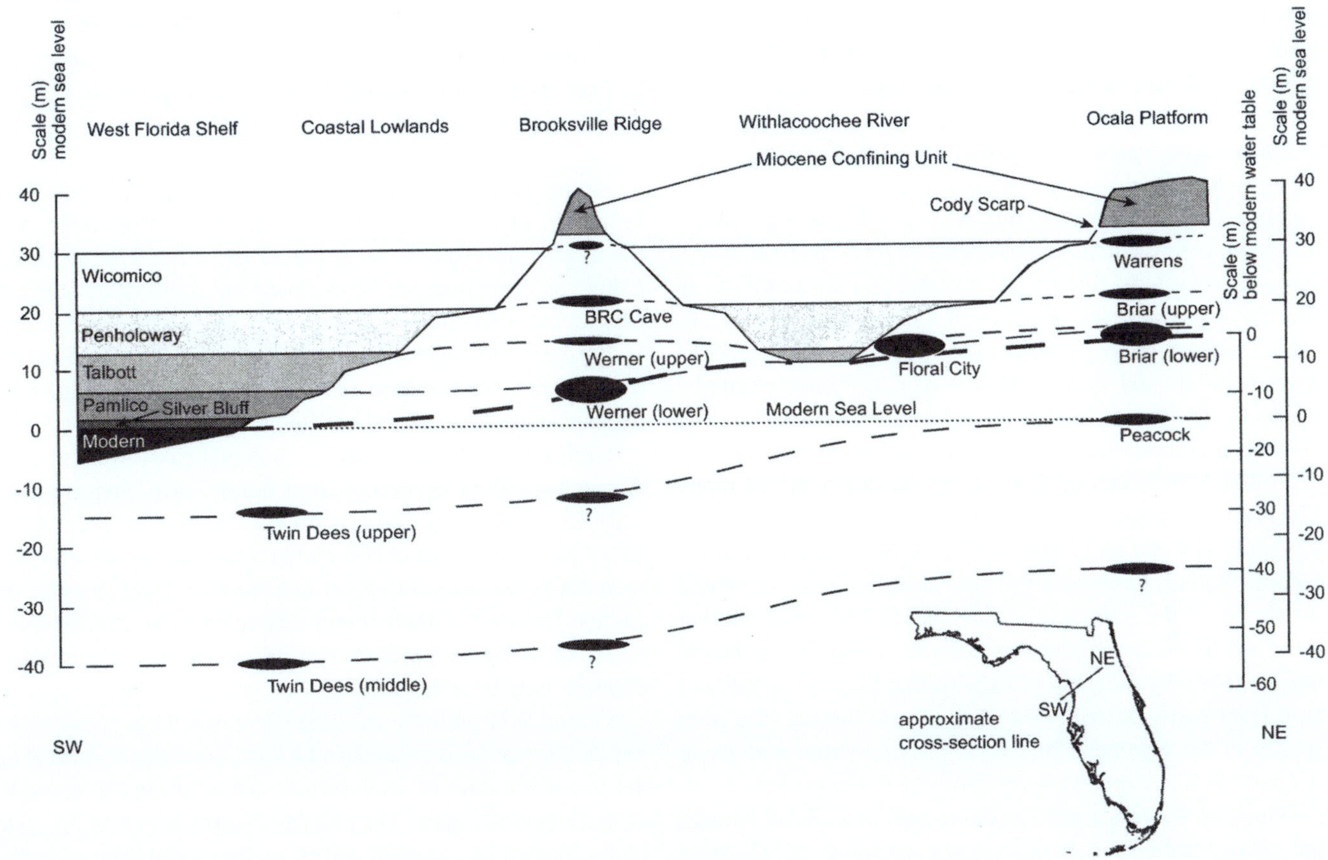


**Fig. 7.8** Elevations of horizontal components of caves in the Brooksville Ridge and Ocala Karst Hills. (From Florea (2006, Fig. 3.9; reproduced with author's permission))

Suwanacoochee Spring (30.3870°N, 83.1718°W) on the west bank of the Withlacoochee River (Fig. 7.4), has been shown by dye tracing to drain the Cathedral-Falmouth cave system by passing under the Suwannee and Withlacoochee rivers. The drainage basin (springshed) of Troy Spring (30.0061°N, 82.9974°W) a first-magnitude spring with discharge that ranges from 3 to almost 6 $m^3/s$, is located on the east side of the Suwannee River while the spring is on the west side (Upchurch et al. 2004; Zwanka 2005; Champion et al. 2005). A large proportion of the discharge from Troy Spring, therefore, discharges from a cave that passes under the Suwannee without significant interaction with the river until reaching the spring vent. Some other springs on the Suwannee and Santa Fe River have drainage basins that extend under the streams, as well.

Gulley (2013) argued that the caves that pass under major streams in contact with the limestone must have formed when the stream beds were dry and not capable of interacting with the underlying groundwater system. That being the case, the origin of the cave system that passes under a stream is likely to have been controlled by dissolution associated with the water table and not leakage from above. It was posited that today's phreatic caves in Florida's eogenetic limestone formed by $CO_2$ enrichment at the water table during low sea stands and water-table conditions, and, as sea level rose, these caves were flooded.

#### 7.2.5.2 Estavelles

Most of the springs on Florida's major streams are either located directly on the shoreline or up a short run from the river. These springs suggest that the cave systems are, at least in part, graded to the modern stream system or that the streams have eroded into, and intercepted pre-existing cave systems.

Because they are at stream level, the caves associated with the springs are subject to back flooding when the streams flood. Springs that experience flow reversals are called estavelles. When streams are in flood, surface water enters the cave system as a form of bank storage. If the surface water is under saturated with respect to the minerals of the host rock either as a result of carbonic acid or humic substances, dissolution and enlargement of the cave is likely to occur while the surface water is stored in the aquifer.

Gulley et al. (2011) explored the mechanisms and consequences of cave enlargement as a result of storage of surface water in the caves associated with estavelles. For example, they found that a single flooding event of the Peacock Spring cave system by the Suwannee River was capable of dissolving up to 3.4 mm of the conduit wall rock.

#### 7.2.5.3 Cave Systems without Swallet Water Sources

Swallets (sinkholes that capture stream flow) can commonly be shown in telogenetic cave systems to form the headwaters for a cave system. Stream-to-sink systems are common in Florida, especially near the toes of karst escarpments. Cave diving and dye tracing have demonstrated the connections of many swallets and springs. Butt et al. (2007), for example, cataloged the fates of water recharging the UFA by way of swallets that drain to the Santa Fe River via caves.

Gulley (2013) argued that the reason that water in many Florida cave systems is so clear is a lack of a swallet system to supply colored and turbid surface water to the limestone aquifer. Rather, he argued, the clear water in caves is related to storage in the porous limestone and in cave systems that have not formed by recharge directly through sinkholes. Observations by the authors suggest that Gulley's argument is at least partly correct. Water movement through the pore spaces in the eogenetic limestone provides time and space for particle filtration and microbial decay of organics.

Many of Florida's cave systems and associated springs are clearly influenced by a complex interplay of (1) direct recharge through sinkholes or swallets combined with (2) storage and long-term discharge from the complex porosity of the limestone matrix (Florea and Vacher 2007; Upchurch 2015).

### 7.2.6 Cave Decorations

Cave decorations, such as stalactites and stalagmites, are familiar because of the worldwide popularity of caves. Florida's caves contain the usual forms of cave decoration, including stalactites, stalagmites, and flowstone. The mechanisms of calcite dissolution and deposition of cave deposits were discussed in Chap. 5. Let us first revisit the chemistry of precipitation of calcite.

#### 7.2.6.1 Causes of Calcite Deposition in Caves

Equation 5.4 is the chemical reaction for the dissolution and precipitation of calcite. In Chap. 5, we were concerned primarily with dissolution and concepts of equilibrium chemistry. This chemical reaction is presented below as Eq. 7.1.

$$H_2O + CO_2 + CaCO_3 \rightleftarrows Ca^{2+} + 2HCO_3^{-}. \qquad (7.1)$$

Assuming that the reaction must remain in equilibrium, there are several ways that calcite can be precipitated. The three options that will drive the reaction to the left and cause precipitation of calcite are:

1. Removal of $H_2O$ from the system by evaporation;
2. Removal of $CO_2$ from the system by degassing; and
3. Increasing the concentrations of $Ca^{2+}$ and/or $HCO_3^{1-}$ in the solution.

Option 1 is a common process in dry caves. In a low humidity, dry cave, Ca-$HCO_3$ facies water evaporates into the cavern atmosphere as it enters the void. The resulting calcite precipitation can be as flowstone, stalactites, or other cave decorations depending on the fate of water movement as it enters the cavern. When evaporation drives precipitation, calcite precipitation is likely to be rapid resulting in deposition of fine-grained calcite crystals.

Option 2, $CO_2$ degassing, is typically a slower process than evaporation, especially in a humid cave environment. With $CO_2$ degassing, there is often time for $Ca^{2+}$ and $CO_3^{2-}$ (from the $HCO_3^{1-}$) to become organized into larger crystals, including coarsely crystalized flowstone and free crystals of calcite. Because of their larger size relative to smaller crystals created by evaporation, larger calcite crystals in cave deposits may be transparent.

Option 3 is unlikely in most scenarios. One scenario in which this option is evident in Florida, however, is related to the transformation of aragonite to calcite during diagenesis or early karstification. The calcite crystals shown in Fig. 6.9 likely formed in a system driven by the dissolution of more soluble aragonite in shells within a humid sediment environment and transport of the Ca-$HCO_3$-enriched water into the shelter porosity of the larger, *Mercenaria* shells where calcite was precipitated.

Many caves in Florida have direct connections to the land surface and are subject to influx of clay and other sediment. This sediment clings to cave deposits with two possible outcomes. First, the finer grained particles may become incorporated in the calcite matrix as it precipitates. When this happens, opaque or discolored layers of calcite result. Second, the presence of siliciclastic clay on mineral surfaces may inhibit calcite crystal growth.

As noted above, vadose caves are likely to have abundant decorations if they have remained above the water table for many years. With the long-term variations in sea level and resulting changes in potentiometric surfaces of carbonate aquifers, it is possible for formerly vadose caverns to become flooded by rising water table. Examples of flooded, heavily decorated caves are known from Mexico, the Bahamas, Yucatan, Bermuda, and other areas. However, decorated, flooded caves are relatively rare in Florida. Noted cave diver Jill Heinerth (2017, pers. com.) reported seeing a poorly developed stalactite in a Florida cave, and stalactites have been observed (Purdy 1991, 2008[3]) at depths of 7.6 and

[3]See Fig. 2.16 in Purdy (2008) for a photograph of the stalactites in Warm Mineral Springs.

about 18 m in Warm Mineral Spring, an unusually deep collapse sinkhole (a cenote, Sect. 7.3.2.5) in Sarasota County.

The lack of abundant cave decorations below the water table in phreatic caves in Florida suggests that either (1) the caverns developed at or near the water table and were flooded before cave decorations could form or (2) they formed well below the water table and have never been exposed to significant vadose zone development.

It is important to understand that rafts of calcite crystals, pools lined with calcite crystals, and other forms of decorations commonly develop at the water table where it is exposed in vadose caves. Hill and Forti (1997) provide excellent discussions of the mechanisms of cave decoration formation.

#### 7.2.6.2 Flowstone and Rimstone

Layered deposits of calcite are often deposited when groundwater flows over carbonate rock and sediment surfaces in air-filled caves. The resulting deposits, known as flowstone, consist of bands of calcite that have precipitated as a result of evaporation or degassing of $CO_2$ from the thin film of water flowing over the flowstone, sediment, or rock surface. The flowstone drapes over the sides and floors of the cavern.

Figure 7.9 illustrates a small rock pedestal that has been capped by flowstone. The small, globular protrusions are tiny stalagmites formed by water dripping on the flowstone. Figure 7.10 is a vertical cross section through a flowstone deposit. Note in Fig. 7.10 that the flowstone has engulfed a small stalactite (ST) and stalagmite (SM) that had formed in a small vug about 70 cm in height. The banding in the flowstone clearly demonstrates the episodic nature of calcite precipitation and flow patterns of the thin film of water flowing over the flowstone surface. The color banding is a result of incorporation of clays in some deposits and not in others. The banding represents alternations of deposition and nondeposition and small-scale changes in the fabric of the calcite crystals. The stalactite (ST, Fig. 7.10) is a single crystal of calcite that has grown in a form termed a soda straw. The low inclusion-content, coarse calcite stalactite and horizontal layers are a result of $CO_2$ degassing. The finer grained, more equant calcite with ferric hydroxide and clay inclusions reflect more rapid crystal nucleation as a result of evaporation. The small openings (AP, Fig. 7.10) in the flowstone were once air or water bubbles that were trapped as the flowstone was deposited.

Figures 7.11 and 7.12 illustrate flowstone draping over bedrock at the base of a stalagmite and several small stalactites, respectively. These draperies and sheets of flowstone are common in vadose-zone caves, where degassing and/or water evaporation contribute to calcite deposition.

Rimstone is flowstone that has accumulated as small ridges and basin rims near actively flowing or dripping water (Fig. 7.13). Rimstone forms small dams that create pools of water on the cave floor. Where agitation is sufficient to move particles within the basins, cave pearls may form (Fig. 7.14).

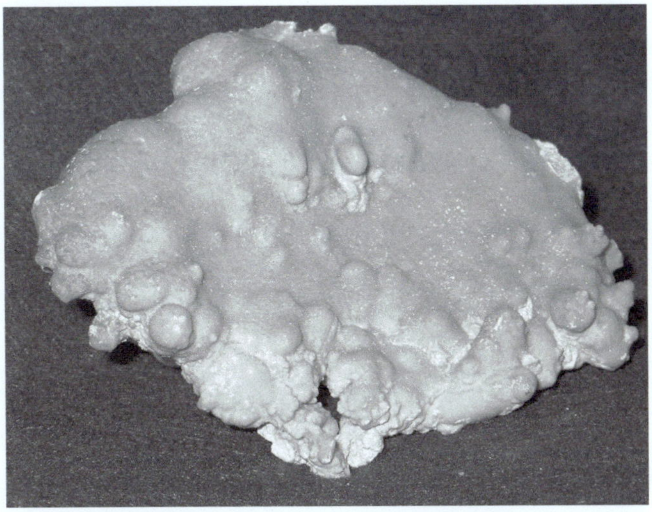

Fig. 7.9 Flowstone draped over a small limestone pedestal from a cave exposed in a quarry near Lecanto, Citrus County. The small, subspherical protrusions are small stalagmites formed by water dripping on the pedestal. Specimen is 8 cm wide

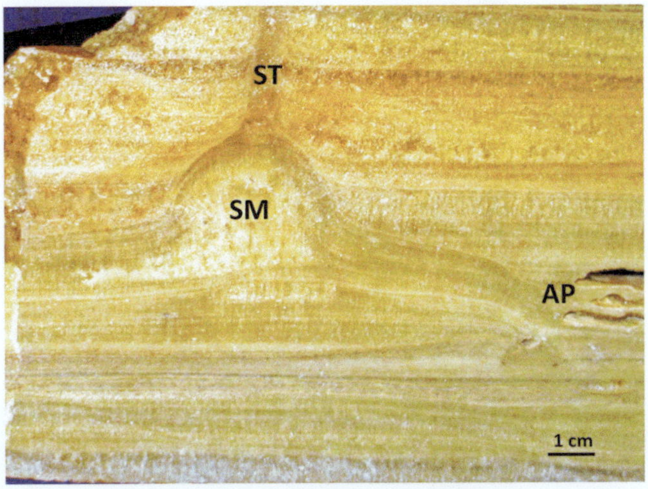


Fig. 7.10 Cross section through a segment of flowstone that has covered a soda-straw stalactite (ST) and stalagmite (SM). The small openings at AP are air or water bubbles that were trapped in the flowstone. Salvaged from a quarry near Lecanto, Citrus County

#### 7.2.6.3 Draperies, Stalactites and Stalagmites

If the ceiling is jointed so that water slowly flows from the crack, sheets of flowstone (draperies) that are suspended from the cavern ceiling are likely to develop. Flowstone draperies are common in caves that have formed in fractured, telogenetic karst systems, where the porosity is mostly related to fracturing and joint development. In Florida, flowstone draperies are less common, but do exist (Fig. 7.15). One commonly observes flattening and alignments of stalac-

**Fig. 7.11** Column of flowstone with small, dripstone stalactites. Florida Caverns State Park (30.8132°N, 85.2309°W), Mariana, Jackson County

**Fig. 7.12** Small stalactites intersecting a sheet of flowstone covering limestone bedrock. Florida Caverns State Park, Mariana, Jackson County

tites, as shown in Fig. 7.15, along the plane of incipient fractures or cracks in the cavern roof.

Where matrix porosity is relatively important, it is common to encounter swarms of small, thin, soda-straw stalactites hanging from the ceiling of the cave (Fig. 7.16). Each stalactite represents a location within the limestone matrix where water has been able to slowly pass through the rock. Many of the soda straws shown in Fig. 7.16 are composed of fine-grained calcite or have ferric hydroxide or clay coatings that make the stalactites appear opaque and outer surfaces

**Fig. 7.13** Rimstone surrounding a drip pool in Florida Caverns State Park, Jackson County. Note the complex fluting and small stalactites on the ceiling

**Fig. 7.14** "Cave pearls" in a drip pool in a cave exposed in a quarry in Washington County. (Photograph courtesy of the Florida Geological Survey)

appear "knobby." However, many of the soda straws in Florida caves are translucent and smooth on the outside. When closely examined, these stalactites consist of one or a few limpid calcite crystals. Figure 7.17 shows some of these stalactites in a north Florida cave. Note the water droplets suspended in the mouths of the central canals that extend along the central axis of each. Note also the globular shaped, clear crystals of calcite with rounded shapes. These stalactites have formed by $CO_2$ degassing accompanied by slow crystallization.

Rarely the calcite crystals grow to the extent that their outer surfaces are hexagonal prisms typical of calcite crystals but with a central canal through which the water passed. Figure 7.18 is an example of the rarest of these calcite crystal stalactites. Locally called a calcite "flower", this specimen is oriented in the figure as it hung from the cave ceiling. At the base is a remnant of the original, cylindrical crystal calcite stalactite. The larger calcite crystals radiating out from the central axis formed at the water table as capillarity drew water up the shaft of the stalactite. Above the "flower" part of the stalactite is a thicker, six-sided prism typical of a single calcite crystal. The central conduit remains intact throughout the specimen.

Stalactites with the more common flowstone-type banding consisting of ferric hydroxide and minor clay inclusions in the calcite with alternating layers within radiating, coarse calcite crystals are also common (white arrow, Fig. 7.19). In these stalactites, deposition of the ferric hydroxide layers (green arrow, Fig. 7.19) was insufficient to "poison" calcite crystal growth, and the radiating crystals maintained crystal

Fig. 7.15 Stalactites and flowstone draperies (D) aligned on a joint in the Suwannee Limestone in a cave exposed in a quarry in Washington County. (Photograph courtesy of the Florida Geological Survey)

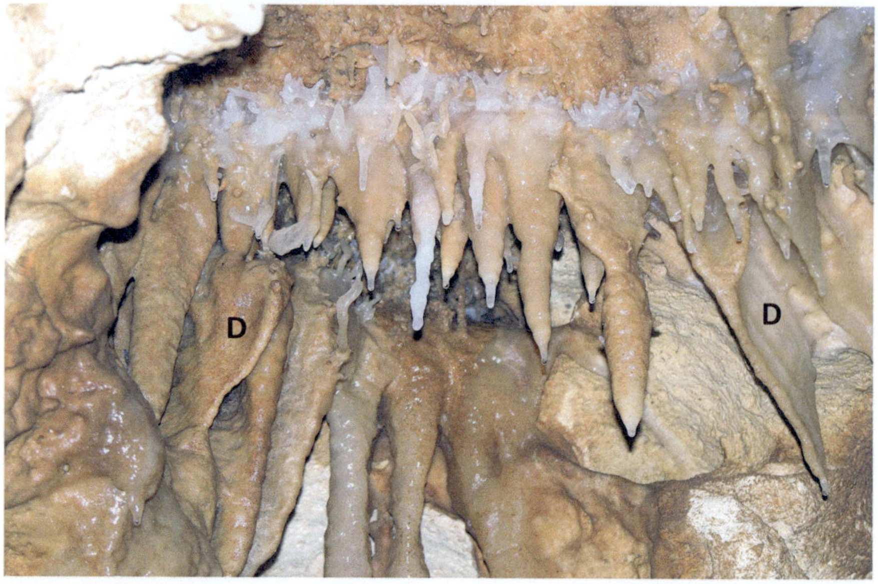


Fig. 7.16 Soda-straw stalactites in Florida Caverns, Marianna, Jackson County. (Photograph courtesy of the Florida Geological Survey)

Fig. 7.17 Soda-straw stalactites in a cave in Washington County. (Photograph courtesy of the Florida Geological Survey)

**Fig. 7.18** A calcite "flower" stalactite shown in growth position. Specimen from a quarry located near Lecanto, Citrus County

**Fig. 7.19** Cross section of a typical stalactite from the Lecanto area, Citrus County. Green arrow points to centripetal ferric-hydroxide-stained banding. White arrow indicates the extent of radial, single calcite crystals. Stalactite is 7.6 cm in diameter

continuity as they grew. Each prism radiating from the center of the stalactite is a single calcite crystal.

Stalagmites (S, Fig. 7.20) show similar, radial crystal forms (single crystal stalagmites are unknown), but the amount of ferric hydroxide and clay may be greater than in stalagmites because of cavern flooding. In some cases, it is clear that these impurities and episodes of flow and sedimentation on the cave floors have interrupted calcite crystallization. When stalactites and stalagmites join, a column (C, Fig. 7.20) is formed.

Finally, helectites (Fig. 7.7) are common in many caves in Florida. Helectites are stalactites that deviate from the vertical and may have branching morphologies. The cause of helectite formation is controversial. Two possible causes are (1) capillarity variations or (2) gentle atmospheric circulation in the cave. The abundance of these exotic stalactites, many of which consist of one or a few calcite crystals, indicates that crystallization was dominated by $CO_2$ degassing.

Water-table calcite is another indicator of environmental conditions in many of Florida's caves. Indicators of former positions of the water table include crystallization of plate-shaped, secondary calcite clusters on the perimeters of stalactites (Fig. 7.21a), calcite "rafts" stranded on crystal clusters (Fig. 7.21b) or on flowstone, and free-standing rafts that consist of radiating crystals (Fig. 7.21c). The stalactite shown in Fig. 7.21a suggests that it was suspended in a pool with water that was over-saturated with respect to calcite, resulting in the small ball of crystals on the tip. Above the tip, a small portion of the original stalactite is exposed. The small shelf at the contact between the original stalactite and the smaller diameter shaft (arrow) to the left suggests the position of the water table.

The rafts shown in Fig. 7.21b, c formed in pools where surface tension allowed them to drift around the water surface as the calcite precipitated. The raft attached to the calcite crystals in Fig. 7.21b consists of more-or-less equant crystals, while the raft shown in Fig. 7.21c shows radial growth of the crystals as the cluster floated in the pool.

#### 7.2.6.4 Speleothem Geochemical Studies

Drip water from active caves has been investigated in order to reconstruct paleoclimate in west-central Florida. Onac et al. (2008) utilized stable isotopes of oxygen and hydrogen to investigate how meteoric water evolves as it travels through the cover materials and limestone and forms stalactites in Florida. Even though issues were noted with sampling, they concluded that that evaporation is the dominant process that affects the oxygen and hydrogen isotopic composition in rainfall. After traveling through cover and limestone, the drip water showed little variability in $\delta^{18}O$ and no correlations between rainfall and drip water compositions were noted. The authors concluded that the speleothems from the two caves investigated cannot document the seasonal cycle in rainfall, but long-term climatic changes can be documented. Lag times of rainfall and drip events of over 2 months were noted. The isotopic ratios indicated that matrix flow was the dominant pathway for the meteoric water in the caves.

**Fig. 7.20** Stalagmites (S) and columns (C) formed where stalactites and stalagmites have joined in the public cave at Florida Caverns State Park, near Marianna, Jackson County

**Fig. 7.21** Examples of water-table calcite forming on a eroded, 6 cm stalactite (**a**), as a small raft of equant crystals (arrow) clinging to 2.5 cm calcite crystals (**b**), and as an isolated 4 cm raft with radiating crystal growth (**c**). Specimens are from various quarries in the Lecanto area, Citrus County

Soto (2005) examined isotopic compositions of four speleothems from two caves (BRC Cave in Hernando County and Briar Cave from Marion County). She utilized uranium-series disequilibrium analyses for chronology of the stalactite layers and $\delta^{18}O$ and $\delta^{13}C$ to identify climatic variability. The $\delta^{18}O$ ratios indicated abrupt changes in precipitation that appeared regional and hemispheric in nature. These changes were related to the Atlantic Multidecadal Oscillation (AMO), el Niño events, and changes in the Intertropical Convergence Zone in the north Atlantic.

The speleothems ranged in maximum radiometric age from 2571 and 29,241 years before present in the BRC Cave stalactites and from 3650 and 4138 years in the Briar Cave samples. Soto determined that the average depositional rates of calcite in the four speleothems ranged from 224 to 540 $mm^3$/yr. in the two speleothems from Briar Cave to 200 to 380 $mm^3$/yr. in the BRC Cave speleothems. Soto identified deposition that extended back into the Pleistocene in one BRC Cave speleothem and determined that the average deposition rate during the "Glacial" interval (21,078 to 29,241 radiometric years before present) was 40 $mm^3$/yr. as compared to 200 $mm^3$/yr. in the "Holocene" record of the same stalactite.

Oxygen and carbon isotopic ratio data from the stalactites resulted in a well-defined pattern suggestive of climatic changes in west-central Florida. Van Beynen et al. (2008)

elaborated on these findings. They concluded that there were periods of increased calcite deposition centered on 2.0, 1.25, and 0.5 ka before present. Comparison of Mg/Ca and Sr/Ca ratios in the speleothems indicated elevated elemental ratios during the periods of enhanced deposition. Correlation of these elevated ratios with calcite nucleation rates suggested an increase of residence times of percolating water during the episodes of increased speleothem growth. These episodes of elevated elemental ratios were inferred to represent dry periods.

#### 7.2.6.5 Why Have Some Cave Decorations Been Partly Dissolved?

Many of the single-crystal calcite stalactites and other cave decorations in Florida caves show clear evidence of dissolution after deposition. This evidence is usually rounding of crystal edges and "polishing" of surfaces as if they were dipped in acid. These evidences of dissolution point to the model proposed by Gulley and his coauthors (Gulley 2013; Gulley and Florea 2016; Gulley and Polk 2017) of enhanced acidity at the water table as a result of $CO_2$ degassing.

## 7.3 Subsidence Features

This section deals with the processes involved with development of depressions on the land surface in Florida. Not all of these depressions are related to karst processes. There is pseudokarst, as well. We begin the section with a detailed discussion of sinkholes and other karst-related depressions.

### 7.3.1 Karst and Pseudokarst Depressions in Florida

A caveat is necessary before discussing sinkholes because of the risk of misapplying the term sinkhole to all closed depressions in Florida. Sinkholes (or dolines) are depressional landforms developed by dissolution of carbonate sediments and contemporaneous or subsequent movement of cave roof or cover materials into the voids created by dissolution. Not all depressions on the land surface in Florida are sinkholes or even karst features (Upchurch et al. 2015). Many of Florida's shallow, land-surface depressions were caused by other processes and are considered pseudokarst (Field 2002; Halladay 2007).

Much emphasis in Florida has been placed on identification of locations where sinkholes are developing and causing property damage. As a result, common depressions, such as are shown in Fig. 7.22, are often attributed to sinkhole development without testing or other forms of confirmation of the origin of the depression. A goal of Sect. 7.3 is to illustrate the mechanisms of sinkhole and pseudokarst development and present characteristics that allow for differentiation of each.

### 7.3.2 Sinkholes (Dolines)

A discussion of sinkholes is included first in this section because they are important as a geologic hazard and dominate the concerns of many Floridians. They are also the most visible karst features in Florida. Note that the term "sinkhole" is not synonymous with "karst" – the words are not interchangeable. Karst is an all-inclusive term that refers to any landform caused by dissolution of carbonate sediments, rocks, or other soluble geologic material. Sinkholes are one of many forms of karst landforms. Sinkholes are depressions in the land surface caused by (1) movement of cover materials or rock collapse into a void in the underlying carbonate sediment or rock or (2) subsidence of cover sediments into a depression caused by dissolution on the upper surface of a carbonate sediment or rock.

The following sections emphasize the processes of sinkhole development, including the initial conditions that exist prior to development of a depression on the land surface. Most of the sinkholes that develop in Florida are relatively harmless and often go undetected or unreported because they are (1) small and do not cause damage or (2) located in rural areas. When they cause significant damage, they receive intense media coverage, both locally and internationally, and cause widespread concern. We also often hear in the media of "sinkholes" that form as a result of water main breaks, pavement wear, or other, non-geologic events. These are not sinkholes according to karst science definitions. It is possible that such processes can accelerate sediment raveling into actual sinkholes, however.

Florida is dotted with ancient circular or near-circular depressions (Fig. 7.23a, b) that reflect a long history of sinkhole development.[4] In many areas of Florida, these ancient depressions are represented by lakes and wetlands that are considered amenities, not hazards, by many Floridians.

The contemporary development of sinkholes in Florida is often unexpected and unpredictable. Structures, particularly residences, are commonly constructed without subsurface testing, so they may be unintentionally placed in harm's way over pre-existing karst features that are likely to develop into a sinkhole. Construction of commercial and public structures normally includes subsurface, geotechnical investigation, and construction standards may be higher than for residential construction. If deleterious, subsurface conditions are investigated and correctly identified; construction of commercial, public, and industrial structures includes correction of these conditions. As a result, commercial and public structures have a lower rate of damage from subsurface activity or subsurface materials as compared to residential structures.

[4]Circular depressions in the landscape are not always ancient sinkholes and should not be considered as confirmed sinkholes without testing. Some other causes of circular depressions include human activity, eolian deflation zones, animal burrows and excavations, weathering and soil formation, and sediment depositional features.

**Fig. 7.22** Closed depressions in the Hardee Upland Province south of Arcadia, Hardee County. Depressions are seasonal wetlands

#### 7.3.2.1 Sinkhole (Doline) Classification

In Florida, there are two major mechanisms that can create a sinkhole. These are (1) collapse of the roof of a cave or cover sediments into a void in the underlying rock and (2) slow migration of cover sediments into a shallow depression created on the upper surface of a shell bed or shallow limestone by dissolution. In the first case, the movement of materials into a void can be caused by raveling or collapse. In the second case, movement of cover material into a depression at the upper surface of a shell bed or rock is called subsidence by soil creep.

The terminology of sinkholes may be confusing and inconsistent. As used in this text, a sinkhole is a landform, or depression at the land surface, created by (1) collapse of the roof of a void, which may be accompanied by raveling of cover materials into the void, (2) raveling of cover materials into a pre-existing void without collapse, or (3) slow dissolution of near-surface carbonate sediment with associated subsidence of cover settlement.

Table 7.3 presents the classifications and terminology used internationally by various authors for sinkholes, or dolines, based on their origin. Note that the term "doline", which is synonymous with sinkhole, is commonly used in Europe, but seldom in the U.S. The table, which is modified from Williams (2004), clearly shows that there is a general concurrence of terminologies among authors, but minor differences may confuse the reader. The classification by Beck and Sinclair (1986) was developed for use in Florida and has been modified to include other terms commonly used in Florida.

This text utilizes a hybrid sinkhole classification scheme (Table 7.4) based on the terminology of Culshaw and Waltham (1987) and Florida studies by Beck and Sinclair (1986), Upchurch and Randazzo (1997), and Tihansky (1999). Many of the differences in terminology used are a result of the limited overlap of study areas of these Florida authors. Most of their work was in west-central Florida, where cover-collapse and cover-subsidence sinkholes are well documented. When the study area is expanded to the entire state, problems with application of these earlier classification schemes may emerge and new modifications of sinkhole development become evident.

Before discussing each sinkhole-development mechanism in detail, it is appropriate to consider the properties of the different mechanisms as suggested by Beck and Sinclair (1986). Table 7.5 presents a summary of the conditions thought by Sinclair et al. (1985), Beck and Sinclair (1986), and Beck and Sayed (1991) to exist for each of their different

**Fig. 7.23** Examples of ancient sinkholes that dot the land surface in Florida. (**a**) Part of the Winter Haven Chain of Lakes in Polk County, Florida. (**b**) These circular depressions are located in northeastern Pinellas County, Florida

**Table 7.3** Classification of sinkholes or dolines based on origin

| Sinkhole- (Doline) forming process | | Ford and Williams (1989) | White (1988) | Jennings (1985) | Bögli (1980) | Sweeting (1973) | Culshaw and Waltham (1987) | Beck and Sinclair (1986) in Florida | Other terms in use |
|---|---|---|---|---|---|---|---|---|---|
| Dissolution | | Solution | Solution | Solution | Solution | Solution | Solution | Solution | |
| Collapse | | Collapse | Collapse | Collapse | Collapse (fast) or subsidence (slow) | Collapse | Collapse | Collapse | |
| Caprock collapse | | | – | Subjacent collapse | | – | – | | Interstratal collapse |
| Dropout | Sub-sidence | | Cover collapse | Subsidence | Alluvial | Alluvial | Subsidence | Cover collapse | |
| Suffosion | | Suffosion | Cover subsidence | | | | | Cover subsidence | Raveled, Shakehole |
| Burial of pre-existing sinkhole | | – | – | – | – | – | Buried | Buried or relict sinkhole | Alluvial, Filled, or Paleo-sinkhole |

Modified from Williams (2004)

**Table 7.4** Sinkhole classification schemes used in Florida and this text

| Sinkhole origin | Sinclair et al. (1985), Beck and Sinclair (1986), Beck and Sayed (1991) | Upchurch and Randazzo (1997) | Tihansky (1999) | This text |
|---|---|---|---|---|
| Dissolution | Solution | Solution | Not addressed | Solution |
| Rock collapse | Collapse | Rock collapse | Not addressed | Rock collapse |
| Cover collapse | Cover collapse | Cover collapse | Cover collapse | Cover collapse |
| Suffosion | Not addressed | Alluvial sinkhole (in part) | Suffosion | Suffosion |
| Carbonate surface dissolution with cover subsidence | Not addressed | Solution | Cover subsidence (in part) | Cover subsidence |
| Reactivation of a pre-existing sinkhole | Not addressed | Raveling sinkhole | Not addressed | Reactivation sinkhole |

sinkhole origins. Their classification scheme was based largely on Culshaw and Waltham's (1987) classification (Table 7.3) and the authors' experiences throughout Florida.

### 7.3.2.2 Solution Sinkholes

There is agreement about the terminology of solution sinkholes (Tables 7.3 and 7.4). Solution sinkholes are depressions in the land surface caused by the dissolution of the upper surface of exposed or bare limestone (or dolostone). As such, development of solution sinkholes is a top-down process that is not associated with undermining and collapse.

As noted in Chap. 6, the rate of dissolution of the limestone is extremely slow, typically requiring many thousands of years to create a solution sinkhole depression of any appreciable size (Table 7.5). There may be minor raveling into small openings in the limestone or the breccia caused by dissolution, but this raveling is not considered the proximal cause of the depression.

Figures 7.24 and 7.25 present some concepts of how solution sinkholes form in Florida. As shown in Fig. 7.24, the upper surface of the limestone[5] is slowly dissolving away above the water table over geologic time, creating a weakened and weathered limestone with volume reduction owing to dissolution. With the volume reduction, a depression in the land surface develops. The weakened and weathered limestone is part of the epikarst.

As suggested by Beck and Sinclair (1986) and discussed in the preceding chapters, meteoric water infiltrating the limestone is initially slightly acidic, so dissolution begins above the water table. Depending on the depth of the water table and the residence time of the water in the vadose zone, meteoric water may fully equilibrate chemically with the limestone before reaching the water table, or dissolution may continue below the water table as suggested by Fig. 7.24.

Figure 7.25 illustrates the possible flow of meteoric water through the limestone through the porous limestone and along fractures, bedding planes, and other openings. Again, Sinclair et al.'s (1985) diagram (Fig. 7.25) suggests that dissolution is limited to the upper weathering horizon, the epikarst, and that sinkhole development is caused by volume loss within the limestone or dolostone surface. The cover materials, if present, are limited in thickness and simply follow the upper surface of the limestone as it is lowered.

Three conditions must exist for solution sinkholes to develop. These are:

- The water table lies below the upper surface of the limestone (Table 7.5);

[5] Circular depressions in thinly covered dolostone are present in west-central Florida.

**Table 7.5** Principal features of the major types of sinkholes in Florida as described by Sinclair et al. (1985), Beck and Sinclair (1986), and Beck and Sayed (1991)

| | Sinkhole type | | | |
|---|---|---|---|---|
| | Limestone solution | Limestone collapse | Cover subsidence | Cover collapse |
| Cover material | Limestone bare or thinly covered | Limestone bare to deeply buried by cover material | Incohesive [*sic*], permeable sand, as much as 15 m (50 feet) or more thick | Generally 1 to 30 m (5 to 100+ feet) thick. Incohesive [*sic*], permeable sand grading downward to clayey sand with relatively cohesive, poorly permeable clay overlying limestone |
| Groundwater levels | Water table generally below top of limestone | | Water table may be in sand or below top of limestone | Water table in cover sand perched above and leaking through poorly permeable clay. Artesian water level in limestone generally below water table but above limestone surface. |
| Sinkhole development | Imperceptibly slow subsidence of land surface by removal of limestone in solution | Cavities develop only at depth beneath limestone surface where they enlarge by upward collapse of the roof, layer by layer, until the limestone roof is too thin to support itself and collapses abruptly; a relatively rare occurrence in terms of landscape development. Sinkhole size controlled by bearing strength of limestone roof and depth of cavity. | Imperceptibly slow subsidence of land surface. Only very small cavities are likely to form at limestone surface because incohesive [*sic*] sand moves continuously or spasmodically downward to occupy space formally held by dissolved limestone. | Solution-enlarged joints at limestone surface are bridged [by] overlying cohesive clay until bearing strength of clay is exceeded. Cavity moves upward by roof spalling until it appears abruptly at land surface. Sinkhole size controlled by bearing strength of clay, usually related to clay thickness. |
| Activities that commonly trigger sinkholes | | Declines in groundwater levels, particularly abrupt declines and fluctuations due to pumping from wells, are the most common factors that trigger collapse.<br>Impoundments, such as reservoirs and holding ponds, may load the land surface and provide a source of water that percolates downward, eroding cover material into solution cavities. Diversion of surface drainage by roads or paving of large areas that concentrates runoff may also accelerate internal erosion.<br>Static loads (buildings) and vibratory and harmonic loads (heavy equipment, trains) may be sufficient to trigger collapse of cavities at shallow depth. | | |
| Effects of above activities on sinkhole development | Slight | Slight. Collapse may be induced where water table is above limestone surface. | Small-scale piping commonly induced. | Collapse commonly induced. |

- Limestone must be at, or near, the land surface; and
- Cover materials must be thin or absent.

Because exposed limestone or dolostone is rare in Florida, solution sinkholes are seldom identified. They do occur on the exposed Pleistocene limestone surface in the Florida Keys and Atlantic Coastal Ridge in the Miami area. Where present, solution sinkholes in southern Florida are typically vertical shafts with diameters that are less than their depths. Typically, the solution sinkholes in southern Florida are about 1–2 m deep and less than a meter in diameter (Fig. 7.26). They often are terminated at the water table or where there is evidence of a former water table. In Florida, these shafts are often termed solution pipes or solution channels.

Since the cover, if present, is thin and the water table is usually below the rock surface, solution sinkholes may be detectable by examination of the rock surface, probing, or hand auger boring, but these tests may lead to incorrect identification of the origin of the depression. Standard penetration testing of solution sinkholes should reveal very shallow limestone (or rarely dolostone), thin cover, accumulations of duricrust, breccia and/or rubble (Fig. 7.26b), and, possibly, a well-developed conduit system within the epikarst in the vicinity of the depression. Individual limestone breccia fragments should show rounding and other evidence of dissolution. Standard penetration test practice in Florida usually involves hand auger boring to a depth of 1.8 m before penetration testing begins. Given that thin or no cover is present in solution sinkholes, standard penetration testing may not

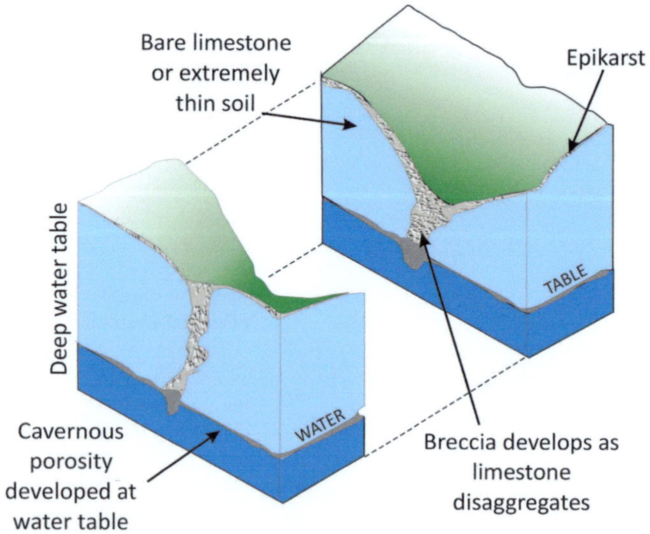


**Fig. 7.24** Formation of a solution sinkhole by erosion of the upper surface of the underlying limestone. (Modified from Beck and Sinclair (1986, Fig. 2))

be a suitable test for identification of solution sinkholes and excavation may well be the only way to document the presence of solution sinkholes (or solution pipes) when covered by fill or sediment.

The origin of solution pipes in southern Florida is inconclusive and subject to debate. Parker (1955, p. 133) concluded that an original cavity was not required for a solution pipe to develop, but that pre-existing cavities hastened the process. He further suggested that some of the vertical solution pipes in southern Florida may have begun by dissolution of limestone or calcareous marl along taproots of trees, but most form by a top-down process initiated by development of small, shallow depressions "resembling raindrop marks in mud" through dissolution on the upper surface of the limestone. These small depressions were thought to gradually deepen until near vertical, shallow tubes result.

Krawiec (1963) noted that the development of the solution pits in the Florida Keys and southern Florida mainland was controlled by the texture and structure of the limestone being dissolved. Coarse-textured lithologies, such as the Pleistocene Key Largo Limestone with its coral heads and boundstones, develop solution pipes that are "V" shaped with irregular margins (Fig. 7.26b). Limestone that is fine-grained and has a uniform texture, such as the oolitic facies of the Miami Limestone, develop "bowl-shaped" depressions with smooth sides. Krawiec (1963) also described the origin and transport of breccia and soils materials into the solution pipes. Sources of sediment included locally derived carbonate, terra rossa soil, and organic materials. Storms were also credited with transporting marine and terrestrial debris into the solution pipes. Finally, Krawiec (1963) concluded that the solution pipes were formed by dissolution by groundwater with localized re-precipitation of calcitic duricrusts on the perimeters of some solution pits. The reddish coloration of the duricrust was thought to have been caused by concentration of iron and other chemicals out of the groundwater.

Perkins (1977) concluded that solution holes were formed by the leaching associated with plant growth, root development, and formation of soil pockets. He further postulated that growth of the solution pit was enhanced by downward percolating groundwater. This interpretation is consistent with the common occurrence of larger, rooted plants within the solution pits.

There is another possible cause of origin of the solution pipes found in the Pleistocene limestones of southern Florida that has been investigated in siliciclastic sand deposits in recent years. This process, known as development of "flow fingers" by wetting surfaces, may be worthy of consideration for southern Florida conditions. When first deposited, the Pleistocene sediment of the Atlantic Coastal Ridge consisted of coarse to medium carbonate sand, silt, and clay containing aragonite and calcite. These sediments initially had open to moderate packing frameworks. The Pleistocene Miami Limestone includes a sand facies composed of ooliths and a facies of muddy, fossiliferous sand (the bryozoan facies). The Key Largo Limestone was largely a boundstone with lenses and interspersed coralgal and molluscan sand. As sea level retreated at the beginning of the Wisconsinian glacial episode, these sediments were exposed to rainfall events and subject to diagenetic alterations, including differential dissolution.

Assuming that the upper surfaces of the sand bodies that became limestone were composed of unsaturated (vadose-zone) carbonate sand at the time that the solution pipes began to develop, the distribution and size of the solution pipes may have been controlled by the pattern of wetting of the unsaturated sand.

Glass et al. (1988, 1989) termed the pattern of wetting of an unsaturated porous medium "flow fingers" after the characteristic pattern of the wetting front as it penetrates the substrate. These flow fingers are thought to be randomly located and related to small-scale heterogeneities in the sandy sediment (Glass et al. 1988, 1989; Pendexter 1991; Pendexter and Furbish 1991). They consist of more-or-less cylindrical wetted zones that initially extend downward as a result of capillarity from the land surface during a precipitation event. Flow fingers developed at the laboratory scale are small, with depth and spacing on the order of centimeters. After development, they continue to function as small-scale zones of increased hydraulic conductivity. De Rooj and Cho (1999) studied the properties of the flow fingers and, among other things, considered the geometries of the flow fingers, including the fact that they tended to terminate at the water table, either by widening or dissipation. Most research on flow fingers has focused on their ability to transport contaminants

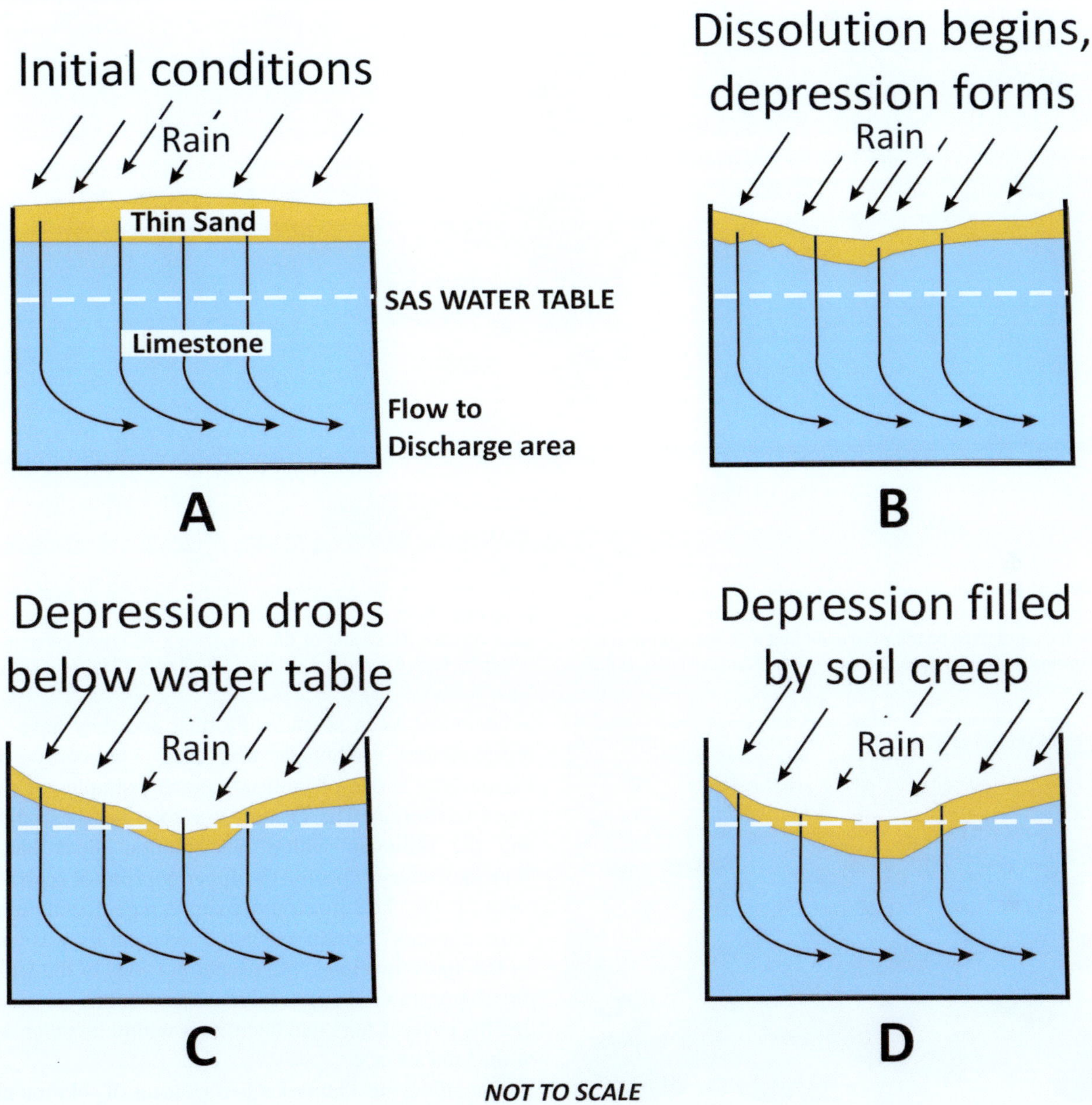


**Fig. 7.25** Mechanisms for formation of a solution sinkhole by dissolution of the underlying limestone surface. (Modified from Sinclair et al. (1985))

into the subsurface or on leaching in the vadose zone. For example, Katz (1993) utilized the concept of leaching of soils within flow fingers to explain how dibromoethane (a.k.a. ethylene dibromide or EDB), a soil fumigant, migrated into the sandy sediments and aquifers under the Lake Wales Ridge. The cylindrical pattern and vertical distribution of flow fingers and the fact that they seem to terminate at the water table seem to be consistent with the patterns of solution pipes observed in sandy sediments in southern Florida. Only the scale differs. Development of the solution pipes may be a result of preferential dissolution within flow fingers. Once pathways developed as a result of flow-finger development, they may have continued to grow by dissolution along the preferential pathways.

In summary, solution pipes have the following properties:

- Formed in poorly to well-lithified rock;
- Nearly vertical shafts;
- Deeper than they are wide, with maximum depths of 1–2 m in many areas of southern Florida;
- Terminate at a common depth which is assumed to have been the position of the water table;
- Often, no apparent connection or obscure connection to karst conduiting below the shaft;
- Walls of the shaft may be bare rock or armored with duricrust and indurated paleosol;

**Fig. 7.26** (**a**) Workers "mucking" out solution pipes in the Miami Limestone, Miami, Dade County. The residual and other soils that originally fill the pipes must be removed and filled with grout prior to foundation construction. (Photograph courtesy of J. Nordqvist). (**b**) Solution pipe exposed in the quarry at the Windley Key Fossil Reef Geological State Park, Monroe County. This solution pipe is "V" shaped and contains breccia. The walls of the pipe have a duricrust colored reddish brown by terra rossa soil. Location: 24.9506°N.5958°W

**Fig. 7.27** Solution pipes in the Buda Mine, Alachua County. (Photograph by W.A. Wisner in 1972)

- Shafts may be open, partly filled with unconsolidated sediment (breccia, soil, organics, and debris) or completely infilled with breccia, paleosol, secondary cements, etc.;
- More recent, marine sediments have been found within the solution pipes;
- Solution pipes may coalesce into larger features; and
- The locations and spacing of solution pipes appears more-or-less random; there does not appear to be any control on location by the structure of the host rock.

In central and northern Florida, solution pipes are commonly developed in Eocene Ocala Limestone. They may be as abundant as in southern Florida, and they have similar morphologies, although they are commonly deeper. Figure 7.27 is one of the best-known examples of solution pipes in Florida. The Ocala Limestone has been quarried, and the sediment within the solution pipes had been "mucked" away exposing the upper portions of each pipe. In contrast, Fig. 7.28 illustrates a single pipe, locally termed a "chimney sink", within which the sediment fill was removed by transport (raveling) into deeper portions of the UFA during a heavy rainfall event. In this example, one can easily see that the cover is thin, and bedrock is within less than a meter of the land surface.

One of the most remarkable exposures of solution pipes is at the Haile Mine near Newberry, Alachua County. Here, the Ocala Limestone is riddled with solution pipes just as at the Buda Pit (Fig. 7.27). Modern quarrying has exposed the solution pipes, so one can easily describe their properties. Where they are undisturbed by mining, the pipes are filled with reddish brown, clayey sandy soil (Fig. 7.29a). At the quarry face, the clayey sandy soil washes out of the solution pits revealing their shapes and extents (Fig. 7.29b). As with the solution pipes in southern Florida, they appear to terminate at a common elevation suggestive of a former position of the water table. Evidence of a former position of the water table (Fig. 7.29b) includes (1) many of the solution pits are interconnected at their bases, and (2) there is often a horizon of small voids where solution pipes are not present as if the voids were created by dissolution at the water table. The solution pipes are typically less than a meter wide, but commonly extend vertically down about 3–4 m. Unlike the

**Fig. 7.28** "Chimney sink" in Chiefland, Levy County. Note the shallow bedrock and nearly circular outline. (Photograph by K. Hill)

**Fig. 7.29** (**a**) Limestone surface showing tops of sediment-filled solution pipes. Haile Quarry near Newberry, Alachua County. (**b**) View of solution pipes in the Haile Quarry near Newberry, Alachua County

solution pipes of southern Florida, the walls are typically smooth and may have a thin (less than 1 cm) rind of recrystallized calcite.

#### 7.3.2.3 Suffosion Sinkholes

The term suffosion is a process synonymous with raveling. The process of suffosion is integral to the formation of many forms of sinkholes and, at some scale occurs during the development of all sinkholes associated with cohesionless cover sediments. Sinkholes formed by simple raveling have been identified worldwide (Table 7.3). Most of the karst scientists who have classified Florida sinkholes deal with suffosion as an integral part of the development of other forms of sinkholes, but Tihansky (1999, Fig. 7.30) suggested that suffosion sinkholes may exist in their own right in Florida.

This form of sinkhole commonly occurs in Florida at locations where a thin sand layer overlies bedrock containing voids. The sand may partly fill a void or it may bridge over the void space. At a later time, water in the sand may (1) induce compaction of the loosely packed sand in the void or (2) break up the bridge over the void. In either case, a suffosion sinkhole forms.

Suffosion sinkholes commonly develop in stormwater basins where the limestone is near the land surface and cohesionless sand rests directly on the limestone. Frequent loading of stormwater in the basin causes sand on the bottom of the basin to ravel into the underlying limestone by way of pre-existing solution pipes.

Figure 7.31a, b show a suffosion sinkhole that developed in a stormwater basin near Newberry, Alachua County. The basin shown in Fig. 7.31a has a history of small- to medium-

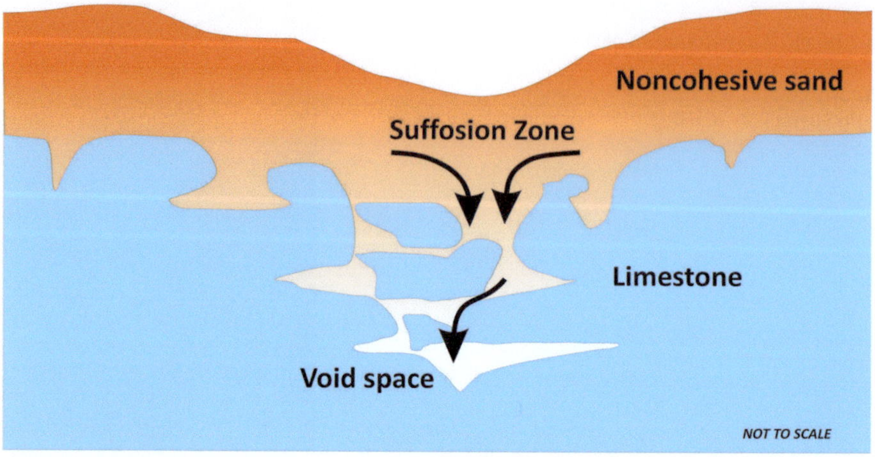


**Fig. 7.30** Suffosion sinkhole. Modified from Tihansky (1999)

**Fig. 7.31** Small suffosion sinkhole developed in a stormwater percolation basin near Newberry, Alachua County. Panel A shows an overall view of the basin and the location of the sinkhole (arrow) depicted in Panel B. Note the thin layer of dark gray silt in the bottom of the normally dry basin and that the sink is located near the perimeter of the silt deposit

sized (0.5–3 m) sinkholes forming during unusual rainfall events. The basin receives stormwater runoff which is temporarily impounded in what is normally a dry basin. Limestone is within 1.5 m of the bottom of the basin.

Dry stormwater basins are common designs in Florida, where the sand layer is intended to serve as a filter for stormwater percolating into the underlying limestone aquifer. These basins frequently develop sinkholes (Fig. 7.31a, b) near the edges of the stormwater pools. The sinkholes often appear to form at the edge of the basin because siltation in the basin floor (note the recent, gray sediment in the bottom of the basin, Fig. 7.31b) hinders percolation through the bottom, while the lack of a silt layer at the water line during unusually high water allows water to move downward more efficiently into openings in the underlying limestone. If the silt and clay sediment layer on the basin bottom is disturbed by burrowing animals, vehicular traffic, or other causes, suffosion sinkhole are likely to develop anywhere within the basin.

The mechanism of formation of the sinkhole shown in Fig. 7.31 appears to be simple suffosion under an unusually high or fluctuating hydraulic gradient. The slightly cohesive, silty sand layer (Fig. 7.31b) remained more-or-less intact as subsidence occurred while the underlying sand raveled into the shallow void.

#### 7.3.2.4 Rock-Collapse Sinkholes

Sometimes termed caprock-collapse sinks (Table 7.3), rock-collapse sinkholes form by the failure of the roof of a cavern or large void within the limestone or dolostone. Figure 7.32 illustrates the mechanism of development of a rock-collapse sinkhole. Note that development of the cavity within the limestone (A, Fig. 7.32) is geologically very slow (Table 7.5), because it is formed at the rate of the kinetics of dissolution in a void with low relative surface area and water residence time. Over time, the roof of the cave is thinned by dissolution until failure begins (B, Fig. 7.32), at which time a breakdown cone, or pile of limestone fragments, begins to form. In Panel

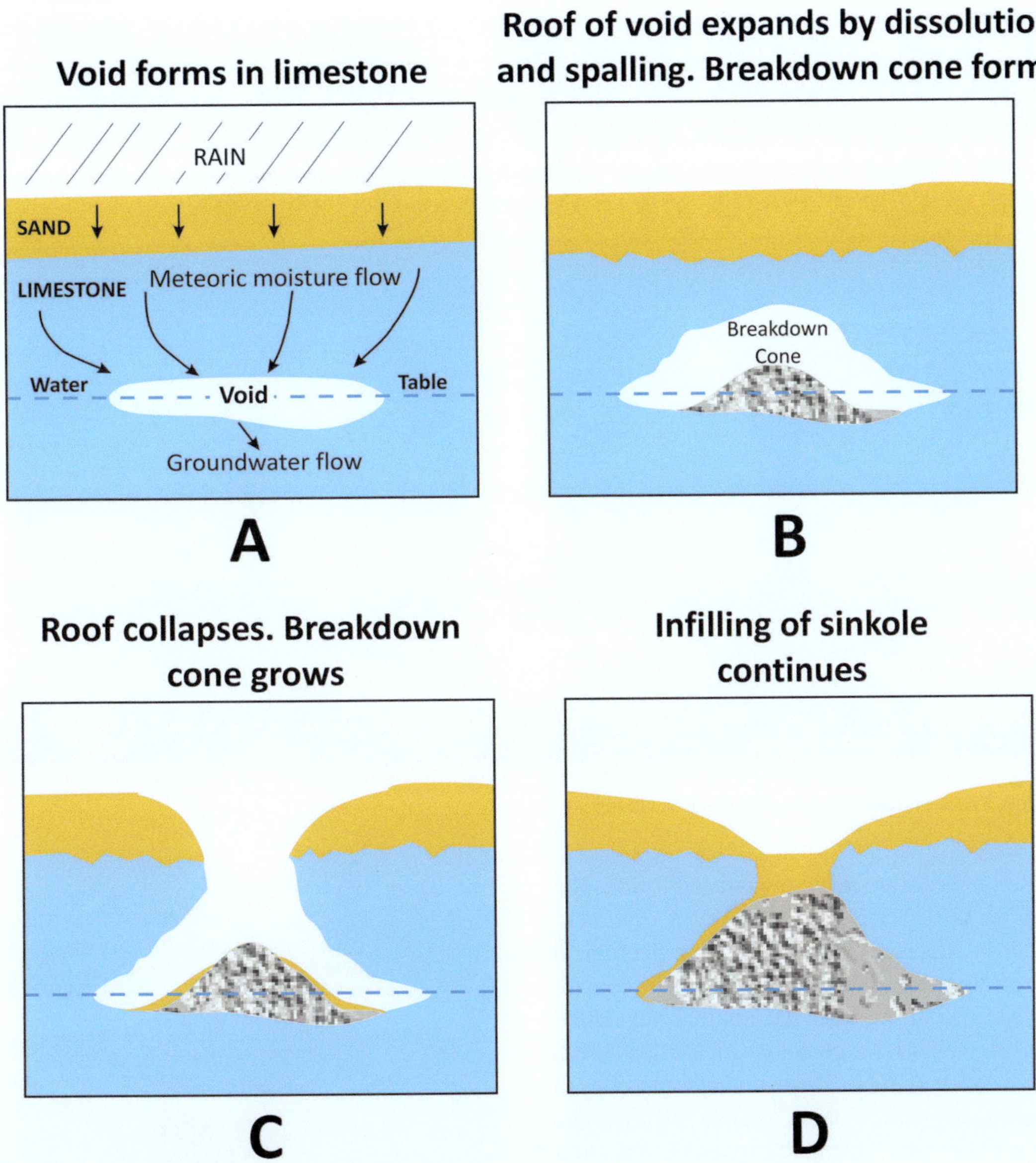


**Fig. 7.32** Mechanism of development of a rock-collapse sinkhole. (Modified from Sinclair et al. (1985))

C (Fig. 7.32) the roof has failed, and a rock-collapse sinkhole has formed. If the water table intersects the opening of the sinkhole, a karst window (window into the aquifer) exists. In areas of covered karst, surficial sediments ultimately migrate laterally into the sinkhole, enlarging its perimeter and making it shallower. Where this late-stage infilling has happened (D, Fig. 7.32), the origin of the sinkhole may be masked.

The actual failure and collapse of the cave ceiling is likely to be rapid. Failure of the roof occurs when the rock roof of the cavity becomes too thin and weak to support itself and any overlying cover material. When failure occurs, a breakdown cone of ceiling rubble and cover material develops on the floor of the void (B and C Fig. 7.32). Bedrock is usually relatively shallow, so it may be observable from the land surface on the sides of the sinkhole (Fig. 6.46).

When rock collapse occurs in a void large enough to cause an obvious sinkhole, the sinkhole often has the following characteristics:

- Steep sides;
- Exposed bedrock on the sides of the aven, or sinkhole throat;
- Deep relative to their areal extent, many are over 60 m in depth;
- Breakdown cone at the bottom of the cavity into which the collapse occurred;
- Little, if any, surface-water drainage into the sink because of poorly developed surface-water drainage basins; and
- Typically water-filled.

Fig. 7.33 Law Sink, a rock-collapse sinkhole in Hamilton County

Rock-collapse sinkholes are relatively rare in Florida. The Devil's Millhopper (29.2071°N, 82.3941°W), a dry rock-collapse sinkhole in Alachua County, is a state geological park with a boardwalk to the bottom. This sinkhole is considered important because of the lush and unusual vegetation that it contains as well as its geology (McClellan 1962; Isphording 1963; Scott 1988). This sinkhole provides a unique opportunity to view a thick section of Miocene Hawthorn Group sediments, which are exposed in the sides of the sink. A similar, accessible rock-collapse sink is present in the Kanapaha Botanical Gardens (29.613°N, 82.405°W) just southwest of Gainesville, Alachua County. Figure 7.33 shows a particularly well exposed wall of Statenville Formation (Hawthorn Group) sediments at Law Sink (Southeastern Geological Society 1981; Lawn 2014), a 24 m-deep, rock-collapse sinkhole in Hamilton County. Note the cross bedding in the Statenville Formation sediments and the small, interstratal cavern just above the head of the person on the left. Water usually cascades down the wall from similar, small caves on the sides of the sink (Fig. 7.33).

### 7.3.2.5 Cenotes

When the large, deep, vertical shafts of rock-collapse sinkholes extend below the water table and are filled with groundwater, they become a special form of karst window known as a cenote. Cenotes in the Yucatan Peninsula of Mexico are

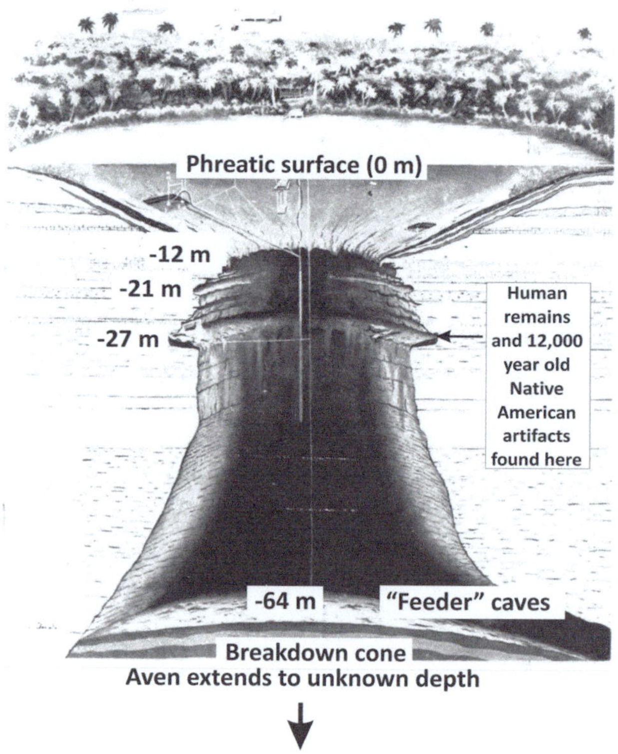


Fig. 7.34 Cross section of Little Salt Spring (27.0749°N, 82.2331°W), Sarasota County. (Modified from a cross section published by Lane (1986) with annotations based on Clausen et al. (1979))

famous because they were used as water sources and ceremonial sites by the Maya civilization. There, the water table is well below the land surface, and the cenotes are spectacular because of the depth of the sinkhole and the clear water at its bottom. Imagine the interest in Florida when cave divers began to explore some of the deep, circular lakes in Florida and discovered that they, too, were cenotes some 12,000 years ago.

Figure 7.34 is a cross section through Little Salt Spring in Sarasota County, Florida. This important rock-collapse sinkhole (a cenote) is 75 m deep with "feeder" conduits that extend from the bottom out and down to even greater depths. During the early Holocene, the water table was significantly lower in this cenote, and early Native Americans utilized it as a water source and ceremonial site. Because of its unique water chemistry (chemically reducing conditions and $H_2S$-rich), human tissues, and other organic human artifacts that dated from 9500 to 7000 years before present were found on the ledge (Fig. 7.34) at 27 m below the water surface. One of the oldest human artifacts found in the United States, a wooden stake $^{14}C$ dated to be about 12,000 years old, was also found in this sink (Clausen et al. 1979). Clearly, the Little Salt Spring rock-collapse sinkhole (note the breakdown cone and sediment fill at the bottom of the cavity; Fig. 7.34) was a cenote utilized by humans for thousands of years before Europeans arrived. An intriguing question about Little Salt Spring is how deep the aven, or conduit, through which collapse occurred extends. Bedrock exposed in the sinkhole belongs to the Arcadia Formation (Miocene Hawthorn Group), but, given the nature of the dolomitic and clayey Arcadia Formation in the vicinity of the spring, it appears improbable that a void large enough to accept the volume of collapsed material developed in the Arcadia. More likely, the original void space into which collapse occurred was in the underlying Ocala Limestone or even deeper.

Warm Mineral Spring (Sect. 7.2.7.1) is located near Little Salt Spring. It, too, had human remains and artifacts that date back 10,000 years. The stalactites at 7.6 and about 18 m (Purdy 2008) below the modern phreatic surface are especially important because they establish that vadose conditions once existed at least below 18 m in the cenote.

There are several similar, deep rock-collapse sinkholes scattered through southern Florida (Fig. 6.38). Upchurch (2017) suggested that they may be related to hypogene karst processes at depth in the UFA or possibly the Lower Floridan Aquifer (LFA). Cunningham et al. (2017) demonstrated the presence of collapse features that extend to the LFA in Broward County.

#### 7.3.2.6 Cover-Collapse Sinkholes

Cover-collapse sinkholes cause more damage (Fig. 7.35) than any other form of sinkhole in Florida. They form when cover (overburden) materials collapse or ravel into a preexisting void at the limestone interface or within the underlying limestone.

**Fig. 7.35** A cover-collapse sinkhole destroyed this residence in Spring Hill, Hernando County. Location: Spring Hill, Hernando County

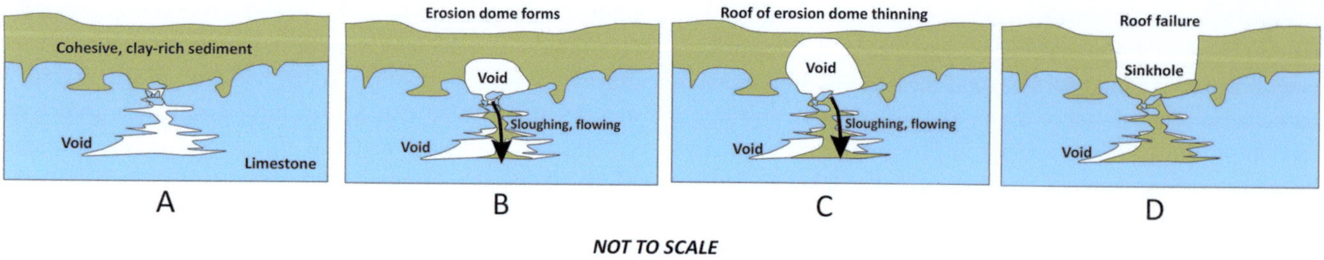


**Fig. 7.36** Development of a cover-collapse sinkhole. Modified from Tihansky (1999)

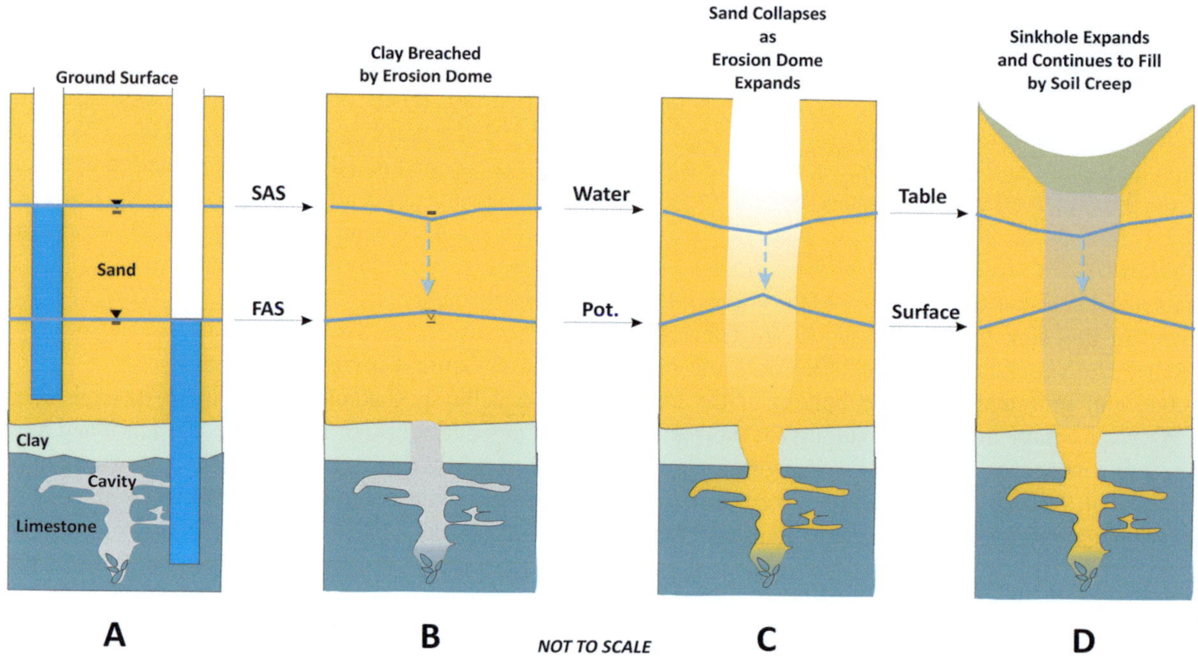


**Fig. 7.37** Sequence of cover-collapse sinkhole development modified from Beck and Sinclair (1986)

In cover-collapse sinkholes, the roof of the void consists of cover material, not bedrock. To form a roof over a void, the cover materials must be cohesive as a result of high clay content or, rarely, well cemented or lithified. Florida's medium- to fine-grained, well sorted sand may temporarily bridge over a small void, but it is considered incapable of bridging over a void large enough that a perceptible sinkhole can form upon failure of the bridge. Note that both of the conceptual models discussed below (Figs. 7.36 and 7.37) begin with a void overlain by cohesive, clay-rich sediment.

The void can be either within the limestone or at the limestone/cover interface. In Florida, the voids at the limestone/cover interface collapse with a failure of the clay-rich sediment that forms the roof of the void. Voids within the limestone body must be connected to the cover by a conduit, which is bridged by cohesive cover materials. When the cover materials can no longer support the load of the overburden, they give way and collapse or ravel into the void. Flowing water, drilling or otherwise puncturing the clay-rich cover may exacerbate failures.

Cover-collapse sinkholes may develop suddenly as the cover fails and can cause significant damage to structures on the property. The damaged structure shown in Fig. 7.35 collapsed during grout installation to stabilize an apparent developing sinkhole. The property was diagnosed as having evidence of the early stages of sinkhole development and subsurface grouting was appropriately prescribed to stabilize the soil and sediment beneath the structure. Penetrating the clay with pipes used to pump cementitious grout into the subsurface and then pumping the grout underground unfortunately resulted in failure of the cover materials and sinkhole development. While this example developed quickly (the grouting contractors were able to save the homeowners' puppies), cover-collapse sinkholes can also develop slowly if there are constraints placed on the process by void size, conduit or aven size, sediment textures, available water, and other factors.

Figures 7.36 and 7.37 illustrate renderings of the process of cover collapse. Figure 7.36 presents Tihansky's (1999) model for the development of a cover-collapse sinkhole. Note that an erosion dome is depicted as developing at the contact between the limestone and cohesive cover materials (B Fig. 7.36). For some period, it may be brief or last many years, the void created by raveling or spalling of cover materials into the underlying limestone persists. The longevity of such a void depends on the shear strengths of the cover materials, their degree of cohesion, water-table position and fluctuations, and other factors. When the roof of the void fails, a cover-collapse sinkhole develops (C and D, Fig. 7.36). Upchurch and Randazzo (1997) and Beck and Sayed (1991) presented similar diagrams to explain cover-collapse sinkhole development.

Beck and Sinclair (1986) published Fig. 7.37 by way of explaining cover-collapse sinkholes and subsequent cover behavior. In Panel A, they show that there is a downward hydraulic gradient from the Surficial Aquifer System (SAS) to the UFA. This gradient is necessary to allow for transport of cover sediment into the underlying void. They also suggested that the void need not be at the top of the limestone or within the epikarst. The only requirement is that the void must have an opening that will allow cover materials to ravel into it. In Panels B and C (Fig. 7.37), failure of the cover materials has begun. Note that the water table is depressed as a result of downward movement of water through the cover and limestone and that the failure is depicted as a cylindrical pipe extending to the land surface.

In general, sinkholes develop vertically through sediments overlying limestone. Whether a cylindrical pipe persists or the sinkhole expands into a cone opening toward the surface is dependent on the clay content of the cover sediment and position of the water table. Once the sinkhole has developed at the land surface, if the sediment is cohesive, a pipe may persist for decades. If the cover above the clay stratum is cohesionless sand, the sinkhole expands through slumping and raveling of sand near the face of the developing sinkhole. This results in a conical sinkhole, as has begun to develop in Fig. 7.37, Panel D. The lateral expansion of the sinkhole in moderately cohesive sediment as a result of slumping (Fig. 7.38) and/or raveling and soil creep in cohesionless sediment is a function of the angle of repose of the sediment. It is important to understand that the movement of the cohesionless, cover sediment toward the aven of the

**Fig. 7.38** Moderate sized cover-collapse sinkhole near Zephyrhills, Pasco County. Note (1) the steep sides that result from the high angle of repose of moist, clayey sand and (2) the incipient slump that is developing on the right-hand side of the sink

**Fig. 7.39** The 1981 Winter Park sinkhole (28.5939 N, 81.3615 W), Orange County. This large cover-collapse sinkhole shows a well-developed aven (A) and peripheral depression caused by lateral soil creep (C). Porsche provides scale. Photograph courtesy of B. Beck, Florida Sinkhole Research Institute

sinkhole represents a queueing problem. Movement is initiated at the wall of the aven, and sediment distant from the aven cannot move toward the aven unless (1) sediment in the queue nearer the aven is moving and (2) the slope vector of sediment motion is sufficient to induce movement and achieve the angle of repose for the material. The sediment that is moving should depict a pattern of declining penetration resistance with depth as a result of dilatation and reorientation of individual particles. Dry (vadose zone) sandy sediments tend to resist movement because of internal friction. Sandy sediments in the phreatic zone may move if the hydraulic gradient is sufficient to lubricate and contribute to movement. The same is true for piping of more cohesive materials toward an aven. Pipes should not develop or persist unless there is a hydraulic vector that is steep and capable of transporting cover sediments toward the aven. Structures at some distance from the active sinkhole, therefore, are not at risk until the developing sinkhole causes loss of support of the structure.

According to Beck and Sinclair (Fig. 7.37), the aven of the sinkhole is a discrete channel through which water and sediment move, creating, in this case, a "plug" of loose sediments that migrates into the underlying void. Panel D shows the cessation of vertical movement down the aven or pipe and the subsequent lateral expansion of the depression by lateral soil creep in the sandy overburden. Lateral soil creep and slumping of the sides of the aven ultimately reduces the slope of the depression walls and partly fills the initial, steep-walled depression.

Finally, the Beck and Sinclair (1986) sketch shows a common consequence of cover collapse. If the volume of the void is less than the volume of sediment that is affected by raveling during the development process, sinkhole formation ceases; leaving a plug of sediment in the aven or pipe. If this plug is eroded away at a later date, the sinkhole may reactivate – a process discussed below. Also, the plug of raveled sediment has very loose, open packing. Later compaction of this poorly packed sediment is likely to result in additional subsidence of the land surface into the sinkhole. Because the aven has higher permeability sediment as a result of open packing of sediment particles and there is a depression at the land surface, the aven is a likely location of preferred water movement, thus accelerating compaction and/or enhancing recharge to the underlying aquifer.

These conceptual models provide the basis for attributing these sinkholes to cover collapse. At some point in their development, a void roofed by cover materials must exist. While many cover-collapse sinkholes fail quickly and dramatically, the rate of development is not a criterion for attribution of collapse to a sinkhole. As shown in Panel D of Fig. 7.37, the queueing of collapse material through the aven is important. If the aven and underlying void are large and can accept the collapsed material quickly, collapse will be rapid. If there are constraints placed by aven or void volume or ability to transport the collapsing material, the sinkhole may form slowly and never reach a large size.

Northern Hillsborough and Pinellas counties and western Pasco, Hernando, and Citrus counties have been collectively termed "sinkhole alley" by the press. Sinkhole development in these areas has driven most of the sinkhole-formation thinking in Florida. Both of the conceptual models presented in Figs. 7.36 and 7.37 were developed based on experiences in west-central Florida. Elsewhere, the sediments of the cover differ in thickness and texture and hydrogeology differs and, as a result, sinkhole risk and development mechanisms may differ somewhat from these models.

Cover-collapse sinkholes can sometimes be identified by the geometry of the young sinkhole. Figure 7.38 illustrates a moderate sized, cover-collapse sinkhole that formed by a process similar to that shown in Fig. 7.37. The subsidence feature has reached the surface (compare with Panel C, Fig. 7.37), but lateral development by creep has not yet begun. A small slump is just beginning on the left-hand side of the sink. In time, slumping and soil creep will broaden and partly fill the depression. However, the steep angle of repose of the moist, clayey sand suggests that (1) slumping and caving of the walls of the sink will cause growth of the upper outline of the sink and (2) creep is not likely to expand significantly away from the walls of the sinkhole.

Figure 7.39 illustrates the 1981 Winter Park sinkhole after it had developed, but before groundwater began to fill it. This large, young sink clearly shows a well-developed aven (A Fig. 7.39), down which the cover collapsed. Lateral sediment creep began to transport cover sediment into the aven as it developed, probably by the rush of groundwater to the aven as the feature drained into the underlying UFA. As a result of lateral creep and slumping, the sinkhole rapidly expanded laterally, finally engulfing a swimming pool and automobile dealership (note the Porsche on the left-hand side of the image). Much of this lateral movement of sediment is creep as shown at the top of the aven in Panel D (Fig. 7.37).

During standard penetration testing or cone penetrometer testing, recognition of cover-collapse, as opposed to other sinkhole-formation mechanisms, requires detailed knowledge of the geology of the site and of soil mechanics. While it may not be possible to find the aven and underlying void, the absence of obvious, widespread bedrock dissolution and presence of systematic weakening of sediment with depth in the cone of lateral and vertical sediment migration, which indicates raveling, should suffice.

#### 7.3.2.7 Volume Constraints of Void Space on Sinkhole Development

When relatively well-consolidated sand is disturbed by separation (dilatation) of the grains during excavation, raveling, or other disruptive processes, there is a volume increase of the material as the degree of packing of the individual grains is reduced. As a result, the volume of a sinkhole depression created by raveling is less than the volume of the sediment that is to ravel into the underlying void. For example, if undisturbed, sandy cover material is more-or-less close-packed with a bulk porosity of 25%, after raveling into a void space it becomes open-packed with a bulk porosity of, say, 35%. After raveling, the raveled sediment occupies 10% more volume than the undisturbed sediment. In other words, 0.9 $m^3$ of undisturbed sediment would fill a 1 $m^3$ void with the hypothesized pore-space increase. The loss of sediment packing and bulk density, therefore, limit the amount of sediment that the void can accept.

The volume of the void and its aven can also affect how large the resulting sinkhole in the cover may be. Figure 7.40 illustrates two scenarios where the void may determine the size of the resulting depression in the cover materials. In scenario A (Fig. 7.40), the void space is large, but accumulation of the cone of raveled sand, which is defined by the stable slope (the angle of repose) of the sand mound, is limited by the height of the cone and the void. Once the cone of accumulated sand intersects the base of the aven and top of the void, the cone of accumulated sand is likely to block additional raveling into the void, the aven fills by soil creep and slumping, and sinkhole development ceases. In other words, the height of the void and cone of accumulated sediment directly limits short-term sinkhole growth in the cover. Later compaction of the cone of accumulated sediment; dispersal of the cone of sediment by groundwater flow, pressure variations, or vibrations; expansion of the void height; or some other process may cause the sinkhole to reactivate as volume is added for additional raveling. This latter process is a common cause of sinkhole reactivation.

Scenario B (Fig. 7.40) illustrates a scenario where the void space available to be filled by raveled sand is limited and, once the void is effectively filled, no additional raveling and growth of the sinkhole in the cover materials can occur. This scenario appears to be a common case in Florida. In this case, the sinkhole may not reactivate for hundreds of years, if ever.

The probability of reactivation of a sinkhole may be related to the size of the void into which overburden has raveled. The larger the sinkhole, the more likely it is to reactivate because there is a higher probability that the host void contains unfilled volume: groundwater flow can move the cone of raveled sediment in the underlying void, or the sediment cone can compact and/or flatten causing opening of additional space.

The volume of the void and thickness of cohesionless sediment available to ravel into the void are both limiting factors. Clearly, for a large sinkhole to form by collapse or raveling, the underlying void must be huge. Most incidences of sinkhole development in Florida result in little or no topographic expression of raveling. When sinkholes do form,

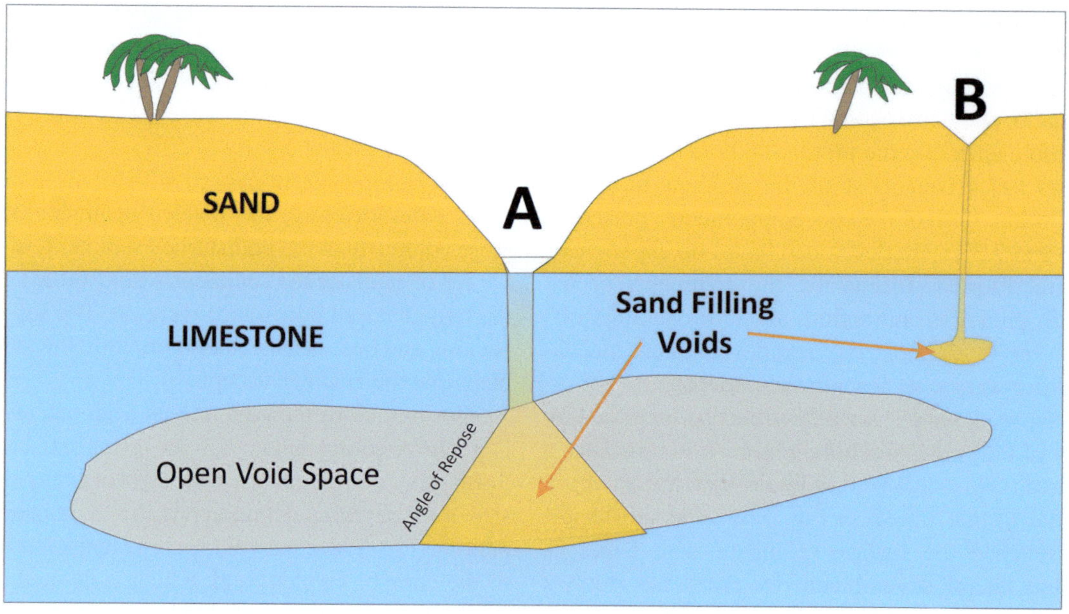


Fig. 7.40 Relationship of void size to the volume of cohesionless sand that can be accepted and the size of the resulting cover-collapse sinkhole

Table 7.6 Summary of the measured subsidence-feature depression sizes reported in the Florida Geological Survey Subsidence Incident Report database as of 2013

| Metric | Diameter in meters | Depth in meters |
|---|---|---|
| Average | 5.2 | 2.8 |
| Median | 2.4 | 1.5 |
| Minimum | 0.08 | <0.10 |
| 25th percentile | 1.1 | 0.76 |
| 75th percentile | 6.1 | 3.0 |
| Maximum | 150 | 61 |
| Number of measurements | 2624 | 2564 |

they are seldom more than 6 m across, and most are much smaller than that (Table 7.6). Given the limitations in sinkhole size[6] reported in Table 7.6, it is clear that thickness of the cohesionless cover materials and void volume are substantial limitations to the size of cover-related sinkholes. Regardless of the thickness of the material available to ravel, a small void can only accept a limited volume of that sediment. The result is likely a pattern of weakening of the cover materials with depth, but little or no probability of continued raveling because there is simply no place for the sediment to go in the limestone.

The Winter Park cover-collapse sinkhole is a unique example of a void that did accept a large volume of collapsed and/or raveled sediment. When the Winter Park sinkhole (Figs. 1.5 and 7.39) developed in 1981, it reached a diameter of approximately 107 m. The sinkhole was developed within a sand column 19 m thick. Below the sand column was about 34 m of sandy clay and clay, and limestone was about 52 m below land surface. The aven, the passage through the sand and clay and into the limestone, was estimated to be 15 m in diameter (Jammal & Associates 1982; Jammal 1984a, b; Jammal and Casper 2009). Assuming that the aven included collapsed sand and clay, a total of approximately 5400 $m^3$ of sand and clay collapsed into the underlying limestone before development of the cone-shaped depression was possible. The volume of the cone-shaped sinkhole, excluding the column of sediments that once occupied the location of the aven, is estimated to have been 61,700 $m^3$, so the total volume of sediment that raveled and/or collapsed into the underlying void was on the order of 67,100 $m^3$. This volume is equivalent to approximately 956 school buses. Assuming that there was some dilatation of the sediment as it settled into the void, this estimated volume is low in terms of the void(s) into which the sinkhole collapsed/raveled. There is no information as to whether the void was a single open space or a series of well interconnected openings.

Given the abundance of void space in Florida limestones and dolostones, it is often futile to try to identify the origins of the voids or determine if specific voids are implicated in development of karst features, such as sinkholes. One can, however, assume that the individual voids are normally relatively small and that a single void or a series of interconnected voids capable of freely transferring sediment from one void to another that is large enough to create a Winter Park sinkhole-sized depression is the exception, not the rule. Most common, large sinkholes are perhaps less than 100–500 $m^3$ in volume, and their sizes are suggestive to the size limitations imposed by the available void volumes.

[6]Care should be taken in interpreting the sinkhole statistics presented in the Florida Geological Survey's subsidence incidence database, because small sinkholes and sinkholes that develop in rural areas are less likely to be reported than larger sinkholes and sinkholes in populated areas.

### 7.3.2.8 Sinkhole Size and Shape

Clearly, the largest modern sinkhole in Florida, the Winter Park Sink (Fig. 7.39), represents a "worst-case scenario." That sinkhole affected approximately 8900 $m^2$. It is important to note that a sinkhole of this size can only exist where limestone is well over 30 m below the land surface – a geological condition that is limited in Florida. Most sinkholes are much smaller than the Winter Park sinkhole, and each is typically only capable of affecting a small area.

When a cover-collapse sinkhole forms, it usually starts as a conduit characterized by vertical movement of sediment. As long as the sides of the sink remain vertical, the diameter of the sinkhole will approximate the diameter of the conduit into which the sediment has raveled. This results in a small-diameter sinkhole. However, sandy, cohesionless sediments cannot support vertical slopes for long periods of time. After the sinkhole initially develops, slumping and spalling or raveling of the sides of the sandy part of the sinkhole begin to dominate and the sinkhole expands in diameter. Lateral soil movement occurs by soil creep, slumping, and other mass wastage processes. Typically, this horizontal migration to the conduit creates a hourglass-shaped cone of soil movement. The diameter of the cone expands until the sandy soils are at their angle of repose.[6] At that point, the sinkhole is mature[7] and unlikely to exhibit additional growth.

Assuming that the angle of repose of cohesionless, sandy sediment is about 45°, a "rule of thumb" states that the radius of a mature, cover-collapse sinkholes is likely to be about twice the depth to the base of the sinkhole cone (Fig. 7.41). Therefore, if a conduit 0.3 m in diameter in limestone occurs under 0.6 m of sand, one would expect the resulting sink to be about 2 to 3 m in diameter when fully mature.

While this is simply a "rule of thumb" and there are many factors that complicate the actual sizes of sinkholes, it is instructive. For example, if limestone or clay[8] is shallow, the sizes of cover-collapse sinkholes are limited. This relationship simply suggests that the thicker the cover of sand over cohesive material, the larger the potential sinkhole is likely to be after it is mature.

The Subsidence Incidence Reports database (Florida Geological Survey 2013) contains data on the sizes of almost 2000 modern sinkholes in Florida. From this database we can obtain some ideas about sizes and shapes of cover-collapse sinkholes at the time of development in Florida.

The sinks averaged 5.2 m in length and 4.9 m in width, and median dimensions were 2.5 m in length and 2.1 m in width (Table 7.4). The large differences in average and median[9] dimensions indicate that the majority of sinkholes are small (75% of all sinkholes were only 6.1 m in width and 4.6 m). The largest sinkhole in the database was 107 m in diameter (the Winter Park Sinkhole).

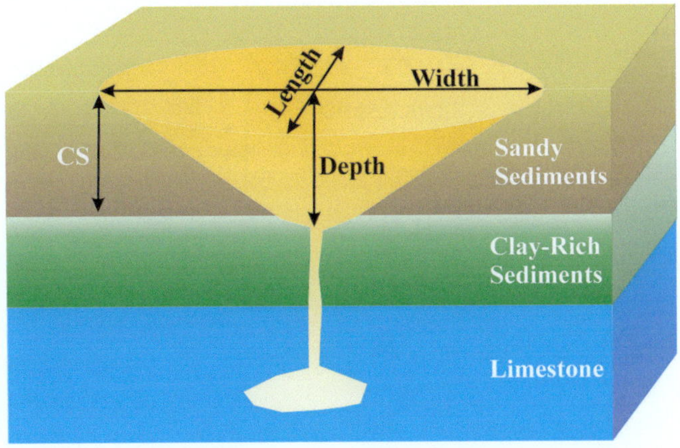


Fig. 7.41 Estimating the size of a fully developed sinkhole based on the thickness of easily raveled, cohesionless sediment (CS)

The average and median length-to-width ratios were 1.1, which indicates that the sinkholes were essentially circular in outline. Small, elongated sinkholes that might preferentially develop in a single direction are reported in the database, but they are rare. Most are very nearly circular.

The average and median width-to-depth ratios were 1.6 and 1.4, respectively. In other words, the sinkholes were about half as deep as they were wide. Most of these sinks were relatively new and probably had not matured, as discussed above.

### 7.3.2.9 Cover-Subsidence Sinkholes

Cover-subsidence sinkholes are the least well understood form of sinkhole in Florida. Cover-subsidence depressions form by the dissolution of the upper limestone/dolostone or carbonate sediment surface, so, like solution sinkholes, the rate of sinkhole development is extraordinarily slow and is limited by the rate of dissolution of the carbonate material. Cover-subsidence sinkholes form as the upper surface of the carbonate rock is dissolved away and the cover materials, which may be relatively thick (Table 7.5), simply sag into the depression as it develops.

In contrast to suffosion sinkholes (Sect. 7.3.2.3), the bedrock or shell deposit that is subjected to dissolution does not form a cavity or void into which thin cover materials ravel. The cover materials in a cover-subsidence sinkhole simply

[7]A "mature" sinkhole is defined as a sinkhole that has grown in size to the point where natural, lateral movement by soil creep and mass wastage is minimized and the slopes are at, or near, the angle of repose for the soil materials involved.

[8]Because it is cohesive, clay does not ravel easily. Therefore, the "hourglass" shape typically develops on top of the clay, not within it (Fig. 7.41).

[9]Averages emphasize large sinkholes while the medians represent the actual frequencies of occurrence. The median is the 50th percentile of the data. In other words, 50% of all sinks were smaller than 2.5 m in length and a few very large sinks caused the average to be about 5.2 m.

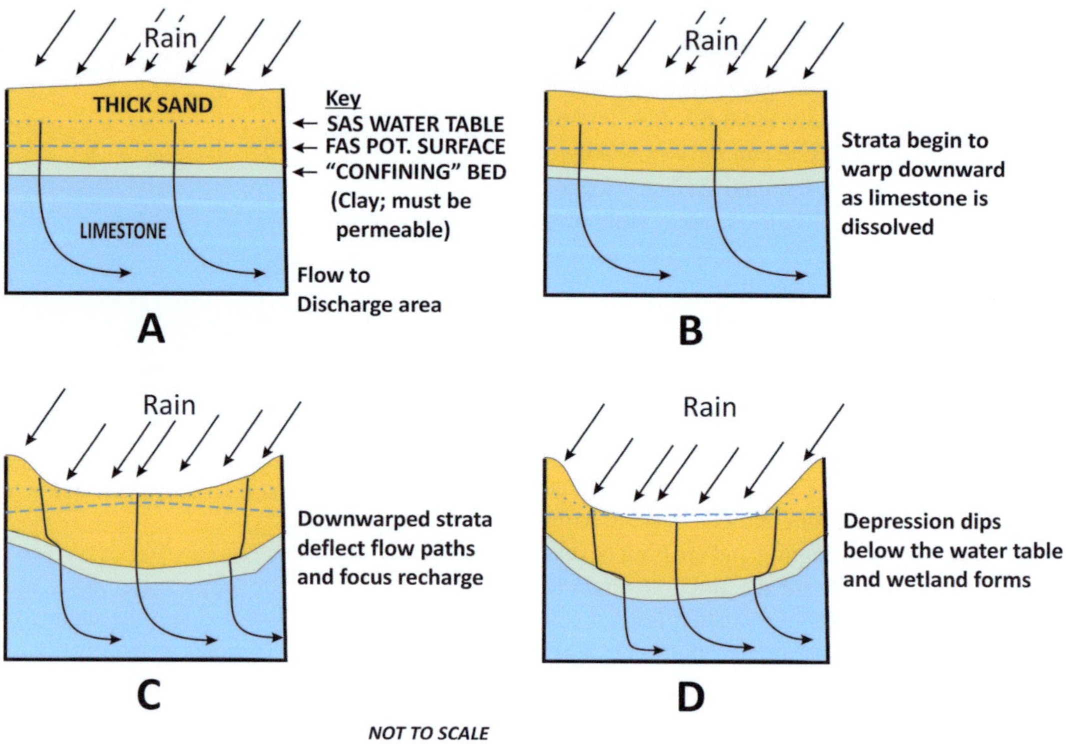


**Fig. 7.42** Development of a cover-subsidence sinkhole according to Sinclair et al. (1985). Modified from Sinclair et al. (1985)

sag as the upper surface of the carbonate stratum is removed by dissolution.

Figure 7.42 illustrates the mechanism of cover-subsidence development as visualized by Sinclair et al. (1985). This diagram illustrates the sequence of events over thousands of years or more that lead to a cover-subsidence sinkhole. Note that the cover is not shown to be raveling or otherwise moving out of the stratum within which it was deposited. Rather, the strata themselves are shown to be sagging over time. This sagging phenomenon often results in water-filled, cover-subsidence sinkholes to be termed "sag" features of "sag ponds" (Fig. 7.22; White 1970).

The model proposed by Sinclair et al. (1985) for development of cover-subsidence sinkholes shows a progression of subsidence through time, and it emphasizes the distributed flow of groundwater in the underlying limestone. There is no obvious void into which the cover materials are migrating. In Panel A (Fig. 7.42), the cover materials consist of sand over a "confining bed" of clay.[10] Both the water table in the SAS and the potentiometric surface of the UFA are within the sand mantle. There is a downward hydraulic gradient from the SAS to the UFA, however. Meteoric and SAS water penetrates the "confining bed", enters the underlying limestone and flows along joints, bedding planes, and other dispersed pathways. In Panel B the cover-subsidence sinkhole is forming as the upper few centimeters of the limestone are dissolved away, and the cover strata cuncurrently sag into the space created by limestone dissolution.

Note that, for chemically aggressive groundwater to dissolve carbonate materials at the top of the limestone (B Fig. 7.42), there should be a permeability contrast between the cover materials and carbonate stratum. Otherwise, it is probable that the water will penetrate the carbonate stratum rather than focusing on dissolution at the upper carbonate-sediment surface. Similarly, if the clay is highly cohesive and resists "relaxing" and sagging, a void is likely to develop at the interface and a cover-collapse sinkhole may result.

Panel C (Fig. 7.42) illustrates the deepening of the depression created by sagging, and Panel D shows the circumstance where the depression has intersected the water table, creating a sag pond. Many of the thousands of cover-subsidence sinkholes in Florida are seasonally flooded as the water table rises to intersect the depression (Fig. 7.22).

Cover-subsidence sinkholes differ from solution sinkholes (Sect. 7.3.2.2) because of the thickness of the cover and difference in mechanism of cover movement. Cover materials must be cohesionless (i.e., sand-rich) and permeable

[10]The concept that the clay constitutes a confining bed comes from the belief that the ICU or its residuum hydraulically confines the UFA in Florida. While this may be true where it is thick, the clay stratum cannot serve as a hydraulic confining bed nor does it prevent development of sinkholes where cover-subsidence sinkholes are present. Sinclair et al. included the clay stratum in deference to the occurrences of clay associated with sinkholes in west-central Florida.

(Table 7.4). The permeability requirement is critical because groundwater must be (1) able to circulate over the top of the carbonate stratum in order to dissolve the carbonate or (2) retained at the top of the carbonate-rich stratum long enough to dissolve the carbonate material. This requirement limits cover-subsidence sinkhole development to coarse-grained sediments and explains why shell materials in low permeability, clayey sand and clay are less likely to develop cover-subsidence features. Because subsidence is contemporaneous with dissolution of the underlying limestone, the rate of cover-subsidence sinkhole development is very slow (Table 7.4) and controlled by the kinetics of carbonate dissolution. Needless to say, the upper surface of the limestone may be pitted, and small cavities may exist, but the key to cover subsidence is overall lowering of the carbonate rock or sediment surface, not raveling into a void.

Florida karst scientists have recognized shallow, cover-subsidence features as sinkholes ever since White's (1970) work, and they are recognized as a form of sinkhole worldwide (Table 7.3). White (1970) and Schmidt and Scott (1984) discussed these features and attributed them to karst processes. In other words, they argued that the preponderance of the shallow, more-or-less circular depressions were likely cover-subsidence sinkholes. Figure 7.22 illustrates how abundant potential cover-subsidence sinkholes can be in southern Florida. Note that not all shallow depressions are a result of cover-subsidence sinkhole processes; they may not even be karst-related (Upchurch et al. 2015).

Investigation of similar, more-or-less circular and linear features at the Tampa Bay Regional Reservoir (Fig. 6.30) in southeastern Hillsborough County by Upchurch et al. (1999) showed that the features, which appear similar to, but larger than, those shown in Fig. 7.22, were related to a variety of possible causes, including sagging related to compaction in the post-Miocene, "leached zone" paleosols, paleosinkholes in the subsurface, and cover-subsidence sinkhole development. Similar, circular depressions can be observed in aerial photographs of the coastal swamps and marshes of Florida, where marine and estuarine sedimentation/erosion patterns develop shallow, circular features. Therefore, one should not assume that all closed wetland depressions in Florida are cover-subsidence sinkholes without adequate subsurface testing.

If the local geology of an area is well understood, recognition of cover-subsidence sinkholes may be relatively simple. The key to recognition is identification of shallow carbonates that have been heavily corroded and subject to dissolution near the top of a carbonate stratum. With the high water-table elevations found in many areas of the state, a preponderance of the cover-subsidence features either are, or were, wetlands. As a result, the sediments that accumulate in the sag depressions often include evidence of the presence of shallow wetlands.

Some characteristics of cover-subsidence sinkholes include, but are not limited to:

- Most commonly, circular or semicircular outline;
- Round or semicircular subsidence feature developed on the upper surface of the carbonate stratum, not at depth within the stratum;
- The carbonate sediment that has undergone dissolution is shallow and shows evidence of extensive dissolution;
- Voids may not be detected in the carbonate substratum;
- Permeable sediments overlying the carbonate stratum;
- Organic-rich sediment accumulations within the depression formed by long-term subsidence (Figs. 7.43 and 7.44);
- Fossils and limestone fragments within the sinkhole cover are eroded and poorly preserved;
- Accumulations of white sand caused by "bleaching" of the sand by humic acids derived from the marsh or swamp vegetation within the wetland (Fig. 7.44);
- "Bowl-shaped" accumulation of sediments developed by both sagging and accumulation of later sediments within the depression; and
- Evidence of seasonal sedimentation, slumping, and bioturbation in the sediments within the depression (Fig. 7.44).

An advantageous way to identify the presence of a cover-subsidence sinkhole in areas where development has obscured the more-or-less circular outline of the sinkhole is through examination of historical photographs that depict the site prior to site development for construction. Since cover-subsidence sinkholes are so slow to develop, there is a high probability that visual evidence of the feature existed prior to development. If there is no evidence of the depression prior to development, it is suggested that excavation of a trench may be necessary to expose enough of the sediment to reveal the sag feature and identify the associated wetland sediments and corroded limestone.

#### 7.3.2.10 Problematic Cover-Subsidence Features

The land surface in Florida is rarely characterized by anomalous circular depressions. They are unusually large and have an unusual profile. The best developed depressions of these depressions are located in the community of Lehigh Acres, Lee County. Figure 7.45 shows an example of one of these depressions. The wetland portion of the depression averages 243 m in diameter and is approximately 1 m deep. Unlike most cover-subsidence sinkholes and sag ponds in Florida, the margins of these depressions are relatively sharp with a 0.5–0.75 m drop over a distance of 10 m or less. In the field and in LIDAR images, these depressions give the impressions of a cylindrical column of sediment that has subsided

**Fig. 7.43** Organic-rich layer with overlying sediment composed of a mixture of iron and organic sand with lumps of light-colored sand and limestone in a cover-subsidence sinkhole near Ft. Drum, Okeechobee County

**Fig. 7.44** Bleached sand with thin beds reflecting annual sediment accumulations. These sediments were exposed in the same pit as Fig. 7.42

as much as a meter. Field testing to determine the origins of these features is lacking at the time of writing of this book.

#### 7.3.2.11 Cover-Subsidence Sinkholes on Shell Deposits

White (1970) discussed the many circular wetlands that occur throughout low areas in southern and central Florida. He identified them as features related to dissolution of the underlying, near-surface shell or limestone and termed them "sag ponds." Schmidt and Scott (1984) also attributed these features to dissolution of shell or limestone. However, the concept that the depressions are karst-related is somewhat problematic because they are developed on near-surface shell beds, not limestone or dolostone and there is no void into which sediments ravel.

It has been incorrectly argued that any stratum with shell content can develop cover-subsidence sinkholes in Florida regardless of (1) the shell bed location within the stratigraphic column, (2) burial depth, or (3) volume proportion of shell material. There is a significant volume and depth

**Fig. 7.45** Houses constructed within a large circular depression that may be a cover subsidence sinkhole. Average diameter of the wetlands part of the depression is 243 m. This site is located in Lehigh Acres, Lee County

problem with this argument. Since the amount of subsidence of the cover materials and land surface is a function of the volume of near-surface shell or limestone material removed (Fig. 7.46) and since disruption of the cover materials as a result of subsidence increases the inter-particle pore space and lowers the density of the cover sediments, a significant amount of carbonate sediment must be removed in order to cause enough subsidence to develop a depression.

For shell deposits to develop cover-subsidence sinkholes, the following conditions control the process:

1. The mantle of cover material must be thin enough that any volume reduction by near-surface shell (or limestone) dissolution is distributed over a small volume of cover.
2. The cover material must be sufficiently permeable for exchange of water across the carbonate-cover sediment interface.
3. Shelly sediment must be shallow; field observations suggest that the shell bed must be within about 3 m of the land surface (Upchurch et al. 2015).
4. For a depression in the land surface to form, the volume of shell removed must be greater than the combined volumes of the depression and additional interstitial pore space introduced in the cover material by dilatation as subsidence occurs (Fig. 7.46).
5. Chemically aggressive water must be able to circulate over and through the upper carbonate sediment for dissolution to occur. The residence time of the water in contact with the shell must be long enough for incremental dissolution of the shell. Chemical equilibration of the water with the shell develops quickly with depth, which means that shells within a short distance below the cover/shelly sediment interface are not likely to be subject to dissolution.

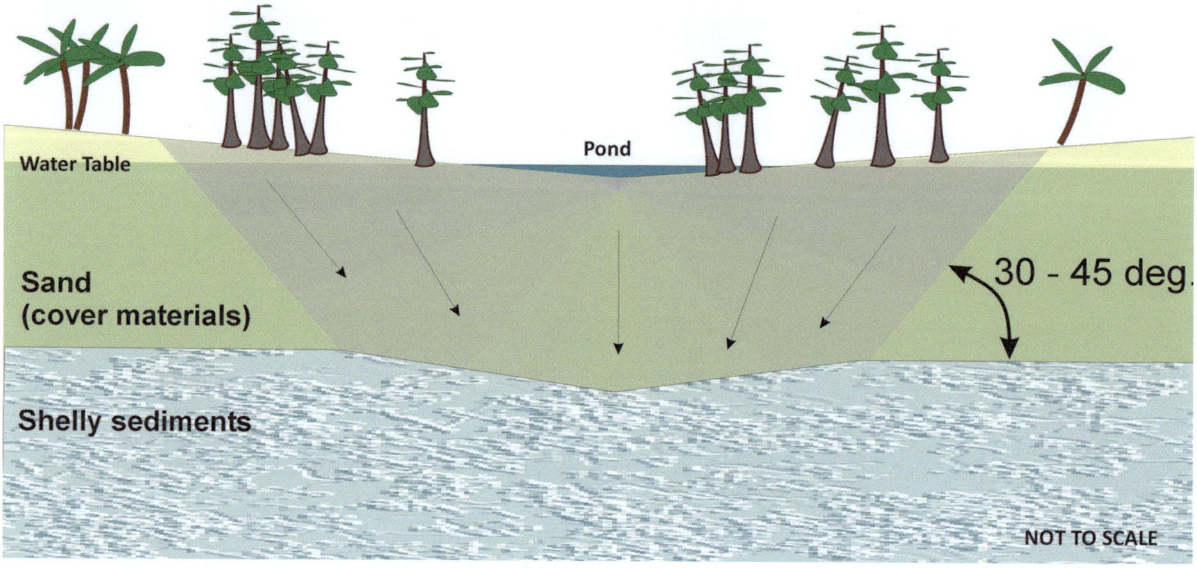


**Fig. 7.46** Hypothetical cross section through a cover-subsidence sinkhole over a shell bed demonstrating the "volume problem" created by the need for the volume of dissolved shell to equal or exceed the volume of subsiding cover materials

6. Any silt- and clay-sized carbonate is most likely to react with the groundwater and cause it to reach equilibrium with respect to the carbonate minerals prior to macro-sized shell dissolution. This is because of the higher surface area of the fine-grained sediment (Chave and Schmalz 1966). Since (1) this silt-and clay-sized material is often interspersed with siliciclastic sediments that can form a grain-supported framework and (2) chemical equilibration is likely to have occurred, dissolution of fine-grained carbonates is significantly less likely to cause development of cover-subsidence sinkholes than is dissolution of shell beds.
7. There must be a high percentage of near-surface shell material in the underlying sediment. Based on field observations, it is estimated that the shell content must be well over 50% of the volume of the shell-rich sediment (Fig. 7.47a) for enough volume to be lost and for subsidence to occur.
8. The dissolution process must be maintained for many hundreds to thousands of years in order for a sufficient volume of shell to be dissolved away and a depression can begin to form.

Recognition of shallow subsidence features developed by dissolution of shallow shell beds often requires petrographic examination of the shells immediately under these depressions in order to determine if the shells and/or shell fragments have undergone substantial dissolution. Note that a very large proportion of the shell (greater than 50%) must be removed in order to create sufficient volume reduction for a depression to form.

Evidence of dissolution of shell fragments includes:

1. Rounding of sharp corners on broken shells;
2. Erosion of shell decorations;
3. Development of shiny surfaces as if the shells were dipped in acid;
4. Development of "punky", earthy surfaces as aragonitic components are selectively removed from the shell by dissolution; and/or
5. Shell remains consist of traces of shell so that only the "ghosts" of collapsed shells and/or molds of shells remain (Fig. 7.47b).

Siliciclastic sediments in soil zones where the shell has been weathered and subjected to dissolution often contain abundant ferric hydroxide, which gives the sediment/soil a reddish hue (Fig. 7.47b). Organic sediments (Fig. 7.43) and crude banding are often present. Since many cover-subsidence sinkholes are either permanent or seasonal wetlands, organic sediments derived from wetland plants are common in cover-subsidence sinks. Also, it is not uncommon to encounter white, "bleached" sand (Fig. 7.44) under the organics. This sand consists of sediment washed into the depression from its periphery and the bleaching is a result of removal of ferric hydroxides and other soluble materials by dissolved humic substances derived from the organic sediments (Fig. 7.43).

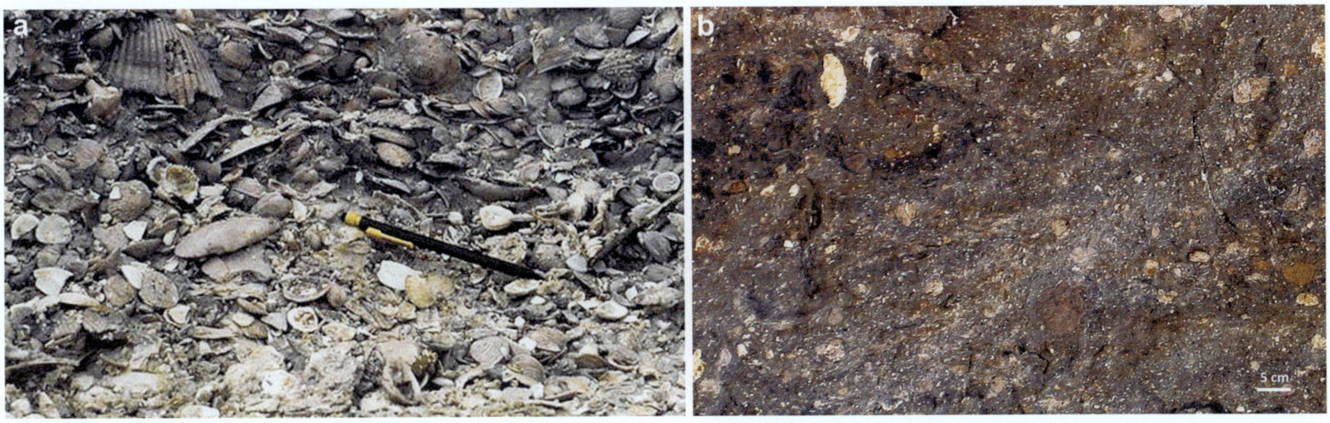


**Fig. 7.47** (**a**) Bed with over 50% shell material. From the Plio-Pleistocene "Bermot beds" in DeSoto County. (Image courtesy of the Florida Geological Survey; reproduced from Upchurch et al. (2015)). (**b**) Weathered shell bed in a cover-subsidence sinkhole developed on a Pleistocene shell bed near Ft. Drum, Okeechobee County. Note the shell fragments, enrichment of the sediment by ferric hydroxide, and scattered organics (black materials). (Image courtesy of the Florida Geological Survey; reproduced from Upchurch et al. (2015))

Because the rate of subsidence is limited by the kinetics of dissolution, cover-subsidence sinkholes pose little threat to structures unless the structure was built over a pre-existing sinkhole that contains deleterious sediments (i.e., peat or muck, woody debris, loose sediments). Development of a new cover-subsidence sinkhole after a structure is placed on the site should have few consequences because the rate of dissolution is extremely slow.

In summary, for shell dissolution to cause development of a cover-subsidence sinkhole, the strata must be true shell beds, not strata with a few shell fragments; the permeability distribution of the strata must create a hydrogeological environment conducive to long-term dissolution at the upper surface of the shell bed; and the cover must be thin enough to subside and create a depression as the shell is removed and siliciclastic sediments dilitate. The shell material slowly dissolves over timeframes of many thousands to millions of years.

#### 7.3.2.12 Rate of Collapse Versus Subsidence during Sinkhole Formation

In Florida, investigators of sinkholes often judge the cause of a modern sinkhole on the basis of the rate of depression development. If the sinkhole develops rapidly (i.e., in hours to days), it is undoubtedly a cover- or rock-collapse sinkhole. However, it is important to understand that collapse-sinkhole development need not be sudden. The final volume and rate of development of a collapse sinkhole is a function of:

1. The available void volume;
2. The mechanism(s) of roof failure (Sect. 7.4);
3. Land-surface slopes surrounding the depression and rate of creep into the developing depression; and
4. Compaction or transport of material in the aven and void.

As a result, many cover-collapse sinkholes take years to develop, and their development may be episodic.

Development of a cover-subsidence sinkhole is controlled by the rate of carbonate sediment dissolution, so it takes thousands of years or more for a depression to form. Similar rates of development may be present as sediment is queued up and transported through the aven in a collapse sinkhole, so the rate of development of a slowly forming sinkhole should not be used to identify the cause of depression formation.

#### 7.3.2.13 Buried or Alluvial Sinkholes and Sinkhole Reactivation

The final class of sinkhole has been termed buried or alluvial sinkholes because pre-existing sinkholes have been filled and plugged by alluvial sediment. Culshaw and Waltham (1987) termed these sinkholes "buried sinks" (Table 7.3). Upchurch and Randazzo (1997) termed them "alluvial sinkholes" (Fig. 7.48). The resulting, inactive sinkhole (A, Fig. 7.48), which often includes a depression in the land surface, may persist for thousands of years. If the sinkhole formed prior to memory, it can be termed a paleosinkhole.

As discussed in Sect. 7.3.2.6, many times the collapse and downward migration of cover sediments into and through the aven leaves a column of cover materials plugging the aven (A, Fig. 7.48). If this plug of sediment fails, the sinkhole is said to have been reactivated.

Reactivated sinkholes are common in Florida, especially near urban areas where groundwater withdrawals and other human activities may weaken the sediment plug in the aven and ultimately lead to failure. In the scenario proposed in Fig. 7.48, the house has been constructed over an ancient, plugged, alluvial paleosinkhole. Groundwater withdrawals (a water well is shown to the right of the house) causes

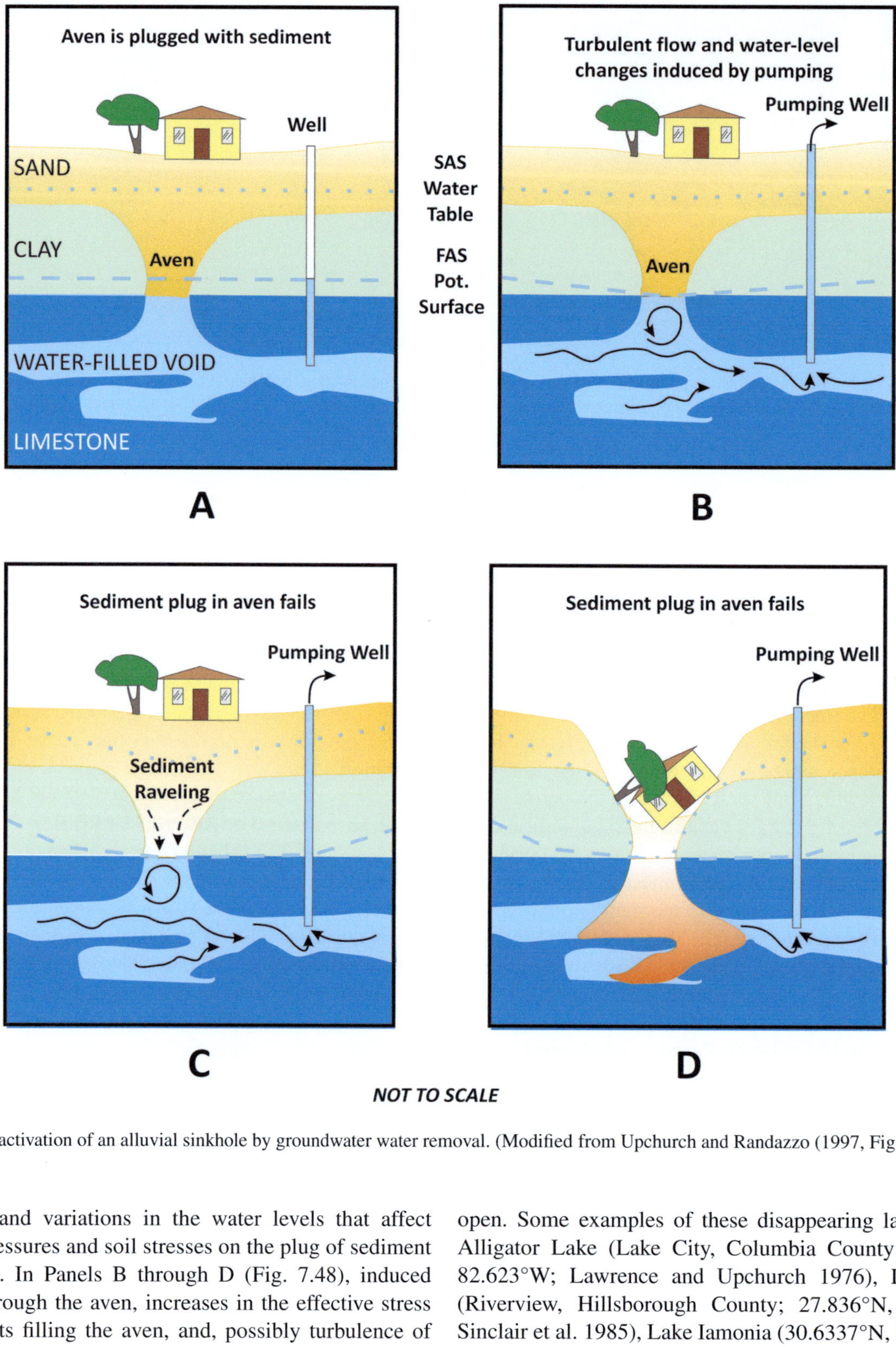


**Fig. 7.48** Reactivation of an alluvial sinkhole by groundwater water removal. (Modified from Upchurch and Randazzo (1997, Fig. 13.11))

turbulence and variations in the water levels that affect effective pressures and soil stresses on the plug of sediment in the aven. In Panels B through D (Fig. 7.48), induced recharge through the aven, increases in the effective stress on sediments filling the aven, and, possibly turbulence of groundwater flow in the underlying conduits as a result of the withdrawals ultimately led to failure of the plug and reactivation of the ancient sinkhole.

Many of Florida's lakes formed as a result of ancient sinkhole activity. Today, several of these lakes are notorious for periodically draining as the sediment plugs in their avens open. Some examples of these disappearing lakes include Alligator Lake (Lake City, Columbia County; 30.170°N, 82.623°W; Lawrence and Upchurch 1976), Lake Grady (Riverview, Hillsborough County; 27.836°N, 82.273°W; Sinclair et al. 1985), Lake Iamonia (30.6337°N, 84.2501°W; Fig. 7.49), Leon County; and Lake Jackson (Leon County; 30.530°N, 84.327°W; Hughes 1967; McGlynn 2006; T. Scott [pers. observation] 1999).

Lake Jackson, which is located on the north side of Tallahassee in Leon County, is the best documented draining lake in Florida. Repeated drought and wet periods have

**Fig. 7.49** Lake Iamonia, a disappearing lake in Leon County. Note the sinkhole in the bottom of the dry lake bed. Location: 30.6337°N, 84.2501°W

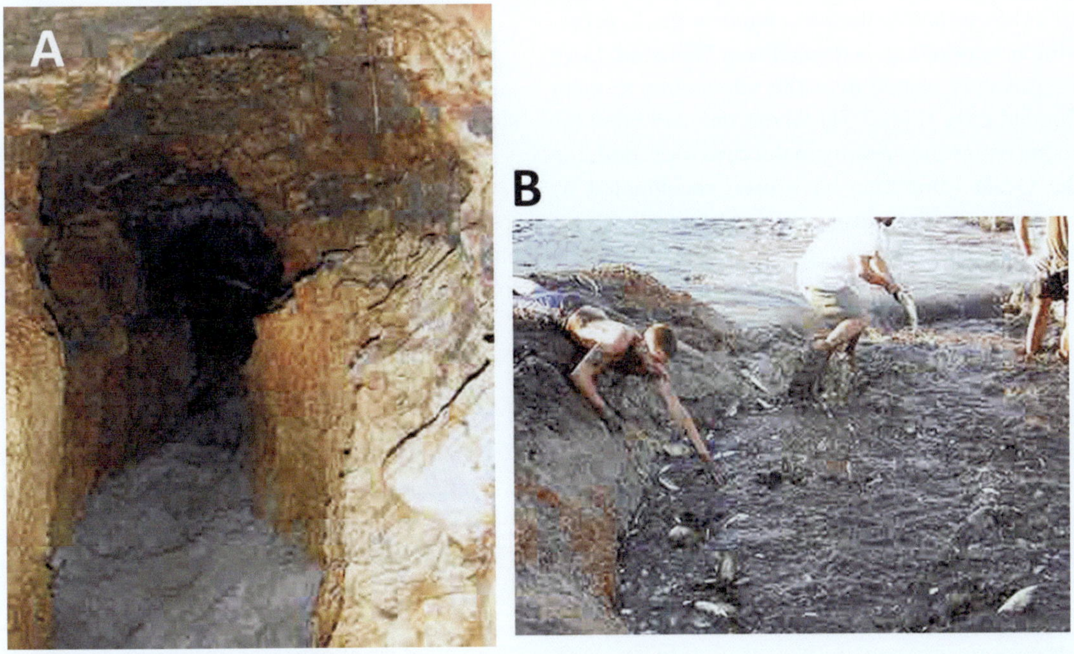


**Fig. 7.50** Scenes from Lake Jackson (30.5309°N, 84.3248°W) after it drained in 1999. (**a**) The aven through which the lake drained. (**b**) Capturing stranded fish in the aven of the sink that drained the lake

caused lake levels to vary dramatically, including drainage of the lake when an aven opens. Informal comparison of lake-level data and episodes of lake drainage does not suggest that drainage events are a result of unusually high rainfall and lake water levels.

One of the authors of this text, T.M. Scott, led a Florida Geological Survey investigation of the lake during and after the 1999 sinkhole reactivation event. The aven penetrated the Pliocene Miccosukee Formation and extended downward through Miocene Hawthorn Group sediments to the underlying limestone (Fig. 7.50a). Figure 7.50b shows men catching fish as the lake was draining.

#### 7.3.2.14 Karst Windows

Karst windows are sinkholes that are typically formed by rock-collapse sinkholes and that intersect a water-filled cave, so that the collapse feature serves as a "window" into the underlying cave and aquifer.

One of the most accessible karst windows in Florida is the feature known as Falmouth Spring (Figs. 7.4 and 7.5) in Suwannee County. This karst window is maintained as a park by the Suwannee River Water Management District. The window is a collapse feature on the Cathedral-Falmouth cave system (Fig. 7.4, Exley 1994). The cave system drains to the Suwannee River, about 5 km west of the window. During normal rainfall conditions, the water in the spring is clear and beautiful, and there is a distinct current to the west, toward the river. When the Suwannee River is high, however, flow in the spring reverses, and the water turns brown as river water replaces groundwater (Scott et al. 2004).

The Devil's Den, a spectacular karst window that is accessible for diving or snorkeling, is located near Williston, Levy County. This privately owned dive site allows one to enter the partly flooded cave (Fig. 7.51) where one can view the breakdown cone below the opening in the cave roof. Bedding planes in the Ocala Limestone have been accentuated by selective dissolution, probably as a result of former locations of the water table.

**Fig. 7.51** The Devil's Den, a privately owned dive center located near Williston, Levy County. Note the rows of voids and dissolution-enhanced water-table positions or bedding planes in the Ocala Limestone. (Photograph courtesy of Mark Long (©marklongphotography.net))

#### 7.3.2.15 Carolina Bays: Hybrid Depressions

In west-central Florida there are at least ten large, shallow depressions that resemble "Carolina bays." Carolina bays are circular to oval wetland depressions that occur on the Atlantic Coastal Plain from New York to Florida. Swarms of the bays have a common orientation (long axis is southwest to northeast), which suggests a common origin, and they often have small, parabolic sand dunes at one end of the elliptical depressions (the presumed down-wind side). The common orientation and dunes suggest that wind, or some other unidirectional transport mechanism played a role in their formation.

The origin of the Carolina bays has been the subject of a long, and sometimes intense, debate. The origins of Carolina bays have been attributed to one, or a combination of, the following processes:

- Meteorite or comet impacts (Prouty 1935, 1952; Wells and Boyce 1953);
- Substrate dissolution and sinkhole development (LeGrand 1953; May and Warne 2004; Willoughby 2007);
- Deflation as a result of eolian erosion and transport (Grant et al. 1998; Ivester et al. 2007); and
- Marine sedimentation patterns (Cooke 1954).

See Eyton and Parkhurst (1975) for a summary of these diverse potential causes. Carolina bays north of Florida are thought to have formed sometime in the period from 10,000

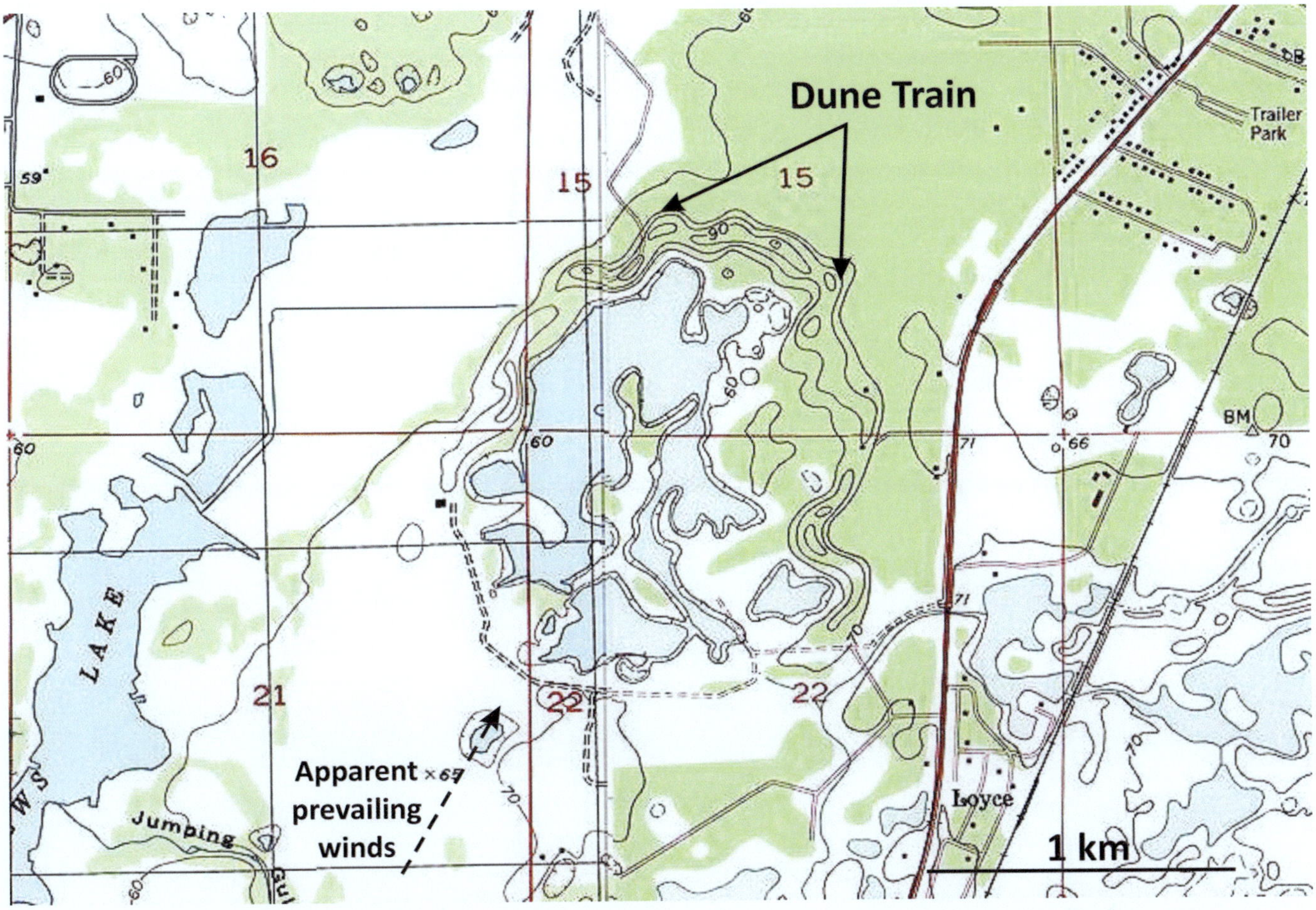


**Fig. 7.52** An example of a Carolina bay-like depression near Loyce, Pasco County. Note the parabolic dune train in the northeastern quadrant and the ephemeral lake/wetland in the center and southwestern quadrant. Location: 28.3935°N, 82.5086°W

to 100,000 years ago (Schalles et al. 1989; Brooks et al. 2001; Ivester et al. 2007).

The depressions in west-central Florida bear a strong resemblance to Carolina bays found in the coastal plain elsewhere in the U.S. They are circular to elliptical in outline, occur in clusters, and the depressions show a common orientation from southwest to northeast (Fig. 7.52, Table 7.7). They have parabolic, aeolian dunes in the northeast quadrant, the apparent down-wind quadrant of the depressions. Most have small ephemeral lakes or wetlands near the center or southwest third of the depression. The interiors of the wetlands are dotted with what appear to be sinkholes, and standard penetration or cone penetrometer testing within the large depressions often presents evidence of the early stages of sinkhole development. Including the dune trains, the depressions average approximately 2.5–3.0 km along the northeast axis and 1–1.5 km across on the northwest axis.

Upchurch et al. (2015) suggested that these depressions formed as a result of localized dewatering of sandy sediments, most likely as a result of sinkhole development during the late Pleistocene. The dry, fine sand was then entrained by prevailing winds and accumulated in down-wind locations to form the parabolic dunes. Therefore, Florida's Carolina bays have a hybrid origin with conditions having been created by late Pleistocene sinkhole development that allowed for large aeolian deflation depressions to form.

### 7.3.3 Pseudokarst Depressions

Many of the shallow land-surface depressions in Florida that have been explored in the subsurface are not true sinkholes formed by collapse or subsidence over a carbonate stratum. The depressions are rarely over a meter in depth. The origins of these depressions are difficult to discern, and, without accurate and appropriate data, the origins of the depressions can be misinterpreted.

Upchurch et al. (2015) investigated some of these depressions using (1) surface geophysics, (2) invasive testing with standard penetration test procedures, (3) development of

**Table 7.7** Summary of the types and properties of pseudokarst depressions observed on the Florida coastal plain

| Depression type | Mode of depression development | Rate of development | Probable evidence | Predominant scale and shape |
|---|---|---|---|---|
| ***"Hybrid depressions"*** | | | | |
| "Carolina bays" | 1. Lowering of water table, possibly as a result of sinkhole development, allows for fine sand to be eroded by wind stress. | Assumed to be formed on the scale of decades | 1. Lake, pond, or wetland depression within larger depression at point of maximum deflation | Long axis of large depression is up to 1000 m. Smaller sinkhole-like depressions within the deflation zone are typically more or less circular and up to 100 m in diameter. Depressions are ovoid with long axis oriented northeast to southwest, the apparent prevailing wind direction |
| | 2. Low basin forms at upwind end of depression and a parabolic dune train accumulates at the downwind end of the depression. | | 2. In Florida, the deepest area within the larger depression is located on the southwest end of the northeast to southwest oriented feature | |
| | 3. Deflation within the "bay" reveals depressions and possible relict sinkholes in the bottom of the larger, deflation-derived depression | | 3. Parabolic dunes developed on upwind, northeastern quarter of the larger depression | |
| ***Features without karst origins (Pseudokarst features)*** | | | | |
| Aeolian deflation depressions | Erosion by wind stresses within dune trains | May be rapid depending on wind stresses and vegetation cover | No subsurface expression; deflation zones typically parallel dune long axes | Depressions are complex and may be elongated, oriented parallel to the long axes of the dune train; the long axes of the depressions are typically less than 100 m |
| Depressions over paleosol and epikarst features | 1. Fine-grained, marine terrace sand deposited over, and infilling, existing depressions developed on the late Miocene to early Pliocene paleosol | Slow (hundreds to thousands of years) | 1. Depression floored by paleosol with no evidence of deeper limestone dissolution | Scale varies with circular depressions up to 200 m and streams kilometers in length. Depressions are circular to linear; infilled stream systems are often occupied by modern streams; |
| | 2. Consolidation and minor, early compaction of the relatively thicker sand within the infilled depressions causes development of depressions in the land surface | | 2. Infilling sediments down warped by compaction; wetland or stream sediments may be included within fill materials | |
| Pedogenetic depressions | Late Miocene to early Pliocene alteration of Miocene smectite to kaolinite, a pedogenetic process, causes volume reduction and land-surface depressions | Very slow (thousands to millions of years) | Clay flooring and bordering depression is kaolinite rich as compared to more distant, smectite-rich sediments | Scale unknown; presumed to be more or less circular |

Modified from Upchurch et al. (2015, Table 1)

cross sections of depressions exposed in excavations, quarries, and borrow pits, and (4) high resolution geologic mapping and categorized these shallow depressions as to origin. Their results indicated that the shallow (less than 1 m deep) depressions have multiple origins (Table 7.7), including pseudokarst (Field 2002; Halladay 2007). Cover-subsidence and suffosion sinkholes, which are true karst features, are discussed above. At least three other, non-karstic mechanisms for formation of shallow depressions were identified (Table 7.7). The three pseudokarst landforms are discussed below.

**Aeolian Depressions** Florida has extensive, Pleistocene and Holocene aeolian dune fields. These are related to the modern coasts and to Plio-Pleistocene marine terraces. As is normal in dune fields, these dunes are characterized by numerous inter-dune and dune-slope depressions. They have more-or-less circular to linear outlines and present on topographic maps and in the field as closed depressions, often with wetlands. The depressions between dune ridges are depositional. The depressions within dunes are deflation features caused by wind erosion.

Because of the closed depressions on topographic maps, they have been mistaken for sinkholes, which are commonly interspersed with the deflation depressions. Only subsurface testing can differentiate them.

**Depressions Over Paleosols** One of the most remarkable features of west-central Florida's Hardee Upland physiographic province can best be appreciated from the air. As one flies over the area, the large number of circular, wetland depressions and streams with segments showing trellis drain-

**Fig. 7.53** Aerial view of depressions and trellised drainage in a portion of the Hardee Upland, Hardee County. Many of the wetlands and streams are developed over somewhat thickened sands in pre-existing low areas developed on the Mio-Pliocene paleosol

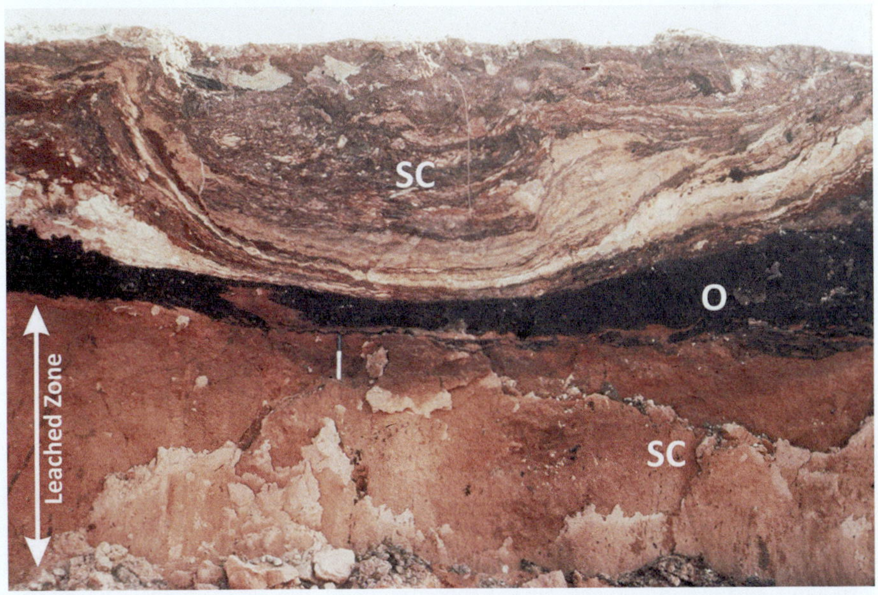


**Fig. 7.54** Sediment fill in a sag feature exposed at the Kingsford phosphate mine near Bradley Junction, Polk County. O = organic sediment, SC = clayey sand. Pen gives scale

age patterns becomes evident (Fig. 7.53). Observation of these depressions in cross section in phosphate mine cuts (Fig. 7.54) and in standard penetration test borings indicates that many of these features are associated with the Mio-Pliocene paleosol (Altschuler et al. 1951; Van Kauwenbergh et al. 1990; Sect. 6.9.4). When the paleosol developed, closed depressions and stream channels were common (Cathcart 1963a, b, c, 1964, 1966). Upon transgression by Plio-Pleistocene seas, these depressions were buried by marine terrace sand. Upon deposition, the terrace sand was somewhat thicker in these depressions relative to elsewhere on the ancient land surface.

During early sediment compaction, the thickness of the marine terrace, sandy sediments dictates the amount of change in thickness of the sediment overlying the paleosol. While the amount of compaction is minor and only changes the relative density of the sediments by a small amount, the result is a shallow depression. For example, if the terrace sand body were 1 m in thickness when it was deposited and shortly after deposition it compacted under its own weight, groundwater percolation, and bioturbation by 5%, the resulting thickness would be 0.95 m, which would likely not be readily discernable. If, however, the sediment were 10 m in thickness when deposited, the post-settlement and compaction thickness would be 9.5 m and a 0.5 m depression would be visible. With a shallow water-table aquifer, this depression would likely be seasonally wet. This process has resulted in the patterns of drainage control and shallow depressions that are so common on the Hardee Upland (Fig. 7.53).

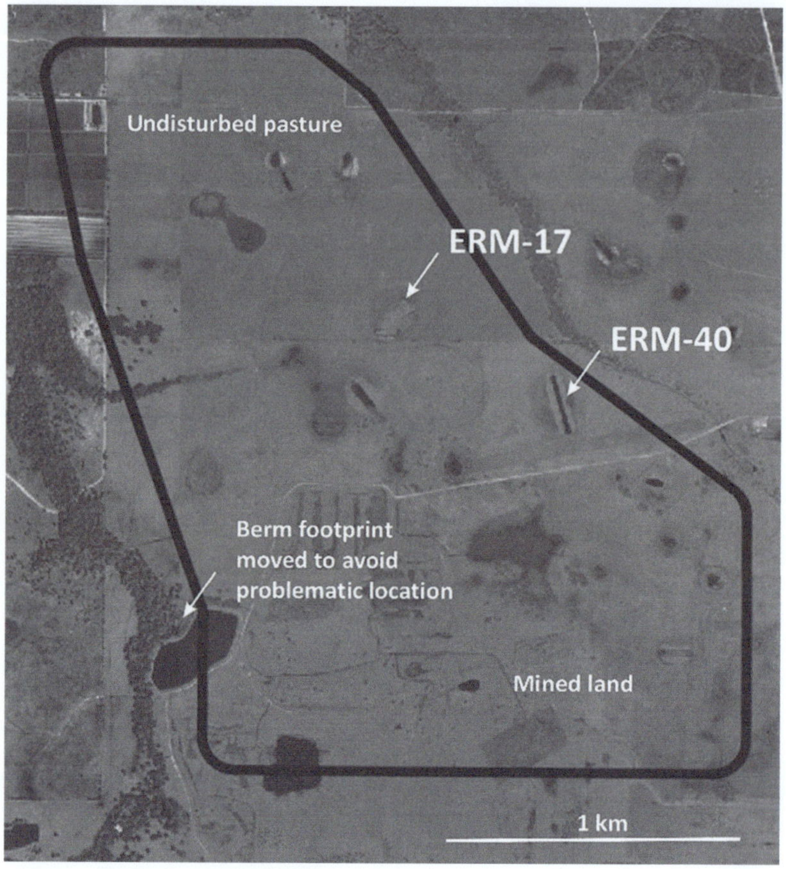


**Fig. 7.55** Aerial view of the C.W. "Bill" Young Reservoir site before construction showing locations of concern. Dark line is the final location of the reservoir berm. Note the small stream channels and circular wetlands

This process was identified on the Hardee Upland by Cathcart (1963b) who noted these features and stated that:

> The subsurface topography of the Hawthorn is similar to the present surface; ancestors of the present surface streams flowed on the surface of the Hawthorn at or close to their present positions. (Cathcart 1963b, p.1).

In other words, the drainage and depressions on the modern land surface mimic the buried topography of the late Miocene/early Pliocene land surface. Post-depositional settlement is greatest where the Plio-Pleistocene sand mantle is locally thickest. With settlement, new drainage systems and wetlands occupy the resulting depressions. Where the drainage was rectilinear or trellis-like (Fig. 7.53), it was probably developed on weak fractures or other nearly linear features developed in the more cohesive and carbonate-rich sediments of the underlying Miocene Hawthorn Group.

The most comprehensive investigation of these depressions was conducted at the site of the C.W. "Bill" Young Regional Reservoir in southeastern Hillsborough County, Florida. Numerous circular wetlands and several small streams crossed the site (Fig. 7.55). A geophysical and stratigraphic study was used to determine the origins and risks of these features to the reservoir (Upchurch et al. 1999; Dobecki and Upchurch 2010; Alfieri et al. 2018). Every photolinear intersection, each depression, and the stream beds were tested by ground penetrating radar, seismic shear-wave analysis, and seismic reflection and refraction and then evaluated with standard penetration testing.

As shown in Fig. 7.55, one problematic cover-collapse sinkhole resulted in moving the berm to avoid the risk. All of the other low areas were determined to reflect depressions in the underlying Mio-Pliocene paleosol or inactive paleosinkholes. Two of the depressions, indicated by the ERM boring designations on Fig. 7.55, were found to be over ancient sinkhole depressions. One of the relict sinkholes was filled with Miocene smectitic clay and the other with Pliocene quartz sand. Neither showed evidence of modern activity and the sediments that infilled these ancient features were competent. The streams and all other wetland depressions were developed over depressions in the upper surface of the paleosol but had no subsurface expression in the underlying Miocene and older strata.

**Pedogenetic Depressions** Finally, Altschuler et al. (1956, 1963) and Isphording (1984) have suggested that sediment volume reductions accompanying alteration of smectite to kaolinite within the Miocene sediments of Florida have caused shallow land-surface depressions (Table 7.7). It is likely that these depressions were the precursors of the

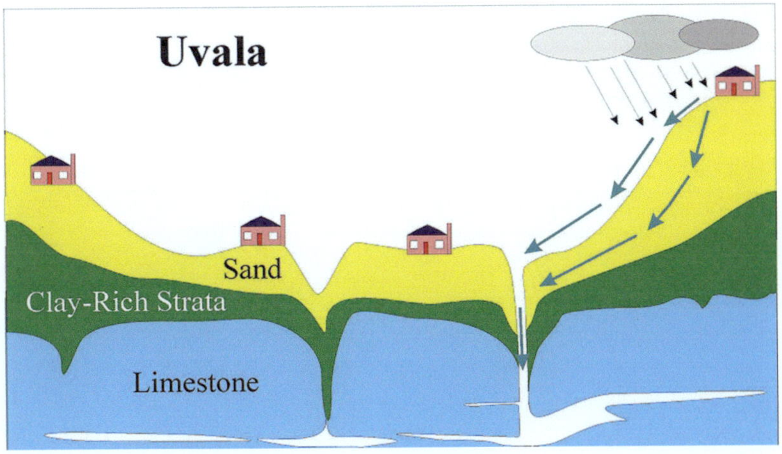


**Fig. 7.56** Cross section showing development of an uvala with coalescent sinkholes and internal drainage. Arrows suggest water pathways

depressions discussed above since they would have developed during the late Miocene/early Pliocene pedogenic event. Small-scale compaction and pedogenic alterations cannot be differentiated as causes of the pseudokarst depressions of the Hardee Upland.

### 7.3.4 Coalescent Sinkholes

Where the cover is thick and sinkhole development widespread, it is inevitable that the depressions formed in the cover will begin to interfere with each other and coalesce. When this happens, large, irregularly shaped, depressions develop. These features may be quite large, depending on the number of sinkholes that have coalesced, the thickness of the cover, and the age of the sinkhole complex. In Florida, there are many such coalesced sinks that exceed 2–3 km$^2$ in area.

In Europe, these features have been given names that reflect their geometries, and they are often located in karst terrains where cover is negligible. Unlike the European examples, Florida's coalescent sinkhole complexes are mantled with cover materials and bare limestone is rarely seen. In spite of this potential difference, the European terminology is utilized in Florida.

#### 7.3.4.1 Uvalas

An uvala is a coalescent sinkhole complex (Fig. 7.56) where the floor of the large basin is irregular with multiple depressions representing sinkholes that have developed within the larger depression. Note that the larger depression is internally drained and that flooding of the depressions within the uvala is common. Some of the sinkholes may be active, while others may be plugged and inactive.

Flooding of uvalas can be problematic in Florida because the cover tends to mute contours and the depressions appear broad and have gentle slopes. As a result, many Florida structures have been constructed within uvalas where they are at risk of flooding and of future sinkhole development or reactivation.

**Fig. 7.57** Uvalas and sinkholes (shaded areas) in a karst terrain near Gainesville, western Alachua County. Black dot is located at 29.6599°N, 82.5150°W. Contour interval is 5 feet (1.5 m)

Figure 7.57 is a topographic map of an area just west of Gainesville in Alachua County. This area is highly karstified with numerous uvalas, individual sinkholes, and cone-shaped hills. It lies near the toe of the Cody Scarp where the cover is

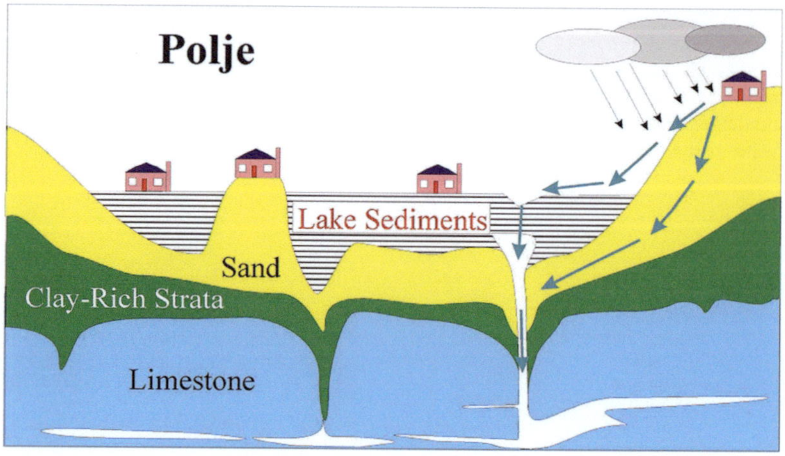


**Fig. 7.58** Cross section showing development of a Florida polje. Arrows indicate possible pathways of water

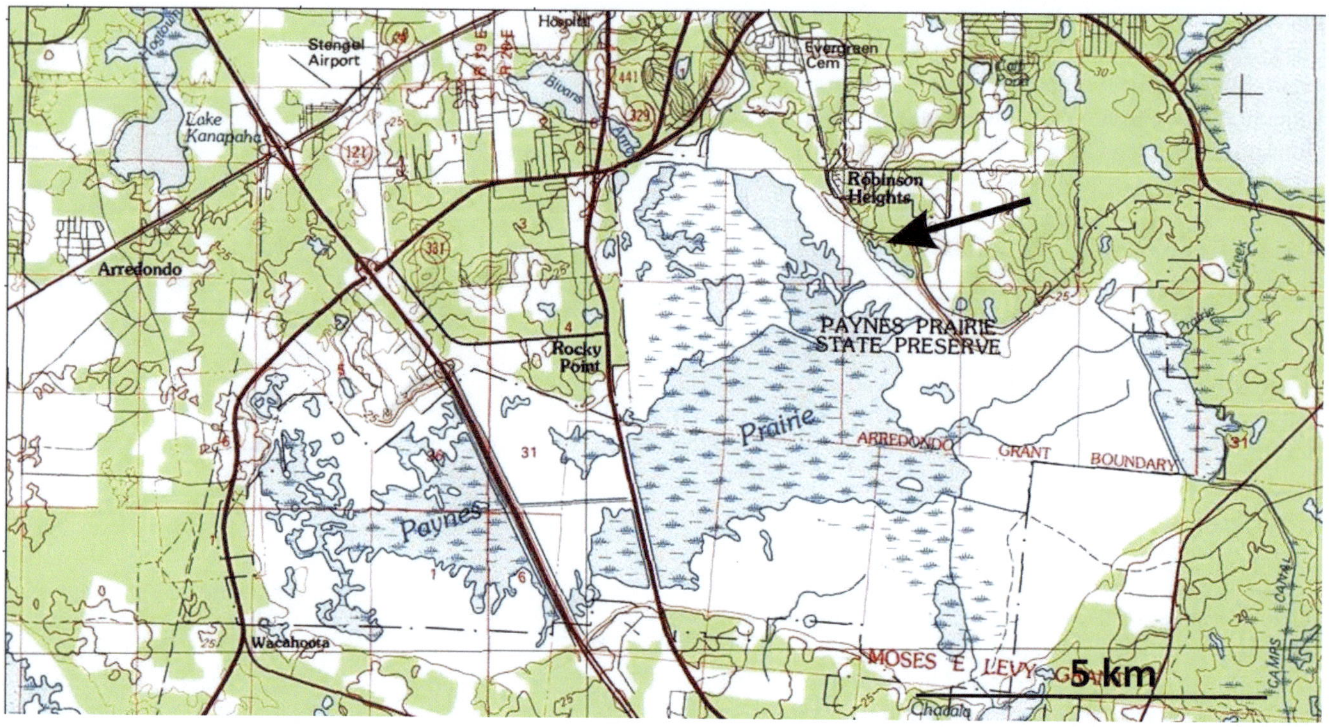


**Fig. 7.59** Topographic map of Paynes Prairie, which is located just south of Gainesville, Alachua County. Arrow indicates the location of the Alachua Sink, the drain for the prairie. Location: 29.581°N, 82.322°W

thick and large sinkholes and uvalas abound. The closed depressions are shaded on the map. Note the many irregularly shaped uvalas. The largest of the uvalas are over a kilometer in maximum dimension and contain as many as ten smaller depressions within the larger uvala depression.

The small, sharply defined depressions with steep sides in the northwest quadrant of Fig. 7.57 are abandoned hard-rock phosphate mines. The phosphate deposits formed in relict sinkholes during the Mio-Pliocene weathering event.

#### 7.3.4.2 Poljes

The definitions of a polje differ (Monroe 1970; Field 2002; Bonacci 2004), but the one factor that all polje definitions have in common is that they are complex, coalesced sinkholes with a flat bottom, usually as a result of post-settlement wetland or lake deposition (Bögli 1980; White 1988; Ford and Williams 1989). Some definitions suggest that a polje must have rocky walls, in many cases as a result of faulting or karst collapse. This is not the case in Florida. With its

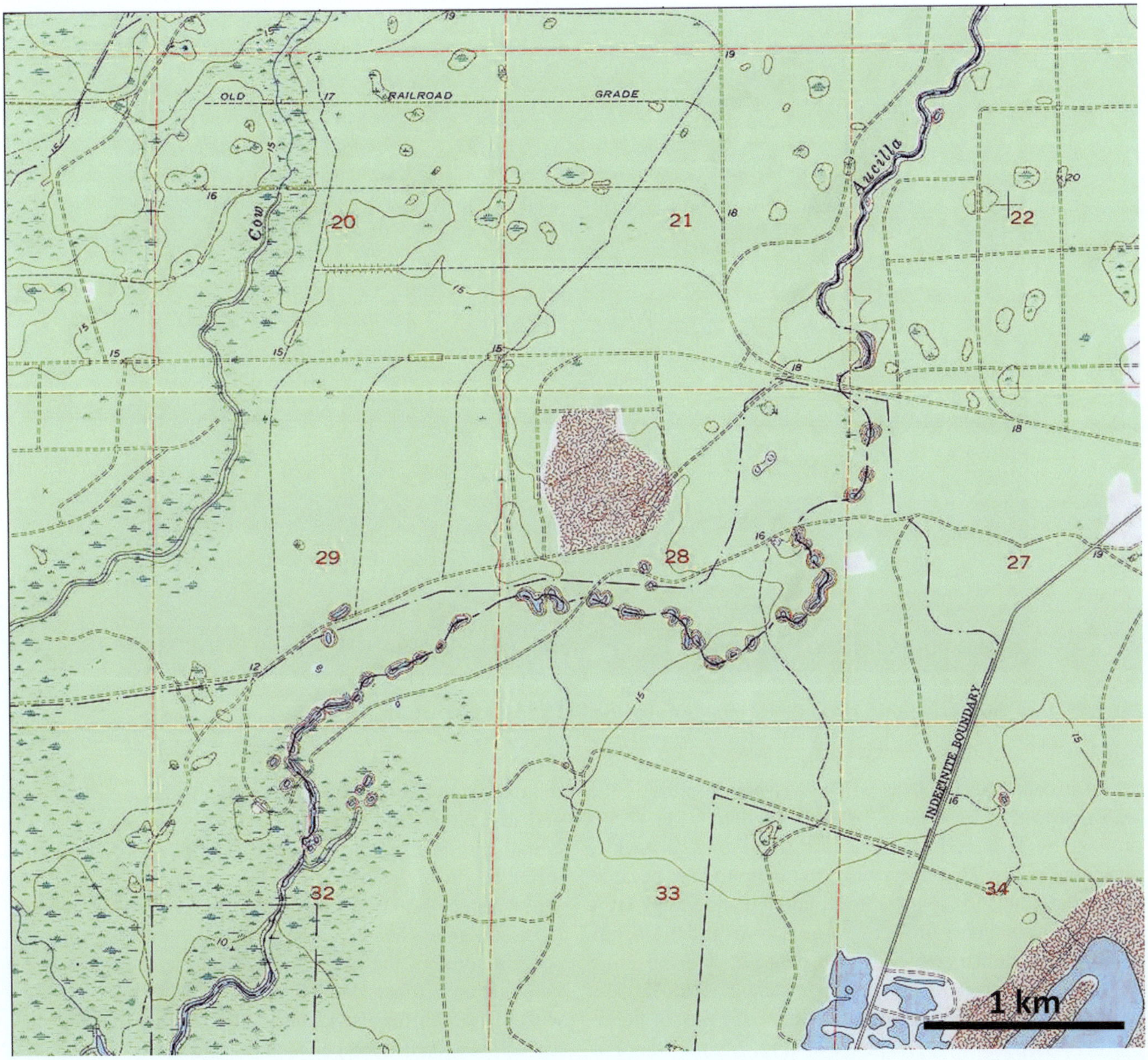


**Fig. 7.60** The string of karst windows on the Aucilla River upstream from Nutall Rise, Taylor County. The river is underground throughout this reach and is exposed in a series of collapse sinkholes (karst windows; Fig. 7.60). The abandoned mine shown in the center of the map is located at 30.1970°N, 83.9365°W

wide-spread sand mantle, most topographic, karst features in Florida have sand draped over them, which causes margins of the features to be rounded with gentle slopes and masks bedrock conditions related to the origin of the depression (Fig. 7.58). Because many important karst depressions in Florida are clearly coalesced sinkhole complexes that have been filled with lake, swamp, or marsh sediments, we have termed them poljes.

With the well-developed cover consisting of cohesionless, marine terrace sand, abundant rainfall, and high water tables, many of the large, coalescent sinkhole complexes in Florida contain lakes and freshwater marshes, which are locally known as wet prairies. As a result of sedimentation in these large depressions, the depressions have flat floors and a blanket of lacustrine (lake) or paludal (marsh) sediment masks the underlying, irregular karst floor of the depression (Fig. 7.58).

Florida poljes are common at the toes of karst escarpments (Chap. 8). Where runoff and streams pass from topographic highlands into the large depressions, lakes and shallow, wet prairies are common. Unlike poljes in bare karst settings, the uplands surrounding Florida's poljes have gentle and rounded slopes.

**Fig. 7.61** Half Mile Rise, a karst window located on the Aucilla River at Half Mile Rise. (Photograph courtesy of J.S. Dunbar)

An excellent example of a lake/wet prairie polje is the well-known Paynes Prairie, a large wetland (Fig. 7.59) crossed by I-75 and US 441 just south of Gainesville, Alachua County. This feature is a wet prairie today because water that drains to the polje flows into the underlying UFA by way of a sinkhole (arrow, Fig. 7.59) known as the Alachua Sink. In the 1800s, the sink was plugged, and the polje became a lake large enough to support steamboat traffic. The sinkhole is maintained today to minimize the risk of flooding. As shown in the topographic map (Fig. 7.59), the floor of Paynes Prairie is flat as a result of accumulations of paludal and lacustrine sediments.

#### 7.3.4.3 Collapsed-Cave Systems

Collapsed cave systems are another form of coalescent subsidence feature found in northern Florida. These features appear to have formed by development of a string of rock-collapse sinkholes over a length of cave. The karst windows ultimately coalesce, resulting in a linear, karst valley. They appear to form preferentially over caves that transport organic-rich surface water that enters the aquifer at a swallet (sinkhole into which a stream drains). These truly underground rivers ultimately discharge at a spring.

Figure 7.60 illustrates a string of karst windows that mark the underground pathway of the Aucilla River (Balfour 2002; Dunbar 2016). Under normal discharge conditions, the river goes underground at a swallet just north of the map and discharges at a large spring/resurgence known as the Nutall Rise just south of the map area. The roof of the cave through which the river passes is slowly coalescing as a series of rock-collapse sinkholes. Several of these rock-collapse sinkholes (i.e., Fig. 7.61) are well known because they contain some of the oldest (~14,500 years before present) Paleoindian artifacts known in North America in context with mammoth fossils (Dunbar 2016). Normally, the shallow depressional trace between the individual sinkholes is relatively dry. However, during extreme runoff events, the infiltration capacity of the sinks is exceeded, and overflow follows the trace of the depressions overland. The string of karst windows on the Aucilla River appears to be in the early stages of collapse of the section where the river flows underground.

Figure 7.62 illustrates an example of a karst feature that represents the final product of the collapse of an underground cave segment. These features occur at several locations along the Suwannee and Santa Fe rivers in Suwannee and Columbia counties. They are more-or-less linear, typically oriented in the northeast-southwest or northwest-southeast orientations of the regional fracture patterns, and consist of elongated, closed depressions with steep sides and occasional piles of rubble that appear to represent breakdown. Given the geometries of the spring runs that exist today on these same rivers,

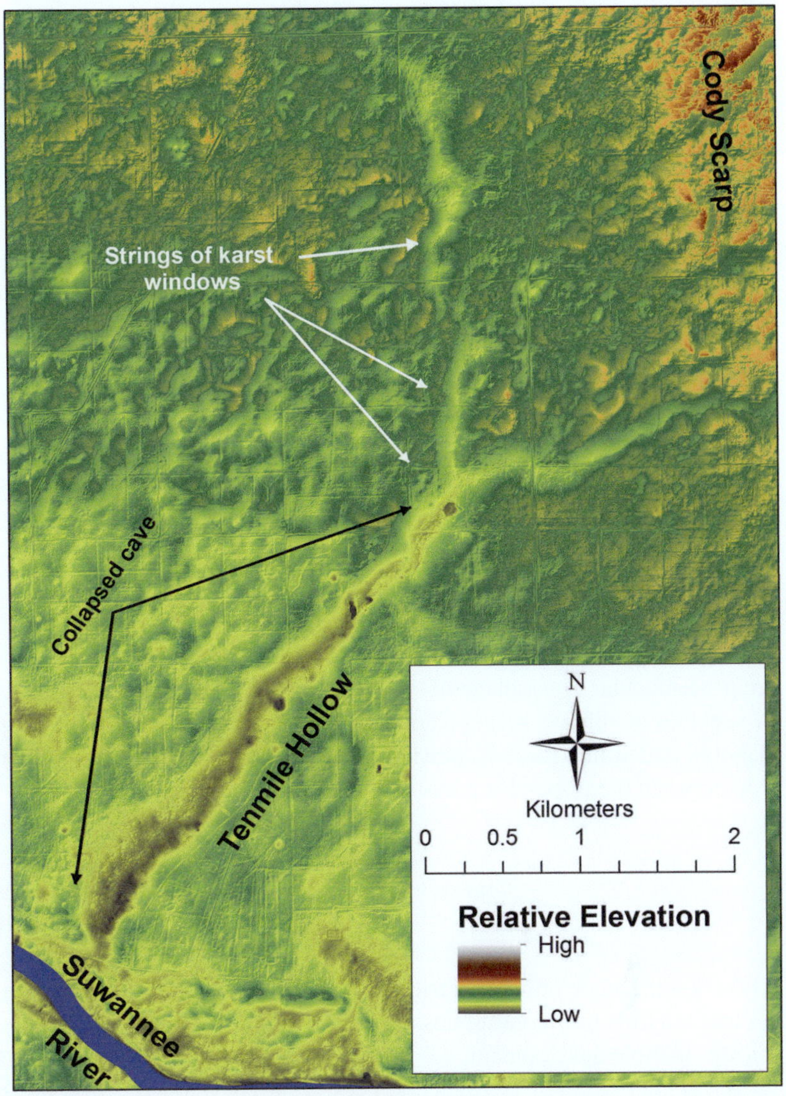


**Fig. 7.62** LIDAR topography of Tenmile Hollow, a collapsed cave system near Luraville, Suwannee County. Location: 30.102°N, 83.085°W

one can surmise that these abandoned sloughs and hollows were once caves that fed spring flow to the rivers. Dissolution of the roofs of the caverns caused extension of the cavern up gradient and collapse down gradient, thus creating a form of headward erosion. Today, these strange valleys are often dry and pockmarked by small sinkholes or they are sloughs with occasional stormwater drainage and wet areas in their floors.

These large, collapsed caves have one thing in common. They, or their drainage systems, originate in the Cody Scarp, and they are connected to swallets that drain streams into the UFA. Water in these streams is typically organic-rich and potentially able to remain chemically aggressive to the surrounding limestone. The underground streams usually retain their coloration from humic substances as they emerge at resurgences downgradient. More will be said about this phenomenon in the following chapter under the discussions of karst bridges and traces.

## 7.4 Mechanisms of Cover Failure

Now that we have discussed the subsidence features common to the Florida landscape, it is appropriate to summarize the mechanisms of cover failure that result in sinkhole development. Recall that the typical stratigraphic section in sinkhole-prone central and northern Florida consists of a thin stratum of clay-rich sediments overlying limestone. This clay-rich stratum may be either intact Hawthorn Group sediments or residuum derived from weathering of the Hawthorn. Plio-Pleistocene quartz sand deposits are usually found above the clay-rich sediments. These cover sediments consist of sand and clay in varying proportions, and there may be alternating layers of sandy and clayey sediments. In this section we discuss the mechanisms that are likely to cause failure of these cover sediments and migration into underlying voids to form sinkholes.

There are four different processes that may contribute to sinkhole development through the cover materials in Florida. These are:

1. Suffosion or raveling of non-cohesive sediments into voids created by dissolution of carbonate materials;
2. Piping as a result of groundwater flow;
3. Settlement of unsupported strata; and
4. Disaggregation or movement of cohesive sediments.

### 7.4.1 Fundamentals

By the 1950s, development of the theory of sediment grain transport was well underway (i.e., Bagnold 1956). This section provides a simple discussion of the process of grain movement and is largely derived from Blatt et al. (1972).

Bagnold's (1956) work showed that cohesionless sediment grains subjected to stress tend to produce forces that are perpendicular to the plane of shearing causing the grains to separate or dilitate perpendicular to the plane of shear. Bagnold called this force dispersive pressure. The relationship between the dispersive pressure (P) and shear stress (T) is

$$\frac{T}{P} = \tan\alpha, \tag{7.2}$$

where $\alpha$ is the angle of internal friction. The angle of internal friction controls how cohesionless, granular sediment ravels during sinkhole development.

Terzaghi (1925) and Terzaghi and Peck (1948) developed the concept that the stress on buried sediment is a function of the mass of sediment bearing at that depth. They also argued that the buoyant effect of pore water at the depth of interest tends to reduce the stress caused by the sediment load. Effective stress (Terzaghi and Peck 1948; Holtz and Kovacs 1981) is the vertical stress applied to materials in the subsurface, adjusted for the buoyancy effects of the pore water within that soil or sediment. That is, the effective stress ($s_e$) is the total lithostatic load ($s_l$; vertical stress caused by the load imposed by overlying soil or sediment) adjusted for the effects of buoyancy on sediment particles caused by immersion in water ($s_w$)

$$s_e = s_l - s_w. \tag{7.3}$$

The effective stress, therefore, increases with depth of burial. This stress is not entirely a result of the vertical load, however. The sediments surrounding the materials of interest are also subjected to pressures related to depth of burial and this pressure field creates confining pressures as well as vertical stresses. Consequently, the concept of effective stress is a simple representation of a complex stress field within which sediment exists at some burial depth.

The effective stress (Eq. 7.3) is an important concept for understanding sinkhole formation in cover materials. The effective stress is important for understanding whether cover sediment movement will occur, and changes in effective stress from fluctuations of the water table caused by rainfall events, groundwater extraction, or changes in drainage can trigger sinkhole formation.

Blatt et al. (1972) related the angle of internal friction to effective stress and shear by the following relationship

$$\tau = c + s_e \tan\phi, \tag{7.4}$$

where $\tau$ is the shear stress on the sediment, c is the cohesion of the sediment, $s_e$ is the effective stress, and $\phi$ is the angle of internal friction.

This important relationship suggests some of the triggering mechanisms for sinkhole formation. Movement of the cover sediment can be exacerbated by:

- Reduction of the cohesion of the sediment;
- Changing the effective stress; or
- Reducing the coefficient of internal friction.

Changes in the coefficient of internal friction develop by such processes as altering the packing of the sediment grains or removal of the organic fraction of the sediment. Increasing the effective stress by increasing the overburden load or lowering the water table is known to cause sinkhole activity. Cohesion is a complicated subject, but sandy sediments with minimal cohesion can be affected by alteration of the hydration state of included clay minerals, oxidation or reduction of iron minerals, and similar processes that alter the fabric and structure of the sediment. Cohesive sediments, such as clays, may lose cohesion during hydration and expansion, dispersion of clay particles, and similar processes.

### 7.4.2 Raveling and Suffosion

#### 7.4.2.1 Definitions of Raveling and Suffosion

As noted above, raveling is the simple movement of cohesionless, granular sediment particles into a void space. The best example of raveling is the movement of sand in an hourglass (Fig. 7.63). In an hourglass, the dry, well-sorted, fine sand moves toward the throat of the hourglass along two pathways: (1) vertical displacement and (2) slope creep (arrows in the upper volume of the hourglass, Fig. 7.63).

In the hourglass, the sand is dry, sub-rounded to rounded, fine and extremely well sorted, so each grain can freely roll and slide toward the aperture of the hourglass. In nature, (1) the sand is less well sorted, (2) the individual grains are more angular, (3) the sediment includes a variety of different grain sizes, and (4) the moisture content ranges from damp to water-saturated, so movement is complicated and much less efficient than in the hourglass. Moving groundwater helps overcome resistance to movement by natural, moist sand movement. Groundwater movement toward a sinkhole aven (represented by the throat in the hourglass) is normally required to lubricate and enable transport of individual sediment particles. One of the conditions that one verifies in determining if a sinkhole is developing is, therefore, the presence of a downward, vertical hydraulic gradient for groundwater flow. Areas with upward or zero groundwater gradients are much less likely to develop sinkholes than regions with downward gradients.

It is important to understand that the process of raveling requires that each particle be more-or-less free to migrate as part of a queue of non-cohesive grains moving under the force of gravity and, normally, water movement. Clay does not ravel because it is cohesive, and the individual particles are electromagnetically attracted to each other.

The term suffosion is synonymous with raveling as described above. This term is used by karst scientists worldwide (Sweeting 1981; White 1988; Ford and Williams 1989; Williams 2004), and its use creates somewhat less confusion than the term "raveling", which is largely used in the United States, especially in Florida.

Many karst textbooks (i.e., Sweeting 1981; White 1988; Ford and Williams 1989; Williams 2004) either state or imply that there are two basic processes by which sinkholes develop. These are suffosion and collapse. This dualistic concept is too general for application to Florida cover materials, which include both cohesionless sand and a wide range of more-or-less cohesive sediments from clayey sand to clay, all with variable carbonate content. It is important to understand the various mechanisms by which cover materials can fail and participate in the subsidence process, especially as the presence of highly cohesive clay may interrupt or modify the subsidence process.

#### 7.4.2.2 Mechanism for Development of a Suffosion Sinkhole

Figure 7.64 illustrates a sequence of sinkhole-development events as a result of suffosion. In this example, we assume that a pathway exists for raveling of the cohesionless sand into a void in the underlying limestone. The mechanisms by which this pathway may develop in the more cohesive sediments (i.e., clayey sand and clay) are discussed below. See Sect. 7.3.2.3 for a discussion of formation of a suffosion sinkhole from the prospective of the underlying limestone.

In Panel A (Fig. 7.64) conditions are favorable for a cover-collapse sinkhole to develop by suffosion, but the raveling process and sinkhole formation have not begun. What will become the sinkhole aven, or raveling pathway, has developed through the cohesive sediments above the limestone.

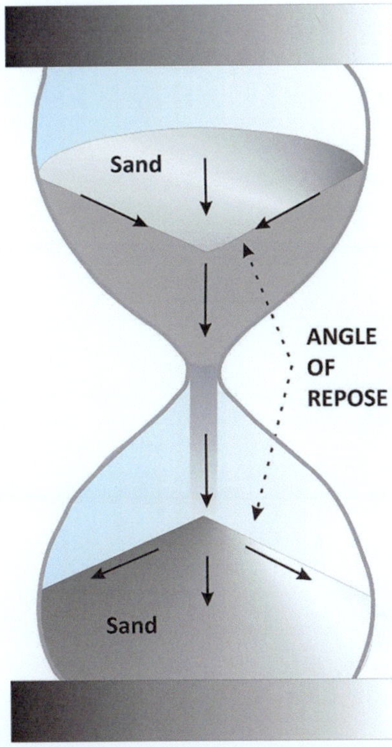


Fig. 7.63 Hourglass showing vertical raveling of sand through the throat and lateral creep in the depression at the top and on the cone at the bottom. Arrows indicate directions of sand movement

Note that the water table elevation is shown in the figure. There is no depression in the water table, which suggests that groundwater flow downward, through the opening has not begun.

In Panel B (Fig. 7.64) raveling is underway with slight loosening, or dilatation (Bolton 1986), of the sand and migration of sand to the aven. Dilatation is the process where the sand grains move apart slightly causing the effective porosity of the sand to increase and packing to decrease. As a result, the strength of the sand begins to decrease. Panel B is included to emphasize that, since the raveling begins at the base of the sand and top of the aven, suffosion is a "bottom-up" process. The loosest sand would be at the base of the ravel zone, near the aven.

Increasing the effective porosity also enhances the efficiency of water to move through and with the sand. Note that there is a depression in the water table. Groundwater, which was originally separated from the underlying limestone aquifer by the confining properties of the clay, is now flowing downward, into the limestone aquifer, under a downward vertical hydraulic gradient. In the process of groundwater flowing downward, the movement of sediment (raveling or suffosion) is enhanced.

In Panel C (Fig. 7.64) raveling is well underway. The sand in the raveling zone is fully dilitated and a cone-shaped ravel zone has developed. If a standard penetration test were

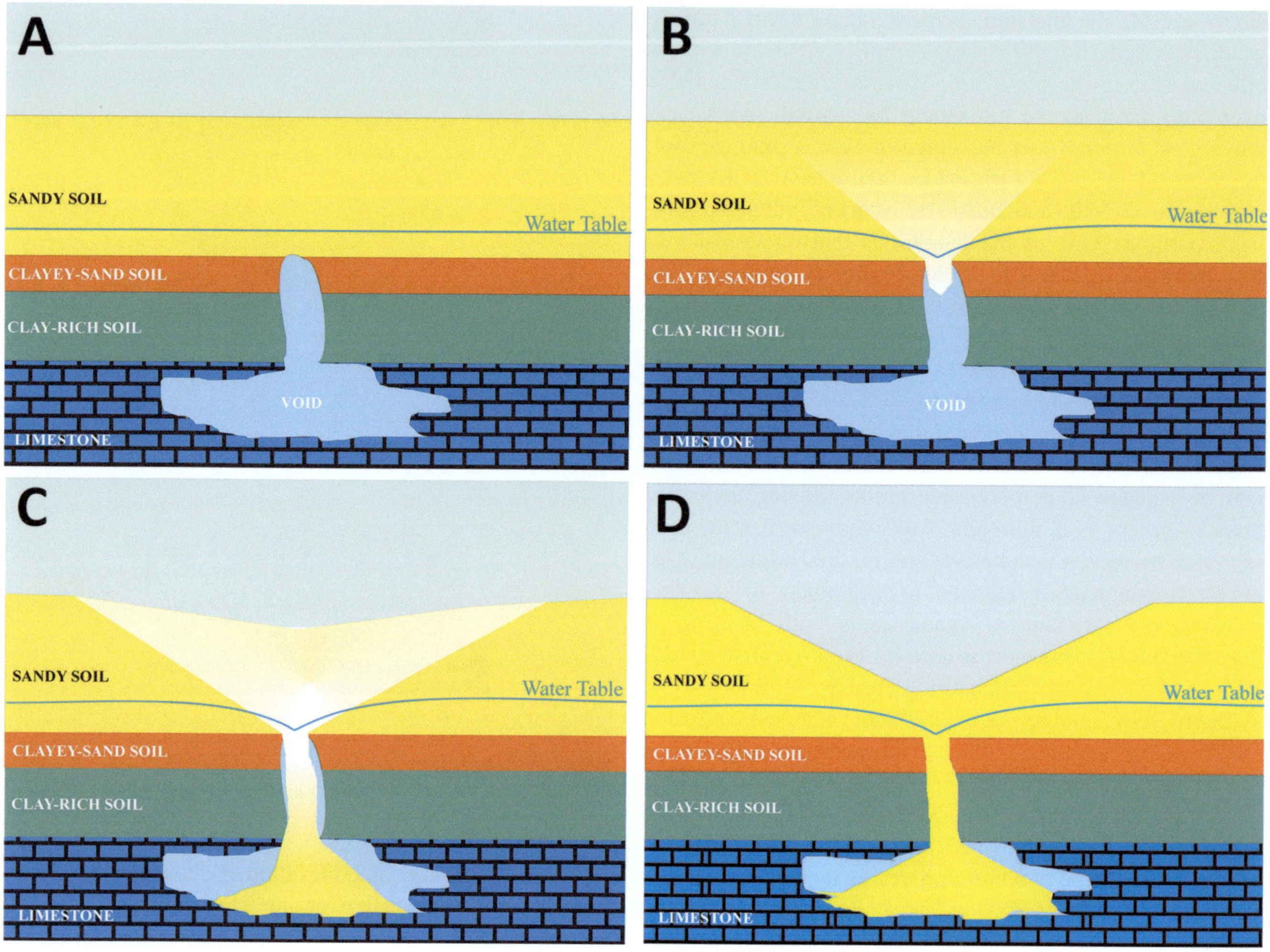


**Fig. 7.64** Sequence of events in development of a cover-collapse sinkhole by suffosion. (**a**) An aven, or pathway, has developed through the cohesive sediments above the limestone. (**b**) Raveling with slight dilatation of the sand is beginning. (**c**) Raveling is underway and the sand in the raveling zone is dilitated. (**d**) Suffosion is complete with the void filled, aven plugged by the cone of raveled sand, and a land surface depression

advanced within the cone of raveled and dilitated sand, there would be a pattern of penetration resistances indicating a decline in relative sediment strength with depth, and, relatively speaking, the weakest sand would be near the aven.

In Panel D (Fig. 7.64) suffosion is nearly complete, and the sinkhole that developed during stages B and C has visibly stopped growing by raveling. The slopes of the sinkhole are at or near the angle of repose of the sediment. In this example, the underlying void is sufficiently large that a mound of raveled sand has accumulated at the bottom of the void. It, too, is at its angle of repose.

Note that the volume of raveled sediment is controlled by two important factors: (1) the volume of the void; and (2) height of the void relative to the height of the cone of sediment accumulated on the bottom of the void. If the top of the cone of accumulated sediment intersects and blocks the aven, the amount of raveled sediment will be limited by the accumulated sediment regardless of the ultimate size of the void.

On Panel D (Fig. 7.64) groundwater flow from the water-table aquifer to the underlying limestone aquifer continues. Inactive sinkholes, such as this, often serve as pathways for long-term recharge and contamination of the underlying aquifer (Chen et al. 1995a, b).

Even through the void has not been completely filled by the raveling sand in Panel D (Fig. 7.63), the mound of raveled sand has plugged the aven, thus arresting significant development of the sink for the time being. If the mound of accumulated sediment settles by compaction or creep or if groundwater flow reduces the volume of accumulated sediment by transport or dissolution, the blocked aven may reopen and sinkhole development may be reinitiated.

At the land surface, the sides of the sinkhole may continue to develop more gradual slopes owing to soil creep,

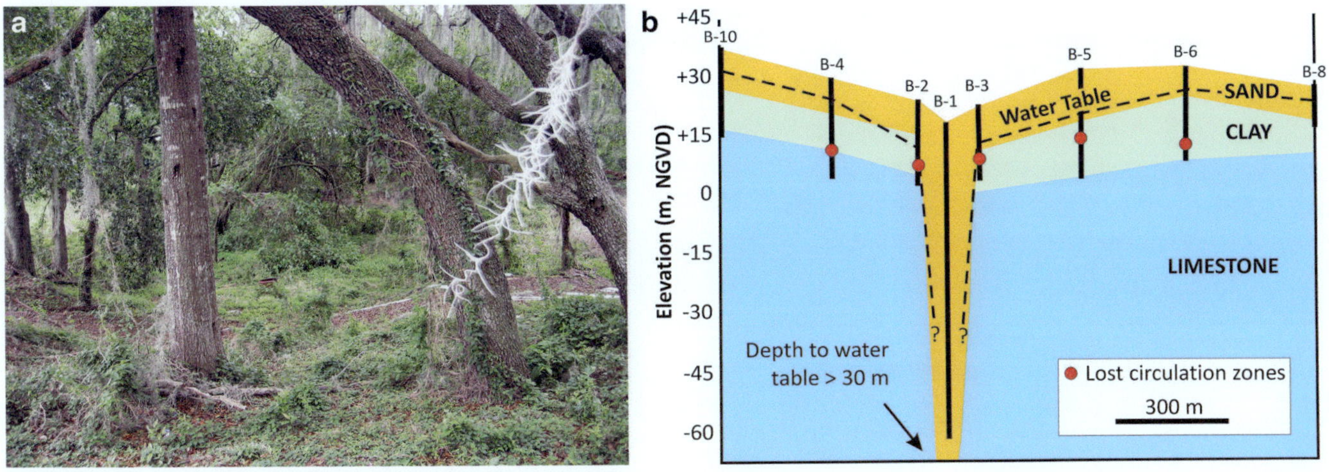


**Fig. 7.65** (**a**) The Mulrennen Road sinkhole, which is approximately 4 m deep. The sinkhole is located in east central Hillsborough County. (**b**) Cross section of the Mulrennen sink. (From Jones and Upchurch (1993))

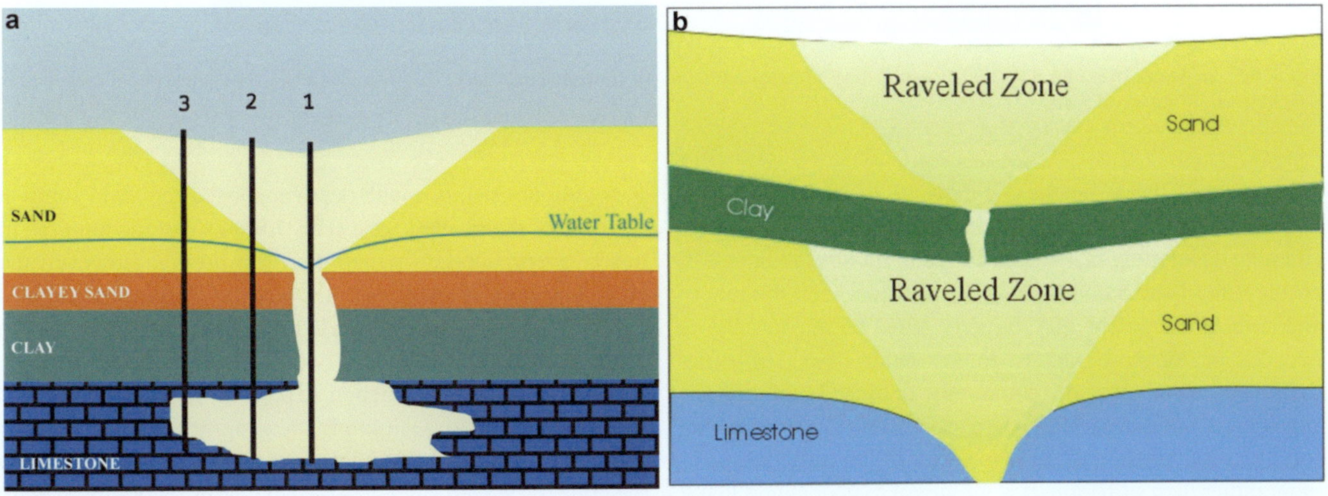


**Fig. 7.66** (**a**) Standard penetration testing in a suffosion-dominated sinkhole. Numbers refer to different testing locations (see text). (**b**) Multiple raveled zones caused by stacked avens

surface-water runoff, and bioturbation, but no further raveling is expected to occur in the scenario as long as the aven is choked with sediment.

#### 7.4.2.3 Rate of Suffosion

As suggested by Fig. 7.63, there are several factors that affect how quickly suffosion may occur and the ultimate extent of the resulting sinkhole. These include:

- Texture (grain size, sorting, and grain angularity) and angle of repose of the cohesionless sand;
- Size of the aven or aperture through which raveling occurs;
- Thickness and imposed load of the sand column;
- Depth of the water table and leakance through an aquitard between overlying and underlying aquifers;
- Timing and magnitude of changes in the vertical hydraulic gradient and effective pressure in the sand column;
- Flow dynamics of groundwater in the void; and
- Size (volume and height at the aven) of the void into which the sand ravels.

With so many variables controlling suffosion and development of the depression at the land surface, the rate of development of the sinkhole is highly variable, spatially and temporally, and relatively unpredictable. If (1) the vertical hydraulic gradient results in groundwater flow sufficient to transport sediment, (2) the aven is of sufficient width, and (3)

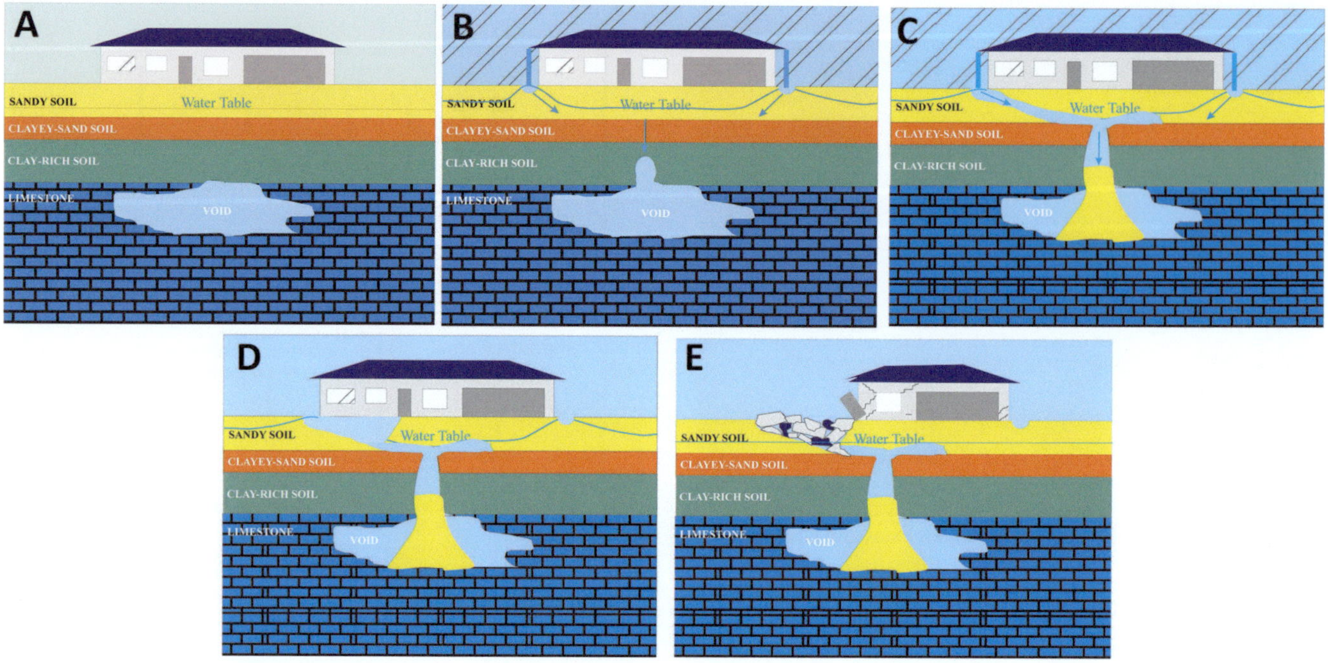


**Fig. 7.67** Sequence of events where a suffosion zone migrates laterally to a concentrated source of water. See text for a discussion of each step in development of the ravel zone

the texture of the sediment is conducive to movement, then suffosion of the cohesionless sediment may be rapid with the sinkhole developing in some instances in minutes to hours. If, on the other hand, conditions limit suffosion, then the rate of sinkhole development may be very slow, taking months to decades or longer to occur. In the latter case, suffosion-dominated sinkholes are commonly confused with cover-subsidence sinkholes, which are also slow to develop, but form by dissolution at the upper surface of the carbonate sediment rather than suffosion. It is important to understand that it is not possible to predict the timing of onset of sinkhole development or the rate at which it progresses. The time frames suggested in this text are simply to suggest rates of sinkhole development, not to provide case-specific rates or processes.

#### 7.4.2.4 Detection of Raveling or Suffosion

Detection of suffosion is normally relatively simple. As shown in Fig. 7.63, the assumption is that raveled, non-cohesive sand will form an hourglass-like depression containing dilitated sand. Furthermore, the relative strength of the sand will decline with depth, often culminating in weight-or-rod/hammer strengths near the base of the raveled zone during standard penetration tests or declines in tip resistance and possible increases in sleeve friction in cone penetrometer test soundings.

Jones and Upchurch (1993, Fig. 11), for example, reported on an extensive drilling program conducted within the Mulrennen Road sinkhole (Fig. 7.65a). Patterns of decreasing penetration resistance with depth and lost circulation zones at the base of the raveled zone (Fig. 7.65b) were encountered in nearly every standard penetration test boring. However, the standard penetration test borings near the margins of the sinkhole only revealed partial raveled zones that were near the top of the sand section.

Care must be taken when exploring cover materials for suffosion-dominated sinkhole development. As shown in Fig. 7.66a, the standard penetration test borings can intersect the cone of raveled sand at different depths. Bloomberg et al. (1988) present examples of the consequences of subsurface testing within the cone of raveled sand, including subsurface tests purposefully located within and peripheral to the aven of the sinkhole. In Fig. 7.66a, boring "1" went down the aven and remained in sand to the top of the limestone within the underlying void. While this rarely happens, the extensive column of very loose to weight-of-rod/hammer strengths coupled with a pattern of declining strength in the upper portions of the sand section during standard penetration test borings and deeper than usual limestone depths are clues that the aven has been penetrated (Boring 1, Fig. 7.66a). The probability of actually penetrating an aven is low because of its small diameter relative to the diameter of the top of the raveled depression. Borings 2 and 3 are common locations for standard penetration testing. In both cases, the cone of raveled sand is completely penetrated, and intact sand was encountered below the raveled zone (Bloomberg et al. 1988). Recognition of the raveled zone may include, but is not limited to, (1) a pattern of systematically[11] declining relative sediment strengths with depth and (2) possible loss of drilling fluid circulation at the base of the raveled zone and above the top of the limestone or any unraveled sediment. Note that the loss of drilling fluid is not evidence of a developing sinkhole because the loss is within the epikarst horizon. The pattern of declining sediment strength in the cover materials above the limestone is a potentially strong suggestion of sinkhole formation if it is local and not a result of depositional processes of the sediment.

Occasionally, one encounters evidence for sinkhole formation that is characterized by "stacked" ravel zones (Fig. 7.67b). In the authors' experience this situation is very rare and would require that (1) a bed of cohesive or lithified sediment occur well above the limestone and that material capable of raveling exists above and below the cohesive stratum and (2) penetrations through the confining bed and the aven in the limestone be more-or-less aligned. If the system is undergoing suffosion in both the upper and lower sands, then stacked ravel zones can develop.

#### 7.4.2.5 Angled Ravel Zones

The only significant, horizontal component to sand movement in a suffosion scenario is the creep of the sand toward the aven from the sides of the zone of raveling (Fig. 7.63). In Florida, there has been much discussion concerning the possibility that conduits of sand movement may occur at low angles or be horizontal, so that sinkhole formation at one location can somehow migrate laterally to damage structures that are not within a conical zone of raveling and creep centered over the aven.

As will be discussed in Sect. 7.4.3, piping failures can migrate laterally over relatively short distances as long as there is sufficient hydraulic gradient for sediment transport through the pipe. Suffosion, however, requires a special situation for lateral migration of a cone of raveling to develop. Specifically, recharge of water to the sand aquifer cannot be spatially uniform, as would occur during a rainfall event over undeveloped ground. The source of water must be concentrated so that groundwater flow is forced to migrate toward the aven with a horizontal component. A recurring scenario where this occurs is shown in Fig. 7.67.

Let us imagine a house constructed over a void in the underlying limestone (A, Fig. 7.67). There is a clay-rich stratum over the limestone. This clay stratum isolates the surficial sand and limestone aquifers and prevents or retards sinkhole development.

[11]The term systematic derives from trend and time-series analyses in statistics. Systematic implies that the preponderance of data demonstrates a more-or-less uniform change over time or distance. In sinkhole exploration, the penetration resistances of sediment within the cone where raveling and/or creep is occurring should more-or-less uniformly decline with a large proportion of depth/test intervals.

During heavy rainfall events, the roof of the structure deflects rainfall to the eaves of the building, where concentrated runoff occurs (B, Fig. 7.67). Since there is no direct, rainfall-related recharge under the footprint of the house, the mound of recharge water that develops under the eaves causes meteoric water to flow laterally under the house, thus raising the water-table elevation.

If there is a zone of weakness, such as a fracture or thin zone in the clay-rich stratum, an aven that penetrates the clay-rich bed may exist or develop (B and C, Fig. 7.67). With the opening developed in the clay bed and with a downward hydraulic gradient, groundwater flow will migrate toward the opening and, thus, into the underlying limestone (C, Fig. 7.67). With the ability of groundwater to flow into the underlying limestone via the opening in the clay-rich stratum, suffosion can occur. In this case, the direction of suffosion will roughly parallel the flow of water and angle toward the aven, most probably as a piping feature. There is no source of water under the house unless there is a broken water or sewer line; the only source of water is runoff from the eaves of the house and general recharge outside of the footprint of the structure.

With water flowing at an angle from the eaves of the house toward the developing aven under the house, the ravel zone begins to develop, first at the base of the ravel zone and then extending at an angle toward the source of water outside the footprint of the structure. In non-cohesive sand, a non-vertical, conical raveling zone develops and undermines the structure's foundation (D through F, Fig. 7.67). Note that cohesive and permeable sediments (e.g., clayey sand and, possibly, sandy clay) are more likely to develop small piping features than a large, conical suffosion zone.

When sinkhole development is expected under a house, we look for "swallow holes" along the foundation. Swallow holes are locations where a piping feature emerges from under a foundation and water from roof runoff passes under the foundation as concentrated flow. Swallow holes represent the stage in sinkhole development shown in Panel D (Fig. 7.67). If the swallow hole is caused by piping, it may remain small.

One must take care in evaluating holes that extend under the foundation. Similar holes can be excavated by burrowing animals, such as by pets and neighborhood "varmints". If water has created the hole by flowing under the foundation, the sediment excavated by erosion of the pipe is transported with the water under the foundation. There should be no significant accumulation of sediment deposited on the outside of the hole. If the origin of the hole is bioturbation, then one would expect to find a mound of dirt on the exterior of the hole.

This event is common in Florida, where sinkholes frequently develop under the eaves and foundations of a structure. In sand with some cohesion, the sinkhole is likely to be

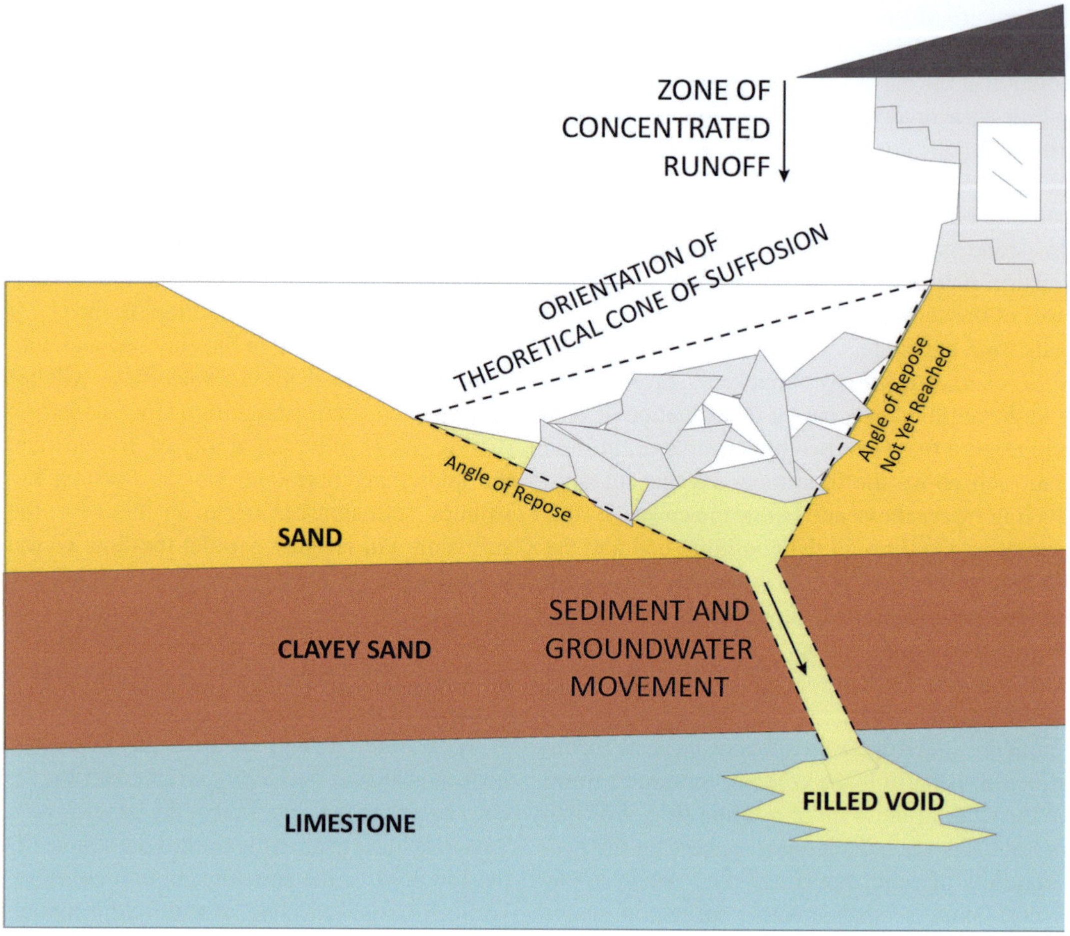


**Fig. 7.68** Diagram showing how an angled cone of raveling or suffosion migrates to the foundation of a house

a simple, tube-like conduit or pipe, but in cohesionless sediments, the resulting sinkhole may be quite large. Where one can follow the zone of raveling under the footprint of the structure, it can be seen to migrate downward toward the center of the house (Fig. 7.68). Unless there is a water source, such as a broken water or sewer line, sinkhole emergence directly under a house is rare.

Figure 7.69 is an example of damage caused by this process. This sinkhole developed in the late 1990s in the community of Mango, eastern Hillsborough County. The duplex was situated on non-cohesive sand overlying clayey sand and limestone. The sinkhole developed under the eave at the front of the house. While the house was cantilevered over the sinkhole, one could see that the pathway of sediment raveling was toward and below the house interior, as suggested by Fig. 7.68.

### 7.4.3 Piping

Piping is erosion caused by the movement of water through soil or sediment that is semi-cohesive to the extent that the sidewalls of a conduit are supported without collapse or dilatation of the host sediment. Unlike suffosion, which causes mass movement of sediment within the cone of raveling, sediment movement during piping is restricted to a specific pathway owing to the cohesion or resistance to erosion of the surrounding host sediment. Erosion in a piping feature is caused by the movement of groundwater that has become localized and concentrated along a pathway (A and B, Fig. 7.70).

The piping failure begins where the water emerges from the slightly cohesive sediment (C, Fig. 7.70). Erosion, therefore, is limited to the face and sides of the piping feature, which migrates upward along the centroid of groundwater flow to the pipe.

The pipe then erodes its way backwards from the discharge point through the host sediment, creating a pipe-like conduit (D and E, Fig. 7.70). Dilatation of the surrounding sediment is limited until the diameter of the pipe is enlarged to the point of collapse. Erosion during piping is generally limited to the head and sides of the pipe.

**Fig. 7.69** A duplex located in Mango (Hillsborough County) that was damaged by sinkhole formation. The sinkhole began under the roof eave. Half of the sinkhole and the aven are under the house and the depression was canted downward, toward the center of the house

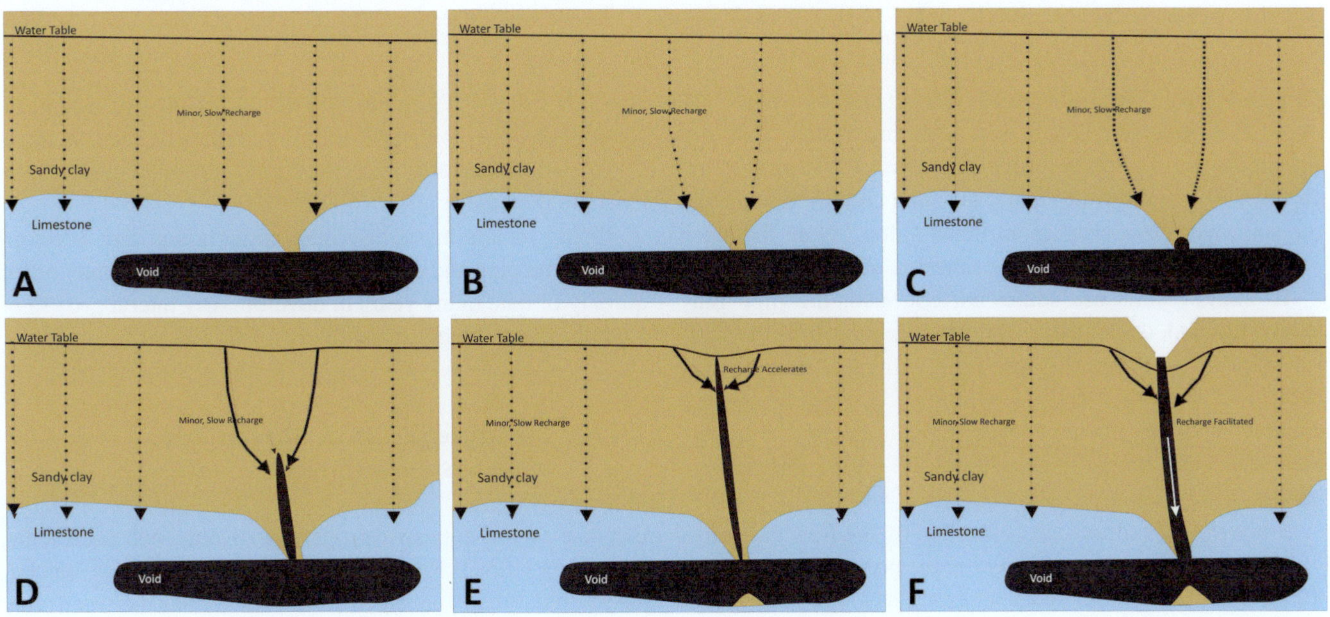


**Fig. 7.70** Sequence of development of a pipe in cohesive sediment. See text for discussion

Once the pipe extends above the water table, depending on the level of cohesion of the sediment and upon time, the pipe may enlarge to a larger depression that is limited in diameter in accordance with the angle of repose of the host sediment (F, Fig. 7.70). The mechanisms of piping are discussed in Holtz and Kovacs (1981) and other geotechnical texts.

Most of the erosion pipes the authors have confirmed have diameters that have been limited to a meter or less. Figure 7.71 is an example of a piping feature observed in

**Fig. 7.71** A cover-collapse sinkhole formed by piping in clayey sand near Tarpon Springs, Pinellas County. The pipe is approximately 1 m in diameter and it is about 75 cm to the top of the sod plug. Bedrock was at 15 m below land surface

Tarpon Springs, Pinellas County. A hand auger boring in the depression indicated that it was a piping feature that had propagated in slightly cohesive clayey sand to near the land surface. Above the water table, the sand collapsed, carrying the sod layer downward as an intact "plug." Note the steep sides of the sinkhole caused by cohesion of the clayey sand.

If shallow soils are somewhat cohesive, the pipe may extend upward, through the sand column, as well. If the shallow sand is non-cohesive, the pipe will most likely break down to a cone of suffosion very quickly. If, however, the shallow sandy sediments contain enough clay to be moderately cohesive, the vertical walls of the piping failure may persist.

Since piping failures are controlled by the flow pathways of water and internal erosion within the pipe, pipes can deviate from the vertical for short distances, as long as a gradient sufficient for water flow and sediment transport within the pipe exists.

If the cover sediment is highly cohesive, piping is less likely. Piping must include a mechanism to overcome any cohesive forces acting on the sediment and for water to move through the sediment. If the cohesive forces are overcome, the mechanism of clay piping may be (1) clay dispersion or (2) failure along fractures, breccia interfaces, or sandy seams within the clay. Where the clay is dehydrated, Sowers (1996) suggested that spalling could account for pipe propagation. Most likely, the piping will follow fractures or occur in areas of the strata that are sandy and have lower cohesion than exists in pure clay.

Piping results in cover-collapse sinkholes. The development of the pipe and any accompanying suffosion may occur quickly or take many years to develop. Any lateral growth of a resulting sinkhole is most likely a result of suffosion.

### 7.4.4 Disaggregation or Movement of Cohesive Sediment

Because of the cohesive properties of clay-rich sediments, they do not ravel. Rather, they fail and subside by mechanisms that break the electromagnetic bonds between clay particles. These processes are discussed in this section.

#### 7.4.4.1 Spalling, Slaking, and Slabbing

Sowers (1996) presented an analysis of the modes of failure of cohesive sediments as a sinkhole develops but did not fully explain how the cohesive forces of the sediment are overcome. Sowers (Fig. 3.9 in 1996) began with formation of an "erosion dome within the clay (A, Fig. 7.72). He implied that the erosion dome was a result of slaking or spalling in

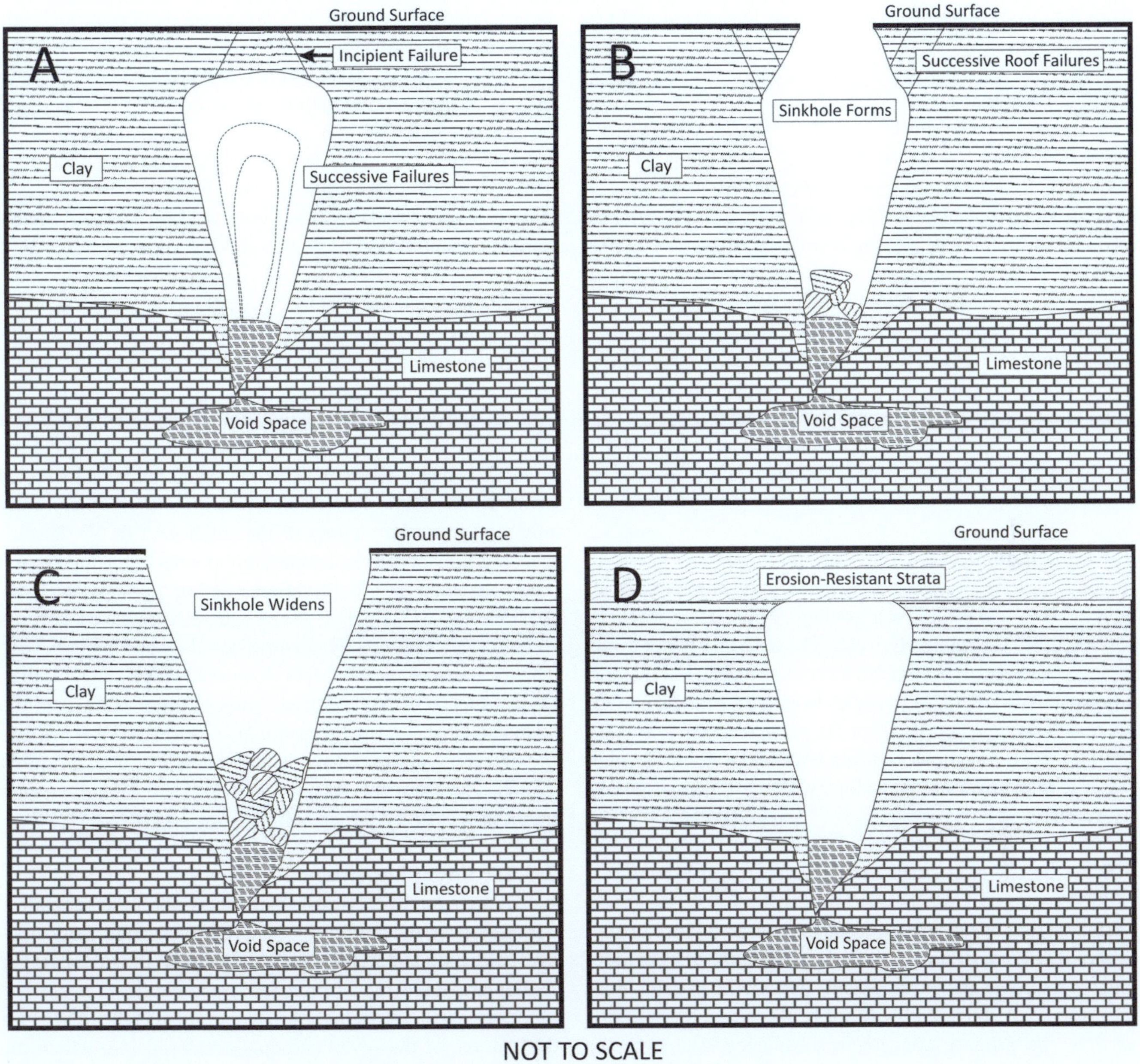


**Fig. 7.72** A proposed mode of failure in cohesive sediment. (Modified from Sowers (1996, Fig. 3.9))

dehydrated clay and that cohesive forces were weakened or negated by repeated dehydration and wetting episodes. When the erosion dome approaches the land surface, the ceiling of the dome fails, and then enlarges (B and C, Fig. 7.72). If a stiff clay layer overlies the erosion dome, the dome may terminate at the base of the more competent stratum (D, Fig. 7.72).

The erosion dome described by Sowers is comparable to the erosion pipe discussed above, but the mechanism proposed by Sowers does not involve flowing water, only periodic wetting and drying of clay-rich sediment. Note how he characterized the successive, headward erosion stages and the final collapse of the upper, unsupported sediment. The only growth of the pipe or erosion dome is by upward migration and then collapse when the remaining sediment cannot be supported over the cavity.

Sowers went on to describe the behavior of the water table in his "erosion dome model." Water was thought to seep downward through fissures and "primary pores" of the overburden and upward from the underlying limestone. When the clay undergoes wetting, Sowers argued that the clay underwent "sudden expansion" and softening. He termed this process slaking. During periodic wetting and drying cycles, the clay would undergo episodes of slaking, which could allow slabs of clay to spall into the cavity or disaggregate.

By this process, Sowers (1996) opined that the erosion dome could grow by spalling of "elongated sheets" of sediment 12–50 mm thick and "5 to 10 times longer" to separate from the cavity walls. This process was termed slabbing by Sowers.

#### 7.4.4.2 Stoping

The upward, or headward, erosion depicted by Sowers (1996) has been likened to the term "stoping" as used by geologists in mining and igneous geology. Stoping is a process by which the upper surfaces of a cavity grow by removal of material. In mining, stoping is removal of rock by human activity. In igneous geology, it involves chemical reactions and melting of ceiling rock in contact with magma to cause growth of a magma chamber. Piping, slaking and spalling, and stoping represent similar upward erosion mechanisms for growth of a cavity in cohesive to semi-cohesive cover materials as a result of repeated wetting and drying. These processes are forms of stoping, a generic term for upward erosion within a pipe or erosion dome where the mechanism is unknown.

#### 7.4.4.3 Dispersion

Sodium-saturated clay has been found to be prone to dispersion in water because the electromagnetic bonds between the clay particles are relatively weak. The authors are not aware of any occurrences of dispersive clay in Florida, but the possibility remains that clay particles in contact with groundwater may simply decouple from the cohesive clay mass as a result of electromagnetic bonding with free water molecules, rather than the hydration envelope within the clay stratum.

### 7.4.5 Settlement of Poorly Supported Strata

In this section, we discuss subsidence caused by loss of support of overlying strata. Since rock collapse is not a phenomenon of the cover, the development of rock-collapse sinkholes was discussed above and will not be discussed here. In this section, the primary concern is subsidence related to the cover/limestone interface.

#### 7.4.5.1 Collapse of Poorly Supported Strata

It is common to encounter weight-of-rod strength materials during standard penetration tests and losses of drilling-fluid circulation at the top of limestone while drilling with mud-rotary drilling techniques. While these features may truly reflect voids (Fig. 7.73), many of these occurrences reflect very soft clay associated with the epikarst.

If there actually is a void at the interface, then there is a concern with (1) the properties of the cover material that can span the top of the void and (2) how large the void can be before the cover material can no longer support the void roof.

Upchurch et al. (2013) documented a 60 m deep, cover-collapse sinkhole (Fig. 1.4) in southeastern Hillsborough County that appears to be the result of failure of sediments over a large void. Figure 7.74 is a cross section through this sinkhole, which developed under the Southeast Hillsborough County Landfill in 2010. The cross section shows the sinkhole after partial remediation. At the time of the subsurface testing, the sinkhole had been filled with grout (CLSM, or controlled, low strength material) and a sand cover had been graded back over the in-filled aven and raveled cone. Most of the aven was, however, filled with landfill debris. A major concern about the origin of this sinkhole was the absence of a significant accumulation of natural (rock) breakdown material on the floor of the gallery below the aven. The laterally equivalent sediment was typically very strong, commonly exhibiting refusal strength sand and limestone during penetration testing. Upchurch et al. (2013) surmised that the bulk of the volume into which the collapse occurred had been removed at some time prior to collapse. Either (1) the cavity was a late Miocene or Pliocene sinkhole that had once been filled with under-consolidated clay, which later migrated from the vicinity of the sinkhole,[12] or (2) dissolution and transport of sediment components had resulted in a cavity spanned by a thin limestone bed or, perhaps, the landfill liner and waste mass. At any rate, a large cavity appears to have been spanned by cover materials that failed in 2010. The aven was about 6–8 m in diameter, which suggests that, under special circumstances, rather large voids can be spanned. Absent a strong spanning material, such as perhaps the limestone in Fig. 7.74, it is unlikely that a roof composed of cohesive sediment can support a void substantially over a meter in maximum dimension.

#### 7.4.5.2 Cover Subsidence

As discussed above, cover-subsidence sinkholes are shallow depressions in the land surface caused by dissolution of shallow carbonate materials. They are normally less than 100 m in diameter and a meter, more-or-less in depth.

Cover-subsidence sinkholes cannot form at depth because the volume of dissolved carbonate material into which the cover sediments migrate is low. Deep dissolution is masked by the minor dilatation of sand and other sediments as they settle. The authors have never encountered a cover-subsidence sinkhole in Florida caused by dissolution at depths greater than 5–10 m.

Figure 7.75 illustrates some typical cover materials within a cover-subsidence sinkhole in southern Florida. This sink was a small "sag pond" with cypress trees prior to mining. The sediment fill was 1.5 m thick. In the figure, the "shell with sand" is the host carbonate sediment for sinkhole development. This sediment consists of over 50% shell with carbonate and siliciclastic sand and minor clay. Dissolution had occurred at the top of the shell bed, creating a shallow

[12] A late Miocene sinkhole filled with clay was documented at a nearby site by Upchurch and his colleagues.

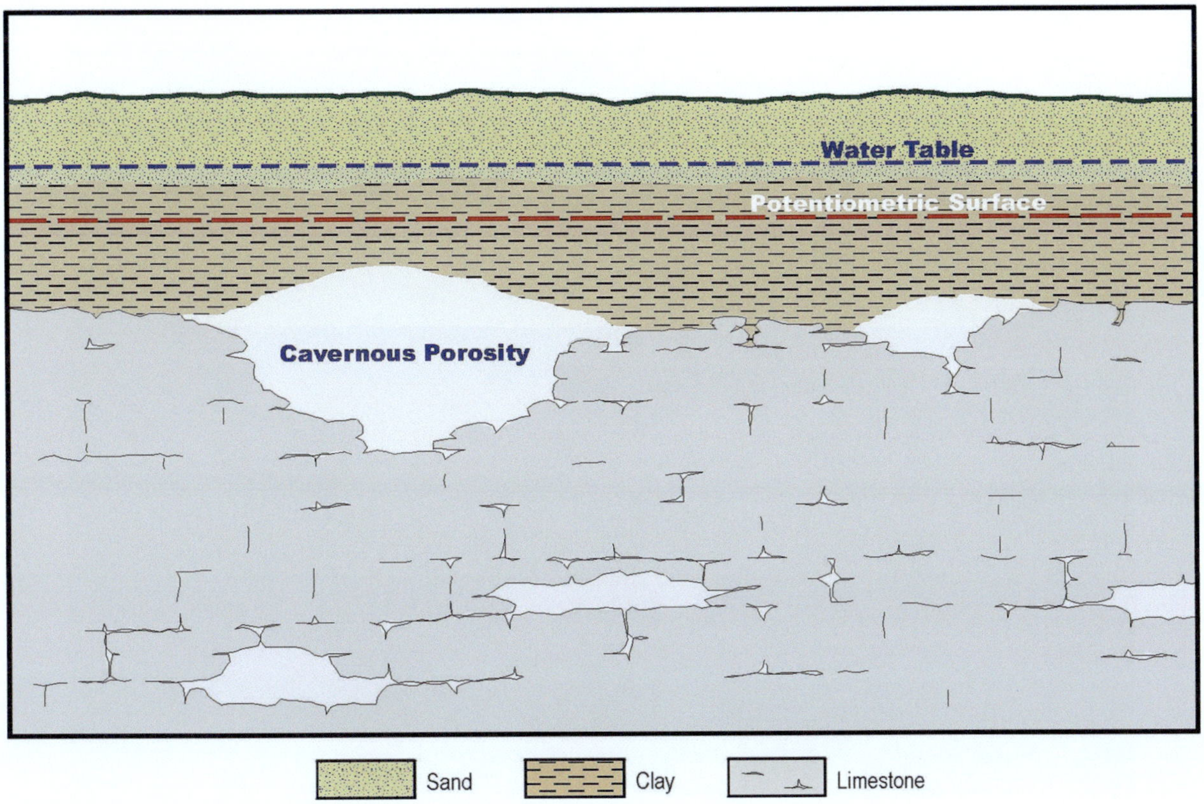


**Fig. 7.73** Cross section through a void with a roof supported by cohesive, clay-rich sediment

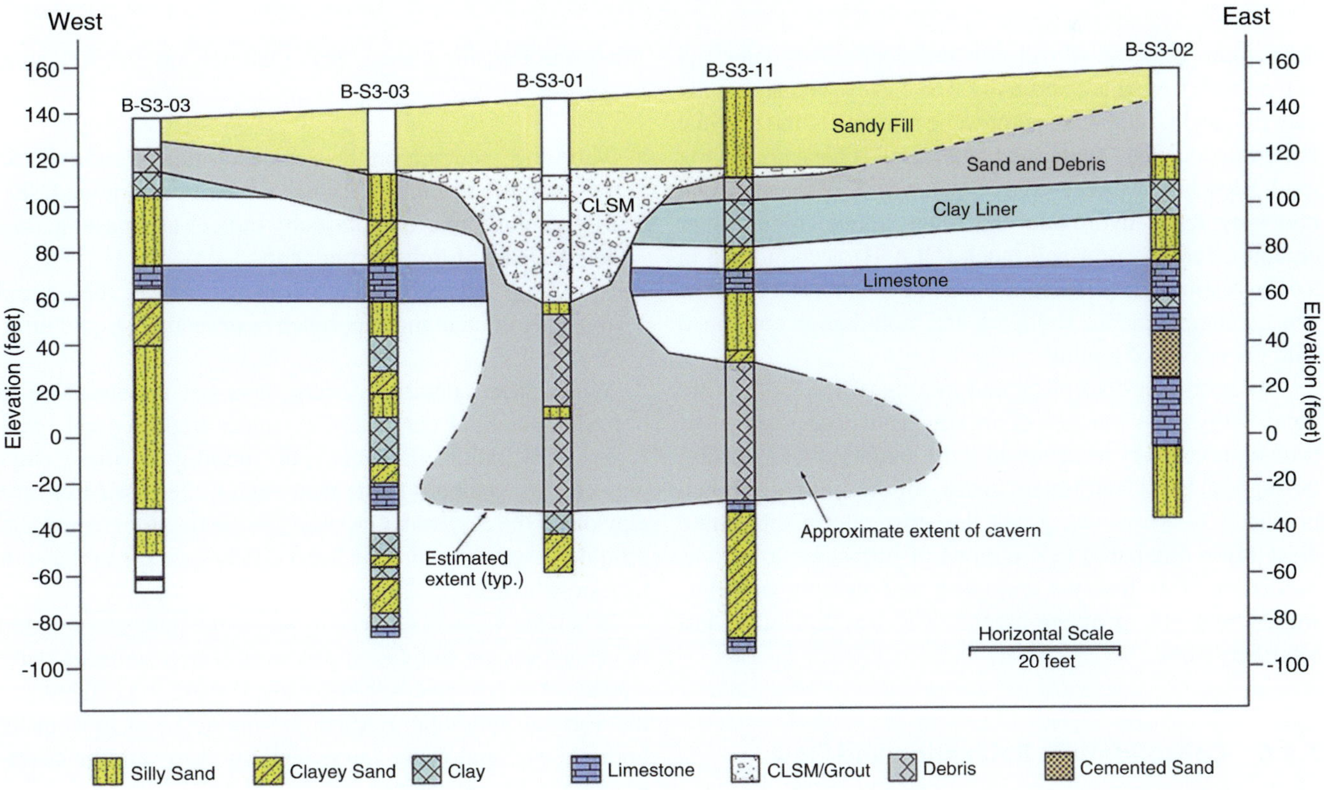


**Fig. 7.74** East-west cross section through the Southeast Hillsborough Landfill sinkhole. See text for a discussion of the mode of collapse of the sinkhole. (From Upchurch et al. (2013))

**Fig. 7.75** Cross section through a cover-subsidence sinkhole showing the sediment types commonly encountered in the sinkhole. Sinkhole was exposed in a shell pit near Ft. Drum, Okeechobee county. See text for a discussion of the sediments that in-fill the sink

depression. Existing siliciclastic sediment slowly subsided as the shells were dissolved away. The clayey sand and sandy clay (SC and CL/CH) constitutes the remains of the original cover material. The white sand labeled SP is bleached quartz sand. Bleaching was caused by organic acids in the sag pond removing ferric hydroxide and other soluble minerals or organics. Finally, peat and muck (OL/OH) accumulated in the depression. The disruption seen in the strata is a result of penecontemporaneous slumping and bioturbation combined with the effects of mining.

The sediment disruption and mixing of lithosomes are noteworthy. The manner of mixing and juxtaposition of lithosomes clearly indicates that the sediment was settling throughout the development history of the sinkhole shown in Fig. 7.75. However, given the slow rate of carbonate dissolution compared to the mode of lithosome preservation, it is clear that the slumping and shifting of lithosomes into the growing depression was episodic and relatively rapid.

### 7.4.6 Consolidation, Relaxation and Flow

It is not uncommon to encounter under-consolidated clay within the Miocene Hawthorn Group and at the cover/limestone transition zone. This material has the following properties:

- Very low penetration resistances (often weight-of-rod or –hammer strengths) during standard penetration tests;
- High liquid limits and plasticity indices as measured during Atterberg Limits testing; and
- Very high natural moisture contents. Often, the natural moisture content may approach or exceed the liquid limit.

When these conditions occur, it is not unreasonable to consider potential clay mobility under high vertical lithologic or hydraulic gradients. The mobility of these clays depends on whether (1) the water can be released from the clay or (2) the clay/water mixture has the freedom to move as a liquid suspension through karst conduits in the epikarst or limestone matrix.

When the water is allowed to vacate the pore space within the clay, the clay simply consolidates, which can cause subsidence of overlying cover materials. If there is a conduit and the vertical hydraulic gradient is conducive to movement, both the clay and water can potentially flow into the downgradient, adjacent conduit.

Upchurch et al. (2013) suggested that these two processes were potential causes of sinkhole development (Fig. 7.76) in

**Fig. 7.76** These sinkholes (arrows) formed along a street in Plant City, Hillsborough County, during the 2010 freeze event

the Dover-Plant City area of Hillsborough County in 2010. A prolonged freeze in the winter of 2009–2010 resulted in extreme over-pumping of the underlying UFA for freeze protection of local strawberry crops, a billion dollar agricultural industry in eastern Hillsborough County. The potentiometric surface of the highly confined UFA declined 18 m, with up to 9 m daily excursions during the period of the freeze (Fig. 7.77).

The authors had the opportunity to test the subsurface of a number of sinkhole sites in the Plant City area shortly after the freeze event. In a majority of sites, there was a thick (25 ± m) clay section overlying Oligocene/Miocene Arcadia Formation limestone or dolostone. At several sites, up to 5 m of the clay was found to have high liquid limits (LL = 90+ percent), plasticity indices (PI = 40+ percent) and high natural moisture contents (NMCs) that frequently exceeded the liquid limits. They concluded that the wet, expansive clay had either collapsed by dewatering under the elevated vertical hydraulic gradient from over-pumping of the UFA or it had migrated as a clay-water suspension into cavities in the underlying limestone/dolostone.

With such high natural moisture contents (NMC $\geq$ 0.8 LL), these clays appear capable of (1) "relaxing" to accommodate the void structure with which they are in contact, (2) consolidating under high vertical hydraulic gradients if dewatering pathways exist, or (3) flowing into adjacent, down gradient void space. In either eventuality, these processes appear to have resulted in multiple, cover-collapse-type sinkholes.

### 7.4.7 Stress-Strain Relationships and Pressure Arches

High penetration resistance, unlithified sand is commonly encountered during subsurface testing of cover materials. The concern is whether or not these sand strata are indicators of pressure arches that have developed over raveled zones or voids. Pressure arches are formed over tunnels and caverns in rock with high compressive strength. As shown in Fig. 7.78, rock overlying a tunnel or void in rock is subjected to the load pressure of the overlying sediment column. On either side of the void and at the highest point in the void, the compressive forces are vertical. However, the sloping sides of the void roof are subjected to stresses[13] that are perpendicular to the sides of the void. These compressional forces are pushing on the void, trying to collapse it. Until the walls of the void either pass through the chord of the void (Fig. 7.78) by deformation or the strain built up in the rock

[13] Stress is the force placed on the rock or sediment and strain is the amount of distortion that results from the stress.

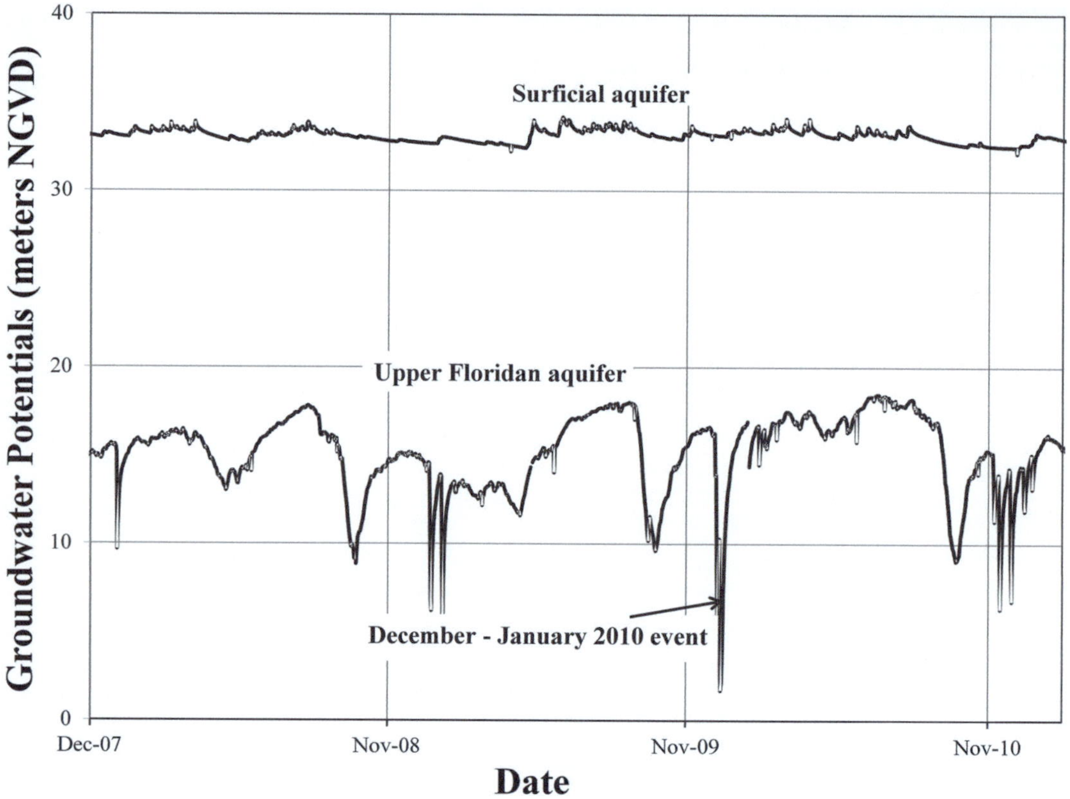


**Fig. 7.77** Water levels in wells near Plant City, Hillsborough County, from October, 1993 to February, 2012. Each extreme drawdown event resulted in sinkhole development in the area. Data source: Southwest Florida Water Management District, 2012; ROMP DV-1 Surficial and Upper Floridan Aquifer wells

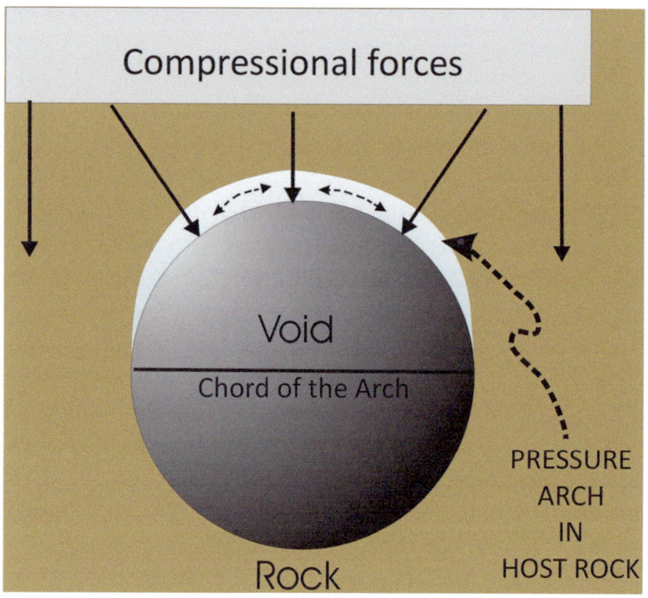


**Fig. 7.78** Development of a pressure arch over a tunnel or cavern

causes failure, the arch of compressed rock will support the roof of the void. This zone of elevated compression that arches over the void is known as a pressure arch.

Figure 7.78 suggests the pressure field on the materials in the ceiling of a cavity. As the materials above and to the sides of the ceiling push downward on the opening, they are under compression. In order to avoid movement, however, the materials must at be compressed without failure. If the materials can withstand the stresses placed on them by the overlying load, the arch is stable and behaves much like a keystone in an arch. If, on the other hand, the materials under compression fail under stress, then the arching materials fragment and collapse.

Caverns in limestone or dolostone develop pressure arches. If the compressive forces increase or if the cavern increases in size through dissolution, the roof of the cavern may fail and slabs and blocks of rock fall to the cavern floor (Fig. 7.79). The ability of the cavern roof to support itself is entirely a result of the stress and strain distribution within the rock.

The ability of the unconsolidated sediment or weak rock to support a pressure arch is dependent on the ability of the material to undergo strain without failure and rupturing. An important concern in soil mechanics is the response of a rock or sediment to stress. Every geologic material has a strain threshold, which, when reached, causes the material to fail and break. While rock may be able to undergo considerable

**Fig. 7.79** Room in Florida Caverns, Marianna, Jackson County, Florida, showing breakdown rock (arrow) that has been partly covered by flowstone

**Fig. 7.80** A drilling rig at the bottom of an induced failure caused by drilling. Only the mast is above the sediment; the engine and tracked vehicle are buried

stress before failing, unconsolidated, well-sorted sand does not (Wang 2005). Non-cohesive, well sorted, and sub-rounded sand has a low stress-strain threshold before failure. As a result, it is not possible for well-developed and strong pressure arches to persist in well-sorted sand over time, although they may develop momentarily during the raveling or suffosion process.

When conducting standard penetration tests in well-sorted sand, there is considerable cyclic loading, and use of drilling fluid injects water into the sediment matrix. Both of these processes are likely to induce failure, sometimes catastrophic failure (Fig. 7.80) of any sand that is temporarily under compression because it overlies a void or actively raveling sand. Rather than encountering high penetration resistances, therefore, the opposite should occur. The sand under compression should either undergo rapid redistribution of the load as it is penetrated or a cave-in of the ravel zone into the void should develop. There is no valid basis for arguing that a high penetration resistance in stable sand represents pressure arches over voids or raveling zones.

## 7.5 Sinkhole Risk in Florida

There have been a number of attempts to characterize sinkhole risk in Florida. Most were map depictions of the locations of sinkholes from the Florida Geological Survey's (FGS) Subsidence Incident Reports database or enumerations of the most probable types of sinkholes on a regional basis based on cover type and thickness.

### 7.5.1 Wright's Sinkhole Map of Florida

The earliest known map that provides information about sinkhole risk for Florida showed types of sinkholes based on (1) cover type and thickness, (2) the relative elevation of the UFA potentiometric surface, and (3) topography. This map, by Wright (1974), proposed risk categories, with the lowest risk in areas where the UFA potentiometric surface is near or above land surface. Higher risk categories were ranked according to overburden (cover) thickness and water use. While the areas where modern sinkhole-claims activity are the greatest were included in Wright's high-risk areas, the high-risk areas are more extensive than current actuarial data suggest. This map is considered inaccurate, largely as a result of improvement of (1) knowledge of stratigraphic conditions in Florida, (2) understanding of karst processes in Florida, and (3) knowledge of changes in hydrogeological conditions since 1974.

### 7.5.2 Sinclair and Stewart's Sinkhole Type Map

Sinclair and Stewart (1985) produced a map (Fig. 7.81) of sinkhole type and distribution based on cover thickness and lithology. The map does not assign risk, but does utilize terms like "numerous" and "very few" to describe how frequently sinkholes were thought to develop in different areas of Florida. As with the Wright map, the Sinclair and Stewart map is small scale, lacks detail, and utilizes limited stratigraphic data. While the map deals with sinkhole type on a regional basis, differentiation of sinkhole type based on cover thickness and lithology is useful for understanding how sinkholes were thought to develop in 1985. Unfortunately, the Sinclair and Stewart map has been widely utilized in recent years in attempts to predict local or site-specific conditions and justify positions in sinkhole-claim litigation. Such use is inappropriate owing to the scale of the map, uncertain boundary locations, and lack of detail in the data presentation.

Neither the Wright (1974) nor the Sinclair and Stewart (1985) map was prepared at a scale that enables site-specific interpretations of sinkhole type, cover thickness, or sinkhole risk. Both maps are now considered out dated and inaccurate. Neither map should be used to support site-specific conclusions relative to either sinkhole risk, occurrence, or type.

### 7.5.3 Quantifying Sinkhole Risk

#### 7.5.3.1 Hillsborough County Sinkhole Risk

The first attempt at actual quantification of sinkhole risk in Florida was an investigation of sinkhole development in Hillsborough County (Figs. 7.82 and 7.83) by Upchurch and Littlefield (1987, 1988). The authors were investigating the possibility of (1) assigning risk and (2) potential for adjusting insurance coverage and rates based on the spatial distribution of sinkhole risk.

Upchurch and Littlefield (1987, 1988) realized that there were issues with the available actuarial data[14] relating to sinkholes that had developed in recent years, including lack of verification of some data and bias towards reportage in urban and suburban areas. Nevertheless, they utilized the Subsidence Incident Report database with local, field verification of the data. Note that, at that time, this database included only sinkhole landforms. The Florida Sinkhole Research Institute (FSRI) data were for the period from 1964 to 1985, a 21-year period. Most of these sinkhole reports, termed "modern" sinkholes by Upchurch and Littlefield,

[14]The available database in 1987 was developed by the Florida Sinkhole Research Institute. It is currently maintained as the FGS' Subsidence Incidence Report database.

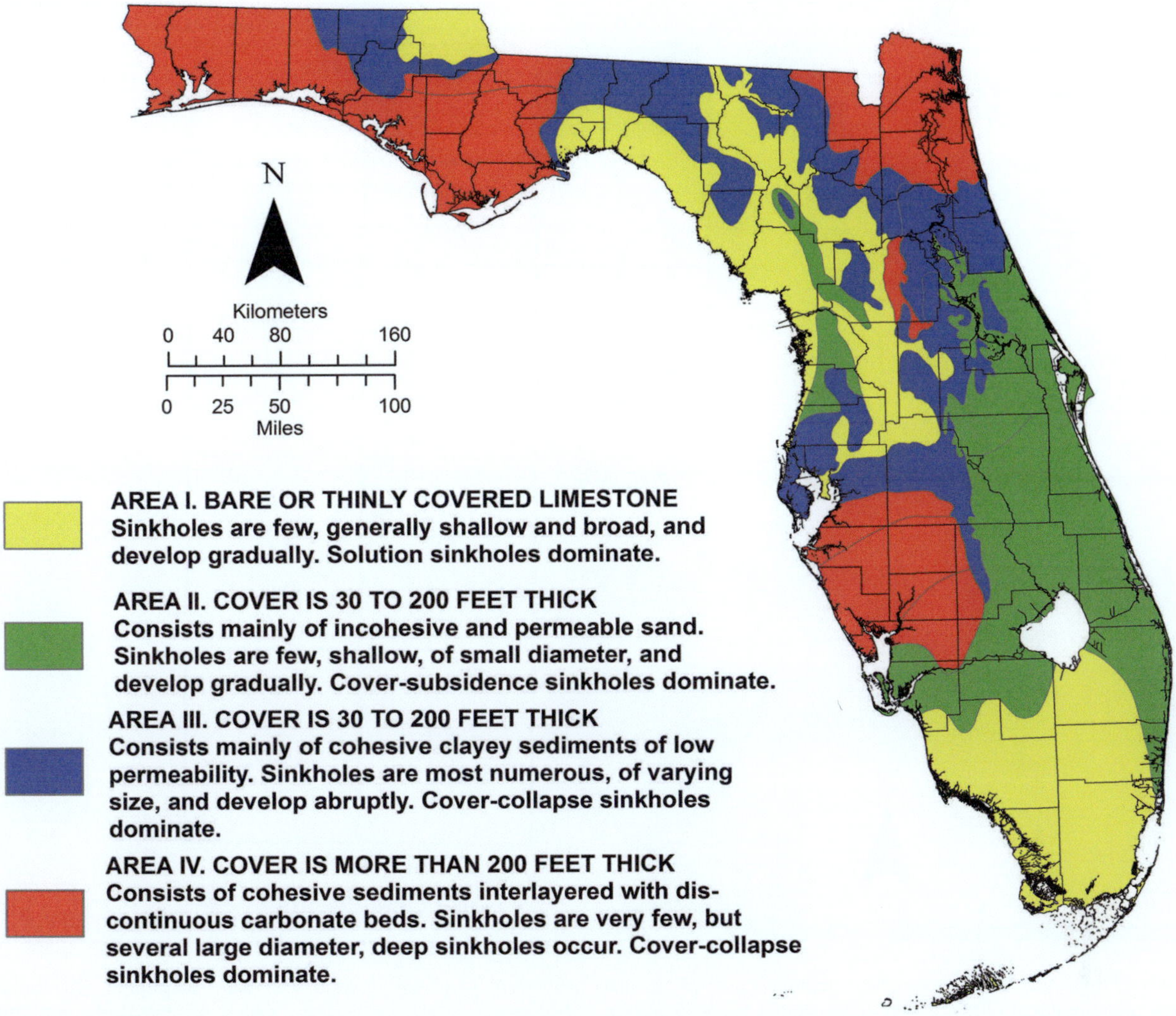


**Fig. 7.81** Map showing sinkhole type, development, and distribution in Florida from Sinclair and Stewart (1985)

were reported in the period between 1981 and 1985, after the FSRI had been created.

To make verification possible, they restricted the investigation to 12 7.5-min, 1:24,000 topographic quadrangle map areas located in Hillsborough County (Fig. 7.82). Where the quadrangle maps crossed county lines, only that portion of the quadrangle map in Hillsborough County was investigated.

They used a surrogate variable, closed depressions as identified on 1:24,000 USGS topographic map coverage (Fig. 7.83), to deal with the issue of spatial bias, i.e., most of the modern sinkhole reports were in densely populated areas or along major transportation corridors. The assumption was made that all closed depressions other than depressions with more-or-less rectangular boundaries represented relict sinkholes. The depressions with angular shapes were field verified as being excavations. Contour maps were prepared using a moving average algorithm for data smoothing to demonstrate the locations and areas of closed depressions (ancient sinkholes) and for the number of modern sinkholes reported on the FSRI database. Statistical reporting was limited to each quadrangle area.

Figure 7.83 illustrates the distribution of ancient, or relict, sinkholes as determined by the spatial density of closed depressions on the topographic maps. There were 2303 discrete, natural closed depressions in the study area. The number of ancient sinkholes (closed depressions) ranged from 0.1 to 3.2/km$^2$. For the entire, 12-quadrangle area, closed depressions averaged 1.1/km$^2$. Within a quadrangle, the area of closed depressions ranged from 0% to 25% of the land surface. The Sulphur Springs Quadrangle (Fig. 7.81) had the most relict sinkholes with 10% of the quadrangle area in closed depressions. The Lithia quadrangle had less than 0.5% of the area within closed contours.

Based on the abundance of closed depressions in the Sulphur Springs area, Upchurch and Littlefield (1987, 1988) termed that area of closed depressions the Sulphur Springs karst terrain (Fig. 7.83). There was also an area of concentrated closed

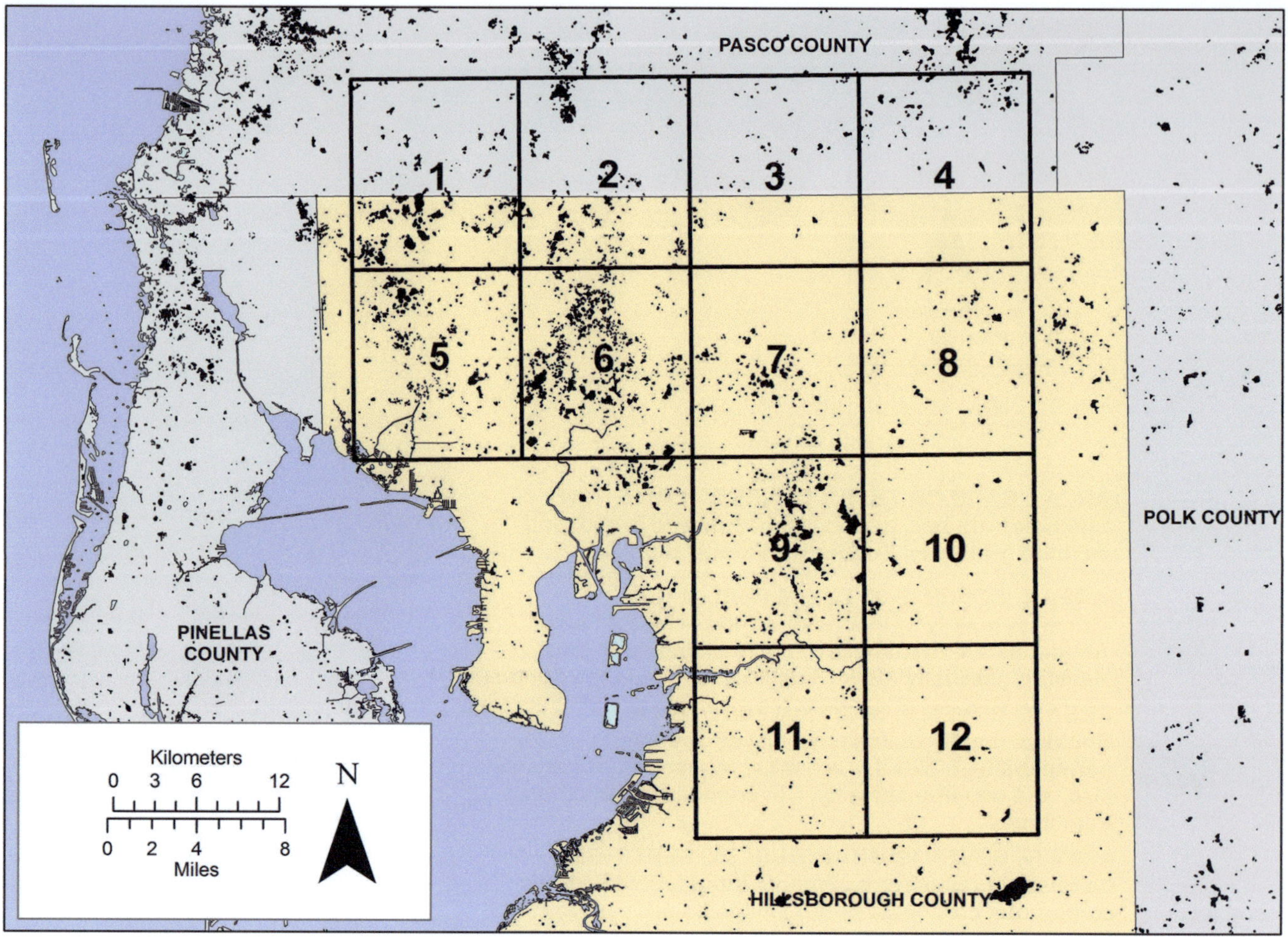


**Fig. 7.82** USGS 1:24,000 topographic quadrangles used in the Upchurch and Littlefield (1987, 1988) Hillsborough county sinkhole risk study. Closed depressions present on the topographic maps are shown as black polygons in the map background. Quadrangles are: 1-Odessa, 2-Lutz, 3-Wesley Chapel, 4-Zephyrhills, 5-Citrus Park, 6-Sulphur Springs, 8-Thonotosassa, 8-Plant City West, 9-Brandon, 10-Dover, 11-Riverview, and 12-Lithia

depressions near the community of Brandon that was termed the Brandon karst terrain. Up to 15% of the Brandon karst terrain was occupied by closed depressions (Fig. 7.83).

There were 179 verified modern sinkholes in the FSRI database for the study area from 1964 to 1985. Temporal-occurrence data (risk as a function of time) for these modern sinkholes were flawed as a result of incomplete reportage. The FSRI had attempted to backfill the database into the 1960s, but early sinkhole data were sparse and largely limited to news reports. The range of occurrence of modern sinkholes was from 0/year in the early years to 29/yr. late in the period of record (see Fig. 7.84 for a portion of the temporal record). The high rate of occurrence is biased by two sinkhole events in 1977 and 1985 in the Dover area (Metcalf and Hall 1984; Bengtsson 1987) when swarms of sinkholes developed as a result of groundwater pumping for freeze protection.

Figure 7.85 presents a moving average analysis map of reported modern sinkholes in the 12 quadrangle study area. The modern-sinkhole data indicated that modern sinkholes were highly localized and ranged from 0 to 1.9/km$^2$, with the highest spatial frequency in the Dover area (Fig. 7.85), where the large number of sinkholes developed in 1977 and 1985 as a result of groundwater extraction for freeze protection.

Because of concerns about under-reporting of modern sinkholes in rural areas and poor recovery of data representing the early parts of the period of record, setting a rate of modern sinkhole development per year was considered problematic in 1987. For the purposes of this text, the FGS Subsidence Incident Report database has been re-examined for the period from 1981 through 1991, the period of highest confidence in the database. Only subsidence events where the dimensions of the depression were reported were considered.

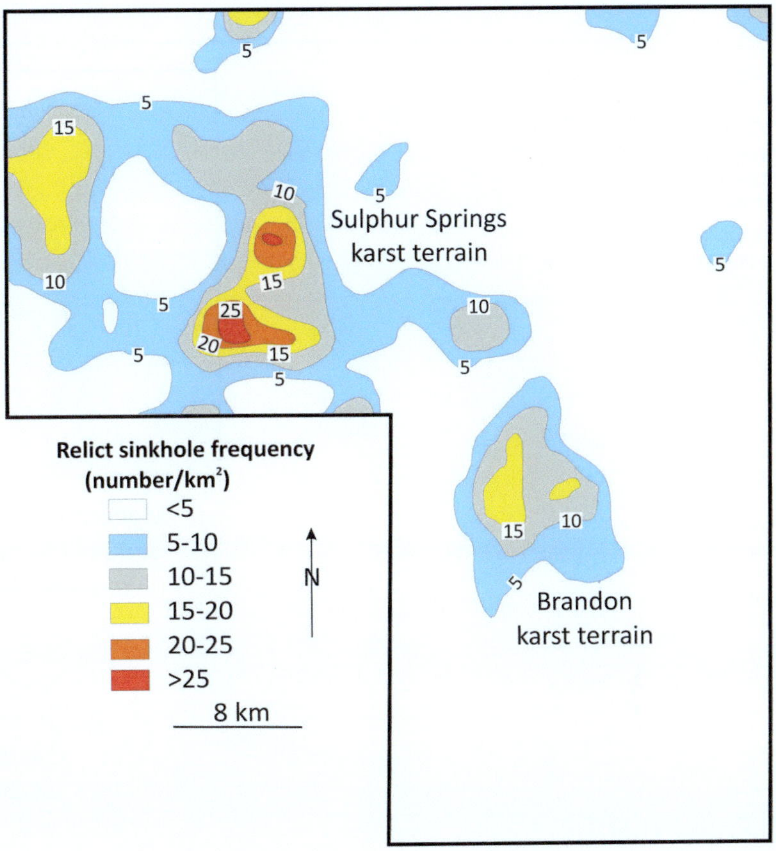


**Fig. 7.83** Relict sinkhole density, based on a moving average algorithm, of natural, closed depressions in the 12-quadrangle area of Hillsborough County depicted in Fig. 7.81. (Source: Upchurch and Littlefield (1988))

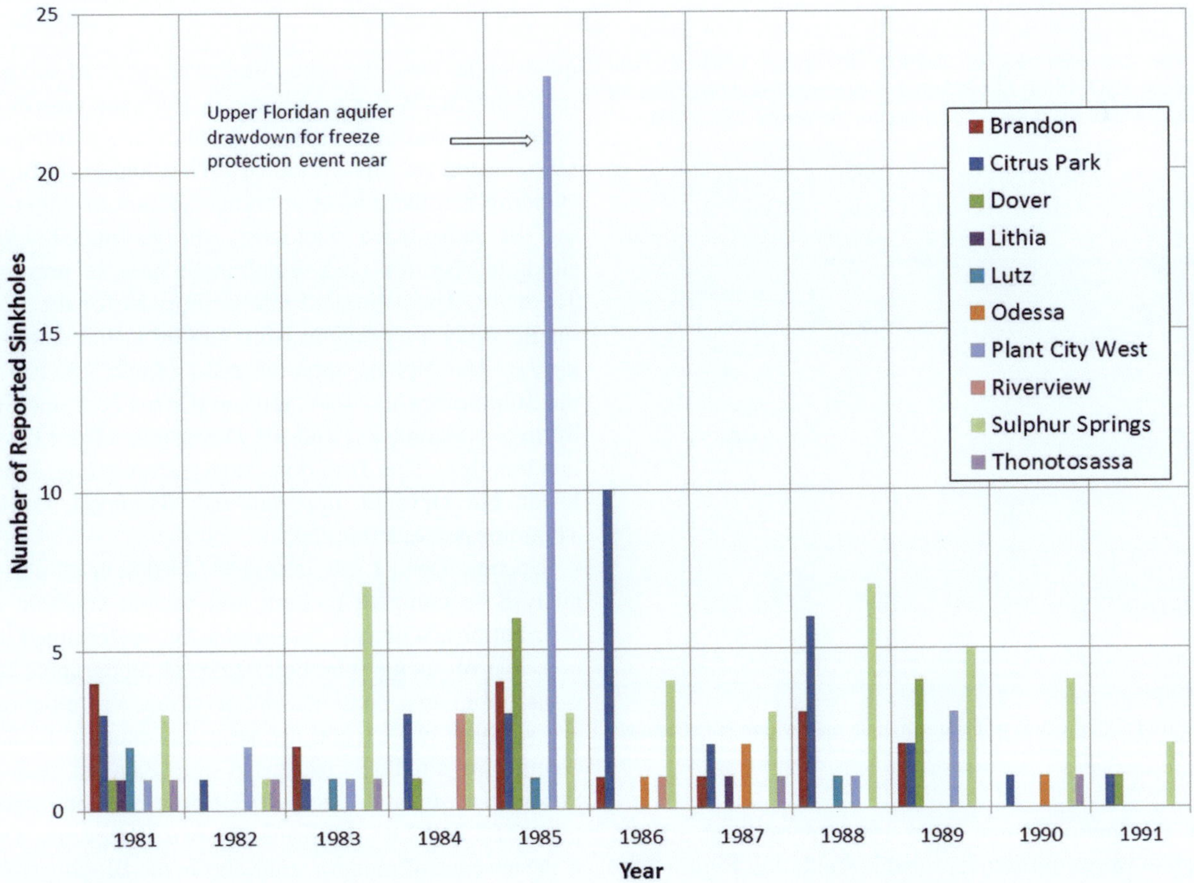


**Fig. 7.84** Probable sinkholes reported to the FGS Subsidence Incident Report database for the period from 1981 to 1991 in the Hillsborough County portions of the 12 quadrangles (Fig. 7.81) investigated by Upchurch and Littlefield (1987, 1988). There were no sinkholes reported in the Wesley Chapel and Zephyrhills quadrangle portions of Hillsborough County in 1981–1991

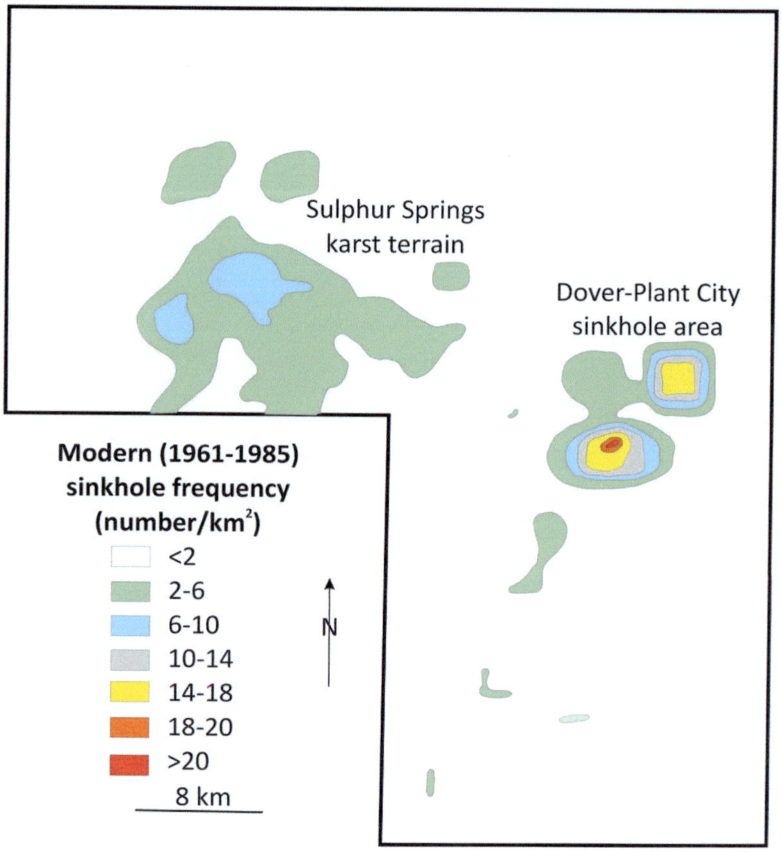


**Fig. 7.85** Density, based on a moving average algorithm, of "modern" (1961–1985) sinkholes in the 12-quadrangle area of Hillsborough County depicted in Fig. 7.81. (Source: Upchurch and Littlefield (1988))

**Table 7.8** Estimated risk of sinkhole activity in Upchurch and Littlefield's (1987, 1988) 12-quadrangle study area based on data from the Subsidence Incident Report database for the period 1981–1991

| Quadrangle | Average number of sinkholes reported per year (rounded to the nearest whole number) | Average number of sinkholes reported per year per square kilometer (rounded to one significant figure) |
|---|---|---|
| Brandon | 2 | 0.0008 |
| Citrus Park | 3 | 0.002 |
| Dover | 1 | 0.0006 |
| Lithia | 0 | 0.0001 |
| Lutz[a] | 1 | 0.0006 |
| Odessa[a] | 0 | 0.0005 |
| Plant City West | 3 | 0.002 |
| Riverview | 0 | 0.0002 |
| Sulphur Springs | 4 | 0.002 |
| Thonotosassa | 1 | 0.0006 |
| Wesley Chapel[a] | 0 | 0.000 |
| Zephyrhills[a] | 0 | 0.000 |

[a]Data restricted to the area of the quadrangle map within Hillsborough County

For this period, there were 156 reported subsidence events that were considered probable sinkholes. Figure 7.85 presents the temporal distribution of sinkholes by quadrangle. Note the high number or reported sinkholes in the Plant City West Quadrangle in 1985, the year of a sinkhole swarm caused by intense groundwater pumpage of the UFA. Based on this period of record and with the understanding that there is an unknown amount of under-reporting of subsidence incidents, the estimated *minimum* sinkhole risk for each quadrangle area is presented in Table 7.8. The estimated minimum risks for the 12-quadrangle study area ranged from 0.0000 to 0.002 sinkholes/km$^2$/yr. The highest apparent risks (Table 7.8) are within the Sulphur Springs karst terrane (Citrus Park and Sulphur Springs quadrangles) and the Dover area (Plant City West quadrangle). The Brandon karst terrane has somewhat lower, but elevated, apparent risk (Brandon, Dover and Thonotosassa quadrangles).

Upchurch and Littlefield (1987, 1988) used chi-square analysis to compare modern and ancient sinkhole spatial distributions. With two exceptions, they determined that the presence of ancient sinkholes (closed topographic depressions) in an area predicted the occurrence of modern sinkholes. These two exceptions point out some of the flaws of quantifying sinkhole risk based on a data set such as the Subsidence Incident Reports database. Comparison of Figs. 7.83 and 7.85 illustrates the two exceptions. The first is the absence of modern sinkholes in the Brandon karst terrane. A small number of sinkholes have developed in this area since 1985, so it appears that under-reporting of sinkhole occurrences from 1964 through 1985 was an issue. The second anomaly is the Dover sinkhole area (Fig. 7.85). This area is in the Hardee Upland Province, an area where karst-related, closed depressions are rare owing to the high degree of confinement of the UFA by the clay-rich sediments of the Miocene Hawthorn Group. These sinkholes developed as a result of intense groundwater extraction from the highly confined aquifer for crop freeze protection in 1977 and 1985 (Metcalf and Hall 1984; Bengtsson 1987; Upchurch et al. 2013; Upchurch 2016). Closed depressions (relict sinkholes) are not common in the area and these sinkholes are human-induced, so the analysis illustrated in Fig. 7.82 did not indicate unusual numbers of closed depressions.

#### 7.5.3.2 Evaluating Sinkhole Reporting Biases

Brinkman et al. (2007a, b) compared modern sinkhole data from the FGS Subsidence Incident Report database with closed depressions mapped at the 1:24,000 scale from USGS topographic maps in Pinellas and Hillsborough counties using Geographic Information System (GIS) techniques. They concluded that the distribution of modern sinkholes differs from the closed depression data, and that processes are causing sinkholes today that were not operative in the past. This conclusion differs from the conclusion that ancient and modern sinkholes are roughly spatially coincident (Upchurch and Littlefield 1987, 1988).

### 7.5.4 Loss of Relict Sinkhole Information

Brinkman et al. (2007a, b) compared historical aerial photographs taken over Pinellas County in a Geographic Information System (GIS) investigation of the fate of closed depressions in an urbanizing environment. They classified features, such as depressions and wetlands, in terms of confidence that the features were ancient, or relict, sinkholes. They then compared the spatial density and distribution of these features in 1926 and 1995. There was a significant loss of features over the 69 years. For example, the percent of land area that was represented by "likely sinkholes" changed from 2.7 to 0.22 over the 69-year time period. Similarly, "possible sinkholes" changed from 3.44% to 0.55% of the land area. Clearly, these data suggest that use of modern aerial photographs to identify ancient depressions or sinkholes in an urban environment is likely to result in a highly censored dataset. One should always utilize the oldest available aerial images for such investigations.

The Brinkman et al. (2007a, b) data also suggest a problematic possibility. That is, buildings and other structures have been constructed in low areas that likely had origins as sinkholes.[15] Combining the "likely" and "possible" sinkhole categories, the percent land surface area covered by sinkholes as interpreted from aerial photographs declined by 5.4%. Does this mean that a little over 5% of the land in the study area has structures and/or fill that is currently at risk because they were constructed in these ancient sinkholes? This is a question that is relevant to all developed areas of Florida and other karst terrains, and Brinkman and his colleagues have demonstrated a possible research approach to this important problem.

### 7.5.5 Comparisons of Modern and Relict Sinkhole Occurrences

While it is possible to assign qualitative risk based on the presence of ancient sinkhole depressions and/or an abundance of actuarial data on modern sinkhole occurrences, assigning quantitative risk is problematic because of the timing and mechanisms of sinkhole development. Post-hoc analysis of Upchurch and Littlefield's (1987, 1988) data demonstrates two of the weaknesses in quantifying sinkhole risk. These weaknesses resulted in the two anomalous areas noted above.

Sinkhole development in the eogenetic karst of Florida is episodic and driven by a variety of causes (Upchurch 2016). Spatial differences in (1) cover material, thickness, and strength, (2) void development, and (3) aquifer potentials, combined with temporal variations in (4) rainfall, (5) aquifer withdrawals, and (6) other human activities results in local differences in the rate of sinkhole development. One cannot assume that there is a steady, year-to-year background of sinkhole development. As shown in Fig. 7.83, there is considerable variability in the year-to-year number of sinkholes that develop in the higher sinkhole-risk areas.

Beck (1991) suggested, for example, that there is a limit of about 5 years during which anecdotal and recorded data can be considered reliable. The lack of modern sinkhole activity reported in the Brandon karst terrane is a good example of this problem. While modern sinkholes have been recorded within this terrane, they were not identified during the Upchurch and Littlefield (1987, 1988) study. Evidently, the period of record was not sufficient to capture adequate subsidence incident data within the Brandon karst terrane.

Furthermore, some sinkhole development in Florida is a result of extreme events, such as pumpage during deep and extended-duration freeze events to protect crops through

[15] The authors' experience with sinkhole investigations in Pinellas County suggests that many of the subsidence features identified by Brinkman et al. were likely cover-subsidence sinkholes or, possibly, pseudokarst, so risk of damage to modern structures may be less than the data suggest.

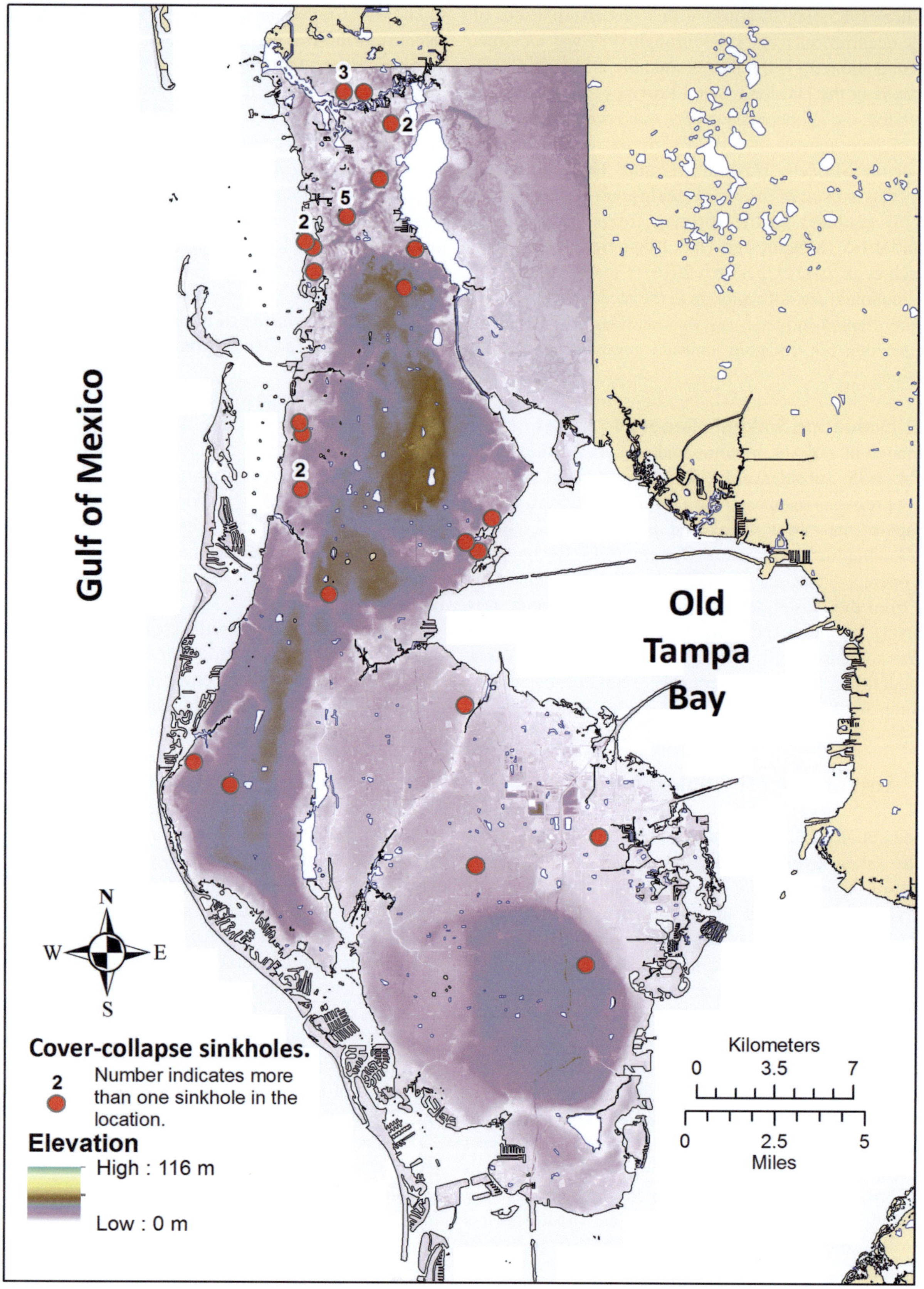


**Fig. 7.86** Cover-collapse sinkholes identified by Beck and Sayed (1991) in Pinellas County

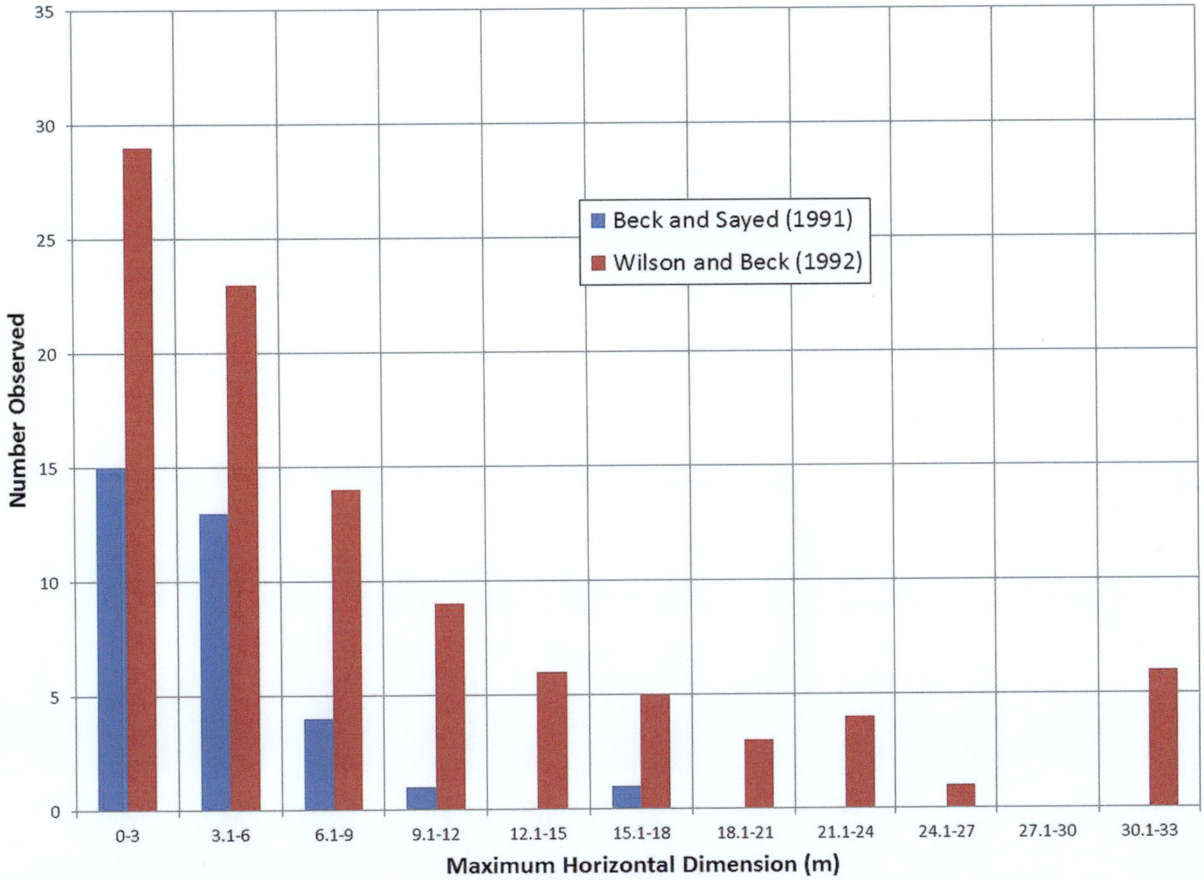


**Fig. 7.87** Maximum diameter of reported cover collapse sinkhole in Pinellas County and the Orlando area. (Modified from Beck and Sayed (1991) and Wilson and Beck (1992))

irrigation (Metcalf and Hall 1984; Bengtsson 1987; Upchurch et al. 2013; Upchurch 2016; Scott 2017b) or high-rainfall tropical storms (Upchurch 2016; Scott 2017a). These events are episodic and related to rapid, intense stresses placed on aquifer pressure fields and/or geologic materials. The 1977, 1985, and 2010 drawdown events in the Dover area of Hillsborough County caused sinkholes in an area where the presence of ancient sinkholes is extremely limited. Here, the absence of ancient sinkholes would not have predicted that the area is at risk, yet excessive pumping of water from the UFA for freeze protection results in the area being amongst the highest for modern sinkhole risk in the study area.

Upchurch and Littlefield's (1987, 1988) conclusion that one can assign general sinkhole risk to identify general areas where relict sinkholes exist is important because it allows one to consider sinkhole risk in rural, undeveloped areas where sinkhole reportage and sinkhole claims have been limited. It also provides a means for corroborating risk based on actuarial data. However, as illustrated by the Dover area sinkholes, one cannot rule out sinkhole development in relatively low risk areas as a result of extreme events.

### 7.5.6 Beck and Sayed's Sinkhole Evaluation of Pinellas County

As part of a comprehensive study of sinkholes in Pinellas County, Beck and Sayed (1991) used the FSRI sinkhole database plus new data obtained through a public request in the media for additional information to characterize sinkhole risk in Pinellas County. All data were verified by the authors as true sinkholes prior to use. Also, the data set purposefully excluded subsidence incidents that were in dispute as to cause.

All totaled, 42 sinkholes were documented in Pinellas County for the period from 1975 to early 1991. Beck and Sayed acknowledged that there was most probably a problem with under-reporting of sinkholes. They also stated that the data reliability was less for the early portion of the period of record, and that it probably improved in the later portions of the record.

The highest concentration of sinkholes occurred in the northern part of the county (Fig. 7.86), where limestone is near the land surface and cover over the limestone ranged from 8 to 15 m in thickness. The extreme northern part of the

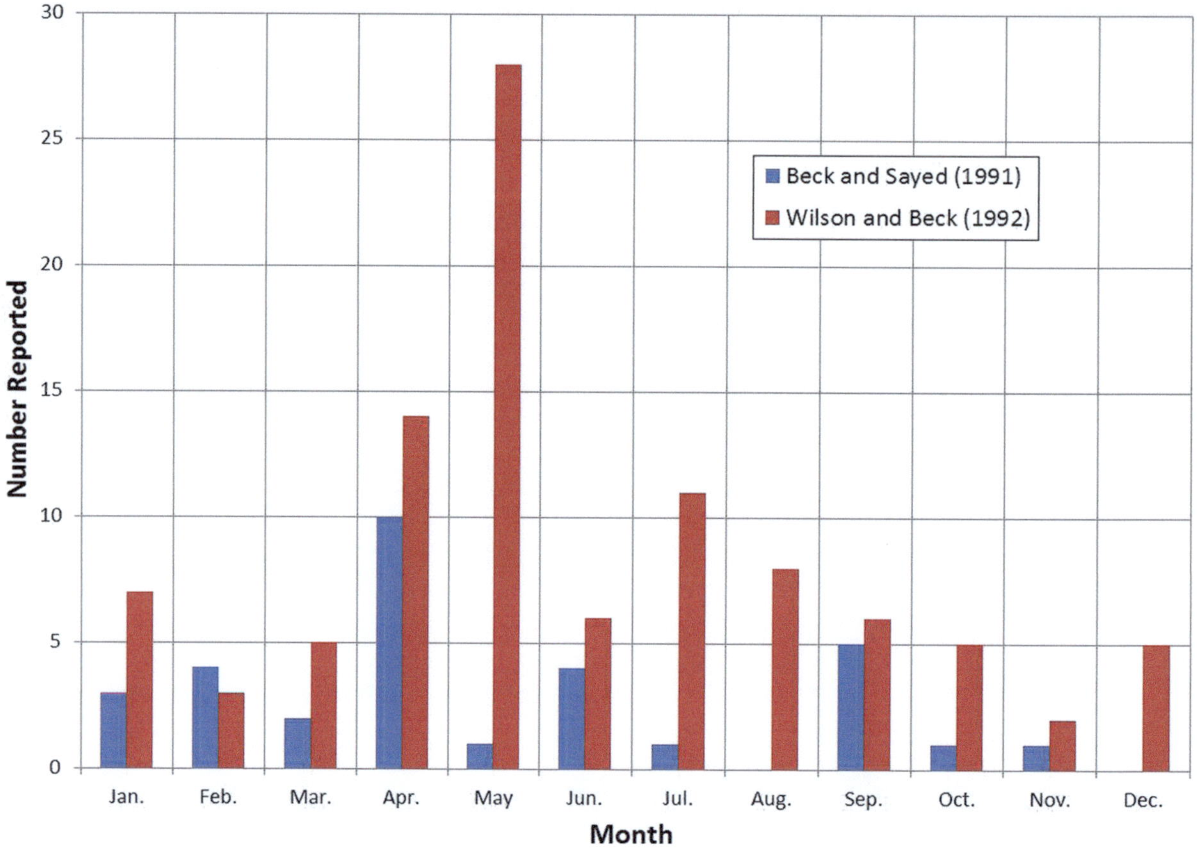


**Fig. 7.88** Frequency of sinkholes reported in Pinellas County from 1975 to 1991 and the Orlando area from 1961 to 1986. (From Beck and Sayed (1991) and Wilson and Beck (1992))

study area had a concentration of 15 cover-collapse sinkholes, and the remainder of the cover-collapse sinkholes was more-or-less randomly scattered about the county. The northern part of the county, an area of approximately 49 km$^2$, experienced a reported 15 sinkholes between 1975 and 1990. Based on these data, Beck and Sayed (1991) estimated the sinkhole risk to be 0.02 sinkholes/km$^2$/yr.

In addition to evaluating sinkhole risk in Pinellas County, Beck and Sayed (1991) examined the seasonal distribution and the dimensions of reported sinkholes. The maximum diameters of the reported sinkholes ranged from less than 3 to 18 m (Fig. 7.87). Modal diameter of the sinkholes was less than 3 m.

April was the month with the most reported sinkholes, and August had the least (Fig. 7.88). The temporal distribution of sinkholes is somewhat problematic. Conventional thinking is that sinkholes develop in wet years and near the onset of the rainy season (late May through June) as a result of hydraulic loading of previously drained sediment and changes in aquifer potentials. April is the height of the dry season in Pinellas County, which may implicate groundwater withdrawals as a cause of sinkhole development. Many of the sinkholes in northern Pinellas County are co-located with penetrations through undifferentiated Hawthorn Group clays that overlie the limestone. Dehydration and shrinkage of the clay during the dry season may enable piping to occur and sinkholes to develop.

### 7.5.7 Wilson and Beck's Investigation of Sinkhole Risk in the Orlando Area

Wilson and Beck (1992) reported on the factors that affect sinkhole development in the Orlando area (Orange and Seminole counties). They investigated sinkhole risk based on the FSRI database. Like previous investigations, they reported an increase in sinkhole reportage beginning in 1981, the year of the Winter Park sinkhole, which was located within the study area.

From 1961 through 1986, the number of new sinkholes averaged 11.2/year. New sinkhole density ranged from 0 to 0.5 sinkholes/km$^2$. They utilized a moving average technique to estimate the spatial distribution of sinkholes throughout the study area (Fig. 7.89). Highest frequency of occurrence was in northern Orange and in Seminole counties, where the limestone is relatively shallow on and near the Mt. Dora Ridge (Fig. 3.14). In these areas the frequency (Fig. 7.89) of sinkholes reached greater than 0.8 sinkholes/km$^2$. Given the

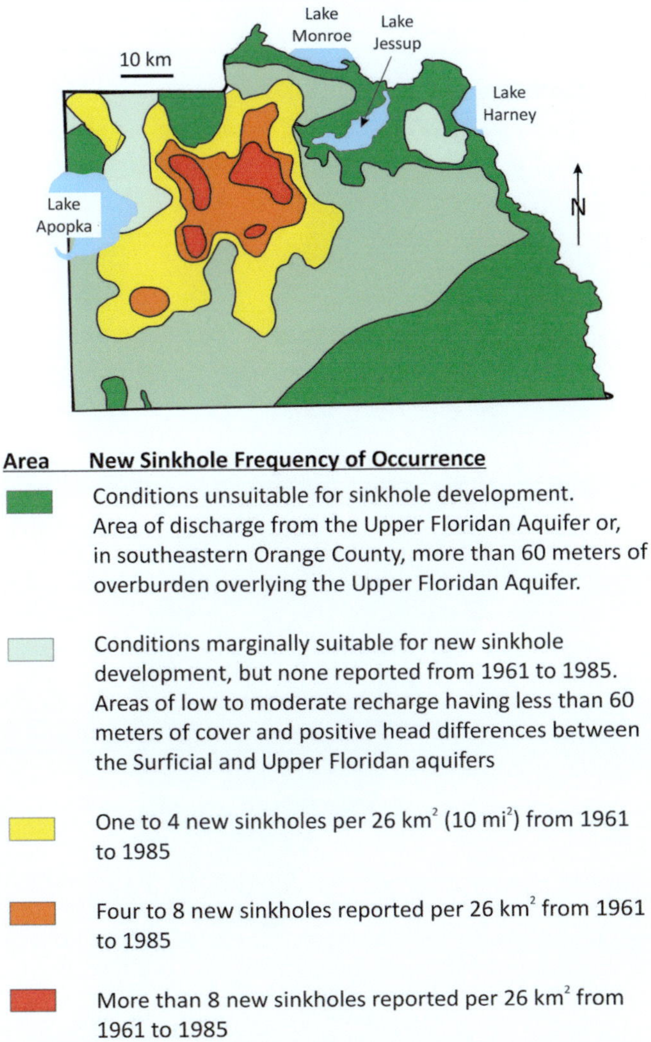


**Fig. 7.89** Density of reported modern sinkholes in Orange and Seminole counties from 1961 to 1985. (Modified from Wilson and Beck (1992))

24 year period of record and issues with under-reporting of sinkholes, especially early in the period of record, the *minimum* sinkhole risk in the high risk areas in Fig. 7.89 is greater than 0.03 sinkholes/km$^2$/yr.

### 7.5.8 Wilson's Northern Peninsular Florida Sinkhole Risk Assessment

Wilson (1995) evaluated sinkhole development in northern peninsular Florida. He examined the densities (occurrence per unit area) of features that were classified into three categories:

- Sinkholes (relict sinkholes or closed depressions represented on 1:24,000 USGS topographic maps; category includes large depressions such as karst valleys, poljes, and uvalas);
- Modern sinkholes from the FSRI database augmented by other data; and
- Buried sinkholes (apparent ancient sinkholes with no topographic expression but observed in ground penetrating radar (GPR) data).

**Table 7.9** Adjustment factors utilized by Wilson (1995) to compensate for under reporting of modern sinkholes

| Area | Source | Estimated amount of under-reporting (adjustment factor) |
|---|---|---|
| Orlando area, Florida | Wilson et al. (1987) | "at least 2.5" |
| Eastern Tennessee | Ketelle et al. (1988) | 5–8.5 |
| Alachua County, Florida | Wilson and McDonald (1988) | 22 |

Wilson's data were derived from investigations undertaken by his company around the state and did not represent a systematic study of the entire region.

According to Wilson, sinkholes (closed topographic depressions) occurred with the following spatial densities (geomorphic provinces are from White 1970):

- Gulf Coastal Lowlands Province: 0–3 sinkholes/km$^2$;
- Lake Wales and Mt. Dora Ridges: 4–7 sinkholes/km$^2$; and
- Western Valley Province: 15–19 sinkholes/km$^2$.

The sinkhole density reported for the Gulf Coastal Lowlands Province compares favorably with the data of Upchurch and Littlefield (1987, 1988) for Hillsborough County, most of which falls within White's (1970) Gulf Coastal Lowlands physiographic province.

Buried sinkholes (apparent, in-filled relict sinkholes without any topographic expression) were reported to occur at densities of 56–250 features/km$^2$ at five upland sites and 460–3400 features/km$^2$ at 10 lowland sites. Modern sinkholes were reported to occur at densities of less than 0.0004 sinkholes/km$^2$/yr. to 1.9 sinkholes/km$^2$/yr.

Wilson (1995) addressed the problem of under-reporting of modern sinkholes through the use of "adjustment factors" (Table 7.9). Based on these previous estimates for the amount of under-reporting, Wilson used the following adjustments for northern peninsular Florida:

- 2.5–5 for urban and dense residential areas;
- 5–10 for light residential areas; and
- 10–20 for rural areas, farmlands, and ranches.

After adjustment, the estimated modern sinkhole occurrence rate was thought to be 0.4 sinkholes/km$^2$/yr. for the

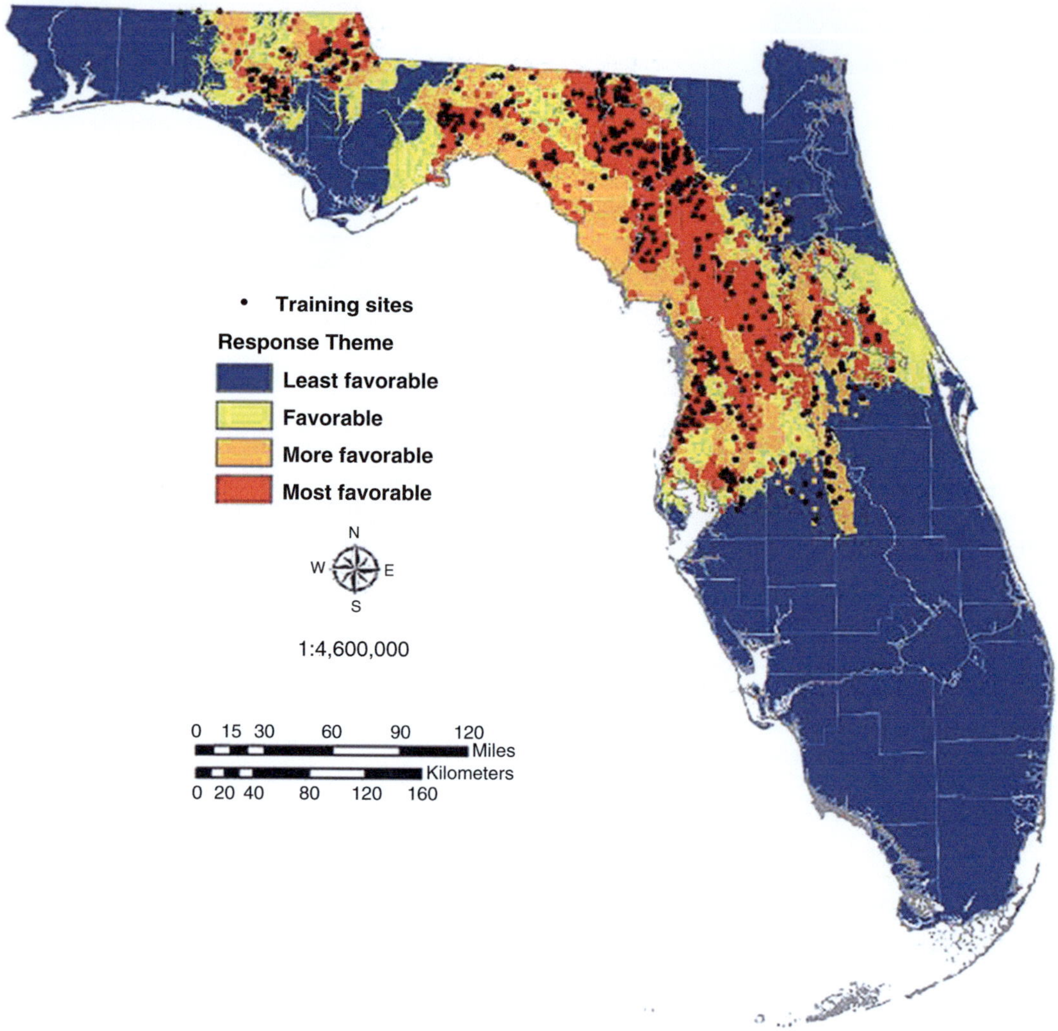


**Fig. 7.90** Sinkhole favorability in Florida. Dots are sinkhole locations used to "train" the weights-of-evidence model to identify sinkhole favorability. (From Kromhout et al. (2017 Fig. 13))

study areas. Use of these adjustment factors has not been well documented or justified. However, it is clear that under reporting of sinkhole incidents is a problem for establishing sinkhole risk.

### 7.5.9 West-Central Florida Sinkhole Risk Study by Scheidt et al.

An interesting investigation by Scheidt et al. (2005) incorporated housing and population densities and other anthropogenic factors to evaluate risk to urban areas in Hillsborough, Pasco, and Pinellas counties. They concluded that Tampa had the highest sinkhole risk when weighted by anthropogenic factors. New Port Richey and Hudson (Pasco County) had the highest risks of communities in the coastal areas of the three-county area. These risks were related to the degree of urbanization and associated groundwater withdrawals.

The data utilized for the analysis consisted of 431 subsidence incidences (assumed sinkholes) from the FGS Subsidence Incident Report database. The authors estimated that annual cost to Tampa property resulting from sinkholes was over \$5 million (M), and each of the other communities had an estimated loss of less than \$0.5 M.

### 7.5.10 Florida Geological Sinkhole Favorability Study

In 2017, the Florida Geological Survey completed a study of sinkhole development potential in Florida based on GIS data and Weights of Evidence (WofE) modeling (Kromhout et al. 2017). WofE modeling utilized Bayesian analysis of 14 spatial databases (model layers) such as (1) top of rock elevations, (2) overburden (cover) thickness, and (3) hydraulic conductivity of soils for comparison to field verified sinkhole locations. Verified

**Table 7.10** Summary of quantitative sinkhole risks estimated for Florida

| Area | Risk estimate (sinkholes/km²/yr.) | Source |
|---|---|---|
| Hillsborough County | 0.0–0.002[a] 0.0–8[b] | Upchurch and Littlefield (1987, 1988) |
| Pinellas County | 0.02[b] | Beck and Sayed (1991) |
| Orange and Seminole counties | >0.03[b] | Wilson and Beck (1992) |
| Multiple areas, northwest peninsular Florida | 0.0004–1.9[b] 0.4[c] | Wilson (1995) |

[a]Based on 7.5-min quadrangle areas
[b]Based on unit area measurements
[c]Based on adjustment factors cited by Wilson (1995)

sinkhole locations were used to "train" the spatial model to identify conditions most conducive to sinkhole development. Sinkholes were classified as (1) collapse (rock- or cover-collapse) if there was "rapid" subsidence and (2) solution (cover subsidence or solution) if subsidence was "slow."

Three (out of 14) spatial databases were found to provide the best results of the modeling. These were (1) overburden thickness, (2) closed depressions on the 1:24,000 scale topographic maps, and (3) vertical distance between the water table and top of limestone. Four risk (favorability categories) were identified: least, favorable, favorable, more favorable, and most favorable. Figure 7.90 illustrates the mapped favorability zones as compared to the "training sites" (field verified sinkholes) identified by the Geological Survey.

### 7.5.11 Summary: Evaluating Sinkhole Risk

In summary, our ability to provide a *qualitative* estimate of actual sinkhole development risk is good. A number of studies have concluded that cover-collapse sinkholes are co-located with ancient or relict sinkholes as represented by closed topographic depressions. It is also possible to conclude that true sinkholes are limited in spatial extent and frequency, and that they tend to develop in areas where others have previously developed. This should not suggest that sinkhole development can be predicted or confirmed by proximity to earlier true sinkholes; only site-specific data can establish the origin of modern land-surface depressions.

*Quantification* of sinkhole risk is highly problematic owing to poor data, under-reporting of sinkhole occurrences, a poor definition and popular perception of sinkholes and sinkhole "activity," and other factors. Table 7.10 summarizes the results of attempts to quantify sinkhole risk. It is clear from these studies that, with the exception of extreme events, such as the Plant City-Dover sinkhole events in 1985 and 2010, the risks of a true sinkhole are low statewide on an annual basis.

**Table 7.11** Fatalities caused by sinkhole development in Florida since 1960

| Year | Incident | Location | Source |
|---|---|---|---|
| 1960 | Driller killed when drilling a test well caused collapse | Keystone Heights, Putnam County | Newton (1987) |
| 1967 | 14-year old man killed in sinkhole | Citrus Park, Hillsborough County | Florida Geological Survey (pers. comm. 2013) |
| 1968 | Bridge over Anclote River collapsed into a sinkhole, one driver was killed | Tarpon Springs, Pinellas County | Newton (1987) |
| 2011 | Well driller killed as sinkhole open under drilling rig | Trenton, Gilchrist County | Gainesville Sun (2011) |
| 2013 | Occupant killed as sinkhole opened under house | Seffner, Hillsborough County | Tampa Bay Times (2013) |

## 7.6 What Are the Costs of Sinkhole Development?

In Florida, we tend to focus on insurance claims and property damage caused by development of sinkholes. The annual costs of property damage varies with the terms of property insurance policies in Florida, but it is in the tens of millions of dollars. However, there are other risks associated with sinkholes in Florida that may affect human welfare and property insurability. These include loss of life and injuries, flooding, and pollution issues.

### 7.6.1 Injuries and Loss of Life

In February, 2011, a gentleman in Seffner, a suburb of Tampa in Hillsborough County, was killed as a sinkhole opened under his home. According to eyewitness accounts, the sinkhole collapsed under his bedroom, and he and the contents of the room were quickly swallowed into the sinkhole. This tragic event was widely reported and frightened many Floridians. The number of sinkhole insurance claims spiked and numerous follow-up stories kept the calamity on our minds for months. This loss of life brings to question how high is the risk of serious injury as a result of a sinkhole "swallowing" a person?

Table 7.11 summarizes the known losses of life as a result of development of sinkholes since 1960. Only five fatalities have been reported in the 54 years for which records can be found. This indicates that the risk of a fatality as a result of a sinkhole opening is vanishingly small (0.1/yr. for the period 1960–2013). Examination of Table 7.11 suggests that one segment of the population, well drillers, is at greatest risk of fatal injury. Forty percent (2) of the fatalities have been to individuals drilling test holes and wells.

**Fig. 7.91** (**a**) September 14, 1984 newspaper article describing flooding in Live Oak (Suwannee County) after Hurricane Dora. (**b**) Flooding in a relict sinkhole in Live Oak (Suwannee County) in March 1980. The building was a pesticide manufacturing facility at the time of the photograph. A stormwater basin and drainage well has since been installed in the depression

No one tracks non-fatal injuries that have resulted from the opening of sinkholes. We often hear of pedestrians, buildings, and cars falling into sinkholes, so the incidence of non-fatal injuries is likely higher than for fatalities.

Finally, use of ancient sinkholes and karst features results in the highest incidence of injury and fatality. Sinkhole lakes and caverns are popular for recreational swimming, open water and cave diving, and boating, and injuries and fatalities

**Fig. 7.92** A water treatment pond associated with a municipal sewage treatment facility in Live Oak, Florida, that had developed sinkholes and drained. (Photo taken in May, 1979)

are common. Clearly, use of existing karst features is much riskier than involvement with a newly developing sinkhole.

### 7.6.2 Flooding and Drainage Wells

Many Florida communities are located in karst terrains where relict sinkholes abound. These depressions represent a hazard because of flooding during extreme precipitation events (Fig. 7.91a, b). These communities typically have flooding problems because the cost of installing and maintaining stormwater lift stations and storm sewers to move stormwater out of the depressional areas at risk is prohibitive given the number and topographic relief of the depressions.

Many communities with internally drained karst basins have "solved" their flooding problems by installing "drainage wells" in the bottoms of the sinks or in engineered stormwater basins to allow drainage of the stormwater directly into the underlying UFA (Kimrey 1978; Kimrey and Fayard 1984; German 1989; German and Bradner 1989; Taylor 1993; Bradner 1996). However, if the drainage wells become plugged or overwhelmed during extreme precipitation events, flooding of the land surface still occurs.

The city of Live Oak (Suwannee County; Fig. 7.91a, b) is an example of a flood-prone karst area where flooding has been a serious issue. Most recently, flooding was widespread in 2013 when Tropical Storm Debbie dumped up to 0.6 m of rain on Live Oak. In addition to the flooding, numerous sinkholes developed in Live Oak as a result of the hydraulic loading within and near the closed depressions (Scott 2017a, b).

### 7.6.3 Groundwater Pollution

Drainage wells have also been a concern, because of their potential for pollution of groundwater in the receiving aquifer. Studies by Hull and Yurewicz (1979), Weber (1981), and Bradner (1991) have shown that there is (1) a high concentration pulse of nutrients, coliform bacteria, fecal material, and metals swept into the aquifer at the onset of a rainstorm and (2) a steady, but lower concentration influx of contaminants thereafter. Situations, such as is depicted in Fig. 7.91b, constitute a liability and hazard, which may have liability implications and certainly will have public health and safety ramifications.

Stormwater basins often develop cover-collapse or suffosion sinkholes because (1) the cover overlying the limestone has been thinned by basin construction and (2) episodic hydraulic loading occurs during stormwater events (Upchurch 2016). These may be sources of introduction of

Fig. 7.93 Pesticide drums disposed of in a sinkhole near Eustis, Florida, in 2005

highway runoff into the receiving aquifer (i.e., German 1989; Kantrowitz and Woodham 1995).

Owing to the costs of installing and maintaining lift stations and sewer mains at industrial sites and in communities with well-developed, large, relict sinkholes, sinkholes have been utilized for gravity-driven water- or effluent-treatment basins in the past. If these basins develop new sinkholes, there is a risk of contamination of the receiving aquifer water. The basin shown in Fig. 7.92 was a membrane-lined "polishing" pond at the City of Live Oak's sewage treatment plant in the 1970s. The pond developed several sinkholes, causing discharge of the partly treated effluent into the UFA and damage to the membrane (the dark surface covering in the photograph). Location of such treatment facilities in sinkhole depressions can constitute a significant contamination risk.

Finally, sinkholes are traditional waste disposal sites in rural areas (Fig. 7.93). Wastes in sinkholes can cause groundwater and soil contamination, a liability and human health concern. Several governmental agencies have funded efforts to identify these waste-filled sinkholes and remove the refuse.

### 7.6.4 Property Damage

Finally, the most obvious and costly sinkhole risk in Florida is damage to structures that overlie the developing sinkhole. In Florida, millions of dollars are expended each year to

Fig. 7.94 Sinkhole damage to Disston Avenue, Tarpon Springs, Pinellas County, in 2011

remediate damages to dwellings and other buildings, roads (Fig. 7.94), and other structures.

Much of the costs of sinkhole remediation are paid by property insurance on the structures. However, remediation of damage to public and many industrial/commercial facilities is typically self-funded.

## 7.7 Summary

This chapter deals with two of the most important aspects of Florida karst: sinkholes and caves. Caves are included in this chapter because they play a role in the development of several forms of sinkholes.

### 7.7.1 Sinkholes

(a) Sinkholes are important in Florida because they result in millions of dollars in damage each year. As a result, they are a subject of great concern to many Floridians and their development is often reported by the press locally. The largest and most destructive receive coverage internationally, as well.

(b) Florida's karst provides important insights into the processes of sinkhole formation in eogenetic karst because the carbonate sediments and rocks are geologically young and many are diagenetically immature.

(c) The common modes of sinkhole formation include

   (i) Solution sinkholes, which form by dissolution of the upper few meters of limestone in areas where cover is thin or non-existent.

      1. In Florida, the most common forms of solution sinkholes are solution pipes or channels, which are cylindrical and deeper than they are wide. Their bottoms appear to occur at, or near, the water table.
      2. Arguments as to how solution sinkholes form vary. There is evidence suggesting that they form (a) around plants or plant roots, (b) by random dissolution at the top of the limestone, or (c) by enlargement of pre-existing locations of enhanced vertical drainage.
      3. Solution sinkholes are common in the Pleistocene Miami and Key Largo limestones of southeastern Florida. As such, they have formed in the last 125,000 years, since the last high sea stand near the end of the Pleistocene.
      4. In central and northern peninsular Florida, they are most common in the Eocene Ocala Limestone. They are often deeper than the southern Florida solution pipes. Fossils as old as the late Miocene have been recovered from the northern solution channels indicating that at least some are several millions of years old.
      5. Solution sinkholes are typically filled with limestone breccia, clay, and sand transported into the solution pipe after formation. In southern Florida, the solution pipes are commonly lined with duricrust; in central and northern Florida they typically have a thin, calcitic rind.

   (ii) Suffosion sinkholes, which are formed when non-cohesive sediments bridge over a void in the underlying limestone.

      1. The term suffosion refers to raveling, the process of free migration of sediment into an underlying cavity. The process of suffosion (or raveling) occurs in many circumstances in Florida, including during cover collapse.
      2. Suffosion sinkholes are commonly observed when human activity results in deposition of thin sediment cover over a pre-existing cavity. They commonly occur in stormwater retention basins and similar areas where the cover is disturbed and thinned. Loading, usually by water, induces washing of the cover into the void.

   (iii) Rock-collapse sinkholes form when the limestone roof of a cave fails and collapses.

      1. Few, modern rock-collapse events have been documented in Florida, but they are relatively common as ancient collapse features.
      2. Karst windows, especially those associated with cavern systems that transport surface water from a swallet (a sinkhole where a stream goes underground) to a spring, are common rock-collapse features.
      3. Cenote, deep water-filled rock-collapse sinkholes, are scattered through central and southern Florida as isolated sinkholes.
         (a) They are unusually deep for Florida sinkholes at over 60 m to the top of the breakdown cone. At least six are known or suspected on the mainland and deep, sediment-filled collapse features have been identified in southeast Florida by geophysical means.
         (b) The depths of the rock-filled avens below the breakdown cones are unknown. The ones on the mainland appear to extend at least as deep as the Ocala Limestone 200+ m below land surface.
         (c) It has been suggested that these deep sinkholes may have a hypogenic origin either related to a freshwater/saltwater mixing zone or dissolution by deep, sulfuric acids

derived from gypsum or anhydrite in the MCU or deeper.

(iv) Cover-collapse sinkholes are formed by the failure of sand and clay cover overlying a void in the underlying limestone.
1. The void space can be either at the cover/limestone interface or within the limestone.
2. The voids in cover-collapse sinkholes are usually bridged by cohesive, clay-rich sediments although cemented sand and other materials may have sufficient strength to bridge voids.
3. The failure of a cover-collapse sinkhole can be rapid, but rate of development is not a valid criterion because size of the void, compaction of the collapsed materials, and other factors can slow or stop downward movement of the raveling sediment.
4. These sinkholes are triggered by a number of factors including (a) seasonal changes in moisture content of the cover, (b) hydraulic loading during rainfall events or changes in surface-water drainage, (c) changes in internal stresses caused by groundwater pumping, especially cyclic pumping, (d) vibrations, and (e) other causes.
5. Because development of cover-collapse sinkholes is often rapid and in locations where the presence of a bridged void is undetected. Cover collapse is the most common, damaging form of sinkhole development in Florida.
6. Collapse is not controlled by the rate of limestone dissolution, voids are pre-existing conditions and the collapse is a failure of the cover materials overlying the void.
7. Extremely large (greater than 100 m in diameter) cover-collapse sinkholes are rare because of limitations in the volumes of voids in the limestone.

(v) Cover-subsidence sinkholes form by the slow dissolution of the upper surface of a carbonate-rich sediment.
1. The depressions created by cover subsidence form by the slow sagging of cover strata into the space created by shell or limestone dissolution. There is no single void into which cover sediments ravel or collapse.
2. Cover-subsidence sinkholes are common throughout Florida where limestone, shell, or, rarely, dolostone is near the land surface.
3. Because it takes thousands of years or more for a cover-subsidence sinkhole to form, subsidence poses minimal hazards unless a structure is constructed over a pre-existing depression with deleterious soils (i.e., peat, expansive clay, poorly consolidated sediment).
4. Cover-subsidence sinkholes can form on unconsolidated sediment, especially shallow (less than 3 m) shell beds. For there to be dissolution of sufficient material to allow for depression formation, it is estimated that shell beds must contain over 50% shell material.
5. Depressions are common in areas where the water table is high. In this circumstance the depressions can be either permanent or ephemeral (seasonal) wetlands.
6. Deeply buried shell, especially in clay-rich sediments, does not form cover-subsidence features.
7. Because of their slow rate of development, cover-subsidence sinkholes can commonly be identified as closed depressions, especially wetland depressions, on early aerial photographs.
8. The weathered, siliciclastic matrix that remains after dissolution of shell is typically (a) sandy clay or clayey sand, (b) stained dark brown to reddish brown by ferric hydroxides, and (c) characterized by weathered and rounded fragments of limestone or shell. Traces of collapsed shells and soft, earthy, weathered shell may be common. Thin lenses or peat or muck are common as a result of the growth of cypress and other plants in the wetlands.
9. Sediments that wash into the depression from the surrounding area may include finely layered sand beds. If there are associated dissolved organics, the sandy soil zones may be "bleached" by the acidic, humic substances in groundwater.

(vi) Alluvial sinkholes are relict sinkholes within which the aven is filled and stable.
1. They may be visible as depressions on the land surface or completely masked by burial under cover sediments.
2. Alluvial sinkholes may be reactivated and undergo additional raveling and subsidence by a variety of processes. Cycling of groundwater levels and circulation may cause short-term changes in stresses on the sinkhole fill and induce additional raveling or sediment compaction, both of which will lead to additional subsidence.
3. Compaction or slumping of a cone of raveled material in the underlying void may provide additional space into which raveling can transport cover materials.

4. Rarely, either by rock collapse or possibly erosion within the void, the space into which a sinkhole has collapsed can increase thus providing space for additional sediment movement.

(vii) Karst windows are sinkholes formed by rock collapse or cover collapse that intersect the water table or potentiometric surface of the carbonate aquifer.
   1. They are located over caves into which the collapse occurs.
   2. In Florida, they form access points to phreatic caves for divers.
   3. A notable occurrence of karst windows is over caverns that transport surface water from a swallet.

(viii) In west-central Florida, there are several closed depressions that appear to be hybrids with karst and aeolian origins.
   1. These depressions are similar to Carolina bays found elsewhere on the Atlantic Coastal Plain.
   2. The may be as large as a kilometer in their longest dimension, which is oriented northeast to southwest.
   3. A parabolic dune train occurs on the northeastern (downwind) margin of the depression.
   4. The center and southwestern portion of each depression is characterized by wetlands and ephemeral lakes.
   5. Sinkholes dot the floors of the wetlands or lakes.
   6. It is thought that they formed in late Pleistocene to early Holocene times when sinkhole development drained the water table and made dry, vegetation-free, cover sand accessible to the prevailing winds. The wind-transported sand accumulated as dunes at the lee margin of the depression.

(ix) Pseudokarst depressions are common in certain areas of Florida, especially in the Hardee Upland of central Florida in Plio-Pleistocene marine terrace sand.
   1. The depressions (ancient sag features, sinkholes, stream beds, etc.) originally existed in the underlying Mio-Pleistocene paleosol.
   2. Minor, differential compaction of the terrace sand after it was deposited on the paleosol resulted in shallow depressions and drainage patterns on the land surface.
   3. Conversion of smectitic to kaolinitic clays during the intense weathering of the Mio-Pliocene paleosol may have contributed to the origin of the ancient depressions.

(x) Uvalas are coalescent sinkholes with complex outlines and multiple sinkholes.
   1. Because of the sand mantle in Florida, their outlines may be rounded by soil creep into the depressions.
   2. Rock exposures are uncommon.
   3. Uvalas can extend kilometers in dimension and they take on the shapes of interfering circular sinks, fractures, and collapsed caves.
   4. Many Florida uvalas have been developed with homes, businesses, and farmland. Because of their size and depressional cross sections, they may flood unless one or more of the sinkholes drain the depression.

(xi) Poljes are coalescent sinkholes (uvalas) that have been filled by lacustrine or paludal sediments.
   1. They have the same properties as uvalas with the exception that sinkholes in the bottoms of poljes may be plugged, thus allowing for deposition of lacustrine or paludal sediments.
   2. The bottoms of poljes are nearly flat, and normally either overlain by standing water or marshes and swamps.

(xii) Collapsed cave systems are the final product of development of a series of rock-collapse sinkholes over a cavern.
   1. The trace of the collapsed cave is either a string of karst windows in an immature system or a linear depression tied to a nearby stream.
   2. The linear depressions are commonly floored with wetlands and/or small streams. As such they are locally termed sloughs.

(d) Sinkhole failure mechanisms during sinkhole collapse are controlled by the texture and cohesiveness of cover materials.

(i) Modes of cover failure include
   1. raveling and suffosion,
   2. piping,
   3. spalling, slaking, and slabbing of clay-rich sediments,
   4. stoping, and
   5. clay dispersion.

(ii) Cohesive, clay-rich sediments may "relax" to conform to open space or flow if the natural moisture content of the clay is near or at the liquid limit. Under-consolidated clays are most susceptible to this process.

(iii) Pressure arches are unlikely to occur and are only temporary in well-sorted sand. The existence of pressure arches in cohesionless sand is problematic because of the nature of subsurface testing, which is likely to cause failure of an arch in sand, if it exists.

(e) Sinkhole risk has been evaluated several times in Florida with varying degrees of success.
   (i) There is no way to predict where sinkholes will develop.
   (ii) Risk is evaluated by either characterizing ancient sinkholes as expressed as closed depressions in the land surface or by modern actuarial data, primarily obtained by "mining" of the Florida Geological Survey's Sinkhole Incidence Database.
      1. This database contains unverified reports and should be carefully conditioned to remove false positive or erroneously located reports.
      2. There is concern that the sinkhole databases are biased because of a preponderance of reportage from urban and developed areas and along transportation corridors.
   (iii) Risk is not uniformly distributed. Areas with increased risk are generally localized to areas where limestone is relatively shallow and human activities disturb the sediment and/or groundwater systems.
   (iv) Causes of sinkhole development and risk include
      1. Groundwater extraction, especially near large wellfields;
      2. Surface water in impounded areas or disposed of on the land surface;
      3. Mining, construction, and related vibration; or
      4. Structures are placed in "harm's way" without knowledge of karst conditions. In this case, the structure may be placed over a developing sinkhole.
   (v) In most cases, the cause of a sinkhole collapse cannot be determined and/or is complex.
   (vi) Forms of sinkhole loss include:
      1. Loss of life as a result of sinkhole development, which is rare. Only 5 fatalities can be identified over the last 50+ years.
      2. Flooding of depressions is a significant hazard. Drainage wells are common in some areas as a means to enhance surface-water drainage and prevent flooding.
      3. Groundwater contamination as a result of waste disposal in sinkholes and use of drainage wells, which can introduce bacteria and trace chemicals to the aquifer.

### 7.7.2 Caves

(a) Both vadose and phreatic caves are common in central and northern Florida, and small caves are known in the Atlantic Coastal Ridge and Rock Glades provinces of southeastern Florida. Cave development has been extensive in the UFA and to a lesser extent in the Biscayne Aquifer.

(b) Horizons of cave development have been attributed to epigenetic processes related to present-day and to former positions of the water table.
   (i) Deep phreatic caves near the coasts appear to have hypogenetic origins related to mixing with freshwater/saltwater transition zones.
   (ii) Because of the high water table or potentiometric surface of the UFA, vadose caves are mostly found on ridges and topographically high areas.
   (iii) In Florida's karstic ridges and scarps, caves are likely to be dry. Groundwater may be found in the caves. On karst plains, the caves are typically phreatic with access by way of springs, karst windows or short vadose passages. Exploration of caves on the karst plains requires cave diving.
   (iv) Cave diving, an extremely dangerous sport requiring a high level of training and vigilance for safety, is a popular sport in Florida's phreatic caves. Cave divers are continually mapping new caves, including some of the longest and deepest known. Several karst windows have been developed as recreational diving centers.

(c) There is an interesting conundrum with respect to the presence of speleothems, such as stalactites and flowstone.
   (i) Vadose caves are commonly decorated with secondary speleothems, such as stalactites and flowstone.
   (ii) Phreatic caves that have been explored rarely contain secondary cave formations. The absence of common cave decorations in phreatic caves can be attributed to a combination of two factors.
      1. First, the caves formed at, or just below, the water table and have not been subjected to long periods of vadose conditions, and
      2. Second, dissolution as a result of $CO_2$ degassing and concentration at the water table encourages dissolution of calcite, not deposition.

(d) Isotopic and elemental analyses of stalactites from caves in central Florida reveal a record of deposition that, at least in one example, extends back into the late Pleistocene. Episodes of deposition can be correlated to major climatic patterns, such as the Atlantic Multidecadal Oscillations (AMO) and el Niño events. Evidence of crystallization through water evaporation was found in $\delta^{18}O$ and $\delta^{13}C$ data.

(e) Speleothems include most of the common forms of cave decorations; however, some appear relatively limited in terms of abundance.
   (i) Flowstone draperies develop under fractures in the overlying limestone. Given the relatively poor development of fractures in doubly porous limestone, they are limited in extent in many caves.
   (ii) Swarms of soda-straw stalactites are common in areas where water moves through the small, solu-

tion enlarged, interstitial openings in the limestone.

(iii) Rafts and clusters of calcite crystals found in many caves indicate water-table positions and modes of crystallization.

(f) There are three processes that can cause deposition of speleothems.

(i) Evaporation of water results in formation of small calcite crystals in speleothems. $\delta^{18}O$ and $\delta^{13}C$ data from stalactites testify to the importance of evaporation as a process in large speleothem formation.

(ii) $CO_2$ degassing results in slow development of larger, and well formed, crystalline calcite. Development of limpid, macro-calcite crystals in Florida caves is an indication of crystallization as a result of degassing in humid cave atmospheres.

Cave walls may be lined by sparry calcite crystals, especially in shallow pools. Rows and clusters of crystals commonly delineate locations of former water-table positions. Soda straws composed of single calcite crystals or limited, multiple crystals are found in some caves.

(iii) Introduction of water that is over saturated with respect to calcite as a result of up-gradient dissolution of aragonite, high-Mg calcite, or some other phase. The presence of calcite crystals filling shelter and moldic porosity in Pleistocene sediments in southern Florida suggest that chemically over-saturated water may have resulted in crystallization.

(iv) Many stalactites consist of radiating, coarse, columnar calcite crystals. These stalactites are usually centrifugally banded with ferric hydroxide, clay, or small calcite crystals. Isotopic and petrographic analyses suggest that these stalactites wither formed by evaporation or had histories of alternating evaporation and $CO_2$ degassing.

(g) The Biscayne aquifer in southern Florida contains small openings that constitute the early stages of cave formation. They occur near the crest of the Atlantic Coastal Ridge (Cutler Ridge) and rarely in the highest elevations of the Rock Glades provinces. These caves are small and do not extend very far underground.

## References


Alfieri MC Upchurch SB Dobecki TL (2018) Photolinears, fractures, and fallacies: A *post hoc* study of photolinears, Hillsborough County, Florida. Proceedings, 15th Multidiscip Conf on Sinkholes and Engineering and Environ Impacts of Karst, Shepherdstown, West Virginia

Altschuler ZS Jaffe EB Dwornik E (1951) The stratigraphy of the upper part of the Bone Valley Formation and its relation to the leached zone. US Geol Surv Trace Elements Memo Rept 237

Altschuler ZS Jaffe EB Cuttitta F (1956) The aluminum phosphate zone of the Bone Valley Formation, Florida, and its uranium deposits. US Geol Surv Prof Pap 300

Altschuler ZS Dwornik EJ Kramer H (1963) Transformation of montmorillonite to kaolinite during weathering. Science, 141(3576):148–152

Bagnold RA (1956) The flow of cohesionless grains in fluids. Royal Soc London, Philosoph Trans, Series A, 249:235–297

Balfour RC III (2002) In search of the Aucilla. Valdosta, Georgia, Colson Printing Co

Beck BF (1991) On calculating the risk of sinkhole collapse. Proc Appalachian Karst Symp, Nat Speleol Soc, pp. 231–236

Beck BF Sayed S (1991) The sinkhole hazard in Pinellas County: A geologic summary for planning purposes. Fla Sinkhole Res Inst Report 90-91-1

Beck BF Sinclair WC (1986) Sinkholes in Florida: An introduction. Orlando, Fla Sinkhole Res Inst Report 85-86-4

Bengtsson TO (1987) The hydrologic effects from intense ground-water pumpage in east-central Hillsborough County, Florida. In: Beck BF Wilson WL (eds.), Karst hydrogeology: Engineering and environmental applications, Proc Second Multidis Conf on Sinkholes and the environmental impacts of karst, Rotterdam, A.A. Balkema, pp. 109–114

Blatt H Middleton G Murray R (1972) Origin of sedimentary rocks. Englewood Cliffs, New Jersey, Prentice-Hall

Bloomberg D Upchurch SB Hayden ML Williams RC (1988) Cone-penetrometer exploration of sinkholes: Stratigraphy and soil properties. Environ Geol and Water Sci 12:99–105

Bögli A (1980) Karst hydrology and physical speleology. Berlin, Springer-Verlag

Bolton MD (1986) The strength and dilatancy of sands. Géotechnique, 36(1):65–78

Bonacci O (2004) Poljes. In: Gunn J (ed.), Encyclopedia of Caves and Science. New York, Fitzroy Dearborn, pp. 599–600

Bradner LA (1991) Water quality in the upper Floridan aquifer in the vicinity of drainage wells, Orlando, Florida. US Geol Surv Water-Resour Invest Report 90-4175

Bradner LA (1996) Estimation of recharge through selected drainage wells and potential effects from well closure, Orange County, Florida. US Geol Surv Open-File Rept 96-316

Brinkman R Wilson K Elko N Seale LD Florea LJ (2007a) Sinkhole distribution based on pre-development mapping in urbanized Pinellas County, Florida, USA. In: Parise M Gunn J (eds.), Natural and anthropogenic hazards in karst areas: Recognition, analysis and mitigation, Geol Soc London, Sp Pub 279:5–11

Brinkman R Parise M Dye D (2007b) Sinkhole distribution in a rapidly developing urban environment: Hillsborough County, Tampa Bay area, Florida. Eng Geol 99:169–184

Brooks MJ Taylor BE Stone PA Gardner LR (2001) Pleistocene encroachment on the Wateree River sand sheet into Big Bay on the middle Coastal Plain of South Carolina. Southeast Geol 40(4):241–257

Butt PL Morris TL Skiles WC (2007) Swallet/resurgence relationships on the lower Santa Fe River, Florida. Contract report prepared by Karst Environmental Services for Water Resource Associates, Tampa, Florida

Cathcart JB (1963a) Economic geology of the Chicora Quadrangle, Florida. US Geol Surv Bull 1962-A

Cathcart JB (1963b) Economic geology of the Keysville Quadrangle, Florida. US Geol Surv Bull 1128

Cathcart JB (1963c) Economic geology of the Plant City Quadrangle, Florida. US Geol Surv Bull 1142-D

Cathcart JB (1964) Economic geology of the Lakeland Quadrangle, Florida. US Geol Surv Bull 1962-G
Cathcart JB (1966) Economic geology of the Fort Meade Quadrangle, Florida. US Geol Surv Bull 1207
Chave KE Schmalz RF (1966) Carbonate-seawater interactions. Geochim et Cosmochim Acta 30:1037-1048
Clausen CJ Cohen AD Emiliani C Holman JA Stipp JJ (1979) Little Salt Spring, Florida; a unique underwater site. Science 203: 609–614
Champion KM Upchurch SB Schneider JC (2005) Ground water chemistry and origin of water discharging from Troy Spring, Lafayette County, Florida. Live Oak, Florida, Suwannee River Water Mgt Dist
Chen J Thomas G Upchurch SB (1995a) Groundwater quality reflects karst development: A case study in central-west Florida. In: Beck BF (ed.), Karst geohazards: Engineering and environmental problems in karst terrane, Rotterdam, A.A. Balkema, pp. 103–109.
Chen J Thomas G Upchurch SB (1995b) A spatial-correlation model for assessment of large-area sinkhole and water quality data. In: Beck BF (ed.), Karst geohazards: Engineering and environmental problems in karst terrane, Rotterdam, A.A. Balkema, pp. 111–116.
Cooke CW (1954) Carolina bays and the shapes of eddies. US Geol Surv Prof Pap 254-I, pp. 195–206
Culshaw MG Waltham AC (1987) Natural and artificial cavities as ground engineering hazards. Quart Jour Eng Geol 20:139–150
Cunningham KJ Kluesner JW Westcott RL Robinson E Walker C Khan SA (2017) Sequence stratigraphy, seismic stratigraphy, and seismic structures of the lower intermediate confining unit and most of the Floridan aquifer system, Broward County, Florida. US Geol Surv Sci Invest Rept 2017-5109
Curl RL (1974) Deducing flow velocity in cave conduits from scallops. Nat Speleo Soc Bull 36:1–5
De Rooj GH Cho H (1999) Modeling solute leaching during fingered flow by integrating and expanding various theoretical and empirical concepts. Jour des Sciences Hydrologiques 44(3):447–465
Dobecki TL Upchurch SB (2010) A multi-level approach to site characterization - C.W. Bill Young Regional Reservoir, Hillsborough County, Florida. Proc Environ and Eng Geophy Soc, Appl Geophy Eng and Enviro Prob (SAGEEP)
Dunbar JS (2016) Paleoindian societies of the coastal southeast. Gainesville, Univ Press Fla
Exley S (1994). Caverns measureless to man. Dayton, Ohio, Cave Books
Eyton JR Parkhurst JI (1975) A re-Evaluation of the extraterrestrial origin of the Carolina bays. Univ Ill at Urbana-Champaign, Occ Pubs Dep Geog Paper Number 9
Field MS (2002a) A lexicon of cave and karst terminology with special reference to environmental karst hydrology. US Environ Prot Agency EPA/600/R-02/003
Florea LJ (2006) The karst of west-central Florida. Ph.D. Diss, Univ South Fla
Florea LJ (ed.) (2008a). Caves and Karst of Florida: A guidebook for the 2008 National Convention of the National Speleological Society. Huntsville, AL, Nat Speleol Soc
Florea LJ (2008b) Geology and hydrology of karst in west-central Florida and north-central Florida. In: Florea LJ (ed.), Caves and karst of Florida: A guidebook for the 2008 National Convention of the National Speleological Society. Huntsville, AL, Nat Speleol Soc, pp. 225–239
Florea LJ (2009) Caves and karst of west-central Florida. In: Palmer AN Palmer MV (eds.), Caves and karst of the USA. Huntsville, AL, Nat Speleol Soc, pp. 189–196
Florea LJ Vacher HL (2006a) Cave levels, marine terraces, paleoshorelines, and the water table in peninsular Florida. In: Onac BP Tamas T Constantin S Persoiu A (eds.), Archives of climate change, Karst Waters Inst Special Pub 10, pp. 188–192
Florea LJ Vacher HL (2006b) Morphologic features of conduits and aquifer response in the unconfined Floridan aquifer system, west central Florida. The 12th Symp Geol Bahamas and other Carbonate Regions, pp. 32–44
Florea LJ Vacher HL (2007) Eogenetic karst hydrology: Insights from the 2004 hurricanes, peninsular Florida. Ground Water, 45(4):439–446
Florea LJ Vacher HL Donahue B Naar D (2007) Quaternary cave levels in peninsular Florida. Quat Sci Rev 26:1344-1363
Florida Geological Survey (2013) Subsidence Incident Reports. Online subsidence incident report database maintained by the Florida Geological Survey, Accessed 11-4-13 at http://www.dep.state.fl.us/geology/gisdatamaps/SIRs_database.htm.
Field MS (2002b) A lexicon of cave and karst terminology with special reference to environmental karst hydrology. US Environ Prot Agy, EPA/600/R-02/003
Ford DC Williams PW (1989) Karst geomorphology and hydrology. New York, John Wiley & Sons
Gainesville Sun (2011) Sinkhole swallows, kills FL well driller. News article accessed 1/26/2014 at http://www.abcactionnews.com/dpp/news/state/sinkhole-swallows,-kills-fl-well-driller
German ER (1989) Quantity and quality of stormwater runoff recharged to the Floridan aquifer system through two drainage wells in the Orlando, Florida, area. US Geol Surv Water Sup Pap 2344
German ER Bradner LA (1989) Artificial recharge to the Floridan aquifer system, Orlando area, Central Florida. Proc Int Symp Artificial Recharge, Amer Soc Civil Eng, pp. 360–366
Glass RJ Steenhuis TS Parlange JY (1988) Wetting front instability as a rapid and far-reaching hydrologic process in the vadose zone. Jour Contam Hydrol 3:207–226
Glass RJ Steenhuis TS Parlange JY (1989) Wetting front instability: 2. Experimental determination of relationships between system parameters and two-dimensional unstable flow field behavior in initially dry porous media. Water Resou Res 25:1195–1207
Grant JA Brooks MJ Taylor BE (1998) New constraints on the evolution of Carolina bays from ground-penetrating radar. Geomorph 22:325–345
Gulden B Coke J (2015) World longest underwater caves. List accessed 11/1/2015 at http://www.caverbob.com/uwcaves.htm
Gulley, J., 2013. On the origin of Florida's underwater caves. Underwater Speleology, 40(4):10–11, 34
Gulley JD Florea LJ (2016) Caves as paleo-water table indicators in the unconfined Upper Floridan aquifer. Florida Scientist, 79(4):239–256
Gulley JD Martin JB Brown A (2016) Organic carbon inputs, common ions, and degassing: Rethinking mixing dissolution in coastal eogenetic carbonate aquifers. Earth Surface Processes and Landforms, 41:2098–2110
Gulley JD Martin JB Moore PE Brown A Spellman PD Ezell J (2015) Heterogeneous distributions of $CO_2$ may be more important for dissolution and karstification than mixing dissolution. Earth Surface Processes and Landforms, 40:1057–1071
Gulley JD Martin JB Moore PJ Murphy J (2013) Formation of phreatic caves in an eogenetic karst aquifer by $CO_2$ enrichment at lower water tables and subsequent flooding by sea level rise. Earth Surface Processes and Landforms, 38:1210–1224
Gulley JD Martin JB Screaton EJ Moore PJ (2011) River reversals into karst springs: A model for cave enlargement in eogenetic karst aquifers. Geol Soc Amer Bull, 123(3/4):457–467.
Gulley JD Polk JS (2017) Hypogene karst influences in the upper Floridan aquifer. In: Klimchouk AB Palmer AN Waela JD Auler AS Audra P (eds.), Hypogene karst regions and caves of the world, Springer Inter Publ, Ch. 50, pp. 745–755
Halladay WR (2007) Pseudokarst in the 21st century. Jour Cave and Karst Stud 69(1):103–113
Hill CA Forti P (1997) Cave minerals of the world. Huntsville, Alabama, Nat Spel Soc, 2nd Ed
Holtz RD Kovacs WD (1981) An introduction to geotechnical engineering. Englewood Cliffs, NJ, Prentice Hall

Hughes GH (1967) Analysis of the water-level fluctuations of Lake Jackson near Tallahassee, Florida. Fla Div Geol Rept Invest 48

Hull RW Yurewicz MC (1979) Quality of storm runoff to drainage wells in Live Oak, Florida, April 4, 1979. US Geol Surv Open-File Rept 79-1073

Isphording WC (1963) A study of the heavy minerals from the Hawthorn Formation and overlying sands exposed at the Devil's Millhopper, Alachua County, Florida, M.S. thesis, Gainesville, Univ Fla

Isphording WC (1984) Sand craters in Gulf Coastal Plain clastic sediments: An extension of the Carolina bays phenomenon? Geol Soc Amer Abst with Programs 16(3):148

Ivester AH Brooks MJ Taylor BE (2007) Sedimentology and ages of Carolina bay sand rims. Geol Soc Amer Abst with Programs 39(2):5

Jammal & Associates, Inc. (1982) The Winter Park sinkhole: A report of the investigation, findings, evaluation and recommendations. Contract report prepared for the City Commission, City of Winter Park, Florida

Jammal SE (1984a) A self-guided field trip to the Winter Park sinkhole. Orlando, Fl Sinkhole Res Inst Report No. 84-85-3

Jammal SE (1984b) Maturation of the Winter Park sinkhole. In: Beck BF (ed.), Sinkholes: Their geology, engineering & environmental impact, Rotterdam, B.A. Balkema, pp. 363–369

Jammal SE Casper JW (2009) The Winter Park sinkhole then and now. Jour Fla Eng Soc, May 2009 issue, pp. 19–20

Jennings JN (1985) Karst geomorphology. Oxford, Blackwell Publishers, 2nd Ed.

Jones GW Upchurch SB (1993) Origin of nutrients in ground-water discharging from the Lithia and Buckhorn Springs. Brooksville, Southwest Fla Water Mgt Dist

Kantrowitz IH Woodham WM (1995) Efficiency of a stormwater detention pond in reducing loads of chemical and physical constituents in urban streamflow, Pinellas County, Florida. US Geol Surv Water-Resour Invest Rept 94-4217

Katz BG (1993) Biochemical and hydrological processes controlling the transport and fate of 1,2-dibromoethane (EDB) in soil and ground water, central Florida. US Geol Surv Water-Sup Pap 2402

Ketelle RH Newton JG Tanner JM (1988) Karst subsidence in east Tennessee. In: Proc Second Conf Environ Prob in Karst Terranes and Their Solutions Conf, Nashville, Tennessee, Dublin, OH, Nat Water Well Assoc, pp. 51–65

Kimrey JO (1978) Preliminary appraisal of the geohydrologic aspects of drainage wells, Orlando area, central Florida. US Geol Surv Water-Resour Invest Rept 78-37

Kimrey JO Fayard LD (1984) Geohydrologic reconnaissance of drainage wells in Florida. US Geol Surv Water-Resour Invest Rept 84-4021

Krause A (2008) Briar Cave. In: Florea LJ (ed.), Caves and karst of Florida: A guidebook for the 2008 National Convention of the National Speleological Society. Huntsville, AL, Nat Speleol Soc pp. 71–74

Krauss, A, 2008. Florida's longest and deepest caves. In: Florea LJ (ed.), Caves and karst of Florida: A guidebook for the 2008 National Convention of the National Speleological Society. Huntsville, AL, Nat Speleol Soc pp. 43–44

Krawiec W (1963) Solution pits and solution breccias of the Florida Keys. Master's thesis, Univ Rochester (conclusions quoted in Multer 1977)

Kromhout C Baker AE Albritton CK Scott TM Cichon JR Miller SR (2017) The favorability of Florida's geology to sinkhole formation. Fla Geol Surv rept.

Lane E (1986) Karst of Florida. Fla Geol Surv Sp Pub 29

Lawrence FW Upchurch SB (1976) Identification of geochemical patterns in ground water by numerical analysis. In: Zaleem EA (ed.), Advances in Groundwater Hydrology, Middleberg, VA, Amer Water Resour Assoc, pp. 199–214

Lawn A (compiler) (2014) Karst hydrogeology of the upper Suwannee River basin, Alapaha River area, Hamilton County, Florida. Southeast Geol Soc Guidebook No. 63

Legrand HE (1953) Streamlining of the Carolina bays. Jour Geol 61(3):263–274

May JH Warne AG (2004) Hydrogeologic and geochemical factors required for the development of Carolina bays along the Atlantic and Gulf of Mexico, coastal plain, USA. Environ & Eng Geosci 5(3):261–270

McClellan GH (1962) Identification of clay minerals from the Hawthorn Formation, Devil's Millhopper, Alachua County, Florida. M.S. thesis, Gainesville, Univ Fla

McGlynn S (2006) Leon County lakes: Lake Jackson watershed. Tallahassee, FL McGlynn Labs, Inc., accessed 9/25/2013 at http://www.mcglynnlabs.com/LakeJackson.pdf

Metcalf SF Hall LE (1984) Sinkhole collapse induced by groundwater pumpage for freeze protection irrigation near Dover, Florida, January, 1997. In: Beck BF (ed.), Sinkholes: Their geology, engineering, and environmental impact. Rotterdam, A.A. Balkema, pp. 3–10

Monroe WH (1970) A glossary of karst terminology. US Geol Surv Water-Sup Pap 1899-K

Newton JG (1987) Development of sinkhole resulting from man's activities in the eastern United States. US Geol Surv Circ 968

Onac BP Pace-Graczyk K Atudirei V (2008) Stable isotope study of precipitation and cave drip water in Florida (USA): Implications for speleothem-based paleoclimate studies. Isotopes in Environ and Health Stud 44(2):149–161

Parker GG (1955) Geomorphology. In: Parker GG Ferguson GE Love SK and others. Water resources of southeastern Florida, US Geol Surv Water Sup Pap 1255, pp. 127–155

Pendexter WS (1996) The influence of media heterogenieties on the development of flow fingers. Ph.D. dissertation, Tallahassee, FL, Fla State Univ

Pendexter WS Furbish DJ (1991) Development of a heterogeneous moisture distribution and its influence on the evolution of preferred pathways of flow in an unsaturated sand soil. Proc Nat Symp on Preferential Flow, Chicago, pp. 104–112

Perkins RD (1977) Part II, Depositional framework of Pleistocene rocks in south Florida. In: Enos P Perkins RD (eds.), Quaternary sedimentation in south Florida, Geol Soc Amer Memoir 147, p. 131–198

Prouty WF (1935) "Carolina bays" and elliptical lake basins, Jour Geol 43(2):200–207

Prouty WF (1952) Carolina bays and their origin. Bull Geol Soc Amer 63(2):167–224

Purdy BA (1991) The art and archaeology of Florida's wetlands. Boca Raton, Telford Press Series, CRC Press

Purdy BA (2008) Florida's people during the last ice age. Gainesville, Univ Press Fla

Schalles JF Sharitz RR Gibbons JW Leversee GJ Knox JN (1989) Carolina bays of the Savannah River Plant, Aiken, South Carolina. Savannah River Plant, Nat Environ Res Park, Aiken, SC

Scheidt J Lerche I Paleologos E (2005) Environmental and economic risks from sinkholes in west-central Florida. Environ Geosci 12:207–217

Schmidt W Scott TM (1984) Florida karst; its relationship to geologic structure and stratigraphy. In: Beck BF (ed.), Sinkholes: Their geology, engineering and environmental impact. Proc First Multidisc Conf on Sinkholes, Orlando, Florida, pp. 11–16

Scott TM (1988) The lithostratigraphy of the Hawthorn Group (Miocene) of Florida. Fla Geol Sur Bull 59

Scott TM (2017a) Case study I: Triggered sinkholes – Tropical Storm Debby in 2012. In: Kromhout C Baker AE Albritton CK Scott TM Cichon JR Miller SR, The favorability of Florida's geology to sinkhole formation. Fla Geol Surv rept., Appendix II

Scott TM (2017b) Case study II: Triggered sinkholes – pumping-related freeze protection, Hillsborough County, January 2010. In: Kromhout C Baker AE Albritton CK Scott TM Cichon JR Miller SR, The favorability of Florida's geology to sinkhole formation. Fla Geol Surv Rept., Appendix III

Scott TM Means GH Meegan RP Means RC Upchurch SB Copeland RE Jones J Roberts T Willet A (2004) Springs of Florida. Fla Geol Sur Bull 66

Sinclair WC Stewart JW (1985) Sinkhole type, development, and distribution in Florida. Fla Bur Geol Map Series 110

Sinclair WC Stewart JW Knutilla RL Gilboy AE Miller RL (1985) Types, features, and occurrence of sinkholes in the karst of west-central Florida. US Geol Surv Water-Resour Invest Rept 85-4126

Soto L (2005) Reconstruction of late Holocene precipitation for central Florida as derived from isotopes in speleothems. Master's thesis, Tampa, Univ South Fla

Southeastern Geological Society (1981) Karst hydrology and Miocene geology of the upper Suwannee River Basin, Hamilton County, Florida. Tallahassee, Southeast Geol Soc Guidebook

Sowers GF (1996) Building on sinkholes. Amer Soc Civil Eng

Sweeting MM (1973) Karst landforms. London, McMillan Press

Sweeting MM (1981) Karst geomorphology. Stroudsburg, PA, Hutchinson Ross Publ Co, Benchmark Paps in Geology, Vol. 59.

Tampa Bay Times (2013) Seffner sinkhole 911 call: "Bedroom floor just collapsed." Accessed 3/2/2013 at http://www.tampabay.com/news/publicsafety/sinkhole-swallows-part-of-seffner-home.r.

Taylor GF (1993) Inventory of drainage wells and potential sources of contaminants to drainage-well inflow in southwest Orlando, Orange County, Florida, US Geol Surv Water-Resour Invest Rept 93-4061

Terzaghi K (1925) Principles of soil mechanics. Eng News Record, 95:742–746, 796–800, 832–836, 874–878

Terzaghi K Peck RB (1948) Soil mechanics in engineering practice. New York, John Wiley & Sons

Tihansky AB (1999) Sinkholes, west-central Florida: A link between surface water and ground water. In: Galloway D Jones DR Ingebritsen SE (eds.), Land subsidence in the United States, US Geol Surv Circ 1182, pp. 121–141

Upchurch SB (2015) Determination of the relationship of nitrate to discharge and flow systems in north Florida springs. In: Doctor DH Land L Stephenson JB (eds.), Sinkholes and the Engineering and Environmental Impacts of Karst, Proc 14th Multidisc Conf, Rochester, MN, Nat Cave Karst Res Inst Symp 5, pp. 347–354

Upchurch SB (2016) The nexus of Florida's groundwater resources and karst processes. Fla Scientist 79(4):208–219

Upchurch SB (2017) Hypogene speleogenesis on the Floridan Platform, U.S.A. In: Klimchouk AB Palmer AN Waela JD Auler AS Audra P (eds.), Hypogene karst regions and caves of the world, Springer International Publ, Ch. 49, pp.735–744

Upchurch SB Champion KM Schneider JC (2004) Geostatistical analysis of water-level and water-quality data for the Troy Spring springshed, Suwannee County, Florida. Live Oak, FL, Suwannee River Water Management District

Upchurch SB Littlefield JR Jr. (1987) Evaluation of data for sinkhole-development risk models. In: Beck BF Wilson WL (eds.), Karst hydrogeology: Engineering and environmental applications, Rotterdam, B.A. Balkema, pp. 359–364

Upchurch SB Littlefield JR Jr. (1988) Evaluation of data for sinkhole-development risk models. Environ Geol and Water Sci 12:135–140

Upchurch SB Dobecki TL Daigle DM (1999) Geological, hydrogeological, and geophysical investigation. In: Law Engineering Services and others, Geotechnical Site Characterization Report – Tampa Bay Regional Reservoir, Volume I, Section 3

Upchurch SB Dobecki TL Scott TM Meiggs SH Fratesi SE Alfieri MC (2013) Development of sinkholes in a thickly covered karst terrane. Proc Thirteenth Multidisc Conf on Sinkholes and the Eng and Environ Impacts of Karst, Carlsbad, NM, Nat Cave and Karst Res Inst, pp. 273–277

Upchurch SB Randazzo AF (1997) Environmental geology of Florida. In: Randazzo AF Smith DL (eds.), Geology of Florida, Gainesville, Univ Press of Fla, Ch 13, pp. 217–249

Upchurch SB Scott TM Alfieri MC Dobecki TL (2015) Shallow depressions in the Florida coastal plain: Karst and pseudokarst. In: Doctor DH Land L Stephenson JB (eds.), Sinkholes and the engineering and environmental impacts of karst, Proc 14th Multidisc Conf, Carlsbad, NM, Nat Cave and Karst Res Inst Symp 5, pp. 231–240

Van Beynen PE Soto L Polk J (2008) Variable calcite deposition rates as proxy for paleo-precipitation determination as derived from speleothems in central Florida, U.S.A. Jour Cave and Karst Stud 70(1):25–34

Van Kauwenbergh SJ Cathcart JB McClellan GH (1990) Mineralogy and alteration of the phosphate deposits of Florida. US Geol Surv Bull 1914

Wang JW (2005) The stress-strain and strength characteristics of Portaway Sand. Ph.D. dissert, Univ Nottingham. (Accessed 9/3/2013 at http://etheses.nottingham.ac.uk/170/1/Thesis_final_PDF.pdf)

Weber KA (1981) Characterization of storm water runoff by numerical analysis. Master's thesis, Univ South Fl

Wells BW Boyce SG (1953) Carolina bays: Additional data on their origin, and history. Jour Mitchell Soc 69(2):119–141

White WA (1970) The geomorphology of the Florida peninsula. Fla Geol Surv Bull 51

White WB (1988) Geomorphology and hydrology of karst terrains. New York, Oxford Univ Press

Williams P (2004) Dolines. In: Gunn J (ed.), Encyclopedia of caves and karst science. New York, Fitzroy Dearborn (Taylor & Francis Books, Inc.), pp. 304–310

Willoughby RH (2007) Carolina bays in Jordan Quadrangle, Clarendon County, central South Carolina are relict survivors from development of sinkholes due to karstic solution (abs), Geol Soc Amer Abst with Prog 39(2), p. 5

Wilson WL (1995) Sinkhole and buried sinkhole densities and new sinkhole frequencies in karsts of northwest peninsular Florida. In: Beck BF (ed.), Karst geohazards, Rotterdam, A.A. Balkema, pp. 79–90

Wilson WL Beck BF (1992) Hydrogeologic factors affecting new sinkhole development in the Orlando area, Florida. Ground Water 30:918–930

Wilson WL McDonald KM (1988) Frequency of sinkhole development in the vicinity of the Southwest Landfill, Alachua County, Florida. Fla Sinkhole Res Inst Rept 87-88-2

Wilson WL McDonald KM Barfus BL Beck BF (1987) Hydrogeologic factors associated with recent sinkhole development in the Orlando area. Fla Sinkhole Res Inst Rept 87-88-4

Wright AP (1974) Environmental geology and hydrology, Tampa area, Florida. Fla Bur Geol Sp Pub 19

Zwanka W (2005) Delineation of the Troy Spring basin using groundwater levels. In: Copeland R (comp), Geomorphic influence of scarps in the Suwannee Basin, Southeastern Geol Soc Field Trip Guidebook 44, p. 53

# Epigene and Hypogene Karst

# 8

**Abstract**

Karren development on the upper surfaces of limestone varies in Florida. Karren forms created by chemically aggressive water flowing over rock surfaces are rare. Grikes, pinnacles, and other karren formed by infiltration of groundwater into porous limestone or along fractures are common in Florida limestones. Vertical shafts (solution pits or channels) are common in some areas. Because they are normally hidden under cover materials, it is unusual to be able to observe karren features.

Duricrusts developed on the upper-most limestone surfaces are composed of laminated crusts, calcite cemented sand, calcitic debris, marine sediments, limestone breccia, terra rossa soils, and thin crusts of recrystallized calcite. While caliche is usually associated with arid environments, common artifacts of caliche development are present in Florida as vadose pisolites and rounded and recemented limestone rubble. Both duricrusts and caliche are formed by repeated wetting and drying of calcium-carbonate sediments.

The Cody Scarp (escarpment) is Florida's best example of the development of fluviokarst. Sinking streams, swallets, traces, and springs characterize the escarpment and karst plains formed by headward erosion and the intersection of streams and limestone near the toe of an escarpment.

Phytokarst developed on rocky shores by dissolution and grazing of algae and fungi by herbivores is found on the rare coastal exposures of limestone in southeastern Florida, and microkarst is well developed in the Rock Glades and Everglades physiographic provinces where limestone is subaerially exposed.

Possible hypogenetic karst includes sulfuric acid or freshwater/saltwater mixing-zone speleogenesis. The Upper Floridan Aquifer (UFA), Lower Floridan Aquifer (LFA), and anhydrite-bound Cretaceous strata contain examples of possible hypogenetic karst.

**Keywords**

Epigenetic karst · Hypogene karst · Pinnacle karst · Solution pipes · Solution channels · Caliche · Karst escarpments · Karst traces · Fluviokarst

## 8.1 Introduction

The previous chapter discussed two of the major landforms, sinkholes and caves that are of interest in Florida. This chapter continues the discussion of landforms and other karst features in Florida. We begin with landforms, such as conical hills, arches and natural bridges, and sandhill lakes. A major emphasis is placed on fluviokarst, including stream-to-swallet-to-emergence systems and the evolution of karst escarpments. Finally, karren and hypogene karst are discussed.

## 8.2 Major Landforms

### 8.2.1 Conical Hills and "Star-Shaped" Valleys

Humid, tropical climates produce several relatively unique landforms, including steep-sided, dome- or conical-shaped hills separated by irregularly shaped ("star-shaped"[1]) sinkhole valleys (Williams 1972a, b; Jennings 1985). The irregularly shaped valleys are created by sinkhole coalescence and interference of sinkhole margins. The mantle of cover materials and low relief make recognition of such features in Florida problematic, but there are examples that suggest origins similar to these classical, tropical karst landforms (Fig. 8.1). These are most evident when associated with thick

[1]Because of the siliciclastic cover sediments in karst terrains, the valleys created by karst processes are rounded and contours softened. As a result, they are seldom truly star-shaped.


S. Upchurch et al., *The Karst Systems of Florida*, Cave and Karst Systems of the World,
https://doi.org/10.1007/978-3-319-69635-5_8

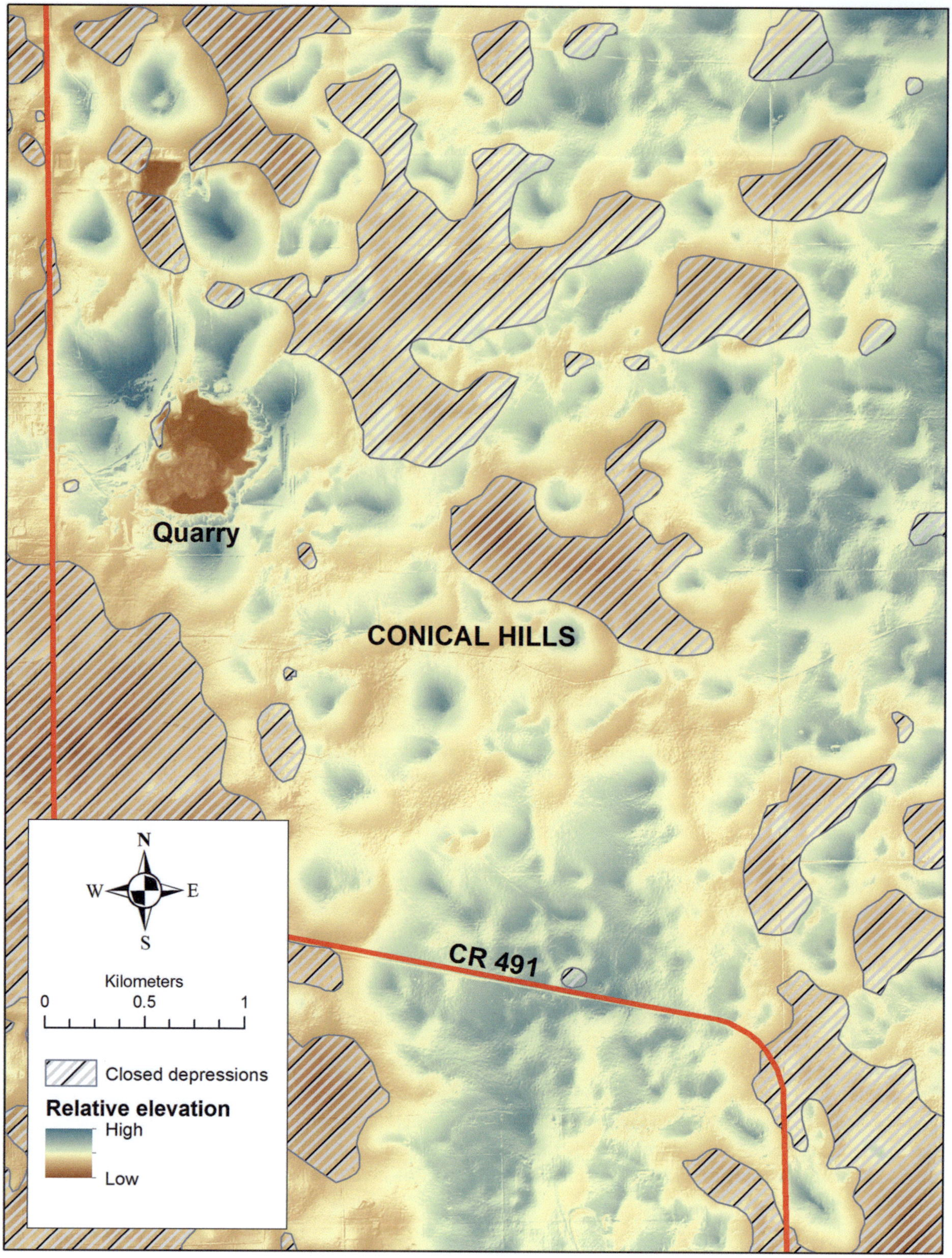


**Fig. 8.1** LIDAR map of conical hills and closed depressions on the Brooksville Ridge just south of Lecanto, Citrus County. Location: 28.841°N, 82.473°W

cover, such as at the margins of karstic ridges and karst escarpments. The thick cover and lack of significant incision into the limestone prevents the hill sides from being highly steepened, but otherwise, many sinkholes are star-shaped because they abut conical- or dome-shaped hills of residual cover materials. Figure 8.1 illustrates several conical hills separated by simple, "star-shaped" sinks and uvalas that are suggestive of tropical karst landforms on the Brooksville Ridge.

## 8.2.2 Arches and Natural Bridges

Limestone arches are also characteristic of karst terrains. They form by the collapse of adjacent sinkholes aligned over a cave system. As discussed in Sect. 7.3.4.3, streams that go underground and flow through caves often develop strings of karst windows (Fig. 7.60). When these sinkholes are widely spaced, we do not think of the intervening spaces as arches. In Florida, the term natural bridge has been used for larger spans between a swallet, or sinking stream, and its resurgence, or point of emergence. For example, Florida's Natural Bridge Battlefield Historic State Park in southern Leon County occupies a span of about 1 km between the swallet and rise of the St. Marks River. Elsewhere the term natural bridge usually refers to shorter arches, but in this text we will call them natural bridges as is common in Florida. In Florida, these arches and natural bridges range from a few meters to over a thousand meters.

### 8.2.2.1 Arches

Arches are often overlooked because the lack of significant relief and high groundwater levels make them small and often unremarkable, especially in comparison with the adjacent karst windows and sinkholes. Figure 8.2a shows a small arch in the run of Charles Spring, a publicly accessible spring on the Suwannee River. Figure 8.2b depicts a 12 m-long arch that once existed in North Miami. This arch collapsed in 1973 and has been reconstructed in Arch Creek Park. Note that both arches are essentially remnants of the limestone roof of a cave separating two karst windows. Figure 8.3 is the more typical arch in Florida with a short span separating two rock-collapse sinkholes.

### 8.2.2.2 Natural Bridges

The term natural bridge is locally used in Florida to include large arches where streams and rivers flow from a swallet underground through conduits and then re-emerge hundreds of meters to kilometers downstream (Cooke 1939). Some of these arches are so large that they cannot be observed as being arches on the ground. In Florida, these long gaps between swallet and resurgence have been called "natural bridges" because they have served historically as dry-land passages over the streams and rivers. The term natural bridge has been adopted by the State of Florida for the bridge at the Natural Bridge Battlefield State Park near Woodville in Leon County. Here, the St. Marks River goes underground and then emerges after a linear distance of approximately 1 km. A Civil War battlefield occupies the space between the swallet and resurgence.

These natural bridges are large enough that there is no indication, other than an ephemeral overflow corridor (trace, Sect. 8.3.4), that the strata that make up the bridge span an underground "river" flowing through a karst conduit. The following section describes one of the largest features described as a "natural bridge" in this very broad interpretation of the term.

### 8.2.2.3 Santa Fe River Natural Bridge

The Santa Fe River Sink and Rise are located in O'Leno State Park and River Rise Preserve State Park, respectively, on the Santa Fe River between Alachua and Columbia counties. As it flows off the Raiford Ridges Province, through the Alachua Karst Hills, and onto the Branford Karst Plain, the Santa Fe River goes underground in a large, impressive swallet. There is a distinct gyre as the river moves underground (Figs. 8.4 and 8.6).

The river rises back to the surface at a resurgence (a spring that discharges surface water from a cave) known as the River Rise 4.6 km southeast of the sink (Figs. 8.5 and 8.6). The land between the River Sink and Rise is a large land bridge that has served early Native Americans and modern occupants as a transportation corridor.

There is a series of wetlands, karst windows, and flood chutes (Fig. 8.6) that form a slough (trace) connecting the Sink and Rise. During floods, the slough serves as a large overflow corridor. The karst windows in this slough area provide access to the cave system that underlies the land bridge and transports water from the Sink to the Rise. The mapped portions of this cavern are known as the Old Bellamy Cave System.

The Old Bellamy Cave System is one of Florida's largest submerged caves. Courbon et al. (2005) reported that the mapped portions of this cave system are 12.7 km in total length, and the cavern extends to a measured depth of 49 m. One of the access points to this cave system is shown as OBCS in Figs. 8.6 and 8.7. Figure 8.7 illustrates a map of the cave system. Cave divers have nearly mapped at least one conduit that connects the Sink and Rise. A somewhat surprising outcome of cave mapping is the development of the cave to the east of a vector from the Sink to the Rise (Fig. 8.7).

The pattern of recharge to the Upper Floridan Aquifer (UFA) in the vicinity of the Santa Fe land bridge is relatively complex. Water Resource Associates et al. (2006) and Brown et al. (2016) investigated stream hydrographs up and downstream from the River Sink-Rise system. Olustee Creek

**Fig. 8.2** (**a**) Small limestone arch spanning the run of Charles Spring (30.167°N, 83.231°W), Suwannee County. Note that the cross section of the opening under the arch is a broad and shallow oval, indicating that the cave that collapsed to form the arch formed at the water table. (**b**) The arch that once existed in Arch Creek Park (25.9010°N, 80.1626°W) in North Miami. Photograph courtesy of the Florida Memory Project. (Image RC09735 by G.W. Romer in 1925)

**Fig. 8.3** Arch developed between two, closely spaced, rock-collapse sinkholes on the north shore of the Suwannee River near the Suwannee River State Park in Hamilton County

(Fig. 8.6), a major tributary to the Santa Fe, loses water through a series of swallets (OS, Fig. 8.6) as it begins to flow from the Raiford Ridges Province, cross the Cody Scarp, and onto the Branford Karst Plain. The fate of this water is unknown. It may emerge at COL61981 (Fig. 8.6), a first-magnitude spring that is located about 5 km northeast of the River Sink on the west bank of the Santa Fe River, or it may contribute to the groundwater passing under the land bridge.

Scientists at the University of Florida have intensely investigated the processes that affect recharge water that enters the Santa Fe River Sink. Special focus has been on the effects of mixing of river water with aquifer water underground. Some of the important findings of these investigations are discussed below.

A dye tracer test found that the River Sink is connected with Sweetwater Lake, a small karst window between the Sink and Rise, and several other karst windows. This tracer was not, however, detected at the River Rise (Hisert 1994). A second dye trace beginning at Sweetwater Lake was detected at the River Rise (Hisert 1994), thus establishing the connection and demonstrating that dilution and/or dispersion may be important processes for reducing concentrations of the dye to less than detection limits. The Hisert investigation was undertaken when the river stage was moderately high (13.5 m above mean sea level). Hisert (1994) also estimated that the rate of travel from the River Sink to Rise was 4.3 km per day.

Dean (1999) and Martin and Dean (1999, 2001) utilized temperature measurements to show that flow rates vary with river stage. At high stage (13.5 m above sea level), travel time from sink to rise was estimated to be less than 6 h. During low stage (10.5 m), the travel time was over 7 h. At "exceptionally low stage" in 2001, the entire flow of the river was captured by a small sink upstream from the River Sink (Smith et al. 2001). After several precipitation events during which discharge pulses moved through the Sink/Rise system, the data obtained indicated the possibility of a complex "plumbing" system controlled by differing elevations of conduits.

Martin and Screaton (2000) used the Sink/Rise system to characterize exchange and mixing of diffuse and conduit flow water. They concluded that only 4% of the water discharging from the River Rise entered the UFA at the River Sink at low river flow while 96% of the water discharging from the Rise was thought to be derived from other sources. The largest source of discharging water was thought to be a conduit system, the Old Bellamy Cave System (Fig. 8.7), which had been mapped by cave divers to extend to the east. They further concluded that this eastern conduit system received little surface water, so most of the discharge was from diffuse flow. At intermediate discharge from the Rise, about 73% of the discharge was from the Sink and the remaining 27% was from other sources. Recently, the University of Florida group has used the Santa Fe Sink and Rise system to explore the importance of $CO_2$ emanating from the water table as a source of dissolution, evolution of groundwater chemistry, and other interactions of groundwater and surface water (see Gulley et al. 2012 and Brown et al. 2016).

The Santa Fe Sink/Rise system is, therefore, a complex system with significant recharge at the Sink and other conduits within the mainstem of the Santa Fe River during moderate to high flow conditions. During droughts, when the river is low, the discharge from the Rise is sustained by movement of water out of the aquifer matrix into the conduit system.

#### 8.2.2.4 Evolution of Landforms at Natural Bridges

Natural bridges form where organic-rich surface water flows into a swallet and then through a well-developed cave to a spring or resurgence some distance away from the swallet. Between the discussions in Chap. 7 and this

Fig. 8.4 The Santa Fe River Sink (29.912°N, 82.573°W), a large swallet that accepts all normal flow of the river

Fig. 8.5 The Santa Fe River Rise (29.874°N, 82.591°W), the resurgence of the Santa Fe River

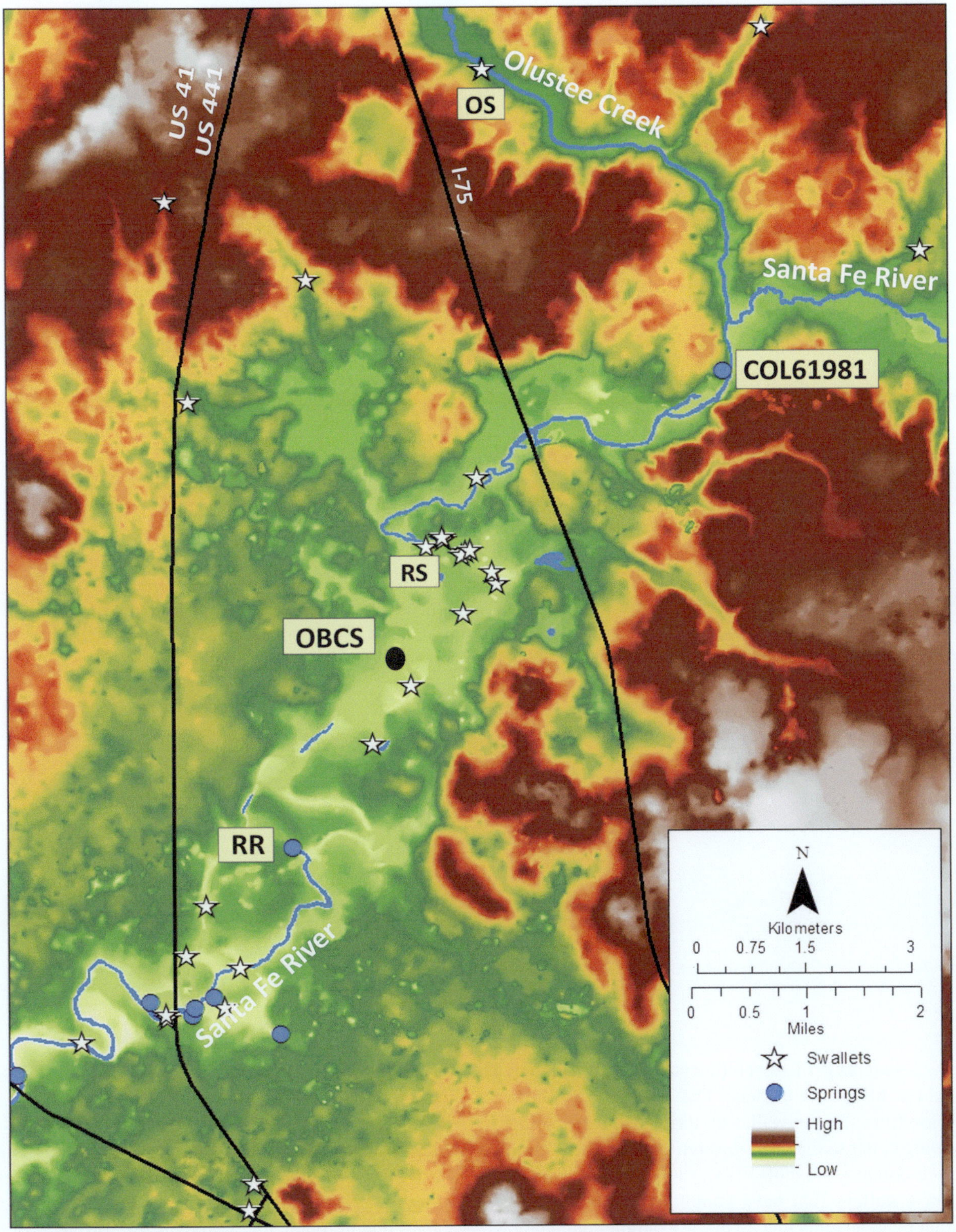


**Fig. 8.6** LIDAR topographic map of the vicinity of the Santa Fe River land bridge. OS-Olustee Creek swallets, COL61981-1st magnitude spring, RS-River Sink, OBCS-Old Bellamy Cave System entrance (karst window), RR-River Rise

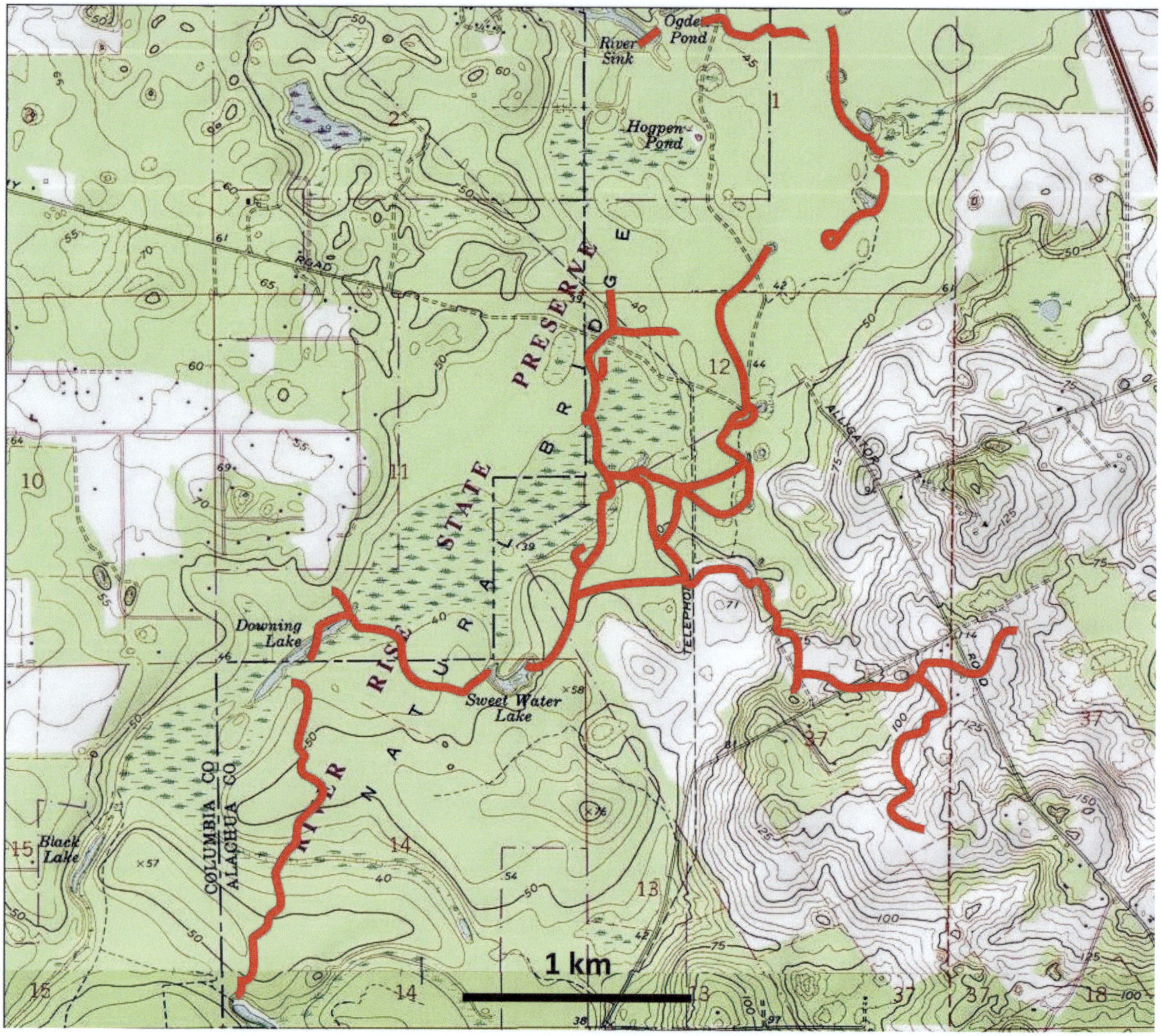


**Fig. 8.7** Topographic map showing the location of the Old Bellamy Cave System. Various sources

chapter, a sequence of developmental events associated with natural bridges is evident. The land bridges show a succession from simple swallet-to-spring pathways with no intervening collapse features, through development of multiple rock-collapse sinkholes (karst windows) and ending with linear valleys that have completely collapsed. It is important to understand that all of the examples cited below are located on streams that originate in the highlands above a karst escarpment. The streams go underground within the escarpment and emerge down-gradient from the escarpment (Sect. 8.5).

The evolution of a natural bridge to a dry valley or slough is demonstrated by examining the examples used in this text. They are mentioned in order of the apparent development of a natural bridge and dry valley or slough.

1. The Alapaha River (Sect. 8.3.6.3) is an example of a long natural bridge with an active overflow stream channel. At low flow (less than 14 $m^3/s$; Upchurch 2014a), the river goes underground for a straight-line distance of about 16 km before emerging at its primary resurgence, a first-magnitude spring. When the river discharge exceeds about 14 $m^3/s$ the river channel floods all the way to the Suwannee River. There are no known rock-collapse sinkholes in the reach from the swallet complex to its resurgence.

2. The natural bridge on the Santa Fe River (Sect. 8.2.2.3) goes into a swallet (River Sink) 4.6 linear kilometers upstream of its resurgence (River Rise). Several rock-collapse sinkholes have developed along the path of the cave that carries the underground Santa Fe River. These karst windows provide access to an extensive, phreatic cave. When the river floods, an overland trace carries the excess flow that cannot be accommodated by the swallets along the trace.
3. The Aucilla River, which was discussed in Sect. 7.3.4.3, also goes underground through a swallet approximately 7.2 linear kilometers above its resurgence (Nutall Rise). The cave that carries the principal flow has been subjected to numerous rock-collapse sinkholes and, today, the karst windows are arrayed like a "string of pearls" along the trace of the river (Fig. 7.60).
4. Finally, the cave system may completely collapse resulting in a dry valley or slough that was termed a collapsed cave (Sect. 7.3.4.3). In Florida, several dry valleys or sloughs can be found on the east bank of the Suwannee River and north bank of the Santa Fe. All of these dry valleys are clearly collapsed caves oriented between the Cody karst escarpment and a river. Tenmile Hollow (Sect. 7.3.4.3, Fig. 7.62) is an example of a collapsed cave system that has apparently undergone the sequence of events described above.

### 8.2.3 Sandhill Lakes

Many of Florida's ridges and larger dune fields have been penetrated by large, single and coalesced sinkholes. They typically occur in areas of high (for Florida) relief, such as on ridges. If these areas (1) have relatively deep water tables and if (2) the surrounding sandy soils and sediments are excessively drained, the sinkhole lakes generate special environmental concerns because of their unusual flora and fauna and resource values. In Florida, these features are termed sandhill lakes (Fig. 8.8).

As can be seen in Fig. 8.8, these lakes range from simple features with circular outlines to complex shapes with numerous depressions in the lake beds. As such, the latter are coalescent, polje-like sinkhole complexes.

The image shown in Fig. 8.8 was taken on January 18, 2012. Lake levels in some lakes were high and there was little shoreline exposed. In other lakes, such as Lake Brooklyn (arrow, Fig. 8.8), only the depressions of individual sinkholes within the lake perimeter contained water, indicating that the lakes respond in different ways hydraulically to rainfall and aquifer water levels. The white margins of the lakes are bare, well-sorted and excessively well-drained sand that has been exposed by lowered lake levels. This sand is reworked from the dune sand of the ridges onto which the lakes have been superimposed. The tan areas between the sandy shoreline and standing water are grassy areas underlain by well-sorted sand with minor clay and organics from lacustrine deposition.

Sandhill lakes are of special concern in Florida because of the following properties.

- Water quality in the lakes is often oligotrophic with low turbidity. Where the lakes have low turbidity and productivity, the white sand bottoms sometimes cause the lakes to appear blue (Fig. 8.9a). Lakes that receive significant water from the organic-rich, surficial aquifer or from surface water are typically colored by humic substances and have variable turbidity. Lakes that receive groundwater from the organic-poor surficial aquifer sand are oligotrophic and have low turbidity.
- The upland, xeric soils that surround the lakes are characterized by unusual and often unique plant and animal communities, including many threatened and endangered species. For example, the carnivorous plants shown in Fig. 8.9b find food from insects as an alternative nutrient source.
- Seepage faces on many sandhill lakes support unique plant communities (Whitney et al. 2004), including such plants as St. Johns-wort (*Hypericum* sp.) and carnivorous plants (Fig. 8.9b).
- Because they are sinkhole lakes situated in geological settings where clay is limited in both terrestrial and lake sediments, plugging of the conduits that connect some sandhill lakes to the underlying UFA may not exist, may be ephemeral, or may be relatively permanent. As a result, water levels in sandhill lakes have a reputation for fluctuating dramatically (Fig. 8.8) with long-term rainfall variability and groundwater withdrawals. Lake Brooklyn (Fig. 8.8) is an excellent example of the range in lake levels typical of many sandhill lakes. The small water bodies shown within Lake Brooklyn are water-filled sinkholes. The larger, brown area is dry lake bed with a thin plant cover. Depending on rainfall, connectivity of the lake and surficial aquifer and the underlying UFA, and water levels in the aquifer, the levels of this lake vary widely.

Figure 8.10 shows a subbottom profile (Dobecki et al. 2013) across Big Blue Lake in Washington County. This is a large sandhill lake with a bean-like shape (inset, Fig. 8.10). Note the smaller, coalescent sinkholes within the lake basin. These sinkholes appear to extend to the UFA, through the surficial sand and underlying semi-confining unit. Also note the evidence of sedimentation (yellow pattern) after the sinkholes formed. This sediment is sand that has slumped into the lake from the sides of the lake basin or been transported into the basin by small tributary streams and sheet wash (Dobecki et al. 2013).

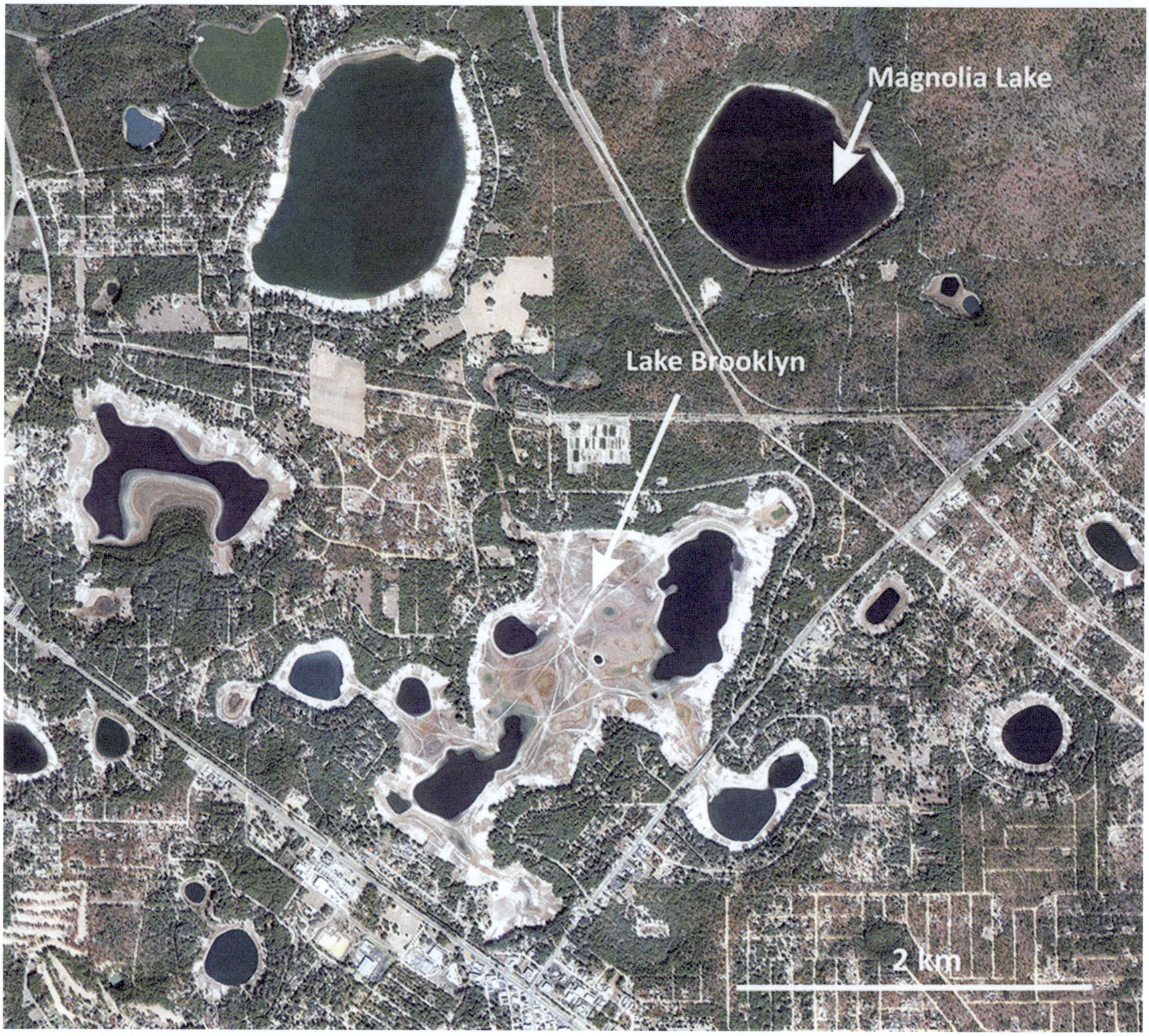


**Fig. 8.8** Sandhill lakes near Keystone Heights, Clay County. Arrow points to the largely dry Lake Brooklyn (29.802°N, 82.028°W), which contains multiple water-filled sinkholes in its floor. Google Earth® image dated 1/18/2012

As a result of such connections, water levels in some sandhill lakes fluctuate greatly because they are correlated with the potentials in the underlying UFA (Yobbi 1996). Water levels in neighboring lakes that remain plugged vary less because the lake levels are a result of variations in rainfall and surficial aquifer water levels only (compare the lake levels in Fig. 8.8). Sacks et al. (1992), Grubbs (1995), Lee (1996), Yobbi (1996) describe the hydrology of sandhill lakes in northern Florida, including the "flow-through" mechanisms in the Surficial Aquifer System (SAS) and interactions with the underlying UFA.

## 8.3 Fluviokarst: Karst Features and the Action of Surface Water

Riverine, or fluvial, processes have had a significant impact on Florida karst, especially in northern Florida where numerous rivers and streams form in the low-recharge uplands of the Okefenokee Basin District (White's (1970) Northern Highlands) and then flow onto karst plains on their way to the Gulf of Mexico. This section describes the most important of the fluviokarst features in Florida.

**Fig. 8.9** (**a**) The shoreline of Big Blue Lake, an oligotrophic sandhill lake in southern Washington County. Note the extremely clear water and sandy bottom. From the air, this lake appears blue because of reflections from the sandy bottom and clarity of the water. (**b**) The seepage-face shoreline of Big Blue Lake. The red plants are sundew plants (*Drosera* sp.), a carnivorous plant that commonly populates nutrient-poor seepage faces on the margins of the sandhill lakes of the panhandle of Florida

### 8.3.1 Swallets

Swallets are sinkholes or other openings into which streams abruptly empty. As such, swallets are significant locations of recharge to the underlying limestone aquifer (Figs. 8.11 and 8.12). Swallets are also considered important point sources for groundwater contamination in Florida because they carry surface-water chemical constituents, including sometimes potentially hazardous chemical constituents, into the Floridan Aquifer System (FAS; Ceryak 1977, Kincaid 1998).

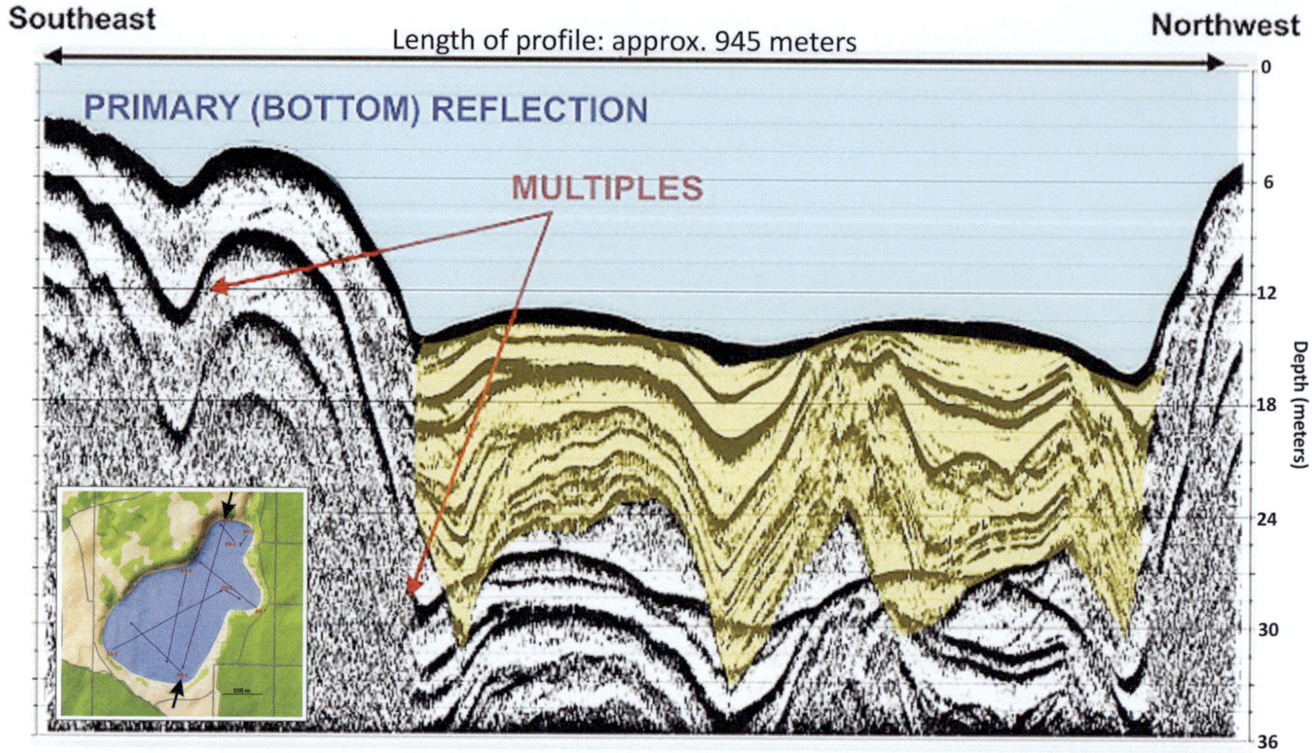


**Fig. 8.10** Acoustic sub-bottom profile across Big Blue Lake, Washington County. The pale blue area in the profile is water and yellow is sand deposited in the lake basin during and after sinkhole development. The features labeled multiples are an artifact of the acoustic signal. The top of the ICU is approximately 24 m below lake level and the top of the UFA is estimated to be at about 36 m below lake level. (Modified from Dobecki et al. 2013)

**Fig. 8.11** Diagram showing a stream originating in a low recharge area and flowing into a karstic terrain where it is captured in a swallet

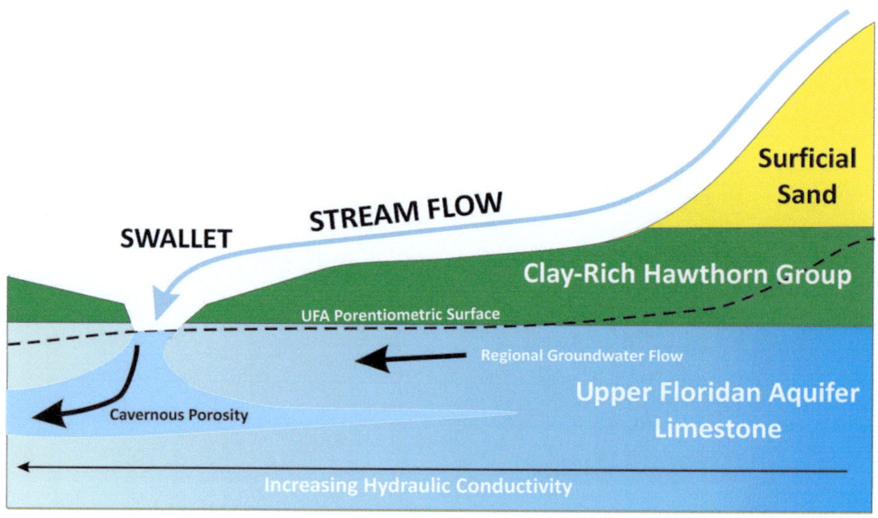


Most of the known swallets in Florida are located in northern and central Florida (Fig. 8.13) where streams drain regions where infiltration of water is limited, such as the Okefenokee Basin geomorphic district, Green Swamp, and San Pedro Bay geomorphic provinces. These areas with limited groundwater recharge allow streams to form and flow onto areas where the UFA is near the land surface and sinkholes have formed in the stream beds (Fig. 8.12). The stream is intercepted by a sinkhole, forming a swallet. Florea (2017, personal communication) reported that several swallets have been found within the Transverse Glades (the former tidal channels between the oolite shoals of the Atlantic Coastal Ridge (Fig. 4.12a, b). These swallets seep water to the underlying Biscayne Aquifer.

In addition to the numerous swallets that capture small streams, several of Florida's major rivers flow underground

Fig. 8.12 Stick Sink, one of the swallets where the Little River (Suwannee County) goes underground

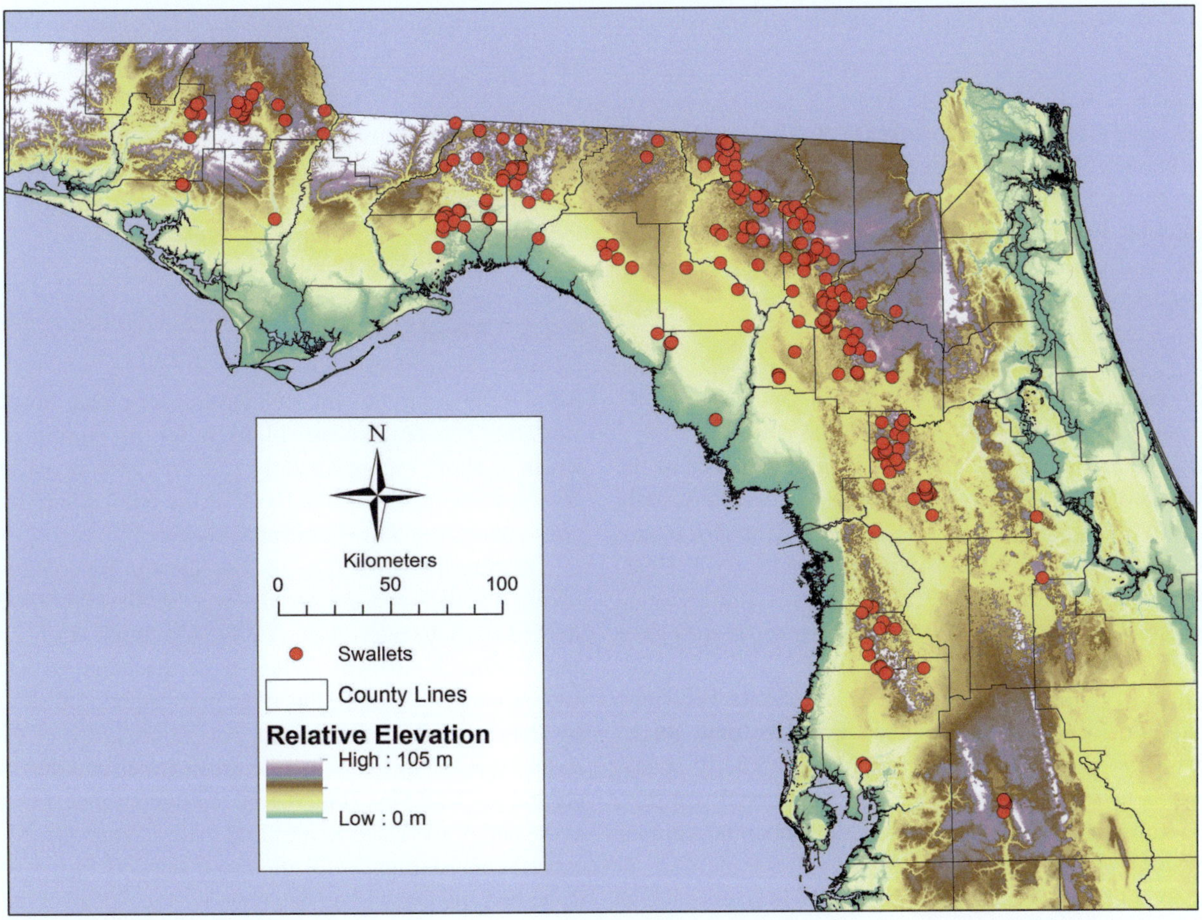


Fig. 8.13 Digital elevation map of Florida showing known swallets in northern and central Florida. Data from the Florida Geological Survey (Means 2005; Chelette and Thurman 2006)

**Fig. 8.14** The "Big Awesome Suck", a siphon that captures portions of the flow of the Santa Fe River, Columbia County downstream from High Springs

through large swallets or river sinks (Sect. 8.2.2), as they are locally known. These rivers, the St. Marks, Aucilla, Alapaha, and the Santa Fe rivers, become, at low-flow stages, truly underground rivers that flow through limestone/dolostone conduits and eventually emerge at large springs known as resurgences or river rises.

### 8.3.2 Siphons

Siphons are in-stream sinkholes that capture portions of the flow of a stream. Unlike swallets, siphons do not normally capture all of a stream's flow. The Santa Fe River, for example, has several known siphons (Hornsby and Ceryak 1998) wherein portions of the river flow passes underground through conduits and then emerges downstream as springs or in-stream boils (underwater springs).

Figure 8.14 illustrates a siphon on the Santa Fe River. This siphon, locally known as the "Big Awesome Suck" (Hornsby and Ceryak 1998) because of its vorticity at low river stage, is typical in that, other than occasional vorticity, there is normally little indication that the siphon is draining river water.

We do not know how many siphons exist in rivers such as the Suwannee, Withlacoochee, or Santa Fe, but one can be assured that many may exist. Like many terrestrial sinkholes, smaller siphons probably become plugged with debris and sediment and then reopen as the plugging material breaks loose.

### 8.3.3 Blind Valleys

Blind valleys are stream valleys that terminate abruptly in swallets. The best known blind valley in Florida is Dead River, a short (approximately 0.8 km) stream valley that diverges from the Alapaha River at the point where the river goes underground in Hamilton County.

The thresholds for flow of the Alapaha River into its in-stream swallets and the Dead River blind valley have been studied by Schneider et al. (2006), Upchurch et al. (2007c), and Upchurch (2014b). When river discharge is less than approximately 14 $m^3/s$, the river goes underground through swallets in the bed of the river. As the discharge of the river nears 14 $m^3/s$, water spills into Dead River and goes underground in Dead Sink (Fig. 8.15a), a large swallet (Fig. 8.15b) at the end of the Dead River blind valley. When the discharge exceeds approximately 14 $m^3/s$, the swallets in the Alapaha River bed and Dead River can no longer accept all of the available water and water continues to flow downstream in the

**Fig. 8.15** (**a**) Dead River Sink, the terminus of Dead River, when the Alapaha River is too low for flow into the blind valley. (**b**) Dead River Sink when the Alapaha River is sufficiently high for Dead River to capture flow

**Fig. 8.16** Siphon Creek Resurgence boil reflecting a location where water from the Siphon Creek siphon re-emerges along a fracture in the bottom of the Santa Fe River. Location: 29.8567°N, 82.7332°W

normally dry river channel (Lawn 1989). Section 8.3.6.3 further describes the relationship of Dead River to the Alapaha River karst system.

### 8.3.4 Resurgences

Resurgences are locations where water derived from surface-water inflow through an in-stream swallet or siphon re-emerges from the underlying aquifer. This term is not used for springs that have known swallets as sources unless the water discharging from the spring is clearly dominated by riverine sources. It is used where rivers and streams go underground through a swallet or siphon and then reemerge as identifiable river water. River water is typically identified by its brown color, which is a result of humic substance content.

Figure 8.16 illustrates a small example of an instream resurgence. Siphon Creek is a small, side channel of the Santa Fe River in Gilchrist County. There is a siphon in the bottom of the creek (Hornsby and Ceryak 1998), and the water taken underground through the siphon re-emerges along a fracture in the limestone in the bottom of the river (Fig. 8.16).

The best-known combination of a swallet and resurgence in Florida is the Santa Fe River Sink and Rise (Sect. 8.2.2.3). Under normal flow conditions, the entire Santa Fe River flows underground through a conduit at the Santa Fe Sink (Fig. 8.4) and then rises at the River Rise (Fig. 8.5). The straight-line distance between River Sink and Rise is about 5 km. During extreme high flow conditions, the sink and rise system are unable to accept all of the river flow and the series of swamps and chutes that lie between the two features (Fig. 8.6) become flooded.

### 8.3.5 Traces

#### 8.3.5.1 Definition

Where streams are captured by swallets, there is often a surface-water over-flow system. During flood events the ability of the swallet(s) in a stream's pathway to capture flow may be overwhelmed. When this occurs, the overflow follows a path that is termed a "trace"[2] in Florida. Traces are characterized by a series of sinkholes, karst windows, and a

[2]The term trace has been widely used by the Florida Department of Environmental Protection's experts on springs and spring protection.

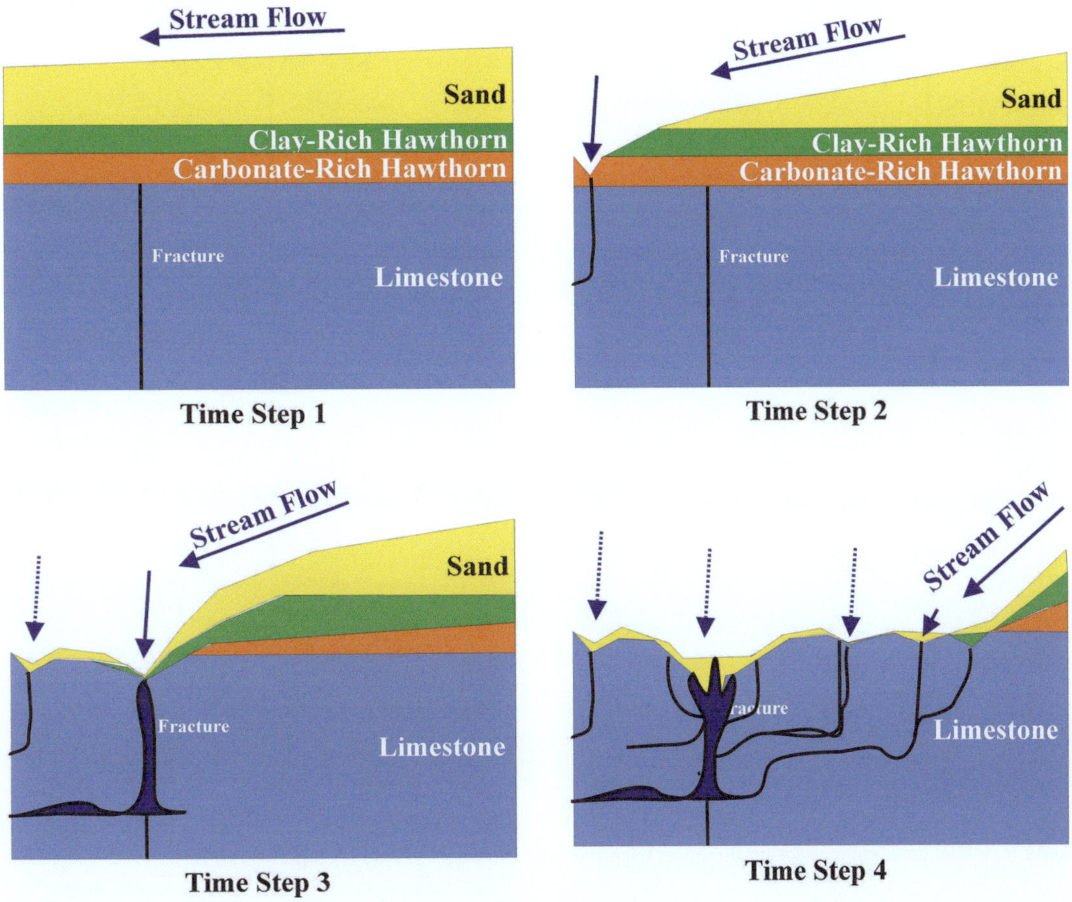


**Fig. 8.17** Sequence of events as a karst escarpment and trace are formed by headward erosion of a stream over a karstic limestone

normally dry stream valley, which floods sequentially downstream during flood events.

The presence of the "string of pearls" array of sinkholes (Fig. 7.60) in a normally dry valley clearly suggests that the traces were formed by retreat along a previously existing stream valley (Fig. 8.17). As the stream erodes headward through strata composed of siliciclastic sediments, sinkholes develop in its bed and capture the runoff when limestone is exposed. In time, headward erosion creates another opportunity for sinkhole development and the former swallet is abandoned as a new, upgradient sinkhole captures the runoff.

The sequence of events shown in Fig. 8.17 is normally seen where a stream enters a karst terrain after flowing over a highland characterized by low groundwater recharge rates. As the scarp at the edge of the highland retreats, a trace with a dry stream valley, "string of pearls" sinkholes, and periodic flooding remains. Excellent examples of traces can be found along the Aucilla River (Fig. 7.60), at Ichetucknee Springs in Columbia and Suwannee counties, and between the River Sink and Rise on the Santa Fe River (Fig. 8.6).

#### 8.3.5.2 Evolution of Traces

The previous sections pointed out many of the attributes of the traces (normally dry valleys) and wet sloughs associated with swallet-to-spring drainage systems. These include

- A primary, linear feature (the trace or slough on the land surface) and a cave in the subsurface that is characterized by a series of shallow to moderate depressions (karst windows, sinkholes) that connect a resurgence to a swallet or swallets up gradient;
- The conduit upon which the karst windows and other sinkholes develop is a well-defined cave that carries the river or stream water, normally without significant dilution or filtration so the water emerging from the resurgence remains colored by humic substances;
- Tributary caves may supply a large proportion of water to the conduit and resurgence, especially during periods of low surface-water flow (Sect. 8.2.2.3);
- Tributary streams and/or dry valleys often terminate in swallets at or near the main trace or slough (Fig. 8.18); and

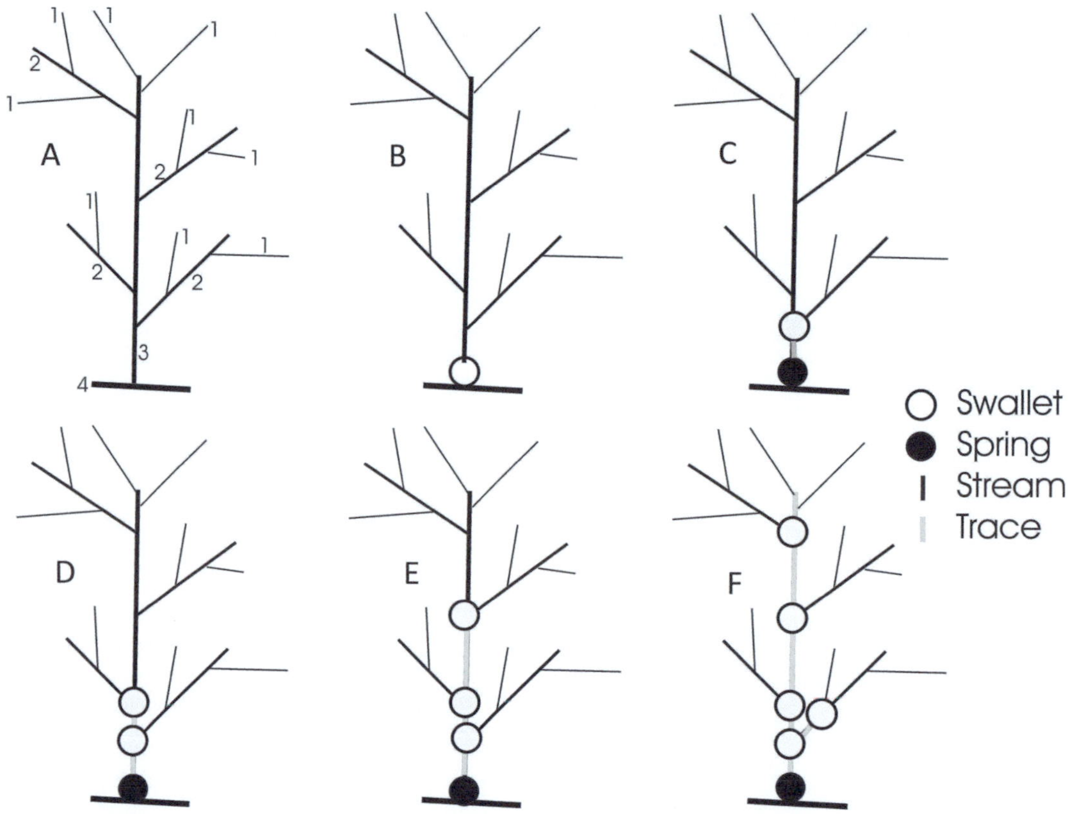


**Fig. 8.18** Model for swallet development during fluviokarst scarp retreat. Numbers in panel A represent Horton's (1936) stream orders

- The traces and sloughs appear to be related to fluviokarst escarpment retreat as they only extend between the rivers and uplands of the Okeefenokee Basin District onto the limestone of the Ocala Karst District.

These observations suggest a model for fluviokarst scarp retreat (Fig. 8.17). Imagine an upland underlain by relatively impervious, clay-rich sediments. This scenario is consistent with the concept that the Miocene Hawthorn Group once extended completely across the Branford Karst Plain in the past (Scott 1981). A drainage network developed under these conditions would likely produce a dendritic pattern such as shown in Fig. 8.18. Horton's (1936) stream orders are identified on Fig. 8.18a for ease of discussion.

As the 3rd-order stream and 4th-order river system erode downward (Fig. 8.17), they encounter limestone and a swallet develops that captures water from the 3rd-order stream (Fig. 8.18b). As headward erosion continues, swallets begin to develop upstream on the 3rd-order stream. Based on the Ichetucknee trace system (Sect. 8.3.6.2), these swallets appear to preferentially develop where major surface-water tributaries enter the master stream (2nd-order streams flowing into the 3rd-order stream in Fig. 8.18d–f).

At some point, headward erosion begins to encourage swallet development on the lower-order streams; again at or near stream intersections (Fig. 8.18f). The low-order streams become fragmented by this process and simply become sinkholes with faint depressions that reflect the earlier, low-order drainage.

### 8.3.6 Examples of Fluviokarst Systems

#### 8.3.6.1 Introduction

Two different fluviokarst systems are discussed in this section. The first is the Ichetucknee River system, which originates at the toe of the Cody karst escarpment in Columbia and Suwannee counties. This springshed and river system demonstrates the development of a trace and interaction of the drainage from highlands to karst lowlands as headward erosion expands the system and swallets, springs, and other karst features evolve in conjunction with local, ephemeral streams.

The second example is the Alapaha River system, which originates in the Okefenokee Basin in southern Georgia and flows across the Cody Scarp to join the Suwannee River. The Alapaha is an example of the development of a fluviokarst system by a perennial river that crosses a karst escarpment.

With the possible exception of the Suwannee River itself, all of the major rivers and streams that flow into the Branford

Karst Plain across the Cody Scarp go underground through swallets and emerge as resurgences. The Withlacoochee River (north[3]) goes underground just north of Valdosta, Georgia at the confluence with Cherry Creek (30.905°N, 83.311°W; McConnell and Hacke 1993). Similarly, the Santa Fe River goes underground at the River Sink (Sect. 8.2.2.3), and the Alapaha River goes underground just south of the Georgia Line.

#### 8.3.6.2 Ichetucknee Trace and River Karst

The complex that consists of the Ichetucknee trace, springs, and river system constitutes one of the best known and most interesting karst systems in Florida because of the Ichetucknee Springs State Park and the location of the Ichetucknee springshed within and adjacent to Lake City, Columbia County. The springs and most of the river are within Ichetucknee Springs State Park, which is considered by many to be the "Crown Jewel" of the Florida State Park System. The Ichetucknee River is one of two main tributaries of the Santa Fe River (the other being Olustee Creek) with karst-controlled interactions with the Santa Fe. The Ichetucknee River begins at the Ichetucknee Head Spring, is fed by several other springs (Fig. 8.19), and then flows into the Santa Fe River approximately 10 km to the southwest.

The river is essentially a large spring run as most of the water in the river under normal conditions is derived from springs at the upper extent of the river. Altogether, the springs have a median discharge greater than 10.1 $m^3/s$ as measured in the river. Therefore, the Ichetucknee spring group constitutes a first magnitude (Meinzer 1927) spring system.

In order to understand the Ichetucknee trace, river, and springs, it is important to understand the entire drainage system, especially the springshed and sources of spring water. Figure 8.20 represents the springshed of the Ichetucknee River and springs. This springshed is well documented owing to a "high-resolution" monitoring program undertaken by the Suwannee River Water Management District in the late 1990s. The large number of monitoring wells within the springshed allowed for a detailed, geostatistically based delineation of springshed boundaries (Champion and Upchurch 2003).

The approximate location of the Cody karst escarpment (Cody Scarp, Puri and Vernon 1964) can be identified in Fig. 8.20 from the pattern of UFA potentiometric-surface contours (Upchurch 2002; Champion and Upchurch 2003). Beneath the uplands of the Alachua Karst Hills and the Northern Okefenokee Basin provinces (e.g., north and east of Lake City), the isopotential contours are relatively widely spaced, which reflects an area of confinement of the UFA by the intermediate aquifer and confining unit system (IAS/ICU). Here, the UFA has relatively low permeability, low horizontal gradients, and receives minimal recharge. The low permeability causes what little water recharges the aquifer to be held within the aquifer, so the potentiometric surface remains high because the water cannot move down gradient efficiently.

The region where the contour lines are close together (i.e., near Lake City; Fig. 8.20) corresponds with a significant change in recharge pattern and permeability in the UFA. Aquifer permeability and recharge increase in the vicinity of the Cody Scarp (Upchurch 2002), so drainage of the aquifer is enhanced and potentials in the aquifer decrease as groundwater flow becomes more efficient.

The lowlands of the Branford Karst Plain form the toe of the Cody Scarp to the south and west in Fig. 8.20. This area is characterized by high aquifer permeability and recharge. In the Branford Karst Plain a more efficient groundwater flow system allows the potentiometric surface of the aquifer to be relatively flat compared to the transition at the Cody Scarp.

As can be seen in Fig. 8.20, portions of the springshed extend into the highlands of the Ocean Pond Plain (Fig. 3.12) near Lake City. This is because the groundwater divide is located in the highly confined, lower permeability portion of the UFA under the highlands. Recharge in this area is low, so rainfall accumulates in streams and flows toward the Cody Scarp. The springshed includes portions of the UFA that are well confined and contribute relatively small amounts of groundwater to the spring system. The functional (efficient) boundary of the groundwater flow system within the springshed is, therefore, near the Cody Scarp, where surface water first enters the groundwater system efficiently.

During the drought of the late 2000s, discharge from the springs declined, the groundwater divide within the UFA potentiometric surface declined, and the potentiometric divide between the Ichetucknee springshed and groundwater basins to the north and east appeared to have shifted westward, reducing the size of the springshed (Grubbs 2011). This shift is a result of the high level of confinement under the Northern Okefenokee Basin and groundwater withdrawals in southeastern Georgia and northeastern Florida. These withdrawals have created a cone-of-depression that extends under the Ocean Pond Plain. With limited recharge in the basin, the cones-of-depression have migrated westward, to locations where they can capture recharge.

There are three flow systems that supply water to the Ichetucknee River and its springs: (1) surface water that enters the aquifer through swallets in the Cody Scarp; (2) overland flow of surface water; and (3) regional groundwater flow.

Alligator Lake in Lake City is the northernmost, known, major source of surface water that enters the groundwater

[3]There are two Withlacoochee Rivers in Florida, one in northern Florida and the other is west-central Florida. To differentiate these rivers, the parenthetic terms north and south are use to indicate the river being discussed.

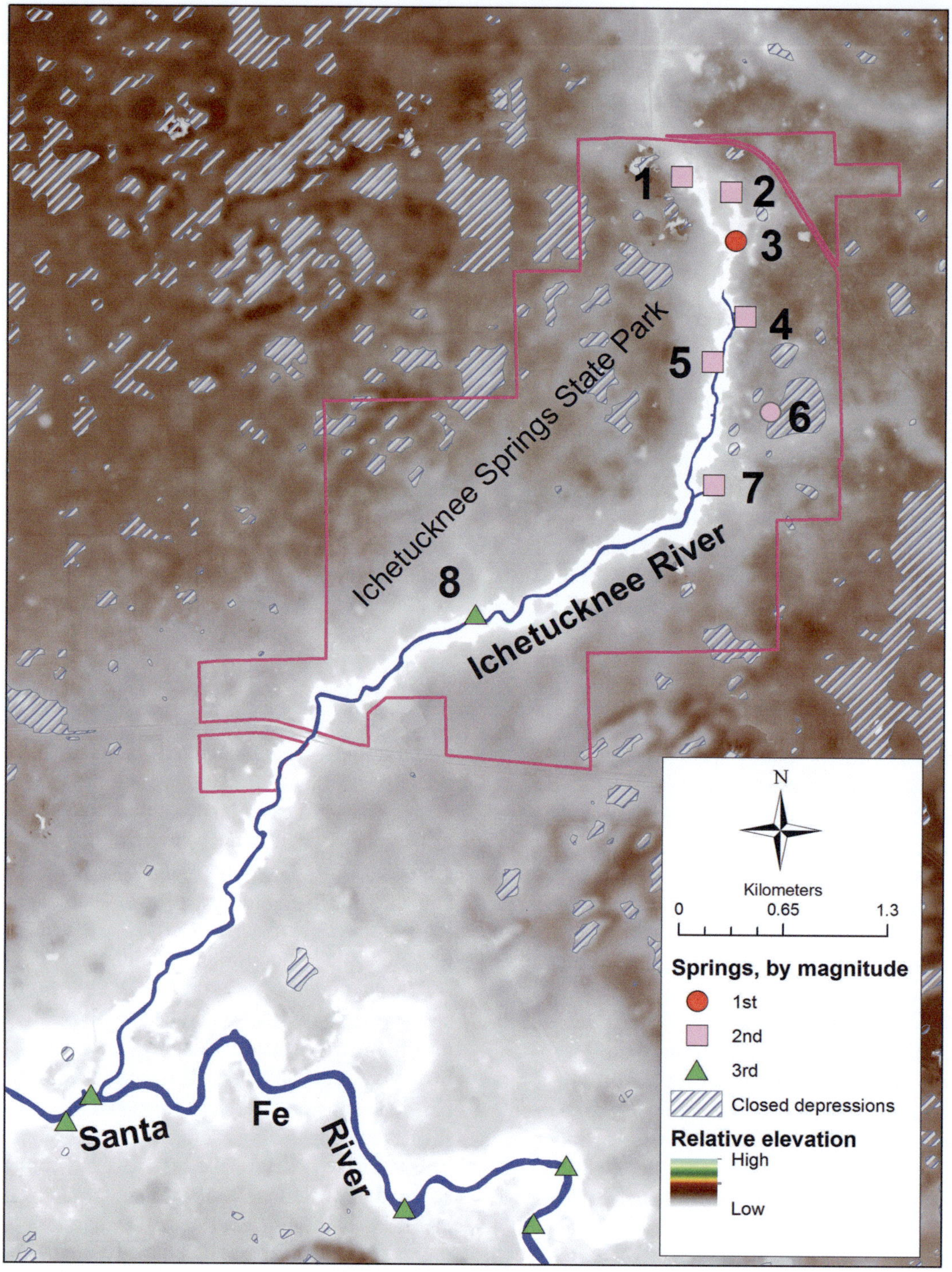


**Fig. 8.19** The Ichetucknee River and springs. 1 Ichetucknee Head Spring, 2 Cedar Head Spring, 3 Blue Hole (Jug) Spring, 4 Mission (Fig) Spring Group, 5 Devil's Eye Spring Group, 6 Grassy Hole Spring, 7 Mill Pond Spring, and 8 Coffee Spring

system. Alligator Lake adds water to the Ichetucknee groundwater system by leakage through its bottom and occasional drainage through sinkholes within the lake basin. Lawrence and Upchurch (1976) studied groundwater quality in the vicinity of Lake City in 1975. At that time Alligator Lake was undergoing one of its periodic draining episodes. They observed a plume of discolored water with organic detritus and high phosphate concentrations in well water from the UFA along a linear trend that reached from the lake to Clay Hole Sink (Figs. 6.23 and 8.21; 30.111°N, 82.683°W), a complex of swallets located 5.6 km southwest of the lake. This observation links the lake as a source of groundwater to the cavern system that ultimately supplies the water discharging from the Ichetucknee springs.

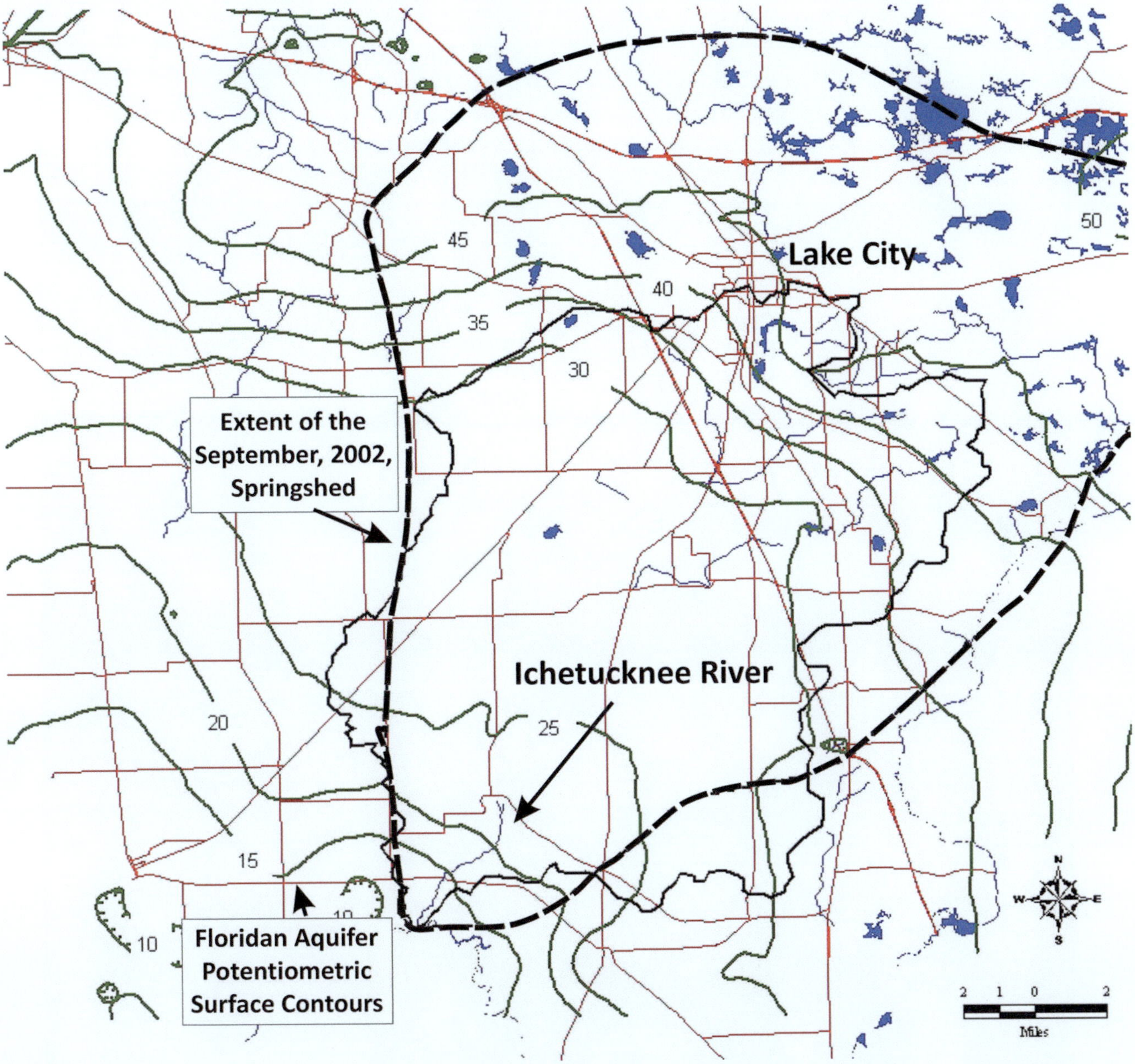


**Fig. 8.20** The Ichetucknee Springshed as identified from September, 2002 UFA potentials. Contour interval is 5 ft (1.5 m). (Data and map from Champion and Upchurch 2003)

When streams flow across the Cody Scarp, they typically go underground through swallets (Upchurch 2014a) at, or near, the first occurrence of limestone in the stream bed. Examples of stream-to-swallet systems that introduce surface water into the Floridan aquifer in the Ichetucknee Basin include (Fig. 8.21):

- Cannon Creek to Cannon Sink (30.120°N, 82.656°W) system (Figs. 8.21 and 8.22);
- Clay Hole Creek to Clay Hole Sink (Fig. 8.21); and
- Rose Creek to Rose Creek Sink (30.069°N, 82.696°W) (Figs. 8.21 and 8.23).

The importance of several of these swallets have been demonstrated by dye-trace studies that established connections of the swallets with the Ichetucknee springs (Karst Environmental Services 1997; Butt and Murphy 2003). These studies and cave-diving explorations have documented a cave system that connects Clay Hole Sink[4] and Rose Creek Swallet (Fig. 8.21) to all of the Ichetucknee springs other than Coffee Spring (Fig. 8.19). Travel times of the dye were as high as 1.6 km/d. These dye traces indicated that

- The cavern system under the sinks and swallets is extensive and directly connected to the Ichetucknee springs;

[4]Dyal and Black Sinks, which were investigated by Butt and Murphy (2003), are located within the larger Clay Hole Sink.

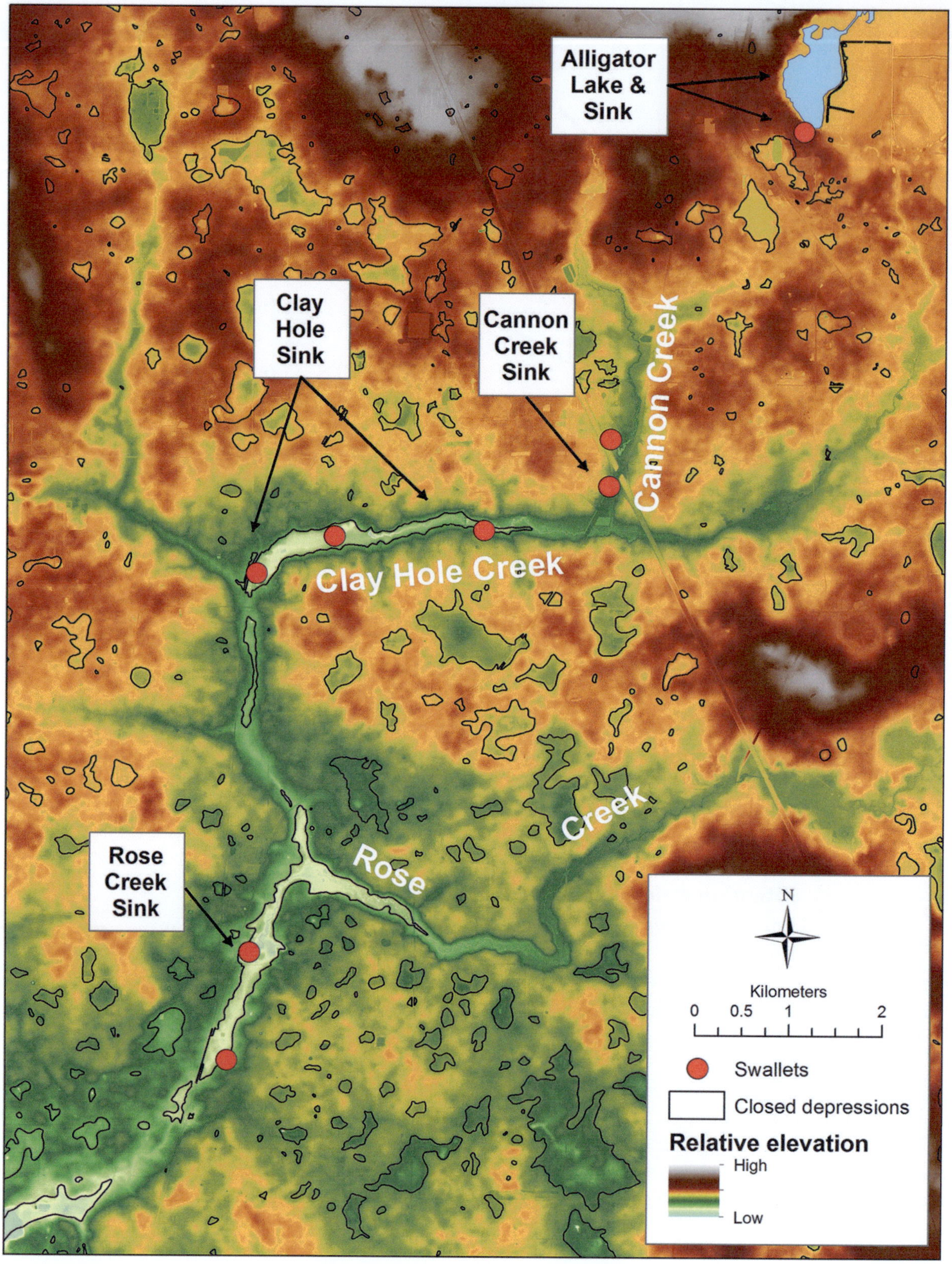


**Fig. 8.21** LIDAR map of the northern part of the Ichetucknee Trace showing the locations of major tributary streams and swallets

**Fig. 8.22** Cannon Sink, which is located in the northeastern quadrant of the intersection of SR 47 and I-75, frequently floods when the ability of the swallet to capture the water in the small stream (Cannon Creek) shown meandering across the sinkhole floor is exceeded.

**Fig. 8.23** Rose Creek Swallet (30.069°N, 82.696°W) is located near Columbia City. This karst window and swallet exposes UFA water, which is free of significant particulates or organics. When Rose Creek (Fig. 8.21) flows, it is captured by this swallet thus introducing surface water with particulates, organics, and other constituents to the aquifer

**Fig. 8.24** Flooding of the Ichetucknee Trace as a result of Hurricane Frances in September, 2004. (Photograph courtesy of the Florida Geological Survey)

- Based on detections in the springs several kilometers from the locations of dye introduction, dilution and dispersion of the dye in the aquifer was limited, indicating that the surface-water drainage to the springs is largely within well-developed conduit systems;
- Travel times in the aquifer are rapid, indicating an efficient flow system; and
- The stream-to-swallet systems make the aquifer highly vulnerable to contamination from human activities in their respective drainage basins.

Note that the swallets that capture flow from these tributary creeks are generally located at or just down gradient from the point of junction of the tributaries and main drainageway. This coincidence appears to indicate that the junction of flow is a likely location for swallet development (Fig. 8.18).

Occasionally, the swallets are overwhelmed by stormwater and cannot recharge all of the surface water flowing through the system. When this happens, flooding follows a sinuous valley, the Ichetucknee Trace, which passes from Cannon Sink (Fig. 8.22) through Rose Creek Sink and near Columbia City, and ends at the head springs of the Ichetucknee River (Fig. 8.19). The Ichetucknee Trace includes a series of abandoned swallets (Upchurch 2014a) that are connected overland by a normally dry, shallow valley (Fig. 8.21) that contains poorly developed stream deposits. The trace is roughly paralleled by the cave complex that normally drains the swallets (Fig. 8.21).

When the Ichetucknee Trace completely floods, surface water drains directly to the Santa Fe River. Normally, the Suwannee and Santa Fe rivers flood at the same time as the Ichetucknee River and Trace, so backwater effects are likely to retard drainage of the flood waters along the river and trace and prolong the duration of the flood.

Water discharging from the Ichetucknee Spring Group consists of a mixture of water gained from the stream-to-swallet systems and regional groundwater flow (Champion and Upchurch 2003). Because the stream-to-swallet drainages are ephemeral, the majority of the discharge from the springs is supplied by regional groundwater flow, which results in clear water with no visible evidence of the inflow of colored, tannin-stained surface water through the stream-to-swallet systems. Some preferential flow was noted, however. For example, there was a pattern of elevated nitrate concentrations that extended from near Clay Hole Sink to the eastern side of the Ichetucknee River near Mission Springs. This pattern was consistent with the Dyal and Black sink dye trace study (Butt and Murphy 2003) which showed dye moving from these sinks within the Clay Hole Creek swallet to

**Fig. 8.25** Ichetucknee Head Spring (29.984°N, 82.762°W; Spring 1, Fig. 8.19)

Mission and other springs on the east side of the Ichetucknee River (Fig. 8.19).

Several of the springs in the Ichetucknee system provide information about the karst in the UFA. The Ichetucknee Head Spring (29.984°N, 82.762°W), which is considered the head of the Ichetucknee River (Fig. 8.19), discharges from a cleft in the limestone (Fig. 8.25). The median discharge, based on 13 measurements from 1917 to 2002, was 0.6 $m^3/s$, so the spring is considered an historic second-magnitude spring. There is an apparent multi-year variability in flow that is related to long-term rainfall patterns. Superimposed on this long-term pattern, there are short-term variations in flow that reflect rainfall and backwater effects from the Santa Fe River. As is typical of all springs in karst terrains, the springs of the Ichetucknee are sensitive to rainfall and water-use patterns. The Ichetucknee Head Spring is the most accessible spring for swimming.

Blue Hole Spring, or Jug Hole Spring, lies at the downstream end of the run from Cedar Head Spring (Fig. 8.19). The name comes from the deep blue color of the spring water (Fig. 8.26), which emerges from a large limestone shaft approximately 3.6 m wide and 7.9 m below the surface (Follman and Buchanan 2015). This spring has a mapped cave system and is a popular diving spot (Hornsby and Ceryak 1998; Scott et al. 2004). There is a short run from Blue Hole Spring to the Ichetucknee River.

Blue Hole Spring, an historic first-magnitude spring, is the largest spring in the Ichetucknee Spring Group. Historic discharge data indicate that the median spring discharge is 2.7 $m^3/s$, but daily measurements since 2002 indicate a much higher flow (3.6$m^3/s$).

Finally, the Mission Spring Group (29.976°N, 82.758°W) (Location 4; Fig. 8.19) consists of at least 8 vents. Flow from these springs is split by a small island (Fig Island) as it passes on its way northwest to the Ichetucknee River. The name comes from a Spanish mission that existed just north of the springs in the 1700s (Weisman 1992, 1993). Prior to discovery of the mission, the springs were known as Fig Spring. The spring on the north side of Fig Island is also known as Roaring Spring (Scott et al. 2004). Roaring Spring, the largest of the Mission Springs Group, discharges from a cave in a ledge of limestone on the north side Fig Island. The spring pool is about 3 m east to west and 4.5 m north to south. Depth near the ledge is 4.5 m (Scott et al. 2004). The smaller spring located south of the island is known as Singing Spring.

The quality of water discharging from the Mission Spring Group is a concern. Eelgrass (*Vallisneria americana*) in the spring run and river just offshore of the mouth of the spring

Fig. 8.26 Blue Hole (Jug) Spring (29.981°N, 82.758°W; Spring 3, Fig. 8.19)

run is covered with brownish periphyton that has degraded the benthic plant and animal community. This periphyton (*Oscillatoria* sp. and other epiphytes) has been attributed to elevated nitrate and other nutrients discharging from the spring. It is thought that these constituents have both local (Butt 2005) and regional sources (Champion and Upchurch 2003). The Mission Springs Group is an historic second magnitude spring group. Monitoring of discharge by the USGS began in 2002, and the median discharge is 28 $m^3/s$ from the daily data.

The floodplain of the Ichetucknee River is relatively narrow, especially near the head springs. Here, the river is either incised in limestone (Fig. 8.27) or it has a narrow floodplain. The middle reach (especially between Mission and Mill Pond springs; Fig. 8.19) of the river has a broader floodplain with a relatively wide marsh dominated by wild rice (*Zizania aquatica*). The lower floodplain downstream from the Devil's Eye Spring run is characterized by scattered marsh and a well-developed swamp with mixed hardwoods and cypress (*Taxodium* sp.). Rock shoals (Fig. 8.28) composed of fractured Eocene Ocala Limestone occur in several locations near the river mouth where the Ichetucknee joins the Santa Fe River.

Figures 8.29 and 8.30 illustrate an interesting, carbonate sedimentation feature of the lower Ichetucknee and Santa Fe rivers. At low to moderate flow, both the Ichetucknee and Santa Fe rivers are essentially spring runs discharging $Ca\text{-}HCO_3$ facies water. During floods, this water is diluted and spread over the river floodplains. As the water evaporates, calcite deposits are formed. Note the white deposits on the bases of the cypress trees in Fig. 8.29. The white deposit on the buttresses of the trees is a residue of fine-grained calcite, which "whitewashes" the lower trunks and buttresses of the trees. Freshwater marls (shelly silt- and clay-sized calcite mud) form natural levees in the flood plains of these rivers (Fig. 8.30). These deposits are a testimony to the amount of calcium and carbonate in the spring water and show that some of it remains as modern-day carbonate sediment deposits within the spring and river systems rather than flowing to the sea.

#### 8.3.6.3 Alapaha River Karst

The Alapaha River system is one of the most interesting and important karst areas in Florida. This section presents an overview of the fluviokarst drainage system of the Alapaha River in Florida and the results of an analysis of stream hydrographs to quantify water capture at a series of swallets in and off the river's mainstem (Fig. 8.31).

The Alapaha River Basin encompasses roughly 259 $km^2$ of central Hamilton County (Conover and Leach 1975). This area represents less than 10% of the Alapaha River drainage basin, which covers approximately 4400 $km^2$ and extends northward nearly 160 km into southern Georgia. The portion of the Alapaha River Basin within Florida contains approximately 15 sub-basins, most of which are internally drained karst basins that do not contribute direct runoff to the Alapaha River except, perhaps, during extreme runoff events. The discharge of groundwater from these closed-basin locations are unknown, but the nearby Alapaha River and several springs along the Suwannee River (Fig. 8.32) are likely candidates.

The mainstem of the Alapaha upstream of the swallets in the river bed and Dead River has perennial flow with a median discharge of 15.4 $m^3/s$ based on gage data from Jennings, Florida. A poorly monitored stream-flow gage downstream of the swallets (Fig. 8.32) indicates that the stream is dry most of the year and is only seasonally flooded. In-stream flow in the Alapaha River reaches the Suwannee River only during periods of high flows and floods, when discharge above the swallets (at Jennings) is about 10–14 $m^3/s$ (Upchurch 2014b). About 50% of the time, the lower river is dry owing to the capture of discharge from the upper river at the swallets. When the stream channel is carrying water, the measured median flow in the normally dry river reach is 1.6 $m^3/s$ (at Jasper).

**Fig. 8.27** Solution pipes and other karren exposed in limestone at Trestle Point near the mouth of the spring run from Blue Hole Springs, Ichetucknee River

**Fig. 8.28** Rock shoals near the mouth of the Ichetucknee River. Note the fractures in the Ocala Limestone

**Fig. 8.29** Carbonate sediment accumulations on cypress tree buttresses near the mouth of the Ichetucknee River. Location: 29.9328°N, 82.7800°W

**Fig. 8.30** The "Santa Fe marl" an accumulation of silt- and clay-sized calcite with pulmonate snails that forms in natural levees on the Santa Fe and lower Ichetucknee rivers

When the lower river is dry, sand bedforms can be seen lying within the rock-bound channel. Lawn (1989) characterized these bedforms and the flow system necessary for their development. There are no known swallets downstream of the entrance to Dead River, but it can be reasoned that sand-filled swallets may exist. If they do, they do not appear to be functional, probably because by the time flow begins in the downstream Alapaha, the potentiometric surface of the Floridan aquifer and Suwannee River levels have risen to the point that the hydraulic gradient hinders flow into the swallets.

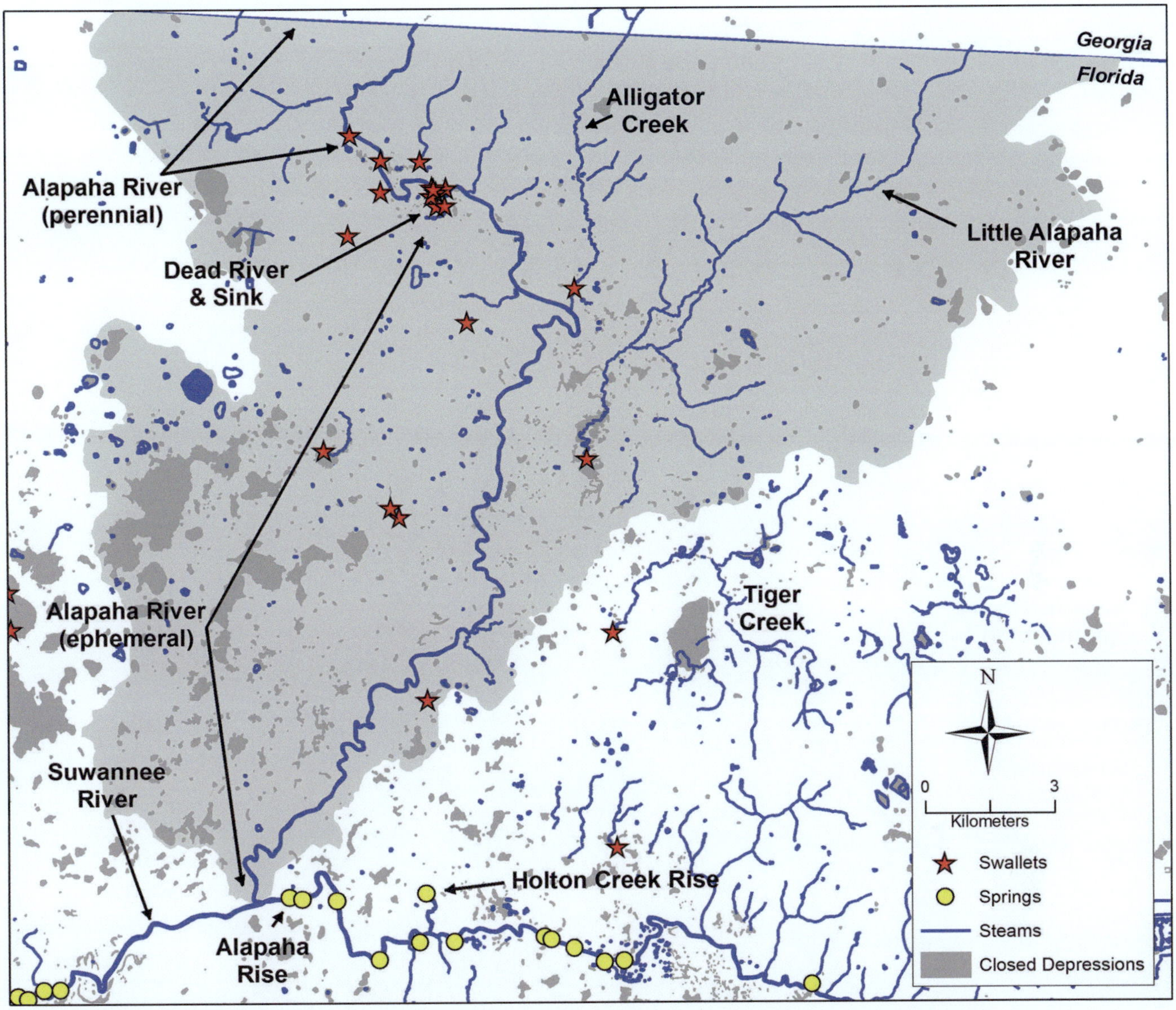


**Fig. 8.31** Locations of karst features in the Florida portion of the Alapaha River drainage basin. (From Upchurch 2014b)

About 5 km north of the swallets, the river is incised into a rock-bound channel with shoals and a relatively narrow floodplain. The shoals are primarily composed of dolostone and sedimentary opal (opal-CT) of the Statenville Formation (Miocene Hawthorn Group; McFadden 1982). Downstream from the shoals, the river enters its karst-dominated reach[5] (Fig. 8.33). In this reach, internally drained basins, blind valleys, and in-stream swallets capture a significant amount of surface-water flow. The development of swallets results from the thinning of the Hawthorn Group semi-confining beds by erosion as the river flows south (Ceryak 1977, 2005; Upchurch 2014b).

[5]There is a karst area along the Alapaha in southern Georgia known as the Valdosta Limesink District (Gore and Witherspoon 2013).

During low flows in the river, water is captured by swallets in the channel of the river. One such swallet (sinkhole) is shown in Fig. 8.34a, b. Exploration of the swallet indicated that it penetrated the Statenville Formation and the upper St. Marks Formation.

As stage in the river increases, the swallets in the river channel are successively flooded. When their capacity to take water is exceeded, river water is diverted into Dead River (Fig. 8.33). A major swallet (Dead River Sink) is located at the terminus of the blind valley (Figs 8.15a, b and 8.35). Dead River and Sink are also incised through the Statenville Formation and into the underlying Miocene St. Marks Formation.

As can be seen in the LIDAR map of the vicinity of the Dead River (Fig. 8.35), Dead River follows a circuitous path,

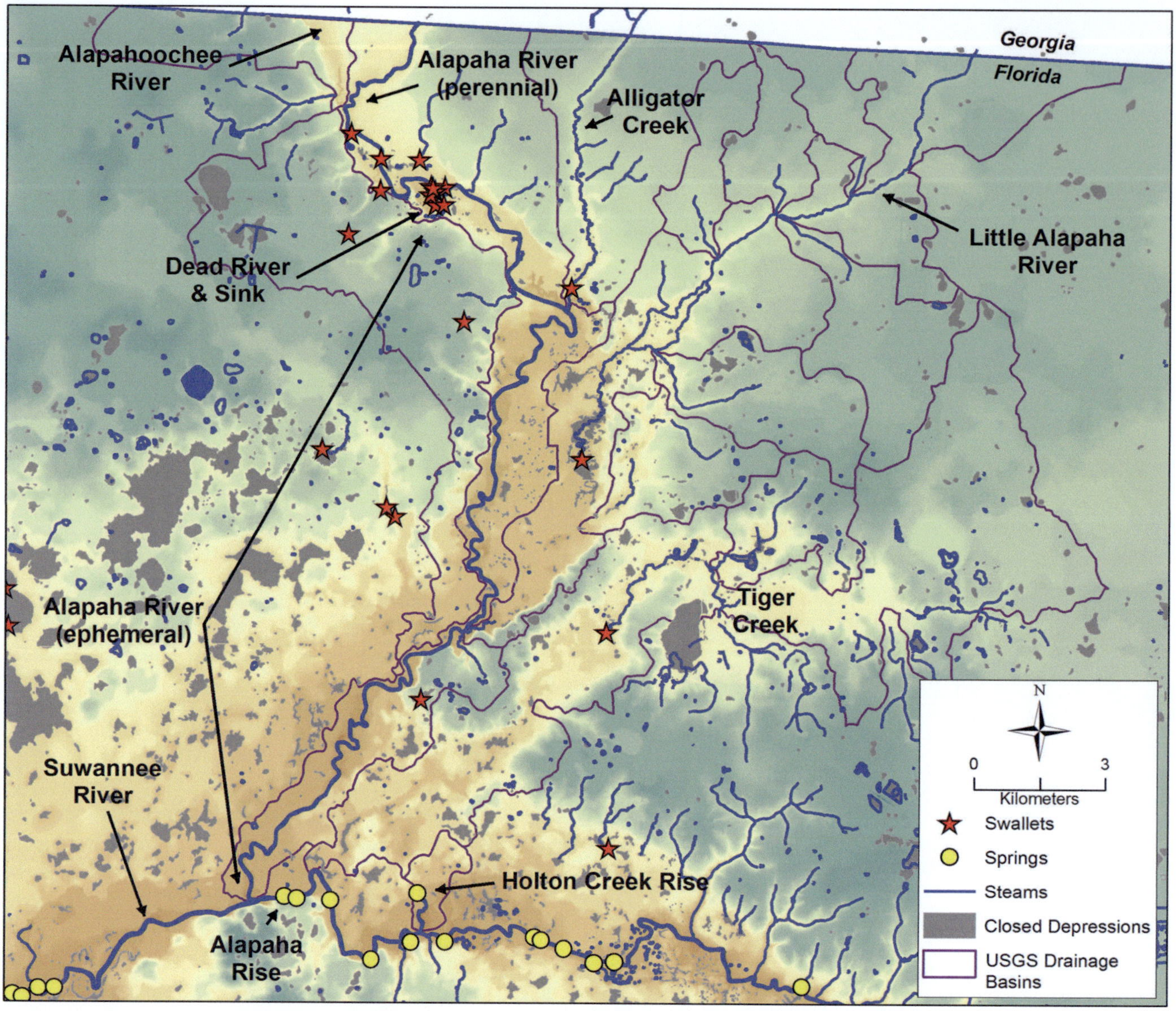


**Fig. 8.32** Digital elevation map of the Alapaha River basin in Florida. Polygons are sub-basins identified by the USGS. These basins show the many, internally drained, stream-to-sink drainage basins in the vicinity of the Alapaha River

apparently following a series of bedrock fractures, on its way to Dead River Sink. Note that the initial reach of Dead River that extends to the west (Fig. 8.35) from the Alapaha River seems to correspond with the base of the broad meander loops to the east and west in the main river valley. Thus, one may hypothesize that the initial reach of Dead River follows the same fracture pattern as does the base of the meander loop.

The LIDAR data (Fig. 8.35) also show the presence of a truncated drainage system within which Dead River now lies. During extreme rainfall events, water flows down this upland drainage until it encounters the series of swallets that includes the Dead River Sink. Unless the Dead River Sink and others upgradient in the truncated stream valley reach their capacity to take water during flood events, water moving down the truncated valley does not reach the Alapaha. This truncated valley and the others shown in Fig. 8.31 provide evidence for the pattern of stream truncation and swallet development shown in Fig. 8.18.

The rock-bound channel is easily seen when river stages are low at the location where the river enters the Suwannee River (Fig. 8.36). Note the incised channel and exposures of limestone throughout the cross section of the dry channel. There is a small ramp of fluvial sand deposited in the Suwannee River at the base of the outfall of the Alapaha. This ramp demonstrates sand transport into and then down the Suwannee.

While there are likely seeps and other discharge points in the Suwannee River, the majority of Alapaha River water that enters the aquifer at Dead River and the swallets in the

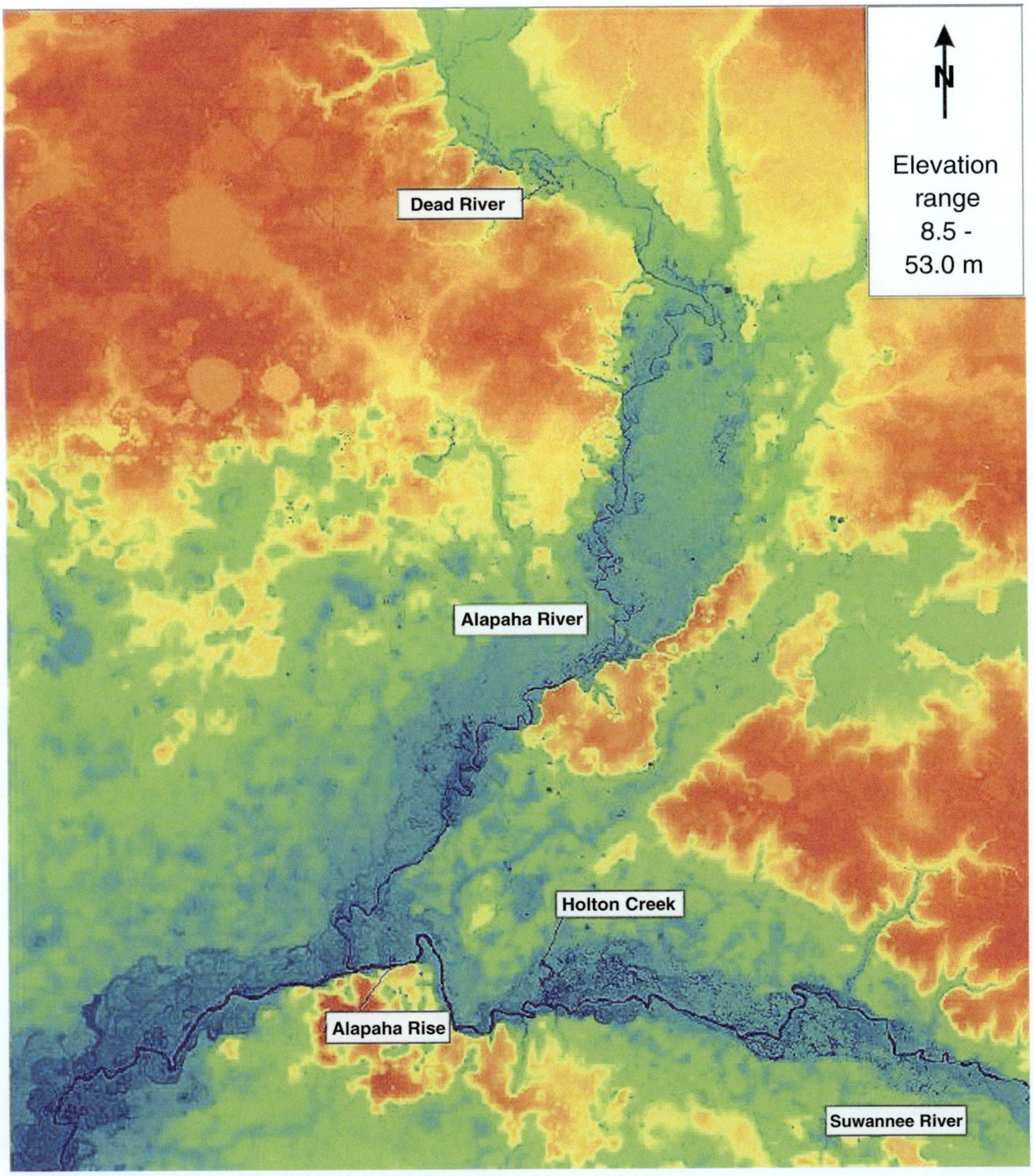


**Fig. 8.33** LIDAR map of the karst area of the Alapaha River basin. (Image from Lawn 2014)

river bed discharges to the Suwannee River through two resurgences (Ceryak 1977; Upchurch 2014b). These resurgences are the Alapaha Rise (Figs. 8.31, 8.32, and 8.37), which drains directly to the Suwannee River through a short run, and Holton Creek Rise (Figs. 8.31 and 8.32), which drains to the Suwannee River through Holton Creek.

The Alapaha Rise is a first magnitude spring with discharge ranging from approximately 5.7 $m^3/s$ to approximately 31 $m^3/s$ (Hornsby and Ceryak 1998). Schneider et al. (2006) studied the lag time between simulated peak flows in the Alapaha Rise and measured peak flows in the Alapaha River. The calculated lag time between a recharge event in the Alapaha River near Dead River and the Alapaha Rise varies from about 2 to 21 days depending on discharge (Upchurch 2014b). This lag reflects a period of time sufficient for water to enter storage in the aquifer and migrate to the resurgences (Alapaha and Holton rises). Discharge from the Alapaha Rise is usually discolored by humic substances as a result of water derived from the Alapaha River. Therefore, it reflects primary capture of river water at the swallets.

Modeling of the response times of the Alapaha Rise by Schneider et al. (2006) found that, when flow in the Alapaha River at the Jennings, Florida gage was less than 14 $m^3/s$, the pulse of water entering the subsurface should arrive in 3–6 days. A dye test was conducted by the Florida Geological Survey and Suwannee River Water Management District in

**Fig. 8.34** (**a**) Sinkhole/swallet in the bed of the Alapaha River near the mouth of the Dead River blind valley. (From Upchurch 2014b). (**b**) Close-up of the swallet shown in Fig. 8.34a. (From Upchurch 2014b)

June 2016. The dye was introduced into Dead Sink, and 6 days later it appeared at both the Alapaha and Holton rises (Greenhalgh et al. 2016, personal communication).

At low flow, discharge from Holton Creek Rise is limited and constitutes relatively clear UFA water. During peak flow episodes, however, the water becomes discolored. As a result, Ceryak (1977) suggested that the Holton Creek Rise was also discharging water captured by the Alapaha River swallets. Analysis of hydrograph data and mass-balance calculations led Schneider et al. (2006) to conclude that the Holton Creek resurgence appears to serve as an "overflow" route for water that is unable to discharge at the Alapaha Rise (Upchurch 2014b). Note that the dye test conducted in 2016 indicated that, at moderate flow, Holton Creek Rise also discharges Alapaha water.

The water that drains into the UFA from the Alapaha can often be identified within the UFA because of color, odor, and iron content. In 1977, Ceryak sampled wells in the vicinity of the Alapaha swallets and river during a period when the swallets were actively capturing water. He showed that this water spreads over an area of about 13 km$^2$ in the aquifer. Note that the area affected by river water in the UFA was not limited to the Alapaha River and its resurgences. Detectable river water was also present along the Suwannee River corridor. This water had apparently entered the aquifer as bank storage during the flood event in the Suwannee.

Screaton et al. (2004) investigated the interaction of stream-derived water in the cavern between the Santa Fe River Sink and Rise and concluded that organic-rich flood waters were chemically undersaturated with respect to calcite, so limestone was dissolved during and following the flood event. The same is undoubtedly occurring in the Alapaha River watershed and Suwannee River corridor. Therefore, development of karst features and enlargement of the conduits that connect the swallets to the rises is on-going, albeit slowly, in fractions of a millimeter each year.

There are other stream-to-swallet systems (Figs. 8.31 and 8.32) of note within the Alapaha drainage basin. These systems are within the lowland surrounding the Alapaha River. The reaches shown in Fig. 8.31 illustrate the fate of two streams, Alligator Creek and the Little Alapaha River, that drain the uplands east of the river. Each stream flows onto the lowland and, then goes underground through a swallet. These streams illustrate the disruption of large drainage systems as headward erosion and development of instream swallets form (Fig. 8.18).

Alligator Creek (Figs. 8.31 and 8.32) drains into a complex of low areas with swallets that resembles a former path of the Alapaha. This low region meanders to the south, where it connects with the low areas related to the Little Alapaha. These low areas are (1) located at the toe of the escarpment that defines the transition from highlands underlain by clay-rich Hawthorn Group sediments to the river corridor incised in the dolomitic Statenville Formation and (2) interconnected in a way that suggests a parallel drainage system to the east of the Alapaha.

The Alapaha River surface-water basin is interesting in that it is asymmetrical (Fig. 8.31). On the east, the Hawthorn Group, which underlies the Ocean Pond Plain (Fig. 3.12), prevents significant groundwater recharge and a dendritic pattern of tributary streams has developed. Note, however, that many of the streams that are considered tributaries of the

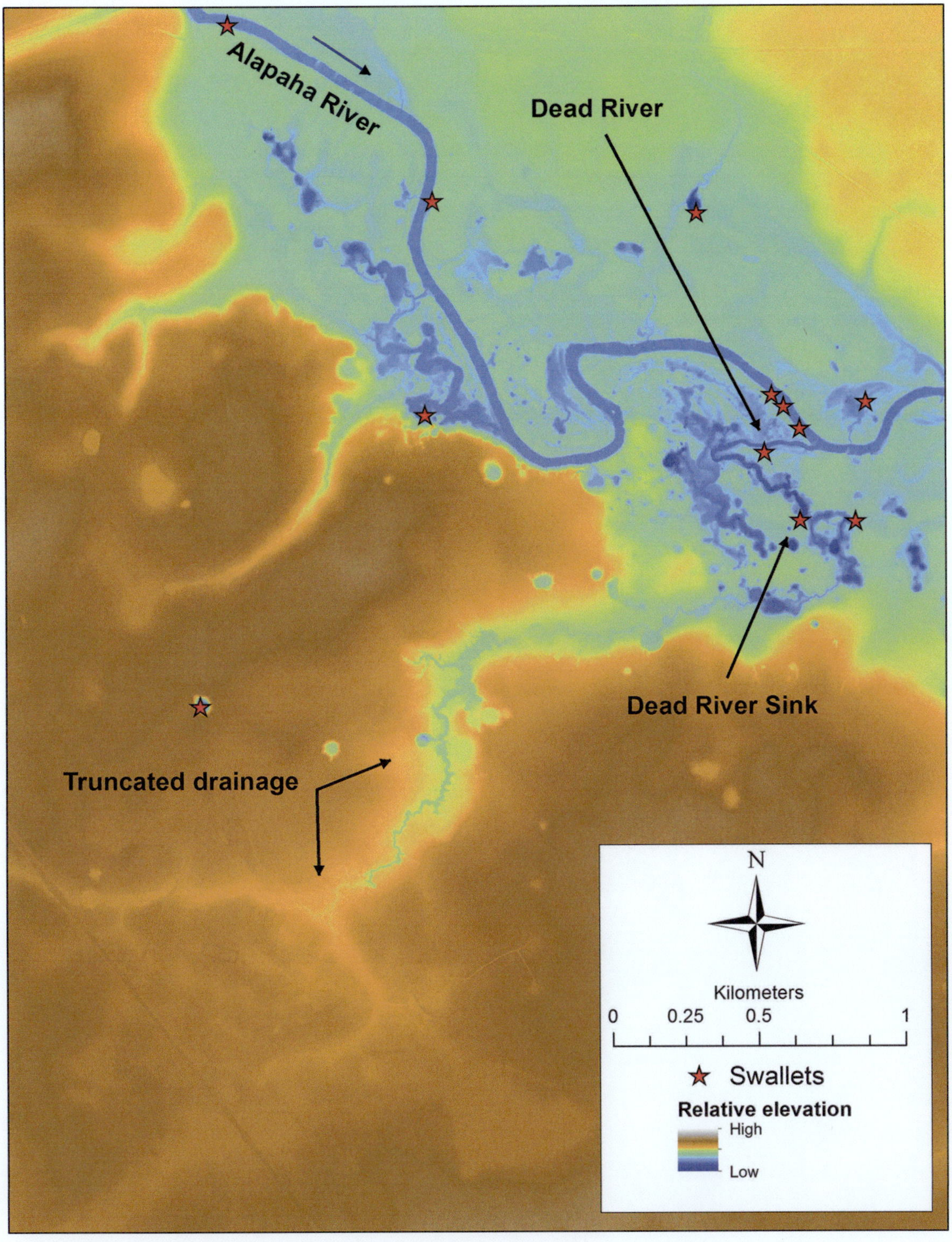


**Fig. 8.35** LIDAR map of the Dead River valley and sink (swallet). Note that the drainage uphill from the swallet has been truncated by the sinkhole

Alapaha, including Alligator Creek and the Little Alapaha River, go underground through swallets before reaching the Alapaha River. Traces indicating that these streams formerly drained to the Alapaha exist between the swallets and the Alapaha during normal flow conditions.

On the west side of the Alapaha River, the Hawthorn Group has been penetrated by numerous sinkholes, and surface-water drainage is almost entirely internal. Only the Alapahoochee River, which lies, in part, within a collapsed cavern system, enters the Alapaha River from the west.

**Fig. 8.36** Mouth of the Alapaha River (30.4366°N, 83.0966°W) as seen from the Suwannee River at moderate low flow

**Fig. 8.37** The Alapaha Rise (30.4388°N, 83.0895°W) on the north bank of the Suwannee River. This first-magnitude spring is a resurgence of Alapaha River water that entered the UFA at the swallets shown in Fig. 8.35

As such, the Alapaha River surface-water drainage basin identified by the U.S. Geological Survey and Florida agencies includes a true surface-water system developed by fluviokarst processes on the east and little direct drainage to the river on the west owing to the existence of a karst plain (Fig. 8.31).

When it enters the Branford Karst Plain, the lower Alapaha River Basin becomes an area of intense karst development, characterized by numerous sinkholes, lack of surface-water drainage, and undulating topography. In karst areas, dissolution has created enlarged cavities along fractures in the limestone, which eventually collapse or reach the surface and form sinkholes. Sinkholes capture surface-water runoff and funnel it underground, which promotes further dissolution of limestone. This dissolution slowly leads to progressive integration of voids beneath the surface over time and allows increasingly larger amounts of water to be transported through the ground-water system.

## 8.4 Karst Escarpments

### 8.4.1 Introduction

The previous discussions about fluviokarst emphasized the processes of interactions between streams flowing off of highlands underlain by relatively low permeability, siliciclastic sediments onto karstic, limestone plains. These processes have been a dominant force in sculpting the Cody Scarp. In this section, we step back and examine the Cody Scarp and other locations in Florida where similar transitions occur. The locations of these transitions from low permeability substrates to karstic, highly permeable limestones are karst escarpments.

Karst escarpments represent the transition areas from topographic *or* hydrologic highlands with well-developed stream networks to an internally drained karst plain (Fig. 8.11). There are several small, karst escarpments (scarps) in Florida (Copeland 2005a, b), but the dominant one is the Cody Scarp, which extends between Leon and Alachua counties along the transition from the Okefenokee Basin and Tifton Upland districts and the karst plains that lie at the toe of the scarp. This transition zone includes the Alachua Karst Hills and portions of the Madison and Tallahassee karst hills.

### 8.4.2 The Cody Scarp

The Cody Scarp was first identified by Puri and Vernon (1964), who considered that the scarp was formed by coastal erosion during a Pleistocene high sea stand. In the vicinity of the community of Cody, Jefferson County, the escarpment is a sharp topographic break (Fig. 8.38). This topographic break and apparent Pleistocene sand bodies that extend to the east from it bear all of the characteristics of a wave-cut scarp. However, elsewhere the scarp is a complex of largely dendritic valleys interspersed with sinkholes, uvalas, and poljes. We now know that the Cody Scarp is a polygenetic escarpment created by fluvial interaction with karst processes at the toe of the scarp and modified by marine-terrace development. The following sections describe the evolution and function of the fluviokarst segments of the Cody Scarp.

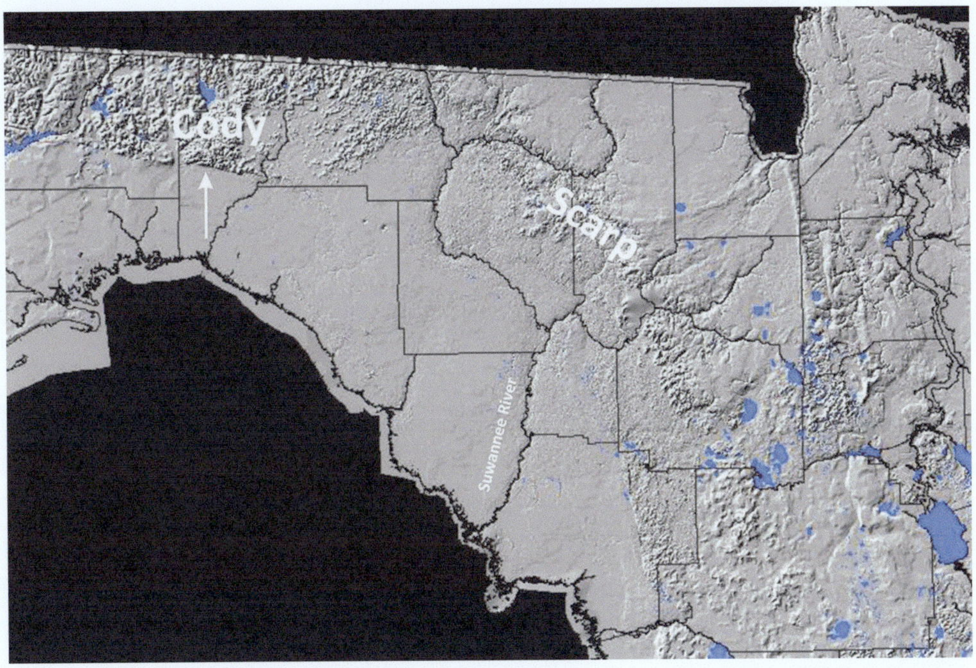


**Fig. 8.38** Digital elevation map of northern Florida showing the Cody Scarp. Note the hilly terrain within the escarpment. Arrow points to the Cody, Florida area, the type location of the Cody Scarp and the area where the scarp appears to be wave cut. Elsewhere, the polygenetic escarpment clearly reflects interaction of fluvial, karst, and marine processes

### 8.4.2.1 Importance

The Cody Scarp includes many excellent examples of a fluviokarst escarpment with numerous poljes, uvalas, sinkholes, sinking streams, siphons, springs, and other karst features along its length.

The Cody Scarp is important to understanding the karst hydrology of the Suwannee River Basin and other areas of north-central Florida because the processes that have formed it greatly affect rivers, groundwater, land forms, and water quality throughout the region (Copeland 2005a, b) (Table 8.2).

In many ways, the karst escarpments of Florida are different from karst escarpments formed in telogenetic karst systems. For example, Table 8.2 compares some of the properties of the Cody Scarp with the well-known Cumberland Escarpment of Tennessee (Crawford 1996).

The most important differences are related to the presence of both primary and secondary porosity in many Florida limestones and the high level of aquifer confinement inland from the crest of the escarpment.

### 8.4.2.2 Geomorphology

The Cody Scarp is a topographic break (Fig. 8.38) with as much as 30 m of relief. The scarp is characterized by sinking

**Table 8.1** Groundwater quality domains in and near the Ichetucknee springshed

| Facies (Groundwater domain) | Chemical characteristics (Major and minor constituents) | Domain location | Springs discharging Facies water (Refer to Fig. 13.43 for spring locations) |
|---|---|---|---|
| I – Recharge domain | $Ca\text{-}Mg\text{-}HCO_3\text{-}SO_4\text{-}NO_3$ | Unconfined springshed from Mission Spring north | Ichetucknee Head Spring Cedar Head Spring Blue Hole Spring |
| I – Recharge domain mixed with IIIa water | $Ca\text{-}Mg\text{-}Na\text{-}HCO_3\text{-}Cl\text{-}SO_4$ | | Mission Spring Devil's Eye Spring |
| II – Uplands domain | $Mg\text{-}Ca\text{-}Na\text{-}HCO_3\text{-}Cl$ | Northern part of springshed where aquifer is confined | No springs identified |
| IIIa – Eastern discharge domain near Santa Fe River | $Na\text{-}Ca\text{-}Mg\text{-}HCO_3\text{-}Cl\text{-}SO_4$ | East of the Ichetucknee River from Mission Spring south | Mill Pond Spring |
| IIIb – Western discharge domain near Santa Fe River | $Ca\text{-}Mg\text{-}HCO_3\text{-}SO_4$ | West of the Ichetucknee river from Mission Spring south | No springs identified |

**Table 8.2** Comparison of karst conditions in the Cumberland plateau scarpment (Crawford 1996) and the Cody Scarp

| Cumberland escarpment | Cody scarp |
|---|---|
| *Crawford's conceptual model* | *This text* |
| [Water] invasion occurs near the contact between impermeable caprock and underlying carbonates as caprock streams, highly aggressive to calcium carbonate, begin to flow through the carbonates to resurgences near the escarpment base. | True caprock is replaced by low-permeability, clay-rich strata. Streams are chemically aggressive to calcite, rarely dolomite. |
| Suspended and bed loads [of siliciclastic sediment] enlarge the conduits by abrasion. | 1. Unverified, but probable. The abundance of quartz sand in stream loads and cavern-fill material suggests a strong possibility of abrasion in high gradient conduits. 2. Streams also contain dissolved ortho-phosphate, which precipitates as water equilibrates with the limestone near swallets. |
| Caves are enlarged almost exclusively by the corrosion (dissolution) and corrasion (abrasion) of swallet streams, particularly during storms. | Spring chemistry suggests that this is true. Most springs show significant swallet (or siphon) flow contributions during high discharge. |
| Percolation input becomes supersaturated on losing carbon dioxide upon entering the cave atmosphere and consequently is not important in cavern development. | With the complex porosity of eogenetic aquifers, meteoric water quickly reaches equilibrium in matrix flow. In conduit flow, equilibration takes longer in terms of both time and transport distance. Cave decorations, where present, usually reflect degassing of $CO_2$ and evaporation of water. |
| Subsurface streams take a "stair-step" route down the Cumberland Plateau Escarpment due to intermittent impermeable strata. | Stratigraphy at the scarp is simple with few, if any, intermittent aquitards. However, there is evidence of stair-stepping of flow within the interstratal karst of the Hawthorn Group. |
| Not all conduit caves result from invasion of caprock streams; many result from domepit input, which is similarly aggressive since its source is the perched caprock aquifer. | Probably true. Deep, rock-collapse sinkholes, such as the Devils Millhopper, suggest that domepits may be developing upgradient from the Cody Scarp. Cenote and other unusually deep rock-collapse sinkholes may reflect hypogene origins. |
| A sinkhole plain, resulting from weathering of carbonates containing less soluble material, follows the retreating escarpment. | Sinkhole plains exist at the toe of the scarps, but they are not caused by reduced solubility. Rather, they appear to be caused by base levels related to modern and ancient sea-level positions. |
| A structural high may result in a stream cutting through the caprock and invading underlying carbonates behind the escarpment. Slope retreat by sapping proceeds away from the initial invasion to form a solution valley. | The absence of significant structure limits formation of solution valleys up-gradient of the escarpment. Isolated karst areas within highlands are possible examples. |

(continued)

**Table 8.2** (continued)

| Cumberland escarpment | Cody scarp |
|---|---|
| *Crawford's conclusions* | *This text* |
| Caprock streams are always [chemically] aggressive to limestone. | Historically true. Widespread use of irrigation with Floridan aquifer water has reduced aggressiveness of some streams. |
| Water entering the swallet and resurging at the rising is virtually always aggressive to calcium carbonate because of the high percentage of aggressive swallet input. (mixing corrosion is also partly responsible for the aggressivity at the rising). | True for many swallets and springs. |

streams, springs, and large sinkholes (uvalas and poljes are common). The sinkholes are large because of the thickness of residual and marine sand and clay cover that remains over the limestone between these large sinkhole features. Plio-Pleistocene marine terrace deposits commonly form elevated ridges at the top of the scarp. Remnants of the Lake City Ridge (a Plio-Pleistocene beach ridge), for example, drape over the scarp in Columbia County (Williams et al. in prep.).

The Cody Scarp is the topographic break between the highlands of the Okefenokee Basin and Tifton Upland districts to the north and east and the karst plains of the Ocala Karst District. The highlands of the Okefenokee Basin and Tifton Upland districts are underlain by a thick sequence of sand, clay, and carbonate sediments of the Miocene Hawthorn Group. Limestone and dolostone of the Eocene Ocala Limestone and Oligocene Suwannee and St. Marks limestones characterize the shallow sediments of the adjacent karst plain lowlands. The scarp is a result of marine, fluvial, and karst-related erosion of the Hawthorn Group sediments.

Figure 8.39 is a map showing the Harris Lake polje located within the Cody Scarp near Lake City, Columbia County. Streams drain to the south and southeast across the highlands where they have well-developed drainage systems. As the streams cross the scarp, they flow into instream swallets, poljes, and uvalas. Recharge of surface water to the UFA in the Harris Lake polje is a source of orthophosphate-rich water found within a karst conduit (cf., Figs. 6.23 and 8.39) that flows to the southeast. The Harris Lake polje and its drainage system provide an excellent example of the effect of karst on drainage within the scarp. The Harris Lake Basin is also an excellent example of one of Florida's poljes. The margins of the polje are steep because (1) the walls consist of relatively erosion-resistant Hawthorn Group sediments, (2) the sinkhole complex is geologically young and rounding of the polje margins has not blurred the sharp margin of the feature, and (3) there is little marine-terrace sand available to infill the margins of the basin.

#### 8.4.2.3 Origin of the Cody Scarp

The Cody karst escarpment developed because of a combination of: (1) headward erosion by streams; (2) dissolution of carbonate rocks by streams and groundwater; and (3) marine coastal processes. The difference between a karst escarpment and any other topographic scarp is that the toe of the scarp is characterized by limestone or dolostone that has been subject to thorough dissolution by the surface- and groundwater as the scarp retreats. Fig. 8.17 illustrates the process of headward erosion on the Cody Scarp.

In Step 1 of fluviokarst scarp retreat (Time-step 1 Fig. 8.17), a stream has developed on the highlands. The stream flows from right to left (east to west) and slowly erodes the surficial sands, clays, and carbonate-rich strata of the Hawthorn Group away. At some point, the stream cuts headward and downward, deep enough to intersect a limestone (or dolostone) stratum. The streams of the highlands and groundwater of the surficial aquifer are typically acidic, so the limestone is attacked by the water as soon as the water and rock are in contact. The water finds a fracture, joint, or other pathway into the limestone and cave development begins as dissolution of the rock occurs (Time-step 2 Fig. 8.17). As the scarp continues to retreat (Time-step 3 Fig. 8.17), the limestone is eroded to form karst landforms with swallets that can capture the surface- or Surficial Aquifer System (SAS) groundwater as it flows off the highlands. This process continues for many thousands of years, resulting in a string of active and inactive swallets (Time-step 4 Fig. 8.17) along the retreating stream valley. Often, a trace, the former path of the stream, can still be seen on the land surface as linear depressions, vegetation and soil tonal patterns, and, of course, the string of sinkhole swallets.

#### 8.4.2.4 Features of the Cody Scarp

Table 8.3 summarizes some of the properties of the Cody Scarp and its adjacent highlands and lowlands. When the sizes and shapes of closed depressions are compared, the Cody Scarp is immediately obvious (Fig. 8.40).

The highlands are relatively flat. There may be small depressions that are either developed during deposition of the Plio-Pleistocene marine terraces or minor sinkholes developed on shallow carbonate strata of the Hawthorn Group. Near the upper margin of the escarpment, the highlands may be penetrated by rock-collapse sinkholes, such as the Devil's Millhopper in Alachua County.

The lowlands (Table 8.3, Fig. 8.40) are characterized by numerous small sinkholes superimposed on a relatively flat platform. The size of the sinkholes is controlled by cover thickness, which is usually thin (less than 3 m). Streams other than the large, allochthonous streams, such as the Suwannee and Santa Fe rivers that originate in highland areas, are rare and often limited to the local sinkhole drainages.

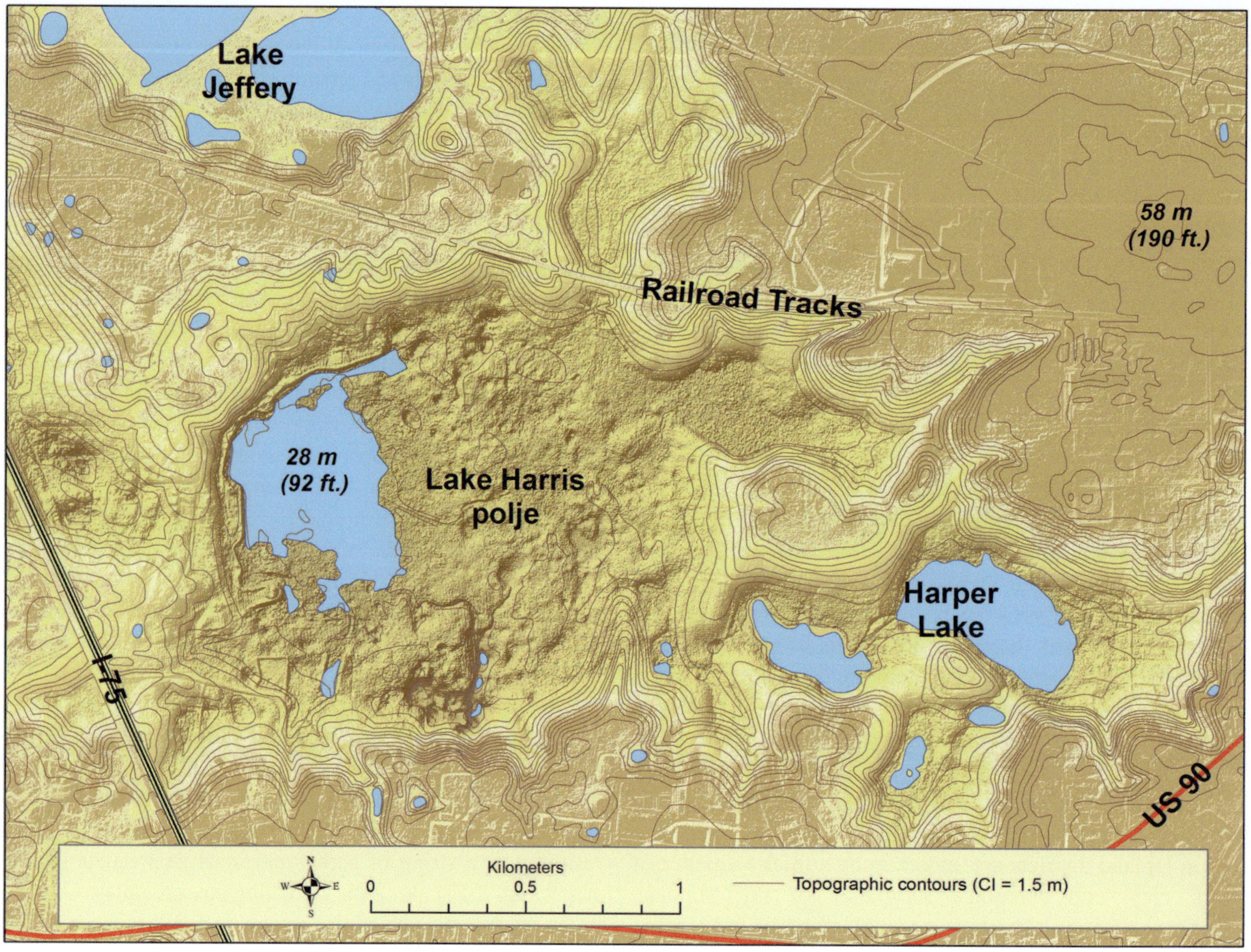


**Fig. 8.39** LIDAR map of the Harris Lake polje near Lake City, Columbia County, Florida. Note that surface water drains from Lake City into the basin via several ephemeral streams and intermediate sinkhole lakes, and that the lake basin drains internally to the UFA. This basin is the origin of the Harris Lake groundwater plume (Fig. 6.23). Lake Harris is located at 30.192°N, 82.684°W

The Cody Scarp is characterized by large sinkholes (including uvalas and poljes) that produce a distinctly hilly appearance to the terrain (Fig. 8.38). Streams that drain the upland areas normally drain into swallets or wet prairies (poljes). The large sizes of sinkholes in the Cody Scarp (Fig. 8.40) are a result of thick cover over the limestone. This cover consists of erosional remnants of Hawthorn Group sediments plus residua and sediment derived from erosion and transport of Hawthorn sediments created by scarp retreat. The thicker the cover, the larger will be the sinkholes.

Karst development is limited in the highlands above the scarp. Limestone and dolostone beds within the Miocene Hawthorn Group may develop localized caves and sinkholes that are limited to the thin Hawthorn carbonate strata. This form of karst (interstratal karst; Wadsworth et al. 1983) results in minor aquifer development, and it has little effect on the hydrology of the Cody Scarp area.

Because of the localized recharge by streams entering swallets along the scarp, karst is very well developed on the Cody Scarp. Recharge tends to result in vertical karst conduits, both under the large sinkholes and in association with swallets and siphons that capture runoff from the highlands. As the groundwater flow turns to move to the south and west, away from the scarp, horizontal passages develop. These processes result in the stream-to-resurgence systems that characterize many of the streams and rivers that originate in or cross the scarp.

#### 8.4.2.5 Effects of the Scarp on Groundwater Quality

Lawrence and Upchurch (1976, 1982) subdivided the Cody Scarp region into three groundwater "domains" (Fig. 8.40) based on groundwater chemistry and development of karst landforms and conduiting. These domains were refined by

**Table 8.3** Some properties of the Cody Scarp and associated physiographic provinces

| Feature | Characteristic |
|---|---|
| **Topography** | |
| Highlands above the scarp | Relatively flat, minor depressions |
| Cody scarp | Hilly, large, closed depressions |
| Lowlands at toe of scarp | Relatively flat but characterized by many, small, closed depressions |
| **Surface drainage** | |
| Highlands above scarp | Well-developed streams, swamps, lakes |
| Cody scarp | Sinking streams, large lakes (poljes) |
| Lowlands at toe of scarp | Internal drainage, few lakes other than a few poljes or streams |
| **Surficial aquifer flow systems** | |
| Highlands above scarp | Local flow systems, aquifer well developed |
| Cody scarp | Generally absent |
| Lowlands at toe of scarp | Mostly absent; the water table aquifer, is part of the Upper Floridan Aquifer |
| **Intermediate aquifer system** | |
| Highlands above scarp | Localized aquifers (carbonate and siliciclastic) poorly to moderately developed; clay-rich strata form aquitards and confine the underlying Floridan aquifer |
| Cody scarp | Aquifers absent or very local, local clay-rich strata are present and may temporarily perch water |
| Lowlands at toe of scarp | Absent |
| **Upper Floridan aquifer** | |
| Highlands above scarp | Porosity development varies; permeability relatively low, flow often sluggish |
| Cody scarp | Secondary porosity well developed by karst processes, flow very dynamic |
| Lowlands at toe of scarp | Secondary porosity well developed by karst processes, flow very dynamic |
| **Recharge to upper Floridan aquifer** | |
| Highlands above scarp | Limited and slow |
| Cody scarp | Rapid, includes runoff from highland streams |
| Lowlands at toe of scarp | Rapid, localized by small catchments of sinkholes |
| **Karst development** | |
| Highlands above scarp | Limited to interstratal karst within Hawthorn Group carbonate strata near the scarp |
| Cody Scarp | Well developed; vertical conduits dominate; large uvalas and poljes present |
| Lowlands at toe of scarp | Well developed; horizontal conduits highly developed; with reduced cover thickness, sinkholes are usually relatively small |

Based on Upchurch (2002)

Upchurch and Lawrence (1984) and Upchurch (2002). The Highland, or Upland, Domain constitutes the highlands above the scarp where streams originate because of low infiltration rates through Hawthorn Group sediment. The Cody Scarp was termed the Scarp Domain and the lowlands at the toe of the scarp constitute the Lowlands Domain (Fig. 8.40). Each domain has different groundwater quality states and forms of karst (Fig. 8.41). The effects of the scarp on groundwater chemistry are discussed in this section.

Water flowing off of the Upland Domain (highlands above the scarp) is typically mildly acidic and contains dissolved phosphate (Odum 1953; Kaufman 1969) derived from weathering of phosphatic sediments. Because of the mineral content of the surficial aquifer and upper strata of the Hawthorn Group (the ICU), there is little chemical buffering to reduce the water acidity, and the mildly acid water moves toward the scarp as either surface water or within the SAS or IAS. Therefore, surface and SAS waters are typically acidic and Na-Cl facies with elevated dissolved orthophosphate.

There is little recharge to the underlying UFA under the Upland Domain, and UFA groundwater is predominantly chemically saturated with respect to calcite. As a result, karst development in the UFA is limited (Fig. 8.41) resulting in low hydraulic conductivities in the aquifer. Flow in the UFA under the Upland Domain is relatively sluggish, and the groundwater facies range from Ca-$HCO_3$ to Ca-Mg-$HCO_3$-$SO_4$.

Upon reaching the vicinity of the Cody Scarp, the acidic, surface and SAS water begin to contact carbonate rocks of the lower Hawthorn Group because of down-cutting in stream valleys (Fig. 8.17). Karst development is initiated within Hawthorn carbonate beds at this stage in scarp retreat. Interaction of the acidic water with the carbonate rocks begins to buffer the water, resulting in an increase in pH. As the lower Hawthorn carbonates are dissolved and physically eroded away, residual sand and clay accumulate and developing sinkholes allow access of the still acidic water to the underlying UFA (Fig. 8.41). Sediments in transport in the streams are swept into the aquifer, and landscape development changes from fluvial to karst near the middle and toe of the scarp.

The SAS is disrupted as scarp development begins, and only remains as sand deposits on the crests of hills and as sandy alluvium draped over the landscape. If there is an intact clay stratum under the larger hills, a perched, ephemeral water table aquifer may remain.

Water quality in the UFA remains as Ca-$HCO_3$ facies, but it becomes saturated with respect to calcite. The water samples analyzed by Lawrence and Upchurch were unfiltered, so colloidal calcite was present in trace quantities. These colloidal particles resulted in water that appeared over-saturated with respect to calcite and hydroxylapatite (Fig. 8.41). Accumulations of clay sized calcite are common on fracture walls and cave sediment in the scarp. Fracture flow of water from swallets and leaking wetlands and lakes is pronounced in the Scarp Domain (Fig. 6.23).

Groundwater of the Scarp Domain in the UFA is highly vulnerable to contamination because of the sinking streams and large drainage basins associated with the large scarp

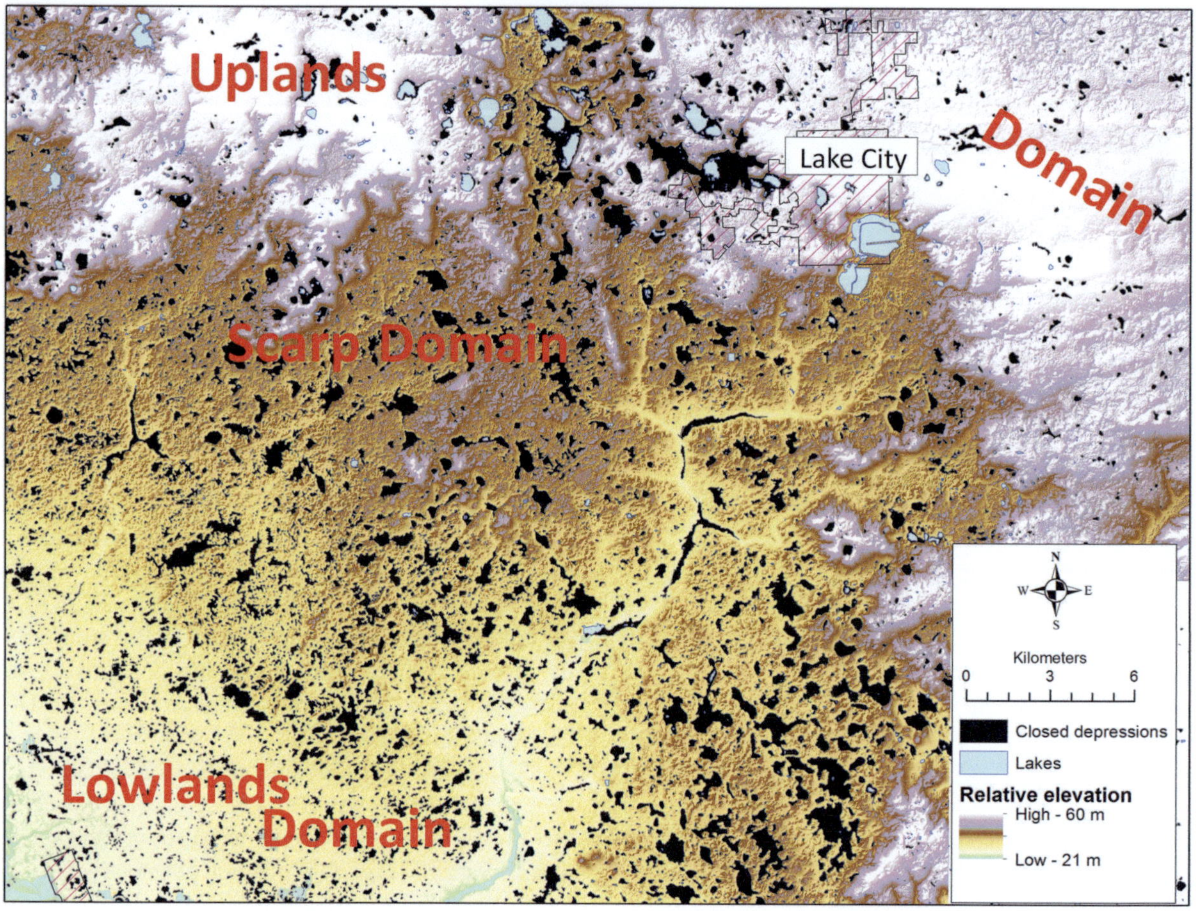


**Fig. 8.40** LIDAR map of the Cody Scarp with its closed depressions near Lake City, Columbia County. Black shapes represent closed depressions (sinkholes, uvalas, poljes), and blue areas are sinkhole lakes. The yellow shaded area that encloses the largest of the closed depressions is the approximate topographic extent of the Cody Scarp in the study area

| GROUND-WATER DOMAINS | LOWLAND DOMAIN | SCARP DOMAIN | UPLAND DOMAIN |
|---|---|---|---|
| **SATURATION STATES** | | **Zone of Colloidal Transport** | |
| Calcite | undersaturated | OVER SATURATED | saturated to sl. undersat. |
| Dolomite | undersaturated | undersaturated | undersat. to saturated |
| Quartz | undersaturated to saturated | OVER SATURATED | SUPERSATURATED |
| Hydroxylapatite | undersaturated | SUPERSATURATED | undersaturated |
| **KARST DEVELOPMENT** | | | |
| Vertical Conduits | | | |
| Horizontal Conduits | | | |
| Intrastratal Karst | | | |
| Solution Doline Development | | | |
| Collapse Sinks | | | |
| Uvala/Polje Development | | | |
| Paleokarst Reactivation | | | |
| **RECHARGE PATTERNS** | DISPERSED RECHARGE | LOCAL, MAJOR RECH. | MINOR RECHARGE |

**Fig. 8.41** Conceptual model of the Cody Scarp with summaries of geologic and geochemical processes in each topographic domain. See text for explanation of the karst development model. (Modified from Upchurch 2002)

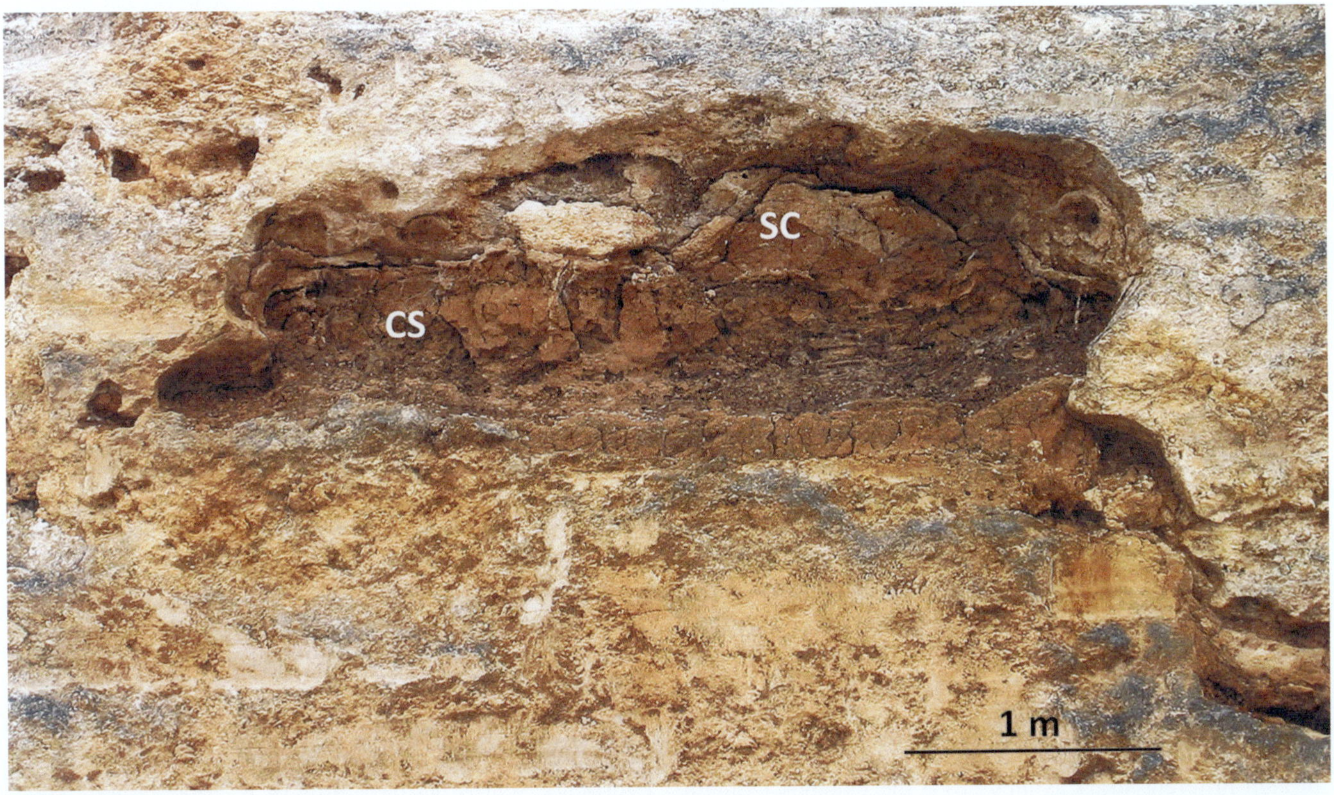


**Fig. 8.42** Sediment filling a cave in the Ocala Limestone at the Haile Quarry near Newberry, Alachua County. CS-sandy clay and SC-clayey sand

sinkholes. Nitrate and other contaminants are likely to first occur in UFA groundwater within the Scarp Domain.

Contact with alkaline water and carbonate rocks in the lower Hawthorn Group and UFA may result in precipitation of phosphate as carbonate hydroxylapatite (Florida's "hard rock phosphate deposits"; Sect. 8.4.3; Upchurch 1992), so orthophosphate concentrations in the UFA (Fig. 8.41) are limited by aquifer chemistry.

The Lowlands Domain remains after the scarp has retreated from an area (Fig. 8.41). The Lowlands domain is pockmarked by relatively small sinkholes as a result of the thin cover. Uvalas are only evident where the cover is thicker. Water quality under the Lowlands Domain is Ca-$HCO_3$ facies. The water varies with respect to calcite saturation state; however, due to local recharge much of the shallow UFA water is undersaturated with respect to calcite.

Soils, which are a combination of residual and marine sands and clayey sands, are highly conducive to agriculture, and contamination by nitrate and related constituents is widespread. Because of high alkalinity, phosphate concentrations are limited in the Lowlands Domain.

#### 8.4.2.6 What Happened to the Siliciclastics?

One of the great conundrums concerning karst development in the Scarp and Lowlands domains is the fate of the sand and clay that once covered the area before scarp retreat. The proposed sequence of development of the scarp (Fig. 8.17) explains the fate of the cover sediments.

1. The initial stages of scarp retreat (Time Step 1, Fig. 8.17) would result in transport of the siliciclastic sediments of the cover to the regional, master streams and, ultimately, to the Gulf of Mexico.
2. However, once a swallet develops (Time Steps 2 through 4; Fig. 8.17), siliciclastic sediments can only be transported to the master streams and sea during extreme floods when the infiltration capacity of the swallets has been exceeded. Most of the insoluble, siliciclastic sediment, therefore, must enter the fractures and karst conduits of the limestone aquifer under normal conditions.
3. The presence of siliciclastic-sediment-filled conduits is not normally evident when drilling test holes because the sediment is lost to the drilling fluid. However, in quarries near the Cody Scarp, sand- and clay-filled caves and other cavities are commonly exposed (Fig. 8.42). Therefore, the UFA may be riddled with karst dissolution features and fractures, but only a small proportion of these openings are likely to function hydraulically at any moment in time. Most of the secondary porosity is apparently plugged by siliciclastic sediments.

Given the closed drainage systems that developed in the Scarp Domain, much of the siliciclastic sediment of the cover must have entered the UFA. The Scarp Domain and virtually all of the Lowlands Domain have only been affected by local fluvial and karst erosion. Trunk streams that might have carried the siliciclastic sediments from the marine terraces and Hawthorn Group to the sea apparently never developed. The Alapaha, Santa Fe, and Suwannee Rivers have only affected narrow corridors through the vast karst plains of the Ocala Karst District. The remainder of the drainage capable of causing erosion and transport of siliciclastic sediments in the Ocala Karst District is from small, stream-to-sink drainages in the Scarp area and by surface runoff and soil creep in the Lowlands. Simple examination of topographic maps of the Scarp Domain where stream-to-sink systems remain suggests that millions of tons of siliciclastic sediments must have entered the UFA.

In the quarries, stratification of the siliciclastic sediment filling caves is often evident with thinly bedded, clay rich sediments interbedded with poorly bedded sandy sediments (Fig. 8.42). Presumably, these different facies of cave fill represent different (1) properties (gradient, drainage basin size, etc.) of the streams and (2) sequences of down cutting through different strata as they erode the Uplands Domain sediments and transport them to swallets.

#### 8.4.2.7 Subsurface Karst near the Scarp

The water-quality data obtained by Lawrence and Upchurch (1976, 1982) and observations in quarries in the Cody Scarp and Lowlands resulted in development of a model (Fig. 8.41) for karst development as the scarp retreats. The horizontal lines at the bottom of Fig. 8.41 represent the approximate locations in the vicinity of the Cody Scarp where the different forms of karst features develop.

In northern Florida, the Hawthorn Group can be roughly subdivided into two strata (Miller 1978; Scott 1988). The upper package of sediments is dominated by siliciclastic sediments with carbonate beds. The lower package is largely carbonate (largely dolostone) with beds of siliciclastics. Under the Upland (Highland) Domain, the carbonate-rich lower stratum develops minor interstratal karst, especially near the scarp.

The UFA experiences minimal epikarst karst development under the Upland Domain (Fig. 8.41). Paleokarst associated with unconformities in the Neogene carbonates exists and produces water to wells. Where fractures exist, connections to the surface may exist and solution enlargement may occur. Fractures enlarged by limestone dissolution occur near the upper margin of the escarpment.

Chemically aggressive water enters the UFA through swallets and sinkholes in the Scarp Domain where dissolution of limestone is evident. Based on exposures in quarries and the general pathways of recharging water, the dominant form of karst conduiting is vertical shafts or solution pits (Figs. 7.27 and 7.29).

Until it equilibrates with the host limestone, chemically aggressive groundwater develops more-or-less horizontal conduits in the Lowlands Domain. Here, Upchurch (2002) reasoned that the sense of conduit development was more horizontal than vertical (Fig. 8.41) in order to comport with the regional flow-net analyses.

Finally, the Scarp and Lowland domains were locations where solution and collapse sinkholes were most likely to develop and to reactivate. Development of uvalas and poljes was most likely in the Scarp Domain because of the thickness of the cover sediments.

### 8.4.3 Other Hydrologic or Topographic Karst Escarpments

There are several other areas in Florida where surface-water streams flow from low recharge regions onto a karst plain. In each case, the stream enters one or more swallets. Some of these areas have topographic expression and are true karst escarpments in the geomorphic sense. Others function as hydrologic scarps but have little or no topographic expression. Other areas have topographic features that suggest that they once were affected by scarp retreat in the manner discussed above, but no longer function as scarps because the scarps have retreated to the extent that there is no longer an upland upon which streams can form. This section discusses three of these areas.

Copeland (2005a, b) argued that surface water draining the San Pedro Bay physiographic province (Fig. 3.13) went underground after exiting the area underlain by clay (the "San Pedro clay"). He posited that the transition from the area underlain by the San Pedro clay to limestone constituted a hydrologic escarpment. The San Pedro Bay physiographic province is a large, swampy area in Taylor County that is underlain by the relatively impermeable San Pedro clay. This clay perches surface water, forming the swamps of the physiographic province. Several prominent streams originate within the San Pedro Bay and drain to the west where they go underground (Fig. 8.43). On the east, small streams can be shown to disappear as they approach the Branford Karst Plain province. On the west, several streams are stream-to-sink systems. The Fenholloway River and Rocky Creek have especially well developed swallets. The swallets where Rocky Creek goes underground can be easily viewed where US 19 crosses the creek at 30.1666°N, 83.6294°W, north of Perry, Taylor County.

One of the criteria used to characterize the Cody Scarp as a karst escarpment was the distribution and size of closed depressions. Figure 8.44 illustrates the distribution of closed depressions within and adjacent to the San Pedro Bay. Note

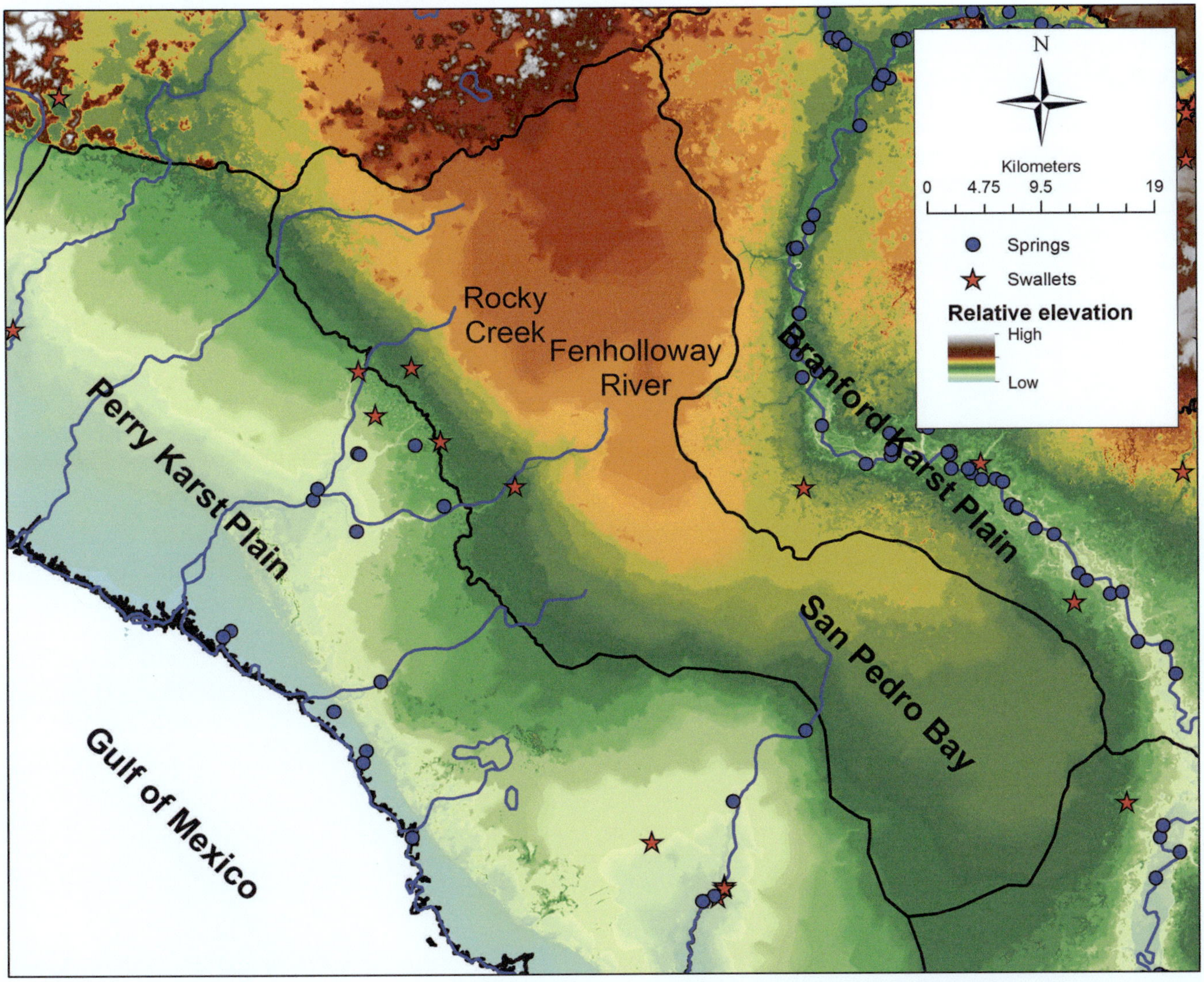


**Fig. 8.43** Digital elevation map of the San Pedro Bay province and surroundings showing stream-to-sink (swallet) systems associated with the Fenholloway River drainage basin

that closed depressions are relatively rare in the center of the province, and that they abound in the surrounding karst plains. Because of the San Pedro clay, karst (sinkhole) development and recharge to the underlying UFA are hindered within the province and enabled within the surrounding karst plains. There is little or no topographic expression for this transition, but the streams and UFA behave in a way that is similar to the Cody Scarp.

Similarly, the transition from the relatively low recharge area of the Green Swamp Province to the surrounding karst areas results in streams forming within the province and going underground through swallets in the adjacent limestone plains. Like the San Pedro Bay, the margin of the Green Swamp appears to be a hydrologic scarp with streams and swamps in the interior of the province and with minimal sinkhole development. Upon leaving the low recharge area, several of the streams enter the UFA by way of swallets (Fig. 4.27). Streams known to enter swallets include the Withlacoochee River (South) near Dade City, Pasco County, and Peace River south of Bartow, Polk County.

### 8.4.4 Hard-Rock Phosphate Deposits: Evidence of Karst Escarpment Evolution

One of the unique artifacts of the retreat of karst escarpments in northern Florida is the presence of "hard-rock" phosphate deposits. The crest of the Ocala Platform was stripped of Hawthorn sediments in Late Miocene, Pliocene, and Quaternary times (Scott 1981). While most of the weathering products of this erosion were swept into adjacent estuaries, rivers, and conduits in the UFA, some of the dissolved phosphate migrated downward into the UFA through sinkholes where it precipitated to form the hard-

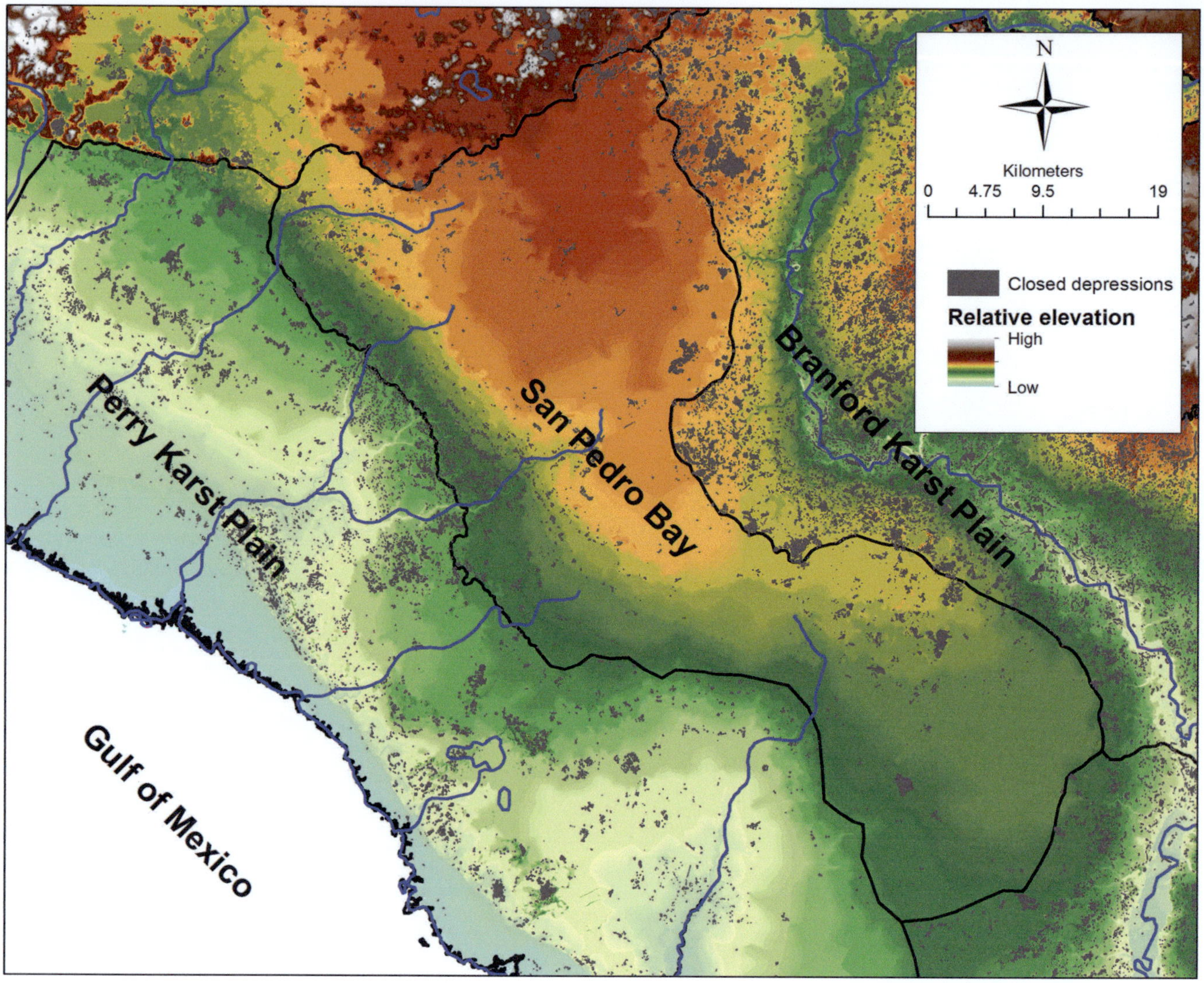


**Fig. 8.44** Digital elevation map of the San Pedro Bay province showing the distribution of closed depressions in the province and adjacent areas

rock phosphate. While the stratigraphy and fossil content of the sinkhole fill indicates that most of the formation of the sinkholes and precipitation of the phosphate occurred in Pliocene time, Upchurch and Lawrence (1984) documented an area in Columbia County where this process is taking place today.

Streams that drain upland areas where the phosphatic Miocene Hawthorn Group is exposed are mildly acidic and likely to contain dissolved phosphate as orthophosphate (o-$PO_4^{3-}$). When these streams crossed an escarpment and entered swallets, the water came in contact with limestone and groundwater rich in bicarbonate ($HCO_3^-$). When the water came in contact with carbonate-rich groundwater or limestone, carbonate hydroxylapatite ($Ca_5(PO_4,CO_3)_3(OH)$) precipitated according to Eq. 8.1.

$$5Ca^{2+} + 1.5H_2CO_3 + 1.5H_3PO_4 + H_2O \rightarrow Ca_5\left(PO_4,CO_3\right)\left(OH\right) + 8.5H^+. \quad (8.1)$$

There were three modes for phosphate preservation in the sinkholes. These were: (1) as colloidal material that settled out in void space; (2) replacement of limestone near the walls of the sinkholes; and (3) crusts that were broken and redeposited as platelets that bridged voids as the sinkhole fill settled (Fig. 8.45). When the deposits of apatite became large enough, the phosphate mineral was mineable as hard-rock phosphate. The mines were concentrated on the east side of the Brooksville Ridge (Sellards 1913a), although a few were located near the toe of the Cody Scarp. This suggests that the Brooksville Ridge and Cody Scarp resulted from scarp retreat associated with phosphate-rich stream-to-sink systems.

These hard-rock phosphate deposits in relict sinkholes provide evidence of scarp retreat and fossils contained in the deposits provide evidence of the timing of sinkhole filling and scarp retreat. Today, most of the known deposits have been mined out and are simply steep-walled depressions that range from a few hectares to half kilometer in maximum dimension.

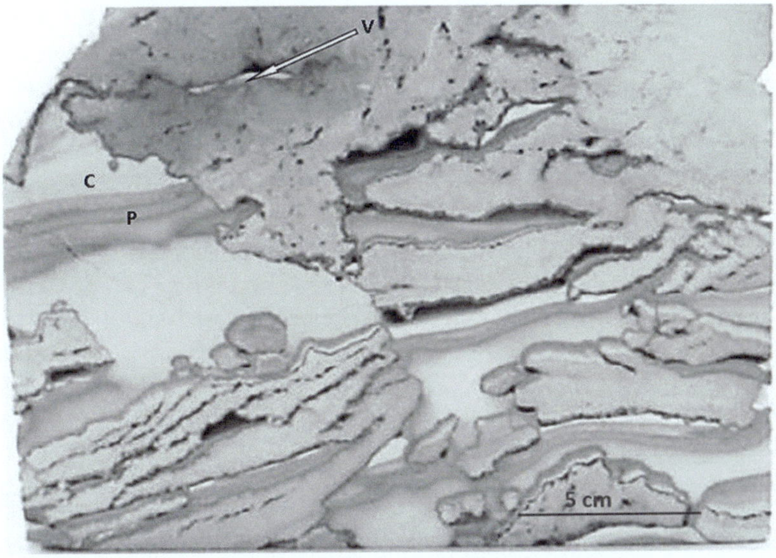


**Fig. 8.45** Slab of hard-rock phosphate (carbonate hydroxylapatite) from an abandoned mine in Hernando, Citrus County. C-colloidal phosphate accumulated as geopetal void filling, P-platelets of replaced limestone and phosphate crusts, V-void space

**Fig. 8.46** Ancient sinkhole filled with sandy clay to clayey sand (S) and massive carbonate-hydroxylapatite deposits (P). Exposure is at the Haile Quarry, near Newberry, Alachua County

The deposits occur in scattered and isolated, ancient swallets (Sellards 1913b; Vernon 1951; Puri and Vernon 1964; Upchurch 1992a, 2002) that were filled by siliciclastic sediments transported by the streams and by deposits of apatite. Fossils found in the deposits included terrestrial vertebrates that lived from the middle Miocene to early Pliocene (Vernon 1951; Puri 1957; Puri and Vernon 1964; MacFadden 1997). The ages of the fossils suggest that the Hawthorn Group sediments were eroded and the phosphate deposits formed during the episode of intense weathering near the end of the Miocene and in early Pliocene, the period of intense weathering and pedogenesis.

The apatite deposits formed massive replacements of limestone (P, Fig. 8.46), clay (colloidal) sized void-filling materials (C, Fig. 8.45), and botryoidal crusts. Sand, clay, and terrestrial fossils were also deposited in these ancient sinkholes (Fig. 8.46). The hard-rock phosphate deposits, therefore, contain important information relative to the evolution of karst escarpments in northern Florida.

## 8.5 Springs

Springs are without doubt the best loved karst features in Florida. With over 700 springs cataloged in 2004 (Scott et al. 2004), and 33 first magnitude springs (Fig. 8.47), Florida has more known springs and more first magnitude springs than any other state in the U.S., and new springs are still being found and documented. A source of concern is the fact that Florida is losing springs as the result of over-permitting of groundwater withdrawals, declines in precipitation, and sea-level rise.

### 8.5.1 Spring Locations

#### 8.5.1.1 Onshore Springs

As shown in Fig. 8.47, most of Florida's cataloged springs are in the central and northern part of the state. Most are located near major streams to which they discharge.

#### 8.5.1.2 Offshore Springs

There are several notable springs offshore in the Atlantic and Gulf of Mexico, a few of which are shown in Fig. 8.47. Notable offshore springs include:

- Springs offshore of Pinellas and Pasco Counties (Hunn and Cherry 1969; Garman 1999, 2010; Reich et al. 2001; Garman et al. 2011),
- Crescent Beach Spring offshore of St. Johns County (Brooks 1961; Stringfield and Cooper 1951; Swarzenski and Reich 2000),
- Warm-water springs offshore of Lee County (Kohout 1974a, b; Kohout et al. 1972, 1979; Breland 1980; Fanning et al. 1979, 1981)
- The Spring Creek Springs offshore of Wakulla County (Lane 2001; Davis and Verdi 2014)
- Springs in Biscayne Bay in Dade County (Kohout 1964, 1966; Langevin 2003), and
- Springs elsewhere in on the continental shelf of the Gulf Coast.

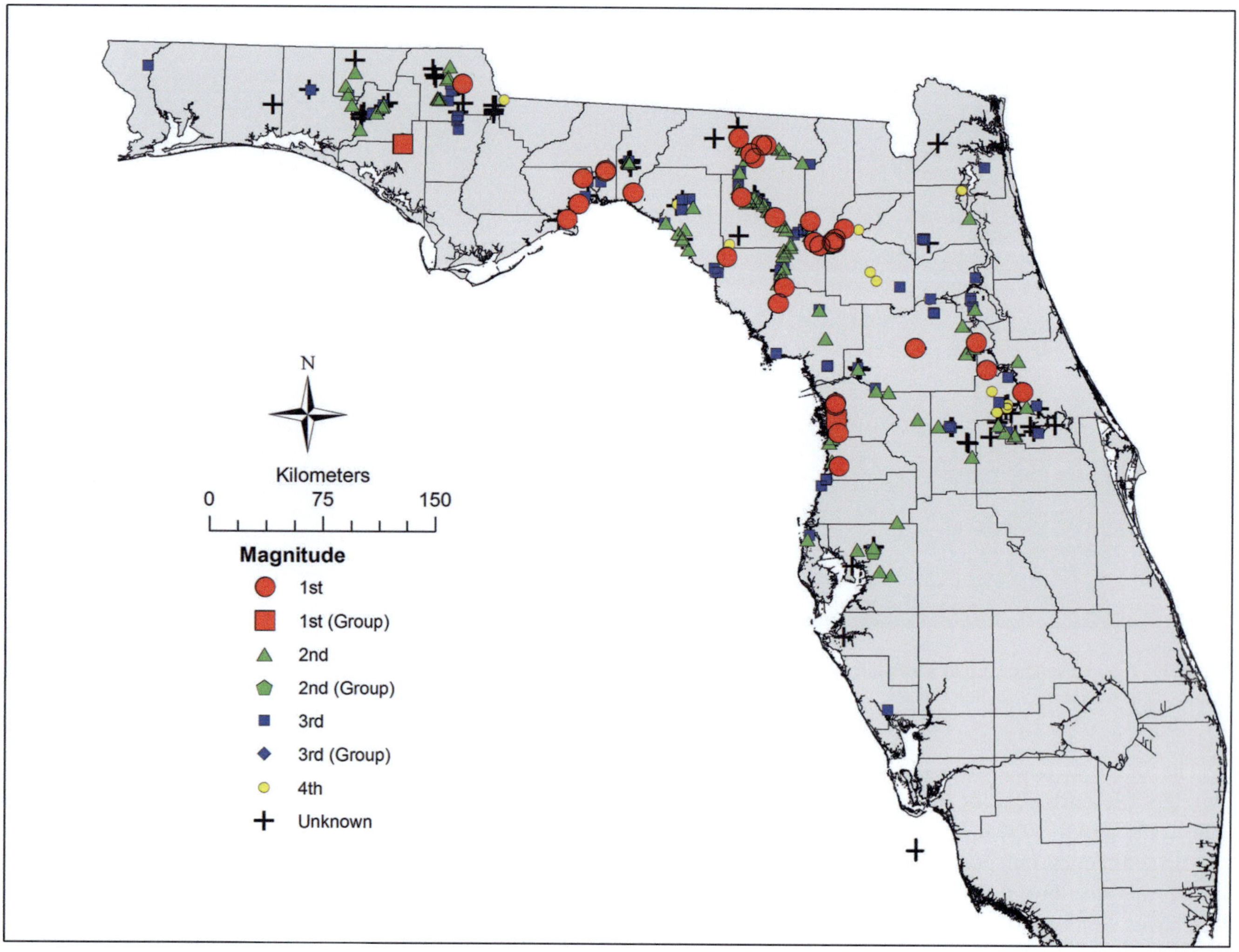


**Fig. 8.47** Springs of Florida, by magnitude. Data from the Florida Geological Survey

### 8.5.2 Importance of Florida Springs

Springs constitute an important economic resource to the local economies of Florida (Bonn and Bell 2003). They are located in popular public and private parks used for swimming, cave diving, camping, and boating. Most of Florida's first magnitude springs (springs with an average discharge greater than 2.8 $m^3/s$) are in state parks. County and private parks are widespread in central and northern Florida, where they serve as swimming and diving centers.

Springs also serve as sources of bottled water (Samek 2004) and public water supply, and they form important wildlife refuge and rehabilitation centers. Sulphur Springs in Tampa, Hillsborough County, a second magnitude spring discharging 1.2 $m^3/s$ (Scott et al. 2004), was once the principle public water source for the City of Tampa and a local tourist attraction with a zoo, water slide, and other amenities. Given its location in the urban area of Tampa (Fig. 8.48; Stewart and Mills 1984, Southwest Florida Water Management District 2004) spring water is contaminated and no longer used for public swimming. The Hillsborough River (Fig. 8.48) serves as the primary water supply for the city, and spring water is only utilized in emergencies by blending with river water and treatment. Millions have been spent to manage storm water and improve water quality in the spring, with minimal success. Sulphur Springs is an excellent example of the need for spring protection.

Many of the offshore springs are of academic interest because they discharge warm water or are associated with cave systems onshore. At least two areas of offshore springs (Crescent Beach Spring offshore of St. Johns County and the springs in Biscayne Bay, Dade County) were historically important because discharge was so fresh that ships could replenish drinking water stores without going ashore.

Groundwater withdrawals have resulted in several examples of springs drying up. The most notorious loss of a spring was Kissingen Spring (Fig. 8.49a, b; 27.8425°N, 81.8111°W) which once drained to the Peace River in Polk County. The third-magnitude spring ceased to flow regularly in the 1950s (Peek 1951) as a result of groundwater withdrawals from the UFA by the phosphate and agricultural industries. Other springs that have ceased to function include White (White Sulphur) Spring in Columbia County, and several small springs in the vicinity of Perry, Taylor County.

Given the economic, ecological, and recreational importance of Florida springs, it is no wonder that much effort has been expended to delineate the sources of water discharging from springs and those areas most likely to cause harm as a result of human activities. We do this by delineating springsheds.

### 8.5.3 Spring and Springshed Definitions

A spring is a location where groundwater emerges from underground. Springs range in size from tiny seeps to large caves discharging millions of liters of water a day (Copeland 2003; Poucher and Copeland 2006).

A springshed is the area that contributes the water that discharges from a spring (Copeland 2003; Scott et al. 2004; Poucher and Copeland 2006). Springsheds include both groundwater recharge areas (the groundwater basin) and any streams that form outside of the recharge area and then flow and sink into the subsurface through siphons or swallets (Fig. 8.50). Note that the springshed may extend under areas where the aquifer is confined as a result of drainage of the confined portions of the aquifer (Fig. 8.20).

### 8.5.4 Spring Magnitude

The magnitude of a spring is a measure of its average (Meinzer 1927) or median (Copeland 2003) discharge. Meinzer (1927) developed the spring-magnitude classification scheme using English units of measurement based on orders of magnitude of spring discharge. A first-magnitude spring is defined as one that averages 2.8 $m^3/s$. Higher spring magnitudes are based on decreasing orders of discharge magnitude (Table 8.4).

As noted by Copeland (2003), the magnitude of a spring is a function of its average or median flow. As a result, multiple measurements of discharge over periods of wet and dry seasons and years should be used to determine magnitude. Unfortunately, many Florida springs, especially the lower magnitude springs, only have one or two historic discharge measurements. As a result, there is no way to evaluate short-term changes in magnitude in order to determine real and significant changes in discharge.

The State of Florida has taken an active role in purchasing and protecting some of Florida's springs, especially the first-magnitude springs. Section 373 of Florida Statutes requires that springs that have been purchased for use as state parks and for conservation purposes must have minimum flows and levels[6] (MFLs) established to protect them from excess permitted withdrawals for water-supply purposes.

This emphasis on the magnitude of the springs has led to much public concern. During deep drought conditions in the

[6]Minimum flows and levels are regulatory levels of discharge or water stage used to manage surface water and groundwater withdrawals in order to protect the resource for water supply and the environmental, recreational, and commercial qualities of the water body (Chapter 373.042 Florida Statutes). Water-body specific MFLs are adopted and are frequently re-evaluated by rule by Florida's water management districts.

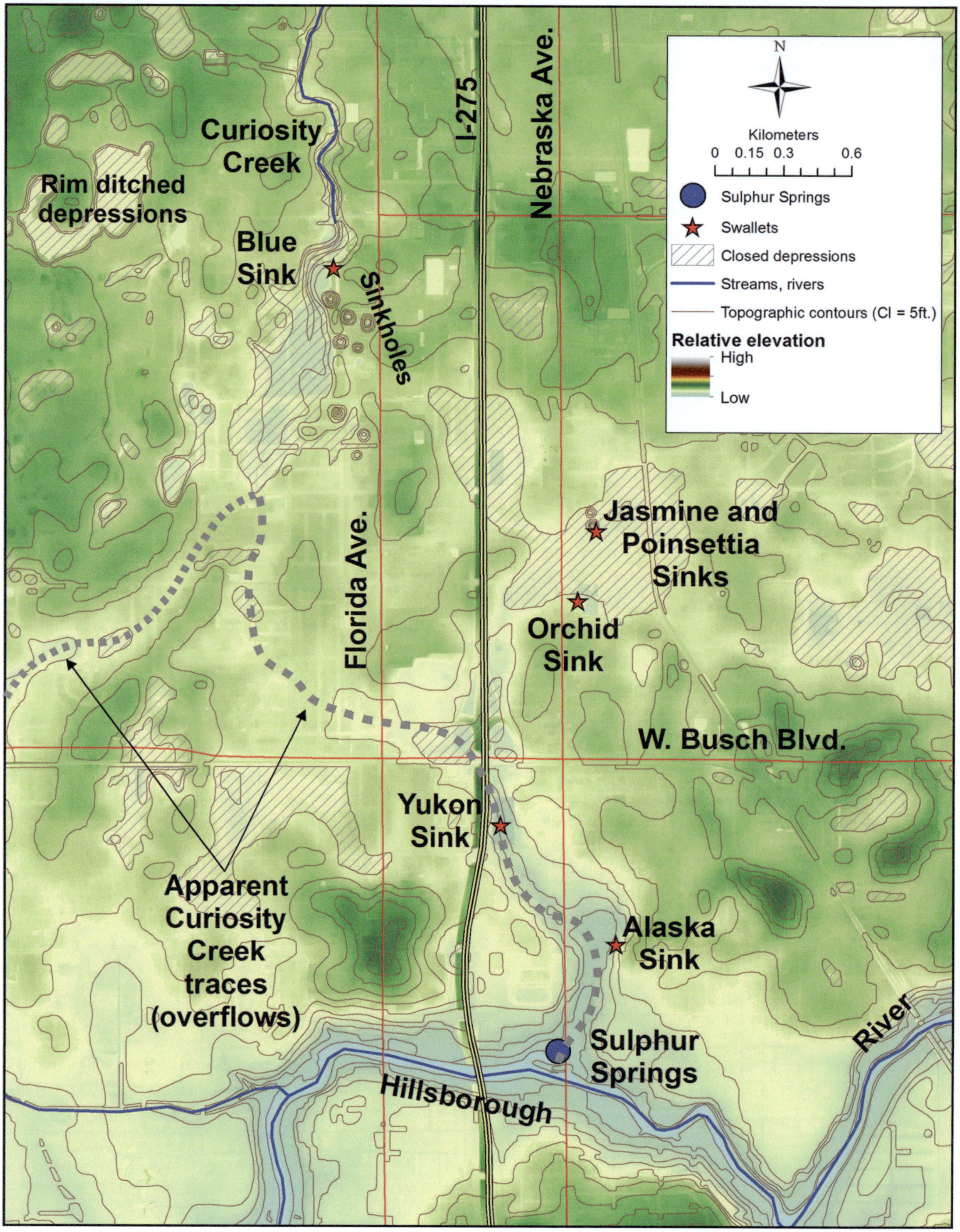


**Fig. 8.48** Digital elevation map of the Sulphur Springs springshed. Sulphur Springs is located at 28.0211°N, 82.4518°W

**Fig. 8.49** (**a**) In 1923 Kissingen Spring was a local spa. Photograph courtesy of Thomas Jackson. (**b**) Today Kissingen Spring is a dry basin that serves as a swallet when the nearby Peace River floods

early 2000s and with increased groundwater consumption, discharge from some of Florida's "first-magnitude" springs has dropped well below their perceived first-magnitude status. There has been a public outcry because of these reductions in spring flow, and efforts have been increased to develop the MFLs needed to protect the discharge regimes of the springs. In many cases, a very small number of discharge measurements were used to determine the magnitude of the spring, so the context for judging real versus perceived changes in magnitude is weak.

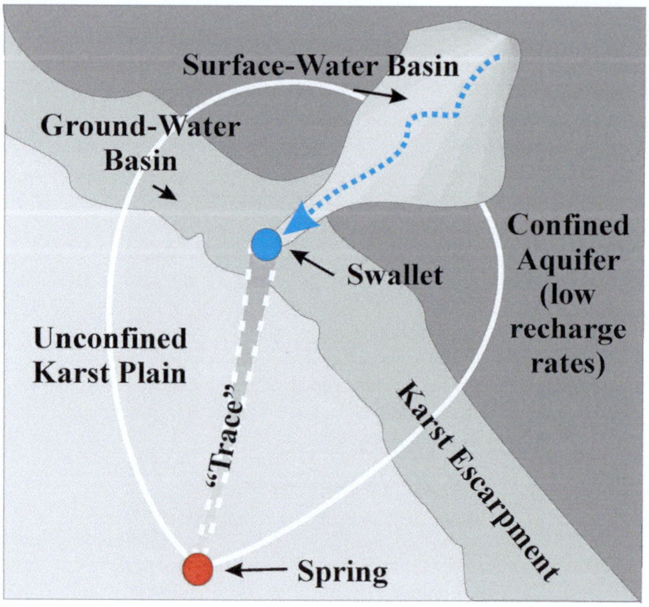


**Fig. 8.50** A springshed, including the groundwater basin and a surface-water basin that drains to the groundwater recharge area. (From Scott et al. 2004)

**Table 8.4** Classification of springs on the basis of average discharge

| | Discharge | |
|---|---|---|
| Spring magnitude | Metric units | English units |
| 1 | ≥2.8 m³/s | ≥ 100 ft³/s |
| 2 | 0.28–2.8 m³/s | 10–100 ft³/s |
| 3 | 0.028–0.28 m³/s | 1–10 ft³/s |
| 4 | 0.0063–0.028 m³/s | 0.1–1 ft³/s |
| 5 | 0.063–6.3 l/s | 0.10–0.1 ft³/s |

From Meinzer (1927)

## 8.5.5 Spring Plumbing and Flow Pathways

### 8.5.5.1 Springs and Spring Groups

Springs are located at the discharge end of complex and poorly known conduit systems that begin in the groundwater basin, often as dispersed recharge areas. Recharge occurs through intergranular porosity, small fractures and openings, sinkholes, and swallets or siphons. As water flows to the spring, it collects into better and better defined conduits until it discharges to the land surface at one or more openings.

Many springs have one, well-defined vent through which all of the water of the springshed system discharges. Others have multiple vents (spring systems with multiple vents are called spring groups**,** Copeland 2003, Poucher and Copeland 2006). Some spring groups have been shown to drain specific subareas within a springshed. For example, Jones et al. (1996) identified six different sub-basins within the Rainbow Springs Group springshed on the basis of groundwater chemistry. Discharge from five of these sub-basins could be attributed to a specific vent within the spring group.

### 8.5.5.2 Conduit Versus Diffuse Flow Systems

The downstream caves of many springs have been explored by cave divers, but the upstream conduits are typically blocked, too far from the spring vent, or too small for exploration. As a result, one can only speculate as to the specific pathways that groundwater follows in its journey from the recharge location to its discharge point. Dye tracing has been used to identify links from swallets and dye-injection points to springs (i.e., Butt and Murphy 2003; Butt et al. 2006, 2007), but detailed pathways are usually unknown. Efforts have been made to relate groundwater flow within a springshed by identification of photolinears or fractures, but such efforts represent informed speculation. Given (1) uncertainties associated with photolineament verification (Alfieri et al. 2018), (2) the large number of conduits that are filled with siliciclastic sediment, and (3) the role of intergranular, diffuse flow, there is no guarantee that conduit flow follows a specific pathway or that the conduits remain open and hydraulically functional through time.

Water-quality data can provide insight concerning the general conduiting of the spring system in terms of residence time and contact surface area at the rock-water interface (White 1988). Diffuse flow systems (Fig. 8.51a) are systems where groundwater is typically moving through small passages where the residence time is long, and the water is in intimate contact with the host rock. Diffuse flow systems move through (1) intergranular pore space and (2) complex networks of small, interconnected voids (i.e., spongework and small conduit mazes), so the water discharging from the spring has relatively high concentrations of calcium, magnesium, bicarbonate, hardness, and other dissolution-related analytes. The relatively high concentrations of these analytes result in elevated specific conductance and pH values consistent with chemical equilibration between the water and rock. Concentrations of these analytes do not vary significantly widely over time or with changes in discharge.

In contrast, conduit flow systems (Fig. 8.51b) are characterized by (1) recharge through swallets, (2) rapid water flow, (3) relatively short residence times, and (4) reduced contact with rock surfaces. As a result, analyte concentrations are typically lower than in diffuse-flow systems and they vary greatly as recharge and drainage events occur.

Spring drainage basins that are dominated by swallets (Fig. 8.51b) typically show conduit flow behavior because of the pulses of surface water that enter the system during runoff events. Springs with relatively stable discharge and water chemistry are dominated by diffuse flow.

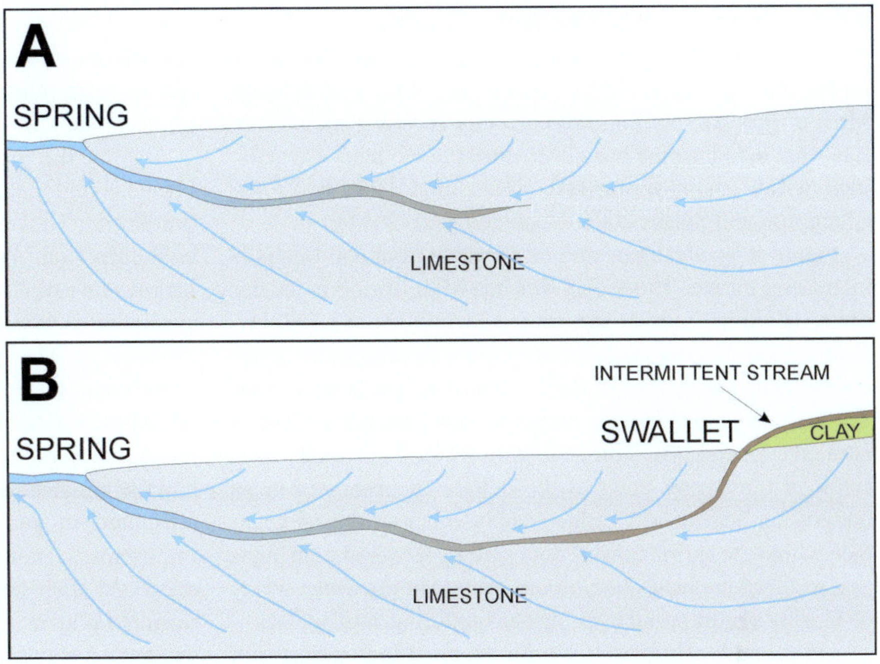


**Fig. 8.51** Simplified examples of spring "plumbing" systems. (**a**) Diffuse flow system; (**b**) Conduit flow system. (From Upchurch et al. 2007b)

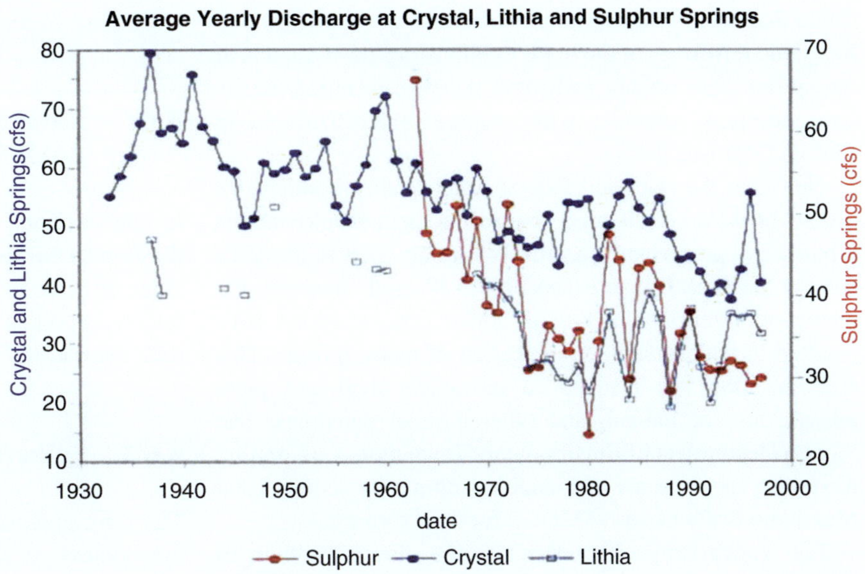


**Fig. 8.52** Historic declines in discharge at Crystal, Lithia, and Sulphur springs in west-central Florida. (Reproduced from Champion and Starks 2001)

## 8.5.6 Springs Are Changing!

### 8.5.6.1 Changes in Spring Discharge and Magnitude

At the present time, there is a consensus that both droughts and groundwater withdrawals by humans have contributed to declines in spring discharge (Fig. 8.52). The specific contributions of each cause are typically unknown, however, and there is much debate as to whether the loss of discharge is a result of the climate or human activity. Much of this debate took place in the early 2000s when a deep, multiyear drought was underway. Many of Florida's springs and rivers experienced record low flows that appeared to be directly related to the drought. With the return of "normal" rainfall, discharge in many of these water bodies has apparently recovered although, in many cases, the discharge remains less than recorded in previous decades (Copeland et al. 2011).

Copeland et al. (2011) looked at both spring discharge and spring water quality trends over time. They concluded that most of the long-term declines in spring discharge appear to be related to human activity. They also identified water-quality trends, such as increases in sulfate concentrations over time, which could be correlated with increased groundwater withdrawals causing upward movement of the sulfate-rich water near the base of the UFA.

#### 8.5.6.2 Changes in Spring Water Quality

A major concern with Florida spring water quality is the widespread increases levels of nitrate ($NO_3^-$) over time. Most of Florida's springs are currently showing such trends (i.e., Fig. 8.53), which have been attributed to land-use activities within their springsheds (Katz and Hornsby 1998; Champion and Starks 2001; Copeland et al. 2011).

Figure 8.54 illustrates one of the most dramatic trends in increasing nitrate. The Weeki Wachee Main spring is located at an old-time tourist attraction (28.5173°N, 82.5732°W; Hollis 2006) in Hernando County. First developed as a privately owned attraction in the 1930's, it is now a state park. Throughout its history, the attraction has featured a team of attractive "mermaids" complete with fish-tail costumes, who swim in the spring bowl to the delight of observers in an underwater auditorium. Unfortunately, much of the rock surface within the spring bowl is now partially covered with algae as a result of the increasing nitrate content of the water. Many of Florida's most popular springs are suffering similar fates.

Jones and Upchurch (1998) investigated the source(s) of the nitrate discharging from Weeki Wachee Springs. They found that the nitrate was largely derived from application of inorganic fertilizers in the many subdivisions developed near the spring. The sudden and rapid increase in nitrate/nitrite concentrations coincides with a surge in fertilizer use in Florida (Hornsby 2007).

Note that the state and federal water-quality standard for nitrate in water is 10 mg/L based on methemoglobinemia ("blue-baby" syndrome) hazards. While only a few of Florida's springs have reached or exceeded the 10 mg/L standard, the increases in nitrate loading at lower concentrations have resulted in widespread eutrophication of many springs. This eutrophication has resulted in unwanted algal and plant growth, loss of habitat, and other adverse outcomes. The Florida Department of Environmental Protection is attempting to reverse these trends of increasing nitrate by setting Total Maximum Daily Loads (TMDLs) for their springsheds.[7]

For comparison with respect to the nitrate problem in Florida, the nitrate concentrations in Alexander Springs (29.0815°N, 81.5759°W; Lake County) can be used as a control. The springshed of Alexander Springs lies within the Ocala National Forest and there is little or no fertilization or waste generation in the springshed. Consequently, nitrate levels in spring water remain at background levels (median concentration is 0.036 mg/L based on 107 samples collected from 1947 to 2015 (Fig. 8.55). This concentration is approximately an order of magnitude lower than that the median nitrate + nitrite concentration (0.22 mg/L, as N) of modern rainfall in Florida (Upchurch 1992 Table 3). The loss of nitrate from rainfall is apparently a result of metabolism by plants.

By far, the greatest concern for spring water quality is nitrate, a regional contaminant based on widespread land-use practices. However, many springs also suffer local contamination. For example, Haber (2003) and his colleagues investigated the presence of priority pollutants in ten springs in the Southwest Florida Water Management District (Table 8.5). Manganese may be a natural or anthropogenic constituent. It is utilized in certain pesticides and gasoline additives. Of the springs tested, Sulphur Springs, which is located within a highly urbanized area of Tampa (Fig. 8.48), had the greatest number of detections, many of which can be attributed to petroleum products, such as gasoline. It is important to note that none of the detected constituents exceeded their water quality standards. Even so, Sulphur Springs, which was once a tourist destination, is no longer available for swimming, and the spring water, which forms a supplemental supply for the City of Tampa, must be blended with other water before use.

[7]TMDLs are the maximum loads of a chemical that be tolerated by a water body before harmful consequences, such as algal growth, begin. Once the TMDL is determined for a water body, the available load of the chemical is apportioned to sources within the watershed or springshed. Chapter 99–223 Florida Statutes establishes the authority and process for TMDL development.

## 8.6 Karst Features Related to In Situ Rock-Surface Erosion

In this section, we discuss the smaller-scale features that develop on the upper surface and upper few meters of the limestone or dolostone because of weathering and erosion. These features, epikarst and karren, are the result of dissolution at the rock interface by meteoric water or shallow groundwater.

### 8.6.1 Epikarst

The term epikarst has been used previously in this book in the context of sinkhole development and timing of karst development in Florida. Because epikarst is so important to understanding Florida karst, the subject is expanded upon in this section. Note that while epikarst may be involved in sinkhole activity, epikarst is <u>not</u> synonymous with a sinkhole or sinkhole development.

Epikarst is, as the name suggests, the uppermost "skin" of a karstic limestone. It forms by chemical and mechanical weathering when the upper surface of limestone (or other carbonate rock) is subjected to subaerial exposure (Klimchouk 2004; Jones 2004; Jones et al. 2004; Mylroie et al. 2004). In west-central Florida, the epikarst on the uppermost limestone began to develop in the late Miocene and early Pliocene and is often covered by Mio-Pliocene or younger clay and marine terrace sand. As such, the uppermost, weathered horizon of the limestone is considered covered epikarst in the sense of

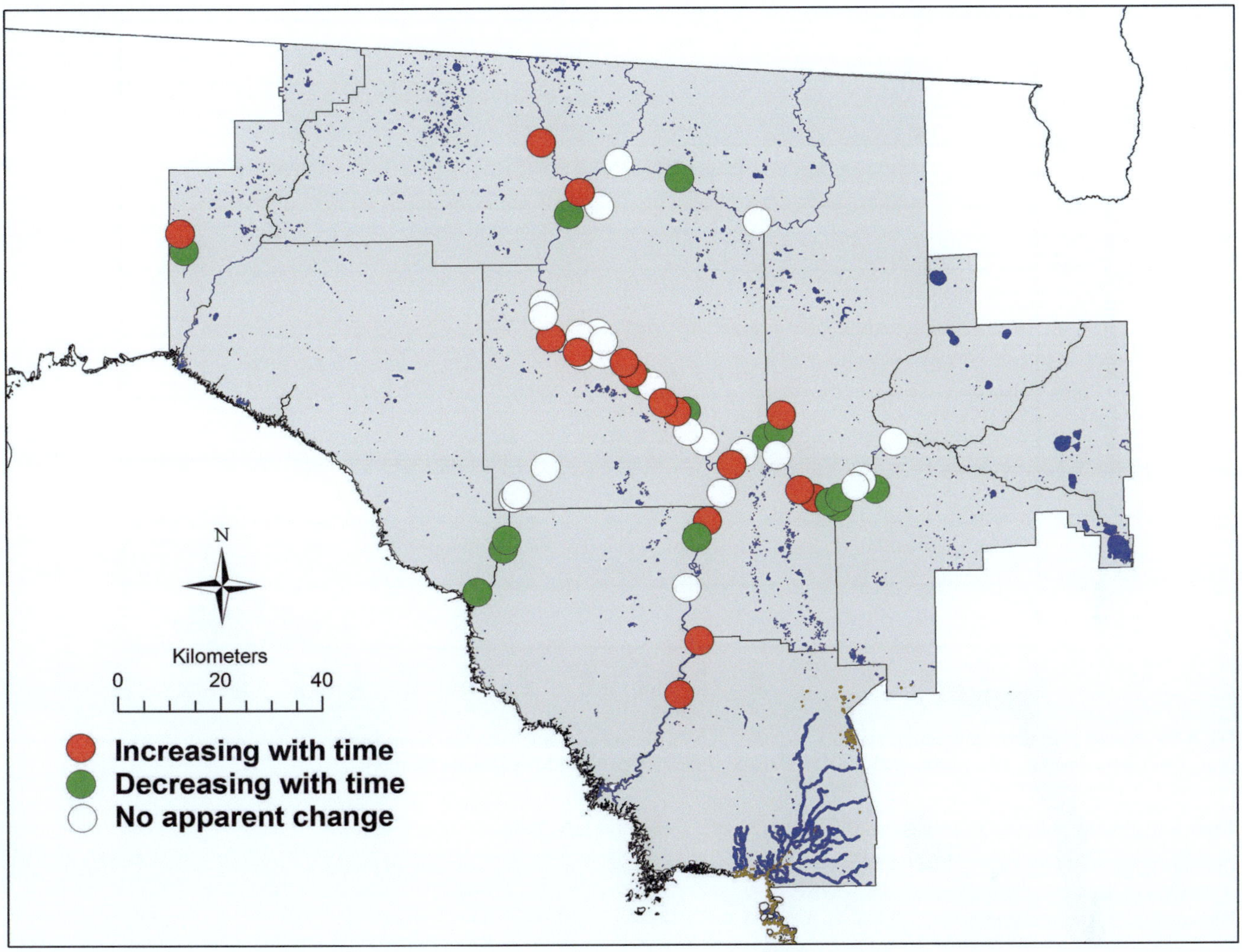


**Fig. 8.53** Temporal trends in nitrate plus nitrite concentrations ($NO_3^- + NO_2^-$, as N) in the springs of the Suwannee River Water Management District. (Modified from Upchurch et al. 2007a)

**Fig. 8.54** Nitrate plus nitrite concentrations ($NO_3^- + NO_2^-$, as N) in water from Weeki Wachee Main Spring (Hernando County) with time. (From Southwest Florida Water Management District 2012)

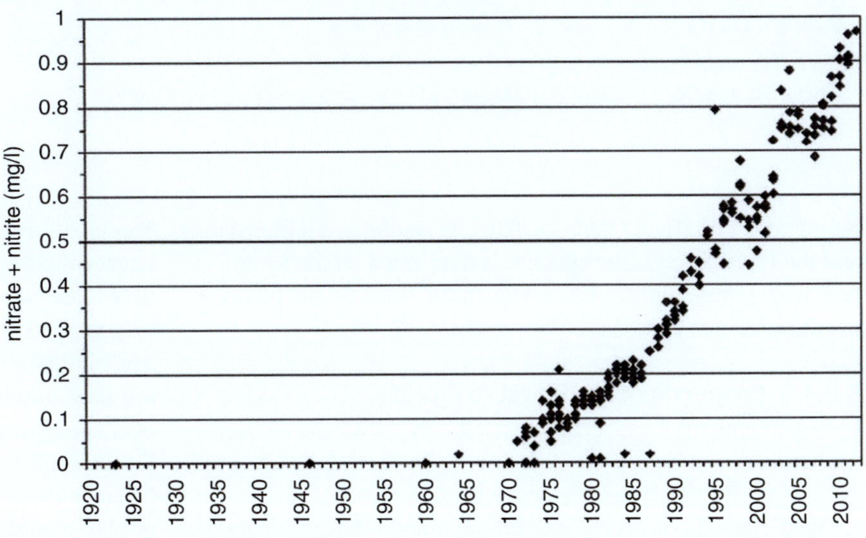

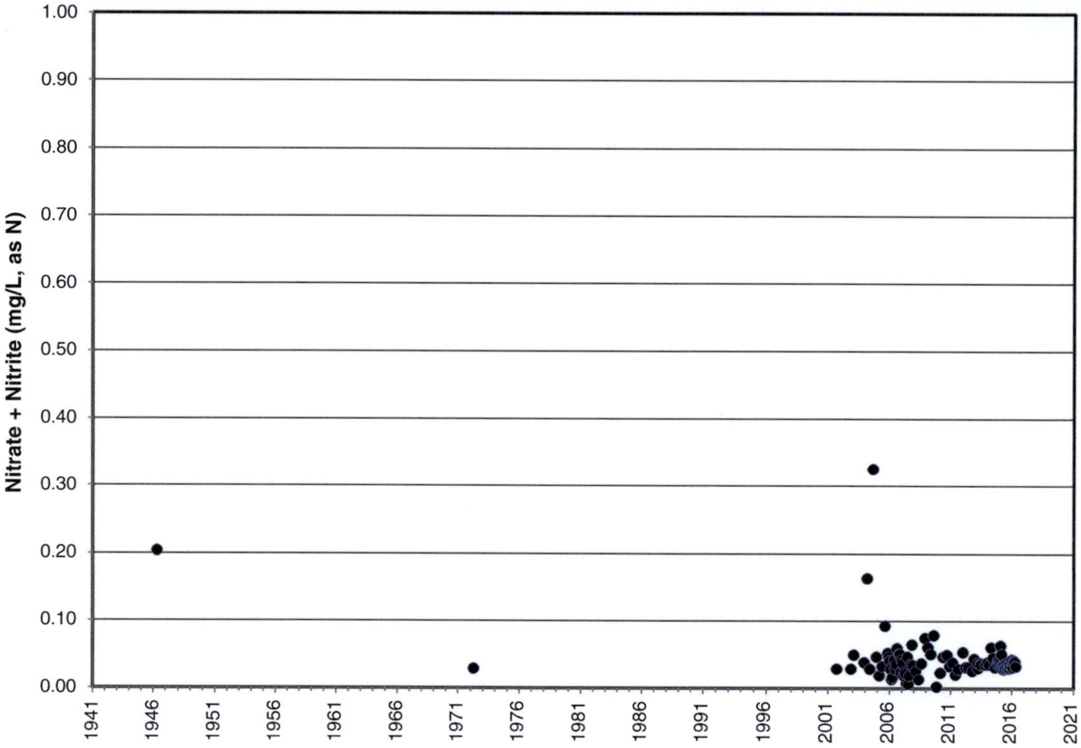


**Fig. 8.55** Nitrate plus nitrite concentrations ($NO_3^- + NO_2^-$, as N) with time in water from Alexander Springs (Lake County). (Data from Ferguson et al. 1947; Rosenau et al. 1977; Scott et al. 2004; and the St. Johns River Water Management District)

**Table 8.5** Priority pollutant screening results in central Florida springs for July, 2002

| Spring | Pollutant | Pollutant concentration (μg/L) | Guidance concentration (water-quality standard; μg/L) |
|---|---|---|---|
| Chassahowitzka main (Citrus County) | Manganese | 4 | 50 |
| Homosassa #3 (Citrus County) | Manganese | 1.1 | 50 |
| | Acetophenone | 4.7 | 700 |
| Hunters (Citrus County) | Chloroform | 0.3 | 6 |
| Sulphur (Hillsborough County) | Nickel | 2.7 | 100 |
| | Benzene | 0.63 | 1 |
| | Manganese | 41.3 | 50 |
| | m,p-Cresols | 250 | 350 |
| | Toluene | 13 | 1000/40[a] |
| Weeki Wachee (Hernando County) | Acetophenone | 2.2 | 700 |

From Haber (2003, Table 3)

[a]Toluene has a primary standard of 1000 μg/L and a secondary standard of 40 μg/L

Klimchouk (2004) and Veress (2016). In southeastern Florida and the Florida Keys, the epikarst is developed on the Miami and Key Largo limestones, which have little or no cover, resulting in bare epikarst.

#### 8.6.1.1 Properties of Epikarst in Florida

Florida epikarst is characterized by the following properties:

- Void space, breccia or rubble, and disaggregated limestone;
- Sand, silt- and clay-sized calcitic sediment caused by disaggregation of poorly lithified limestone, micritization of limestone allochems, and repeated wetting and drying episodes;
- Mechanically mixed carbonate fragments with the overlying clay and/or sand as a result of marine transgression over the epikarst and bioturbation;
- Siliciclastic sand- and clay-filled voids;
- Chert nodules "floating" in the weathered carbonate and/or clay matrix;

**Fig. 8.56** Epikarst development in the Suwannee Limestone near Brooksville, Hernando County. Photograph courtesy of the Florida Geological Survey

- Irregularities in the depth to limestone and to "competent"[8] limestone, with well-developed pinnacles and cutters (low areas between the pinnacles);
- Highly variable thickness of the weathered zone;
- Enhanced groundwater flow as a result of relatively high permeability of the weathered limestone (Klimchouk 2004).
- Wide variation in relative strength during standard penetration testing with weight-of-rod/-hammer strength sediments adjacent to split-spoon refusal-strength rock; and
- Wide variation in tip resistances during cone penetrometer test sounding, with minimal to refusal-strength resistances.

With Florida's many transgressions and regressions of the sea over the Neogene and Holocene, limestone strata in the subsurface often have highly porous and permeable horizons at unconformities. These paleo-epikarst horizons are common targets for water-well development.

As a result of the complex weathering and post-depositional history of the epikarst horizon, the epikarst can present a wide variety of conditions in three-dimensions and over very short distances (a few meters).

For example, Fig. 8.56 illustrates epikarst in the Vulcan Mine near Brooksville, Hernando County. This quarry is developed in the Oligocene Suwannee Limestone, with some Eocene Ocala Limestone exposed in the deeper portions of the quarry and outside of the view of the quarry face. The quarry face shown in Fig. 8.56 includes many of the features that typically occur within the epikarst in central and northern Florida. The upper surface of the limestone, which had small (less than 1 m), well-developed pinnacles, and the cover materials had been removed in preparation for quarrying when the image was captured. Features to note in Fig. 8.56 include:

- An upper horizon of rubble with hard and soft zones depending on the amount of disaggregation or cementation of the limestone;

[8]The term "competent" refers to the penetration resistance of the rock. Competent strata have refusal strength penetration resistance; in other words, the penetration test device is unable to pass through the stratum being tested. Strata that have high penetration resistances, but do not prevent penetration testing, may be suitable for construction even they do not have the sobriquet competent.

- Zones within the rubble zone that have admixed, green clay derived by migration of Miocene Hawthorn Group clay into the epikarst, apparently during subsequent marine transgressions, particularly during the late Miocene or early Pliocene[9];
- A clay-filled solution pipe (vertical shaft with a circular outline) filled with reddish-brown, clayey sand that apparently entered the epikarst after having been weathered, so it is more recent than the green clay discussed above;
- Two horizons of small voids that apparently represent former positions of the water table within the epikarst (NB, presence of former positions of the water table may not be an epikarst criterion);
- A zone of intact, but weakened, Oligocene Suwannee Limestone caused by initial disaggregation and micritization of the allochems in the limestone;
- Chert nodules included within the weakened and partly altered Oligocene Suwannee Limestone; and
- A relatively large cave filled with laminated, green clay and sand.

Clearly, there has been a long history of weathering and epikarst formation at this location prior to exposure by mining. The greenish clay filling the cavern and rubbly zone indicates that karst features, including caves and epikarst, were developed by the time of deposition and or initial weathering of the Miocene Hawthorn Group sediments that once overlaid the Oligocene Suwannee Limestone. Multiple zones of small voids resulted from dissolution of the Oligocene Suwannee Limestone at various positions of the water table. The exposed rock face is approximately 8 m, of which 5 m has been affected by epikarst development.

#### 8.6.1.2 Timing of Epikarst Development

Where epikarst is near the land surface and exposed or covered by permeable sand, weathering and epikarst formation is likely continuing at the present time. In contrast, where there is plastic, clay mantle over much of the epikarst, modern weathering is likely minimal. The clay is either intact Miocene Hawthorn Group sediment deposited directly over the limestone or residual Hawthorn sediment that remains after deposition, weathering, and several Plio-Pleistocene marine transgressions. See Figs. 2.25 and 2.26 for a comparison of weathered and intact Miocene sediment and Sect. 2.6.2.5 for a discussion of sediment textures and of methods for recognition of intact versus weathered clay-rich cover. The clay mantle can be used to establish a time of formation of much of the covered epikarst. As noted by Carr and Alverson (1959), the majority of the epikarst formation on top of the Tertiary limestones in central and northern Florida is late Miocene to early Pliocene in age.

When drilling through the limestones of central and northern Florida, one typically encounters voids, rubble, and enhanced groundwater flow at the unconformities between formations. These horizons of enhanced flow are ancient epikarst systems that were developed when the older strata were exposed to subaerial weathering.

#### 8.6.1.3 Origin of Epikarst

Epikarst forms by differential chemical and physical weathering of the uppermost surface of the limestone when it is exposed to acidic groundwater and the weather. The slightly acidic, meteoric water dissolves the limestone, but not as a uniform surface. Weathering is differential, resulting in an irregular surface with bare rock, rubble, disaggregated carbonate mud, pinnacles and depressions forming over short horizontal distances (Fig. 8.57).

In much of central and northern Florida, Plio-Pleistocene transgressions of the sea resulted in deposition of sand, clay, and shell beds that buried and mostly filled in the epikarst surface created during the Mio-Pliocene weathering event. Geologists call these buried weathering surfaces erosional unconformities. In this situation, the epikarst is essentially a buried, paleosol horizon or weathering surface at the top of the limestone and below the sand and/or clay.

When exposed to view, the most obvious features of the epikarst are karren.

### 8.6.2 Karren

#### 8.6.2.1 Definition of Karren

Karst scientists have long been fascinated by the fluting and sculpting of exposed limestone surfaces (Fig. 8.58). These often-exotic and beautiful features are known as karren (Sweeting 1973; Bögli 1980; Veress 2010). Karren are created when slightly acidic water encounters limestone or other soluble rock. As the water flows over or penetrates the rock surface, karren form by dissolution and/or mechanical erosion. Karren can form on bare rock or on covered limestone surfaces. The nature of the formation of karren depends on:

- Porosity and permeability of the limestone, including the presence of fractures, intergranular porosity, and other pathways for water;
- Dip (inclination) of the limestone stratum;
- Soil cover; and
- Residence time of the water in contact with the limestone.

[9]The green tint of the clay is typical of Miocene, ferrous-iron-rich smectite clays (see Chap. 2). The fact that the clay has not been substantially weathered and the iron oxidized to the ferric state, as is the iron in the overburden clay, indicates that the clay entered the epikarst during or shortly after deposition at the exposure.

**Fig. 8.57** Rubble and brecciated pinnacles developed in the epikarst of the Suwannee Limestone near Brooksville, Hernando County

**Fig. 8.58** Karren developed on the Ocala Limestone near Ozello (Citrus County)

Certain forms of karren are well-developed in Florida. These forms are typically varieties that develop in covered karst rather than bare, or free, karst. Because of the relatively high primary porosity of much of Florida's flat-lying limestone, karren, where present, often represent differential erosion caused by infiltrating meteoric water, not by water flowing over the surface of the low-permeability rock.

#### 8.6.2.2 Classification of Karren

Many karst scientists have attempted to classify karren structures by their form and others by the processes by which they formed. Table 8.6 presents portions of the classification system developed by Sweeting (1973) based on the size and geometry of the karren. Sweeting used the German terms for the common karren (the English equivalents are given in the table). White (1988), Veress (2010), and others have taken a more genetic approach to classification of karren. Table 8.7 presents a version of White's (1988) genetic classification modified to fit the possible forms of karren in Florida. Note that the three controlling factors are cover, type of runoff over the surface of the limestone, and structure of the limestone. In Florida, cover is certainly a major factor, but, owing to the presence of intergranular porosity of the limestone in many areas, runoff over the surface of the limestone may be limited. Also, the structure of the limestone limits the diversity of karren forms in Florida. Florida limestone typically has little or no dip, and, while fracturing is present, other, non-structural pathways of water movement, such as

**Table 8.6** A classification of small karren structures

| Term used in this text (Table 8.7) | Synonyms | Typical size (depth or height, length) | Occurrence: Free (bare) or covered karst? | Occurrence: Horizontal or inclined surfaces? | Sharp or smoothed crests | Known to exist in Florida karst? |
|---|---|---|---|---|---|---|
| Rill karren | Rillenkarren, Solution flutes | 1–2 cm, ≤ 50 cm | Free | Inclined | Sharp | Yes (rare) |
| Step karren | Trittkarren, Heel-print karren | 3–50 cm, 20–100 cm | Free | Horizontal | Sharp | No |
| Solution grooves | Rinnenkarren | 50 cm, ≤20 cm | Free and half-free | Inclined | Sharp crests, sometimes slightly rounded bases | Yes (rare) |
| Pinnacle karren | Spitzkarren | 50 cm, 50 cm | Free | Inclined crests and roof-like surfaces | Both | Yes |
| Meandering karren | Meanderkarren | 50 cm, ≤20 cm | Free and half free | Only slightly inclined | Sharp crests, slightly rounded bases | Possible |
| Round karren | Rundkarren | 12–50 cm, ≤15 cm | Covered | Inclined | Smoothed | Yes |
| Solution basins, solution pans | Kamenitzas | ≤50 cm, few cm to over 3 m diameter | Both free and covered according to type | Horizontal | Sides fretted when free karren, and smoothed when covered, bases normally smoothed | Yes |
| Grikes | Kluftkarren | Few cm to 4 m, ≤4 m | Both free and covered | Development along joints or fractures | Sides sharp when free and smoothed when covered, bases normally smoothed | Yes |
| Undercut solution runnels | Hohlkarren or Mohrkarren | 60 cm – 1 m, 50 cm | Covered – Formed under peat | Inclined or slightly inclined | Smoothed sides and bases | Possible |
| Root karren | Deckenkarren | Normally a few mm to cm deep | Covered – Formed by direct action of plants | Both inclined and horizontal | Smoothed but can be sharp | Yes (subject to some debate) |

Dimensions and descriptions modified from Sweeting (1973). See Bögli (1980), White (1988), and Ford and Williams (1989) for illustrations of these karren structures

burrows, may be present to interrupt the flow of water. The following sections discuss those forms of karren that are known to exist in Florida.

Because of the presence of cover materials, lack of significant slope or original relief on the limestone surfaces, and geologically young ages of the karst surfaces, karren development in Florida is usually limited and subdued. It is also often difficult to assign the various classification schemes to the shapes developed on exposed, Florida limestone surfaces. For example, Fig. 8.59 illustrates an exposed karren surface on the Oligocene Suwannee Limestone at Suwannee Springs, Suwannee County. The surface includes poorly developed fluting reminiscent of solution flutes or grooves and step karren superimposed on a horizontal surface that appears smoothed by dissolution and reprecipitation of calcite.

Note that meandering, round karren, undercut solution runnels, and root karren (Table 8.6) have not been observed by the authors in Florida.

**Table 8.7** Some karren terminology based on origin

| | Hydraulic forms: Sheet flow | Hydraulic forms: Channel flow | Etched forms: Structural weakness | Etched forms: Massive bedrock |
|---|---|---|---|---|
| Bare bedrock surface | Rill karren | Wall runnels | Solution grooves | Rain pits |
| | Stepped karren | Pit and tunnel karren | | |
| | Solution basins or pans (Kaminitza) | Meandering runnels | Cleft karren | Pinnacle karren |
| | | Solution channels (pipes, wells) | | |
| Covered surface | (not applicable) | Rounded runnels | Ledge pendants | Irregular rounded pinnacles |

Modified from White (1988, Table 2.4) to emphasize the forms of karren observed or possible in Florida

### 8.6.2.3 Solution Flutes, Heel-Print or Step Karren, and Solution Grooves

Features that resemble solution flutes, heel-print or step karren, and solution grooves (Tables 8.6 and 8.7) are rare on the limestone surfaces in Florida. Step karren were thought by Sweeting (1973) to form on horizontal surfaces, while the other forms of karren were thought to have formed by runoff on inclined surfaces. Figure 8.59 illustrates what passes for solution flutes in Florida, but they are located on essentially horizontal surfaces that have undergone dissolution and reprecipitation. Where features that resemble these types of karren exist in Florida, they are poorly developed and on horizontal surfaces. Therefore, one should not assume that these flutes are a result of mechanical erosion by surface-water runoff. The features shown in Fig. 8.59 are located on

**Fig. 8.59** Karren developed on the Suwannee Limestone at Suwannee Springs, Suwannee County

a surface that, unlike surrounding surfaces, is smooth and lacks molds and other irregularities. The absence of irregularities suggests that the surface may have developed as a result of dissolution and reprecipitation of calcite during weathering cycles.

Low permeability, well lithified "caprock" is locally common in the epikarst of the Eocene and Oligocene limestone of central and northern Florida and to Paleocene limestone of southwestern Florida. However, the upper surfaces of this dense limestone are rarely seen. One cannot rule out the presence of similar karren, particularly solution grooves, on the surfaces of the limestones, and there are a few exposures along the Gulf of Mexico coast where cover is thin or nonexistent and these forms of karren can be found.

Duricrust surfaces developed on the upper surface of the Pleistocene Miami and Key Largo limestones in the Keys (Fig 7.26b) and Miami area often exhibit relatively low permeabilities owing to recrystallization and cementation of the limestone surface. As a result of runoff of rain water falling on the bare rock surfaces of the duricrust, small-scale fluting is common. While they are nothing like the classical karren of bare, telogenetic limestones, it is intriguing to consider that these features may reflect miniature solution flutes or grooves and step karren.

#### 8.6.2.4 Pinnacle Karren, Covered Pinnacle Landforms, and Cutters

Small pinnacles are commonly developed in the epikarst of both covered and bare karst throughout central and northern Florida (Fig. 8.60). There is a continuum from small pinnacles that are clearly part of karren surfaces (Fig. 8.57) to very large pinnacles that, while buried under thick cover sediments, would qualify as landforms if exposed. Unlike Sweeting's (1973) description (Table 8.6), the small, karren pinnacles are commonly a meter or more in height and occur in complex galleries.

Large pinnacles that exceed 10 m in height (Fig. 8.61) are commonly found near and within Florida's ridges (Vernon 1951; Puri and Vernon 1964). These pinnacles are covered by marine terrace and aeolian dune sand and are not visible on the natural land surface but they are commonly exposed in quarries and abandoned, hard-rock phosphate mines. If these large pinnacles were exposed or constituted positive topographic features, they would likely be considered tropical landforms, such as small mogote.

The low areas between the pinnacles are termed cutters. Cutters form by dissolution of limestone along pathways of preferential water movement. These pathways may be joints and vertical fractures that have been enlarged or they may be isolated shafts in the bedrock. Their presence is an indication of karst processes, but not necessarily sinkhole development that involves raveling or subsidence of overlying cover materials. While cutters may be related to sinkhole development, many appear to be related to the movement of water through fractures in the underlying rock or, in the case of small pinnacles, along the cover/limestone interface without concomitant raveling or subsidence of the cover.

Where cutters are exposed in quarries, the cover materials that fill the cutters often appear undisturbed and show evidence of pedogenesis (Fig. 8.62). The cutters may have developed when the limestone bedrock was exposed, in which case they may represent solution-enlarged fractures

**Fig. 8.60** Pinnacles approximately 1 m in height in the Suwannee Limestone at the Vulcan Mine, Brooksville, Hernando County

**Fig. 8.61** Pinnacles as high as approximately 10 m in the Suwannee Limestone at the Vulcan Mine, Brooksville, Hernando County

**Fig. 8.62** A depression (cutter) between pinnacles whose tops have been cut off in preparation for mining at the Vulcan Mine near Brooksville, Hernando County. Note the horizontal soil horizons developed in the cover materials that fill the bedrock depression

that have since been filled by cover materials. It is also possible that they developed after having been buried under cover materials.

### 8.6.2.5 Solution Basins and Pans

Where limestone is exposed in Florida, such as in the Florida Keys and Miami area and the Gulf Coast of Hernando and Citrus counties, small solution pans are common. Figure 8.58 illustrates several small solution pans superimposed on the Ocala Limestone near Ozello, Citrus County.

### 8.6.2.6 Grikes

Grikes are solution-enlarged fractures or joints in the limestone (Figs. 8.63 and 8.64). They are common in areas where the upper surface of the limestone is exposed in central and northern Florida. When drilling and conducting geophysical surveys, linear features that may reflect grikes are also commonly encountered (Stewart and Wood 1991).

### 8.6.2.7 Vertical Shafts

Vertical shafts (i.e., solution pits, solution channels, solution pipes; Sect. 7.3.2.2) are common karst landforms in certain areas of Florida. The most common type of vertical karst shaft in Florida has very nearly circular in outline (Figs. 7.26a and 8.65) and a depth that is greater than the diameter of the opening.

White's (1988) geometrical classification of karst landforms suggests that these features can be collectively termed vertical shafts. Vertical shafts are, according to White (1988, "right circular cylinders" that cut bedding regardless of its orientation. White concluded that vertical shafts are formed by "fast-moving water films." White also opined that they occur where there is a "solution-resistant caprock" that retards infiltration and directs water to specific locations where water can enter and dissolve the limestone.

Three versions of this geometry are notable. In southeastern Florida, the Pleistocene Miami and Key Largo limestones commonly have circular depressions that are 1 m or less in diameter and 1–3 m in depth (Fig. 7.26a, b). In central and northern Florida, the Oligocene Suwannee and Eocene Ocala limestones have similar diameter vertical shafts with depths of 2–10 m (Fig. 8.65). In a few situations, vertical shafts can be quite large. For example, the shaft into which Florida's highest waterfall cascades is approximately 30 m deep and 6 m in diameter. This shaft is one of several impressive shafts and sinkholes located at Falling Waters State Park near Chipola (Washington County) The shaft is located at 30.7249°N, 82.5306°W.

**Fig. 8.63** Conjugate fractures forming grikes (arrow) at the Ozello boat ramp, Citrus County

**Fig. 8.64** Close up of one of the Ozello grikes and other karen developed on the Ocala Limestone. (Photograph courtesy of the School of Geological Sciences, University of South Florida)

Vertical shafts often can be shown to terminate at or near the present or a former water-table elevation and the sides of the shafts may (1) be bare rock, (2) have a thin, calcite crust or rind, or (3) a duricrust with calcite-cemented, mixed calcite and siliciclastic sediment lining.

As is discussed in Sect. 7.3.2.2, there are several models for how Florida's vertical shafts have formed. The solution pipes that are most common in the Pleistocene Miami and Key Largo limestones of southern Florida have been attributed to dissolution around plant taproots (Parker et al. 1955; Perkins 1977), random dissolution (Parker et al. 1955), and other causes. The larger, chimney-like shafts in the Oligocene Suwannee and Eocene Ocala limestones have been attributed to vertical recharge patterns at karst escarpments, such as in the Cody Scarp (Upchurch and Lawrence 1984; Upchurch 2002, 2005). The large shafts at Falling Waters State Park are located on a small hill that is an erosion remnant. It is a rock-collapse sinkhole (Rupert and Lane 1992).

### 8.6.3 Duricrusts

Duricrusts are rinds or hard layers formed by cementation on the upper surfaces of soil or sediment. They may include the host sediment/rock and/or soils that rest upon the host. Cements of duricrusts include calcium carbonate, silica (opal-CT and quartz), ferric hydroxides, and other minerals.

In Florida, duricrusts are typically composed of calcite-cemented soil and host rock materials. They often show two or three distinct layers: (1) an upper, zone of cemented soil (S, Fig. 8.66), (2) a central zone of laminated and fine-grained

**Fig. 8.65** Vertical chimney-like shafts in the Ocala Limestone, Haile Quarry near Newberry, Alachua County. These shafts range from 3 to 10 m in depth

crust (LC, Fig. 8.66), and (3) a lower breccia composed of fragments of the host sediment or rock (B, Fig. 8.66). Duricrusts are commonly laminated as a result of different episodes of wetting and drying (Fig. 8.67).

It is thought that Florida duricrusts form by repeated wetting and drying of carbonate rock in the epikarst, with small-scale dissolution when the material is wet and local precipitation of calcite cement when the material dries out (Multer and Hoffmeister 1968; Perkins 1977; Robbin and Stipp 1979). Multer and Hoffmeister (1968) obtained radiocarbon ($^{14}C$) dates for five duricrust samples from the Florida Keys. The dates suggested formation within the last 4395 ± 90 years, at which time the present-day Keys were clearly subjected to subaerial exposure. Using radiocarbon dating techniques on samples from Big Pine Key and the mainland, Robbin and Stipp (1979) concluded that the duricrusts ranged uniformly in age from 5680 years B.P. at the base to 400 years B.P. the top. From these studies, it is clear that the laminated crusts in southern Florida have formed within the last 5000–6000 years. Duricrusts lining solution channels and within the epikarst in the Florida Keys and southeastern Florida are commonly stained reddish brown (Fig 7.26b) because of incorporation of *terra rossa* derived from deposition of extra-continental airborne dust (Prospero 1999; Prospero et al. 1987, 2001, 2010; Shinn 2001).

Duricrusts are found at disconformities in the Tertiary limestones and dolostones, including the Miocene Hawthorn Group and Oligocene Suwannee (Fig. 8.66) and Eocene Ocala limestones. These ancient duricrusts represent past episodes of subaerial exposure and exposure to wetting and drying episodes.

### 8.6.4 Caprock

In karst science, the term caprock is often used to describe non-carbonate strata that overlie a cave and that may collapse into a cave to form a "caprock collapse sinkhole" (Williams 2004). For example, in karst systems in Kentucky and Tennessee, beds of quartz sandstone (caprock) may overlie and form part of the roof a cave. This caprock is important because it helps (1) support cave roofs and (2) distribute water seepage into the cave. This situation is unknown in Florida.

In Florida, "caprock" is a term used to describe a hard, well-lithified horizon located at the upper surface of limestone and other calcareous sediments. Florida's caprock is a dense, lithified rock that forms by repeated wetting and drying. The process of dissolution and reprecipitation of calcite results in thorough cementation with superimposed karst

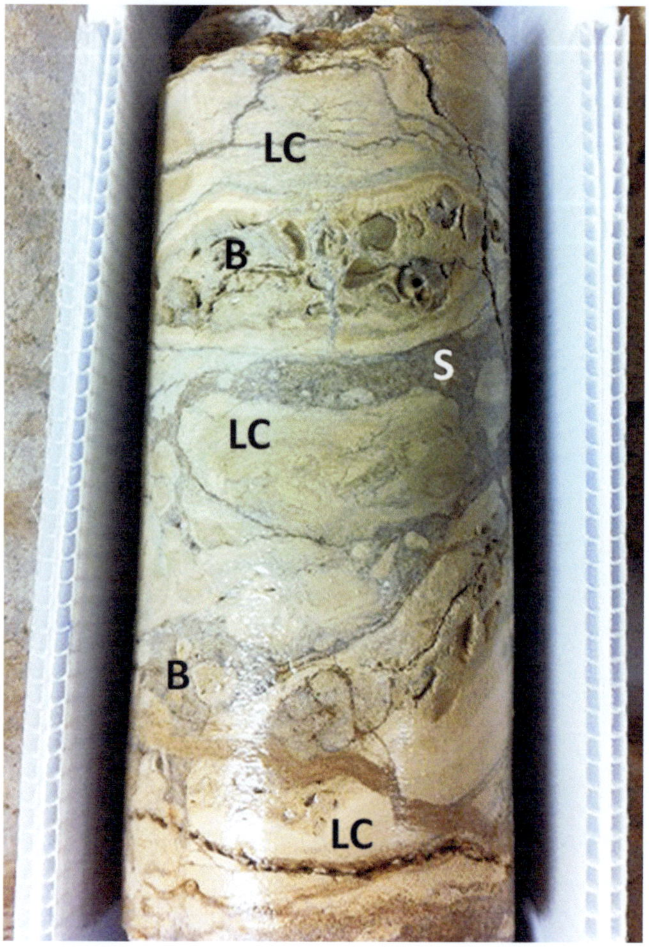


Fig. 8.66 Duricrust in a core from the Suwannee Limestone in Clearwater, Pinellas County. Note the laminated crust (LC), cemented soil (S), and breccia (B) zones

features. The caprock of southwestern Florida is notorious because it is so well lithified and difficult to penetrate or excavate.

While caprock has similar origins to duricrusts, it is a thick well-cemented horizon that can include breccias, duricrusts, cemented clayey sand, and vadose nodules of calcite (Fig. 8.68). Caprock is significantly thicker than duricrust; it is frequently as thick as 0.5–1 m, and it lies on less well cemented and, therefore, softer sediments. Like duricrusts, caprock is thought to form by repeated wetting and drying during seasonal water-table variations (Duever et al. 1985; Weedman et al. 1997; Kruse et al. 2000).

Caprock is best developed in the Plio-Pleistocene upper Fort Thompson and Tamiami formations in southwestern Florida where carbonate strata fall within the zone of water-table fluctuation (Weedman et al. 1997; Fig. 8.69). The caprock is commonly less than 1 m in thickness and it often has small (less than 1 m) openings (solution pits; Fig. 8.70) through which plant roots grow and water infiltrates to the underlying sediments. The origin of these openings has not been proven, but Duever et al. (1985) have suggested that the openings are caused by the plant roots that penetrate the caprock.

In southern Florida, there are typically two to three zones of the caprock layer. The uppermost horizon is a stained, sometimes laminated crust that resembles duricrust. This horizon is not always present. The next horizon down is ubiquitous, massive, well lithified, hard, sandy limestone to clayey, calcareous sand. This horizon is commonly represented by a disjointed layer of "pillow-shaped" limestone boulders (Fig. 8.69). The limestone was originally shelly

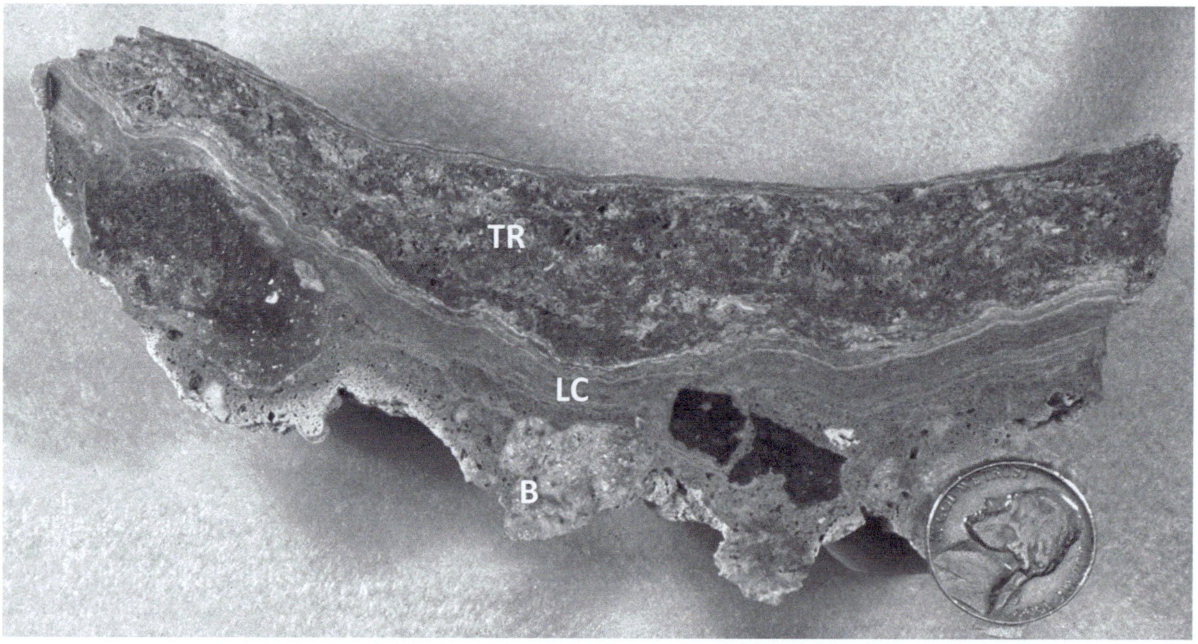


Fig. 8.67 Duricrust developed on the Key Largo Limestone at Key Largo, Monroe County. TR cemented *terra rossa* soil, LC laminated crust, and B limestone breccia

**Fig. 8.68** Hand sample of caprock composed of cemented clayey sand with rounded, calcitic, duricrust nodules from East Hinson Island, Collier County. Specimen is 8 cm across

grainstone to packstone with varying amounts of included quartz sand. The fabric of this limestone has been altered by micritization and cementation, which has resulted in a very low permeability, dense limestone. Water passes through the caprock horizon through solution pipes that commonly penetrate the limestone (Fig. 8.70) and by by-passing individual masses of caprock (Fig. 8.69). The third horizon is common and consists of partly cemented limestone breccia with varying amounts of quartz sand (Fig. 8.70). The clasts of the breccia are fragments of the host limestone that have been partly cemented into the massive basal zone.

Depressions in the caprock have been attributed to karst processes, including "collapse"[10] (Duever et al. 1985). These depressions are locations where cypress domes and wetlands, including sag ponds, form (Geologic Resources Division 2008). Figure 8.71 shows an area of intensive sag pond development north of I-75 in Collier County. Caprock is found underlying these features. Note that many of the sag ponds are aligned suggesting (1) control by jointing, (2) linear separations in the caprock, or (3) differential erosion along water flow-paths. The origin of these depressions has not been investigated in detail. These are the sag ponds that have been termed cover-subsidence sinkholes by White (1970), but this origin has not been confirmed through geological testing.

The term caprock is also used to describe thoroughly lithified limestone that immediately underlies thin cover and the land surface in northern Florida. This hard limestone layer is most obvious and best developed near major streams (Fig. 6.43). Thin sections of this limestone indicate cementation with microcrystalline calcite which suggests that repeated wetting and drying cycles have caused its formation.

### 8.6.5 Rubble Zones

One of the most distinctive features of many epikarst horizons is a zone of poorly sorted rubble near the top of the limestone (Fig. 8.72). Rubble consists of accumulations of loose, angular to sub-rounded breccia fragments of bedrock that have not been transported away from their point of origin. Fragments in the rubble zone of the epikarst can range in size from cobble- to clay-sizes. The nearly ubiquitous rubble zone of the epikarst is the reason epikarst horizons can conduct groundwater so well (Klimchouk 2004).

Figures 8.57 and 8.72 illustrate rubble zones overlying limestone in a quarry in Hernando County. Note the small, light-colored pinnacles and the laterally equivalent rubble. The rubble was formed by repeated wetting and drying of the limestone surface during weathering. Evidence of this wetting and drying includes minor rounding of sharp protrusions on the clasts and growth of weak calcite cements and fibrous calcite nodules, all of which give the rubble zones the appearance of poorly developed caliche.

### 8.6.6 Caliche and Vadose Concretions

Caliche is typically considered to be a soil zone of arid environments. It forms by the dissolution of calcium carbonate-containing minerals during rainfall events and then re-precipitation of carbonate during the arid intervals. As a result of this process, classic caliche has a rubbly texture and includes boulder- to clay-sized particles. Calcium-carbonate precipitates cement soil particles and form crusts in the soil.

Owing to distinctive wet and dry seasons, upper surfaces of limestone deposits in Florida often develop caliche-like characteristics, including duricrusts, rubble zones, and cemented breccia clasts and nodules. These features have been discussed in previous sections of this chapter. In this section we focus on calcite nodules that are found within the caliche-like rubble zones and that are clearly a result of precipitation of calcite in the vadose zone.

[10] The term collapse is not appropriate in the context of the sinkhole classification scheme discussed above. If they are related to dissolution of the caprock rather than failure of the caprock to form, the depressions are cover-subsidence features, which White (1970) termed sag ponds or sag features.

Fig. 8.69 Pliocene Tamiami Formation caprock exposed along a canal on US 41 near Monroe Station, Collier County. Note that the upper surfaces of the "pillow-like" slabs are relatively smooth

Fig. 8.70 A slab of caprock from the Tamiami Formation in Jerome, Collier County. The slab has been turned on end to show the solution pipes penetrating the thoroughly cemented limestone with quartz sand and shell fragments of the caprock. This view is of the bottom of the caprock showing the basal, well-cemented limestone breccia

These vadose concretions show fibrous internal structures that are similar to vadose pisolites found in arid soils (Scholle 1978). The fibers may form rod-like sheaths of crystals, sprays that mimic large crystals (Fig. 8.73), and more-or-less spherical pisolites.

### 8.6.7 Phytokarst

Phytokarst (Folk et al. 1973; Veress 2010) is a form of karren created by dissolution and mechanical abrasion of limestone exposed along coasts. In intertidal and wave-

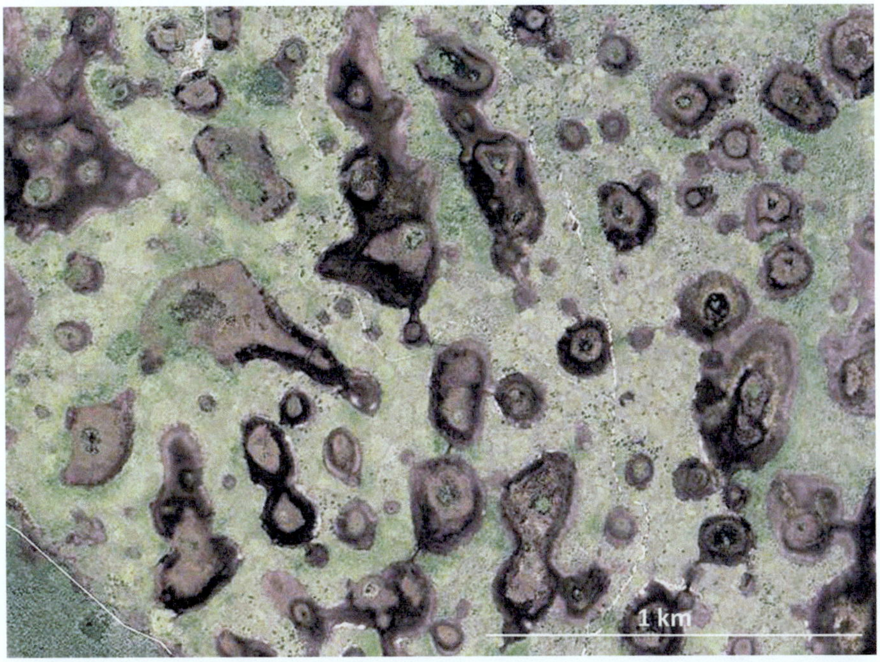


**Fig. 8.71** Sag ponds (wetlands) caused by depressions in the caprock on East Hinson Island in the Big Cypress Swamp, Collier County

**Fig. 8.72** Rubble zone transitioning into a pinnacle in epikarst of the Suwannee Limestone in the road cut shown in Fig. 8.72. The arrow points to iron-stained calcitic silt and clay caused by weathering and disaggregation of the limestone

washed rocky coasts, algae and fungi penetrate the carbonate minerals and any interstices between sediment particles, forming organic mats on the rock surface. The growth of the holdfasts and hyphae of the algae and fungi penetrate the rock by dissolving the carbonate material and wedging between grains. Wave impacts, dissolution by acids secreted by the algae and fungi, and grazing by marine herbivores erode the rock and form this distinctive form of karren.

Unlike karren formed by runoff over or infiltration into the limestone, phytokarst forms random, hackly, sharp, protrusions (arrows, Fig. 8.74) on the limestone surface. Where grazing is intense, the hackly surface may be smoothed and non-existent.

Other forms of karren structures can be found on rocky, limestone coasts (Taboroši and Kázmé 2013). Examination of Florida's rare rocky coastlines will undoubtedly reveal other forms of karren.

### 8.6.8 Microkarst

Microkarst is a term used in this text for small-scale morphological features formed by dissolution of the limestone surface. Like phytokarst, microkarst surfaces are characterized by small, sharp, more-or-less random irregularities, often giving the surface a hackly appearance. Microkarst surfaces are common in the Everglades, where organic acids and chemical complexing in standing or slowly moving water erode the limestone surface.

**Fig. 8.73** Vadose concretion found in the rubble zone of the epikarst at a quarry near Lecanto, Citrus County. The crystal-like shapes are bundles of fibers, aggregated in more-or-less random clusters. Sample is 10 cm in maximum dimension

Figure 8.75 shows microkarst developed on the Pleistocene Miami Limestone along the Pineland Trail in Everglades National Park. The highly irregular upper surface and abundant solution pipes are characteristic of the epikarstic limestone surface of the swamps and marshes of southern Florida.

### 8.6.9 Collapse Breccias: Recognizing Sinkhole Fill and Cave Collapse

Now that we have introduced rubble and breccias formed by subaerial exposure and/or weathering of the epikarst, it is appropriate to revisit the breccias and other sediments related to collapse of sinkholes. Breakdown cones in caves represent collapse of the limestone roof. Unless the breakdown cone is flooded by surface water containing sediments from the surface, the breccia consists of angular fragments of limestone with little or no sand or clay matrix (Fig. 8.76). The breakdown debris from cover-collapse sinkholes, in contrast, typically has little in the way of limestone fragments unless epikarst rubble is involved. Therefore, if breccia is encountered in the subsurface, there is a model that allows for differentiation of epikarst rubble, cover-collapse debris, and cave roof collapse:

**Fig. 8.74** Phytokarst (arrows) developed on the Pleistocene Anastasia Formation at St. Lucie Rocks, House of Refuge historic site, Stuart, Martin County. Location: 27.199°N, 80.166°W

Fig. 8.75 Microkarst superimposed on the Miami Limestone in Everglades National Park. The limestone has been penetrated by numerous solution pipes, as well. Location: 25.423° N, 80.680°W

Fig. 8.76 Collapse breccia cemented by a thin rim of calcite from a quarry near Lecanto, Citrus County

- Epikarst rubble – rounded fragments, weathered appearance of the limestone fragments, mixed with sand and clay derived from the cover sediments;
- Cover-collapse debris – sand and clay, often reddish brown to yellowish brown in color; and
- Cave-roof breakdown breccia – little or no sand and clay, angular fragments, wide variety of sizes from clay to blocks over 1 m in diameter.

## 8.7 Hypogene Karst in Florida

### 8.7.1 Introduction

Section 6.5.4.4 and other sections of this book briefly addressed hypogene karst in Florida. This section provides more detail and some possible examples from the Lower Floridan Aquifer (LFA) and deeply buried, Cretaceous strata of the South Florida Basin (Fig. 6.38).

In the latter part of the Twentieth Century, karst scientists began to realize that not all caves were a result of epigenetic speleogenesis. That is, not all dissolution features in the subsurface were a result of carbonic acid in meteoric or shallow ground water dissolving away limestone as it percolates downward.

The previous sections of this chapter deal with epigenic processes wherein mildly acidic water enters limestone from the land surface. Because epigenetic caves and karst form by the action of meteoric water, dissolution, and cavern formation are a "top-down" process. The predominant dissolution processes are in the vadose and shallow phreatic zones. Caves are typically best developed at shallow depths where chemically aggressive water is available.

In the last 30 years, many caves have been identified worldwide to have been formed by dissolution of limestone or dolostone by waters rising from below (Klimchouk et al.

2017). These waters gain their chemical aggressiveness through two major sets of processes: (1) formation of dilute sulfuric acid through oxidation of hydrogen sulfide in groundwater and (2) mixing of two water masses with different chemical compositions on a freshwater/saltwater transition zone. In 1991, Palmer utilized the term hypogene (or hypogenetic) speleogenesis for these processes. The term hypogene refers to the origin of the chemically aggressive water from below as opposed to from above the limestone stratum undergoing dissolution. Since hypogene speleogenesis is a "bottom-up" process, the resulting karst features tend to be deep seated and independent of the water table or meteoric processes. Both forms of hypogene speleogenesis are likely operating in Florida although documenting the origins of hypogenetic caves is problematic because the caves are underwater.

## 8.7.2 Speleogenesis by Groundwater Mixing

### 8.7.2.1 Introduction

Mixing of groundwater with different chemical compositions was discussed in Sect. 5.2.2. To summarize, mixing of two water masses with different chemical compositions, each of which may be chemically equilibrated with respect to calcite, can result in a solution that is undersaturated with respect to calcite (Bögli 1964a; Plummer 1975; Wigley and Plummer 1976; Back et al. 1986; Starks 1986; Sanford 1987; Smart et al. 1988; Sanford and Konikow 1989; Wicks and Herman 1996; Melim et al. 2004; Rezaei et al. 2005). In addition to dissolution with respect to calcite, these mixtures may also be responsible for precipitation of dolomite (Runnels 1969; Cander 1994; Gaswirth et al. 2007). Dolomitization as a result of groundwater mixing is discussed in Sect. 5.12.

Several of the benchmark papers dealing with groundwater mixing as a cause of speleogenesis have utilized Florida water-quality data in thermodynamic calculations to identify the effects of mixing (i.e., Plummer 1975; Back et al. 1986; Starks 1986). Typically, these papers deal with mixing of saline groundwater (Na-Cl facies) with fresh groundwater (i.e. Ca-$HCO_3$ facies water), both of which may be chemically saturated with respect to calcite. Within certain ranges of mixing, these waters are shown to result in water that is undersaturated with respect to calcite and, therefore, capable of hypogene speleogenesis.

### 8.7.2.2 Mixing-Zone Cave Development in the Upper Floridan Aquifer

The result of this mixing, which takes place within the freshwater/saltwater transition zone, is the development of caves that are similar to flank-margin caves that develop on carbonate islands with a freshwater lens (Mylroie and Carew 1990). These caves are:

- Related to the coastal position of the freshwater/saltwater transition zone;
- Likely to develop conduits that should parallel the transition zone and gently or imperceptibly dip landward;
- Likely to extend to greater depths than epigenetic karst because the dissolution process is in the zone of mixing at depth; and
- Subject to changes in sea level over time.

Few studies have actually documented hydrogeochemical conditions conducive to cave development as a result of groundwater mixing in Florida. Back and Hanshaw's (1970) investigation of saturation states of UFA groundwater with respect to calcite indicated that the coastal waters were typically saturated to oversaturated, not undersaturated, with respect to calcite. Katz (1992) examined the saturation states with respect to calcite of water samples from 787 monitoring wells that tap the UFA throughout Florida (Fig. 5.27). He related geochemical facies relationships to dissolution reactions and groundwater mixing. Katz identified coastal, UFA waters that were undersaturated with respect to calcite where the UFA is confined. Coastal and inland waters were commonly undersaturated with respect to calcite in unconfined areas of the aquifer. Based on groundwater chemistry measurements, a coastal mixing zone where hypogenic speleogenesis could occur was not detected. It should be noted that a large part of the data set that Katz utilized (the Florida Department of Environmental Protection's Background Water Quality Network) purposefully avoided areas of known saline-water intrusion (Maddox 1992).

So why does theory suggest that groundwater in the transition zone should be aggressive to calcite, yet field data are inconclusive as to the occurrence of aggressive water on the mixing zone? The answer is two-fold. First, most of the UFA groundwater samples studied in Florida were purposefully taken from the potable portions of the aquifer, not from the mixing zone. Second, water in the mixing zone travels slowly toward the coast and is in intimate contact with limestone and dolostone of the UFA. It would appear that the rate of equilibration of the initially aggressive, mixing zone water is rapid relative to water velocity. As a result, one might not expect to sample undersaturated water in the mixing zone, chemical undersaturation may be too transient. There has been one study that captured mixing-zone water in the unconfined UFA, however.

Starks (1986) studied mixing-zone water quality in the vicinity of the coastal springs in Citrus County with the idea of identifying the processes of mixing-zone cave development. He found that regional, fresh groundwater was undersaturated with respect to calcite. This water is locally recharged, in part, and clearly has not had opportunity to chemically equilibrate with the host limestone minerals. Based on modeling with PHREEQE (Parkhurst et al. 1980) and direct observations, Starks concluded that water near the coastal springs remains undersaturated with respect to calcite until the mixture contains about 12% sea water. At 12% sea water, the solution becomes supersaturated with respect to dolomite, but remains undersaturated with respect to calcite. At about 45% sea water the solution becomes saturated with respect to both minerals. Therefore, Starks concluded that dissolution of limestone would continue within the transition zone until the mixed solution was 45% sea water. Dolomitization, perhaps as a replacement for calcite in the host limestone, would take place in mixtures from 12% to 45% sea water.

In areas where the UFA is confined, contributions of new freshwater to the regional freshwater lens are limited and the hydrochemical facies that exist along the freshwater/saline water transition zone are better developed. Portions of the UFA that are well confined and have deep flow systems may, as indicated in Fig. 4.29, have two transition-zone horizons where conditions suitable for speleogenesis may exist. Fig. 4.29 illustrated the groundwater quality with depth in a coastal well in Sarasota County. This well taps the UFA where confinement is excellent. Note that there is a wedge of sulfate-rich water that begins at about 80 m below land surface. The onset of Ca-$HCO_3$-$SO_4$ water in this well is in the lower Hawthorn Group. Chloride concentrations do not increase until about 220 m below land surface. The freshwater/saline-water transition zone in this region of Florida is, therefore, a complex zone with an upper layer of sulfate-rich water derived from deep UFA water flowing along the upper surface of the middle confining unit (MCU) and picking up sulfate by dissolution of the gypsum or anhydrite in the underlying confining unit. If conditions are suitable (i.e., reducing conditions caused by microbial activity) then $H_2S$ may form from this $SO_4$-rich water (Rightmire et al. 1974; Rye et al. 1981; Sacks 1995; Sacks and Tihansky 1996; Sacks et al. 2012). Mixing of this water with seawater or shallow freshwater may result in speleogenesis.

A limited number of vadose and phreatic caves have been attributed to groundwater mixing in Florida (Polk and Brinkman 2013). In most studies where cave origin has been discussed, the caves are considered epigenetic (e.g., Florea and Vacher 2006a, b; Florea et al. 2003, 2007). Where a mixing-zone origin has been suggested, it is usually a result of (1) proximity to modern or ancient coastlines (2) modern water-quality conditions, and/or (3), erosional patterns (e.g. maze-work caves and cupolas) in phreatic caverns (Polk and Brinkman 2013). Given the many Plio-Pleistocene excursions of sea level, it is likely that many epigenetic caves have been modified by hypogene processes related to past positions of the freshwater/saltwater transition zone.

There are several coastal cave systems that are currently being explored by deep, technical diving procedures. The best known of these deep caverns are the Twin Dees/Weeki Wachee spring system (A, Fig. 6.38) and Diepolder II (B, Fig. 6.38). Both cave systems have been touted as the deepest phreatic caves in the United States. The Weeki Wachee Springs cave has been explored to a depth of at least 124 m in a cavern that has been explored to 1400 m in length (Neill and Anderson 2009). The Diepolder II cavern has been explored to a depth of 110 m at 469 m from the entrance sinkhole (Clesi and King 1991; Poucher 2008).

These coastal cavern systems are somewhat unique because of their depths and the large rooms that exist at depth. For example, the "Ballroom" in the Diepolder II cave system (Figs. 6.39, 6.40; Clesi and King 1991) is over 30 m high at an average depth of 64 m below the water table and, based on a cave profile published by Clesi and King (1991), about 160 m long. "Middle Earth", a room in the Twin Dees/Weeki Wachee cave system, is considered one of the largest rooms in a Florida cave. It is 107 m long, 61 m wide, and 24 m high. This room is located between 55 m and 79 m below the phreatic surface (Karst Underwater Research, Inc. pers. comm. 2017). In both cases, the phreatic surface is at approximately +6 m NGVD. The profiles of the caves are above the modern freshwater/saltwater transition zone, but they approximately parallel the mixing zone.

Starks (1986) and Katz (1992) concluded that unconfined, shallow UFA water in the vicinity of these coastal caves is chemically undersaturated with respect to calcite. Therefore, dissolution of the host limestone is predicted regardless of the current mixing zone location. These unusually large and deep caves show evidence of dissolution, such as development of small alcoves and cupolas in the rock surface (Fig. 6.40).[11]

A notable feature of these caves is the large amount of breakdown debris (Fig. 6.40), which suggests that a significant mode of cavern development is breakdown. Recall that the limestone, in this case the Eocene Ocala Limestone, is poorly to moderately cemented and fractured. Bedding planes are evident in the caves, and the breakdown material typically consists of large slabs with bedding plane surfaces. One can conjecture that the weakening of the cave ceilings is a result of either (1) chemically aggressive water penetrating the less cemented limestone enabling dissolution and detachment of large blocks of better cemented limestone when sea level was higher and the freshwater/saltwater transition zone intersected the strata or (2) dissolution in the vadose environment when sea level was significantly lower than present.

[11] For an excellent collection of cave diving videos, largely within the Twin Dees/Weeki Wachee cave system, see Pitkin (2016).

Little else is known about the architecture of the caves. However, the depth and large room sizes of these caves make these coastal cave systems candidates for hypogene speleogenesis by groundwater mixing in Florida.

#### 8.7.2.3 Mixing-Zone Speleogenesis in the Lower Floridan Aquifer

As noted in Sect. 4.5.4, confusion exists about the importance of karst processes in the boulder zones of the LFA. This is because of the limited data available for evaluation of the artifacts of speleogenesis. Our knowledge of the boulder zones of the LFA is entirely based on borehole cores and videos and on drillers' logs.

Based on our knowledge of groundwater quality and circulation in the LFA, it is probable that mixing-zone, hypogene speleogenesis has contributed to the high transmissivity zones of the aquifer (Herbert and Upchurch 2016; Upchurch 2017). High hydraulic-conductivity zones in the aquifer may be a result of mixing on a transition zone powered by geothermally driven groundwater circulation. Sproul (1977) described geothermal gradients in south Florida and identified anomalous areas of high and low heat flow. The warm-water anomalies were thought to reflect upwelling of warmer-than-normal saline water near the west coast of Florida. A few warm-water springs, such as the cenotes Warm Mineral and Little Salt springs (Cockrell 1973, 1975; Clausen et al. 1975, 1979; C Fig. 6.38) and submarine Mud Spring (Breland 1980; Fanning et al. 1979, 1981; E Fig. 6.38) can be found in the areas of these anomalies. These springs are thought to be discharging deep, saline water, possibly from the LFA. The cenotes are examples of deep (<50 m) rock-collapse sinkholes. A small number of similar collapse features exists elsewhere in south Florida (Fig. 6.38).

The Kohout-type circulation system in the LFA (Sect. 4.5.4.3) presents opportunities for mixing zone speleogenesis associated with seawater and native Ca-$HCO_3$ facies water. Given the gypsum/anhydrite strata that confine the LFA, the Kohout circulation model also suggests opportunities for hypogene speleogenesis in association with sulfide- and sulfate-rich water.

### 8.7.3 Speleogenesis by Sulfuric Acid Solutions

The LFA and Cretaceous strata that underlie the FAS include several strata sequences that consist of carbonate rocks bounded above and below by anhydrite. These strata sequences meet the conditions for sulfuric-acid generation of Klimchouk (2009, 2014) for hypogene speleogenesis.

#### 8.7.3.1 Lower Floridan Aquifer

The stratigraphy of the LFA compares favorably with Klimchouk's (2007, 2009) geological conditions for sulfuric-acid development and hypogene speleogenesis. The upper and lower confining strata include gypsum or anhydrite, and there is a well-developed groundwater circulation system associated with breccias and fractures (the boulder zones) that may have been at least enlarged by dissolution.

There is minimal direct evidence at this time to suggest that on-going or former hypogenesis has enhanced the porosity of the boulder zones. Others have suggested its influence based on geophysical or geochemical studies that reveal evidence of deep seated dissolution, however. Randazzo (1997) speculated that hypogene speleogenesis had enhanced porosity in the LFA based on sulfate mineralization in aquifer water. Cunningham and Walker (2009), Cunningham (2013, 2015) and Cunningham et al. (2017) utilized seismic methods to identify deep-seated collapse zones associated with sinkhole-like depressions on- and offshore of Dade and Broward counties of southern Florida (D and E, Fig. 6.38). They attributed these collapse features to hypogene processes within the deep FAS. While Popenoe et al. (1984) did not suggest hypogene origins of collapsed zones, they also noted apparently deep, collapse features offshore along the northern, Atlantic shelf of Florida in a seismic survey.

Onshore, the deep cenotes and collapse sinks in southern Florida may have a similar origin. Little Salt Spring (C Fig. 6.38) discharges anoxic, $H_2S$-rich groundwater from a vent at 73 m depth (Yang et al. 2008). This water is mixing with shallow, oxic water resulting in $H_2SO_4$. While the cenote is 75 m deep, the bottom is a debris cone, and no one knows how deep the breakdown-filled aven extends. The presence of the vent opens the possibility that the cenote is a deep-seated collapse rather than a feature limited to the UFA, within which the cenote's known extent lies. Similar cenotes in the Yucatan Peninsula of Mexico have been attributed to sulfur reduction-oxidation origins (Stoessell et al. 1993; Schmitter-Soto et al. 2002; Socki et al. 2002).

There are numerous on- and offshore, deep sinkholes, including cenote with abnormal geothermal heat flow (Fig. 6.38) that may reflect these hypogene processes and offshore sinkholes that are currently at elevations too deep to have formed on a modern saline-water transition zone. For example, the 200+ km long Pourtales and Miami terraces (F to G Fig. 6.38; Jordan 1954; Jordan and Stewart Jr 1961; Jordan et al. 1964; Land et al. 1995; Land and Paull 2000) offshore of southeastern Florida have sinkholes at water depths of 180–450 m. Either the terraces (1) subsided to these depths after karstification or (2) deep-seated hypogene speleogenesis has caused their development.

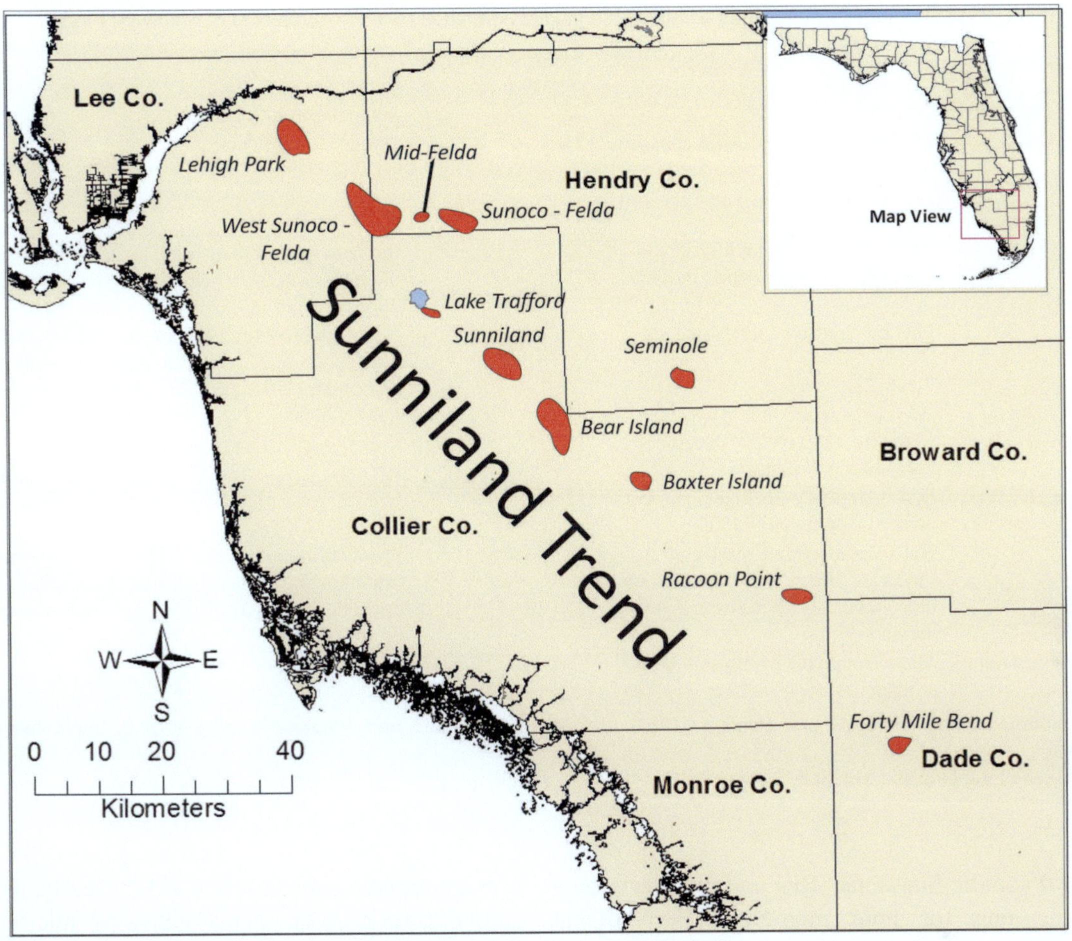


**Fig. 8.77** Oil fields within the Sunniland Trend in southwest Florida. (Figure from Herbert and Upchurch 2016)

#### 8.7.3.2 The Lower Cretaceous Sunniland Formation

The eleven oil fields of the Sunniland Trend (Fig. 8.77, Table 8.8) are important petroleum-producers in southwestern Florida. The host strata, which make up the Aptian (Cretaceous) Sunniland Formation, are deeply buried, having subsided as much as 3500 m below sea level.

The individual fields occur in a northwest-southeast linear trend that corresponds with a series of rudistid knolls (Mitchell-Tapping and Mitchell-Tapping 2003; Herbert and Upchurch 2016) that developed along a hinge line in the South Florida Basin (Fig. 6.38). Southwest of the Trend, lower Cretaceous strata dip to the southwest. Oil production has been sustained since the 1940s, and many of the fields have produced over a million barrels of oil from within the small structures.

The Sunniland Formation is underlain by the Lower Cretaceous Punta Gorda Formation (Table 8.8), a massive limestone and anhydrite formation that serves as a lower confining unit for the Sunniland Formation (Herbert and Upchurch 2016). The reservoir seal across the Trend is the Lake Trafford Formation, a similar, massive sequence of anhydrite and dense limestone.

The Sunniland Formation is up to 75 m in thickness (Huddlestun et al. 1988). The unit has been subdivided into five informal stratigraphic units (Table 8.8; Applegate and Pontigo 1984; Mitchell-Tapping and Mitchell-Tapping 2003). It is within these five facies that hypogene speleogenesis is likely to have enhanced formation porosity and hydraulic conductivity. The basal unit consists of two major transgressive marine lithosomes (Table 8.8). The lower, basal facies is a fossiliferous limestone with abundant fractures and breccia zones. This horizon, locally known as the "rubble zone", is petroliferous. The second facies, which overlies the "rubble zone", is informally known as the "black shale." This unit is a dense, argillaceous limestone with abundant organics that are thought to be the source of the oil, either from within or adjacent to the Trend (Applegate and Pontigo 1984; Palacas 1978a, b; Palacas et al. 1981, 1984a). The facies that overlies the organic facies is a low porosity

**Table 8.8** Stratigraphic relationships of lithosomes within the Sunniland Formation (Lower Cretaceous) of southwest Florida

| Formation | Facies[1] | Lithology | Average thickness (meters) | Function |
|---|---|---|---|---|
| Lake Trafford Formation | | Limestone, anhydrite | 46 | Upper confining unit |
| Sunniland Formation | 1 | Molluscan limestone with anhydrite blebs in uppermost part, partly oolitic, bioturbated; regressive phase deposition | 25 | Primary oil production zone |
| | 2 | Intertidal limestone with rudistid fragments forming debris mounds | 20 | Phreatic environment with subaerial "leaching" near end of deposition (Mitchell-Tapping and Mitchell-Tapping 2003); primary oil production zone in west-Felda field |
| | 3 | Shallow, intertidal limestone | 10 | According to Mitchell-Tapping and Mitchell-Tapping (2003) phreatic and hypersaline conditions existed near end of deposition; primary oil production zone in the Felda fields |
| | 4 | Light brown to cream limestone, occasionally dolomitic | 15 | Low porosity, low permeability; possible confining stratum |
| | 5 | Dense, argillaceous limestone; dark colored; "organically rich" | 12 | Considered a possible source rock for Sunniland oil |
| | | Fossiliferous, transgressive limestone; fractured with common breccia masses; petroliferous odors common | 15 | Production in the Lake Trafford field; termed the "rubble zone" because of fractures and rubble encountered during drilling; possible source rock for Sunniland oil |
| Punta Gorda Formation | | Massive anhydrite with salt stringers | 180–250 | Lower confining unit |

Descriptions modified from Applegate and Pontigo (1984) and Mitchell-Tapping and Mitchell-Tapping (2003). Table from Herbert and Upchurch (2016)

[a]Facies from Mitchell-Tapping and Mitchell-Tapping 2003

limestone to dolomitic limestone. This unit may serve as a semi-confining unit for fluid migration (Herbert and Upchurch 2016). Next is a shallow, intertidal limestone (Mitchell-Tapping and Mitchell-Tapping 2003). Mitchell-Tapping and Mitchell-Tapping (2003) reported evidence of hypersaline solutions associated with "phreatic" conditions in this facies. Herbert and Upchurch (2016) interpreted the term "phreatic conditions" to represent evidence of dissolution and brecciation. Above the intertidal limestone is a fragmental, intertidal limestone with abundant rudistid debris mounds (Mitchell-Tapping 1987; Mitchell-Tapping and Mitchell-Tapping 2003). Mitchell-Tapping and Mitchell-Tapping (2003) also reported evidence of phreatic conditions in this facies. This facies is a major oil-production horizon. Finally, the primary oil producer in the Sunniland Trend is a fossiliferous limestone deposited under marine regression conditions according to Mitchell-Tapping and Mitchell-Tapping (2003). The facies shows evidence of bioturbation and deposition of secondary anhydrite. The unit that overlies the Sunniland Formation is the Lake Trafford Formation (Table 8.8).

Sunniland petroleum has been studied extensively (Palacas 1978a, b; Palacas et al. 1981, 1984). Applegate and Pontigo (1984) characterized the petroleum from the Sunniland as "moderately immature" to "moderately mature." Thermal maturity increases with depth, as does the total organic carbon content. Palacas (1978b) noted that the average organic content was 0.41%. Clearly, there is ample carbon to serve as an energy source for microbial transformation of sulfur compounds and petroleum. Mitchell-Tapping (2002) argued that the oil is too mature for current bottom-hole temperatures; bottom-hole temperatures are less than the published minimum temperatures for hydrocarbon maturation.

Migration of the petroleum may have been widespread. Palacas et al. (1984b) noted that lower Cretaceous limestone in the eastern Gulf of Mexico had staining and residual asphalt that appeared to represent a "post-migration alteration product of a mature oil that has migrated from source rocks deeper in the section, or from stratigraphically equivalent … [strata]…down-dip…." The petroleum was found to be compositionally similar to Sunniland petroleum. If the maturity and organic-carbon content constraints are valid, there appears to be an unaccounted for petroleum-migration path operative in the Sunniland Trend oil fields.

As noted by the production figures (Herbert and Upchurch 2016), the amount of recoverable oil and ease of extraction seem high relative to the size and distribution of the fields within the Sunniland Trend, particularly from low-relief structural features. Detailed examination of the production and distribution of wells within the fields suggests that some form of conduit flow and secondary stratigraphic trap development exists. Within fields in the Trend, closely spaced wells result in a pattern of highly variable production with

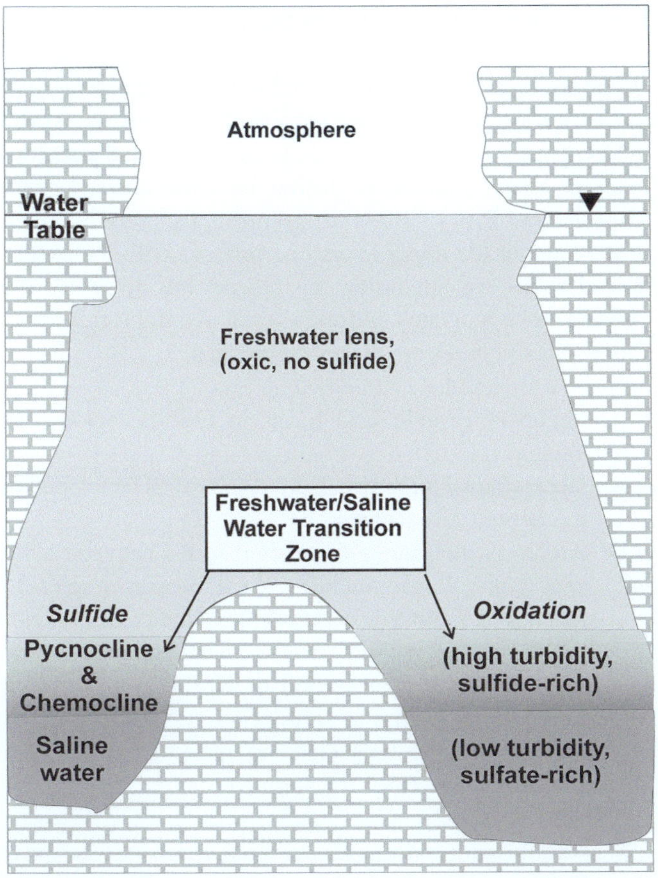


**Fig. 8.78** Sketch of the hydrochemical facies relationships in an anchialine spring or sinkhole

wells that have produced $10^8$ L of oil in close proximity to dry or low-recovery wells. Within the fields, intervals of no core recovery or weight-of-drill string gaps are commonly described in drilling logs. This information, combined with the concerns about source strata and geothermal temperatures, suggests a more complicated system than has previously been described. Simply considering the entrained fluids as a karst-facilitated circulation system may account for the lower than expected geothermal gradient.

Sunniland petroleum is "sour." That is, it contains abundant $H_2S$. Hill (1990) argued that the $H_2SO_4$ that formed Carlsbad Cavern was derived from $H_2S$ gases that evolved from the petroleum in underlying strata. The abundance of $H_2S$ in Sunniland oil may be a similar source in the Florida strata.

Abundant brecciated horizons in three of the facies of the Sunniland Formation (Table 8.8) are consistent with collapse as conduiting developed. Reports of evidence of "phreatic conditions" may also suggest dissolution in a hypogene system. Alternatively, syndepositional brecciation may have facilitated later speleogenetic enhancement of the breccias and conduits.

Finally, there is evidence of fluids seeping along the base of the continental shelf. Paull and Dillon (1980), Paull and Neumann (1987), Commeau et al. (1987), Chanton et al. (1993), Robb (1990), and Hine (1997) discuss brine seeps near the base of the continental slope off Florida in both the Atlantic Ocean and Gulf of Mexico. It has also been suggested that the shelf margin is forming by sapping and karst development near the seeps (Paull and Neumann 1987; Paull et al. 1990; Paull and Dillon 1980). The seeps offshore of southwestern Florida are at about 3270 m below sea level and are in Cretaceous limestone. They are characterized by relatively cool water. These seeps appear to occur at about the same stratigraphic position as the Sunniland. According to these authors, the seeps appear to be locations of developing karst as a result of dissolution, causing sapping at the base of the continental margin.

Chanton et al. (1993) characterized limestone dissolution at the seeps as being a result of mixing of seawater and sulfate-depleted brine derived from the Cretaceous sediments. Two sulfide sources were suggested to explain the isotopic compositions of sedimentary pyrite at the seeps. These are (1) a $^{34}S$-enriched source derived from the Florida Platform and (2) a $^{34}S$-depleted component derived from microbial reduction of seawater sulfate. As a result of these microbial-driven reactions, the seeps support populations of chemotrophic organisms.

In summary, there is evidence of sulfur-driven geochemical reactions in pore or conduit waters within the Cretaceous sediments in the vicinity of the Sunniland Trend. This, combined with stratigraphic sequencing of anhydrite and fractured limestone strata consistent with Klimchouk's hypogene speleogenesis model, evidence of collapse and brecciation, and a distribution of oil production suggestive of large conduits, suggests that hypogene speleogenesis may be responsible for development of the "plumbing" system that has resulted in migration of oil into the structures forming the Sunniland fields. If so, the speleogenesis in the Sunniland Trend most likely occurred prior to subsidence to present depths.

## 8.8 Coastal Anchialine Caves

Anchialine caves consist of openings in the host limestone wherein the freshwater lens is in contact with seawater within the cave (Fig. 8.78; Bishop et al. 2015). The contact between the two water masses is transitional. The thickness of the mixing zone is driven by diffusion or minor advective mixing.

Anchialine sinkholes are found near the Gulf of Mexico coastline and in shallow, nearshore Gulf waters in Florida. Cave divers commonly report highly turbid water as they pass from the freshwater lens to the underlying saline water.

They also report odors and irritation of the eyes, lips, etc. in this turbid layer as a result of exposure to $H_2S$. The presence of a turbid, pycnocline layer indicates an anchialine system. It has been suggested (Garman 2010; Garman and Garey 2005; Garman et al. 2011) that the sulfur reduction-oxidation reactions within the pycnocline may cause dissolution of limestone and cause modification of the caverns or sinkholes.

The freshwater lens (Fig. 8.78) is exposed to the atmosphere and the water is oxic. Any sulfur in the freshwater is dissolved $SO_4^{2-}$ or suspended particulates. The freshwater/saltwater transition zone is characterized by a well-developed chemocline and pycnocline, commonly in conjunction with high turbidity. The density of the water increases as total dissolved solids increase with depth. Below the pycnocline is saline water with low turbidity and elevated $SO_4^{2-}$ concentrations. Anchialine sinkholes and springs are characterized by circulating fresh and saline groundwater. The circulation must not be strong enough to mix the entire water column, however.

The zone of density change traps the particulate organics, microbes, pyrite framboids, and clay minerals (Garman 2010; Garman et al. 2011) resulting in the turbid, low visibility pycnocline layer. Microbial activity chemically reduces $SO_4^{2-}$ obtained from the underlying water forming $H_2S$ and colloidal to clay-sized $FeS_2$ framboids.

Based on Hydrolab® data obtained by divers and fixed station measurements, pH, temperature, ammonia, dissolved oxygen, and other analyte data were obtained. In addition, $\delta^{18}O$ ratio data were obtained. The position of the pycnocline was found to migrate with seasons and other stresses. Based on pH profiles, Garman (2010) suggested that in some situations, pH lowering by $H_2SO_4$ from oxidation of $H_2S$ within or just above the turbid zone may cause speleogenesis.

If a pH minimum exists because of $H_2SO_4$ production within the zone wherein $H_2S$ is oxidized and if calcite dissolution occurs, the tides, migrations of the pycnocline caused by cold-water intrusions, and sea-level fluctuations can be expected to "smear" the zone of calcite dissolution over some vertical and horizontal distance.

Investigations of anchialine caves have mostly been by biologists because of interest in microbial activities in these unusual chemical systems. Chemoautotrophic microbes and other unusual organisms are commonly found in these systems. To our knowledge, no one has examined these features for evidence of cave wall architectures that may reflect hypogene speleogenesis. Garman and Garey (2005), Garman (2010), Garman et al. (2011), Menning et al. (2015), and Rubelmann and Garey (2016) describe the microbial populations and metabolic dynamics of Florida's anchialine and anoxic sinkholes and caves.

## 8.9 Summary and Conclusions

While the forms of karst observed in Florida are diverse and, in many ways, parallel karst terrains throughout the world, the geological youth of Florida karst has limited development of many landforms. Burial by sand and clay-rich Miocene and Plio-Pleistocene sediments has masked much of the karst at the upper limestone surface. With the exception of caves and sinkholes, this chapter has discussed the epigene and hypogene landforms developed in Florida karst. Caves and sinkholes are discussed in Chap. 7.

- Common epigenetic land forms in Florida include the following:
  - Cone-shaped hills, which are locally well developed in the Brooksville Ridge.
  - Arches, which range from short spans between adjacent rock-collapse sinkholes along the axes of caves to reaches of a kilometer or more along the paths of major subterranean streams.
- Arches are found near several of Florida's springs and scattered along swallet-to-spring cave systems.
- Arches that bridge long (100s of meters) reaches of a swallet-to-spring cave are termed natural bridges in Florida.
  - Natural bridges form where organic-rich surface water flows into a swallet and then through a well-developed cave to a spring.
  - Natural bridges are often accompanied by traces of the former pathway of the surface-water drainage on the land surface.
  - The land bridges show a succession of development from simple swallet-to-spring pathways with no intervening collapse features, through development of multiple rock-collapse sinkholes (karst windows) and ending with linear valleys that have completely collapsed.
    - The Alapaha River is an example of a long natural bridge with an overflow stream channel. There are no known rock-collapse sinkholes in the reach from the swallet complex to its resurgence.
    - The natural bridge on the Santa Fe River goes into a swallet (River Sink) 4.6 linear kilometers upstream of its resurgence (River Rise). Several rock-collapse sinkholes have developed along the path of the cave that carries the underground Santa Fe River. When the river floods, an overland trace carries the flow.
    - The Aucilla River, which was discussed in Chap. 7, also goes underground through a swallet approximately 7.2 linear kilometers above its resurgence (Nutall Rise). The cave that carries the principal flow has been subjected to numerous rock-collapse

sinkholes and, today, the karst windows are arrayed like a "sting of pearls" along the trace of the river.
    - Finally, the cave system may completely collapse resulting in a dry valley or slough, such as Tenmile Hollow (Chap. 7).

- Sandhill lakes are commonly found on ridges where there is a thick cover of well-drained sand.
  - Sandhill lakes formed as cover-collapse sinkholes.
  - Because their primary source of water is rainfall and surficial aquifer flow, they are typically oligotrophic and have low turbidity. Also, water levels in the lakes are highly sensitive to rainfall.
  - Because their oligotrophic status, plant and animal communities are limited.
  - Seepage faces on the margins of the lakes may contain unusual, and often threatened, biota.
  - The sand from the surrounding uplands covers the bottoms of the lakes as a result of creep on slopes above lake level and, perhaps, slope failures underwater.
  - If the lakes are connected with the UFA, as is commonly the case if the sand has not plugged the avens of the sinkholes within the lake bottoms, water levels in the lakes fluctuate with potentials in the UFA, rendering them sensitive to groundwater withdawals.
- Fluviokarst is well developed on, and near, karst escarpments.
  - As streams exit uplands where recharge to underlying aquifers is limited because of the low hydraulic conductivities of strata that underlie the uplands, streams enter swallets.
    - In a topographic escarpment, headward erosion of the low permeability strata results in development of swallets upon first encountering soluble, carbonate rocks.
    - Headward erosion continues and, at some point, a new swallet develops and the older one is abandoned.
    - This sequence progresses resulting in a trace of the original stream that is dotted by karst windows that formed earlier as swallets. The trace is occasionally flooded during high rainfall events and after the swallets within the trace can no longer accept flood waters.
- Siphons are underwater swallets that partially capture stream flow on the toe of the escarpment.
- As headward erosion continues, the stream may be truncated, with swallets on its former tributaries.
- Blind valleys develop if, after truncation, the river gradient is reversed by either dissolution of carbonate rock or during flood episodes.
- Hard-rock phosphate deposits are found in relict sinkholes in the eastern flank of the Brooksville Ridge and on the Cody Scarp.
  - These are evidence of scarp retreat and precipitation carbonate hydroxylapatite from phosphate-rich, mildly acidic water as it encountered Ca-$HCO_3$ water and limestone.
  - Fossils found in the sinkhole fill are predominantly late Miocene or early Pliocene, which dates a major episode of scarp retreat to the Mio-Pliocene weathering episode.
- There are also hydrologic escarpments in Florida where surface water flows from lower permeability strata onto higher permeability, more soluble strata with little or no topographic expression. The low permeability strata can be clay, as in the San Pedro Bay example, or from one limestone to another, as is the Green Swamp example.
- Karren development on the upper surfaces of limestone varies in Florida.
  - Because of the presence of (1) low dips of the upper surfaces of limestone, (2) double porosity in some limestones, and (3) extensive, clay-rich cover, karren forms created by chemically aggressive water flowing over rock surfaces are rare.
  - Solution flutes and solution basins are rarely seen.
  - Grikes are commonly encountered in quarries and locations where bedrock is exposed, but they are seldom identified in the subsurface where cover materials mask the limestone surface.
- Vertical shafts (solution pits or channels) are common in the Pleistocene limestones of southern Florida and in certain Eocene limestones in northern and central Florida.
- Duricrusts developed on the upper most limestones are common in southern Florida and rarely in central and northern Florida.
  - Duricrusts in southern Florida are typically composed of laminated, calcite cemented sand, calcitic debris, marine sediments, limestone breccia, and *terra rossa* soils.
  - In northern and central Florida, they are often thin crusts of recrystallized calcite. Lamination is rare and incorporation of soil materials unusual.
  - Duricrusts are commonly found in the Oligocene and Eocene strata at or near unconformities in the subsurface.
  - Duricrusts form by repeated wetting and drying, which is typical of subtropical environments.
  - They are seldom over 10 cm in thickness.
- Caprock is a relatively thick (greater than 10 cm) densely recrystallized limestone or calcite-cemented sand and clayey sand.
  - Like duricrusts, caprock appears to have been formed by repeated wetting and drying.
  - Caprock in southern Florida is associated with limestone breccia, which the dense limestone often incorporates and overlies.

  - Where the caprock is calcite-cemented sand or clayey sand, clasts of calcite and shell may be present and the caprock overlies calcitic siliciclastic deposits.
- While caliche is usually associated with arid environments, it is present in Florida as vadose pisolites and rounded and recemented limestone rubble.
- Phytokarst developed on rocky shores by dissolution and grazing of algae and fungi by herbivores is found on the rare coastal exposures of limestone in southeastern Florida.
- Microkarst development is well developed in the Rock Glades and Everglades provinces where limestone is subaerially exposed.
- Documentation of possible hypogene karst in Florida includes evidence from equilibrium chemistry, possible caves and high permeability zones in Florida's Tertiary aquifers, and older sequences in the Cretaceous oil fields of southwestern Florida.
  - Speleogenesis as a result of groundwater mixing and generation of sulfuric acid is likely in Florida.
  - As noted in Chap. 5, calculation of chemical equilibration states for mixtures of fresh and saline groundwater from the FAS suggests that mixtures within the freshwater/saltwater transition are undersaturated with respect to calcite and are forming caves.
  - A likely candidate for development of caves that have developed as a result of mixing-zone speleogenesis can be found in the unusually deep, phreatic caves near the Gulf of Mexico coastline.
  - Mixing-zone speleogenesis may also be responsible for enhanced porosity and permeability in the boulder zones of the LFA. The LFA is also a location where generation of sulfuric acid derived from the gypsum and/or anhydrite has possibly enhanced porosity of the aquifer.
  - Breccias and unusual porosity in Cretaceous limestone of the Sunniland Trend of oil fields in southwestern Florida may reflect relict porosity developed by sulfuric acid derived from anhydrite. The depth of the Sunniland Formation below the land surface suggests that the hypogene karst, if present, originated below subsidence.
  - Deep, rock-collapse sinkholes that have been documented on the Atlantic continental shelf and in the interior of Florida appear to have originated by collapse into void spaces deep in the FAS as a result of hypogene speleogenesis.
  - Springs on the deep continental slope of the Gulf of Mexico and Atlantic offshore are probable indicators of sulfuric-acid derived speleogenesis in the Cretaceous strata of southern Florida.

## References


Alfieri MC Upchurch SB Dobecki TL (2018). Photolinears, fractures, and fallacies: A *post hoc* study of photolineaments, Hillsborough County, Florida. Proceedings, 15th Multidisc Conf on Sinkholes and the Eng and Environ Impacts of Karst, Shepherdstown, WV, Carlsbad, NM, Nat Cave Karst Res Inst, pp. 77–88

Applegate AV Pontigo FA (1984) Stratigraphy and oil potential of the Lower Cretaceous Sunniland Formation on south Florida. Fla Geol Surv Rept Invest 89

Back W Hanshaw BB (1970) Comparison of chemical hydrogeology of the carbonate peninsulas of Florida and Yucatan. Jour Hydrol 10:330–368

Back W Hanshaw BB Herman JS Van Driel JN (1986) Differential dissolution of a Pleistocene reef in the ground-water mixing zone of coastal Yucatan, Mexico. Geology 14:137–140

Bishop RE Humphreys WF Cukov N Žic V Boxshall GA Cukov M Iliffe TM Kršinić F Moore WS Pohlman JW Sket B (2015) "Anchialine" redefined as a subterranean estuary in a crevicular or cavernous geological setting. Jour Crustacean Biol 35(4):511–514

Bögli A (1964) Érosion par Melange des Eaux. Internat Jour Speleology 1:61–70. (Translated as Mixed-water corrosion in Sweeting MM (1981) Karst geomorphology, Benchmark Papers in Geology 59:90–99)

Bögli A (1980) Karst hydrology and physical speleology. Berlin, Springer-Verlag

Bonn MA Bell FW (2003) Economic impact of selected Florida springs on surrounding local areas. Tallahassee, Fla State Univ

Breland JA (1980) Chemical and physical characteristics of a saline geothermal submarine spring off Florida's southwestern coast. Master's thesis, St. Petersburg, Univ South Fla

Brooks HK (1961) The submarine springs off Crescent Beach, Florida. Fla Scien 24:122–134

Brown A Young C Martin JB (2016) Groundwater-surface water interactions in the Suwannee River basin. Fla Scien 79(4):220–238

Butt PL (2005) Dye trace: Potential infrastructure impact to Ichetucknee Springs State Park. High Springs, FL, Karst Environmental Services

Butt PL Murphy GJ (2003) Dyal and Black Sinks Dye Trace, Columbia County, Florida, May-September, 2003. High Springs, FL, Karst Environmental Services, Inc., available at http://www.dep.state.fl.us/springs/reports/index.htm

Butt PL Boyes S Morris TL (2006) Mill Creek and Lee Sinks Dye Trace, Alachua County, Florida, July-December, 2005. High Springs, FL, Karst Environmental Services, Inc.

Butt PL Morris TL Skiles WC (2007) Swallet/resurgence relationships on the lower Santa Fe River, Florida. Contract report prepared by Karst Environmental Services for Water Resource Associates, Tampa, Florida

Cander HS (1994) An example of mixing zone dolomite, Middle Eocene Avon Park Formation, Floridan aquifer system. Jour Sed Res A64:615–629

Carr WJ Alverson DC (1959) Stratigraphy of Middle Tertiary rocks in part of west-central Florida. US Geol Surv Bull 1092

Ceryak R (1977) Hydrogeology of a river basin in karst terrain, Alapaha River, Hamilton County, Florida. Suwannee River Water Mgt Dist Info Circ 5

Ceryak R (2005) Significance of the Cody Scarp on the hydrogeology of the Suwannee River Water Management District. In: Copeland R (comp), Geomorphic influence of scarps in the Suwannee Basin, Southeastern Geol Soc Field Trip Guidebook 44, p. 37

Champion KM Starks R (2001) The hydrology and water quality of springs in west-central Florida. Tampa, Southwest Fla Water Mgt Dist

Champion KM Upchurch SB (2003) Delineation of spring-water source areas in the Ichetucknee Springshed. Tallahassee, Fla Dept Environ Prot, Div State Lands

Chanton JP Martens CS Paull CK Coston JA (1993) Sulfur isotope and porewater geochemistry of Florida escarpment seep sediments. Geochim Cosmochim Acta 57(6):1253–1266

Chelette A Thurman D (2006) Report of progress on locating, cataloging and describing Florida's swallets. Fla Geol Surv

Clausen CS Brooks HK Wesolowsky AB (1975) Florida spring confirmed as 10,000 year old early man site. Gainesville, Fla Anthropol Soc, Pub 7, Fla Anthropologist 28(3, part 2):1–38

Clausen CJ Cohen AD Emiliani C Holman JA Stipp JJ (1979) Little Salt Spring, Florida; a unique underwater site. Science 203: 609–614

Clesi D King J (1991) Team Diepolder '91: Our story. Underwater Speleol 18(3):7–12

Cockrell WA (1973) Remains of early man recovered from spring cave. Archives and History News, vol 4, no 2

Cockrell WA (1975) Warm Mineral Springs, 1975: A multi-disciplinary approach to a 10,000 BP archeological site (abs.). Fla Scien 38, Sup 1, p 19

Commeau RF Paull JA Commeau JA Poppe LJ (1987) Chemistry and mineralogy of pyrite-enriched sediments at a passive margin sulfide brine seep: abyssal Gulf of Mexico. Earth and Planet Lett 82(1–2):62–74

Conover CS Leach SD (1975) River basin and hydrologic unit map of Florida. Fla Geol Surv, Map Ser 72

Cooke CW (1939) Scenery in Florida Interpreted by a Geologist. Fla Geol Surv, Bull 17

Copeland R. (comp.) (2003) Florida spring classification system and spring glossary. Fla Geol Surv, Sp Pub 52

Copeland R (2005a) An overview of the influence of scarps on a variety of topics within the Suwannee River Basin of Florida. In: Copeland R (comp.), Geomorphic influence of scarps in the Suwannee Basin, Southeastern Geol Soc Field Trip Guidebook 44, pp. 1–35

Copeland R (comp.) (2005b) Geomorphic influence of scarps in the Suwannee River Basin. Tallahassee, Southeastern Geol Soc Field Trip Guidebook 44

Copeland R Dorian NA White AJ Upchurch SB (2011) Regional and statewide trends in Florida's spring and well groundwater quality (1991–2003). Fla Geol Surv Bull 69 (Revised)

Courbon P Chabert C Bosted P Lindsley K (2005) Atlas of great caves of the world. Dayton, OH, Cave Books

Crawford NC (1996) The karst hydrogeology of the Cumberland Plateau Escarpment of Tennessee. Tenn Div Geol Rept Invest 44, Part IV

Cunningham KJ (2013) Marine seismic-reflection data from the southeastern Florida Platform: A case for hypogenic karst (abs.). Int Assoc Hydrogeol, Proc Int Sym on Hierarchical Flow Systems in Karst Regions, p. 59

Cunningham KJ (2015) Seismic-sequence stratigraphy and geologic structure of the Floridan aquifer system near "boulder zone" deep wells in Miami-Dade County, Florida. US Geol Surv Sci Invest Rept 2015-5013

Cunningham KJ Kluesner JW Westcott RL Robinson E Walker C Khan SA (2017) Sequence stratigraphy, seismic stratigraphy, and seismic structures of the lower intermediate confining unit and most of the Floridan aquifer system, Broward County, Florida. US Geol Surv Sci Invest Rept 2017-5109

Cunningham KJ Walker C (2009) Seismic-sag structural systems in Tertiary carbonate rocks beneath southeastern Florida, USA: Evidence for hypogenic speleogenesis? In: Klimchouk A Ford D (eds.), Hypogene speleogenesis and karst hydrogeology of artesian basins, Ukrainian Inst Speleol and Karst Sp Pap 1, pp. 151–158

Davis H Verdi R (2014) Groundwater flow cycling between a submarine spring and an inland fresh water spring. Ground Water,52(5):705–716

Dean RW (1999) Surface and ground water mixing in a karst aquifer: An example from the Floridan aquifer. Master's thesis, Gainesville, Univ Fla

Dobecki TL Upchurch SB Scott TM Fratesi B Alfieri MC (2013) Subbottom profiling investigation of sinkhole lake structure in Bay and Washington counties, Florida. Proc Thirteenth Multidisc Conf Sinkholes and the Engin and Environ Impacts of Karst, Carlsbad, NM, Nat Cave and Karst Res Inst, pp. 207–211

Duever MJ Carlson JE Meeder JF Duever LC Gunderson LH Riopelle LA Alexander TR Myders RL Spangler DP (1985) The Big Cypress National Preserve. Nat Audubon Soc, Res Rept 8, Chap 3, pp.13–35

Fanning KA Betzer PB Byrne RH Breland JA Jolley RR (1979) Characteristics of a submarine geothermal spring on the west Florida shelf. Fla Scien 42 Sup 1, p. 21

Fanning KA Byrne RH Breland JA Betzer PB Moore WS Elsinger RJ Pyle TE (1981) Geothermal springs of the west Florida continental shelf: Evidence for dolomitization and radionuclide enrichment. Earth and Planet Lett 52:345–354

Ferguson GE Lingham CW Love SK Vernon RO (1947) Springs of Florida. Fla Geol Surv Bull 31

Florea LJ Vacher HL (2006a) Cave levels, marine terraces, paleoshorelines, and the water table in peninsular Florida. In: Onac BP Tamas T Constantin S Persoiu A (eds.), Archives of climate change, Karst Waters Inst Sp Pub 10, pp. 188–192

Florea LJ Vacher HL (2006b) Morphologic features of conduits and aquifer response in the unconfined Floridan aquifer system, west central Florida. 12th Symp Geol Bahamas and other Carbonate Regions, pp. 32–44

Florea LJ Vacher HL Oches EA (2003) Karst studies in west central Florida: USF seminar in karst environments. Brooksville, Southwest Fla Water Mgt Dist, accessed 9/4/2013 at http://uweb.cas.usf.edu/geokarst/Publications/SWFWMD2003/intro.pdf

Florea LJ Vacher HL Donahue B Naar D (2007) Quaternary cave levels in peninsular Florida. Quat Sci Rev 26:1344–1363

Folk RL Roberts HH Moore CH (1973) Black phytokarst from Hell, Cayman Islands, British West Indies. Geol Soc Amer Bull, 84:2351–2360

Follman J Buchanan R (2015) Springs fever: A field & recreation guide to 500 Florida springs. 2nd ed, accessed 11/16/2015 at http://springsfever.org/Springbook/index.html

Ford DC Williams PW (1989) Karst geomorphology and hydrology. New York, Chapman and Hall

Garman KM (1999) New tunnel discoveries in the Crystal Beach Spring cave system. Nat Speleol Soc News, Jun, 1999, pp. 164–168

Garman KM (2010) The biogeochemistry of submerged coastal karst features in west central Florida. Ph.D. dissert, Tampa, Univ South Fla

Garman KM Garey JR (2005) The transition of a freshwater karst aquifer to an anoxic marine system. Estuaries 28(5):686–693

Garman KM Rubelmann H Karlen DJ Garey JR (2011) Comparison of an inactive submarine spring with an active nearshore anchialine spring in Florida. Hydrobiologia, 677:67–87

Gaswirth SB Budd DA Farmer LG (2007) The role and impact of regionally extensive freshwater-seawater mixing zones in the maturation of regional dolomite bodies within the proto-Floridan aquifer. Sedimentology, 54:1065–1092

Geologic Resources Division (2008) Big Cypress National Preserve: Geologic resource evaluation report. U.S. Dept Interior, Nat Resour Rpt NPS/NRPC/GRD/NRR-2008/021

Gore PJ Witherspoon W (2013) Roadside geology of Georgia. Missoula, MT, Mountain Press

Grubbs JW (1995) Evaluations of ground-water flow and hydrologic budget for Lake Five-O, a seepage lake in northwestern Florida. US Geol Surv Water-Resour Invest Rept 94-4145

Grubbs JW (2011) Analysis of long-term trends in flow from a large spring complex in northern Florida. In: Kuniansky EL (ed.), US

Geol Surv Karst Interest Grp Proc, Fayetteville, AR, US Geol Surv Sci Invest Rept 2011-5031, pp. 160–167

Gulley J Martin JB Spellman P Moore PJ Screaton EJ (2012) Influence of partial confinement and Holocene river formation on groundwater flow and dissolution in the Florida carbonate platform. Hydrol Proc, 28:705–717

Haber JD (2003) Priority pollutant screening at select springs within the Southwest Florida Water Management District. Brooksville, Southwest Fla Water Mgt Dist

Herbert TA Upchurch SB (2016) The potential role of hypogene speleogenesis in the lower Floridan aquifer and Sunniland Oil Trend, south Florida, U.S.A. In: Chavez T Reehling P (eds.), Proc DeepKarst 2016: Origins, resources, and management of hypogene karst, Carlsbad, NM, Nat Cave and Karst Res Inst, Symp 6, pp. 119–129

Hill CA (1990) Sulfuric acid speleogenesis of Carlsbad Cavern and its relationship to hydrocarbons, Delaware Basin, New Mexico and Texas. Bull Amer Assoc Petrol Geol, 74(11):1685–1694

Hine AC (1997) Structural and paleoceanographic evolution of the margins of the Florida Platform. In: Randazzo AF Jones DS (eds.). The geology of Florida. Gainesville, Univ Press Fla, Ch 11, pp 169–216

Hisert RA (1994) A multiple tracer approach to determine the ground and surface water relationships in the western Santa Fe River, Columbia County, Florida. Ph.D. Dissertation, Gainesville, Univ Fla

Hollis T (2006) Glass bottom boats and mermaid tails: Florida's tourist springs. Mechanicsburg, PA, Stackpole Books

Hornsby D Ceryak R (1998) Springs of the Suwannee River basin in Florida. Suwannee River Water Mgt Dist, WR99-02

Hornsby DH (2007) Influences on the distribution and occurrence of nitrate-nitrogen and total phosphorus in the water resources of the Suwannee River Water Management District. Ph.D. Dissertation, Gainesville, Univ Fla

Horton RE (1936) Erosional development of streams and their drainage basins: Hydrophysical approach to quantitative morphology. Geol Soc Amer Bull 56:275–370

Huddlestun PF Hunter ME Scott T Chowns TM Lloyd JM (1988) Southern Florida, CSD 140, Col. 31. In: Lindberg FA, editor. Gulf Coast Region, Correlation of Stratigraphic Units of North America (COSUNA) Project. Tulsa, OK, Amer Assoc Petrol Geol

Hunn JD Cherry RN (1969) Remote sensing of offshore springs and spring discharge along the Gulf Coast of central Florida. In: Nat Aeron Space Admin, 2nd Ann Earth Resour Aircraft Program Status Review, Hydrology and Oceanography, 3:39–1 – 39–7

Jennings JN (1985) Karst geomorphology. Oxford, Blackwell Publishers, 2nd ed

Jones GW Upchurch SB (1998) Increasing nitrate levels in the major coastal springs in Citrus and Hernando Counties, Florida. In: Darling SH Harper HH (eds.), Integrating Water Resources and Growth into the 21st Century, Proc Seventh Ann Southeast Lakes Mgt Conf, Fla Lake Mgt Soc, pp. A121–A122

Jones GW Upchurch SB Champion KM (1996) Origin of nutrients in ground-water discharging from Rainbow Springs. Brooksville, Southwest Fla Water Mgt Dist

Jones WK (2004) Introduction to epikarst. In: Jones WK Culver DC Herman JS (eds.), Epikarst. Karst Waters Inst Sp Pub 9, pp. 3–10

Jones WK Culver DC Herman JS (eds.) (2004) Epikarst. Karst Waters Inst Sp Pub 9

Jordan GF (1954) Large sinkholes in Straits of Florida. Amer Assoc Petrol Geol Bull 38:1810–1817

Jordan GF Stewart H Jr (1961) Bathymetry of western Straits of Florida. Geol Soc Amer Bull 72:1051–1058

Jordan GF Malloy RJ Kofoed JW (1964). Bathymetry and geology of the Pourtales Terrace, Florida. Marine Geology 1:259–287

Karst Environmental Services (1997) Ichetucknee Springs Water Quality Working Group Cooperative Dye Trace: Rose Creek Swallet-Ichetucknee Springs Group. High Springs, FL, Karst Environ Serv

Katz BG (1992) Hydrochemistry of the Upper Floridan Aquifer, Florida. US Geol Surv Water-Resour Invest Rept 91-4196

Katz BG Hornsby HD (1998) A preliminary assessment of sources of nitrate in springwaters, Suwannee River Basin, Florida. US Geol Surv Open File Rept 98-0069

Kaufman MI (1969) Generalized distribution and concentration of orthophosphate in Florida streams. Fla Geol Surv Map Series 33

Kincaid TR (1998) River water intrusion to the unconfined Floridan aquifer. Environ Eng Geosci IV(3):361–374

Klimchouk A (2004) Towards defining, delimiting and classifying epikarst: Its origin, processes and variants of geomorphic evolution. Speleogen Evol Karst Aquifers 2(1):1–13

Klimchouk A (2009a) Principal characteristics of hypogene speleogenesis. In: Stafford KW Land L Veni G (eds.), NCKRI Symp 1 Adv Hypogene Karst Stud, p. 1–11

Klimchouk A (2014) The methodological strength of the hydrogeological approach to distinguishing hypogene speleogenesis. In: Klimchouk A Sasowsky ID Mylroie J Engel SA Engel AS (eds.), Hypogene Cave Morphologies. Leesburg VA, Karst Waters Inst Sp Pub 18, pp. 4–12

Klimchouk A (2007) Hypogene speleogenesis: Hydrogeological and morphogenetic perspective. Carlsbad, New Mexico., Nat Cave Karst Res Inst.

Klimchouk A (2009b) Principal characteristics of hypogene speleogenesis. In: Stafford KW Land L Veni G (eds.), Carlsbad, NM, Nat Cave Karst Res Inst Symp 1, Advances in Hypogene Karst Studies, p. 1–11.

Klimchouk AB Palmer AN Waela JD Auler AS Audra P (eds.) (2017) Hypogene karst regions and caves of the world, Springer Science Publishers

Kohout FA (1964) The flow of fresh water and salt water in the Biscayne aquifer of the Miami area, Florida. US Geol Surv Water-Sup Pap 1613–C, pp. 12–32

Kohout FA (1966) Submarine springs, a neglected phenomenon of coastal hydrology. Central Treaty Org Symp Hydrol and Water Resour Devel, pp. 391-41.

Kohout FA (1974a) Geothermal spring off Florida's west coast. Geol Soc Amer Abstracts with Progs, 6:829

Kohout FA (1974b) Hydrologic significance of a geothermal submarine spring off Florida west coast. Fla Scient 37(Supplement 1):13

Kohout FA Munson RC Turner RM Royal WR (1979) Satellite observations of a geothermal submarine spring off Florida west coast. In: Deutch M Wiesnet DR Rango A (eds.), Satellite Hydrology, Proc Fifth Ann William T. Pecora Mem Symp Remote Sensing, Amer Water Resour Assoc, pp. 570–578

Kohout FA Dill RF Royal WR Benjamin GJ Hill RE (1972) Ocean-ground-water interface; sinkholes, blue holes, and submarine springs - geologic-hydrologic windows in the Florida-Bahama platform. Geol Soc Amer Abstracts with Progs, 4(7):566

Kruse SE Schneider JC Campagna DJ Inman JA Hickey TD (2000) Ground penetrating radar imaging of cap rock, caliche and carbonate strata. Jour Appl Geophy 43:239–249

Land LA Paull CK (2000) Submarine karst belt rimming the continental slope in the Straits of Florida. Geo-Marine Lett 20:123–132

Land LA Paull CK Hobson B (1995) Genesis of a submarine sinkhole without subaerial exposure: Straits of Florida. Geology 23:949–951

Lane E (2001) The Spring Creek submarine springs group, Wakulla County, Florida. Fla Geol Surv Sp Pub 47

Langevin CD (2003) Simulation of submarine ground water discharge to a marine estuary: Biscayne Bay, Florida. Ground Water 41(6): 758–771

Lawn AM (1989) Sedimentology of a low-sinuosity meander within the Alapaha River, Hamilton County, Florida. Master's thesis, Tampa, Univ South Fla

Lawn A (compiler) (2014) Karst hydrogeology of the upper Suwannee River basin, Alapaha River area, Hamilton County, Florida. Southeastern Geol Soc Guidebook 63

Lawrence FW Upchurch SB (1976) Identification of geochemical patterns in ground water by numerical analysis. In: Zaleem EA (ed.), Advances in groundwater hydrology, Amer Water Resour Assoc, pp. 199–214

Lawrence FW Upchurch SB (1982) Identification of recharge areas using geochemical factor analysis. Ground Water 20:680–687

Lee TM (1996) Hydrogeologic controls on the groundwater interactions with an acidic lake in karst terrain, Lake Barco, Florida. Water Resour Res, 32:831–844

MacFadden BJ (1997) Fossil mammals of Florida. In: Randazzo AF Smith DL (eds.), Geology of Florida, Gainesville, Univ Press Fla, Ch 8, pp. 119–137

Maddox GL (1992) Data collection and management methods. In: Maddox GL Lloyd JM Scott TM Upchurch SB Copeland R (eds.), Florida ground water quality monitoring program -- Vol 2, Background hydrogeochemistry, Fla Geol Surv Sp Pub 34, Ch. II, pp. 3–5

Martin JB Screaton EJ (2000) Exchange of matrix and conduit water with examples from the Floridan aquifer. US Geol Surv Water-Resour Invest Rept 01-4011, pp. 38–44

Martin JB Dean RW (1999) Temperature as a natural tracer of short residence times for ground water in karst aquifers. In: Palmer AN Palmer MV Sasowsky ID (eds.), Karst modeling. Karst Waters Inst Sp Pub 5, pp. 236–242

Martin JB Dean RW (2001) Exchange of water between conduits and matrix in the Floridan aquifer. Chem Geol 179:145–165

McConnell JB Hacke CM (1993) Hydrogeology, water quality, and water-resources development potential of the upper Floridan aquifer in the Valdosta area, south-central Georgia. US Geol Surv Water-Resour Invest Rept 93-4044

McFadden M (1982) Petrology of porcellanites in the Hawthorn Formation, Hamilton County, Florida. Master's thesis, Tampa, Univ South Fla

Means GH (2005) The Florida Geological Survey swallet mapping project. In: Copeland R (comp.), Geomorphic influence of scarps in the Suwannee basin, Southeastern Geol Soc Field Trip Guidebook 44, pp. 44–46

Meinzer OE (1927) Large springs in the United States: US Geol Surv Water Sup Pap 557

Melim LA Swart PK Eberli GB (2004) Mixing-zone diagenesis in the subsurface of Florida and the Bahamas. Jour Sed Res 74(6):904–913

Menning DM Wynn JC Garey JR (2015) Karst estuaries are governed by interactions between inland hydrological conditions and sea level. Jour Hydrol 527:718–733

Miller JA (1978) Geologic and geophysical data from Osceola National Forest, Florida. US Geol Surv Open-File Rept 78-799

Mitchell-Tapping HJ (1987) Application of the tidal mudflat model to the Sunniland Formation of south Florida. Trans Gulf Coast Assoc Geol Soc 37:415–426

Mitchell-Tapping HJ (2002) Exploration analysis of basin maturity in the South Florida sub-basin. Trans Gulf Coast Assoc Geol Soc 52:753–764

Mitchell-Tapping HJ Mitchell-Tapping AM (2003) Exploration of the Sunniland Formation of southern Florida. Trans Gulf Coast Assoc Geol Soc 53:599–609

Multer HG Hoffmeister JE (1968) Subaerial laminated crusts of the Florida Keys. Geol Soc Amer Bull 79: 183–192

Mylroie JE Carew JL (1990) The flank margin model for dissolution cave development in carbonate platforms. Earth Surf Processes and Landforms 15:413–424

Mylroie JE Jenson JW Mylroie JR Carew JL (2004) Epikarst in eogenetic rocks. In: Jones WK Culver DC Herman JS (eds.), Epikarst. Charles Town, WV, Karst Waters Inst, Sp Pub 9, pp. 132–134

Neill L Anderson J (2009) Cave divers explore deepest parts of Weeki Wachee Springs. Tampa Bay Times, April 19, 2009 (http://www.tampabay.com/news/humaninterest/cave-divers-explore-deepest-parts-of-weeki-wachee springs)

Odum HT (1953) Dissolved phosphorus in Florida waters. In: Miscellaneous Studies, Fla Geol Surv, Rept Invest 9, Part 1, pp. 1–41

Palacas JG (1978a) Preliminary assessment of organic carbon content and petroleum source rock potential of Cretaceous and lower Tertiary carbonates, South Florida Basin. Trans Gulf Coast Assoc Geol Soc 28:357–381

Palacas JG (1978b) Distribution of organic carbon and petroleum source rock potential of Cretaceous and lower Tertiary carbonates, South Florida Basin. US Geol Surv Open-File Rept 78-140

Palacas JG Anders DE King JD (1984a) South Florida Basin – prime example of carbonate source rocks of petroleum. In: Palacas JG (ed.), Petroleum geochemistry and source rock potential of carbonate rocks, Tulsa, OK, Amer Assoc Petrol Geol Studies in Geology 18:71–96

Palacas JG Daws TA Applegate AV (1981) Preliminary petroleum source rock assessment of pre-Punta Gorda rocks (lowermost Cretaceous – Jurassic?) in south Florida. Trans Gulf Coast Assoc Geol Soc 31:369–376

Palacas JG King JD Claypool GE (1984) Origin of asphalt and adjacent oil stains in lower Cretaceous fractured limestones, Deep Sea Drilling Project Leg 77. In: Buffler RT Schlager W and others (eds.), Initial Repts Deep Sea Drilling Proj 77:477–488

Parker GG Ferguson GE Love SK, and others (1955) Water resources of southeastern Florida, US Geol Surv Water Supply Pap 1255

Parkhurst DL Thorstenson DC Plummer LN (1980) PHREEQE – A computer program for geochemical calculations. US Geol Surv Water-Resour Invest Rept 80-96

Paull CK Neumann AC (1987) Continental margin brine seeps: Their geological consequences. Geology 15:545–548

Paull CK Dillon WP (1980) Erosional origin of the Blake Escarpment: An alternative hypothesis. Geology 8:538–542

Paull CK Spiess FN Curray JR Twichell DC (1990) Origin of Florida Canyon and the role of spring sapping on the formation of submarine box canyons. Geol Soc Amer Bull 102:502–515

Peek HM (1951) Cessation of flow of Kissengen Springs in Polk County, Florida. Fla Geol Surv, Rept Invest 7, pp. 73–82

Perkins RD (1977) Part II, Depositional framework of Pleistocene rocks in south Florida. In: Enos P Perkins RD (eds.), Quaternary Sedimentation in South Florida, Geol Soc Amer, Mem 147, pp. 131–198

Pitkin A (2016) Andrew Pitkin's Videos. Accessed 6/6/2016 at https://vimeo.com/apitkin/videos

Plummer LN (1975) Mixing of sea water with calcium carbonate ground water. In: Whitten EHT (ed.), Quantitative studies in geological sciences, Geol Soc Amer Mem 142, pp. 219–236

Polk J Brinkman R (2013) Climatic influences on coastal cave and karst development in Florida. In: Lace MJ Mylroie JE (eds.), Coastal karst landforms, Dordrecht, Springer Science, Coastal Research Library, Book 5, Chap 15, pp. 317–345

Popenoe P Kohout FA Manheim FT (1984) Seismic-reflection studies of sinkhole and limestone dissolution features on the northeastern Florida shelf. In: Beck BF (ed.), Sinkholes: Their geology, engineering and environmental impact, Proc First Multidisc Conf on Sinkholes, Rotterdam, A.A. Balkema, pp. 43–57

Poucher M (2008) Diepolder Cave System. In: Florea LJ (ed.), Caves and karst of Florida: A guidebook for the 2008 National Convention of the National Speleological Society. Huntsville, AL, Nat Speleol Soc, pp. 205–209

Poucher S Copeland R (comp.) (2006) Speleological and karst glossary of Florida and the Caribbean. Gainesville, Univ Press Fla

Prospero JM (1999) Long-range transport of mineral dust in the global atmosphere: Impact of African dust on the environment of the southeastern United States. Proc Nat Acad Sci, USA, 96:3396–3403

Prospero JM Landing WM Schulz M (2010) African dust deposition to Florida: Temporal and spatial variability and comparisons to models, Jour Geophy Res 115, D13304, doi:https://doi.org/10.1029/2009JD012773

Prospero JM Nees RT Uematsu M (1987) Deposition rate of particulate and dissolved aluminum derived from Saharan dust in precipitation at Miami, Florida. Jour Geophy Res 92:14,723–14,731

Prospero JM Olmez I Ames M (2001) Al and Fe in PM 2.5 and PM 10 suspended particles in south-central Florida: The impact of the long range transport of African mineral dust, Water, Air, and Soil Poll 125:291–317

Puri HS (1957) Stratigraphy and zonation of the Ocala Group. Fla Geol Surv Bull 38

Puri HS Vernon RO (1964) Summary of the geology of Florida and a guidebook to the classic exposures. Fla Geol Surv Sp Pub 5 (revised)

Randazzo AF (1997) The sedimentary platform of Florida: Mesozoic to Cenozoic. In: Randazzo AF Jones DS (eds.), The geology of Florida, Gainesville, Univ Press Fla, pp. 39–56

Reich CD Swarzenski PW Kindinger JD Flocks JG Hickey TD Spechler RM (2001) Direct linkages between onshore karst aquifers and offshore marine environments: Crescent Beach Spring, Florida. In: Kuniansky EL (ed.), US Geol Surv Karst Interest Grp Proc, Water-Resour Invest Rept 01-4011, p. 106

Rezaei M Sanz E Raeisi E Ayora C Vázquez-Sune E Carrera J (2005) Reactive transport modeling of calcite dissolution in the fresh–salt water mixing zone. Jour Hydrol 311(1–4): 282–298

Rightmire CT Pearson FJ Jr. Back W Rye RO Hanshaw BB (1974) Distribution of sulfur isotopes in ground waters from the principal artesian aquifer of Florida and the Edwards aquifer of Texas, United States of America. Isotope Tech Groundwater Hydrol, v. 2, Vienna, Int Atomic Energy Agency, pp. 191–207

Robb JM (1990) Groundwater processes in the submarine environment. In: Higgins CG Coates DR (eds.), Groundwater geomorphology: The role of subsurface water in earth-surface processes and landforms, Geol Soc Amer Sp Pap 252, Ch 12, pp. 267–281

Robbin DM Stipp JJ (1979) Depositional rate of laminated soilstone crust, Florida Keys. Jour Sed Pet 49:175–180

Rosenau JC Faulkner GL Hendry CW Jr. Hull RW (1977) Springs of Florida. Fla Bur Geol Bull 31 (Revised)

Rubelmann H Garey JR (2016) Biotic changes in a deep sulfidic offshore sinkhole (abs.). In: Chavez T Reehling P (eds.), Proc DeepKarst 2016: Origins, resources, and management of hypogene karst, Carlsbad, NM, Nat Cave and Karst Res Inst, Symp 6, p. 159

Runnels DD (1969) Diagenesis, chemical sediments, and the mixing of natural waters. Jour Sed Pet 39:1188–1201

Rupert FR Lane E (1992) The geology of Falling Waters State Recreation Area. Fla Geol Surv Leaflet 16

Rye RO Back W Hanshaw BB Rightmire CT Pearson FJ Jr (1981) The origin and isotopic composition of dissolved sulfide in groundwater from carbonate aquifers in Florida and Texas. Geochim Cosmochim Acta 45(10): 1941–1950

Sacks LA (1995) Geochemical and isotopic composition of ground water with emphasis on sources of sulfate in the upper Floridan aquifer in parts of Marion, Sumter, and Citrus counties, Florida. US Geol Surv Water-Resour Invest Rept 95–4251

Sacks LA Lee TM Tihansky AB (1992) Hydrogeologic setting and preliminary data analysis for the hydrologic-budget assessment of Lake Barco, an acidic seepage lake in Putnam County, Florida. US Geol Surv Water-Resour Invest Rept 91-4180

Sacks LA Tihansky AB (1996) Geochemical and isotopic composition of ground water, with emphasis on sources of sulfate, in the upper Floridan aquifer and intermediate aquifer system in southwest Florida. US Geol Surv Water-Resour Invest Rept 96-4146

Sacks LA Herman JS Kauffman SJ (2012) Controls on high sulfate concentrations in the upper Floridan aquifer in southwest Florida. Water Resour Res 31(10):2541–2551

Samek K (2004) Unknown quality: The bottled water industry and Florida's springs. Jour Land Use 19(2):569–595

Sanford WE (1987) Assessing the potential for calcite dissolution in coastal salt-water mixing zones. Ph.D. thesis, University Park, Penn State Univ

Sanford WE Konikow LF (1989) Porosity development in coastal carbonate aquifers. Geology 17:249–252

Schmitter-Soto JJ Comin FA Escobar-Briones E Herrera-Silveira J Alcocer J Suárez-Morales E Elías-Gutíerrez M Díaz-Arce V Marín LE Steinich B (2002) Hydrogeochemical and biological characteristics of cenotes in the Yucatan Peninsula (SE Mexico). Hydrobiol 467:215–228

Schneider JC Upchurch SB Champion KM (2006) Stream/aquifer interactions in a karstic river basin, Alapaha River, Florida. Geol Soc Amer, Abs with Progs, 38(3)

Scholle PA (1978) A color illustrated guide to carbonate rock constituents, textures, cements, and porosities. Amer Assoc Petrol Geol Mem 27

Scott TM (1981) The paleoextent of the Miocene Hawthorn Formation in peninsular Florida (abs.). Fla Scien 44(Supplement 1)

Scott TM (1988) The lithostratigraphy of the Hawthorn Group (Miocene) of Florida. Fla Geol Surv Bull 59

Scott TM Means GH Meegan RP Means RC Upchurch SB Copeland RE Jones J Roberts T Willet A (2004). Springs of Florida. Fla Geol Surv, Bull 66

Screaton E Martin JB Ginn B Smith L (2004) Conduit properties and karstification in the unconfined Floridan aquifer. Ground Water 42(3):338–346

Sellards EH (1913a) Map of Florida showing topography, hard rock and land pebble phosphate deposits, and areas of artesian flow. Fla Geol Surv

Sellards EH (1913b) Origin of the hard rock phosphates of Florida. Fla Geol Surv, 5th Ann Rept, pp. 23–80

Shinn EA (2001) African dust causes widespread environmental distress. US Geol Surv, Open-File Rept 01–246

Smart PL Dawans JM Whitaker F (1988) Carbonate dissolution in a modern mixing zone. Nature 335:811–813

Smith LA Martin JB Screaton EJ (2001) Surface water control of gradients in the Floridan Aquifer: Observations from the Santa Fe River Sink-Rise system. In: Martin JB Wicks CM Sasowsky ID (eds.), Hydrogeology and biology of post-Paleozoic carbonate aquifers, Charlestown, WV, Karst Waters Inst Sp Pub 7, pp. 44–48

Socki RA Perry EC Romanek CS (2002) Stable isotope systematics of two cenotes from the northern Yucatan Peninsula, Mexico. Limnol Ocean 47:1808–1818

Southwest Florida Water Management District (2004) The determination of minimum flows for Sulphur Springs, Tampa, Florida. Brooksville, FL, Southwest Fla Water Mgt Dist

Southwest Florida Water Management District (2012) The impact of nutrients on the Weeki Wachee. Presentation to the Hernando Co Bd County Comm, accessed 5/3/2016 at https://www.swfwmd.state.fl.us/files/database/site_file_sets/2874/Weeki_Wachee_Nutrients_-_BCC_Final.pdf

Sproul CR (1977) Spatial distribution of ground water temperature in south Florida. In: Smith DL Griffin GM (eds.), The geothermal nature of the Floridan Plateau. Fla Bur Geol Sp Pub 21, pp. 65–89

Starks MJ (1986) Mixing-zone diagenesis in a carbonate aquifer, west-central Florida. Master's thesis, Tampa, Univ South Fla

Stewart JW Mills LR (1984) Hydrogeology of the Sulphur Springs area, Tampa, Florida. US Geol Surv Water-Resour Invest Rept 83-4085

Stewart MT Wood J (1991) Geologic and geophysical character of fracture zones in a Tertiary carbonate aquifer, Florida. In: Ward SH (ed.), Geotechnical and Environmental Geophysics, 2:235–244

Stoessell RK Moore YH Coke JG (1993) The occurrence and effect of sulfate reduction and sulfide oxidation on coastal limestone dissolution in Yucatan cenotes, Ground Water 31(6): 566–575

Stringfield VT Cooper HH Jr (1951) Geology and hydrologic features of an artesian submarine spring east of Florida: Water Resour Stud, 1951, pt. II, Fla Geol Surv Rept Invest 7, pp. 57–72

Swarzenski PW Reich CD (2000) Re-examining the submarine spring off Crescent Beach, Florida. US Geol Surv Open-File Rept 00-158

Sweeting MM (1973) Karst landforms. London, McMillan Press

Taboroši D Kázmé M (2013) Erosional and depositional textures and structures in coastal karst landscapes. In: Lace MJ Mylroie JE (eds.), Coastal karst landforms, Dordrecht, Springer Science, Coastal Research Library, Book 5, Ch 2, pp. 15–57

Upchurch SB (1992) Quality of waters in Florida's aquifers. In: Maddox GL Lloyd JM Scott TM Upchurch SB Copeland R (eds.), Florida Ground Water Quality Monitoring Program -- Volume 2, Background Hydrogeochemistry, Fla Geol Surv Sp Pub 34, Ch. IV, pp. 12–52, 64–84, 90–347

Upchurch SB (2002) Hydrogeochemistry of a karst escarpment. In: Martin JB Wicks CM Sasowsky ID (eds.), Hydrogeology and biology of post-Paleozoic carbonate aquifers. Charles Town, WV, Karst Waters Inst Sp Pub 7, pp. 73–75

Upchurch SB (2005) Geochemical and geomorphic evolution of a karst escarpment. In: Copeland R (comp.), Geomorphic influence of scarps in the Suwannee River Basin. Tallahassee, Southeastern Geological Society, Field Trip Guidebook 44, pp. 38–42.

Upchurch SB (2014a) An introduction to the Cody Escarpment, north-central Florida. In: Lawn A (comp.), Karst hydrogeology of the upper Suwannee River basin, Alapaha River area, Hamilton County, Florida. Tallahassee, Southeastern Geol Soc Field Trip Guidebook 63, pp. 13–28

Upchurch SB (2014b) Hydrogeology of the swallet and resurgence system in the Alapaha River, Hamilton County, Florida. In: Lawn A (comp.), Karst hydrogeology of the upper Suwannee River basin, Alapaha River area, Hamilton County, Florida. Tallahassee, Southeastern Geol Soc Field Trip Guidebook 63, pp. 30–45

Upchurch SB (2017). Hypogene speleogenesis on the Floridan Platform, U.S.A. In: Klimchouk AB Palmer AN Waela JD Auler AS Audra P (eds.), Hypogene karst regions and caves of the world, Springer Science Publishers, Ch. 49, pp. 735–744

Upchurch SB Lawrence FW (1984) Impact of ground-water chemistry on sinkhole development along a retreating scarp. In: Beck BF (ed.), Sinkholes: Their geology, engineering & environmental impact. Rotterdam, A.A. Balkema, pp. 23–28

Upchurch SB Chen J Cain CR (2007a) Trends of nitrate concentrations in waters of the Suwannee River Water Management District, 2007. Live Oak, FL, Suwannee River Water Mgt Dist

Upchurch SB Chen J Cain CR (2007b) Relationships of nitrate to flow in springs of the Suwannee River Water Management District. Live Oak, FL, Suwannee River Water Mgt Dist

Upchurch SB Schneider J Champion K (2007c) Stream-aquifer interaction in two karstic river basins, north Florida, USA. Nat Ground Water Assoc, 4th Conf on Hydrogeology, Ecology, Monitoring, and Management of Ground Water in Karst Terrains, Safety Harbor, Florida

Veress M (2010) Karst environments: Karren formation in high mountains. New York, Springer Science

Veress M (2016) Covered karsts. Dordrecht, Springer Science

Vernon RO (1951) Geology of Citrus and Levy counties, Florida. Fla Geol Surv Bull 33

Wadsworth JR Brook GA Carver RE (1983) Surface expression of heavily mantled interstratal karst bordering Okefenokee Swamp, Georgia. Tech Paps, 49th Ann Meet, Amer Soc Photogrammetry, pp. 463–470

Water Resource Associates, SDII Global Corporation, and Janicki Environmental, Inc. (2006) MFL Establishment for the Upper Santa Fe River. Tech Rept, Live Oak, FL, Suwannee River Water Mgt Dist

Weedman SD Paillet FL Means GH Scott TM (1997) Lithology and geophysics of the surficial aquifer system in western Collier County, Florida. US Geol Surv Open-File Rep 97-436

Weisman BR *(1992) Excavations on the Franciscan frontier: Archaeology at the Fig Springs Mission. Gainesville, Univ Press Fla*

Weisman BR (1993) Archaeology of Fig Springs Mission, Ichetucknee Springs State Park. In: McEwan BG (ed.), The Spanish missions of La Florida. Gainesville, Univ Press Fla, Ch. 7, pp. 165–192

White WA (1970) The geomorphology of the Florida peninsula. Fla Geol Surv Bull 51

White WB (1988) Geomorphology and hydrology of karst terrains. New York, Oxford University Press

Whitney E Means DB Rudloe A (2004) Priceless Florida: Natural ecosystems and native species. Sarasota, Pineapple Press

Wicks CM Herman JS (1996) Regional hydrogeochemistry of a modern coastal mixing zone. Water Resour Res 32(2):401–407

Wigley TML Plummer LN (1976) Mixing of carbonate waters. Geochim Cosmochim Acta, 40:989–995

Williams CP Scott TM Upchurch SB Paul DT Hannon LM (in prep) Florida Geomorphology. Fla Geol Surv Bull 72

Williams P (2004) Dolines. In: Gunn J (ed.), Encyclopedia of caves and karst science. New York, Fitzroy Dearborn (Taylor & Francis Books, Inc.), pp. 304–310

Williams PW (1972a) Morphometric analysis of polygonal karst in New Guinea. Geol Soc Amer, Bull 83:761–796

Williams PW (1972b) The analysis of spatial characteristics of karst terrains. In: Chorley RJ (ed.), Spatial analysis in geomorphology, London, Methuen, pp. 136–163

Yang Y Schaperdoth I Albrecht H Freeman KH Macalady JL (2008) Geomicrobiology and hopanoid content of a sulfidic subsurface vent biofilm, Little Salt Spring, Florida. Amer Geophy Union Abst 1:0517

Yobbi DK (1996) Analysis and simulation of ground-water flow in Lake Wales Ridge and adjacent areas of central Florida. US Geol Surv Water-Resour Invest Rept 94-4256

# Index




S. Upchurch et al., *The Karst Systems of Florida*, Cave and Karst Systems of the World,
https://doi.org/10.1007/978-3-319-69635-5

## R



## S

MIX
Papier aus verantwortungsvollen Quellen
Paper from responsible sources
FSC® C105338

If you have any concerns about our products,
you can contact us on
**ProductSafety@springernature.com**

In case Publisher is established outside the EU,
the EU authorized representative is:
**Springer Nature Customer Service Center GmbH**
**Europaplatz 3, 69115 Heidelberg, Germany**

Printed by Libri Plureos GmbH
in Hamburg, Germany